D1258954

Readings in Marine Ecology

Readings in Marine Ecology

Edited by
James W. Nybakken
California State College at Hayward
and the Moss Landing Marine Laboratories

Harper & Row, Publishers
New York, Evanston, San Francisco, London

Readings in
Marine Ecology

For information address Harper & Row, Publishers, Inc., 49 East 33rd Street, New York, N.Y. 10016.

Standard Book Number: 06-044868-7

Library of Congress Catalog Card Number: 75–148936

Contents

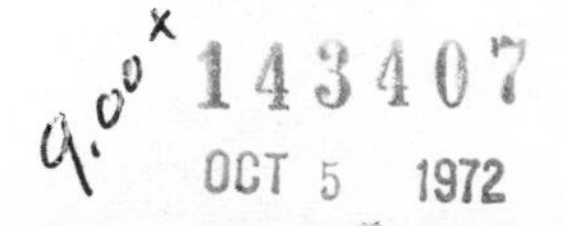

DEEP-SEA ECOLOGY

INTERTIDAL ECOLOGY

TROPICAL ECOLOGY

CONCEPTS IN MARINE ECOLOGY

Preface

I was first motivated to assemble this collection of papers during a post-doctoral year spent at the University of Washington when I became aware of the need such a book might fill in this field, a field in which there are few recent textbooks but considerable current interest. The book has a dual purpose—first, to serve as a supplement to an existing text in marine ecology or biology by furnishing an introduction to some of the more current and classic literature of marine ecology, and secondly, to serve itself as a text for advanced courses and seminars at the graduate level.

The papers which are included here represent as wide a coverage of the whole of what can be called marine ecology as was possible given the limitations of the size of the volume and the length of paper that it was possible to include. As is usual in books such as this, for each paper included many outstanding ones had to be omitted, and it is most probable that another ecologist would choose different papers, or at least a different arrangement of the existing papers. Since the existing choice represents my own bias, I can only hope that those who use this volume will help to reduce that bias through criticism and suggestion.

These papers were selected with the upper division and graduate student in mind, and many have been used in my courses in marine ecology at the Moss Landing Marine Laboratories.

I have chosen not to abridge the papers because it seems that students often feel a certain sense of frustration when they are unable to read the whole paper. I have not felt the same way concerning appendixes, which I have sometimes removed.

I am indebted to my colleagues at the University of Washington, particularly Alan Kohn, Gordon Orians, Robert Paine, Karl Banse, and Ulf Lie, each of whom contributed in some way to my own education in marine ecology and indirectly to this volume. I remain, however, responsible for any omissions.

I also owe a word of thanks to my students and colleagues at the Moss Landing Marine Laboratories whose suggestions and ideas have more often than not been useful in the development of this volume.

Finally, I would like to express my appreciation to the publishers of the journals and the authors of the papers through whose permissions this volume has been made possible.

James W. Nybakken

Editor's Introduction

Coinciding with the recent resurgence of the discipline of ecology at all levels in human society there has been a corresponding increase in interest in that part of ecology known as marine ecology. This interest in marine ecology has been aided by the recent establishment of the Sea Grant Program at the federal level, the many recent instances of oceanic pollution, and the interest of many nations, under the lengthening shadow of portending protein famine, in the use or manipulation of the sea for food resources for an ever-increasing human population. This new awareness of marine ecology also comes at a time when many new approaches and methods of analysis are being employed on marine organisms and communities and new concepts proposed to answer some long puzzling problems in the field.

The selections in this volume represent an attempt to explore some of these recent approaches, methods, and concepts. They do not represent an attempt to "cover" marine ecology. There is a notable gap in that no papers on plankton productivity are included because of my unfamiliarity with that field coupled with a serious space limitation. Nekton also receive short shrift, a situation which at least partly reflects the lack of papers here and the great difficultiy of work with open ocean organisms. A few papers are old, such as the Riley and Doty papers, but are included here because they have become classics or because they represent approaches which were ahead of their time and indicaitve of trends to be developed later—Riley for statistical analysis and mathematical modeling and Doty for the experimental approach. The level of difficulty of the papers varies, and some, especially those that employ advanced statistical analyses and computer methods such as Cassie and Michael and Fager and McGowan, will be too difficult for many undergraduates to understand fully. However, they are indicative of the current methods employed by practicing ecologists and, I hope, may stimulate the serious student to undertake the study of advanced statistics and computer methods to obtain this now necessary mathematical background. At the least the student will be warned about the necessary skills to work in the field.

I have grouped the papers under seven headings: Introduction, Plankton, Benthic Ecology, Deep-Sea Ecology, Intertidal Tropical Ecology, and Concepts in Marine Ecology. This arbitrary division of the papers is due mainly to the dual purpose of the book, and the papers could be arranged differently under other circumstances. The Redfield paper serves as a fine introduction to the marine environment, and hence it is set aside in an Introduction. The remaining six sections represent a convenient, if not always natural, division of the communities of the sea which corresponds well to the divisions in texts in marine ecology and biology. Hence, this arrangement is a convenience for those who will use this book as a supplementary text. Advanced courses may prefer to consider the papers in quite different sequences. I have not necessarily placed the papers to be read sequentially except that the Sverdrup paper should be read before the Parsons *et al.* paper in the Plankton section.

Within each of the six main headings I have attempted to include a set of papers that will balance each other somewhat in terms of coverage of major ecological principles, organisms and habitats, and recent approaches and trends thus allowing the book to stand by itself as a text. The intertidal section reflects this attempt by covering population dynamics (Frank), competition (Connell), predation (Connell), energy flow (Teal), and interrelationships (Cassie and Michael) and at the same time providing coverage of the two major groups of intertidal habitats: rock (Connell, Frank, Doty) and soft bottoms (Odum and Smalley, Teal, Cassie and Michael). Similarly, I have attempted to equalize coverage in the plankton section between zooplankton (Cushing, Fager and McGowan, Steeman-Nielsen) and phytoplankton (Riley, Sverdrup, Parsons *et al.*) while also considering two of the basic problems in plankton communities, the relationship of the phytoplankton and zooplankton (Cushing, Steeman-Nielsen, Steele) and the conditions which cause the spring blooms (Sverdrup, Parsons *et al.*). Finally, within this framework I feel that the papers are outstanding examples of recent and modern approaches to ecological questions.

I have included a section on tropical communities because the tropics contain two unique communities or ecosystems, mangrove swamps and coral reefs, which cover extensive areas and have a considerable importance in the general economy of the tropical seas.

I have included a final section on Concepts in Marine Ecology to allow the inclusion of theoretical or speculative papers that challenge some established concepts or suggest new approaches to understanding the construction and functioning of marine ecosystems. They also serve to direct attention to some general ecological questions, such as: What are the determinants of stability in marine communities and ecosystems? What is the relationship, if any, between species diversity and stability? I feel that future work in marine ecology will be directed to testing the ideas or approaches suggested by these authors. Reading these selections should help the student gain an insight into a few of the many yet unsolved problems in marine ecology. I have also included in this section two papers by Ryther which speak directly to the problem of potential food from the sea.

Two papers are included because they demonstrate the use of two modern mathematical techniques for analyzing complex ecological data. These are the Fager and McGowan paper, which uses factor analysis to sort out species associations in zooplankton, and the Cassie and Michael paper, which treats benthic data to a multivariate analysis. Both of these techniques generally require computers, and both are becoming more widely used. Hence I feel it is desirable to include them to illustrate the trends of future analysis.

I must conclude by admitting my bias in favor of intertidal ecology and benthos, the fields with which I am most familiar. The 32 papers here are certainly not the only examples which could be used and still accomplish the purpose of the book. Undoubtedly several other sets of papers would cover the field as well; I can only hope that those assembled here are representative and equally as useful as any other set.

Readings in
Marine Ecology

Introduction

THE BIOLOGICAL CONTROL OF CHEMICAL FACTORS IN THE ENVIRONMENT[1]

By ALFRED C. REDFIELD

IT IS a recognized principle of ecology that the interactions of organisms and environment are reciprocal. The environment not only determines the conditions under which life exists, but the organisms influence the conditions prevailing in their environment. The examples on which this principle rests are usually difficult to describe in quantitative terms and are frequently local in their application. In the ocean the principal interactions between organisms and environment are chemical. Because of its unity, its fluid nature, and the intensity of the mixing to which the water is subject their relations can be examined statistically and expressed in quantitative terms.

The purpose of this essay is to discuss the relations between the statistical proportions in which certain elements enter into the biochemical cycle in the sea, and their relative availability in the water. These relations suggest not only that the nitrate present in sea water and the oxygen of the atmosphere have been produced in large part by organic activity, but also that their quantities are determined by the requirements of the biochemical cycle. The argument is not simple and in order that it may be understood the nature of the biochemical cycle and the circulation of the elements involved are reviewed.

The Biochemical Cycle

The production of organic matter in the sea is due to the photosynthetic activity of microscopic floating plants, the phytoplankton, and is limited to the surface layers where sufficient light is available. The formation of organic matter in the autotrophic zone requires all the elements in protoplasm, of which carbon, nitrogen, and phosphorus are

[1] A Sigma Xi National Lecture for 1957–58. Contribution No. 976 Woods Hole Oceanographic Institution.

Reprinted from AMERICAN SCIENTIST, Vol. 46, No. 3, Sept., 1958

of particular concern. These are drawn from the carbonate, nitrate, and phosphate of the water. Following the death of the plants the organic matter is destroyed, either by the metabolism of animals or the action of microorganisms. Normally, decomposition is completed by oxidation so that carbon, nitrogen, and phosphorus are returned to the sea water as carbonate, nitrate, and phosphate, while requisite quantities of free oxygen are withdrawn from the water.

The autotrophic zone has a depth of 200 meters at most and includes less than five per cent of the volume of the ocean. Below this zone, life depends on organic matter carried down by organisms sinking from above or by the vertical migrations of animals back and forth between the depths. Although the greater part of the nutrient chemicals absorbed in the autotrophic zone complete the cycle in this layer, the portion which sinks as organic matter tends to deplete the surface layers of these chemicals and, with the decomposition of the organic matter in the depths, to enrich this heterotrophic zone with the products of decomposition.

The existence of the vast reservoir of deep water in which organic matter may accumulate and decay out of reach of autotrophic resynthesis is a distinctive feature of the oceanic environment which enables one to separate, in observation and thinking, the constructive and destructive phases of the biochemical cycle.

The synthesis of organic matter is a highly selective process which results in products having specific composition. Differences in the composition of various species or individuals exist, of course, but the similarities on the whole are greater than the differences. When the population of a large region of the sea is considered it seems reasonable that its composition will be uniform in a statistical sense, and will be reflected in the changes in the chemistry of the water from which its materials are drawn or to which they are returned. In the decomposition of a given amount of organic matter the quantity of oxygen consumed must be determined exactly by the quantities of carbon, nitrogen, etc., to be oxidized, and the relative changes in the quantity of oxygen, phosphate, nitrate, and carbonate in the water must depend exactly on the elementary composition of the plankton [1].

The validity of this concept is shown by a comparison of the proportions in which the elements exist in the plankton and the proportions in which they vary in samples of water from the open sea. The analysis of many samples of plankton, taken in a variety of places with nets designed to take organisms of different size indicates that atoms of phosphorus, nitrogen, and carbon are present on the average in the ratios: 1:16:106 [2]. The oxidation of this material is estimated to require 276 atoms of oxygen. In comparison, data on the phosphate, nitrate, and carbonate content of sea water collected from various depths in the several oceans show that the available phosphorus, nitrogen, and carbon vary from sample to sample in the atomic ratios: 1:15:105 and that about 235 atoms of oxygen disappear for each addition of one phosphorus atom, assuming the water to have been in equilibrium with the oxygen of the atmosphere when it was last at the sea surface [3–5]. It is as though various quantities of material containing phosphorus, nitrogen, and carbon in these ratios had decomposed in the water samples. The correspondence of these ratios with those obtained from the analysis of plankton, as shown in Table 1, leaves little doubt that these elements vary in sea water almost entirely as the result of the synthesis or decomposition of organic matter [1].

This conclusion permits one to approach the biochemical cycle in the sea in much the same way as the physiologist examines the general metabolism of an individual organism. It defines the quantitative relations of the cycles of the separate elements which are involved in the biochemical cycle as a whole. It provides a quantitative criterion, similar to the principle of combining proportions in chemistry, for examining many problems in marine ecology. Consider, for example, the question of the relative availability in the sea of the various substances required for the growth of organisms. In accordance with Liebig's law of the minimum, that constituent of the sea water present in smallest quantity relative to the requirement for growth of organisms will become the limiting factor. The ratios shown in Table 1 define precisely what these relative requirements are. When the composition of sea water is examined it is found that phosphorus and nitrogen are usually available in about the proportions, 1:15, required for the formation of the plankton. Carbon, as carbonate, in contrast, is present in great excess, the phosphorus-carbon ratio being about 1:1000 whereas the required ratio is 1:105. Consequently, carbonate never becomes a limiting factor. In a similar way sulfur, an important plant nutrient, is available in great quantities relative to phosphorus and nitrogen, and the same is true of calcium, magnesium, and potassium.

TABLE 1

ATOMIC RATIOS OF ELEMENTS IN THE BIOCHEMICAL CYCLE

	P	*N*	*C*	*O*
Analyses of plankton	1	16	106	−276
Changes in sea water	1	15	105	−235
Available in sea water	1	15	1000	200–300

Phosphorus and nitrogen thus appear to be the constituents of sea water present in limiting quantities. It was pointed out by Harvey in 1926 that it is a remarkable fact that in the English Channel plant growth should strip sea water of both nitrates and phosphates at about the same time [6]. To this coincidence I will return. Data for the oceans indicate that this is a world-wide occurrence in the surface waters of tropical and temperate seas. This would not be the case were other substances, such as the various trace elements, generally limiting factors for growth in the sea.

It should be recognized that in coastal waters and in the surface of the open ocean the ratios of the elements under discussion frequently depart widely from those which are here attributed to the oceans as a whole. This is due, in part at least, to the fact that where the elements are substantially depleted by the growth of plants small unused residues of one or another element may greatly alter the ratios, and also because in the process of decomposition phosphorus tends to be regenerated more rapidly than nitrogen. It is believed, however, that in the great bulk of the open sea the ratios presented represent the statistical composition of the water and of the organisms living in it sufficiently well to serve for the argument to be developed. This argument depends upon the consideration that the physiology of marine plants places definite restrictions on the proportions in which certain elements participate in the biochemical cycle, rather than on the exactness with which these proportions can be stated at present.

The Biochemical Circulation

The withdrawal of nutrients from the autotrophic zone would soon exhaust its fertility and bring the cycle to a close if means did not exist to restore them to the surface. This is done by purely physical processes, the vertical mixing of water by turbulence and by direct upwelling of deeper water to the surface—frequently after transport for great distances by currents in the depths of the sea [7].

The production of organic matter in the autotrophic zone at the surface of the sea depends upon how fast the biochemical cycle runs its course. Many factors, such as the temperature and light intensity, may intervene, but it is clear that the cycle is limited by the return of nutrient materials to the surface by the motion of the water. This, in turn, depends in part upon the quantities of nutrients available in the subsurface waters. We may consequently inquire with interest into the concentrations of nutrients in the deeper layers of the sea as they are distributed throughout the world.

The sea water of the oceans may be considered to be thoroughly mixed. The major ions of sea water occur so nearly in the same proportions that by this criterion one cannot distinguish water from one part of the ocean from another. The principal variables used by oceanographers to distinguish different waters are the temperature and salinity. Changes in these properties are produced at the sea surface, and while the effects may be relatively large they influence only a small part of the total mass of water. Ninety per cent of the water in the oceans differs by less than 3 per cent from the average in its salinity or its heat content.

In contrast, the principal elements involved in the biochemical cycle differ greatly in their abundance in different parts of the sea. It was first shown by the cruises of the Danish research vessel "Dana" and the "Carnegie" that phosphate and nitrate were much more abundant in the Pacific and Indian Oceans than in the Atlantic [8, 9]. A chart showing the quantities of phosphorus present in the world oceans at a depth of

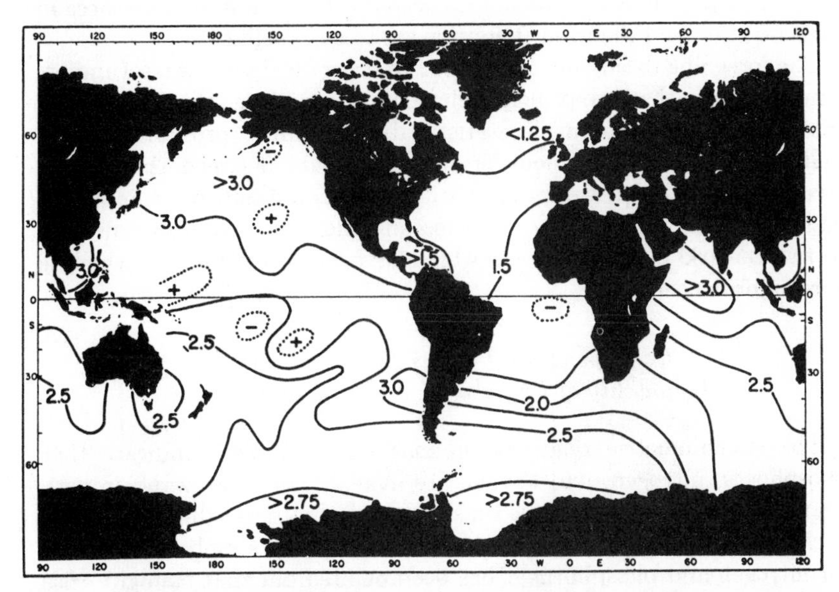

FIG. 1. Distribution of phosphorus at depth of 2000 meters in Oceans of World. Contour interval 0.25 mg atoms/m³. This diagram is based on some 1600 measurements between depths of 1900 and 2100 meters. Data are available from about 75 per cent of the 10° squares. About half the observations fall within the designated contour interval, the remainder within the adjacent intervals.

2000 meters, which represents adequately the concentrations present in the great reservoir of deep water, is shown in Figure 1. The variation in phosphorus content is widespread and systematic, the values increasing from less than 1.25 mg atoms per cubic meter in the North Atlantic to more than 3.0 of these units in the North Pacific and Indian Oceans. In the Mediterranean Sea comparable values are less than 0.5 unit, so that there is a variation of more than sixfold in the potential fertility of the deep reservoirs of the major seas.

It is evident that there is some world-wide mechanism at work producing this distribution. Since the elements involved are those which enter the biochemical cycle this mechanism must be biological.

Since the distribution differs from that of the biologically inactive components of the water there must be a biochemical circulation which is different from, though dependent upon, the physical circulation of the water [10]. How can this difference in behavior be explained?

The presumptive agency of fractionation which separates the biochemical from the inactive elements in sea water is the selective absorption of the former in the synthesis of protoplasm near the surface, followed by the sinking of the organized matter to greater depths prior to its decomposition. In addition the differential motions of the water at different depths must be taken into account if substantial differences in concentration and their redistribution are to be explained.

The foregoing discussion is intended to bring out two points of importance, namely, that the principal elementary constituents of protoplasm enter the biochemical cycle statistically in definite proportions, and that the cycle of movement of these elements between the surface layers and the deep waters runs with sufficient intensity to influence markedly their distribution on an ocean-wide scale. We may now turn to two remarkable coincidences which have led to the inquiry which is our major concern.

Correspondence Between Requirement and Availability of Phosphorus, Nitrogen, and Oxygen

The stoichiometric relations summarized in Table 1 indicate that phosphorus, nitrogen, and oxygen are available in ocean water in very nearly the same proportions as those in which they enter the biochemical cycle. In discussing the remarkable coincidence in the supply and demand for nitrogen and phosphorus it has been pointed out that it might arise from: (1) a coincidence dependent on the accidents of geochemical history; (2) adaptation on the part of the organisms; or (3) organic processes which tend in some way to control the proportions of these elements in the water [1].

Of the first alternative not much can be said except that the probability that the ratio in the sea be what it is rather than any other is obviously small. That the coincidence applies to the oxygen as well as to the nutrient elements compounds the improbability.

For the second alternative, it may be said that the phytoplankton do have some ability to vary their elementary composition when one element or another is deficient in the medium in which they grow. Such physiology might account for the coincidence in the nitrogen-phosphorus ratios. However, it is not evident how adaptation could determine the oxygen relation since this depends more on the quantity than the quality of the organic matter formed, and the oxygen requirement is felt only after the death of the living plant.

For these reasons the third alternative deserves serious consideration. Mechanisms should be examined by which organic processes may have tended to control the proportions of phosphorus, nitrogen, and oxygen available for life in the sea.

The Phosphorus-Nitrogen Ratio

When the coincidence between the elementary composition of marine plankton and the proportions of available nitrogen and phosphorus in sea water was first noted, it was suggested that it might have been brought about by the activity of those microorganisms which form nitrogenous compounds from atmospheric nitrogen, or liberate nitrogen gas in the course of their metabolism. The composition of such organisms must be more or less fixed in regard to their relative phosphorus and nitrogen content. When living in an environment containing a deficiency of nitrate relative to phosphate, the growth and assimilation of the nitrogen-fixing organisms might tend continually to bring the proportions of nitrogen and phosphorus nearer to that characteristic of their own substance. Thus, in the case of Azotobacter, it has been found that for every atom of phosphorus available in the medium about 10 atoms of nitrogen are fixed or assimilated into microbial protein [11]. In an environment populated by organisms of this type, the relative proportion of phosphate and nitrate must tend to approach that characteristic of their protoplasm. Given time enough, and the absence of other disturbances, a relation between phosphate and nitrate such as observed in the sea may well have arisen through the action of such organisms.

Nitrogen fixation is employed practically in agriculture whenever leguminous plants are used to enrich the soil. It is not unreasonable to assume that the same process has been effective on a larger scale in nature. Hutchinson has estimated that nitrogen is being fixed on the earth's surface at the annual rate of 0.0034 to 0.017 mg/cm^2 [12]. At the lesser value it would require only 40,000 years to fix the 7×10^{14} kilograms of nitrogen estimated to be available as nitrate in the ocean.

Nitrogen fixation is so active that there is no difficulty in assuming that it might serve in adjusting the phosphorus nitrogen ratio in the sea. The difficulty is rather in explaining why there is not a great excess of nitrate nitrogen in the water. The ratio of nitrogen to phosphorus in fresh waters is higher than that in ocean water, while the ratio in sedimentary rocks is very much lower. Consequently the ratio in the sea must tend to increase, unless some process is returning nitrogen to the atmosphere. Denitrifying bacteria might operate in this sense, in which case the phosphorus-nitrogen ratio is fixed by a complex balance.

Biological mechanisms adequate to influence the phosphorus-nitrogen ratio in sea water are known. Whether they do in fact operate in a regulatory sense is a subject for future investigation.

The Phosphorus-Oxygen Ratio

The relation between the quantity of phosphorus present in sea water and the amount of oxygen available for the decomposition of organic matter is less obvious than the relation of phosphorus and nitrogen. The quantity of oxygen dissolved in sea water when it is at the sea surface appears to be fixed by equilibrium with the atmosphere, which contains about 21 per cent oxygen. Consider what would happen if a unit volume of water, containing its characteristic quantities of plant nutrients, was brought from the depths of the ocean to the surface. There, under the influence of light, photosynthesis would convert the available nutrients to organic matter until all the phosphate is exhausted. At the same time the water is saturated with oxygen by equilibration with the atmosphere. If the unit volume of water is now returned to the depths where the organic matter is decomposed and oxidized, will the

TABLE 2

UTILIZATION OF OXYGEN IN OCEAN WATER

	Average S. W.	*North Atlantic*	*North Pacific*
Phosphorus content mg atoms/m^3	2.3	1.25	3.0
Equivalent oxygen Utilization* ml/liter	6.9	3.75	9.0
Oxygen content of saturated S. W.	7.1	6.7	7.5
Excess oxygen	0.2	2.95	−1.5
Oxygen in "minimum oxygen layer"		3.0	0.01

* Calculated allowing 276 atoms of oxygen to be used in oxidizing a quantity of organic matter containing one atom of phosphorus in which case 1 mg atom P/m^3 is equivalent to 3 milliliters O_2/liter.

dissolved oxygen present be sufficient for the purpose, will there be a large excess of oxygen, or will the oxygen be deficient?

The answer can be given only within approximate limits, for two reasons. First, the quantities of nutrients are different in the deep waters of the several oceans, as we have seen. We can consider the extreme cases of the North Atlantic containing phosphorus in the concentration of 1.25 mg atoms per m^3, the North Pacific containing 3.0 mg atoms per m^3, and an "average" sea water containing, say, 2.3 mg atoms per m^3. Second, the quantity of oxygen dissolved in the water when it is at the surface will vary depending upon the temperature, and to a lesser degree upon the salinity.

The lowest oxygen concentrations are not found at the bottom of the oceans, but at some intermediate depth, where also maximum concentrations of phosphate and nitrate occur. The temperature at this depth in the North Atlantic is about 8°C. and when at the surface at this temperature the water would have taken up 6.7 milliliters of oxygen per liter. The corresponding temperature of the minimum oxygen layer in the North Pacific is about 3°C. and the oxygen solubility 7.5 milliliters per liter.

From these numbers Table 2 has been prepared to show the excess oxygen which might be expected to remain in the minimum oxygen layer if all the phosphorus present has been derived from the oxidation of organic matter. In the case of the "average" sea water it appears that the quantity of organic matter which can be formed from the nutrients in a unit of volume of "average" sea water is almost exactly that which can be completely oxidized by the oxygen dissolved at the surface. In the North Atlantic, where the phosphorus content is reduced, only about one half of the dissolved oxygen would be consumed and the excess oxygen corresponds well with the quantities observed to remain in the zone of minimum oxygen content. In the North Pacific, the phosphate content of the deep water is so great that the dissolved oxygen would be more than exhausted if the process went to completion as postulated.

Actually the Pacific Ocean does not appear to be anaerobic at any depth. However, large volumes of water at intermediate depths contain only small traces of oxygen [4]. In high latitudes where the deep water of the oceans originates the photosynthetic processes do not convert all of the available nutrients into organic matter before the water sinks. Consequently, the deep water contains phosphates which have not been liberated by oxidation during the preceding turn of the cycle. Presumably this effect accounts for traces of oxygen which remain in the minimum oxygen layer of the North Pacific.

Although the oxygenation of the oceans appears to be adequate to oxidize the products of the biochemical cycle at present the margin is not large. Actually anaerobic conditions exist in a number of isolated areas, such as the Black Sea and the Sea of Azov, Kaoe Bay in the East Indies, the Cariaco Trench, and numerous fjords in Norway and elsewhere [4]. A decrease in the oxygen content of the atmosphere, an increase in surface ocean temperatures or reduction in the vertical circulation of the water might lead to the extension of anaerobic conditions over much wider areas.

It is widely held among geochemists that the primitive atmosphere was devoid of oxygen, or at least contained very much less oxygen than at present. During the course of geological history atmospheric oxygen is thought to have been produced by the photochemical dissociation of water in the upper atmosphere and by the photosynthetic reduction of carbon dioxide, previously present in much greater quantities [13].

Estimates of the quantity of reduced carbon present in the earth's crust as coal and petroleum indicate that photosynthetic processes have been much more than adequate to produce the present oxygen content of the atmosphere. It has not been suggested, to my knowledge, why this process has proceeded just so far as it has; that is, why there is 21 per cent of oxygen in the atmosphere at present, no more or no less. It is, however, on this fact that the quantity of oxygen dissolved in the sea depends.

My supposition is that the actual quantities of oxygen present in the sea may have been regulated by the activities of sulfate-reducing bacteria. This group of bacteria are known to have the ability to use sulfates as a source of oxygen when free oxygen is absent and organic compounds are present to supply a source of energy [14]. The over-all reaction is

$$SO_4^{=} \longrightarrow S^{=} + 2O_2$$

The process should be broken down into two steps, each of which takes place in the sea under different environmental conditions, (1) Sulfate Reduction, $SO_4^{=} + 2C \rightarrow 2CO_2 + S^{=}$ which occurs at depth under anaerobic conditions, and (2) Photosynthesis $2CO_2 \rightarrow 2C + O_2$ which occurs near the surface in the presence of light.

In these equations C represents the reduced carbon present in organic matter. The decomposition of this material by sulfate-reducing bacteria according to the first step also liberates a corresponding quantity of nitrogen and phosphorus, which permit the CO_2 formed to re-enter the biological cycle when the second step comes into play. The CO_2 produced in this way can thus contribute to the production of oxygen in a way in which the excess carbonate normally present in sea water cannot.

The first step, which depends upon the presence of organic matter in excess of the free oxygen required to complete its decomposition, will initiate a mechanism which will tend to increase the oxygen when, and only when, the quantity of available free oxygen is deficient. If the total mechanism has operated on a large enough scale in the course of geochemical history, it may have kept the supply of oxygen available in the sea adjusted to the requirements of the biochemical cycle.

There is very good evidence that sulfate reduction does operate on a large scale in the sea wherever anaerobic conditions exist. Figure 2 shows the relations between the phosphorus content and the combined utilization of oxygen and sulfate in the waters of the Black Sea and Cariaco Trench, both anaerobic basins. The phosphorus content may be taken as an indication of the quantity of organic matter which has decomposed in the water. The combined utilization is the quantity of free oxygen which has disappeared from the water since it left the surface, plus *four times* the quantity of sulfide present. That is to say, it is assumed that for each ion of sulfide liberated in the water four atoms of oxygen have been produced by sulfate reduction and utilized in oxidizing organic matter. The figure shows that as increasing quantities of organic matter decompose in the water the utilization of oxygen increases approximately in the ratio of 235 atoms for each atom of phosphorus set free [3]. After the phosphorus content reaches a little more than 2 mg atoms per m^3 the free oxygen is exhausted and hydrogen sulfide appears

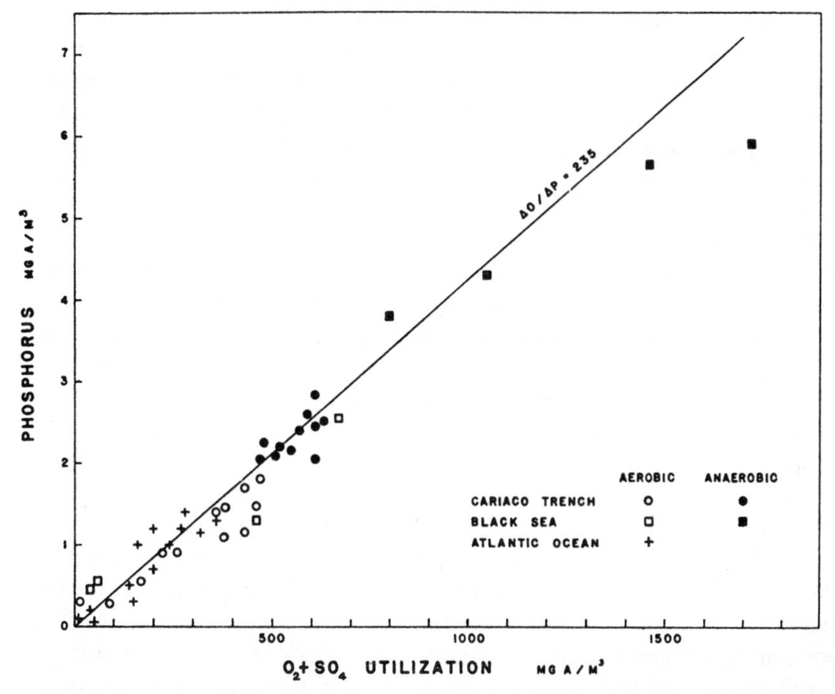

FIG. 2. Relation between the inorganic phosphorus and the combined utilization of oxygen and sulfate in waters of Cariaco Trench, Black Sea, and western Atlantic Ocean. After Richards and Vaccaro [3].

in the water. The increase in hydrogen sulfide continues as though it were produced by the reduction of the quantity of sulfate required to oxidize the accumulated organic matter. In the depths of the Black Sea more than two thirds of the phosphorus present may be attributed to the decomposition of organic matter through sulfate reduction. If the products of such decomposition find their way back to the surface, then the oxygen derived from sulfate will be liberated by the photosynthetic process and will be added to the supply of free oxygen in the environment.

At the present time, anaerobic basins such as the Black Sea and the Cariaco Trench are rarities and the greater part of the ocean is well oxygenated. Extensive areas of marsh and estuary present anaerobic conditions, as do quite generally marine muds. At present the production of oxygen through sulfate reduction in such situations may not be more than sufficient to balance the losses due to the oxidation of eroding terrestrial surfaces. If, however, in the past the oxygen of the atmosphere were lower than at present, anaerobic conditions may have been much more prevalent. The reduction of sulfates may then have served to bring the oxygen content of the atmosphere and sea into correspondence with the requirements set by the quantities of phosphorus available.

In considering the influence of the biochemical cycle on the environment it should be remembered that the cycle can, and apparently for the most part does, run its course without adding more to the environment at one phase than it withdraws at another. Photosynthesis cannot increase the oxygen content of the environment unless the products of photosynthesis are in some way withdrawn from reoxidation. The reduction of sulfates will not make more oxygen available in the sea if the sulfides produced are reoxidized in the sea water. Fractionating mechanisms are required which separate the products of the cycle, so that they cannot re-enter it. Such mechanisms exist at the sea bottom, where materials may be buried in the sediments, and at the surface where volatiles may pass into the atmosphere.

Consider the fate of the sulfide and oxygen which are produced in anaerobic basins by the activity of sulfate-reducing organisms. When the water containing these products is brought to the surface layers of the sea by mixing processes the sulfide will be reoxidized to sulfate, consuming a quantity of free oxygen equivalent to that produced by the original reduction. No change in the total oxygen in the water column would result if this were the complete picture. However, if a portion of the sulfide forms insoluble compounds, such as iron sulfide, which settle into the bottom sediments, or if a portion of the oxygen liberated near the sea surface passes into the atmosphere, these portions cannot re-enter the cycle immediately and a net increase in the oxygen of the environment will be produced.

Conditions on the land, in contrast with the sea, appear to be much less favorable for producing changes in the free oxygen of the atmosphere, either by photosynthesis or in response to anaerobic conditions through sulfate reduction. The predominance of erosion does not permit the permanent entrapment of large quantities of organic matter on land. Soils are usually well aerated so that organic carbon does not remain long out of circulation. Anaerobic conditions do exist in swamps, and many coal deposits must represent reduced carbon formed in such places which have contributed oxygen to the atmosphere, but such withdrawals must be small compared to the quantities of organic matter accumulated in marine sediments.

The exchange of oxygen across the sea surface must be important in this connection. The solubility of oxygen in water is such that under equilibrium conditions more than twenty times as much oxygen occurs in a unit volume of air than in a comparable volume of water at the sea surface. Of any increase in the quantity of oxygen in the water, whether produced by the reduction of sulfates or the photosynthetic reduction of carbonates, only a small portion would remain in the water since a larger fraction would pass into the atmosphere.* In this way the biochemical cycle of the sea can continue to add to the oxygen content of the environment until a new equilibrium is established between the atmosphere, the sea, and the available nutrient materials.

It appears, then, that the biochemical cycle in the sea may have produced the major amount of oxygen in the atmosphere and that conditions in the sea have adjusted its level to that which occurs at present. We think of the atmosphere as determining the oxygen content of the sea. This is because the atmosphere is the great reservoir of oxygen on the earth's surface and because the motion of the air gives it a conveniently constant composition. Perhaps it is more correct, however, to think of the sea, and particularly of its nutrient content, as determining the composition of the atmosphere.

Geochemical Considerations

As a final check on these speculations, we can look at the relative availability of the principal materials of the biochemical cycle on the earth's surface, to see if they conform to the postulates. In Table 3 estimates are given of the total quantities of the pertinent elements in the atmosphere, the ocean, and the sedimentary rocks. The content of the atmosphere is quite accurately known; that of the ocean can be approximated since the greater part of its water is of relatively uniform composition. The values for the sedimentary rocks are quite uncertain, partly because of the variability in their composition and the incompleteness in sampling, and partly because the volumes in question are arbitrarily chosen. However, we are concerned with orders of magnitude so the numbers are useful.

The first column of numbers in Table 3 are the estimated weights of each element on the earth's surface. Divided by the phosphorus content of the ocean and adjusted for atomic weights, the second column gives the numbers of atoms relative to the atoms of phosphorus in the ocean.

* The intensity of the exchange of oxygen across the sea surface is indicated by the observations that in the Gulf of Maine a quantity of oxygen sufficient to form a layer 0.9 meters in depth enters the atmosphere each spring as the result of photosynthetic activity by phytoplankton. This is about 15 per cent of the oxygen locally present to a depth of 200 m. This quantity of oxygen returns to the water during the following fall and winter to replace that used in the oxidation of the summer crop of organic matter [15].

So reduced, the relative quantities are indicated in Figure 3 which presents in diagrammatic form the biochemical cycle as described.

TABLE 3

QUANTITIES AND PROPORTIONS OF ELEMENTS AT THE EARTH'S SURFACE*

Ocean	P	1×10^{14}	1
	N as NO_3	7×10^{14}	15
	as N_2	2.3×10^{16}	510
	C	4×10^{16}	1,000
	S	1×10^{18}	10,000
	O	1.4×10^{16}	270
Atmosphere	N as N_2	3.8×10^{18}	62,000
	C	6.4×10^{14}	16
	O	1.2×10^{18}	23,000
Earth's crust			
Sedimentary rocks	P	4×10^{18}	40,000
	N	4.4×10^{17}	10,000
	S	1×10^{18}	10,000
	C	1.5×10^{19}	400,000
Coal and petroleum	C	6×10^{18}	160,000

* Estimates are based on following sources: Atmosphere [19]; Ocean, volume 1.37 $\times$ 10^9 Km^3 [20], P 71 $\times$ 10^{-6} gr/Kg(original), N as NO_3 15 $\times$ P, N as N_2 17.3 $\times$ 10^{-3} gr/Kg assuming saturation at 2°C. and 34.33 % S [21], O as O_2, including consumption in biochemical cycle, 10.7 $\times$ 10^{-3} gr/Kg assuming saturation at 2°C. and 34.5 % S [22], C as $H\bar{C}O_3$ 0.028 gr/Kg [20] S as SO_4 0.884 gr/Kg [20]; Earth's crust, area 5.1 $\times$ $10^{18}/cm^2$, Sedimentary rocks 170 Kg/cm^2, P 0.46 per cent (arbitrary weighting of analyses by Stokes and Steiger) N 0.051 per cent, S = 0.12 per cent (arbitrary weighting of analyses by Stokes), C 3 Kg/cm^2, and coal and petroleum 1.26 Kg/cm^2 after Kalle [19].

The diagram shows phosphate, nitrate, and carbonate entering the organic phase of the biochemical cycle near the sea surface, through the process of photosynthesis. Phosphorus, nitrogen, and carbon are selected by the synthetic process in the proportions of 1:15:105. This is the step which coordinates the cycles of the several elements in a unique way and gives meaning to the comparisons. The elements are carried in these proportions to the point of decomposition where they are oxidized to their original state as phosphate, nitrate, and carbonate. The oxygen required is just that set free by photosynthesis. Such a cycle could run indefinitely in an otherwise closed system so long as light is supplied.

To account for the correspondence in the ratios of phosphorus and nitrogen in the organic phase of the cycle and in the inorganic environment, bacterial processes of nitrogen fixation and denitrification are indicated at the upper right and, similarly, the sulfate reduction process is shown at the lower left. This latter is assumed to operate effectively only when the environment becomes anaerobic. Finally, the exchanges with the atmosphere and the sediments of the sea bottom are shown. If these processes are operative it is necessary that supplies are adequate and that their products exist in suitable quantities.

Considering first nitrogen, there exist in sea water for each atom of phosphorus 15 atoms of nitrogen available as NO_3 and a reserve of 510 atoms of nitrogen as dissolved N_2 which may be drawn on by nitrogen-fixing bacteria. In addition, there is a reserve of nitrogen in the atmosphere equivalent to 86,000 atoms of phosphorus, which is available to replace that dissolved in the sea were it to be drawn on. The nitrogen of the sedimentary rocks is about one-sixth that in the atmosphere and twenty times that in the ocean. More than four-fifths of this is fossil nitrogen which may be assumed to be derived from organic matter [12]. Consequently, large quantities of nitrogen have passed through the biochemical cycle in its passage from the atmospheric reserves to be deposited in sediments at the sea bottom. The quantity withdrawn in this way is small, however, in comparison to the reserve in the atmosphere. Clearly, the nitrogen supply is adequate.

Sulfate is one of the most abundant ions in sea water. In this form there is present sulfur equivalent to 10,000 atoms of phosphorus. It would be capable of supplying oxygen equivalent to 40,000 atoms of phosphorus on reduction. Clearly, the sulfate reduction mechanism could continue to operate for a long time. If it has operated as postulated in the past much sulfide may have been removed from the sea. Sedimentary rocks are estimated to contain sulfur equivalent to 10,000 atoms of oceanic phosphorus. If this were all the product of sulfate reduction, this would have produced oxygen equivalent to 40,000 atoms of oceanic phosphorus, which is almost twice that present in the atmosphere. It is not clear how much of the sulfur in sedimentary rocks is present as sulfides, but much of it is. Clearly, much oxygen can have been produced in the past by sulfate reduction and possibly this process has contributed to an important degree in producing the oxygen of the atmosphere.

Carbon is present in the sea, chiefly as carbonate ions, in about ten times the quantity required for the biochemical cycle. Much of the large deposits of carbon in the sedimentary rocks is present as carbonates and cannot have contributed to the production of free oxygen. The estimated carbon present as coal and petroleum, equivalent to 160,000 atoms of oceanic phosphorus, is sufficient to yield oxygen on reduction equivalent to 320,000 atoms of oceanic phosphorus, which is more than ten times the present content of the atmosphere.

The known facts of geochemistry do not appear to contradict the suppositions presented on the mechanism which may have controlled the relative availability of phosphate, nitrate, and oxygen in the sea. Sources of nitrogen and sulfate are available in great excess and the by-products of the reactions can be adequately accounted for. According to these suppositions, phosphorus is the master element which controls the availability of the others. How about phosphorus itself?

In the sedimentary rocks, formed for the most part by marine deposition, there are present some 40,000 atoms of phosphorus for every atom in the sea water. Evidently large quantities of phosphorus, carried into the sea from the rivers in past times, have passed through the sea to contribute to these deposits. The concentration of phosphorus in average sea water is very small, amounting to less than one part in 10 million. Probably this represents a condition determined by the low solubility of phosphates under the conditions existing in sea water at the sea bottom. Dietz, Emery, and Shepard [16] have suggested that sea water deeper than a few hundred meters is essentially saturated with $Ca_3(PO_4)_2$. It is not possible to be exact in this regard because the solubility will vary with the conditions, and the physical constants are not well known. However, they have described the presence of phosphorite deposits on the floor of the North Pacific ocean which they explain in this way. The high ratio of phosphorus to nitrogen in sedimentary rocks, 4:1, when compared to the ratio characteristic of the plankton, 1:15, also suggests that phosphorus has been introduced into the sediments by some mechanism other than the entrapment of organic matter.

If the argument presented is sound it may be concluded that the quantity of nitrate in the sea, and the partial pressure of oxygen in the atmosphere are determined through the requirements of the biochemical cycle, by the solubility of phosphate in the ocean. This is a physical property of a unique chemical compound and as such is not subject to change except in so far as alterations in conditions may influence the activity coefficients of the ions involved. It follows then that the nutrient supplies in sea water, and the oxygen content of the atmosphere have been about as at present for a long time in the past and will remain at much the same level into the future. This argument may then be added to those reviewed by Rubey [17] that the composition of sea water and atmosphere has varied surprisingly little at least since early geologic time.

Speculations about geochemical history are obviously not subject to proof in the ordinary sense. The best one can do is to develop circumstantial evidence which shows that known processes occur which are adequate to produce the effects postulated, and to demonstrate that the quantities of matter and its distribution as presently observed are concordant with the hypotheses. The present hypothesis appears to be supported by evidence of this sort. What is lacking is direct evidence that the postulated processes do, in fact, act in a regulatory way so as to bring the atomic ratios of phosphate nitrate and oxygen in the

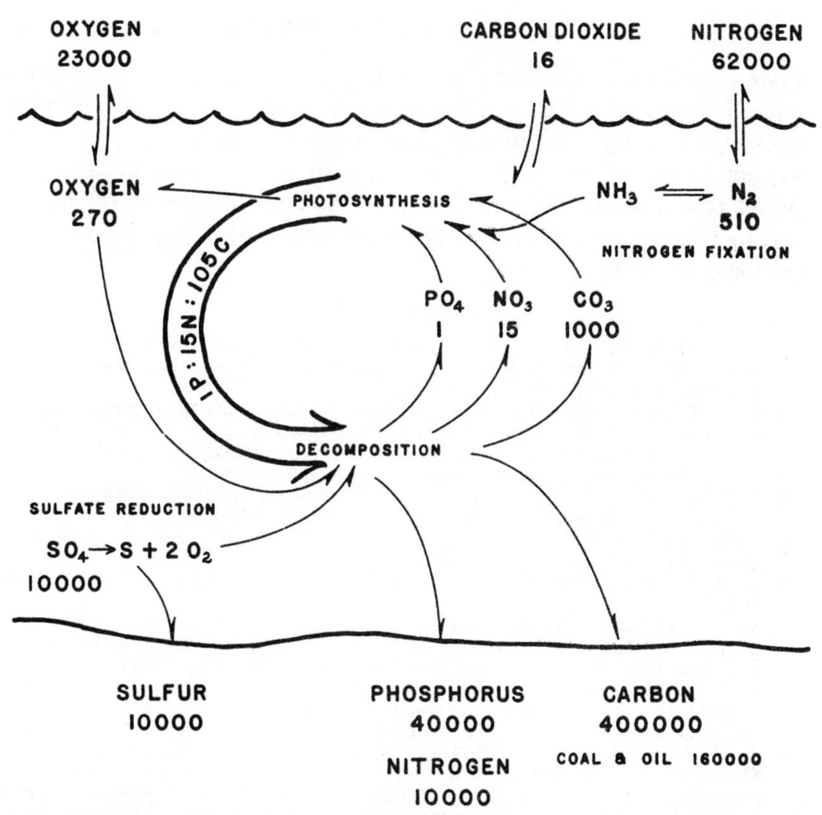

FIG. 3. The Biochemical Cycle. Numbers represent quantities of respective elements present in the atmosphere, the ocean, and the sedimentary rocks, relative to the number of atoms of phosphorus in the ocean.

environment into conformity with the requirements of the biochemical cycle. It is desirable, and would seem quite possible, to set up controlled microcosms in the laboratory, such as von Brand, Rakestraw, and Renn [18] employed to demonstrate the steps in the nitrogen cycle, which would test whether the physiology of the nitrogen and sulfur manipulating bacteria is such as to provide the homeostatic regulation which has been assumed.

REFERENCES

1. A. C. Redfield, "On the proportions of organic derivatives in sea water and their relation to the composition of plankton," James Johnstone Memorial Volume, pp. 176–192, Liverpool (1934).
2. R. H. Fleming, "The composition of plankton and units for reporting populations and production," *Proc. Sixth Pacific Sci. Congress, 3*, pp. 535–540 (1940).
3. F. A. Richards and R. F. Vaccaro, "The Cariaco Trench, an anaerobic basin in the Caribbean Sea," *Deep-Sea Research*, 3, pp. 214–228 (1956).
4. F. A. Richards, "Oxygen in the Ocean," Treatise on Marine Ecology and Paleoecology, J. W. Hedgpeth, Ed., Vol. 1, Chapter 9, pp. 185–238 (1957).
5. L. H. N. Cooper, "On the ratio of nitrogen to phosphorus in the sea," *Jour. Marine Biol. Assn., 22*, pp. 177-182 (1937).
6. H. W. Harvey, "Nitrate in the Sea," *Jour. Marine Biol. Assn., n.s. 14*, pp. 71–88 (1926).
7. A. C. Redfield, "The processes determining the concentration of oxygen, phosphate and other organic derivatives within the depths of the Atlantic Ocean," *Papers in Physical Oceanography and Meteorology, 9(2)* 22 pp. (1942).
8. H. Thomsen, "Nitrate and phosphate contents of Mediterranean Water," Report on the Danish Oceanographical Expeditions 1908–1910 to the Mediterranean and Adjacent Seas. No. 10 (3) pp. 1–11 (1931).
9. H. W. Graham and E. G. Moberg, "Chemical results of the last cruise of the Carnegie," Sci. Results of Cruise VII of the Carnegie during 1928–29 under Command of Captain J. P. Ault, Chemistry I, v plus 58 pp., Washington (1944).
10. A. C. Redfield, "The hydrography of the Gulf of Venezuela," Papers in Marine Biology and Oceanography, Supplement to *Deep-Sea Research*, 3, pp. 115–133 (1955).
11. J. Stoklasa, "Biochemischer Kreislauf des Phosphat-Ions im Boden," *Centrbl. Bakt.* II, 29; 385–419 (1911).
12. G. E. Hutchinson, "Nitrogen in the biogeochemistry of the atmosphere," *Am. Scientist, 32*, pp. 178–195 (1944).
13. J. H. J. Poole, "The evolution of the atmosphere," *Royal Dublin Soc., Sci. Proc., 22* (n.s.), pp. 345–365 (1941).
14. S. A. Waksman, "Principles of soil microbiology," Ed. 2, 894 pp., Baltimore (1932).
15. A. C. Redfield, "The exchange of oxygen across the sea surface," *Jour. Marine Research*, 7, pp. 347–361 (1948).
16. R. S. Dietz, K. O. Emery, and F. P. Shepard, "Phosphorite deposits on the sea floor off southern California," *Bull. Geol. Soc. America, 53*, pp. 815–847 (1942).
17. W. W. Rubey, "Geologic history of sea water," *Bull. Geol. Soc. America, 62*, pp. 1111–1147 (1951).
18. T. von Brand, N. W. Rakestraw, and C. E. Renn, "Further experiments on the decomposition and regeneration of nitrogenous organic matter in sea water," *Biol. Bull.*, 77, 285–296 (1939).
19. K. Rankama and T. G. Sahama, "Geochemistry," 912 pp., Chicago (1950).
20. H. U. Sverdrup, M. W. Johnson, and R. H. Fleming, "The Oceans," 1087 pp., New York (1942).
21. N. W. Rakestraw and V. M. Emmel, "The solubility of nitrogen and argon in sea water," *Jour. Phys. Chem., 42*, pp. 1211–1215 (1938).
22. F. A. Richards and N. Corwin, "Some oceanographic applications of recent determinations of the solubility of oxygen in sea water," *Limnology and Oceanography*, 1, pp. 263–267 (1956).

Plankton

Vol. XCV, No. 882 The American Naturalist May–June, 1961

THE PARADOX OF THE PLANKTON*

G. E. HUTCHINSON

Osborn Zoological Laboratory, New Haven, Connecticut

The problem that I wish to discuss in the present contribution is raised by the very paradoxical situation of the plankton, particularly the phytoplankton, of relatively large bodies of water.

We know from laboratory experiments conducted by many workers over a long period of time (summary in Provasoli and Pintner, 1960) that most members of the phytoplankton are phototrophs, able to reproduce and build up populations in inorganic media containing a source of CO_2, inorganic nitrogen, sulphur, and phosphorus compounds and a considerable number of other elements (Na, K, Mg, Ca, Si, Fe, Mn, B, Cl, Cu, Zn, Mo, Co and V) most of which are required in small concentrations and not all of which are known to be required by all groups. In addition, a number of species are known which require one or more vitamins, namely thiamin, the cobalamines (B_{12} or related compounds), or biotin.

The problem that is presented by the phytoplankton is essentially how it is possible for a number of species to coexist in a relatively isotropic or unstructured environment all competing for the same sorts of materials. The problem is particularly acute because there is adequate evidence from enrichment experiments that natural waters, at least in the summer, present an environment of striking nutrient deficiency, so that competition is likely to be extremely severe.

According to the principle of *competitive exclusion* (Hardin, 1960) known by many names and developed over a long period of time by many investigators (see Rand, 1952; Udvardy, 1959; and Hardin, 1960, for historic reviews), we should expect that one species alone would outcompete all the others so that in a final equilibrium situation the assemblage would reduce to a population of a single species.

The principle of competitive exclusion has recently been under attack from a number of quarters. Since the principle can be deduced mathematically from a relatively simple series of postulates, which with the ordinary postulates of mathematics can be regarded as forming an axiom system, it follows that if the objections to the principle in any cases are valid, some or all the biological axioms introduced are in these cases incorrect. Most objections to the principle appear to imply the belief that equilibrium under a given set of environmental conditions is never in practice obtained. Since the deduction of the principle implies an equilibrium system, if such sys-

*Contribution to a symposium on Modern Aspects of Population Biology. Presented at the meeting of the American Society of Naturalists, cosponsored by the American Society of Zoologists, Ecological Society of America and the Society for the Study of Evolution. American Association for the Advancement of Science, New York, N. Y., December 27, 1960.

tems are rarely if ever approached, the principle though analytically true, is at first sight of little empirical interest.

The mathematical procedure for demonstrating the truth of the principle involves, in the elementary theory, abstraction from time. It does, however, provide in any given case a series of possible integral paths that the populations can follow, one relative to the other, and also paths that they cannot follow under a defined set of conditions. If the conditions change the integral paths change. Mere failure to obtain equilibrium owing to external variation in the environment does not mean that the kinds of competition described mathematically in the theory of competitive exclusion are not occuring continuously in nature.

Twenty years ago in a Naturalists' Symposium, I put (Hutchinson, 1941) forward the idea that the diversity of the phytoplankton was explicable primarily by a permanent failure to achieve equilibrium as the relevant external factors changed. I later pointed out that equilibrium would never be expected in nature whenever organisms had reproductive rates of such a kind that under constant conditions virtually complete competitive replacement of one species by another occurred in a time (t_c), of the same order, as the time (t_e) taken for a significant seasonal change in the environment. Note that in any theory involving continuity, the changes are asymptotic to complete replacement. Thus ideally we may have three classes of cases:

1. $t_c \ll t_e$, competitive exclusion at equilibrium complete before the environment changes significantly.
2. $t_c \simeq t_e$, no equilibrium achieved.
3. $t_c \gg t_e$, competitive exclusion occurring in a changing environment to the full range of which individual competitors would have to be adapted to live alone.

The first case applies to laboratory animals in controlled conditions, and conceivably to fast breeding bacteria under fairly constant conditions in nature. The second case applies to most organisms with a generation time approximately measured in days or weeks, and so may be expected to occur in the plankton and in the case of populations of multivoltine insects. The third case applies to animals with a life span of several years, such as birds and mammals.

Very slow and very fast breeders thus are likely to compete under conditions in which an approach to equilibrium is possible; organisms of intermediate rates of reproduction may not do so. This point of view was made clear in an earlier paper (Hutchinson, 1953), but the distribution of that paper was somewhat limited and it seems desirable to emphasize the matter again briefly.

It is probably no accident that the great proponents of the type of theory involved in competitive exclusion have been laboratory workers on the one hand (for example, Gause, 1934, 1935; Crombie, 1947; and by implication Nicholson, 1933, 1957) and vertebrate field zoologists (for example, Grinnell, 1904; Lack, 1954) on the other. The major critics of this type of approach, notably Andrewartha and Birch (1954), have largely worked with insects in the field, often under conditions considerably disturbed by human activity.

DISTRIBUTION OF SPECIES AND INDIVIDUALS

MacArthur (1957, 1960) has shown that by making certain reasonable assumptions as to the nature of niche diversification in homogeneously diversified[1] biotopes of large extent, the distribution of species at equilibrium follows a law such that the rth rarest species in a population of S_s species and N_s individuals may be expected to be

$$\frac{N_s}{S_s} \sum_{i=1}^{r} \frac{1}{S_s - i + 1} .$$

This distribution, which is conveniently designated as type I, holds remarkably well for birds in homogeneously diverse biotopes (MacArthur, 1957, 1960), for molluscs of the genus Conus (Kohn, 1959, 1960) and for at least one mammal population (J. Armstrong, personal communication). It does not hold for bird faunas in heterogeneously diverse biotopes, nor for diatoms settling on slides (Patrick in MacArthur, 1960) nor for the arthropods of soil (Hairston, 1959). Using Foged's (1954) data for the occurrence of planktonic diatoms in Braendegård Sø on the Danish island of Funen, it is also apparent (figure 1) that the type I distribution does not hold for such assemblages of diatom populations under quite natural conditions either.

MacArthur (1957, 1960) has deduced two other types of distribution (type II and type III) corresponding to different kinds of biological hypotheses. These distributions, unlike type I, do not imply competitive exclusion. So far in nature only type I distributions and a kind of empirical distribution which I shall designate type IV are known. The type IV distribution given by diatoms on slides, in the plankton and in the littoral of Braendegård Sø, as well as by soil arthropods, differs from the type I in having its commonest species commoner and all other species rarer. It could be explained as due to heterogeneous diversity, for if the biotope consisted of patches in each one of which the ratio of species to individuals differed, then the sum of the assemblages gives such a curve. This is essentially the same as Hairston's (1959) idea of a more structured community in the case of soil arthropods than in that of birds. It could probably arise if the environment changed in favoring temporarily a particular species at the expense of other species before equilibrium is achieved. This is, in fact, a sort of temporal analogue to

[1]A biotope is said to be *homogeneously diverse* relative to a group of organisms if the elements of the environmental mosaic relevant to the organism are small compared to the mean range of the organisms. A *heterogeneously diverse biotope* is divided into elements at least some of which are large compared to the ranges of the organisms. An area of woodland is homogeneously diverse relative to most birds, a large tract of stands of woodland in open country is heterogeneously diverse (Hutchinson, 1957, 1959).

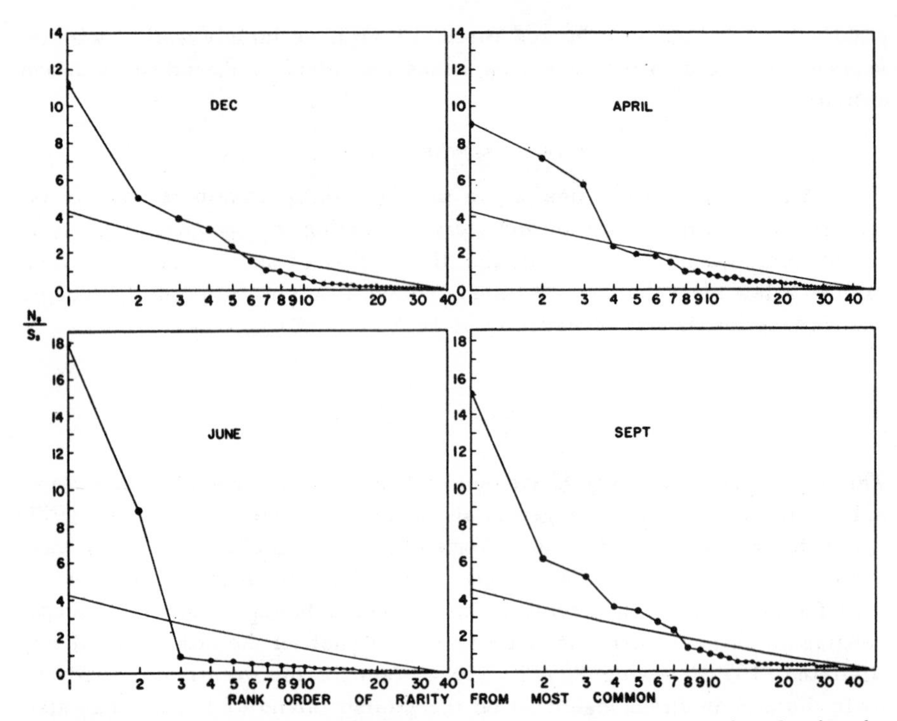

FIGURE 1. Abundance of individual species plotted against rank order for the planktonic diatoms of Braendegård Sø, for the four seasons, from Foged's data, showing type IV distributions. The unmarked line gives the type I distribution for a like number of species and individuals. The unit of population for each species is the ratio of total number of individuals (N_S) to total number of species (S_S).

heterogeneous diversity. Existence of the type IV distribution does not necessarily imply non-equilibrium, but if we assume niches are separated out of the niche-hyperspace with any boundary as probable as any other, we may conclude that either non-equilibrium in time or unexpected diversity in space are likely to underlie this type of distribution.

APPLICATION TO THE PLANKTON

Before proceeding to inquire how far plankton associations are either never in equilibrium in time or approach heterogeneous diversity in space in a rather subtle way, it is desirable to inquire how far ordinary homogeneous niche diversification may be involved. The presence of a light gradient in all epigean waters by day does imply a certain diversification, but in the epilimnia of lakes the chances of any organism remaining permanently in a particular narrow range of intensities is small in turbulent water. By day the stability of the epilimnion may well never be zero, but since what has to be explained is the presence of many species of competitors in a small volume of water, the role of small vertical variations is probably insignificant. A few organisms may be favored by peculiar chemical conditions at the surface film, but again this hardly seems an adequate explanation. The Langmuir spirals in the wind drift might also separate motile from non-motile forms or organisms of different densities to some extent but again the effect is likely to be small and transitory. It is hard to believe that in turbulent open water many physical opportunities for niche diversification exist.

SYMBIOSIS AND COMMENSALISM

The mathematical theory of competition permits the treatment of commensal and symbiotic relations by a simple change in sign of one or both of the competition functions. It can be shown (Gause and Witt, 1935) that under some conditions commensal or symbiotic species can occupy the same niche. There is a little evidence that occasionally water in which one alga has been growing may be stimulatory to another species (Lefèvre, Jacob and Nisbet, 1952; see also Hartman, 1960) though it is far more likely to be inhibitory. Since some phytoplankters require vitamins and others do not, a more generally efficient species, requiring vitamins produced in excess by an otherwise less efficient species not requiring such compounds, can produce a mixed equilibrium population. It is reasonably certain that this type of situation occurs in the phytoplankton. It is interesting to note that many vitamin-requiring algae are small and that the groups characteristically needing them (Euglenophyta, Crytophyceae, Chrysophyceae, and Dinophyceae) tend to be motile. The motility would give such organisms an advantage in meeting rare nutrient molecules, inorganic or organic. This type of advantage can be obtained by non-motile forms only by sinking in a turbulent medium (Munk and Riley, 1952) which is much more dangerous than even random swimming.

ROLE OF PREDATION

It can be shown theoretically, as Dr. MacArthur and I have developed in conversation, that if one of two competing species is limited by a predator, while the other is either not so limited or is fed on by a different predator, co-existence of the two prey species may in some cases be possible. This should permit some diversification of both prey and predator in a homogeneous habit.

RESULTS OF NON-EQUILIBRIUM CONDITIONS

The possibility of synergistic phenomena on the one hand and of specific predation on the other would probably permit the development of a somewhat diversified equilibrium plankton even in an environment that was essentially boundaryless and isotropic. It may, however, be doubted that such phenomena would ever permit assemblages of the order of magnitude of tens of species to co-occur. At least in homogeneous water in the open ocean there would seem to be no other alternative to a non-equilibrium, or as MacArthur (1960) would term it, an opportunistic community.

The great difficulty inherent in the opportunistic hypothesis is that since, if many species are present in a really variable environment which is controlling their competition, chance extinction is likely to be an important aspect of the process.[2] That this is not an important aspect of the problem, at least in some cases, is shown by the continual presence of certain dominant species of planktonic diatoms as microfossils in sediments laid down under fairly uniform conditions over periods of centuries or millenia. This is, for instance, clear from Patrick's (1943) study of the diatoms of Linsley Pond, in which locality *Stephanodiscus astrea*, *Melosira ambigua* and certain species of Cyclotella must have co-occurred commonly for long periods of time. It is always possible to suppose that the persistent species were continually reintroduced from outside whenever they became extinct locally, but this does not seem a reasonable explanation of the observed regularity.

IS THE PHYTOPLANKTON A VALID CONCEPT?

In view of the paradoxical nature of the phytoplankton, perhaps it is justifiable to inquire to what extent the concept itself has validity. In the ocean it is reasonably certain that the community is a self-perpetuating one, but in lakes it has long been regarded as largely an evolutionary derivative of the littoral benthos (for example, Wesenberg-Lund, 1908, pp. 323-325) and in recent years much evidence has accumulated to suggest that the derivation in some cases is not an evolutionary process in the ordinary sense of the word, but a process occurring annually, some individuals of a benthic flora moving at times into plankton. The remarkable work of Lund (1954, 1955) on Melosira indicates that the planktonic species of this genus become benthic, though probably in a non-reproductive condition, when turbulence is inadequate to keep them afloat. Brook (1959) believes that some of the supposed planktonic varieties of littoral-benthic desmids are non-genetic modifications exhibited by populations annually derived from the littoral. If most of the phytoplankton consisted of species with well-defined, if somewhat restricted, benthic littoral niches, from which at times large cultures in the open water were developed but perhaps left no descendants, much of our paradox would disappear. In the sea we should still apparently have to rely on synergism, predation and opportunism or failure to achieve equilibrium, but in fresh waters we might get still more diversity from transitory invasions of species which in the benthos probably occupy a heterogeneously diverse biotope like the soil fauna studied by Hairston (1959).

[2]The chance of extinction is always finite even in the absence of competition, but for the kind of population under consideration the arguments adduced, for instance, by Cole (1960) appear to the writer to be unrealistic. In a lake of area 1 km^2 or 10^6 m^2, in a layer of water only one meter deep, any organism present at a concentration of one individual per litre, which would be almost undetectibly rare to the planktologist using ordinary methods, would have a population N_o of 10^9 individuals. If the individuals divided and the two fission products had equal chances of death or reproduction, so that in the expected case the population remained stable, the probability of random extinction (Skellam, 1955) is given by $p_e = [t/(1+t)]^{N_o}$ where t is measured in generations. For large values of N_o and t we may approximate by $t = -N_o/\ln p_e$. In the lake in question p_e would reach a value of 0.01 in 2.2×10^8 generations which for most phytoplankters would be a period of over a million years. Less than half a dozen lakes are as old as this, and all these are vastly larger than the hypothetical lake of area 1 km^2.

The available data appear to indicate that in a given lake district there is no correlation between the area of a lake and the number of species comprising its phytoplankton. This is apparent from Järnefelt's (1956) monumental study of the lakes of Finland, and also from Ruttner's (1952) fifteen Indonesian lakes. In the latter case, the correlation coefficient of the logarithm of the numbers of phytoplankton species on the logarithm of the area (the appropriate quantities to use in such a case), is -0.019, obviously not significantly different from zero.

It is obvious that something is happening in such cases that is quite different from the phenomena of species distribution of terrestrial animals on small islands, so illuminatingly discussed by Dr. E. O. Wilson in another contribution to this symposium. At first sight the apparent independence indicated in the limnological data also may appear not to be in accord with the position taken in the present contribution. If, however, we may suppose that the influence of the littoral on the species composition decreases as the area of the lake increases, while the diversity of the littoral flora that might appear in the plankton increases as the length of the littoral, and so its chances of diversification, increases, then we might expect much less effect of area than would initially appear reasonable. The lack of an observed relationship is, therefore, not at all inconsistent with the point of view here developed.

CONCLUSION

Apart from providing a few thoughts on what is to me a fascinating, if somewhat specialized subject, my main purpose has been to show how a certain theory, namely, that of competitive exclusion, can be used to examine a situation where its main conclusions seem to be empirically false. Just because the theory is analytically true and in a certain sense tautological, we can trust it in the work of trying to find out what has happened to cause its empirical falsification. It is, of course, possible that some people with greater insight might have seen further into the problem of the plankton without the theory than I have with it, but for the moment I am content that its use has demonstrated possible ways of looking at the problem and, I hope, of presenting that problem to you.

LITERATURE CITED

Andrewartha, H. G., and L. C. Birch, 1954, The distribution and abundance of animals. XV. 782 pp. Univ. of Chicago Press, Chicago, Ill.

Brook, A. H., 1959, The status of desmids in the plankton and the determination of phytoplankton quotients. J. Ecol. 47: 429–445.

Cole, L. C., 1960, Competitive exclusion. Science 132: 348.

Crombie, A. C., 1947, Interspecific competition. J. Animal Ecol. 16: 44-73.

Feller, W., 1939, Die Grundlagen der Volterraschen Theorie des Kampfes von Dasein in Wahrscheirlichkeitstheoretischer Behandlung. Acta Biotheoret. 5: 11–40.

Foged, N., 1954, On the diatom flora of some Funen Lakes. Folia Limnol. Scand. 5: 1–75.

Gause, G. F., 1934, The struggle for existence. IX. 163 pp. Williams and Wilkins, Baltimore, Md.

1935, Vérifications expérimentales de la théorie mathematique de la lutte pour la vie. Actual. scient. indust. 277: 1-62.

Gause, G. F., and A. A. Witt, 1935, Behavior of mixed populations and the problem of natural selections. Amer. Nat. 69: 596-609.

Grinnell, J., 1904, The origin and distribution of the chestnut-backed Chickadee. Auk 21: 364-382.

Hairston, N. G., 1959, Species abundance and community organization. Ecology 40: 404-416.

Hardin, G., 1960, The competitive exclusion principle. Science 131: 1292-1298.

Hartman, R. T., 1960, Algae and metabolites of natural waters. *In* The Pymatuning symposia in ecology: the ecology of algae. Pymatuning Laboratory of Field Biology, Univ. Pittsburgh Special Publ. No. 2: 38-55.

Hutchinson, G. E., 1941, Ecological aspects of succession in natural populations. Amer. Nat. 75: 406-418.

1953, The concept of pattern in ecology. Proc. Acad. Nat. Sci. Phila. 105: 1-12.

1957, Concluding remarks. Cold Spring Harbor Symp. Quant. Biol. 22: 415-427.

1959, Il concetto moderno di nicchia ecologica. Mem. Ist. Ital. Idrobiol. 11: 9-22.

Järnefelt, H., 1956, Zur Limnologie einiger Gewässer Finnlands. XVI. Mit besonderer Berücksichtigung des Planktons. Ann. Zool. Soc. "Vancimo" 17(1): 1-201.

Kohn, A. J., 1959, The ecology of *Conus* in Hawaii. Ecol. Monogr. 29: 47-90.

1960, Ecological notes on *Conus* (Mollusca: Gastropoda) in the Trincomalee region of Ceylon. Ann. Mag. Nat. Hist. 13(2): 309-320.

Lack, D., 1954, The natural regulation of animal numbers. VIII. 343 pp. Clarendon Press, Oxford, England.

Lefèvre, M., H. Jakob and M. Nisbet, 1952, Auto- et heteroantagonisme chez les algues d'eau dounce. Ann. Stat. Centr. Hydrobiol. Appl. 4: 5-197.

Lund, J. W. G., 1954, The seasonal cycle of the plankton diatom, *Melosira italica* (Ehr.) Kütz. subsp. *subarctica* O. Müll. J. Ecol. 42: 151-179.

1955, Further observations on the seasonal cycle of *Melosira italica* (Ehr.) Kütz. subsp. *subarctica* O. Müll. J. Ecol. 43: 90-102.

MacArthur, R. H., 1957, On the relative abundance of bird species. Proc. Natl. Acad. Sci. 45: 293-295.

1960, On the relative abundance of species. Amer. Nat. 94: 25-36.

Munk, W. H., and G. A. Riley, 1952, Absorption of nutrients by aquatic plants. J. Mar. Research 11: 215-240.

Nicholson, A. J., 1933, The balance of animal populations. J. Animal Ecol. 2(suppl.): 132-178.

1957, The self-adjustment of populations to change. Cold Spring Harbor Symp. Quant. Biol. 22: 153-173.

Patrick, R., 1943, The diatoms of Linsley Pond, Connecticut. Proc. Acad. Nat. Sci. Phila. 95: 53–110.

Provasoli, L., and I. J. Pintner, 1960, Artificial media for fresh-water algae: problems and suggestions. *In* The Pymatuning symposia in ecology: the ecology of algae. Pymatuning Laboratory of Field Biology, Univ. Pittsburgh Special Publ. No. 2: 84–96.

Rand, A. L., 1952, Secondary sexual characters and ecological competition. Fieldiana Zool. 34: 65–70.

Ruttner, F., 1952, Plankton studien der Deutschen Limnologischen Sunda-Expedition. Arch. Hydrobiol. Suppl. 21: 1-274.

Skellam, J. G., 1955, The mathematical approach to population dynamics. *In* The numbers of man and animals, ed. by J. G. Cragg and N. W. Pirie. Pp. 31–46. Oliver and Boyd (for the Institute of Biology), Edinburgh, Scotland.

Udvardy, M. F. D., 1959, Notes on the ecological concepts of habitat, biotope and niche. Ecology 40: 725–728.

Wesenberg-Lund, C., 1908, Plankton investigations of the Danish lakes. General part: the Baltic freshwater plankton, its origin and variation. 389 pp. Gyldendalske Boghandel, Copenhagen, Denmark.

Species Diversity in Net Phytoplankton of Raritan Bay[1]

Bernard C. Patten

Virginia Institute of Marine Science
College of William and Mary
Gloucester Point, Virginia

ABSTRACT

The annual diversity cycle in Raritan Bay net phytoplankton is described, using entropy-related diversity indices. A theory of diversity change during succession is developed. Mean diversity levels in the estuary increased downbay in association with diminishing pollution, and the spatial pattern was strikingly related to general patterns of water mass circulation.

Introduction. An *ecological community* may be defined simply as any aggregation of mixed taxa, and particular attention to its composition may raise the question: how are the individuals distributed amongst the species? This is the *diversity* problem. The present study is concerned with the annual cycle of species diversity in the net planktonic flora of Raritan Bay.

The numbers of species (m) and the total individuals (N) determine the range of diversity available to a community. The extremes are (i) all individuals belong to one species ($m = 1$); (ii) every individual belongs to a different species ($m = N$). Between these lies a number of possibilities, determined by the values of m and N.

A number of approaches to the diversity problem have been proposed, for example, by Gleason (1922), Fisher *et al.* (1943), Preston (1948), Goodall (1952), Williams (1952), and Koch (1957). Margalef (1956, 1957, 1958a, 1958b) has advocated using an entropy measure, and this is the approach taken here. The following derivation of such a diversity index is after Branson (1953) and Brillouin (1956).

A mixed species population in a biotope space consists of $n_1, n_2, \ldots, n_i, \ldots, n_m$ individuals of m different types, the total number being

[1] From a dissertation presented to the Graduate Faculty of Rutgers University in candidacy for the degree of Doctor of Philosophy, 1959.

$$N = \sum_{i=1}^{m} n_i . \tag{1}$$

The probability of selecting in sampling a species of i-th type is

$$p_i = n_i/N , \tag{2}$$

where

$$\sum_{i=1}^{m} p_i = 1 .$$

The permutations of N objects having n_1 elements alike of one kind, n_2 elements alike of a second kind, and so on for m kinds, are

$$NP\, n_1, n_2, \ldots, n_m = \frac{N!}{n_1!\, n_2! \ldots n_m!} = \frac{N!}{\prod_{i=1}^{m} n_i!} . \tag{3}$$

If each permutation is equally probable, then the Boltzmann equation from statistical mechanics gives the entropy of the aggregation as

$$H = k \log P , \tag{4}$$

or, from (3),

$$H = k\,[\log N! - \sum_{i=1}^{m} \log n_i!] . \tag{5}$$

Assuming a reasonably large sample, the logarithms of the Γ functions may be approximated by Stirling's formula:

$$\log \Gamma(z+1) = \log z! = z\,(\log z - 1) . \tag{6}$$

Eq. (5) then becomes

$$\begin{aligned} H &= k\,[N(\log N - 1) - \sum_{i=1}^{m} n_i (\log n_i - 1)] \\ &= k\,[N \log N - N - \sum_{i=1}^{m} n_i \log n_i + \sum_{i=1}^{m} n_i] \\ &= -k \left[\sum_{i=1}^{m} n_i \log n_i - N \log N \right] \\ &= -k \left[\sum_{i=1}^{m} n_i \log \frac{n_i}{N} \right] , \end{aligned} \tag{7}$$

and the mean entropy per individual is

$$\overline{H} = -k\left[\sum_{i=1}^{m} \frac{n_i}{N} \log \frac{n_i}{N}\right], \tag{8}$$

or, from (2)

$$\overline{H} = -k \sum_{i=1}^{m} p_i \log p_i \tag{9}$$

(Shannon, 1948; Wiener, 1948). If base 2 logarithms are used, the entropy is expressed as binary digits (*bits*), one bit representing the information required to make a choice between two equally probable alternatives.

The variables H and $\overline{H}$ possess a number of properties which make them reasonable measures of species diversity. If all the p_i's except one are zero ($m = 1$), the outcome of sampling is certain and H vanishes ($H_{min} = 0$). At the other extreme ($m = N$), where uncertainty is greatest, H is maximal ($H_{max} = \log N!$). Least diversity when $m > 1$ corresponds to the situation where all individuals except ($m - 1$) belong to a single species, and the remainder are distributed one each to the other species:

$$H_{min} = k\{\log N! - \log[N - (m-1)]!\}. \tag{10}$$

Any change toward equalization of the p_i's increases H, resulting in maximum diversity whenever $n_i \geq 1$ when the individuals are equally apportioned among the species, thereby minimizing the second term of

$$H_{max} = k[\log N! - m \log (N/m)!]. \tag{11}$$

The position of H in the range between H_{max} and H_{min} is denoted by the *redundancy*, which we define here to be

$$R = \frac{H_{max} - H}{H_{max} - H_{min}}. \tag{12}$$

The following table illustrates some of the properties of $\overline{H}$ and R which make them satisfactory extensive and intensive (respectively) expressions for diversity.

Communities
($N = 6$)

Species	*A*	*B*	*C*	*D*	*E*	*F*	*G*	*H*	*I*	*J*	*K*
s_1	1	2	2	3	2	3	4	3	4	5	6
s_2	1	1	2	1	2	2	1	3	2	1	–
s_3	1	1	1	1	2	1	1	–	–	–	–
s_4	1	1	1	1	–	–	–	–	–	–	–
s_5	1	1	–	–	–	–	–	–	–	–	–
s_6	1	–	–	–	–	–	–	–	–	–	–
$\overline{H}$ (bits)....	2.58	2.25	1.93	1.79	1.61	1.47	1.25	1.00	0.92	0.65	0.00
R	0.00	0.13	0.25	0.30	0.38	0.43	0.52	0.61	0.64	0.75	1.00

Note how $\overline{H}$ vanishes as the probability of selecting a particular species becomes a certainty, and increases the more uncertain the choice becomes. R, on the other hand, is maximal when no choice exists and disappears when there is most choice. The fact that one bit constitutes a binary choice between two equally probable alternatives is illustrated by community H.

Diversity indices permit large amounts of information about numbers and kinds of organisms to be succinctly, if implicitly, summarized. More than this, they allow direct study of communities *at* the community level. Detailed analysis at the species level of cumbersome lists which summarize the same information more explicitly is obviated. If details are required, however, they can be evinced readily by traditional methods. It is the purpose of this paper to illustrate how H and R may be employed as *bona fide* ecological variables to describe changes in community composition and to denote successional status precisely and unambiguously.

The study is based on net phytoplankton collections made in Raritan Bay from July 1957 through September 1958. Since the emphasis was to be more methodological than descriptive, the total phytoplankton were partitioned for convenience in a repeatable, if arbitrary, manner to obtain a study unit which was discrete, if also arbitrary. This unit, the *net phytoplankton*, is defined to include all recognizable forms which remained in a preserved sample.

Methods. Six sampling stations were established in Raritan Bay (Fig. 1). Hydrographic observations included temperature, salinity, turbidity and dis-

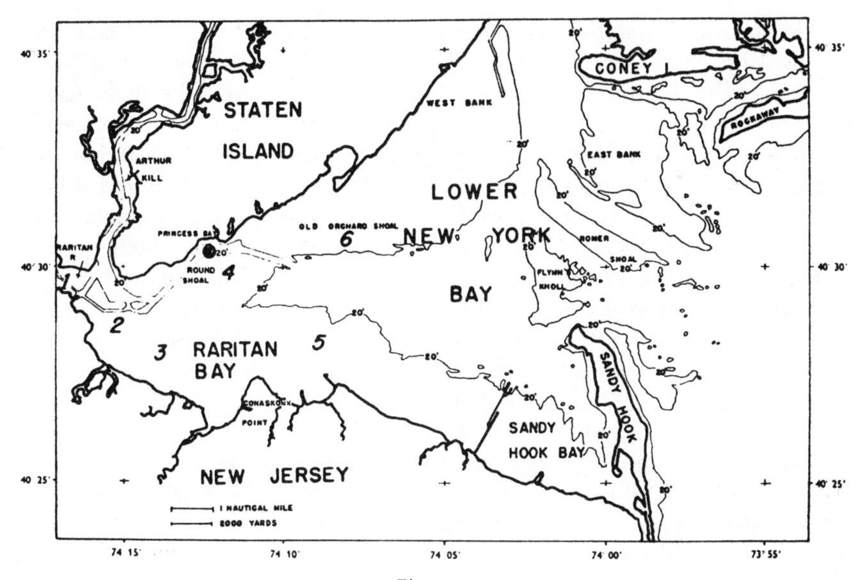

Figure 1.

solved oxygen. Nitrate and orthophosphate were determined. The net phytoplankton were sampled by oblique hauls of 90 seconds duration with a Clarke-Bumpus sampler to which a no. 20 net was attached. The samples were preserved in approximately 3 °/o unneutralized formalin. Physical units (chains, colonies or individual cells) were counted in a Sedgwick-Rafter chamber and results tabulated as *units*, n_i, of each species per liter. Diversity indices were computed from equations (5), (10), (11), and (12) with an IBM Type 650 Data Processing Machine at the Rutgers University Computation Center.

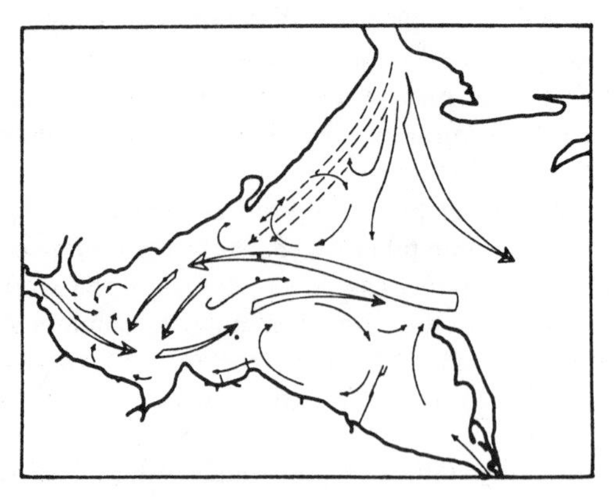

Figure 2.

The General Circulation. Water mass movements in the estuary have been discussed by Ayers *et al.* (1949), and hydrographic and nutrient details have been given by Jeffries (1959, 1961) and Patten (1959). The main source of fresh water is the Raritan River (highly polluted before the opening of a trunk sewer in 1959), exchange between the Hudson River and Raritan Bay being impeded by a sluggish eddy at Old Orchard Shoal. Sea water, diluted by discharge of the Hudson into Lower Bay, enters the Raritan between Sandy Hook and Romer Shoal. Moving westward along the Staten Island shore, it thrusts southwestwardly in the vicinity of Princess Bay to a point just west of Conaskonk Point. This tongue is presumed to accelerate the seaward movement of fresher water along the south shore while damming back low salinity water accumulated at the head of the Bay. These general relationships are diagrammed in Fig. 2.

The Flora. During this study 133 taxa were identified in the net phytoplankton. (A list with record of monthly occurrence is available from the author upon request.) In Table I, the numbers of taxa (m) recorded on each sampling date are given. Total units (N) per liter are listed in Table II. Table I indicates (i) generally fewer species downbay, presumably due to mixtures of

TABLE I. Record of the number of taxa (m) recorded in the net phytoplankton at Sts. 1–6.

Date	Stations 1	2	3	4	5	6
1957						
Jul 5	–	3	7	–	–	–
11	14	13	9	6	8	6
18	11	12	7	8	–	–
26	5	–	9	6	9	9
Aug 1	15	–	9	–	10	13
8	11	11	18	15	17	18
15	18	17	15	16	–	–
21	10	12	13	16	16	9
29	18	11	8	16	11	11
Sep 14	14	11	5	5	7	6
24	20	17	10	6	9	13
Oct 8	17	14	10	4	11	13
29	8	7	5	4	4	15
Nov 11	18	–	10	8	9	9
26	–	–	5	–	5	–
Dec 17	–	–	14	–	15	–
1958						
Jan 20	18	25	24	21	22	25
Feb 6	24	23	23	21	20	19
Mar 1	27	24	24	13	18	18
11	21	21	17	17	16	18
25	23	22	18	11	19	18
Apr 10	26	20	20	16	20	20
24	25	20	12	–	14	11
May 10	15	16	8	14	12	10
30	23	22	16	9	12	14
Jun 23	18	14	13	14	17	15
Jul 7	10	8	10	11	13	15
24	10	11	9	12	12	10
Aug 11	9	10	8	8	7	–
21	6	9	7	10	11	16
Sep 11	22	19	20	12	–	–

brackish and salt water populations in the headwaters, (ii) maximum values of *m* during early spring, with a secondary pulse in late summer, and (iii) smallest species numbers in autumn following the summer bloom, and in early summer following the spring bloom (Table II).

The net flora may be grouped into four associations based on seasonal distribution patterns:

1) *Constants*, collected the year round. The three major species were *Coscinodiscus asteromphalus* Ehrenberg, *C. subtilis* Ehrenberg and *Lithodesmium undulatum* Ehrenberg, all of which have similar habitat preferences: temperate or south temperate, and neritic, although *C. subtilis* is a low-salinity form.

TABLE II. Record of the total number of counting units (N) per liter of all net phytoplankton taxa at sts. 1–6.

Date	Stations 1	2	3	4	5	6
1957						
Jul 5......	–	3,771	4,494	–	–	–
11......	5,660	9,501	7,057	4,458	3,531	5,967
18......	27,730	23,104	19,646	6,690	–	–
26......	56,511	–	21,134	39,180	12,588	5,727
Aug 1......	170,922	–	16,382	–	19,615	3,877
8......	13,108	2,362	7,636	6,328	73,846	7,563
15......	10,146	4,780	2,933	9,371	–	–
21......	40,028	73,674	76,118	43,738	118,163	12,571
29......	22,617	42,984	31,129	2,608	29,382	255,718
Sep 14......	30,830	10,729	3,551	3,369	949	5,569
24......	18,817	15,510	40,107	15,920	10,592	15,120
Oct 8......	16,473	25,875	25,292	20,847	16,611	5,250
29......	3,208	1,740	621	335	175	1,595
Nov 11......	3,875	–	1,139	239	1,098	125
26......	–	–	240	–	757	–
Dec 17......	–	–	495	–	788	–
1958						
Jan 20......	5,986	12,101	11,604	16,245	7,649	17,998
Feb 6......	7,289	8,631	17,225	13,055	22,101	10,387
Mar 1......	2,950	6,427	11,570	27,064	10,152	16,243
11......	94,290	193,764	236,128	296,942	331,594	288,843
25......	46,041	61,161	105,141	326,162	303,609	360,260
Apr 10......	86,115	125,129	84,056	141,710	133,855	134,032
24......	117,046	1,289,403	2,377,179	–	1,392,188	1,360,673
May 10......	603	466	167	228	360	298
30......	3,050	1,530	1,586	6,347	6,046	6,018
Jun 23......	15,903	2,718	15,240	6,303	18,490	3,454
Jul 7......	33,408	41,912	30,174	114,887	100,903	93,395
24......	21,004	36,401	25,443	50,999	44,209	80,946
Aug 11......	42,064	73,484	88,325	50,501	56,514	–
21......	1,548	4,999	7,995	7,200	2,785	23,499
Sep 11......	11,815	21,118	3,851	5,649	–	–

2) *Vernal-serotinal species*, present bimodally in spring and late summer. The four most important species were *Skeletonema costatum* (Greville) Cleve, *Thalassiosira gravida* Cleve, *Chaetoceros decipiens* Cleve, and *Gyrosigma acuminatum* (Kützing) Cleve, which range in preferences from fresh water (*Gyrosigma*) to oceanic (*Chaetoceros*); most are species of the northern seas, except *Gyrosigma*, a temperate form.

3) *Serotinal species*, collected in late summer. The most abundant were *Nannochloris atomus* Butcher, *Prorocentrum micans* Ehrenberg, *Peridinium trochoideum* (Stein) Lemmermann, *P. breve* Paulsen, and *P. divaricatum* Meunier, which, with the exception of *Nannochloris*, are dinoflagellates.

These are neritic or estuarine forms with preference for low salinities, excepting *Prorocentrum*, which tolerates oceanic salinity.

4) *Hiemal species*, dominant in winter. The winter flora is diatomaceous, the most important species being *Nitzschia seriata* Cleve, *Leptocylindricus danicus* Cleve, *Rhizosolenia setigera* Brightwell, *R. imbricata* Cleve, *R. alata* Brightwell, *Asterionella japonica* Cleve, *Thalassionema nitzschioides* Grunow, *Guinardia flaccida* (Castracane) Peragallo, *Melosira sulcata* (Ehrenberg) Kützing, and *Actinoptychus undulatus* (Bailey) Ralfs, which are neritic or oceanic and typical of northern areas, excepting *Asterionella*, a south temperate form.

The obvious seasonal gaps in the foregoing classification were occupied by two nannoplanktonic associations: 1) *Aestival species* of early summer, dominated by the small dinoflagellate *Massartia rotundata* (Lohmann) Schiller; and 2) *Autumnal species*, also dominated by a dinoflagellate, the colorless *Oxyrrhis marina* Dujardin.

The two most significant species in the estuary were *Skeletonema costatum* and *Nannochloris atomus*. The former dominated the spring bloom, achieving a maximum of 17.3×10^6 units/L on March 11 at st. 5. Conover (1956) has recorded 35×10^6 cells/L for Long Island Sound. Since a *Skeletonema* unit, as used here, probably averages 6–10 cells or more, it seems that the Raritan maximum is at least 3–5 times greater than that of Long Island Sound. With respect to sheer numbers of cells, and possibly also with respect to contribution to total productivity of the estuary, the chlorophyte *Nannochloris* is probably the most important single phytoplanktont. The highest concentration recorded was 57.3×10^7 units/L (unfiltered, fresh sample) on August 11, 1958, more than half of Ryther's (1954) record of 10^9 "small forms" (*Nannochloris* and *Stichococcus*) in nearby Moriches Bay.

Summarizing the annual succession of communities, diatoms dominated the cold-water floras while phytoflagellates and *Nannochloris* were dominant during the warmer seasons. Hence, from termination of the spring diatom maximum until midautumn, the most important forms were nannoplanktonic flagellates and *Nannochloris*. From autumn until termination of the spring flowering, diatoms prevailed.

Annual Diversity Cycle. In describing the annual cycle of diversity, both H and R will be reported together in a decimal notation. In 1133.41, for instance, the digits to the left of the point (1133.) denote H in bits/L while those to the right (.41) represent the redundancy. Data from the six Raritan sampling stations are reported in this manner in Table III, which shows that, during the study period, H varied from 87 bits/L in November to 487,220 in April while R ranged from 0.99 in October to 0.17 in May. The annual cycle may be divided into four periods.

TABLE III. Diversity and redundancy of net phytoplankton at sts. 1–6. The digits to the left of the decimal point represent H in bits/L; those to the right denote R.

Date	Stations					
1957	1	2	3	4	5	6
Jul 5......	–	309.95	262.98	–	–	–
11......	4,952.78	1,597.96	3,004.87	2,455.79	904.92	869.95
18......	2,550.97	1,569.98	3,941.93	3,991.80	–	–
26......	4,057.97	–	3,097.96	8,461.92	1,796.96	8,944.51
Aug 1......	30,379.95	–	5,220.90	–	10,389.84	7,959.45
8......	9,202.80	4,965.39	13,530.52	12,346.50	159,045.47	20,890.34
15......	10,345.76	10,055.49	8,481.26	24,156.46	–	–
21......	41,712.69	131,544.50	138,640.51	58,601.67	108,023.77	12,183.70
29......	46,251.51	34,119.77	12,592.87	5,345.49	19,118.81	51,181.94
Sep 14......	4,857.96	2,322.94	2,082.75	2,798.65	656.77	6,774.53
24......	5,190.94	6,082.91	3,102.98	1,960.95	2,068.94	5,862.90
Oct 8......	3,051.96	2,642.98	2,823.97	505.99	6,334.89	8,293.58
29......	1,041.90	732.86	203.88	287.59	101.75	2,788.56
Nov 11......	1,420.92	–	196.97	151.85	279.94	128.77
26......	–	–	87.89	–	118.95	–
Dec 17......	–	–	1,133.41	–	1,686.46	–
1958						
Jan 20......	5,636.78	18,169.68	8,754.84	15,376.79	7,438.79	21,435.75
Feb 6......	17,918.47	19,515.50	31,628.60	27,662.52	59,140.38	26,701.40
Mar 1......	7,434.48	10,214.66	19,284.64	13,414.87	9,136.79	11,242.84
11......	57,335.86	82,941.90	68,141.93	90,917.93	76,314.94	145,848.88
25......	37,239.82	28,478.90	37,266.92	65,394.94	116,141.91	157,551.90
Apr 10......	53,028.87	52,315.90	32,290.91	51,673.91	66,254.89	63,805.89
24......	45,099.92	350,744.94	487,220.94	–	448,701.92	268,233.94
May 10......	1,171.52	1,531.17	216.61	497.45	715.46	328.71
30......	5,827.59	4,321.37	3,015.53	2,993.86	4,483.80	8,593.63
Jun 23......	20,407.69	4,273.59	18,901.67	12,695.47	26,686.65	4,726.66
Jul 7......	18,640.83	19,162.85	11,350.89	23,074.94	23,925.94	26,721.93
24......	11,439.84	9,232.93	4,468.95	9,708.95	14,802.91	5,805.98
Aug 11......	10,065.93	8,912.96	8,731.97	5,557.96	6,658.96	–
21......	1,865.54	5,229.67	13,365.40	14,099.41	3,351.46	28,625.70
Sep 11......	30,424.42	47,585.47	4,954.71	13,348.34	–	–

July 5–August 21, 1957 (Serotinal). On July 5, only sts. 2 and 3 were sampled, with low diversities and high redundancies prevailing: 309.95 and 262.98, respectively. *Nannochloris atomus* was the only significant species. Subsequently, by August 21, diversity at both stations had increased steadily to peaks of 131,544.50 at st. 2 and of 138,640.51 at st. 3, associated with increased importance of several other species (*Skeletonema costatum*, *Coscinodiscus asteromphalus*, and *Peridinium trochoideum*). At st. 1 a similar increase to 41,712.69 occurred (and, on August 29, 46,251.51). The increase at st. 4 was virtually exponential, from 2455.79 on July 11 to 58,601.67 on August 21. At st. 5 a steep increase from 904.92 to 108,023.77 occurred during this

period, reaching a maximum of 159,045.47 on August 8. At st. 6, alternate increases and decreases prevailed, with a net increase during the period from 869.95 to 51,181.94. Thus, highest diversity levels during this period were encountered along the south shore of the estuary at sts. 2, 3 and 5, the lowest at sts. 1, 4 and 6.

Unlike H, R did not show any uniform trends during this period, except at st. 1, where the values decreased on each successive sampling date. In spite of the uneven rates of change, the general pattern for R was similar at all stations: decreases to minima followed by increases (Table III). The curves in Table III are out of phase along the time axis, minimum values being obtained in early August downbay, in mid-August at the midbay stations, and in late August at st. 1. Thus, the tendency to realize higher proportions of the available diversity developed first in the lower reaches of the estuary, where oceanic influences were greatest, and then spread upbay. The lowest redundancy obtained at each of sts. 1–6 was, respectively, .51, .39, .26, .36, .47, .34.

August 21–November 26, 1957 (Autumnal). This period was marked by a general diversity decrease throughout the Bay, although small isolated increases occurred at various times and places. St. 4 seems to represent the general trend; in the week between August 21 and 29, diversity here decreased from 58,601.67 to 5345.49, and the decline continued steadily to its low for the year, 151.85 on November 11. The lowest values of H (bits/L) recorded for sts. 1–6 were 1041, 732, 87, 151, 101, and 128, indicating that the stations upbay maintained a considerably higher diversity level than those downbay. These generally low diversities of the autumnal period were associated with dominance by one form, *Nannochloris*. Further downbay (especially at st. 6), several other species, *e.g.*, *Coscinodiscus asteromphalus* and *Chaetoceros decipiens*, maintained some degree of importance.

The situation at st. 5 was slightly different from the others. In the three weeks between August 21 and September 14, diversity declined precipitously from 108,023.77 to 656.77 (well below the other stations), and in the next month it rose again tenfold to 6334.89 (October 8) before falling to its lowest value of 101.75 on October 29, where it remained, approximately, throughout November. The same pattern of abrupt decline to a low on September 14, followed by subsequent increase before the final decline, was also expressed, though less markedly, at south-shore sts. 2 and 3, and very slightly at st. 6. These patterns presumably reflected local environmental conditions in the estuary at the time. Less than a week before September 14, high northeasterly winds and heavy seas produced great turbidity throughout the Bay on September 9. Examination of samples collected on this date revealed much silt and many fragments of lysed cells and diatom frustules. Even *Nannochloris* was reduced to only 842 units/L at st. 5 compared to 30,367 at the relatively protected st. 1. Even more sensitive to this environmental disturbance than H was H_{max}. Although H remained fairly unaltered at sts. 2, 3, 4, and 6, H_{max} was depressed throughout the Bay on September 14 and increased sharply in the weeks following.

The general trend of H_{max} during the autumnal period was downward. The behavior of H_{min} was more erratic, with a net decline at some stations and an increase at others. Counting units (Table II) remained quite high during September and early October, while number of species (Table I) was generally low. The net result of decreasing H_{max} while leaving H_{min} relatively unaffected was to diminish the range of available diversity to a very narrow spread in November. This would be expected, *a priori*, to depress the values of R. But Table III indicates very high redundancies. The lowest values recorded after decline of the summer bloom were, for sts. 1–6, .90, .86, .75, .59, .75, and .53, markedly higher upbay and along the south shore. The highest redundancies obtained were .96, .98, .98, .99, .94, and .90, higher at the head than downbay. Thus, in spite of the small spreads between the limits H_{max} and H_{min}, the realized diversity was so low during this period that it constituted a very small proportion of the total diversity available.

NOVEMBER 26, 1957–MAY 10, 1958 (HIEMAL). This extended period was characterized at all stations by a gradual rise in diversity accompanied by an increase in the spread between H_{max} and H_{min}. These trends were least pronounced at sts. 1 and 2, but greater and fairly similar for the remaining stations downbay. The long diversity ascent (Table III) culminated in maxima at all stations on April 24, except st. 4 which was not sampled then, and st. 1 which had a higher diversity (53,028.87) on April 10. The diversity levels recorded on April 24 were 45,099.92, 350,744.94, 487,220.94, – – –, 448,701.92, and 268,233.94 at sts. 1–3, 5, 6. The corresponding number of species diminished downbay (25, 20, 12, —, 14, 11), and the number of units per liter was greatest at st. 3 and least at st. 1 (117,046; 1,289,403; 2,377,179; – – –; 1,392,188; 1,360,673). Redundancies were high throughout the Bay, as indicated.

The most important species during the spring flowering were *Skeletonema costatum*, *Leptocylindricus danicus*, *Thalassiosira gravida*, *Goniaulax* sp. Diesing, *Rhizosolenia setigera*, *Tropidoneis lepidoptera* (Gregory) Cleve, and *Asterionella japonica*. Although each of these species had high cell counts, resulting in high total diversity, *Skeletonema* was abundant enough to produce high redundancy. This unusual condition has special significance which will be discussed later.

Although generally high redundancies prevailed throughout the spring and at the peak of the diatom bloom, the lower values obtained during the winter months through early March (Table III) indicate that no species dominated. In January, *Nitzschia seriata* became more important than the other members, and in February the distribution of increased total numbers among many species produced higher diversities and lowered redundancies.

A slight interruption of the general upward diversity trend resulted on March 1 from several days of heavy rainfall that produced a heavy silt load and the highest turbidity levels encountered during the study. The resultant depression of diversity was followed by rapid recovery to former levels, which led eventually to the spring maximum.

Spring freshets with their entrained nutrients and growth factors presumably made the spring diatom flowering possible. The cause of the rapid diversity loss following the flowering peak is uncertain. Nutrient levels were very high on May 10, and high N/P ratios were recorded. This leaves as possible factors (i) growth factor, silicate, and/or trace metal impoverishment, (ii) build-up of toxic "exocrines", or (iii) zooplankton grazing (record concentrations of *Eurytemora* were present during this period). Regardless of the cause, on May 10 very low diversity levels prevailed throughout the estuary, which was in striking contrast to the annual maxima recorded on April 24. In the upper reaches, at st. 1 in particular, the rate of decline was not quite so cataclysmic, partly because communities here were less diverse during the spring maximum than those at other stations. St. 1 declined from 45,099.92 to 1171.52, whereas, st. 3, for example, decreased from 487,220.94 to 216.61.

The general pattern of redundancy change from November through April was from prevailing low values in winter (.47, .50, .60, .52, .38, .40 on February 6) to extremely high values during the spring bloom (.92, .94, .94, – –, .92, .94 on April 24) reflecting *Skeletonema* dominance. Disappearance of the diatom pulse was followed on May 10 by low redundancies (.52, .17, .61, .45, .46, .71).

May 10–September 11, 1958 (Vernal-Aestival-Serotinal). This period may be regarded as the counterpart in 1958 of the first period (July 5–August 21) in 1957, but it differed from that period in several respects. The low diversity levels of early May were followed by subsequent increases to moderate maxima in early July, and these levels were maintained through August and into September. The highest diversity recorded during the summer of 1958 was 47,585.47 on September 11, only one-third of the maximum levels of 1957 (159,045.47 on August 8, 1957). Redundancies increased during this period from lowest values following the spring flowering to highest in July and early August (Table III) associated with increased dominance by *Nannochloris*. Low values prevailed again in late August and early September, corresponding to similar troughs in the preceding year.

Summarizing, each of the four periods distinguished displays describable trends in net phytoplankton diversity at each station considered. Gross disturbances in the general trend and subsequent recovery were recorded on two different occasions, these being associated with corresponding environmental changes.

Diversity and Succession Theory. Alterations in the structure of a phytoplankton community may occur through (i) differential reproduction and selective elimination of component species, and (ii) water mass transfer and consequent dispersal or concentration (Gran and Braarud, 1935). The resultant diversity changes, although more or less cyclic over many years, give in a single year the appearance of ecological succession (Margalef, 1958 a). Margalef construes a phytoplankton succession as a gradual, irreversible change in community composition from loosely organized systems of smaller organisms with rapid dynamics and high productivity/biomass ratios to stabler communities having larger organisms with a slower thermodynamic output as well as lower productivity/biomass relationships. Margalef (personal communication) also conceives that niche structure develops in succession and that each niche tends to become occupied by a single species in accordance with the principle of competitive exclusion (Volterra, 1926; Winsor, 1934; Gause, 1934, 1935; Hardin, 1960; Cole, 1960). Since niche theory is critical to a consideration of diversity trends in succession, Hutchinson's formal theory (1957) is briefly outlined below, slightly modified.

In Hutchinson's model, a *fundamental niche*, v_i, of a species, s_i, is a phase space whose dimensions correspond to the range of each significant environmental variable permitting s_i to survive under all conditions of variable interaction. Each point in this space corresponds to an environmental state at which s_i can exist. Now, considering the ordinary n-space of the physical biotope, β, it is clear from the definition of niche that $v_i \simeq \beta$ (the niche space is isomorphic to the biotope space), so that any point in v_i can correspond to one or more points in β. If there are no such 1–1 correspondences, then v_i is not approximately equal to β, and β is regarded as *incomplete* with respect to s_i. β may also be *complete* or only *partially complete* with respect to s_i.

The niches of different species in a biotope may be *separate* (v_i is not $\cap\, v_j$) or *intersecting* ($v_i \cap v_j$). In the first case there is no basis for interspecific interactions; such phenomena occur only in the intersection subset ($v_i \cap v_j$). In Hutchinson's view, competitive exclusion posits that for any small element of $v_1 \cap v_2$ there do not [at equilibrium] exist in β corresponding small parts, some inhabited by s_1 and others by s_2. Hence, a biocoenose at equilibrium may be construed as an aggregation of mutually exclusive species occupying mutually exclusive niches in a perfectly partitioned biotope. This obviously never occurs, but it is a helpful ideal to have in mind. Several cases of interspecific relationships are distinguishable: (i) $v_1 = v_2$, but $s_1 \neq s_2$ is so unlikely that it may be disregarded [One wonders, however, whether Hutchinson has not nominally provided an ecological definition of species such that $v_i \equiv s_i$]; (ii) $v_2 \subset v_1$, in which competition may lead to survival of s_1 throughout, or survival of s_2 in $(v_1 \cap v_2)$ and of s_1 in $(v_1 \sim v_2)$; (iii) $(v_1 \cap v_2) \subset v_1, v_2$, leading to survival of s_1 in $v_1 \sim v_2$, of s_2 in $v_2 \sim v_1$, and of s_1 or s_2 (but not both) at any point in the β-space where $v_1 \cap v_2$. If the *realized niche*, v_1', is defined as $(v_1 \sim v_2) \cup [(v_1 \cap v_2) \supset s_1]$, and v_2' as $(v_2 \sim v_1) \cup [(v_1 \cap v_2) \supset s_2]$, then the competitive exclusion principle implies that $v_1' \cap v_2' = \varphi$, where φ is the null set.

In a β-space containing an equilibrium community of m species $s_1, s_2, \ldots, s_m$, each represented by $n_1, n_2, \ldots, n_m$ individuals, it is possible to identify for each s_i $(i = 1, 2, \ldots, m)$ a number of elements in β corresponding to a whole or part of v_i'. Thus, if at any given instant, each such element, $\Delta_i\beta$, is occupied by one individual of s_i, then the total partition of β by s_i will be $n_i \Delta_i\beta$, the *specific biotope* of s_i. Since, at equilibrium, the biotope universe will be perfectly partitioned into nonoverlapping subsets by the m species present, it follows that

$$\beta = \sum_{i=1}^{m} n_i \Delta_i \beta . \tag{13}$$

From this we get

$$\sum_{i=1}^{m} n_i = \frac{\beta}{\sum_{i=1}^{m} \Delta_i \beta},$$

which may be substituted into (7) to obtain

$$\begin{aligned} H &= -k \left[\sum_{i=1}^{m} \frac{\beta}{\Delta_i \beta} \log \frac{n_i}{N}\right] \\ &= -k\beta \left[\sum_{i=1}^{m} \frac{1}{\Delta_i \beta} \log \frac{n_i}{N}\right], \end{aligned} \tag{14}$$

which displays (i) a direct relationship between community diversity and magnitude of the biotope space, and (ii) an inverse relationship between diversity and $\sum_{i=1}^{m} \Delta_i \beta$, the sum of the sizes of each element of β occupied by an individual of s_i. Hence, "larger" organisms (in the sense of occupying more of the biotope phase space) lead to a reduction of diversity independently of their numbers n_i, all else being equal.

This applies to communities at equilibrium. As mentioned above, true equilibrium could hardly ever be expected to prevail in nature. Nonequilibrial conditions fall into two basic classes: (i) early in succession when the community is inchoate and the biotope incompletely utilized:

$$\beta > \sum_{i=1}^{m} n_i \Delta_i \beta ; \tag{15}$$

and (ii) later in the sere when biotope resources are competed for:

$$\beta < \sum_{i=1}^{m} n_i \Delta_i \beta . \tag{16}$$

In the first case equilibrium can be approached by (i) increase in number of species, m, (ii) increase in n_i, and (iii) increase in $\sum_{i=1}^{m} \Delta_i \beta$, which, if m is held constant, can only be accomplished by increase in some or all of the $\Delta_i \beta$. In general, all three of these changes are characteristic of ecological succession, which may defined by the transformation

$$S: \left| \begin{array}{l} \beta > \sum_{i=1}^{m} n_i \Delta_i \beta \\ \downarrow \\ \beta = \sum_{i=1}^{m} n_i \Delta_i \beta , \end{array} \right. \tag{17}$$

with the implied diversity changes, which cannot be specified without actual data, since n_i and $\Delta_i \beta$ affect H in opposite ways (eq. 14). Therefore, although diversity within a niche must decline in succession, the total diversity of the community is a function of niche development, and may therefore either decrease or increase at various stages. Hence it appears unrealistic to try to assert categorically that a climax community will have a higher or lower diversity than a particular seral stage since it would depend upon whether that stage were characterized by (15) or by (16). Hence, Margalef's (1958 b) statement that successions proceed from high to low diversity would seem to be theoretically untenable, although such may often be the case in actuality.

Based on the idea of diversity decline in succession, Margalef advocated H as a succession index. The present data indicate that redundancy rather than diversity reflects successional status better. Although no formal basis can yet be provided for using R to measure succession, we proceed on strictly empirical grounds on the assumption that in phytoplankton, where biotope monotony seriously restricts niche diversification, redundancy tends to increase and is maximal at climax. Diversity and redundancy relationships for st. 3 are graphed in Fig. 3, where suggested seres are indicated by the heavy, broken, vertical lines. During each period designated, the communities evolved from a condition of low to one of high redundancy. Three successions are indicated for August 1957 to August 1958. The other stations had similar patterns (Table III), excepting st. 6, which had considerably less redundant communities during autumn 1957.

In the first succession from August to November 1957 (Fig. 3), diversity declined steadily in accordance with Margalef's (1958 b) thesis. The terminal phase may be regarded as a "climax" since it persisted for three months, indicating considerable stability and approach to the condition of (13). In the second sere, from November 1957 through April 1958, both diversity and redundancy increased. The two variables were simultaneously maximal at the height of the spring flowering. Apparently this is a function of large N since, by (12), H and R would appear inversely related. When N is very large, H_{max} must grow faster than H, producing redundancy increase coincident with diversity increase. Such a condition is construed to correspond to equation

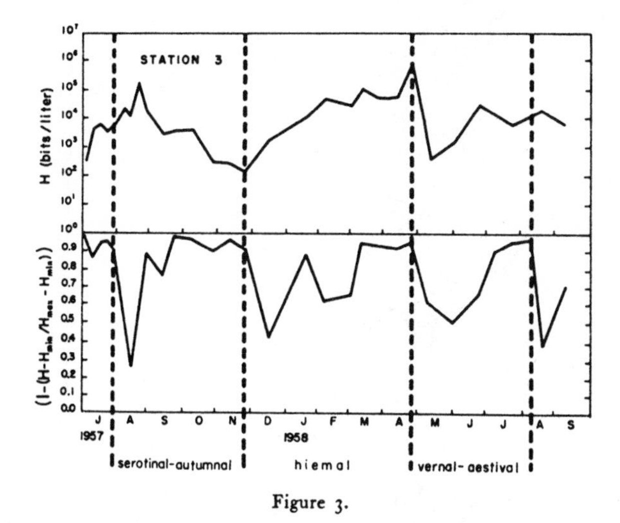

Figure 3.

(16) and must be marked by extreme exploitation of the biotope. The ultimate consequence, when some essential resource becomes impoverished, must be a precipitous return of the community to the condition in (15). This is expressed in nature as the catastrophic disappearance of the spring diatom bloom, to be followed subsequently by a new succession (17).

It seems clear from these results, viewed against the theoretical background developed, that diversity-redundancy relationships offer considerable promise for development of a formal theory of compositional changes during succession, and that the methods explored here could lead to rigorous criteria for denoting the successional status of any ecological community at any time. Further development will be necessary, however, before such an ideal can be realized.

Conclusion. Mean redundancy levels in the Raritan net phytoplankton during the term of study were .85, .81, .86, .83, .85, .85 for sts. 1–6, a remarkably consistent set of values. If high redundancy signifies late successional stages in plankton, then these figures indicate the Raritan net flora to be, on the average, in essentially long term steady state (climax). If this is true, then by (14) the mean diversity at a given location logically reflects the quality of the biotope there, greater β-spaces being capable of sustaining higher diversity communities. The mean diversity (in bits $\times 10^{-3}$/L) at each station, based only on dates for which data from all six stations were complete, was 17.7, 22.0, 21.2, 20.6, 35.2, 30.5; these relationships are displayed spatially in Fig. 4. The similarity between Fig. 4 and Fig. 2 depicting the general circulation of the Raritan is striking and constitutes an effective testimony to the basic efficacy of the method.

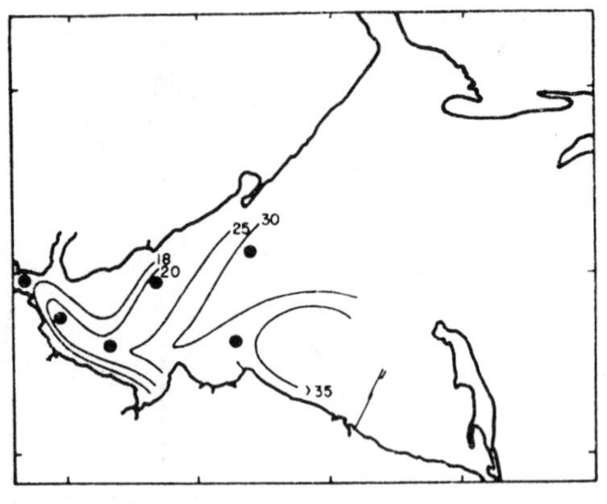

Figure 4.

As Fig. 4 indicates, higher mean diversity levels prevailed in the lower estuary, signifying higher biotope quality proceeding toward Lower Bay. Progressive diminution of diversity upbay indicates a more unsatisfactory biotope, undoubtedly a reflection of the gross pollution originating at the head of the estuary. Consequently, the Raritan is interpreted to be a generally poor quality ecosystem with respect to its capacity for developing a niche structure adequate to maintain high diversity steady states, at least relative to its lower segment. This interpretation is supported by additional data from a series of production experiments performed off Princess Bay during summer 1959 which indicated a mean daily energy loss in the water column of 0.37 g cal/cm^2 (Patten, 1961). The extent to which the established diversity relationships will be altered by the lifting of pollution from the Raritan River should provide an interesting area for investigation against the background presented here.

Acknowledgments. This study was supported by NIH grant 5278, H. H. Haskin and E. T. Moul, Rutgers University, principal investigators. The field program was jointly carried out with H. P. Jeffries, who concurrently studied the zooplankton for his doctoral dissertation. David Trend graciously volunteered the use of a power boat without which sampling could not have continued through cold weather. F. G. Fender of the Rutgers Computation Center wrote the program for the IBM 650 Computer. The diversity-succession model was developed to its current state at the author's present institution, and represents Contribution No. 111 from the Virginia Institute of Marine Science.

REFERENCES

AYERS, J. C., B. H. KETCHUM, and A. C. REDFIELD

1949. Hydrographic considerations relative to the location of sewer outfalls in Raritan Bay. Woods Hole Oceanog. Inst., Ref. No. 29-13.

BRANSON, H. R.

1953. A definition of information from the thermodynamics of irreversible processes and its application to chemical communication in biological systems; *in* Information theory in biology, H. Quastler, (ed.) Univ. Illinois Press, Urbana, pp. 25-40.

BRILLOUIN, L.

1956. Science and information theory. Academic Press, New York. 320 pp.

COLE, L. C.

1960. Competitive exclusion. Science, *132*: 348-349.

CONOVER, S. A. M.

1956. Oceanography of Long Island Sound, 1952-1954. IV. Phytoplankton. Bull. Bingham Oceanogr. Coll., *15*: 62-112.

FISHER, R. A., A. S. CORBET, and C. B. WILLIAMS

1943. The relation between the number of species and the number of individuals in a random sample of an animal population. J. Anim. Ecol., *12*: 42-58.

GAUSE, G. F.

1934. The struggle for existence. Williams and Wilkins, Baltimore. 163 pp.

1935. Verifications experimentales de la theorie matehmatique de la lutte pour la vie. Actual. Sci. Industr., *277*. 61 pp.

GLEASON, H. A.

1922. On the relation between species and area. Ecology, *3*: 156-162.

GOODALL, D. W.

1952. Quantitative aspects of plant distribution. Biol. Rev., *27*: 194-245.

GRAN, N. H., and T. BRAARUD

1935. A quantitative study of the phytoplankton in the Bay of Fundy and the Gulf of Maine. J. biol. Bd. Canada, *1*: 279-467.

HARDIN, G.

1960. The competitive exclusion principle. Science, *131*: 1292-1297.

HUTCHINSON, G. E.

1957. Concluding remarks. Cold Spring Harbor Symp. Quant. Biol., *22*: 415-427.

JEFFRIES, H. P.

1959. The plankton biology of Raritan Bay. Ph. D. Thesis. Rutgers Univ., New Brunswick, N.J.

1961. Physical-chemical studies on Raritan Bay, a polluted estuary, *ms*.

KOCH, L. F.

1957. Index of biotic dispersity. Ecology, *38*: 145-148.

Margalef, D. R.
1956. Información y diversidad específica en las comunidades de organismos. Invest. Pesquera, *3*: 99–106.
1957. La teoría de la información en ecología. Mem. R. Acad. Cien. Artes, *32*: 373–449.
1958a. Temporal succession and spatial heterogeneity in phytoplankton; *in* Perspectives in marine biology. A. A. Buzzati-Traverso (ed.). Univ. California Press, Berkeley.
1958b. Information theory in ecology. Gen. Systems, *3*: 36–71.

Patten, B. C.
1959. The diversity of species in net phytoplankton of the Raritan estuary. Ph. D. Thesis, Rutgers Univ., New Brunswick, N.J.
1961. Plankton energetics of Raritan Bay. Limnol. Oceanogr., *6*: 369–387.

Preston, F. W.
1948. The commonness, and rarity, of species. Ecology, *29*: 254–283.

Ryther, J. H.
1954. The ecology of phytoplankton blooms in Moriches Bay and Great South Bay, Long Island, New York. Biol. Bull. Woods Hole, *106*: 198–209.

Shannon, C. E.
1948. A mathematical theory of communication. Bell. Syst. Tech. J., *27*: 379–423, 623–656.

Volterra, V.
1926. Vartazioni e fluttuazioni del numero d'individui in specie animali conviventi. Mem. R. Acad. Lincei, (6) *2*: 181–187.

Wiener, N.
1948. Cybernetics. Wiley, New York. 194 pp.

Williams, C. B.
1952. Diversity as a measurable character of an animal or plant population. Colloque intern. C.N.R.S. sur l'Ecol., Paris, 1950: 129–144.

Winsor, C. P.
1934. Mathematical analysis of the growth of mixed populations. Cold Spring Harbor Symp. Quant. Biol., *2*: 181–187.

FACTORS CONTROLLING PHYTOPLANKTON POPULATIONS ON GEORGES BANK[1]

By

GORDON A. RILEY

Woods Hole Oceanographic Institution

and

Bingham Oceanographic Laboratory

A complex field such as oceanography tends to be subject to two opposite approaches. The first is the descriptive, in which several quantities are measured simultaneously and their inter-relationships derived by some sort of statistical method. The other approach is the synthetic one, in which a few reasonable although perhaps over-simplified assumptions are laid down, these serving as a basis for mathematical derivation of relationships.

Each approach has obvious virtues and faults. Neither is very profitable by itself; each requires the assistance of the other. Statistical analyses check the accuracy of the assumptioms of the theorists, and the latter lend meaning to the empirical method. Unfortunately, however, in many cases there is no chance for mutual profit because the two approaches have no common ground. Until such contact has been established no branch of oceanography can quite be said to have come of age. In this respect physical oceanography, one of the youngest branches in actual years, is more mature than the much older study of marine biology. This is perhaps partly due to the complexities of the material. More important, however, is the fact that physical oceanography has aroused the interest of a number of men of considerable mathematical ability, while on the other hand marine biologists have been largely unaware of the growing field of bio-mathematics, or at least they have felt that the synthetic approach will be unprofitable until it is more firmly backed by experimental data.

However valid the latter objection may be, the present paper will attempt, in the limited field of plankton biology, to establish continuity between some purely descriptive studies that have been made and mathematical concepts based on what seem to be logical assumptions about plankton physiology. The need for such an attempt has

[1] Contribution No. 353 from the Woods Hole Oceanographic Institution.

become apparent during the course of several plankton surveys in which the data were analyzed statistically with the idea of correlating plankton populations and their rate of growth with various environmental factors such as solar radiation, temperature, dissolved nutrient salts, *etc.* In each survey a reasonably high degree of correlation was found, but the empirical nature of the relationships was often confusing. For example, temperature affects plankton in several different ways, and the relative importance of these effects varies from time to time and from place to place. The statistical relationship of temperature and plankton represents an average of these different effects. Therefore, it may happen in examining sets of data for particular areas and times that the temperature constant varies widely from set to set, and study of the values of their constants does not lead easily to a universally applicable theory.

Furthermore, there is no good reason for assuming that the variations of plankton with environment are always linear. To treat them as such may introduce an error. To evolve nonlinear relationships on a purely empirical basis is possible, but this generally requires a larger set of observations than is readily available.

These limitations of the statistical method will become apparent in the pages that follow. The only way to avoid them is by the opposite approach—that of developing the mathematical relationships on theoretical grounds and then testing them statistically by applying them to observed cases of growth in the natural environment. At present this can be done only tentatively, with over-simplification of theory and without the preciseness of mathematical treatment that might be desired. It is not expected that any marine biologist, including the writer, would fully believe all the arbitrary assumptions that will be introduced. However, the purpose of the paper is not to arrive at exact results but rather to describe promising techniques that warrant further study and development.

ACKNOWLEDGMENTS

My best thanks are due Dean F. Bumpus, who kindly provided me with the necessary zooplankton data, and Henry Stommel, who gave me valuable advice on some of the mathematical treatment.

STATISTICAL SECTION

Phytoplankton studies in the Georges Bank area of the western North Atlantic during the period from 1939 to 1941 have been described in a series of publications (Riley, 1941, 1942, 1943; Riley and Bumpus, 1946). In the first of these papers it was noted that part of the variations that occurred in the distribution of phytoplankton from one part of the bank to another and from one month to the next could be correlated with such factors as the depth of water, temperature and dissolved phosphate and nitrate. Since that time the study of the zooplankton collections has been completed and examination of the data has shown that grazing by zooplankton is important in controlling the size of the phytoplankton population. With the inclusion of the zooplankton material, it is now possible to develop a relatively complete statistical treatment of the ecological relationships of the Georges Bank plankton.

Observations. The original observations made during the 1939–1940 cruises were listed in the papers cited above. This material is briefly summarized in Table I in the form of means and standard deviations for each cruise. Correlations of phytoplankton with its various environmental factors have also been published. These have been

TABLE I. MEANS AND STANDARD DEVIATIONS OF GEORGES BANK PLANKTON AND ENVIRONMENTAL FACTORS

		Sept.	*Jan.*	*Mar.*	*Apr.*	*May*	*June*
Mean	*Depth of water in meters*	247	209	135	209	82	206
	Mean temperature, upper 30 m.	15.24	4.61	2.60	3.81	5.14	9.66
	Mg-atoms phosphate P per m², upper 30 m.	14.4	33.7	34.7	21.9	16.6	19.2
	Mg-atoms nitrate N per m², upper 30 m.	153	209	172	129	285	155
	Number of animals, thousands per m²	135	14	24	32	106	103
	Plant pigments, thousands of Harvey units per m²	560	118	828	2303	871	478
σ	*Depth*	540	371	163	530	357	458
	Temperature	2.14	0.71	0.33	0.37	0.39	2.58
	P	7.2	11.1	8.1	8.9	3.5	8.7
	N	108	209	68	129	79	155
	Zooplankton	126	13	18	38	77	85
	Plant pigments	195	67	781	827	522	233

used to develop multiple correlation equations by which the variations in horizontal distribution of plant pigments during each cruise are calculated according to the variations in environmental factors. Comparison of calculated values with the actual determinations for plant pigments shows an average error on the different cruises of 20–40%. In other words 60–80% of the variations in phytoplankton on Georges Bank can be accounted for on the basis of variations in depth, temperature, phosphate, nitrate and zooplankton. The multiple correlation equations for each cruise are as follows:

$$\text{Sept.}: PP = -.011D - 23.8t + 5.20P + .371N - .26Z + 829$$
$$\text{Jan.}: PP = .191D - 61.2t + 7.96P + .956N + .47Z - 115$$
$$\text{Mar.}: PP = -D - 770t - 23.18P - 6.98N - .62Z + 4989$$
$$\text{Apr.}: PP = .469D - 331t - 61.4P + .197N - 5.31Z + 4954$$
$$\text{May}: PP = .007D + 236t + 15.1P - 2.316N - 3.50Z + 437$$
$$\text{June}\ PP = -.066D - 48.1t - 15.5P + .070N - .55Z + 1300$$

PP is thousands of Harvey units of plant pigments per m^2, D is depth of water, t is temperature, P is mg-atoms of phosphate P per m^2 in the upper 30 meters, N is mg-atoms of nitrate N, and Z is thousands of animals per m^2.

Discussion of the effect of environmental factors on the horizontal distribution of phytoplankton. The equations show, within the limits of error stated above, the amount of variation in the phytoplankton crop that is obtained by varying any one or all of the environmental factors. For example, at a particular station of the September cruise, if all the factors were found to have exactly the mean values as stated in Table I, then the plant pigments would be expected to have the mean value for September. If phosphate were increased one milligram-atom, it would increase the calculated value for plant pigments by the amount of the phosphate constant, or 5.2 thousands of units. If the phosphate varied a "normal" amount, as indicated by the limits of its standard deviation, the plant pigments would be changed $\pm$ 6.5%. Use of the standard deviation in this way is a convenient method of rating the importance of a given factor, and in Table II it is applied to all the variables in the equations. Although the standard deviation is a positive or negative variation around the mean, it serves a useful purpose to give the values in the table the same sign as the constant in the equation. The figures then represent the change in plant pigments produced by raising each factor from its mean to the upper limit of its standard deviation.

Table II shows that although one particular factor may be of outstanding significance, such as nitrate during the March cruise, phos-

TABLE II. Percentage Change in the Phytoplankton Crop Produced by Increasing the Value of Each Environmental Factor from Its Mean to the Limit of Its Standard Deviation

	Sept.	*Jan.*	*Mar.*	*Apr.*	*May*	*June*
Depth	−1	60	−20	11	0	− 1
Temperature	−9	−37	−31	− 5	10	−26
P	7	74	−23	−24	6	−28
N	7	41	−57	1	−21	1
Zooplankton	−6	5	− 1	− 9	−31	−10

phate in April, and zooplankton in May, there is no indication of complete control of the phytoplankton crop by a particular factor. It appears to be a highly complex relationship in which one factor after another gains momentary dominance. This table also shows that each variable has a vastly different significance at different times of the year, which is in accord with our present knowledge of phytoplankton ecology.

It has been shown (Riley, 1942) that the depth of water plays a significant role in the inception of the spring diatom flowering, and it is reasonable to find a strong negative relationship in March. This effect disappears later in the season when radiation becomes strong enough so that vertical turbulence is no longer able to prevent growth by dissipating the surface crop.

Temperature is generally supposed to have a negative effect on phytoplankton because increased respiration uses up part of the store of energy that would otherwise be used in the production of new plant material. The predominantly negative relationship shown in Table II is therefore expected. The positive relation in May is anomalous and not readily explainable.

The nutrient-phytoplankton relationship is one in which cause and effect are not clearly separable; although a large quantity of available nutrients is likely to stimulate growth, the growth-utilization process will reduce the quantity of nutrients so that the relationship becomes negative. The diversified results in Table II come from the complexities of this inter-relation. Probably the relationships are of three main types. First, in January, when the quantity of available nutrients was large and growth was slight because of low light intensity, the observed positive relationship is indicative of a slight stimulating effect by nutrients, which to a slight degree counteracted the inhibiting effect of insufficient light. Second, in spring, when radiation increased and growth became more abundant, negative relationships of the growth-utilization type were established. However, they did not become progressively stronger with the advance of the season, leading to complete exhaustion of nutrients; whether or not the observed partial exhaustion had an inhibiting effect on growth cannot be determined from these data. Third, a situation was established which was particularly apparent in September but which probably began early in the summer; in this situation the total quantity of plant pigments was fairly uniform all over the bank, but with certain localized areas of slightly higher crops accompanied by larger quantities of nutrients. These were sufficiently important to provide a direct relationship of a moderately low order between plankton and nutrients. They included some shallow water stations as well as four in deeper water (50 to 100 m.) on the northern and western edges of Georges Bank. In all cases the vertical distribution of temperature, as well as nutrients, was more nearly uniform than at near-by stations that had smaller crops. It is concluded, therefore, that the observed relationship was due to localized turbulence and upwelling of the nutrient-rich lower waters. Probably such conditions were of transient nature, for it seems likely that a degree of regeneration strong enough to maintain a positive nutrient-phytoplankton relationship for any length of time would lead to a much larger growth than was observed.

The phytoplankton-zooplankton relationship has been discussed in some detail in a previous paper (Riley, 1946) and need be only briefly summarized here. It was concluded that the predominantly negative relationship was due to grazing. The quantities of animals and plants were such as to indicate that the observed relationship could have been established in a very short time, possibly in a day or in a few days. A theory was postulated that tidal currents and turbulent motion of the Georges Bank waters tended continually to destroy the horizontal gradients in phytoplankton, but that the zooplankton, because of their habit of vertical migration, would be absent part of the time from the surface waters where mixing processes are strongest, and hence they would not be so readily dispersed. Therefore, they would tend constantly to reestablish the phytoplankton gradients by their grazing activity.

Effect of Environmental Factors on the Seasonal Cycle of Phytoplankton. It is apparent from the preceding discussion that the relationship of a particular environmental factor with the horizontal distribution of phytoplankton may differ from one month to another both in quantity and in kind. Nevertheless, there are seasonal trends in these factors which are related with plankton variations, as can be observed by inspection of the data in Table I. Thus the seasonal cycle of phytoplankton can be correlated with its environment with a fair degree of accuracy, even though such treatment makes no allowance for special effects that are operative only at particular times during the year.

A multiple correlation equation was developed from the cruise averages in Table I. It is

$$PP = -153t - 120P - 7.3N - 9.1Z + 6713.$$

The symbols are the same as in previous equations. The relation between the observed cruise averages and the values determined from the equation is shown in Fig. 14. The average error between observed and calculated values is 20%, which is slightly less than that obtained in the treatment of individual cruises. Probably the errors caused by

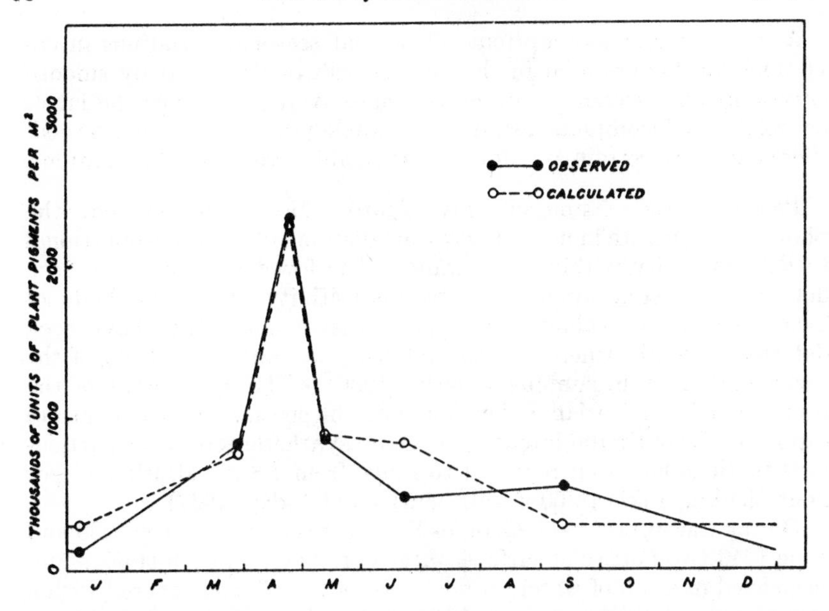

Figure 14. Comparison of the observed phytoplankton population with the calculated population as determined by a multiple correlation analysis of the relationship between phytoplankton and environmental factors.

ignoring certain special seasonal effects are more than counterbalanced by the reduction in analytical errors when averages are used for the calculation.

THEORETICAL SECTION

The rate of change of the phytoplankton population is determined by the difference in reaction rates between the process of accumulation of energy by the population and the processes of energy dissipation. It will be assumed that all the important reaction rates are included in the equation

$$\frac{dP}{dt} = P\,(P_h - R - G)\,, \tag{1}$$

in which the rate of change of the population P in respect to time is determined by the photosynthetic rate per unit of population (P_h), the rate of phytoplankton respiration (R), and the rate of grazing by zooplankton (G). Each of these rates is subject to environmental influences and therefore is continually changing with the seasons. In order to arrive at a practical solution of the equation it is necessary to examine each of its component parts in the light of present day plankton physiology.

A second major assumption will be that seasonal variations in environmental factors used in the analysis can be expressed by smooth curves drawn through the observed cruise averages. Since the latter are relatively incomplete, as far as the whole yearly cycle is concerned, there must be a certain amount of unavoidable error in the calculations.

Photosynthesis. Numerous investigators have reported that the photosynthetic rate in actively growing diatom cultures is proportional to light intensity within wide limits. The lower limit has not been determined accurately due to the insensitivity of the methods of measurement, but values of the right order of magnitude have been detected at depths where the light intensity was about 0.1–1.0% of the surface intensity in summer (Clarke, 1936). The upper limit of the proportionality is variable, depending on the species and the length of exposure; the optimum intensity for photosynthetic activity in particular situations has been reported to range from 1.8 g. cal. per cm^2 per hour (Jenkin, 1937) to 60 g. cal. (Curtis and Juday, 1937).

During the six cruises to Georges Bank between September 1939 and June 1940 two bottles of surface water were taken at each station and suspended in a tub of water on deck, one of them being covered with a bag of several thicknesses of dark cloth. After twenty-four hours' exposure the oxygen in the two bottles was measured and the difference in their oxygen content was used as a rough estimate of the photosynthetic activity of the surface plankton. The inset in Fig. 15 shows the average photosynthetic rate[2] obtained on each cruise plotted against the average incident solar radiation in the area at that time of year. Radiation values are obtained from data published by Kimball (1928) and reproduced in Sverdrup, Johnson and Fleming (1942: 103). The plotted points for the January, March, April and May cruises approximate a linear relationship in which the photosynthetic rate equals 2.5 times the incident radiation in g. cal. per cm^2 per minute. The June and September rates are lower, but it will be shown later that this can be explained as a result of nutrient depletion.

With these facts at hand the mean photosynthetic rate of the population will be estimated according to the following assumptions:

1. When nutrient depletion does not limit photosynthesis,

$$P_h = pI \tag{2}$$

[2] The photosynthetic rate is expressed as grams of carbon produced per day per gram of carbon in the surface phytoplankton crop, using the formula photosynthetic rate = .375 × oxygen production in g/m^3/day ÷ 17 × 10^{-6} Harvey units of plant pigments/m^3. The conversion factor is based on analyses described in a previous paper on the plankton of Georges Bank (Riley, 1941).

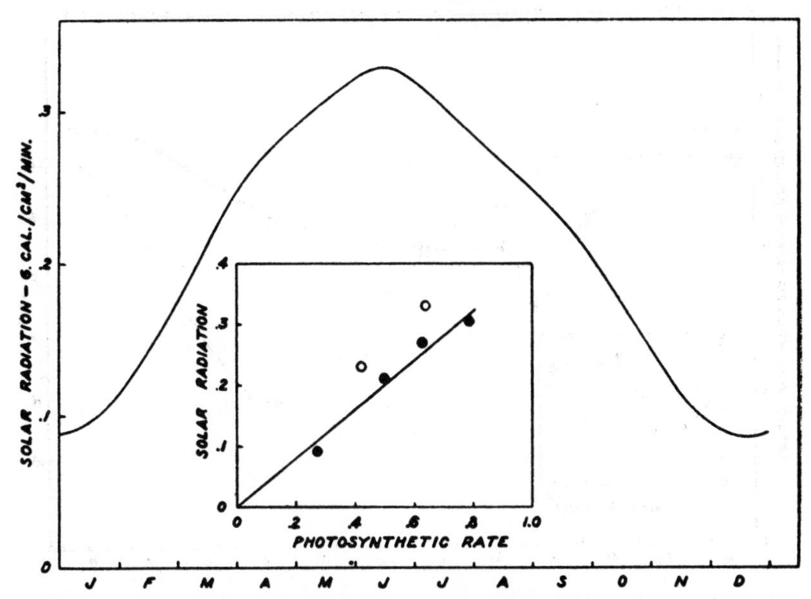

Figure 15. Seasonal variations in average incident solar radiation in the Georges Bank area. Inset shows average observed surface photosynthetic rate plotted against incident radiation. Dots are averages for January, March, April and May; circles are June and September.

in which P_h is the photosynthetic rate, I is radiation in g. cal./cm²/ minute at the depth of the photosynthesizing plankton, and p, the photosynthetic constant, is 2.5.

2. The intensity at the surface, I_0, may be determined for any time in the year from the curve in Fig. 15, which is based on Kimball's data, cited above.

3. The intensity at any other depth z is determined from the formula

$$I_z = I_0 e^{-kz}$$

and therefore

$$P_{hz} = pI_0 e^{-kz}. \qquad (3)$$

In these formulas the extinction coefficient k is defined as 1.7 divided by the depth of the Secchi disc reading, a rough conversion factor suggested by Poole and Atkins (1929). Secchi disc determinations for the Georges Bank area are shown in Fig. 16.

4. According to equation (3) the photosynthetic rate approaches zero as the depth approaches infinity. But the depth at which a measurable and significant amount of photosynthesis occurs is limited, and so is the depth at which viable phytoplankton can be found.

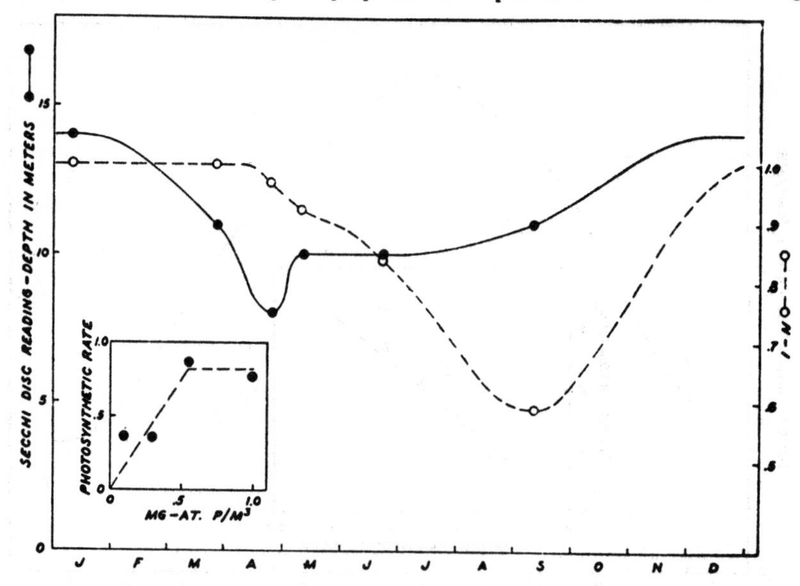

Figure 16. Solid curve is transparency as determined by Secchi disc readings. Dotted line is the estimated reduction in photosynthetic rate caused by nutrient depletion. Inset shows averages of photosynthetic rate plotted against phosphate concentration during June and September.

Therefore, it is convenient to set an arbitrary limit to the depth of the euphotic zone. This depth will be called z_1 and will be defined as the depth at which the light intensity has a value of 0.0015 g. cal./cm²/ minute. This approximates the intensity at the maximum depth of photosynthesis as reported by Clarke (1936). Calculated values for z_1 are shown in Fig. 17.

5. To find the mean photosynthetic rate in the euphotic zone, equation (3) is integrated from the surface to z_1, and divided by z_1:

$$\overline{P_h} = \frac{pI_o \int_o^{z_1} e^{-kz}\, dz}{z_1} = \frac{\frac{pI_o}{-k} e^{-kz} \Big|_o^z}{z_1}$$

$$\underset{o \rightarrow z_1}{\overline{P_h}} = \frac{pI_o}{kz_1} (1 - e^{-kz_1}) . \qquad (2)$$

6. It was postulated that the proportionality between photosynthesis and radiation holds only when nutrients are abundant. The fact noted previously that the ratio between the photosynthetic rate and light intensity was reduced in June and September led to an in-

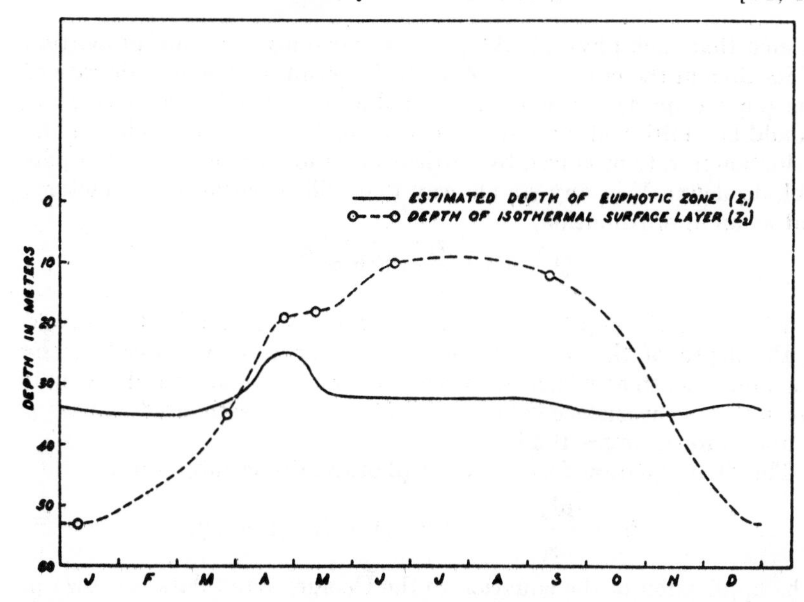

Figure 17. Estimated depth of the layer in which a measurable amount of photosynthesis occurs (z_1), and the depth of the virtually isothermal surface layer (z_2).

vestigation of nutrients as a possible cause. Large individual variations were found, and the correlation between photosynthetic rate and phosphate concentration was poor; nevertheless, when the rates were averaged for different ranges of phosphate concentration, there was a pronounced reduction in the average rate when the phosphate fell below about 0.5 to 0.6 mg-atom of P per m³ (Fig. 16, inset). Ketchum (1939) reported a decrease in the growth rate of experimental cultures of *Nitschia closterium* when the phosphate concentration was less than 50 gamma of PO_4 per liter (0.55 mg-atom per m³). Therefore it seems reasonable to assume that the mean photosynthetic rate as determined in equation (4) should be multiplied by a factor $(1 - N)$, in which N is the reduction in rate due to nutrient depletion. According to the facts above,

$$N = \frac{0.55 - mg\text{-}at.\ P/m^3}{0.55} \qquad \text{when } P \leqq 0.55.$$

Mean values for $(1 - N)$ are shown in Fig. 16.

7. Several investigators have pointed out the importance of vertical turbulence in reducing phytoplankton crops by carrying the breeding stock below the euphotic zone. That this is an important phenomenon on Georges Bank has been demonstrated (Riley, 1942). If turbulence

is such that each phytoplankton cell spends only a certain proportion of its time in the euphotic zone, then the mean photosynthetic rate of the population as a whole will be reduced. Therefore equation (4) should be multiplied by still another factor $(1 - V)$, in which V is the reduction in rate produced by vertical water movements. It is impossible to define V in any simple way that will be entirely satisfactory, but as an approximation,

$$(1 - V) = \frac{z_1}{z_2} \text{ when } z_1 \leqq z_2,$$

in which z_1 is the depth of the euphotic zone as previously defined and z_2 the depth of the mixed layer, which is arbitrarily defined as the maximum depth at which the density is no more than 0.02 of a σ_t unit greater than the surface value. Fig. 17 shows the estimated values for z_1 and z_2 for Georges Bank.

The final equation for the mean photosynthetic rate is now

$$P_h = \frac{pI_0}{kz_1}(1 - e^{-kz_1})(1 - N)(1 - V). \tag{5}$$

The application of the equation to the Georges Bank data is shown in Fig. 18. The upper curve shows the primary calculation of the photo-

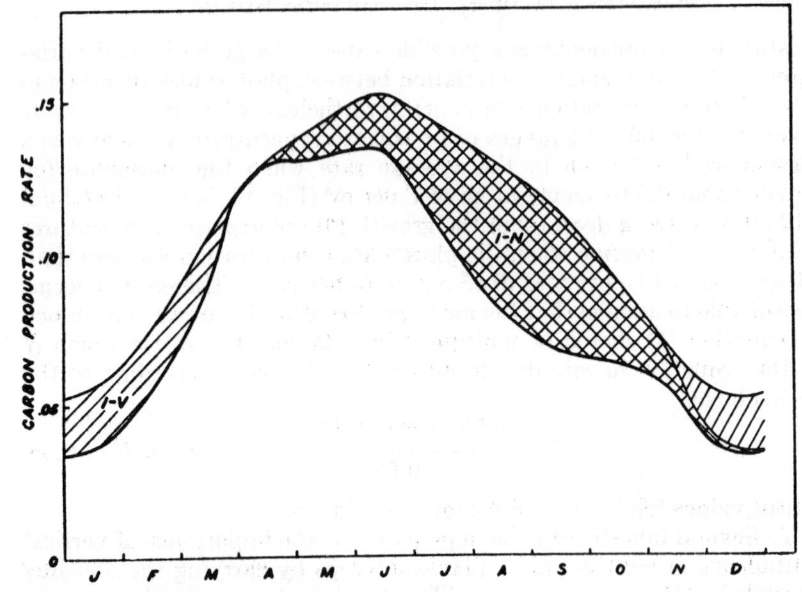

Figure 18. Estimated mean photosynthetic rate. Upper curve is the maximum possible rate as determined by incident radiation and transparency. Lower curve is the estimate obtained by introducing correction factors for the effects of vertical turbulence and nutrient depletion.

synthetic rate based on light intensity. The reduction obtained by introducing $(1 - N)$ and $(1 - V)$ is indicated by hatched areas. The heavy lower curve is the final estimate of the mean photosynthetic rate.

Phytoplankton respiration. The few available measurements of the respiration of pure diatom cultures have not yielded precise results. Observed rates have varied from one species to another as well as during different stages of growth of the same culture. The recorded values differ by a factor of 10 to 20, and there are not enough of them to draw a good average.

No direct measurements have been made of the respiration of a natural phytoplankton population, since the measured oxygen consumption also includes zooplankton and bacterial respiration. Statistical estimates have been made by the writer on the basis of the observed correlation between phytoplankton and total oxygen consumption. The best of these, judging by quantity and homogeneity of data, was obtained in Long Island Sound (Riley, 1941a). It was estimated that in winter (average temperature 2.05° C.) the respiration rate was 0.024 ± 0.012 mg. of carbon consumed per day per mg. of phytoplankton carbon. Calculations based on summer observations suggested that the respiration increased with higher temperatures (0.110 ± 0.007 at 17.87° C.), the rate being approximately doubled by a 10° increase in temperature.

On the basis of these rather scanty data it will be assumed that:

1. The temperature effect can be stated as

$$R_T = R_o e^{rT} \tag{6}$$

in which R_T is the respiratory rate at any temperature T, R_o is the rate at 0° C. and r is a constant expressing the rate of change of the respiratory rate with temperature. The value of r is 0.069 when the rate is doubled by a 10° increase in temperature. The seasonal cycle of Georges Bank surface temperatures used in computing respiratory rates is shown in Fig. 19.

2. The value chosen for R_o will be 0.0175. This is the mean of the two estimates derived from the Long Island Sound data mentioned above, in which the calculated values of R_o for winter and for summer are respectively 0.020 and 0.015.

Grazing. The greater part of the zooplankton population consists of filter-feeding organisms which tend to strain a relatively constant volume of water in a given time irrespective of the quantity of food material in it. Therefore a fixed proportion of the phytoplankton population will be consumed in successive units of time. This is stated as

$$G = gZ\,, \tag{7}$$

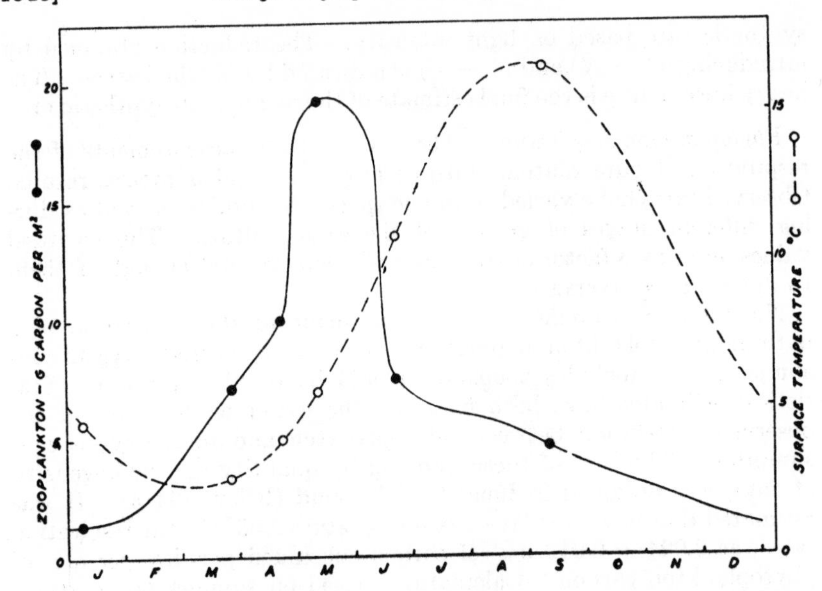

Figure 19. Solid line is the seasonal cycle of zooplankton. Measurements of zooplankton volume by the displacement method are treated by a conversion factor (wt. in g. = 12.5% × vol. in cc.) to derive a rough estimate of the carbon content. Dotted line is the mean surface temperature.

in which G is the rate of grazing, g is the rate of reduction of phytoplankton by a unit quantity of animals and Z is the quantity of zooplankton in grams of carbon per m^2.

There is some question as to whether g is nearly constant over long periods of time or undergoes a marked seasonal change. On the one hand there are the experiments of Marshall, Nicholls and Orr (1935) which showed that the respiratory rate of *Calanus finmarchicus* increases with increasing temperature, implying a greater food requirement at higher temperatures and possibly an increased filtering rate. On the other hand, feeding experiments by Fuller (1937) showed variations of a more complex nature. The grazing rate of *Calanus finmarchicus* was greater at 8° C. than at 3° or 13° C., and animals captured in the late summer, when the natural breeding stock was declining, had lower feeding rates than those studied earlier in the summer. Thus a factor that can be called "depressed physiological state" for want of a more precise term appears to counterbalance the expected effect of high temperature late in the summer. It is clear that the whole process of zooplankton feeding requires much more thorough study. However, lacking the necessary information to describe the process accurately, it is believed that it can best be approximated by the simple form of equation (7).

The value of g must be of an order of magnitude which will at least satisfy the minimum respiratory requirements of the zooplankton population at times when the latter is stable. According to Marshall, Nicholls and Orr (1935), the daily food requirement of *Calanus* in winter (5° C.) lies between 1.3 and 3.6% of the carbon content of the animals. This estimate applied to the January plankton on Georges Bank yields a grazing constant of 0.0091 to 0.0252. In summer (15° C.), these authors suggest a food requirement of 1.7–7.6%, for which the corresponding values of g are 0.0084 to 0.0374 in September on Georges Bank. The latter estimates are perhaps too high, since the zooplankton was decreasing at a rate of 0.5% per day and therefore probably was not getting enough food to satisfy the minimum respiratory requirements. If it is assumed that the food intake equaled the food requirement minus the rate of population decrease, then the food intake was 1.2–7.1% of the animals' carbon content per day, and the corresponding values for g are 0.0059 and 0.0350.

Within these wide limits it is difficult to choose a correct value for the grazing constant, and again the need for more experimental work is apparent. On a purely empirical basis, a good fit for the data is obtained by using the average of the minimum values of g for the September and January cruises, namely 0.0075. This factor, multiplied by the quantity of zooplankton, shown in Fig. 19, estimates the Georges Bank grazing rate.

Conclusions. The original equation

$$\frac{dP}{dt} = P\,(P_h - R - G)$$

can now be expanded by substituting the right hand terms of equations (5), (6) and (7):

$$\frac{dP}{dt} = P\left[\frac{pI_0}{kz_1}\,(1 - e^{-kz_1})\,(1 - N)\,(1 - V) - R_0e^{rT} - gZ\right]. \quad (8)$$

The rate of change of the population is dependent on six ecological variables: the incident solar radiation, transparency of the water, the quantity of phosphate, the depth of the mixed layer, the surface temperature and the quantity of zooplankton. The results of the application of equation (8) to the Georges Bank data are shown graphically in Fig. 20. The curve at the top is the photosynthetic rate previously illustrated (Fig. 18). The second curve is the photosynthetic rate minus the respiratory rate of the phytoplankton, or in other words the

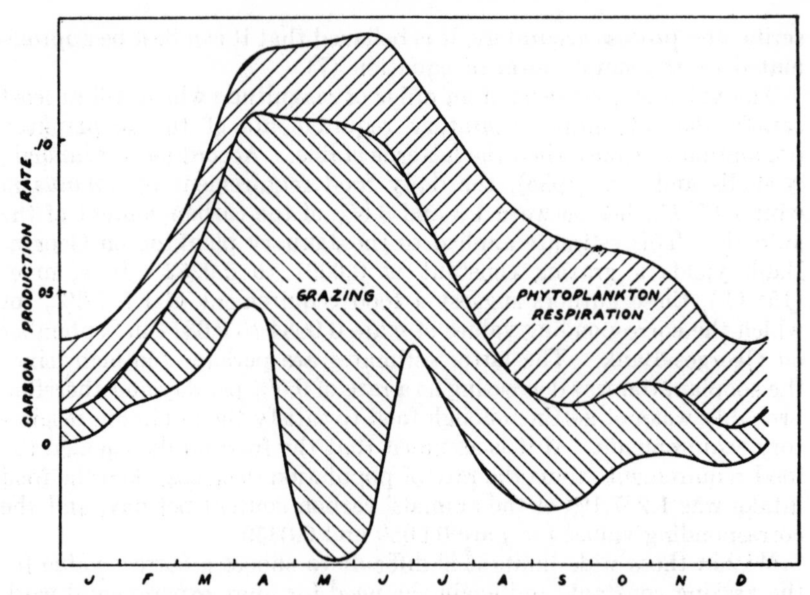

Figure 20. Estimated rates of production and consumption of carbon. Curve at top is the photosynthetic rate. By subtracting the respiratory rate the second curve is obtained, which is the phytoplankton production rate. From this is subtracted the zooplankton grazing rate, yielding the curve at the bottom, which is the estimated rate of change of the phytoplankton.

phytoplankton production rate. By subtracting the grazing rate from the production rate, the rate of change of the population is obtained. This is the bottom curve of Fig. 20. Numerical values used in drawing these curves are shown in the appendix.

Equation (8) cannot be integrated readily, but an approximation is obtained by integrating over successive short intervals of time and assuming for each variable a constant, average value during that time. Thus the change in population in the time interval 0 to t is determined by

$$lnP_t - lnP_o = \overline{P_h} - \overline{R} - \overline{G}.$$

Therefore, by a series of integrations the quantity $lnP_t - lnP_o$ can be approximated for the whole seasonal cycle. This quantity indicates the relative size of the population from one part of the cycle to another. To convert to absolute terms requires evaluation of the integration constant P_o (the size of the population at the minimum point in the cycle), which is readily obtained if the quantity of plankton (P_t) is measured at one or more times during the year. In the present case, in which six cruise averages are available, P_o was statistically determined so as to give the best fit for the data. The results are shown in Fig. 21, in which the theoretical population cycle is shown by a smooth curve, and the observed cruise averages are indicated by dots. The average error is 27%.

It is now apparent that a few simple and commonly measured environmental factors can be used with a fair degree of accuracy to evaluate the quantitative aspects of the seasonal cycle of phytoplank-

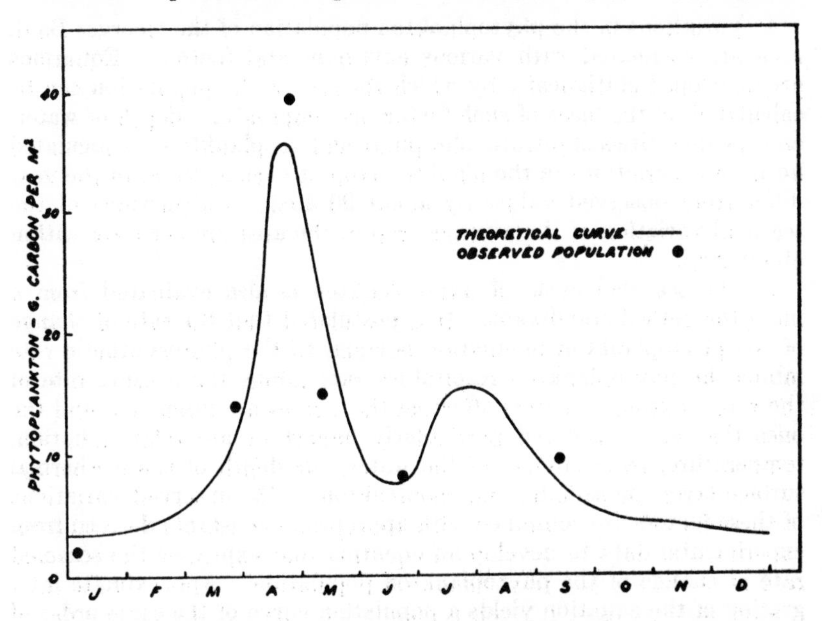

Figure 21. Curve shows the calculated seasonal cycle of phytoplankton, determined by approximate integration of the equation for the rate of change of the population. For comparison the observed quantities of phytoplankton are shown as dots.

ton. Furthermore, this can be done in two different ways: first by statistical comparison of the simultaneous variations of phytoplankton and environmental characteristics in a particular locality; second, by theoretical evaluation of the way in which changes in the environment might affect growth as evidenced by the results of various physiological experiments. Each of these methods has something to contribute to plankton biology. The statistical method is useful in determining whether a particular factor is significant; the theoretical method carries on from there, discriminating between cause and effect and helping to establish certain quantitative relationships that are not likely to be derived empirically.

While these methods are obviously crude at the present time and need to be developed further, both by examination of other areas and by better experimental evaluation of constants, it does not seem too much to hope that they will eventually solve some of the problems of seasonal and regional variations that puzzle marine biologists today.

SUMMARY

1. Variations in the phytoplankton population of the Georges Bank area are correlated with various environmental factors. Equations are developed statistically by which the size of the population can be calculated on the basis of such factors as temperature, depth of water, and the quantities of nitrate, phosphate and zooplankton. Calculated horizontal variations in the plankton crop at various times in the year differ from observed values by about 20–40%. Calculations of the seasonal variation of the average crop in the area are accurate within about 20%.

2. The seasonal cycle of phytoplankton is also evaluated from a more theoretical standpoint. It is postulated that the rate of change of the phytoplankton population is equal to the photosynthetic rate minus the phytoplankton respiratory rate minus the grazing rate of the zooplankton. Factors affecting these rates are discussed, and the ones that are considered particularly important are solar radiation, temperature, transparency of the water, the depth of the isothermal surface layer, phosphate, and zooplankton. The observed variations of these factors are combined with appropriate constants derived from experimental data to develop an equation that expresses the seasonal rate of change of the phytoplankton population. Approximate integration of the equation yields a population curve of the same order of accuracy as the statistical estimate.

REFERENCES

CLARKE, G. L.

1936. Light penetration in the western North Atlantic and its application to biological problems. Rapp. Cons. Explor. Mer, *101*, Pt. 2: 1–14.

CURTIS, J. T. AND C. JUDAY

1937. Photosynthesis of algae in Wisconsin lakes. III. Observations of 1935. Int. Rev. Hydrobiol., *35:* 122–133.

FULLER, J. L.

1937. Feeding rate of *Calanus finmarchicus* in relation to environmental conditions. Biol. Bull. Wood's Hole, *72:* 233–246.

JENKIN, P. M.

1937. Oxygen production by the diatom *Coscinodiscus excentricus* Ehr. in relation to submarine illumination in the English Channel. J. Mar. biol. Ass. U. K., (N. S.) *22:* 301–342.

KETCHUM, B. H.

1939. The absorption of phosphate and nitrate by illuminated cultures of *Nitzschia closterium.* Amer. J. Bot., *26:* 399–407.

KIMBALL, H. H.

1928. Amount of solar radiation that reaches the surface of the earth on the land and on the sea, and methods by which it is measured. Mon. Weath. Rev. Wash., *56:* 393–398.

MARSHALL, S. M., A. G. NICHOLLS AND A. P. ORR

1935. On the biology of *Calanus finmarchicus.* Part VI. Oxygen consumption in relation to environmental conditions. J. Mar. biol. Ass. U. K., (N. S.) *20:* 1–27.

POOLE, H. H. AND W. R. G. ATKINS

1929. Photo-electric measurements of submarine illumination throughout the year. J. Mar. biol. Ass. U. K., (N. S.) *16:* 297–324.

RILEY, G. A.

1941. Plankton studies. IV. Georges Bank. Bull. Bingham oceanogr. Coll., *7* (4)*:* 1–73.

1941a. Plankton studies. III. Long Island Sound. Bull. Bingham oceanogr. Coll., *7* (3)*:* 1–93.

1942. The relationship of vertical turbulence and spring diatom flowerings. J. Mar. Res., *5* (1)*:* 67–87.

1943. Physiological aspects of spring diatom flowerings. Bull. Bingham oceanogr. Coll., *8* (*4*)*:* 1–53.

RILEY, G. A. AND D. F. BUMPUS

1946. Phytoplankton-zooplankton relationships on Georges Bank. J. Mar. Res., *6* (1)*:* 33–47.

SVERDRUP, H. U., M. W. JOHNSON AND R. H. FLEMING

1942. The oceans. Prentice-Hall, Inc., New York. 1060 pp.

APPENDIX

NUMERICAL VALUE OF QUANTITIES USED IN THE CALCULATION OF THE SEASONAL RATE OF CHANGE OF THE GEORGES BANK PHYTOPLANKTON POPULATION AS DEVELOPED IN THE THEORETICAL SECTION OF THIS PAPER

Date	I_o	k	z_1	$1-N$	z_2	$1-V$	Ph	T	R_T	Z	G	dP/dt	$lnP_t - lnP_o$	lnP_t	P_t
1/1	.088	.121	34	1.00	53	.64	.034	5.2	.024	1.3	.010	.000	.000	1.217	3.4
1/15	.094	.121	34	1.00	53	.64	.036	4.1	.023	1.4	.011	.002	.030	1.247	3.5
2/1	.112	.124	35	1.00	51	.69	.044	3.2	.021	1.7	.013	.010	.180	1.397	4.0
2/15	.138	.128	35	1.00	48	.73	.055	2.7	.021	2.6	.020	.014	.390	1.607	5.0
3/1	.174	.136	35	1.00	45	.78	.071	2.4	.021	4.2	.031	.019	.675	1.897	6.7
3/15	.212	.145	34	1.00	40	.85	.091	2.5	.021	5.8	.043	.027	1.080	2.297	10.0
4/1	.247	.159	32	1.00	32	1.00	.120	2.7	.021	7.5	.056	.043	1.725	2.942	19.1
4/15	.272	.200	26	1.00	23	1.00	.130	3.4	.021	8.9	.067	.042	2.355	3.572	35.5
5/1	.290	.205	25	.95	19	1.00	.131	4.5	.024	17.2	.129	−.022	2.025	3.242	25.7
5/15	.306	.170	31	.92	18	1.00	.132	5.9	.026	19.3	.145	−.039	1.440	2.657	14.2
6/1	.321	.170	32	.90	15	1.00	.134	7.6	.030	18.8	.141	−.037	.885	2.102	8.2
6/15	.329	.170	32	.88	11	1.00	.134	9.7	.035	14.0	.105	−.006	.795	2.012	7.5
7/1	.319	.170	32	.82	10	1.00	.122	11.8	.038	6.9	.052	.032	1.275	2.492	12.1
7/15	.302	.170	31	.76	9	1.00	.108	13.9	.045	6.2	.047	.016	1.515	2.732	15.4
8/1	.284	.165	32	.69	9	1.00	.093	15.5	.051	6.0	.045	−.003	1.470	2.687	14.8
8/15	.267	.162	32	.63	10	1.00	.081	16.3	.054	5.7	.043	−.016	1.230	2.447	11.5
9/1	.250	.159	32	.60	11	1.00	.073	16.6	.056	5.1	.038	−.021	.915	2.132	8.4
9/15	.230	.154	33	.59	13	1.00	.067	16.4	.054	4.5	.034	−.021	.600	1.817	6.1
10/1	204	.145	34	.63	16	1.00	.065	15.5	.051	3.9	.029	−.015	.375	1.592	4.9
10/15	.174	.138	34	.69	20	1.00	.063	14.2	.045	3.3	.025	−.007	.270	1.487	4.4
11/1	.144	.131	35	.77	27	1.00	.060	12.4	.040	3.2	.024	−.004	.210	1.427	4.2
11/15	.115	.126	35	.85	37	.95	.053	10.5	.037	2.6	.020	−.004	.150	1.367	3.9
12/1	.094	.121	34	.92	45	.76	.039	8.5	.031	2.0	.015	−.007	.045	1.262	3.5
12/15	.086	.121	33	.97	50	.66	.033	6.7	.028	1.6	.012	−.007	−.060	1.157	3.2

I_o = incident solar radiation in g. cal./cm² 1 /cm²/minute

k = extinction coefficient = 1.7/Secchi disc reading in meters

z_1 = depth of euphotic zone in meters, defined as depth where the light intensity is .0015 g. cal.

$1 - N$ = correction factor for nutrient depletion = mg-at. phosphate P/0.55 when $P \leqq 0.55$.

z_2 = depth of mixed layer = maximum depth at which $\sigma_{tz} - \sigma_{t_0} \leqq 0.02$.

$1 - V$ = correction factor for vertical turbulence = z_1/z_2 when $z_1 \leqq z_2$.

Ph = estimated mean photosynthetic rate according to equation (5) in text.

T = mean surface temperature.

R_T = estimated phytoplankton respiration according to equation (6).

Z = g. of zooplankton carbon/m², estimated on the assumption that the weight of carbon in grams = 12.5% of the volume (by displacement) in cubic centimeters.

G = grazing rate = $0.0075Z$

dP/dt = rate of change of the phytoplankton population = $Ph - R - G$.

$lnP_t - lnP_o$ = the summation of 15 × dP/dt (since the rate determinations are at approximately 15-day intervals)

lnP_t = $lnP_t - lnP_o + 1.217$. The latter is a value for P_o determined statistically as the best fit for the observed population data.

P_t = estimated population in g. of phytoplankton carbon/m² (considered equivalent to 17 × Harvey units of plant pigments × 10^{-6}/m²).

On Conditions for the Vernal Blooming of Phytoplankton.

By

H. U. Sverdrup,
Norsk Polarinstitutt, Oslo.

IN order that the vernal blooming of phytoplankton shall begin it is necessary that in the surface layer the production of organic matter by photosynthesis exceeds the destruction by respiration. The destruction of organic matter by respiration goes on continuously wherever there are plants or animals, but photosynthesis can take place only in the presence of light, carbon dioxide, and nutrient salt such as nitrates, phosphates, and other minor constituents. Gran and Braarud (1935) have pointed out that production cannot exceed destruction if there exists a deep mixed top layer. Their reasoning is that within a well mixed layer the plankton organisms are about evenly distributed, but a net production takes place only above the compensation depth, whereas below the compensation depth there is a net loss of organic matter. The total population cannot increase if this loss exceeds the net production.

The reasoning is illustrated in Figure 1, where the curves dp and dr show the increase of organic matter by photosynthesis and the decrease by respiration, respectively, as functions of depth. Both increase and decrease apply to unit volume and unit time. The production is supposed to decrease logarithmically with depth, corresponding to the logarithmic decrease of light intensity, but the destruction is supposed to be independent of depth in agreement with the assumption that the organisms are evenly distributed in the mixed layer. At the compensation depth, D_c (Fig. 1), gain and loss balance each other, $dp = dr$. The condition for an increase of the total population is that the total production P must exceed the total destruction by respiration, R, or in our representation, that on an average for 24 hours the area *acd* must be greater than the area *abcd*. This implies that there must exist a *critical depth* such that blooming can occur only if the depth of the mixed layer is *less* than the critical value.

Gran and Braarud concluded that the critical depth was 5 to 10 times the compensation depth. Riley (1942) has carried some of these ideas further and has shown that on Georges Bank there existed in the spring of 1941 a relation between plankton and stability. We will attempt a more precise description of the conditions that are shown in Figure 1 and, on certain assumptions, we will derive an analytical expression for the critical depth.

Our assumptions are:—

1. There exists a thoroughly mixed top layer of thickness D.
2. Within the top layer the turbulence is strong enough to distribute the plankton organisms evenly through the layer.
3. Within the top layer the production is not limited by lack of plant-nutrition salts.

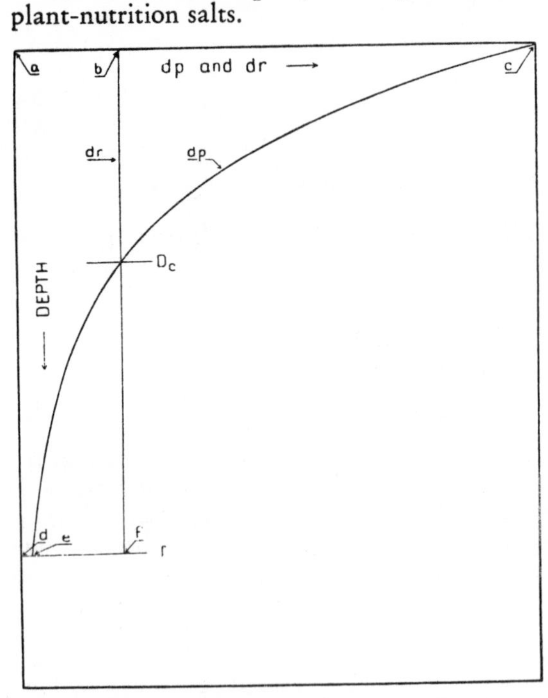

Figure 1. Schematic representation of the variation with depth of the increase of organic matter by photosynthesis, *dp*, and the decrease by respiration, *dr*. Increase and decrease apply to unit volume and unit time.

4. Within the top layer the extinction coefficient, k, of radiation energy is constant. This assumption is not quite correct even if the transparency of the water is independent of depth because, owing to selective absorption, the spectrum of the radiation changes with depth. The wave-lengths for which the extinction coefficient is smallest penetrate to the greatest depths, for which reason the extinction coefficient for total energy decreases with depth (Sverdrup *et al.*, 1942, p. 107; Jerlov, 1951). However, below a depth of a few metres the decrease is so small that the introduction of a constant value of the coefficient does not lead to a serious error, provided that the transparency of the water does not change greatly with depth.

5. When dealing with photosynthesis it is not necessary to consider the energy at wave-lengths shorter than 4200 or longer than 5600 Å. This assumption is justified because for the shorter and the longer wave-lengths the extinction coefficients are so large that the energy associated with these wave-lengths is absorbed in the upper one metre or less. The energy of the spectrum in the wave-length range 4200 to 5600 Å represents about 20 per cent. of the total energy of the incoming radiation, but of the latter a fraction, a, is reflected from the surface.

6. The production of organic matter by photosynthesis is proportional to the energy of the radiation at the level under consideration. According to Jenkins (1937) this assumption is correct if the energy flux is less than 1·8 g. cal. cm.$^{-2}$ hour^{-1}. In spring this condition may be expected to be fulfilled below a depth of a few metres.

7. The energy, I_c, at the compensation depth is known. The compensation depth is defined as the depth at which the energy intensity is such that the production by photosynthesis balances destruction by respiration. This energy level may depend upon the temperature because photosynthesis and respiration may not stand in the same relation to temperature, and it must depend upon the composition of the plankton. It must, for instance, lie higher for a mixed population of phyto- and zooplankton than for a pure phytoplankton population. For the latter it may depend on the species of which the population is composed. By experiments in the English Channel with the diatom *Coscinodiscus excentricus* Jenkins (1937) found the energy at the compensation depth equal to 0·13 g. cal. cm.$^{-2}$ hour^{-1}. Pettersson *et al.* (1934) found a value of 400 luxes in Gullmar Fjord, Sweden, working with a mixed population. Using the same conversion factor as Jenkins this light intensity corresponds to an energy of 0·15 g. cal. cm.$^{-2}$ hour^{-1}.

On the basis of these assumptions we can derive an expression for the critical depth. It will be shown that the critical value depends upon the amount of incoming radiation, the transparency of the water as expressed by the extinction coefficient for total energy, and the energy level at the compensation depth.

Let the total incoming energy from the sun and the sky, expressed in g. cal. cm.$^{-2}$ hour^{-1}, be I_0. Of this amount a fraction, a, is lost by reflection from the sea surface. The energy that passes through the sea surface, I_w, is then $I_w = (1—a)\, I_0$. With a clear sky the fraction a depends upon the altitude of the sun as follows:—

Altitude of sun (°)	5	10	20	30	40	50 to 90
Fraction reflected	·40	·25	·12	·06	·04	·03

With a completely overcast sky the fraction is assumed to have the constant value $a = {·}08$.

Of the amount of energy that passes the sea surface only 20 per cent. need be considered when dealing with photosynthesis because the major part is absorbed in the upper one metre (assumption 5). The "effective" energy that passes the sea surface is, therefore, $I_e = 0{·}2\, I_w$. With a constant coefficient of extinction, k, the energy that reaches a depth z is then, taking the z-axis position upwards:—

$$I_z = I_e e^{kz} = 0.2\ (1 - a)\ I_0 e^{kz} \tag{1}$$

The production at the depth z is supposed to be proportional to the available energy. The production in the time interval dt is then $dp = m\, I_z\, dt$, whereas the destruction in the same time interval is simply $dr = n\, dt$. Here m and n are factors which are independent of z, but depend upon the character of the population and the temperature of the water. By definition $dp = dr$ at the compensation depth where $I_z = I_c$. Therefore, $n/m = I_c$.

In the time, T, the total production by photosynthesis between the surface and the depth $z = -D$ is:—

$$P = m \int_0^T \int_{-D}^0 I_e e^{kz}\ dtdz = \frac{m}{k}\ (1 - e^{-kD}) \int_0^T I_e dt \tag{2}$$

The total destruction is:—

$$R = n \int_0^T \int_{-D}^0 = nTD \tag{3}$$

The condition for an increase of the phytoplankton population is:—

$$P > R \tag{4}$$

Introducing the value for I_e from (2), putting

$$\overline{I_e} = \frac{0.2}{T} \int_0^T (1 - a)\ I_0\ dt \tag{5}$$

and remembering that $n/m = I_c$, we find that the critical depth, D_{cr}, is defined by the equation

$$\frac{D_{cr}}{1 - e^{-kD_{cr}}} = \frac{1}{k}\ \frac{\overline{I_e}}{I_c} \tag{6}$$

So far we have been concerned only with the critical thickness of the mixed top layer, which must not be exceeded if the production in the layer shall be greater than the destruction. It is, however, possible to add a few observations. In the first place it may be pointed out that on the general assumptions made here the rapidity with which a given population may grow depends upon how much the thickness of the top layer deviates from the critical value. If it is only slightly smaller the increase of the total population must be slow, but if it is much smaller, the increase may be very rapid. These conclusions may be greatly modified if grazers are present (Fleming, 1939). In that case the phytoplankton population may remain small in spite of heavy production.

In the second place a phytoplankton population may increase independently of the thickness of the mixed layer if the turbulence is moderate. In this case the plankton may be unevenly distributed at the end of the daylight hours, with greater concentration above the compensation depth where production has taken place. During the night hours mixing may not be complete and when daylight again makes photosynthesis possible, the concentration of plankton may still be greater near the surface. In these circumstances the production will be greater than according to our assumptions, and the population may increase as long as the conditions prevail. A similar development may take place even with strong turbulence if the phytoplankton displays a positive phototaxis.

Returning to our main topic we may interpret our results in terms of "stabilization". In middle and high latitudes there generally exists a deep mixed layer at the end of the winter, but as the season advances, there develops a shallow mixed layer below which the density increases so rapidly with depth that turbulence is suppressed. By "stabilization" we understand the development of such a layer. This development may be caused by spring heating of the surface layer or by lowering of the salinity of the top layer, related to spring run-off.

Table 1. Thickness of top mixed layer according to observations at Weather Ship "M", 66°N., 2°E.Gr.

Date 1949	Thickness m.	Date 1949	Thickness m.	Date 1949	Thickness m.	Date 1949	Thickness m.
3. March	150	4. April	300	28. April	100—300	16. May	25
7. „	100—150	6. „	50	29. „	100	18. „	100
10. „	100	8. „	100	2. May	100	19. „	100
17. „	100	11. „	50—100	4. „	>150	21. „	0
21. „	150	15. „	200—400	5. „	150	23. „	0
23. „	150	16. „	75—100	6. „	100	25. „	25
26. „	200	22. „	50	11. „	75	27. „	0
30. „	100	25. „	100	12. „	100	28. „	50
2. April	400	27. „	>100	14. „	75	31. „	0

Observations carried out in the spring of 1949 at Weather Ship "M" in the Norwegian Sea (66°N., 2°E. Gr.) make possible a test of our conclusions. The observations comprise measurements of the thickness of the upper mixed layer, counts of the number per litre of phytoplankton organisms at one or more depths near the surface, and counts of the copepods and nauplii as collected by vertical hauls from 100 m. to 50 m. and from 50 m. to the surface. We will deal only with the observations from the months of March, April, and May.

The observations of the thickness of the upper mixed layer were made several times a week. They are based on temperature and salinity observations at fixed depths and on continuous temperature records with the Mosby thermosonde. The latter has a very small scale, for which reason the thickness of the layer can be determined with an accuracy of only ± 25 m. The observed values as communicated by Mr. O. S æ l e n are entered in Table 1.

The counts of phytoplankton, copepods, and nauplii are entered in Tables 2 and 3. If observations were made at more than one depth the

Table 2. Phytoplankton organisms per litre in the surface layer according to observations at Weather Ship "M", 66°N., 2°E.Gr.

Date 1949	Numbers Diatomaceae	Coccolitho-phoridae	Dinoflagel-latae	Total
10. March	500	1,000	0	1,500
17. „ *)	120			120
28. „ *)	160	2,500	0	2,500
6. April	9,240	4,500	8,500	22,240
11. „	4,300	9,500	1,020	14,820
22. „ *)	3,100	8,500	1,000	12,600
28. „ *)	6,100	10,000	1,000	17,100
5. May*)	3,060	10,260	2,520	15,840
12. „	8,500	28,000	40	36,540
19. „ *)	7,540	22,000	4,000	33,540

*) An asterisk indicates observations from 0 m., otherwise the numbers pertain to depth of maximum concentration.

data apply to the depth of maximum population. In the case of the copepods and the nauplii the values from the two vertical hauls above 100 m. have been combined. The phytoplankton data have been placed at my disposal by cand. real. Per Haller-Nielsen, the zooplankton data by mag. sc. Ole J. Østvedt. I wish to express my thanks to these two and to Mr. Sælen for their kindness.

No observations of the incoming radiation were made at the Weather Ship, but approximate daily values have been computed by means of Mosby's formula:—

$$\bar{I}_0 = 0{\cdot}026\,(1-0{\cdot}075\,\bar{C})\,\bar{h}\ (\text{g. cal. cm.}^{-2}\ \text{min.}^{-1})$$

where $\bar{C}$ is the average cloudiness on the scale 0 to 10 and $\bar{h}$ is the average altitude of the sun. The latter can readily be computed for every day for the latitude of the Weather Ship, and the cloudiness is obtained from the meteorological observations. From the values of I_0 that have been determined in this manner daily values of D_{cr} have been found and have been smoothed by forming overlapping 5-day averages.

Measurements of transparency were not made, but at an oceanic

Table 3. Numbers of copepods and nauplii from vertical hauls between 100 m. and the surface at Weather Ship "M", 66°N., 2°E.Gr.

Date 1949	Numbers Copepods	Nauplii	Total	Date 1949	Numbers Copepods	Nauplii	Total
2. March	1,319	0	1,319	23. April	11,220	10,190	21,410
9. „	2,339	0	2,339	27. „	3,674	1,005	4,679
16. „	2,758	0	2,758	4. May	2,313	2,730	5,043
7. April	1,644	85	1,729	18. „	39,131	10,550	49,681
14. „	4,606	3,480	8,08	25. „	42,103	9,370	51,473

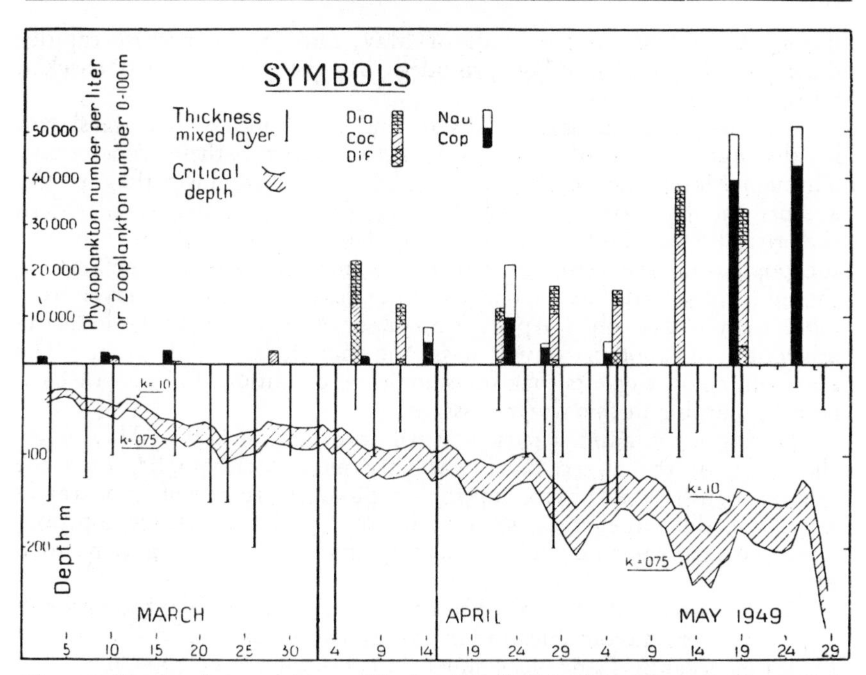

Figure 2. Results of observations at Weather Ship "M" (66°N., 2°E. Gr.). The symbols are explained in the graph, where the following abbreviations have been used:—Dia, Diatomaceae; Coc, Coccolithophoridae; Dif, Dinoflagellatae; Nau, Nauplii; and Cop, Copepods.

locality the coefficients of extinction of total energy may be assumed to lie between 0·075 and 0·10 (Jerlov, 1951; Sverdrup *et al.*, 1942, p. 107). According to Jerlov the value 0·075 is probably the more correct. It should, however, be borne in mind that the value may vary from day to day, and that it probably increases as the season advances.

All variables are presented graphically in Figure 2. In the lower part of the figure the critical depth is shown, which is supposed to lie between the curves marked $k = 0{\cdot}10$ and $k = 0{\cdot}075$. These curves are based on 5-day means of D_{cr}. It is seen that the critical depth increases more or less regularly from about 30 to 40 m. at the beginning of March to nearly 300 m. at the end of May. The observed values of the depth of the mixed layer are also entered. It is seen that stabilization leading to the development of a very shallow mixed layer does not take place before the middle of May.

The graphs bring out two striking features. (1) Until the final week of April the depth of the mixed layer is greater than the critical depth, and (2) after the middle of May the depth of the mixed layer is very much smaller than the critical depth. This implies, if our reasoning is correct, (1) that until the beginning of April the phytoplankton population must remain very small, (2) that it should be expected to increase

2*

during April and the first half of May, and (3) to increase rapidly during the last half of May, provided that the increase is not checked by the presence of grazers.

The observed amounts of phyto- and zooplankton that are shown in the upper part of the graph fully confirm these conclusions. Through March the amounts of plankton remained insignificant. On 4. April the depth of the mixed layer was for the first time smaller than the critical depth, and on the following day an appreciable phytoplankton population was recorded, but only a small number of copepods. It should be observed that the change in the depth of the mixed layer from more than 300 m. on 2. April to 50 m. on 4. April probably indicates advection of another water mass and not local stabilization. The plankton count suggests that for some time conditions for growth have been favourable in this water mass.

During the remaining part of April and the first half of May, when the depth of the mixed layer was only moderately smaller than the critical depth, the phytoplankton population remained moderately large, and did not increase systematically. At the same time copepods appeared in greater numbers, indicating that considerable grazing took place.

In the latter half of May when conditions for phytoplankton growth should be very favourable, the counts show larger numbers of phytoplankton organisms and very much larger numbers of copepods. The latter feature suggests that the development of the phytoplankton is effectively suppressed by a heavy grazing. Thus, the entire sequence of the development of the plankton of the surface layer appears to be consistent with our conclusions.

It may be pointed out that in the spring of 1949 the first increase in phytoplankton was not associated with stabilization, but with the seasonal increase of the compensation depth. Stabilization led to a second increase in population after the middle of May.

In the introduction it was mentioned that G r a n and B r a a r u d estimated the critical depth to be about 5 to 10 times the compensation depth. Since $D_{cr} \doteq I_e/I_c k$ (Equation 6) and $I_c = I_e e^{-kD_c}$ we obtain

$$\frac{D_{cr}}{D_c} = \frac{e^{kD_c}}{kD_c} \tag{7}$$

Applying this equation to the data that are shown in Figure 2 we find that the ratio D_{cr}/D_c increases as the season advances and exceeds the value 5 after the beginning of May. A curve showing the values of the ratio could have been entered on the graph but would have complicated the presentation. It is sufficient to state that on 15. March, 15. April, and 15. May the values of D_{cr}/D_c were 2·9, 4·0, and 5·5 respectively, meaning that the results arrived at here appear to confirm the conclusions by G r a n and B r a a r u d.

The data shown in Figure 2 represent the only available series of fairly complete records which make possible an analysis of certain features of the productivity of the surface layer as related to available light and thickness of mixed layer. It is, therefore, not advisable to place too great emphasis on the agreement between theory and observations. It is hoped, though, that the results presented here may encourage further systematic observations and collections, not only at Weather Ship "M", but at other Weather Ships as well.

Summary.

On certain assumptions a "critical depth" is defined. The depth of a mixed surface layer must be less than this critical depth if the phytoplankton population of the mixed layer shall increase. The results are applied to observations made at Weather Ship "M" (66°N., 2°E. Gr.) in March, April, and May 1949. A striking agreement is found between the hydrographic conditions and the development of the plankton communities.

References.

Fleming, R. H., 1939. "The control of diatom populations by grazing." Cons. Perm. Internat. Explor. Mer, Journ. du Cons., **14**, 2, pp. 210—227.

Gran, H. H., & T. Braarud, 1935. "A quantitative study of the Phytoplankton in the Bay of Fundy and the Gulf of Maine." Journ. Biol. Board Canada, **1**, 5, pp. 279—467.

Jenkins, Penelope M., 1937. "Oxygen production by the diatom *Coscinodiscus excentricus* Ehr. in relation to submarine illumination in the English Channel." Journ. Mar. Biol. Assoc. U. K., **22**, pp . 301—342.

Jerlov, N. G., 1951. "Optical Studies of Ocean Waters." Rep. Swedish Deep-Sea Exp. 1947—1948, **3**, 1, pp. 1—69.

Pettersson, H., H. Höglund, & S. Landberg, 1934." Submarine daylight and the photosynthesis of phytoplankton." Oceanogr. Inst. Göteborg, Meddelanden, No. 10, 17 pp. (Göteborgs K. Vetensk. Handl., 5. ser. B.)

Riley, G. H., 1942. "The relationship of vertical turbulence and spring diatom flowering." Journ. Mar. Res., **5**, pp. 67—87.

Sverdrup, H. U., M. W. Johnson, & R. H. Flemming, 1942. "The Oceans." Prentice Hall, New York. 1087 pp.

The Advent of the Spring Bloom in the Eastern Subarctic Pacific Ocean[1]

By T. R. Parsons, L. F. Giovando, and R. J. LeBrasseur

Fisheries Research Board of Canada
Pacific Oceanographic Group, Nanaimo, B.C.

ABSTRACT

The spring phytoplankton bloom in the eastern subarctic Pacific Ocean was described from estimations of the critical depth and the depth of the mixed layer. The results suggested that the spring bloom begins during February in the area south of 45°N and east of 135°W. During March the bloom area advances in a northwesterly direction to 50°N at 125°W and 45°N at 135°W. A net increase in primary production is also possible during March near 55°N and 155°W. During April, the spring bloom is generally well established throughout the region except in a central area where suitable conditions are not firmly established until May. This description is supported by the distribution of copepods in the region during April.

INTRODUCTION

Conditions for the onset of the spring phytoplankton bloom in temperate latitudes have been shown to be largely dependent both on the stability of the water mass and on the penetration of sufficient light to cause a net increase in production. One approach to an examination of these conditions was suggested by Gran and Braarud (1935) and developed into a prediction model by Sverdrup (1953). Studies using the model have been carried out by Marshall (1958), Semina (1960), and Cushing (1962). A similar type of study was carried out by Riley (1957).

The approach described by Sverdrup (1953) is based on a comparison between the depth of the mixed layer and the depth at which light conditions (radiation and transparency) are sufficient to allow a net increase in the production of a water column. The latter depth is known as the "critical depth" and is defined as the depth at which the total production beneath a unit surface is equal to the total respiration. It follows that, if the critical depth is greater than the depth of mixing, a net increase in production can take place. Sverdrup (1953) determined a mathematical expression for the critical depth as follows:

$$\frac{D_{cr}}{1 - e^{-k_e D_{cr}}} = \frac{I_e}{I_c k_e} \tag{1}$$

where D_{cr} is the critical depth in metres, k_e is the extinction coefficient (m^{-1}), I_e is the average energy which passes the sea surface per unit time and is available for photosynthesis, and I_c is the energy at the compensation depth.

[1]Received for publication August 6, 1965.

J. Fish. Res. Bd. Canada, 23(4), 1966.
Printed in Canada.

(The compensation depth is defined as the depth at which the energy intensity is such that production by photosynthesis balances destruction by respiration.) I_e and I_c have been expressed in langleys (ly) per hour, one langley being equal to one gram calorie per square centimetre.

Since the term $1 - e^{-k_e D_{cr}}$ is little different from 1 the equation can be simplified to:

$$D_{cr} \doteqdot \frac{I_e}{I_c k_e} \quad (2)$$

In the following discussion the model described above was applied to the northeast Pacific Ocean. This area was arbitrarily defined as lying between 40 and 60°N lat from 160°W long to the coast of North America and thus practically corresponds to the eastern subarctic Pacific. Within this area special reference is made to Ocean Station P (50°N lat, 145°W long).

The conditions for a net increase in primary production described in this discussion were compared with the distribution of copepods at Ocean Station P and at several other points in the northeast Pacific. The latter data were employed here for two reasons. Firstly, 6 years of chlorophyll *a* data collected at Station P indicated that throughout the year very little change is observed in the standing stock of phytoplankton per cubic metre (Parsons, MS, 1965). There is good evidence that this absence of a marked change is due to intensive zooplankton grazing (McAllister et al., 1960), which must commence simultaneously with a net increase in the primary production during the spring. (Another example of this suppression of changes in the standing stock of phytoplankton per cubic metre is discussed by Heinrich (1962).) Thus, studies at Station P suggested that, in the eastern subarctic Pacific, changes in the standing stock of copepods (as herbivorous feeders) may be used to indicate a net increase in primary production during the spring.

The second reason for employing these data is that they represent the most geographically comprehensive data collected in this area during the spring. In contrast, C-14 primary production data are lacking except at Station P, where some seasonal changes have been reported and are discussed at the end of this presentation.

As far as available data permitted, results reported in this presentation were averaged over several years in order to obtain a general assessment of the occurrence of the spring bloom from which future anomalies can be observed.

METHODS

The critical depth was determined using expression 2 above. The extinction coefficient, k_e (m^{-1}), was calculated for maximum and minimum values for each month from Secchi disc data accumulated during 1957 through 1962 at Ocean Station P (Parsons, MS, 1965). These values are entered in Table

I as extinction coefficients for blue light as derived from the formula given by Poole and Atkins (1929):

$$k_e = \frac{1.7}{D},$$

where D is the Secchi disc depth in metres. For the remainder of the eastern subarctic Pacific, Utterback and Jorgensen's (1934) oceanic average extinction for the area, 0.073 m^{-1} at 4800 A, was employed. This value is representative of minimum values at Station P during the spring. Maximum extinctions during the spring at Station P also were included in calculations of the critical depth in the remainder of the northeast Pacific by using the maximum extinction for March of 0.13 m^{-1} (Table I).

The value used by Sverdrup (1953) for I_c, the energy at the compensation depth, was taken from Jenkin (1937) as being 0.13 ly/hr. This value was determined for a day length of 16 hr, which suggests that a slightly larger value might be considered for the spring when the day length is less. However, in the absence of sufficient information on the effect of light and dark periods on the compensation light intensity, the value of 0.13 ly/hr was employed

TABLE I. Estimations of the critical depth at Ocean Weather Station P (50°N, 145°W).

Month	Mean radiation[a] (*ly/hr*)	Mean radiation corrected for reflection (*ly/hr*)	PAR[b] (*ly/hr*)	Compensation light intensity (*ly/hr*)	Extinction coefficient k_{max} (m^{-1})	Extinction coefficient k_{min} (m^{-1})	Critical depth (*m*) Max	Critical depth (*m*) Min
Jan.	2.91	2.30	0.46	0.13	0.11	0.065	54	32
Feb.	4.73	4.11	0.82	0.13	0.12	0.068	93	53
Mar.	8.50	7.82	1.56	0.13	0.13	0.074	162	98
Apr.	11.2	10.6	2.12	0.13	0.15	0.081	200	108
May	15.3	14.7	2.94	0.13	0.18	0.089	253	126
June	15.3	14.7	2.94	0.13	0.20	0.095	237	113
July	14.4	13.8	2.76	0.13	0.23	0.10	211	92
Aug.	12.5	11.9	2.38	0.13	0.24	0.097	189	77
Sept.	10.3	9.6	1.92	0.13	0.23	0.089	165	64
Oct.	7.25	6.45	1.29	0.13	0.19	0.083	121	52
Nov.	3.71	3.15	0.63	0.13	0.15	0.076	62	32
Dec.	2.16	1.68	0.34	0.13	0.13	0.071	37	20

[a]Average radiation for each month determined from January 1960 to February 1964 (Monthly Radiation Summary, Meteorological Branch, Dept. Transport, Canada).
[b]Photosynthetically active radiation.

here for comparability with other authors (Marshall, 1958; Sverdrup, 1953; Cushing, 1962).

The average energy available for photosynthesis per unit time, I_e, was determined at Station P from the total solar radiation measured with an Epply pyrheliometer. These values, averaged from January 1960 to February 1964, were expressed as the mean hourly radiation and corrected for reflection losses by determining the mean sun altitude for each month (Sverdrup, 1953). The amount of energy available for photosynthesis was determined by reducing the total radiation by a factor of 0.2 to allow for absorption of nonphotosynthetic energy in the first metre of sea water (Sverdrup, 1953).

For the rest of the eastern subarctic Pacific, solar radiation estimates computed by the U.S. Bureau of Commercial Fisheries, San Diego (Marine Weather Observation Summary for the Pacific Ocean), were utilized. These data, which were available by 5° squares of latitude and longitude for the years 1962 through 1964, were averaged and expressed as the hourly photosynthetic radiation, corrected for cloud cover and for reflection, for each month.

Estimates of the depth of the mixed layer during February for the years 1947 through 1963 were taken from Giovando and Robinson (MS, 1965). For the areas 55–60°N, 150–160°W, and 40–45°N, 150–160°W, where no statistical estimation of the depth for the mixed layer has been made by these authors, an approximation of the mean mixed layer depth plus or minus one standard deviation was made from bathythermograph data (available at the Pacific Oceanographic Group, Nanaimo) for January to April during the years 1957 and 1962 through 1964. Data on the mixed layer depth obtained by Giovando and Robinson (MS, 1965) are represented in Fig. 1 for each month

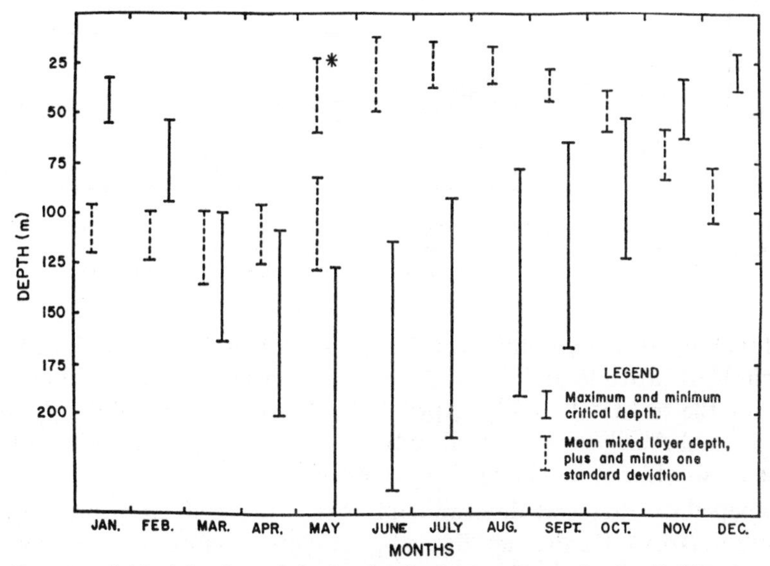

FIG. 1. Critical depths and the depths of mixing at Ocean Station P. *(During the years 1956–59 and 1962–63 the new seasonal thermocline was formed near the middle of May.)

of the year. In Fig. 2 are shown values of the mixed layer depth from north to south along lines of longitude, at 10° intervals, from 125 to 155°W. In this manner, estimates of the critical depth, which were made by 5° intervals of latitude and 10° intervals of longitude, can be compared with estimates of the mixed layer depth, which have been reported by areas composed of 5° intervals of latitude and 2° intervals of longitude (Giovando and Robinson, MS, 1965). In reporting the latter results, where two areas of latitude and longitude are adjacent (east and west) to a common line of longitude, the larger maximum and the smaller minimum values for the mixed layer depth are entered in Fig. 2.

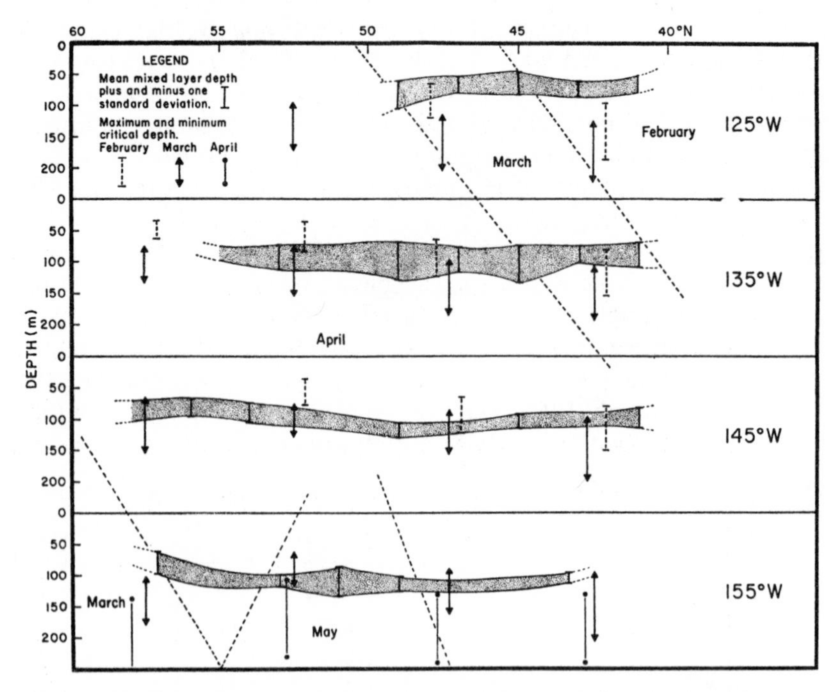

FIG. 2. Critical depths and the depths of mixing in the eastern subarctic Pacific Ocean, February–April. (Dotted lines separate areas in which conditions for a net increase in primary production became established in the same month.)

Maximum and minimum critical depths for March are also shown in Fig. 2 for each 5° of latitude and 10° of longitude. Critical depths for February are shown for 125 to 145°W and for April at 155°W. Other critical depth values for February and April were omitted only to clarify the figure and the following arguments would not be changed by their inclusion.

Copepod weights were determined from samples collected during April with a standard North Pacific net hauled vertically from 150 m to the surface. Total wet weight of copepods averaged by areas composed of 2° of latitude and 10° of longitude for the years 1962 and 1963 are shown in Fig. 3. Total copepod wet weights at Station P averaged by month for the years 1956 through 1964 are reported as an inset to Fig. 3.

RESULTS AND DISCUSSION

From December to February the maximum critical depth at Station P was less than the minimum depth of mixing (Fig. 1). This would indicate that there could be no net increase in primary production at Station P during this period. On the other hand, from May through September the minimum critical depth was equal to or greater than the maximum depth of mixing, indicating that this is the principal period in which a net increase in primary production

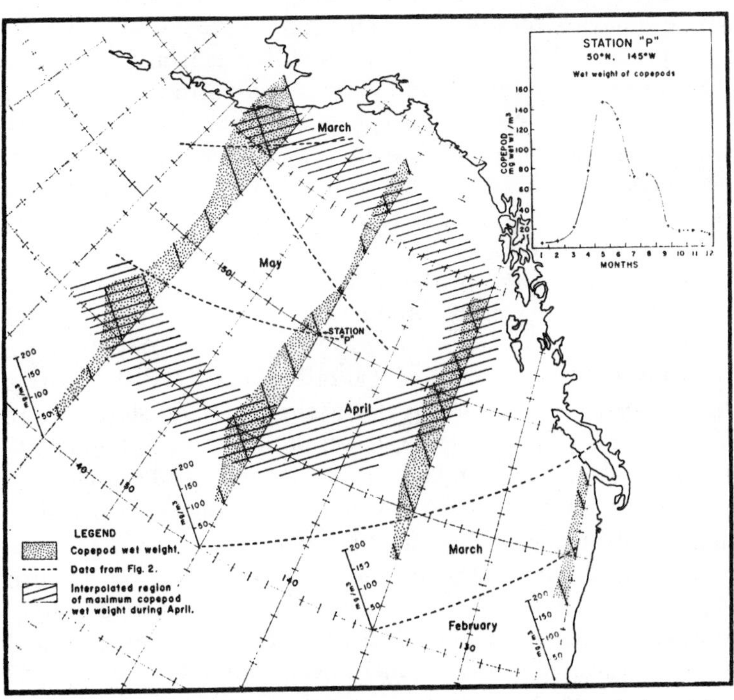

FIG. 3. Copepod wet weights during April and occurrence of the spring bloom in the eastern subarctic Pacific Ocean, February–May.

can occur. For 2 months during the spring, March and April, and 2 months during the fall, October and November, the conditions for a net increase in primary production were such as to indicate that some increase may take place under favourable conditions of radiation and stability.

In contrast to Fig. 1, it may be seen in Fig. 2 that, throughout a small area south of 45°N and east of 135°W, conditions for a net increase in primary production were possible during February. During March this area extended to 50°N at 125°W, and to about 45°N at 135°W. Another small area in which a net increase in primary production was possible developed during March in the vicinity of 55°N and 155°W. During April, critical depths exceeded the depth of the mixed layer over the whole eastern subarctic region except between about 50° and 55°N at 155°W. Included also in this area (but not apparent from Fig. 2) should be the results for Station P (Fig. 1), for which the minimum critical depth was still above the maximum depth of mixing for April.

Dotted lines on Fig. 2, drawn to separate areas for which conditions for a net increase in primary production become established in the same month, were transposed to Fig. 3 where average copepod wet weights are shown for the month of April. An area of approximate copepod maxima was drawn as an interpolation of the results of copepod wet weights shown in this figure. Comparison of these results shows that maximum copepod weights were encountered in areas north and south of Station P in a semi-circle approximating the areas in which a net increase in primary production occurred during March to April. Minimum copepod weights were encountered in a central portion of the northeast Pacific between a wide area at 155°W and a narrower area extending east beyond Station P. This area of low copepod weight approximated the area in which conditions for a net increase in primary production were not firmly established until May. These results are further substantiated by the inset to Fig. 3 which shows that maximum copepod weights at Station P do not occur until the period May–June.

Other evidence supporting the description of the spring bloom discussed here can be found in primary production data for Station P (McAllister, 1962). These data show values of up to 1100 mg C/m^2/day during the period May–June compared with values of 200 or less mg C/m^2/day during March–April. Data reported by Stefansson and Richards (1964) show that in the area of 40–45°N and 125–130°W nitrate depletion starts in February and the nutrient becomes exhausted from the surface layers by May. These results are also in keeping with the sequence of events shown in Fig. 2 and 3.

Semina (1960) has described conditions for the onset of the spring bloom in the northwestern part of the Pacific Ocean, north of 55°N. Although the exact timing of events is not reported, the early onset of a spring bloom is indicated as far north as the Gulf of Anadyr. This is reported to be due to the high stability in the area caused by melting ice and river discharge. The former stabilizing influence is also reported in studies by Marshall (1958) in the Barents Sea. In this area, the onset of the spring bloom was reported to occur in April whereas, in Atlantic waters further south, sufficient stability was not achieved until June of the same year. In contrast to these results the early establishment of favourable conditions for a spring bloom in the northwestern corner of the Gulf of Alaska (Fig. 3) appears to be influenced more by solar radiation than by fresh water. Thus the mean total monthly radiation corrected for cloud cover and reflection during the period March–May in the region 55–60°N and 155°W is 300 ly/month compared with 207 ly/month in the region 45–55°N and 155°W (U.S. Bureau of Commercial Fisheries, San Diego, Marine Weather Observation Summary for the Pacific Ocean).

In conclusion, timing of the spring bloom in the northeast Pacific can be inferred from a knowledge of the critical depth and the mixed layer depth during the 4-month period, February–May. However, the description given here applies to a protracted period in which other factors, such as currents, may also be significant in determining distribution of plankton. Thus while the general shape of the curves showing the distribution of copepods in Fig. 3 is in agreement with the description of the spring phytoplankton bloom, the absolute quantities of copepods at 155°W are greater than along other lines of longitude. An explanation for this type of variation is currently under investigation but it appears initially that one solution is to be found in separating growth rate and recruitment of the copepod stock. Preliminary results indicate that only the former is influenced by the timing of the spring bloom.

ACKNOWLEDGMENTS

The authors are grateful to the U.S. Bureau of Commercial Fisheries, San Diego, for making available the computed solar radiation data corrected for cloud cover and reflection in the northeast Pacific Ocean.

REFERENCES

CUSHING, D. H. 1962. An alternative method of estimating the critical depth. J. Conseil, Conseil Perm. Intern. Exploration Mer, **27**: 131–140.

GRAN, H. H., AND T. BRAARUD. 1935. A quantitative study of the phytoplankton in the Bay of Fundy and the Gulf of Maine. J. Biol. Bd. Canada, **1**: 210–227.

GIOVANDO, L. F., AND M. K. ROBINSON. MS, 1966. Characteristics of the surface layer in the northeast Pacific Ocean. Fish. Res. Bd. Canada, Oceanogr. Limnol. MS Rep. (In preparation).

HEINRICH, A. K. 1962. The life histories of plankton animals and seasonal cycles of plankton communities in the Oceans. J. Conseil, Conseil Perm. Int. Exploration Mer, **27**: 15–24.

JENKIN, P. M. 1937. Oxygen production by the diatom *Coscinodiscus excentricus* Ehr. in relation to submarine illumination in the English Channel. J. Mar. Biol. Ass. U.K., **22**: 301–343.

MCALLISTER, C. D. MS, 1962. Photosynthesis and chlorophyll *a* measurements at Ocean Weather Station "P," July 1959 to November 1961. Data record. Fish. Res. Bd. Canada, Oceanogr. Limnol. MS Rep., No. 126, 14 p.

MCALLISTER, C. D., T. R. PARSONS, AND J. D. H. STRICKLAND. 1960. Primary productivity and fertility at Station "P" in the northeast Pacific Ocean. J. Conseil, Conseil Perm. Intern. Exploration Mer, **25**: 240–259.

MARSHALL, P. T. 1958. Primary production in the Arctic. Ibid., **23**: 173–177.

PARSONS, T. R. MS, 1965. A general description of some factors governing primary production in the Strait of Georgia, Hecate Strait and Queen Charlotte Sound, and the N.E. Pacific Ocean. Fish. Res. Bd. Canada, Oceanogr. Limnol. MS Rep., No. 193, 34 p.

POOLE, H. H., AND W. R. G. ATKINS. 1929. Photoelectric measurements of submarine illumination throughout the year. J. Mar. Biol. Ass. U.K., **16**: 297–324.

SEMINA, J. H. 1960. The influence of vertical circulation on the phytoplankton of the Bering Sea. Int. Revue Ges. Hydrobiol., **45**: 1–10.

STEFANSSON, V., AND F. A. RICHARDS. 1964. Distribution of dissolved oxygen, density and nutrients off the Washington and Oregon coasts. Deep-Sea Res., **11**: 353–380.

SVERDRUP, H. U. 1953. On conditions for the vernal blooming of phytoplankton. J. Conseil, Conseil Perm. Intern. Exploration Mer, **18**: 287–295.

RILEY, G. A. 1957. Phytoplankton of the north central Sargasso Sea. Limnol. Oceanogr., **2**: 252–270.

UTTERBACK, C. L., AND W. JORGENSEN. 1934. Absorption of daylight in the North Pacific Ocean. J. Conseil, Conseil Perm. Intern. Exploration Mer, **9**: 197–209.

THE QUANTITATIVE ECOLOGY OF MARINE PHYTOPLANKTON

By JOHN H. STEELE

Scottish Home Department Marine Laboratory, Torry, Aberdeen

(*Received* 14 *February* 1958)

CONTENTS

I. INTRODUCTION

The study of organic production in the sea has a recent history rather like that of a backward country into which new techniques have been introduced. It has advanced very quickly, perhaps too quickly. Although the broad relationships between the water, the phytoplankton and the zooplankton are known, and have been fully described in *Biological Reviews* by Harvey (1942), many of the details, chemical and biological, are still unknown. Yet such details may be fundamental in the attempts which are being made to fix numerical values for the primary production of organic matter in the sea, and to explain these values in terms of the environment.

One reason for this rapid development is that large numbers of microscopic plants drifting passively in the sea form very suitable material for the mathematical methods used in population studies. Further, for the mathematics to be manageable, the postulates must be simple; thus a comparative lack of knowledge is almost an asset, since the investigator is not distracted by a plethora of facts. Yet this mathematical approach to phytoplankton populations has been moderately successful, which suggests that the factors which control production in the sea are indeed simple. This conclusion is, however, very doubtful because of the weakness in our knowledge of details of the basic relations. This uncertainty is increased by the present inability to relate the results to work in other branches of the subject: for example, to work with laboratory cultures on vitamin B_{12} requirements.

The purpose of this article, then, is to describe these mathematical studies, but since the mathematics are merely a logical structure, the discussion is really concerned with the information used to set up the postulates and to test the conclusions. By implication it also has to be concerned with the information which is ignored.

Since Harvey has reviewed the general developments in this subject up to 1940, the work discussed here will be limited mainly to more recent studies.

II. DESCRIPTION OF THE PROBLEM

Although mathematical equations are only a means of determining the logical consequences of a set of postulates, by their nature they impose restrictions on the type of postulates which can be used in this way. To show how this has affected the study of plant populations in the sea, it is convenient to begin with the first mathematical 'model' which was developed.

Fleming (1939) studied a well-known feature of the phytoplankton cycle—the spring outburst. In the spring the number of plants increases rapidly and then dies down again. The plant species involved may change during this period and as these species can vary greatly in size, the usual method of cell counts was not an adequate estimate of the quantity of organic matter present. For this reason Fleming turned to measurements made in the English Channel by Harvey, Cooper, Lebour & Russell (1935) of concentration of plant pigments as a factor common to all species.

The initial rapid growth of this plant population can be explained by the increasing light during the spring, and the simplest assumption to make about this growth, as Fleming did, is that its rate is constant. The ultimate decrease of the population must, however, be explained by some other factor, and for this Harvey *et al.* had given evidence that the spring outburst was limited in time and quantity by the grazing of the herbivorous plankton animals. Since a constant rate of grazing by a fixed size of zooplankton population would merely produce a slower increase in the plants, Fleming postulated a linear increase in the animal population.

With these postulates he set up an equation (see Appendix) which produced the bell-shaped curve shown in Fig. 1 and which very roughly fitted the data. He thus showed that the spring outburst could be largely explained by two simple postulates: a constant growth of plants and a constantly increasing amount of grazing. From his equation he could deduce the production of organic matter if he knew the relation between plant pigment and organic matter in phytoplankton.

It can be seen that the species of plant-eating animals, like those of the plants, is not directly relevant. What is required is an estimate of grazing in terms of some measure of the 'amount' of herbivores present, independent of their species composition: a simple figure to put into an equation rather than a more complex, and arithmetically cumbersome, description of the population. It is in this way that the mathematics select the appropriate data.

It is obvious that Fleming's model is inadequate for further development, since, in fact, the zooplankton population will not go on increasing. Thus it is now necessary to consider some of the factors which Fleming neglected. There is the possibility that the plants are not neutrally buoyant and by sinking will leave the euphotic zone where photosynthesis exceeds respiration. There is the fact that nutrients are not plentiful and as the plants remove these nutrients the low concentrations may limit further growth. There are changes in photosynthesis due to variation in light intensity. Lastly, the rate of vertical mixing between the euphotic zone and the lower waters will affect both the supply of nutrients and the concentration of plants.

All these features have been given the necessary exact formulations for use in the later mathematical models. Is the evidence sufficiently good for such exactitude?

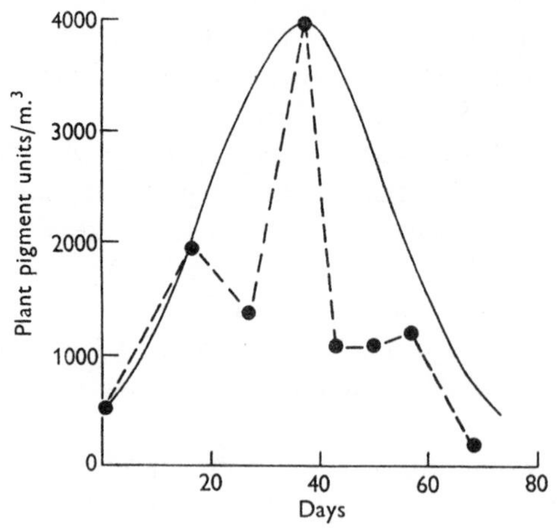

Fig. 1. Observed (– – ● – –) and calculated (——) plant pigments in the English Channel.

III. THE POSTULATES

(1) *Plant population*

As their index of plant population Harvey *et al.* (1935) used a visual estimate of the colour density of the plant pigments of plankton extracted with 80% acetone. This approach had advantages in simplicity and generality, and it was developed first into a colorometric estimate of chlorophyll (Krey, 1939) and then as a spectroscopic analysis of the chlorophylls and other pigments found in algae (Richards, 1952; Richards & Thompson, 1952).

These developments, though they add to our knowledge of the absorption spectra of the pigments and the variations of these spectra, do not solve the main problem in the use of chlorophyll in population studies. This work requires a conversion factor from chlorophyll to some more general measure of living matter, such as carbon content, and it has become apparent that there is no constant factor. Tropical phytoplankton in general probably contains less chlorophyll per unit of carbon than that of temperate waters (Riley, Stommel & Bumpus, 1949) and this may be due to adaptation similar to the light-shade effects found in land plants (Rabinowitch, 1945, p. 419). There may also be species differences, since Gillbricht

9-2

(1952) gives evidence to suggest that dinoflagellates have only one-third the content compared with diatoms. Finally, not all the chlorophyll may be in living cells but partly associated with plant detritus. Gillbricht estimates that for the turbid waters of Kiel Bay only one-quarter of the chlorophyll concentration is derived from living plants.

These difficulties are part of the reason for the great variability in conversion factors, but a further reason is that these values are found by indirect methods, statistical or chemical. An idea of the range is shown by the following ratios of carbon to chlorophyll by weight:

phytoplankton in temperate waters (Riley *et al.* 1949)	57
phytoplankton in tropical waters (Riley *et al.* 1949)	150
phytoplankton in the English Channel (Harvey, 1950)	26
diatoms in Kiel Bay (Gillbricht, 1952)	4
peridinians in Kiel Bay (Gillbricht, 1952)	12

On this basis any value is open to question.

Thus the original aim of an index of plant population easily measured and of general application has not been fulfilled. Yet no other methods are at present available, although Krey (1951) has suggested that estimates of protein may provide a better indication of living matter. The most that may be said for chlorophyll is that for a given offshore region where detritus in suspension should be relatively unimportant, it may provide a rough quantitative measure of plant population.

(2) *Light intensity, photosynthesis and respiration*

Since the light intensity at any depth in the sea can be found from a knowledge of the surface illumination and of the transparency of the water, the information needed is the relation between the rate of photosynthesis of the plants and light intensity. The simplest assumption is that this dependence is a strict proportionality (Riley, 1946), but although this holds for the lower intensities there is a level at which the rate is at a maximum and above which photosynthesis tends to be inhibited. This effect has been demonstrated in the sea (Jenkin, 1937; Steeman Nielsen & Jensen, 1957) and many times in the laboratory (Rabinowitch, 1951, ch. 28). From Rabinowitch's review it is evident that the relationship between light intensity and photosynthetic rate is not fixed even for a given species, but depends partly on the previous 'light history' of the plants. There is evidence, however, from experiments by Ryther (1956) that the inhibition effects show similarities within broad groups, chlorophytes, diatoms, dinoflagellates.

This similarity has led Ryther (1956; Ryther & Yentsch, 1957) to construct a formula from which photosynthetic rates can be deduced in terms of light intensity at the surface, the extinction coefficient of the water, and the chlorophyll concentration. From the light/photosynthesis curves for the three groups of plants, a mean curve is constructed for general phytoplankton: the rate for any light intensity being given in terms of the maximum rate of carbon assimilation. For the ratio of this maximum rate to chlorophyll *a* content Ryther and Yentsch take a value of 3·7, which is the mean of a number of observations, mainly original but including three quoted from other workers. On this basis, using chlorophyll data, and light and transparency readings, it is possible to calculate the total photosynthesis in a water column during any period of time. When applied to data from the sea Ryther and Yentsch found good agreement with observations of photosynthetic rate which varied by a factor of about ten. An essentially similar approach has been used by Talling (1957) to compute the photosynthesis of freshwater algae.

As Ryther and Yentsch say, this method of measuring photosynthesis has the advantage of simplicity compared with the more direct but more elaborate experimental techniques (see §IV). It shares, however, one difficulty with them; the most interesting value is the net production of organic matter during 24 hours and to derive this an estimate of the loss due to plant respiration is required. A value may be obtained by assuming that the ratio of maximum photosynthesis to respiration is 10:1 and this has been used by Ryther (1956) and Steeman Nielsen & Jensen (1957). Kok (1952), however, as reported by Ryther, has shown that *Chlorella* can have a ratio varying from 20:1 to 1·4:1 depending on environmental conditions.

Another, and perhaps more natural, way of estimating respiration is to assume that respiration per unit of plant carbon depends only on temperature. In their mathematical work Riley *et al.* (1949) used a formula based on this supposition, which they constructed from experimental data and statistical analysis of natural phytoplankton associations. Unfortunately the practical use of this formula in conjunction with chlorophyll data involves the problem of carbon/chlorophyll ratios mentioned in the preceding section.

An interesting example of the interrelations between these factors is given by Fleischer (1935) who showed that chlorotic cells of *Chlorella* could be produced by nitrate deficiency; yet, over a wide range of chlorophyll content per cell, gross photosynthesis was proportional to chlorophyll. This supports the assumption of a constant assimilation number, but indicates that the chlorophyll/carbon ratio will be very variable, and thus also, probably, the chlorophyll/respiration ratio.

So, finally, it is not possible to pass from chlorophyll to production in terms of light intensity alone and so avoid the complications raised by other variations in the environment, in particular the problem of nutrient limitation.

(3) *Nutrient requirements of plants*

It is only in terms of the supply of nutrients that the great differences in productivity of different areas can be explained. There are many examples of low nutrient values corresponding to sparse vegetation, and, on the other hand, in regions where nutrients well up from deeper waters, the plant populations are dense. The problem is to find data sufficiently exact and general to provide a basis for the quantitative formulation of this limiting effect of nutriments.

In sea water the two main nutrients which limit production are phosphorus and nitrogen compounds, but, owing to difficulties of estimating the forms in which nitrogen occurs (Harvey, 1955) and the comparative ease of making inorganic phosphate measurements, it is the phosphates on which most work has been done, both in the laboratory and in the sea.

Ketchum (1939*a*) found for cultures of *Phaeodactylum tricornutum* (originally called *Nitzschia closterium* forma *minutissima*; see Hendey, 1954) that there appears to be a decrease in the growth rate as the phosphate concentration falls below about 0·55 μg. at. P/l. Comparable results had been obtained by Harvey (1933) using the same species. This, together with data implying similar changes in the sea, led Riley (1946) to make the simple assumption that the relation of nutrients to photosynthesis was a linear decrease below 0·5 μg. at. P/l. For other species, however, other values have been found: Barker (1935) gives evidence that the limiting concentration for the dinoflagellate *Prorocentrum micans* was below 0·30 μg. at. P/l. For the marine diatom *Asterionella japonica* Goldberg, Walker & Whisenand (1951), using the isotope ^{32}P, found inhibition of growth at about 0·25 μg. at. P/l.

There are, however, many complications. Ketchum (1939*b*) has shown that plant cells grown in a non-nutrient medium contain only one-third of the normal phosphorus and one-half of the normal nitrogen; these deficiencies can be eliminated later by absorption from a nutrient medium both in dark and in light. The freshwater diatom *Asterionella formosa* can concentrate phosphorus from a medium containing less than 0·03 μg. at. P/l. (Mackereth, 1953) and the phosphorus is sufficient for considerable growth during which the cell phosphorus is decreased. On this basis Mackereth deduced that in Windermere silica and not phosphorus limits the growth of *Asterionella*, even though the phosphorus concentration in the water is never above 0·06 μg. at. P/l.

Another complexity in fresh water is shown by the work of Rodhe (1948) and was confirmed by Mackereth in Windermere. Rodhe found that *Asterionella* from Lake Erken, in Sweden, grown in an artificial medium needed 32 μg. at. P/l. for maximal growth, whereas in sterilized lake water only 0·1 μg. at. P/l. was required. Rodhe concludes that lake water contains one or more factors facilitating the use of phosphorus by *Asterionella* at very low concentrations. For *Scenedesmus quadricauda*, on the other hand, Rodhe found that lake water permitted no reduction from the high maximal phosphorus concentration.

A different problem concerns the use of organic forms of nutrients. Chu (1946) showed that bacteria-free algal cultures could use some organic phosphorus compounds for growth; for example, *Skeletonema costatum* grew as well with phytin as with orthophosphate. Harvey (1953) found that phosphorus-deficient cells could take up organic phosphorus place in the dark. In the sea organic phosphorus is often a significant proportion of the total (Redfield, Smith & Ketchum, 1937) and sometimes plentiful when inorganic phosphate is scarce (Harvey, 1950). The nature of the compounds present and whether they are used directly in the sea is not known.

In general, the interpretation of laboratory experiments in terms of an index for nutrient limitation is very difficult owing to the artificial conditions involved. The concentrations of plants have usually been very high and the initial levels of nutrients may be a hundred times that found in the sea. With these concentrations it has been observed, using the isotope ^{32}P, that the phosphorus content of the plants varies with the concentration in the medium (Gest & Kamen, 1948; Goldberg *et al.* 1951) Much of this phosphorus, however, is labile as it can be removed by washing the cells; in high concentrations about 50–70% can be removed in this way. Further, algae which have been washed free of this excess phosphate appear to retain normal photosynthetic activity (Kamen & Spiegelman, 1948).

Another artificiality of culture experiments is that high concentrations of plants and nutrients give rates of change of these factors which are much greater than the rates found in natural conditions. For example, the limiting value for phosphate of 0·25 μg. at. P/l. in the experiment of Goldberg *et al.* is derived from a culture in which the phosphate in the medium, starting at 2·7 μg. at. P/l., dropped from 1·8 to 0·25 μg. at. P/l. in one day, after which exponential growth ceased. Because of the large changes in such a short time, the value of 0·25 may be too low as a result of the plants using phosphorus in a previously mentioned 'labile' form taken up originally at the high concentrations.

For the other main limiting nutrient, nitrogen, very little is known about limiting values, although Ketchum (1939 *a*) found no effect down to 3·4 μg. at. N/l. for *Phaeodactylum*. Thus the question is whether phosphorus can be used as a general nutrient index, and this is linked to the problem of phosphorus/nitrogen ratios in the sea and in plankton. The original work, reviewed by Sverdrup, Johnson & Fleming (1942), showed that in general, and independent of the particular concentrations in the sea, the P:N ratio was roughly constant at about 1 : 15 atoms both in the sea and in plankton. Many anomalous ratios have, however, been found for certain bodies of water, although these are often enclosed areas, such as the Mediterranean with 1:30 (Cooper, 1937), or Moriches Bay on the east coast of America which is affected by pollution (Ryther, 1954*a*). In a detailed study of Long Island Sound, Riley & Conover (1956) found that in mid-winter the ratio was only 1:8 and in summer nitrate was virtually absent although phosphorus was still comparatively high at 0·5 μg. at. P/l.; in other words, nitrogen not phosphorus was then the limiting factor.

The conclusion to be drawn from all these studies is that, although there is general agreement that nutrients limit plant production, the way in which this happens and the levels at which deficiencies become effective cannot be fixed. The results described here show merely the external effects of complex metabolic changes resulting from lack of nutrients. Phosphorus is essential for the photosynthetic reaction itself and also for nearly all other metabolic processes (Glass, 1951, 1952). Nitrogen deficiency produces large changes in the protein content (Spoehr & Milner, 1949). At present, however, these kinds of results cannot be used for the quantitative description of the relation between nutrient deficiency and production of organic matter. Yet until this is possible no truly adequate formulae are likely to be obtained.

(4) *Regeneration of nutrients*

The other main factor affecting quantitative studies of the nutrient cycle is the rate at which the nutrients return to soluble inorganic form as the result of excretion and of the decomposition of organic matter.

To measure the excretion of inorganic phosphorus by zooplankton Gardiner (1937) kept net hauls of plankton in jars for periods of 3 hr. For seven collections of about 200–400 animals, mainly copepods, each in 200 ml. of water the phosphate content of the water increased by $0{\cdot}82 \times 10^{-3}$ μg. at. P/l./animal or about 0·0013 μg. at. P/animal/day. Cushing (1954), using a similar method with *Calanus finmarchicus* Stage V, found a rate of 0·01 μg. at. P/animal/day. An indirect idea of expected excretion can be obtained by assuming that zooplankton excretes phosphorus in proportion to its respiration of carbon. Using the ratio for carbon to phosphorus normally found in plankton (Sverdrup *et al.* 1942) and data on the respiration of *Calanus* Stage V from Marshall & Orr (1955*a*), a value of the order 0·0016 μg. at. P/animal/day is obtained, agreeing with Gardiner.

This divergence is related to disagreement concerning the assimilation of the plant material eaten by herbivores. During the spring outburst Harvey *et al.* (1935) noted that the faecal pellets contained large amounts of plant pigments and assumed that in dense plant concentrations the animals only digested part of their food, excreting most of it in a partially decomposed form which allowed very rapid regeneration of nutrients. This might be used to explain the different excretion rates of Gardiner and Cushing since Gardiner's experiments were done in the autumn when plant life might be slight, whereas Cushing's took place in the spring (personal communication). Recent work, however, by Marshall & Orr (1955*b*, *c*), using phytoplankton cultures labelled with ^{32}P and ^{14}C as food for *Calanus*, has shown that even in dense cultures assimilation is as high as 80%.

There may also be regeneration of phosphorus due to excretion by the plants themselves, and this has been studied using the isotope ^{32}P. Pratt (1950), with a mixed algal population grown in concrete tanks, found that when the assimilation rate was 0·36 μg. at. P/l./day, the regeneration rate had the comparatively high value of 0·13 μg. at. P/l./day. On the other hand, Goldberg *et al.* (1951) kept a culture in the dark at the end of the exponential growth phase and found that the release of phosphorus was extremely slow; only about 5% of the phosphorus originally in solution was returned in 20 days. Neither this culture nor Pratt's tanks was bacteria-free.

The third mode of regeneration is from the decomposition of dead plankton. Seiwell & Seiwell (1938) found a very rapid liberation of phosphate from zooplankton during the first 24 hr. but a comparatively slow rate thereafter. Similar results have been obtained by Cooper (1935) and Hoffman (1956). For phytoplankton Cooper found a considerably slower rate. A correspondingly rapid liberation of nitrogen as ammonia has been shown by von Brandt, Rakestraw & Renn (1937). The ammonia is then changed to nitrite and the nitrite in turn to nitrate. These changes are mainly due to bacteria, which show a rapid development. The rapidity of the phosphorus changes is also probably due to bacteria which grow very quickly when sea water is enclosed in bottles (Waksman & Renn, 1936). Thus these experiments cannot be used to indicate the rates of regeneration in the sea.

From investigation of the phosphorus cycle in the Gulf of Maine, Redfield *et al.* (1937) suggested that the organic matter produced during the spring outburst sank below the productive upper layers before decomposing. From phosphate data for the North Sea, I inferred that regeneration of phosphorus took place 2 months after its assimilation and was found mainly below the euphotic zone (Steele, 1956). These types of studies would not reveal rapid release and re-assimilation of phosphorus in the euphotic zone, but I found that during both spring and summer plant production estimates from phosphate data were not lower than estimates made by use of the carbon isotope ^{14}C (Steele, 1957*a* and see §IV). This implies that for the area studied such rapid recycling does not occur to any great extent.

But again, apparently different results have been obtained in fresh water where, using ^{32}P added to small lakes, it has been shown that there is rapid exchange between the water, the plant life and the bottom (Hutchison & Bowen, 1950; Hayes & Coffin, 1951). It was found by Rigler (1956) that the turnover time for phosphate in the surface waters of such lakes was less than 1 hr. and he deduced that this was due almost entirely to bacteria. Since these bacterial populations are perhaps a thousand times greater than those of the open sea, these results may not be relevant; yet as with the other results from fresh water, it is not possible to ignore them completely.

These data together show a range of values from very slow to very fast regeneration, and although laboratory experiments can always be dismissed either because bacteria were present, or because they were not, interpretations of natural changes are very indirect. Slow rates of decay seem rather more probable but there is little certainty about this.

(5) *Sinking of plants*

Gran (1915) pointed out that the vertical distribution of plants in the sea could be explained only by assuming that the plants were sinking. From the gradual increase in depth of a particular population Allen (1932) suggested a rate of about 4 m./day. When, however, Allen observed diatoms in the laboratory the rate was up to 20–30 m./day, yet Riley (1943) with *Nitzschia* found only 0·06 m./day.

The difficulty in the sea is that the plants as well as sinking are also being grazed on and are subject to the effects of vertical mixing of the water. To combine these factors requires a quantitative formulation of the way in which they are interrelated. Thus the values for sinking rates in the sea are part of the results from the use of mathematical models. Riley *et al.* (1949), from the study of vertical distributions of chlorophyll, deduced that the rate must lie between 2·6 and 3·6 m./day; from changes with time of chlorophyll distribution, I found (1956, 1957*a*) that the rates were nearly always between 1·4 and 6·0 m./day; K. P. Andersen (in Steeman Nielsen & Jensen, 1957) from cell counts gave a value of 11 m./day. Considering the wide range of areas studied, these results are in comparatively good agreement.

The problem raised by such results is of a different kind. What exactly is meant by a 'sinking rate' deduced in such an indirect manner from a series of numerical operations? In what way does it refer to a physical fact; to what extent is it merely a term defined by its place in a set of equations? Gross & Zeuthen (1948) suggested that healthy plants do not sink because they can use energy to reduce their density by maintaining low concentrations of divalent ions. Munk & Riley (1952) have pointed out, however, that in order to maintain a nutrient supply, plants must change their immediate environment and, except for flagellates, this can only be done by sinking. Thus it is uncertain whether the 'sinking rate' applies generally to the population or mainly to less viable parts of it, or perhaps even to dead material.

(6) *Grazing by herbivores*

Indirect mathematical estimates of grazing have also been made and the results are stated as the rates at which water is filtered by unit weights of zooplankton. The mean values for a year given by Riley *et al.* (1949) for different areas range from 1·3 to 3·0 m.^{3}/g. zooplankton carbon. Nine out of ten values for the North Sea calculated at the same time as the sinking rates vary from 1·0 to 7·7 m.^{3}/g. (Steele, 1956, 1957*a*). Again this is comparatively good agreement, but once more the indirect approach hides many difficulties. The use of a unit such as g. zooplankton carbon gives an appearance of generality but neglects the great variations in the structure of the animal population. At certain times the animals may be mainly herbivorous copepods but on other occasions the carnivorous *Pleurobrachia*, for example, may predominate. Further, there is the question whether the simple concept of filtration adequately explains the way in which herbivores feed.

Fortunately there has been considerable work on the behaviour of copepods, mainly by Marshall & Orr (1955*a*), who have studied among other factors the grazing and respiration of *Calanus finmarchicus*. In a review of their own and others' work with *Calanus* in the laboratory they imply that feeding is mainly by filtration rather than selection, but unfortunately the rates of filtration, ranging from 1 to 240 c.c./animal/day, are much too low to provide adequate feeding for *Calanus* in the sea as estimated from respiration rates and plant densities. In their natural environment Cushing (1954) has calculated that *Calanus* would need to filter between 1 and 3 l. per day. Cushing puts forward a possible explanation of this discrepancy in an 'encounter' theory of grazing. From the assumption that copepods sweep clear the area covered by the hairs projecting from the antennules, and using speeds taken from the literature on vertical migration, he calculates that the volume filtered per day could be as great as 3 l. (This work will be fully described in a future publication by Dr Cushing.) Cushing (1957) suggests that the low values found in experimental work are due to the smallness of the volumes of water used, since the highest values of 240 c.c. (Harvey, 1937) were obtained with half-litre beakers which were much larger than the vessels used by other workers. (In Harvey's experiments this held for only one species used as food, *Ditylium brightwelli*, results for other species being much lower.) To confirm this suggestion, Cushing fed *Temora longicornis* on *Skeletonema costatum* in bottles ranging from 5 to 500 c.c. and found that the filtering rates varied with the size of bottle from 6 to 150 c.c. per day.

It would seem that the anomalous experimental results may be explained in this way, but at present only the mathematical interpretations are available as estimates.

(7) *Vertical mixing*

The complex movements of water in the sea can be separated formally into transport and mixing processes. The large-scale transport of water in currents such as the Gulf Stream is the best known form of lateral movement and the largest vertical motion of water masses is found in those regions where the process known as 'upwelling' occurs (Sverdrup *et al.* 1942). Turbulent mixing, however, both lateral and vertical, is always present, especially in the surface layers of the sea. Thus, although lateral movements disperse and intermix plankton populations, and in special regions the upwelling of nutrients gives massive plant production, it is vertical turbulent mixing which generally controls the production of phytoplankton. This control operates through the restriction of production in the euphotic zone because of the limited supply of nutrients and the downward transport of plants: an increase in vertical mixing will carry up supplies of nutrients from the richer lower waters, permitting faster growth, yet at the same time greater mixing will remove more plants from the region of light and so tend to decrease production.

To estimate the balance of these effects, a measure of the vertical mixing must be obtained. This can be done by studying the changes in some conservative property such as temperature or salinity. These changes can provide a value for the coefficient of eddy diffusivity by their use in a mathematical formulation of turbulent mixing (Riley, 1946; Riley *et al.* 1949; Steele, 1956). The mathematical form used (see Appendix) is open to question, however, since there is no evidence that it can be applied to fluids where the density is varying (Goldstein, 1938).

IV. MEASURE OF PRODUCTION

These then are the various factors which are postulated to control plant life in the sea and which may be used in a mathematical model to describe variations in the distribution of plants. But since a main reason for this work is the attempt to determine the causes of changes in primary production, direct measures of this production would form the most valuable part of the available evidence.

Once more, the simplest and most general unit is carbon, but as already mentioned in §III(2) there are two definitions of production in terms of carbon converted from inorganic to organic form: total production given by the amount of carbon taken into the plants by photosynthesis, and net production which is the balance during any time interval between the gain due to photosynthesis and the loss due to respiration. Since the latter is more interesting, it is a pity that, of the two experimental techniques, the light-dark bottle method (Gaarder & Gran, 1927; Riley, 1939) measures total and the ^{14}C method (Steeman Nielsen & Jensen, 1957) gives an estimate somewhere between total and net production.

In the former method one part of a water sample is exposed to the light and the other kept in the dark. It is assumed that the respiration of plants, animals and bacteria will be the same in both bottles, and so, after a period of time, when the oxygen concentrations in the two bottles are measured the difference will be a measure of the total production. The drawback of this method is that when the photosynthetic rate is low, the samples have to be exposed to the light for a considerable time before there is significant difference between the oxygen concentrations: 3 days in the Sargasso (Riley, 1939). During such a long exposure extensive unnatural changes can be expected to occur in the bottles, especially through the large growths of bacteria discussed in §III(4). Beyond this, Steeman Nielsen (1955) has suggested that the unexpectedly high values found by Riley for supposedly barren areas are due to a smaller growth, and thus smaller oxygen consumption, by the bacteria in the light bottle owing to the production of an antibiotic by phytoplankton growing in the light.

A much more sensitive method has been developed by Steeman Nielsen (1952) using the radioactive isotope of carbon ^{14}C, to estimate the uptake of carbon by plants during a period of 4–6 hr. The full results from the world cruise of the *Galathea* are given by Steeman Nielsen & Jensen (1957). On this cruise simultaneous use of the ^{14}C and oxygen methods led to estimates where the 'oxygen' values for production were 200 times the ^{14}C values, the latter being corrected to 'total' production by adding 10% (Steeman Nielsen, 1954). An argument has arisen about whether such a small correction is sufficient in waters which are very low in nutrients. Ryther (1954*b*) gave experimental evidence that in nutrient-deficient cultures the uptake by photosynthesis was used almost immediately for respiration, so that the ^{14}C results were only one-twentieth of the oxygen estimates of photosynthesis. Steeman Nielsen & Al Kholy (1956) did not obtain this result and concluded that Ryther's experiments were not satisfactory. Once more it is only the external effects which are observed and the existence of this problem is due to ignorance of the changes in metabolism produced by nutrient deficiency. Thus although the ^{14}C method is extremely sensitive and is being widely used, the exact interpretation of observations is still doubtful.

An entirely different approach has been used by Cushing (1955) to measure the production rate of diatom populations. Owing to the mode of reproduction of diatoms the width of their cells decreases at each division, and by using an electron microscope it is possible to measure the thickness of the cell wall and the transapical width. These measurements made for each species in a population permit the number of divisions to be calculated. Knowing the average volume of the cells of each species it is possible to give a production rate for a mixed diatom population in terms of total cell volume. This method has the advantage that it is still related to the type and number of species present and is not abstract as are estimates in terms of carbon. Its disadvantages are the time and trouble required in making the measurements, and also the fact that its use is confined to the spring outburst when diatoms are the dominant forms.

Other and less direct estimates of production can be derived from observations of relevant quantities in the sea. The use of chlorophyll and light intensity has already been described in §III(2), but the main attempts have been through estimations of changes in oxygen and phosphate. These variations in concentration with time and depth will be due partly to water movement and partly to 'biological change'. If it is assumed that the hydrographic part is caused by vertical mixing and this can be measured from temperature distributions, then by suitable calculation its effect can be removed to provide an estimate of the 'biological change' alone.

In one way oxygen is more suitable than phosphate since it can be more directly related to production, but it has one major drawback—exchange through the sea surface. It has not been possible to provide accurate measures of this and so the value of the 'biological change' is doubtful (Redfield, 1948; Riley, 1956; Steele, 1957*b*). The problem with phosphate, on the other hand, is to relate a phosphate decrease in the upper layers to an increase in organic carbon. The simplest relation is that given by chemical analysis of plankton (Sverdrup *et al.* 1942), but all the difficulties mentioned in §III arise in this connexion, especially that of how much organic phosphorus is regenerated in the euphotic zone. The results show that there is general agreement between phosphate, oxygen and ^{14}C estimates (Riley, 1956; Steele, 1957*a*), although the details of depth distribution of phosphate uptake are anomalous. This implies that phosphate uptake and carbon production are not exactly equivalent in time, but that extraneous sources of phosphate, regeneration and organic phosphorus, are negligible.

These methods of using oceanographic data are very indirect compared with the ^{14}C technique, which although difficult to interpret gives immediately a value in terms of carbon uptake. The ^{14}C results do not, however, in themselves provide clues to explain why the production is large or small. It is for this reason that the phosphate method is complementary; it requires several postulates before providing an estimate of production, but this estimate is then related to two of the controlling factors, nutrient deficiency and vertical mixing. In this sense the phosphate method is already a mathematical model.

V. MODELS OF THE POPULATION CYCLE

The various aspects of the environment which have been listed here are those used in the mathematical models to be discussed. Their choice for these models is based on the extensive work which has been done to show that these factors are relevant to the changing plant populations observed in the sea. Thus the models are a continuation of previous studies and depend on their general conclusions. This earlier work, however, indicates at most that the variations of certain factors are similar and suggests the reasons for this similarity. Thus, Gran & Braarud (1935), working in the Bay of Fundy and the Gulf of Maine, found that though nutrients were more plentiful in the former area, the density of the plant populations was not greater. So, although nutrient limitation could explain the conditions offshore in the Gulf of Maine, it appeared that in the Bay of Fundy the turbidity and excessive turbulence of the water were the cause of low populations. Similarly, Sverdrup (1953), studying conditions in the Atlantic, showed that the bursts of phytoplankton depended on the ratio of the compensation depth (at which respiration equals photosynthesis) to the depth of the surface mixed layer, and so these bursts could be explained in terms of light, water mixing and turbidity.

The density, however, of plant population will be simultaneously controlled by the other factors discussed, namely photosynthetic rate, grazing, sinking and regeneration. The general pattern can be shown in broad outline, as Harvey (1950) has done, in summing up the long-term investigations in the English Channel, but for a detailed quantitative study of their interrelated effects it is finally necessary to use a mathematical model, where they can be considered together.

The change from one approach to the other is shown in the work of Riley (1946). In the first part of this paper he considered the statistical correlation between sets of observations collected during a year in an offshore area known as Georges Bank. He showed that about 70% of the variations in phytoplankton could be accounted for statistically by variations in depth, temperature, phosphate, nitrate and zooplankton. It could be seen that 'one factor after another gained momentary dominance so that each variable has a vastly different significance at different times of the year'. For this area it was known that depth is related to vertical mixing; also temperature can change the respiration rate of the plants.

In the second part of this paper Riley set up exact formulations of each of these effects. Photosynthetic rate, nutrient limitation, respiration and grazing were defined along the lines indicated in the preceding sections. Turbulence was introduced in terms of the ratio between the depth of the euphotic zone and the depth of the surface mixed layer. These were united in an equation for the plant population under unit surface (see Appendix). The solution of this equation gives the yearly cycle of production of the plant population which can be compared with observations (Fig. 2*a*) and the success of the mathematical approach is about the same as that from the statistical analysis.

As a continuation of the study of the Georges Bank material, Riley (1947) developed an equation for the herbivores (see Appendix). For this the grazing term of the previous equation provided the relationship for zooplankton feeding. Respiration was taken from the *Calanus* data of Marshall, Nicholls & Orr (1935). The predation on herbivores was supposed to be proportional to the number of *Sagitta elegans*, and this factor of proportionality, together with a constant natural death-rate, was determined statistically to give the best fit. This fit is shown in Fig. 2*b*.

In this work Riley studied changes with time of the total populations. The other aspect of the problem is variation with depth and to study this Riley *et al.* (1949) chose areas and times where the temporal changes could be considered as very slow and thus neglected. Thus any one set of observations could be analysed separately. Equations were developed for plants and phosphate-phosphorus, which described the concentrations at any depth and included terms for mixing and sinking of plants (see Appendix). For the herbivores it was assumed that they spent

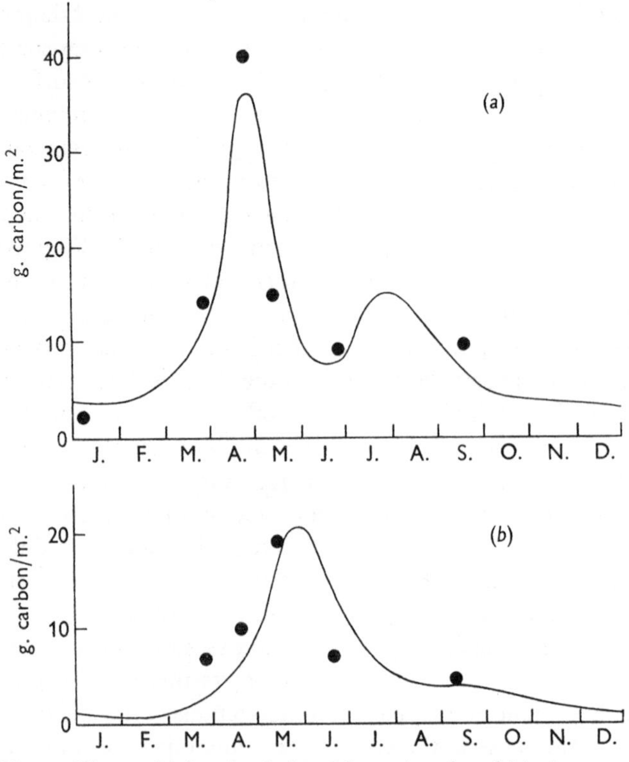

Fig. 2. Observed (●) and calculated (——) cycles of (*a*) plants, and (*b*) herbivores on Georges Bank.

equal times at all the depths considered, so only the total quantity of herbivores need be known and an expression for this was developed. On this basis, by calculation, profiles of the vertical distribution of phosphorus and plant concentration could be constructed from 'physiological' coefficients for respiration, filtering and sinking, using data for the physical factors, temperature, light, transparency and deep-water nutrient concentration. In Fig. 3 three examples of the plant profiles are given for areas providing very different types of environment.

These two methods used by Riley, considered separately, change with time and change with depth. Using data from the North Sea and the same form of general equation as Riley, I calculated the expected plant profiles at the end of successive time intervals from the profiles at the beginning together with temperature data and production estimates based on phosphate measurements (Steele, 1956; see Appendix). The correspondence between observed and calculated profiles had about the same measure of success as Riley's calculations.

These models suffer from one type of drawback; there is an element of circularity in the calculations. Both Riley *et al.* (1949) and I, in calculating sinking rates,

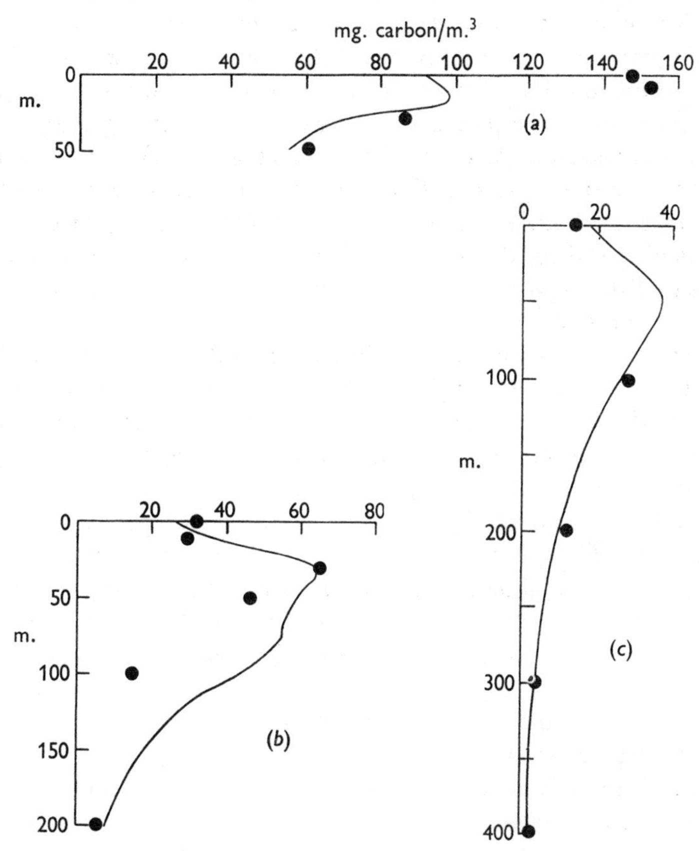

Fig. 3. Observed (●) and calculated (——) profiles of plant population at (*a*) Georges Bank, (*b*) Slope Water, (*c*) Sargasso Sea.

used the same data and the same form of equation that were used later to derive the expected plant distributions (see Appendix). Because of this the results are not true predictions and the success of the method must be considered in terms of the general consistency of the results. This, however, is comparatively good when the wide range of environments is taken into account, yet there are occasions when the method apparently breaks down, and then, unfortunately, and because of the lack of detailed knowledge, it is always easy to provide probable explanations—for example in terms of large-scale lateral water movements.

The biological variables used in these models have all been related to carbon content, through chlorophyll estimates for plants and dry-weight estimates for herbivores. The gain in generality is balanced by the rather artificial abstraction of the units which makes it difficult to have any insight into the effects of species changes and of the various stages in the development of the animals. It is for this reason that Cushing's studies are of special interest since he has used the more realistic information of diatom and copepod counts. His method for estimating the production by diatoms from decrease in cell sizes and his 'encounter' theory of grazing by copepods has already been mentioned. In a paper (Cushing 1959) which Dr Cushing has allowed me to see in manuscript these are incorporated in a model of the events in the central North Sea during the spring of 1949. The production rate, derived from data given by Lund (1949) for freshwater *Asterionella*, is a division rate-energy relationship similar to the photosynthesis-chlorophyll formula of Ryther. Nutrient limitation and respiration are ignored, but this is probably justified during the spring outburst. Turbulence is introduced in the same way as by Riley (1946). The results of these estimates are in general agreement with the cell-decrease data.

The herbivores at this time are nearly all copepods, particularly *Calanus*, and so the extensive work on *Calanus* (Marshall & Orr, 1955*a*) can be used to make estimates of reproduction rates, growth rates (in terms of copepodite stages) and 'natural' mortalities for this species. These data are also applied to the other copepods termed 'non-*Calanus*'. On this basis an equation similar to Riley's (1946) can be solved in the same step-by-step manner, and the curves for algae and herbivores are shown in Fig. 4 in terms of wet weight. There are not many observations for comparison; for the algae a sample taken at about the time of the peak of the curve is three times greater than the theoretical maximum. In the model the numbers of *Calanus* and non-*Calanus* increase towards 0·5 and 4·0 per litre respectively; the peak numbers for this area over a period of years were 0·4 and 2·2 respectively. The final increase in herbivores and decrease in algae are, however, rather greater than expected.

In this way Cushing's model is a more intimate picture of the spring outburst, especially of copepod development. Its drawback, apart from the great amount of work required in measuring and counting plants and animals, is that it can be checked only for the period when the plankton consists almost entirely of diatoms and copepods.

It has already been stated that the advantages of these models lies in the ability to combine several features of the environment which are varying simultaneously. Once this has been done, however, the problem is to try to discover the relative importance of the variations of each factor; in particular their individual effects on the cycle of production which they together control. For this type of study a more theoretical approach is required, in which any particular factor can be varied while the others are kept constant. I have attempted this for conditions that are generally applicable to the northern North Sea (Steele, 1958). In this area the thermocline usually occurs between 30 and 50 m. and the euphotic zone has about the same depth range, although not necessarily at the same times. In this comparatively small area the most important environmental variable is the rate of mixing through the thermocline. On this basis a model was constructed of a two-layered sea divided at 40 m., a depth which also corresponded to the euphotic zone; nutrient limitation was introduced into the photosynthetic term in the same way as by Riley (1946) except that a lower initial level was used; zooplankton respiration

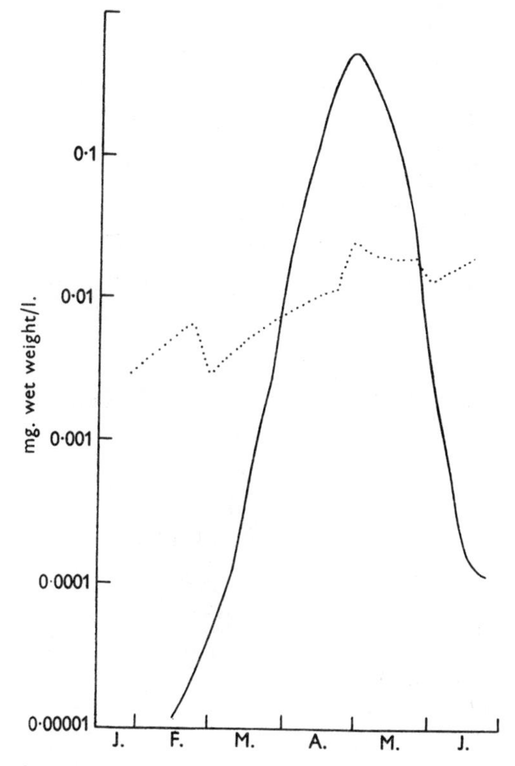

Fig. 4. Calculated values of plants (——) and herbivores (····) in the central North Sea.

was ignored in the euphotic zone. This defined two equations, for plants and for nutrients. For herbivores an empirical equation was constructed from data for the North Sea. There were thus three equations for three variables, nutrients, plants and herbivores, and, by choosing values of sinking, grazing, mixing and photosynthetic rate (excluding nutrient effect), a yearly cycle of the three variables could be estimated and also the corresponding cycle of production. In this way changes in any coefficient could be studied in terms of the differing cycles produced (see Appendix).

As an example in Fig. 5*a* the curves of production are given for four mixing rates, showing that increased mixing can be expected to decrease the speed and the maximum of the spring outburst, but give a higher production in the summer; in Fig. 5*b* an appropriate theoretical curve for the Fladen Ground is compared with values for production from phosphate data and ^{14}C experiments. The results of variations in mixing rate were also used to make a chart of the expected production over the northern North Sea which showed moderate agreement with estimates from phosphate data. To this extent the model, although it has many idealized features, is not unrealistic.

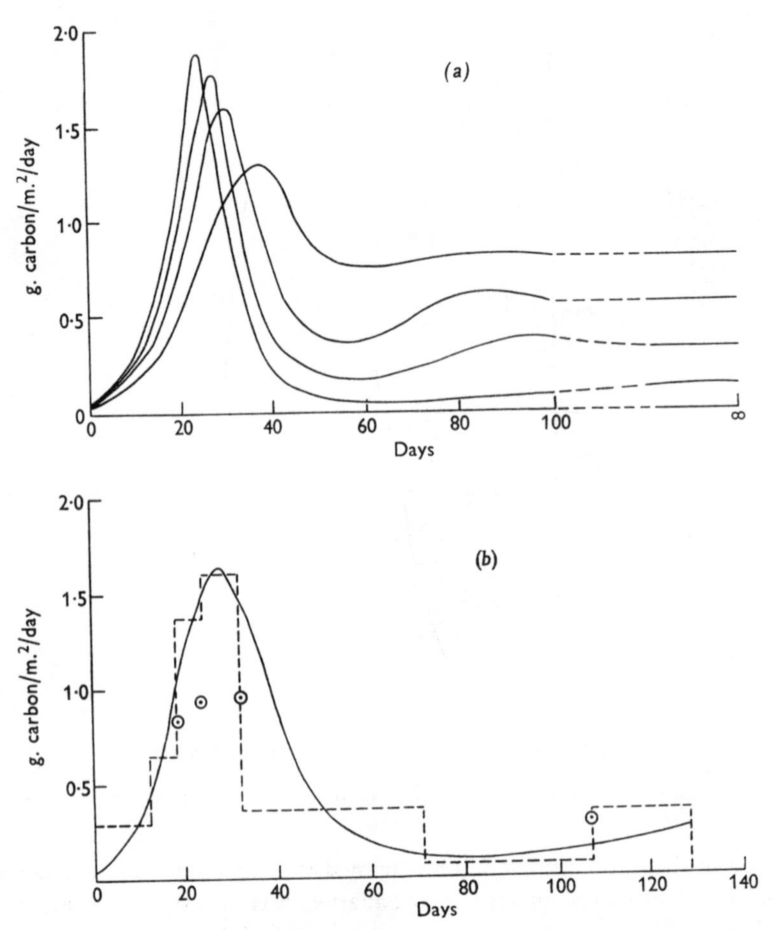

Fig. 5. (*a*) Theoretical cycles of production after winter conditions for four mixing rates (in arbitrary units). (*b*) Production on the Fladen Ground in 1956: theoretical (—); phosphate estimates (- - - -); ^{14}C experiments (⊙).

VI. RESULTS OF THE MATHEMATICAL APPROACH

Each of these models forms part of the investigation of a particular region and so provides insight into individual problems about the distribution of plants and the quantities of organic matter produced. But beyond this the ability of similar models to provide fairly good agreement with observations varying both with time and with depth over a wide range of environments incites general conclusions about the factors controlling plant production.

It has been shown that the models are extremely simple mathematically; that the observables, phosphate, chlorophyll, dry weight, are only very rough indices of the variables, nutrient, plant carbon, herbivore carbon; that the coefficients, photosynthetic rate, sinking rate, grazing rate, have very wide limits, especially in experimental studies in the laboratory. Yet the fact that these bases are sufficient for adequate descriptions of the main features of the changing plant life in the sea implies that the factors controlling this life must indeed be simple in character and few in number.

From this point of view the apparent complexities of the succession of plant species can be thought of merely as a means for achieving optimum production under a few limiting physical conditions. In a sense this may be regarded as the expected result of adaptation. Similarly, the changes in herbivore species and their grazing abilities would have the same type of function. For natural land vegetation Pearsall (1954) has pointed out that over a wide range of plant types the yearly production values are similar.

Simplicity of this kind, as Poincaré (1905) has said, is only an intermediate stage between previous and succeeding complexities. Its general use is to provide direction for further developments, since attempts to determine the various factors more exactly and to apply the methods to new situations will show up the critical points at which the simplicity breaks down. The study of nutrients already displays this feature. By using phosphate as a nutrient index the assumption is made that all nutrients become limiting at the same time. Again, in terms of adaptation, this is an attractive hypothesis, but the study of phosphate/nitrate ratios shows that it does not always hold and although in practice some variability can be allowed, a more complex postulate appears necessary. Further, it seems likely that broad groups of plants may have different reactions to light intensity and nutrient deficiency, and the accuracy of the models might be increased by introducing parameters for these effects.

Yet, apart from attempts to construct more accurate models, there remain many serious difficulties which go back to the foundations on which these models were erected. The range of values for regeneration of nutrients would permit either the supposition that regeneration in the euphotic layer is negligible compared with the role of vertical mixing or that it is the dominating factor in controlling the population. The apparent range in the ratio of photosynthesis to respiration is another unknown where different hypotheses could radically alter the interpretations of production estimates. The fact that these and other problems can be raised indicates the weaknesses which result from the rapid advance to general conclusions on very shaky postulates. The generality is possible because of the abstract nature of the variables, but this at the same time tends to hide the basic problems.

Also the areas which have been studied in detail by these methods are carefully chosen to emphasize certain types of changes; temporal changes on Georges Bank; stationary plant populations in offshore semi-tropical waters; lack of lateral movement on the Fladen Ground. It is not impossible that such choices avoid areas and times where other and unused factors become critical and control the population. This possibility is raised acutely by the results found in another rapidly developing branch of phytoplankton research.

VII. UNUSED EVIDENCE

Historically the present detailed laboratory work which cannot be assimilated directly into the mathematical models developed from the same problems. One feature of the plant-animal relationship which concerned Harvey and Fleming was the fact that high concentrations of both are not usually found together; Fleming's model was partly intended to show how this could be explained by grazing. Another explanation had been put forward by Hardy (1935); he suggested that animals deliberately avoided dense concentrations of plants for reasons at that time unknown. This concept of 'animal exclusion' was used to explain herring distributions in the southern North Sea, and although these explanations are no longer accepted by Cushing (1956), they led to an interest in the possible unknown factors. The term 'external metabolite' was used by Lucas (1947) for any substance released into sea water by plants or animals which could have a metabolic or behavioural effect on the same or other species. By transferring the work to laboratory cultures fruitful results have been obtained. In several papers the needs for and effects of particular metabolites, especially B_{12}, have been reviewed (Provasoli & Pintner, 1953; Lucas, 1955; Droop, 1957*a*). The metabolites released by one species into the water can affect the growth both of this species itself and of other species. It is suggested that such results will ultimately explain the phenomena of plant species succession.

As a result of this work natural sea waters may have been examined for metabolites. Johnston (1955) has shown that extracts of organic matter from sea water can promote or inhibit the growth of algal cultures. Cowey (1956) has found large variations in the B_{12} content of sea water, changes which in time and relative order of magnitude are similar to the changes observed in phosphate. On the other hand, Droop (1957*b*) has shown that, for several culture organisms, there would be more than sufficient B_{12} to produce the size of crops normally encountered; the organisms used by Droop are, however, virtually never found in the open sea.

These types of data about changes in the sea can be fitted theoretically but indirectly into the mathematical picture, either by supposing that metabolites determine species succession without affecting the total production, or by including the more general factors (such perhaps as B_{12}) in the concept of a group of nutrients which become simultaneously limiting. Either approach requires the extra condition that circumstances are 'normal', since it becomes possible that under 'abnormal' conditions the simplicity of the model would be upset by lack of some unknown micro-constituent in the same way that low nitrate/phosphate ratios can upset a model based on phosphorus as the index of deficiency. But, for these circumstances, the problem of defining 'normal' and 'abnormal' conditions in the sea is almost impossible and so the work on metabolites raises doubts about a set of simple basic hypotheses. Unfortunately the species used in the experimental work have with rare exceptions (Droop, 1955), not been common marine organisms. Thus, although the problem exists, the laboratory studies have at present no relevance to particular cycles of production in the sea.

In any investigation, however, the problem of 'normality' can to a limited extent be discussed by studying the changing distributions of particular species in the plankton. These 'indicator species' delineate roughly the general types of environment—oceanic, neritic, etc. (Russell, 1935; Fraser, 1937). They provide an index for the effects of the varying environment and their study is perhaps the most useful, if still rather second-hand, method of introducing species composition into the quantitative study of production.

VIII. PRODUCTION AND THE FISHERIES

This discussion has been concerned with only a few steps of the trophic levels—water, plants, herbivores—but such studies are intended to be, biologically and administratively, part of the general chain which extends to the commercial fisheries.

It is in the study of these fisheries that the other, and more practically important, development of population models has taken place (Beverton & Holt, 1956). Yet this work is also in a similar position to quantitative plankton studies. It makes little use of detailed biological studies of fish, its knowledge of the effects of food limitation are slight and it cannot be used to explain adequately the large variations in recruitment of young fish. It is these factors which are linked with the lower trophic levels and the problem of production studies in this respect is whether the type of results being obtained has, or can have in the future, any relevance to fishery problems.

It has been suggested that by mapping out the basic productivity of the actual and possible fishing grounds, an index could be obtained of their ability to support commercial fisheries (Kesteven & Laevastu, 1957). This sort of procedure can sometimes be successful (Sette, 1955), but since there are fundamental disagreements about the postulates and occasionally about the conclusions, studies in new areas are at present more likely to be useful in unravelling basic problems than as ready-made information for the fishery biologist.

Further, most of the production studies have been made in areas where the fishing possibilities are adequately known, yet even in these areas it has not been possible to relate causally the production to the fishing. The only work so far has dealt with statistical correspondence (Cushing, 1955).

From another aspect the total yearly primary production may not be a reliable index, since the level of fish population maintained may depend partly on the food available at some critical period, such as the end of the yolk sac stage. It may even be that such a dependence is on a particular species of plant, in which case a mathematical model of the type discussed here would be useless. For these kinds of reasons, although a high level of production is necessary, it is not a sufficient condition and thus, for example, the English Channel appears to produce more marketable fish than Long Island Sound despite the high basic production of the latter (Riley, 1955).

Such speculations show once more our inadequate knowledge of the fundamental relationships between different biological levels. It is because of this lack that the study of plankton production cannot yet be regarded as the satisfactory instrument which is needed to link the fisheries with their environment. Attempts to do this can form a useful background and a source of ideas, but to make this the main aim of research at present would be to treat theories about plankton ecology as firm rules, whereas they are still rather tentative hypotheses.

IX. SUMMARY

1. Microscopic algae are suited for the use of mathematical techniques in population studies.

2. These techniques require exact formulation of the basic factors, such as photosynthesis, respiration, sinking, nutrient requirements, grazing and the vertical mixing of the water. The available information is, however, vague and often contradictory.

3. Yet simple mathematical models with a suitable choice of postulates can explain the main features of populations inhabiting a wide range of environments.

4. The conclusion of this approach is that marine phytoplankton populations can be defined by a few simple physiological characteristics, and their general abundance in the sea is controlled by a small number of physical factors.

5. Looked at in this way the total population and its production of organic matter appear to be largely independent of the variations in species composition. Estimates of the plants in abstract terms, such as carbon content, provide more insight into their ecology than detailed studies of the species present in any sample.

6. The weakness in the postulates, however, and the choice, for the detailed studies, of ecologically simple regions, make the general application of these conclusions doubtful. The work on the external metabolites of phytoplankton, although it might be included theoretically in the bases of the models, also raises many practical and fundamental difficulties.

7. Because of these doubts about the foundations of this type of work, phytoplankton ecology cannot yet provide links between the environment and commercial fisheries.

My thanks are due to Drs H. W. Harvey, B. H. Ketchum, C. E. Lucas and G. A. Riley who have kindly read and commented on this article in manuscript.

X. APPENDIX: THE MATHEMATICAL FORMULAE

List of symbols used

p = Plant concentration in unit volume of water at depth z.
P = Total plant concentration in a vertical water column.
P^* = Mean plant concentration below euphotic zone.
P_h = Photosynthetic rate for unit plant concentration.
G = Grazing rate of animals.
g = Filtering rate of unit quantity of animals.
Z = Total quantity of herbivores in unit volume of water.
R = Respiration rate of unit of plants = $R_0 e^{rT}$, where R_0 and r are constants.
R' = Respiration rate of unit of herbivores.
T = Temperature.
γ = Grazing rate of unit of *Sagitta*.
S = Mean number of *Sagitta* in unit volume of water.
D = Constant natural death rate of unit of herbivores.
n = Phosphate-phosphorus concentration at depth z.
N = Mean phosphate-phosphorus concentration in euphotic zone.
N^* = Mean phosphate concentration below euphotic zone.
I_0 = Light intensity at surface.
k = Extinction coefficient of water.
z_0 = Depth to which model operates.
z_1 = Depth of euphotic zone.
z_2 = Depth of surface mixed layer.
v = Sinking rate of plants at zero temperature.
μ_0 = Viscosity of water at zero temperature.
μ_T = Viscosity of water at temperature T.
A_Q = Coefficient of eddy diffusivity for any quantity Q.
t = Time.
ρ = Density.
ν = Phosphorus/carbon ratio.
m = Idealized mixing coefficient.

Equations used

For mean conditions in a body of water, the change with time of plants can be expressed by

$$\frac{dP}{dt} = (P_h - R - G)\,P. \tag{1}$$

Fleming (1939) took $P_h - R = a$,

$$G = b + ct,$$

where a, b and c are positive constants. The solution for P at time t is then

$$P_t = P_0 \exp\left[(a-b)\,t - \tfrac{1}{2}ct^2\right],$$

and the constants can be determined from the observed data in Fig. 1 by choosing the starting-point at the first observation and the maximum of P_t at the largest observed value, to give the curve in Fig. 1. Riley (1946) obtained a mean value for photosynthetic rate in the euphotic zone by integrating,

$$P_h = \alpha I_0 e^{-kz}, \quad \text{where } \alpha \text{ is a constant} \tag{2}$$

to give

$$P_h = \frac{\alpha I_0}{kz_1}(1-e^{-kz_1}). \tag{3}$$

This applies when there is no limitation due to nutrients or to mixing. The first is introduced by a term $(0{\cdot}55-N)/0{\cdot}55$ and the second by z_1/z_2 when $z_1 < z_2$.

Respiration is given by $R_0 e^{rT}$ and the grazing $G = gZ$, with g chosen somewhat empirically to provide a good fit. Thus,

$$\frac{1}{P}\frac{dP}{dt} = \frac{I_0}{kz_2}(1-e^{-kz_1})\,\frac{0{\cdot}55-N}{0{\cdot}55} - R_0 e^{rT} - gZ. \tag{4}$$

This is solved by a step-by-step process from

$$P_{t+1} = P_t \exp\overline{(P_h - R - G)}, \tag{5}$$

where the initial value, P_0, is the concentration of plants in the winter.

For the herbivores Riley (1947) used a similar equation,

$$\frac{dZ}{dt} = (gP - R' - \gamma S - D)\,Z \tag{6}$$

solved in a similar manner except that the constants γ and D were chosen statistically to provide the best fit.

Riley *et al.* (1949) set up the equations for plants, phosphate and animals as

$$\frac{\partial p}{\partial t} = (P_h - R - gZ)\,p + \frac{\partial}{\partial z}\left(\frac{A_p}{\rho}\frac{\partial p}{\partial z}\right) - \frac{v\mu_0}{\mu_T}\frac{\partial p}{\partial z}, \tag{7}$$

$$\frac{\partial n}{\partial t} = \nu[R'Z - p(P_h - R)] + \frac{\partial}{\partial z}\left(\frac{A_n}{\rho}\frac{\partial n}{\partial z}\right), \tag{8}$$

$$Z = \frac{\displaystyle\int_0^{z_0} p(P_h - R)\,dz - \frac{v\mu_0}{\mu_T}\overline{(p)_{z_0}}}{\displaystyle g\int_0^{z_0} p\,dz}. \tag{9}$$

They solved them for steady state conditions, i.e.

$$\frac{\partial p}{\partial t} = \frac{\partial n}{\partial t} = 0.$$

In equation (7), assuming $A_p = A$ is a constant

$$(P_h - R - gZ)\, p = \frac{\mu_0 v}{\mu_T}\frac{\partial p}{\partial z} - \frac{A}{\rho}\frac{\partial^2 p}{\partial z^2}. \tag{10}$$

By dividing the water column into two vertical layers and with estimated values for the other factors, (10) can be used to solve for A and v. From solutions for various areas a mean value for v is found. Using this value in (7), (8) and (9) expressed as finite difference equations the vertical distributions of p and n can be estimated by relaxation methods assuming $A_p = A_n = A$ in the mixing terms. A is calculated from

$$\frac{\partial T}{\partial t} = \frac{\partial}{\partial z}\left(\frac{A}{\rho}\frac{\partial T}{\partial z}\right)$$

by integrating to depth z from z_0

$$\int_z^{z_0} \frac{\partial T}{\partial t}\, dz = \frac{A}{\rho}\frac{\partial T}{\partial z} \tag{11}$$

Using (11) to compute A and neglecting $R'Z$ above the level z_1, I obtained (Steele, 1956) values for net production during any time interval from (8) in the form

$$\int_0^{z_1} p(P_h - R)\, dz = \nu^{-1}\left[\frac{A}{\rho}\frac{\partial n}{\partial t} - \int_0^{z_1}\frac{\partial n}{\partial t}\, dz\right]. \tag{12}$$

Integrating (7) for the layers $O - z_1$ and $z_1 - z_0$ and using successive profiles of plant population, mean values of g and v are solved for each interval. Using these values, (7) can then be applied to predict a profile at the end of an interval in terms of the profile at the beginning.

Cushing (1959) used an equation of the form

$$\frac{dP}{dt} = (P_h - G - M)\, P, \tag{13}$$

where M, the natural mortality, was estimated from the number of empty diatom frustules and a sinking rate of 10–15 m./day. On this basis M was found to be only 1 % of the other factors and thus was ignored. The equation was solved by the usual step-by-step process.

For a two-layered sea divided at 40 m., I developed an idealized form of (7) and (8) (Steele, 1958), in which the terms for sinking and mixing take a simple form

$$\frac{dP}{dt} = (P_h - R - gZ) - \frac{vP}{40} - m(P - P^*),$$

$$\frac{dN}{dt} = -\nu(P_h - R)\, P - m(N - N^*),$$

where P, N and P^*, N^* are the mean concentrations above and below 40 m.; P^* is taken as zero, N^* is taken as constant; m is a mixing coefficient.

For zooplankton, the empirical form

$$\frac{dZ}{dt} = aP - bZ^2$$

was found to give values of a and b which fitted the spring maximum and summer minimum of zooplankton data for the Fladen Ground.

XI. REFERENCES

ALLEN, W. E. (1932). Problems of flotation and deposition of marine plankton diatoms *Trans. Amer. micr. Soc.* **51**, 1.

BARKER, H. A. (1935). Culture and physiology of the marine dinoflagellates. *Arch. Mikrobiol.* **6**, 157.

BEVERTON, R. J. H. & HOLT, S. J. (1956). The theory of fishing. In *Sea fisheries*, ed. Michael Graham. London: Arnold.

BRANDT, T. VON, RAKESTRAW, N. W. & RENN, C. E. (1937). The experimental decomposition and regeneration of nitrogenous organic matter in sea water. *Biol. Bull., Woods Hole*, **72**, 165.

CHU, S. P. (1946). Utilization of organic phosphorus by phytoplankton. *J. mar. biol Ass. U. K.* **26**, 285.

COOPER, L. H. N. (1935). The rate of liberation of phosphate in sea water by the breakdown of plankton organisms. *J. mar. biol. Ass. U.K.* **20**, 197.

COOPER, L. H. N. (1937). On the ratio of nitrogen to phosphorus in the sea. *J. mar. biol. Ass. U.K.* **22**, 177.

COWEY, C. B. (1956). A preliminary investigation of the variation of vitamin B_{12} in oceanic and coastal waters. *J. mar. biol. Ass. U.K.* **35**, 609.

CUSHING, D. H. (1954). Some problems in the production of oceanic plankton. Document VIII presented to Commonwealth Oceanographic Conference, 1954.

CUSHING, D. H. (1955). Production and a pelagic fishery. *Fish. Invest., Lond.* (2), **18**, no. 7.

CUSHING, D. H. (1956). Phytoplankton and the herring. V. *Fish. Invest., Lond.* (2), **20**, no. 4.

CUSHING, D. H. (1958). The effect of grazing in reducing the primary production. *Rapp. Cons. Explor. Mer*, **144**, 149.

CUSHING, D. H. (1959). On the nature of production in the sea. *Fish. Invest., Lond.* (2), [in Press].

DROOP, M. R. (1955). A pelagic marine diatom requiring cobalamin. *J. mar. biol. Ass. U.K.* **34**, 229.

DROOP, M. R. (1957*a*). Auxotrophy and organic compounds in the nutrition of marine phytoplankton. *J. gen. Microbiol.* **16**, 286.

DROOP, M. R. (1957*b*). Vitamin B_{12} in marine ecology. *Nature, Lond.*, **180**, 1041.

FLEISCHER, W. E. (1935). The relation between chlorophyll content and rate of photosynthesis. *J. gen. Physiol.* **18**, 573.

FLEMING, R. H. (1939). The control of diatom populations by grazing. *J. Cons. int. Explor. Mer.* **14**, 210.

FRASER, J. H. (1937). The distribution of Chaetognatha in Scottish waters during 1936, with notes on the Scottish indicator species. *J. Cons. int. Explor. Mer*, **12**, 311.

GAARDER, T. C. & GRAN, H. H. (1927). Investigation of the production of plankton in the Oslo Fjord. *Rapp. Cons. Explor. Mer*, **42**, 3.

GARDINER, A. C. (1937). Phosphate production by planktonic animals. *J. Cons. int. Explor. Mer*, **12**, 144.

GEST, H. & KAMEN, M. D. (1948). Studies of the phosphorus metabolism of green algae and purple bacteria in relation to photosynthesis. *J. biol. Chem.* **176**, 299.

GILLBRICHT, M. (1952). Untersuchungen zur Produktionsbiologie des Planktons in der Kieler Bucht. I. *Kieler Meeresforsch.* **8**, 173.

GLASS, B. (1951). Summary of the symposium. *A symposium on phosphorus metabolism. I.* Baltimore: Johns Hopkins Press.

GLASS, B. (1952). Summary of the symposium. *A symposium on phosphorus metabolism. II.* Baltimore: Johns Hopkins Press.

GOLDBERG, E. D., WALKER, T. J. & WHISENAND, A. (1951). Phosphate utilization by diatoms. *Biol. Bull., Woods Hole*, **101**, 274.

GOLDSTEIN, S. (1938). *Modern developments in fluid dynamics*. Oxford University Press.
GRAN, H. H. (1915). The plankton production in the north European waters in the spring of 1912. *Bull. plankt., Copenh.*, 1912, 7.
GRAN, H. H. & BRAARUD, T. (1935). A quantitative study of the phytoplankton in the Bay of Fundy and the Gulf of Maine. *J. biol. Bd Can.* **1**, 279.
GROSS, F. & ZEUTHEN, E. (1948). The buoyancy of plankton diatoms: a problem of cell physiology. *Proc. roy. Soc.* B, **135**, 382.
HARDY, A. C. (1935). The plankton community, the whale fisheries, and the hypothesis of animal exclusion. *'Discovery' Rep.* **11**, 273.
HARVEY, H. W. (1933). On the rate of diatom growth. *J. mar. biol. Ass. U.K.* **19**, 253.
HARVEY, H. W. (1937). Note on selective feeding by *Calanus*. *J. mar. biol. Ass. U.K.* **22**, 97.
HARVEY, H. W. (1942). Production of life in the sea. *Biol. Rev.* **17**, 221.
HARVEY, H. W. (1950). On the production of living matter in the sea off Plymouth. *J. mar. biol. Ass. U.K.* **29**, 97.
HARVEY, H. W. (1953). Note on the absorption of organic phosphorus compounds by *Nitzschia closterium* in the dark. *J. mar. biol. Ass. U.K.* **31**, 475.
HARVEY, H. W. (1955). *The chemistry and fertility of sea waters*. Cambridge University Press.
HARVEY, H. W., COOPER, L. H. N., LEBOUR, M. V. & RUSSELL, F. S. (1935). Plankton production and its control. *J. mar. biol. Ass. U.K.* **20**, 407.
HAYES, F. R. & COFFIN, C. C. (1951). Radioactive phosphorus and exchange of lake nutrients. *Endeavour*, **10**, 78.
HENDEY, N. I. (1954). Note on the Plymouth '*Nitzschia*' culture. *J. mar. biol. Ass. U.K.* **33**, 335.
HOFFMAN, C. (1956). Untersuchungen über die Remineralisation des Phosphors im Plankton. *Kieler Meeresforsch.* **12**, 25.
HUTCHINSON, G. E. & BOWEN, V. T. (1950). Limnological studies in Connecticut. IX. A quantitative radio-chemical study of the phosphorus cycle in Linsley Pond. *Ecology*, **31**, 194.
JENKIN, P. M. (1937). Oxygen production by the diatom *Coscinodiscus excentricus* Ehr. in relation to submarine illumination in the English Channel. *J. mar. biol. Ass. U.K.* **22**, 301.
JOHNSTON, R. (1955). Biologically active compounds in the sea. *J. mar. biol. Ass. U.K.* **34**, 185.
KAMEN, M. D. & SPIEGELMAN, S. (1948). Studies on the phosphate metabolism of some unicellular organisms. *Cold Spr. Harb. Symp. quant. Biol.* **13**, 151.
KESTEVEN, G. L. & LAEVASTU, T. (1957). The oceanographical conditions for life and abundance of phytoplankton considered with respect to fisheries. *F.A.O. Fish. Div. Biol. Branch* (mimeogr.).
KETCHUM, B. H. (1939*a*). The absorption of phosphate and nitrate by illuminated cultures of *Nitzschia closterium*. *Amer. J. Bot.* **26**, 399.
KETCHUM, B. H. (1939*b*). The development and restoration of deficiencies in the phosphorus and nitrogen composition of unicellular plants. *J. cell. comp. Physiol.* **13**, 373.
KOK, B. (1952). On the efficiency of *Chlorella* growth. *Acta Bot. Meerl.* **1**, 445.
KREY, J. (1939). Bestimmung des Chlorophylls. *J. Cons. int. Explor. Mer*, **14**, 201.
KREY, J. (1951). Quantitative Bestimmung von Eiweiss im Plankton mittels der Biuretreaktion. *Kieler Meeresforsch.* **8**, 16.
LUCAS, C. E. (1947). The ecological effects of external metabolites. *Biol. Rev.* **22**, 270.
LUCAS, C. E. (1955). External metabolites in the sea. *Pap. mar. Biol. and Oceanogr. Deep-sea Res.* **3** (suppl.), 139.
LUND, J. W. G. (1949). Studies on *Asterionella*. I. The origin and nature of the cells producing seasonal maxima. *J. Ecol.* **37**, 389.
MACKERETH, F. J. (1953). Phosphorus utilization by *Asterionella formosa* Hass. *J. exp. Bot.* **4**, 296.
MARSHALL, S. M., NICHOLLS, A. G. & ORR, A. P. (1935). On the biology of *Calanus finmarchicus*. VI. Oxygen consumption in relation to environmental conditions. *J. mar. biol. Ass. U.K.* **20**, 1.
MARSHALL, S. M. & ORR, A. P. (1955*a*). *Biology of a marine copepod*. Edinburgh: Oliver & Boyd.
MARSHALL, S. M. & ORR, A. P. (1955*b*). On the biology of *Calanus finmarchicus*. 8. Food uptake, assimilation and excretion in adult and stage V *Calanus*. *J. mar. biol. Ass. U.K.* **34**, 495.
MARSHALL, S. M. & ORR, A. P. (1955*c*). Experimental feeding of the copepod *Calanus finmarchicus* (Gunner) on phytoplankton cultures labelled with radio-active carbon (^{14}C). *Pap. mar. Biol. and Oceanogr., Deep-sea Res.*, **3** (suppl.), 110.
MUNK, W. H. & RILEY, G. A. (1952). Absorption of nutrients by aquatic plants. *J. mar. Res.* **11**, 215.
PEARSALL, W. H. (1954). Growth and production. *Advanc. Sci., Lond.*, **11**, 232.
POINCARÉ, H. (1905). *Science and hypothesis*. London: Walter Scott.
PRATT, D. M. (1950). Experimental study of the phosphorus cycle in fertilized salt water. *J. mar. Res.*, **9**, 29.

PROVASOLI, L. & PINTNER, I. J. (1953). Ecological implications of *in vitro* nutritional requirements of algal flagellates. *Ann. N.Y. Acad. Sci.* **56**, 839.
RABINOWITCH, E. I. (1945). *Photosynthesis and related processes. I.* New York: Interscience Publishers.
RABINOWITCH, E. I. (1951). *Photosynthesis and related processes. II. Part I.* New York: Interscience Publishers.
REDFIELD, A. C. (1948). The exchange of oxygen across the sea surface. *J. mar. Res.* **7**, 347.
REDFIELD, A. C., SMITH, H. P. & KETCHUM, B. H. (1937). The cycle of organic phosphorus in the Gulf of Maine. *Biol. Bull., Woods Hole*, **73**, 421.
RICHARDS, F. A. (1952). The estimation and characterisation of plankton populations by pigment analyses. I. *J. mar. Res.* **11**, 147.
RICHARDS, F. A. & THOMPSON, T. G. (1952). The estimation and characterization of plankton populations by pigment analyses. II. *J. mar. Res.* **11**, 156.
RIGLER, F. H. (1956). A tracer study of the phosphorus cycle in lake water. *Ecology*, **37**, 550.
RILEY, G. A. (1939). Plankton studies. II. The western North Atlantic, May–June, 1939. *J. mar. Res.* **2**, 145.
RILEY, G. A. (1943). Physiological aspects of spring diatom flowerings. *Bull. Bingham oceangr. Coll.* **8**, Art. 4.
RILEY, G. A. (1946). Factors controlling phytoplankton populations on Georges Bank. *J. mar. Res.* **6**, 54.
RILEY, G. A. (1947). A theoretical analysis of the zooplankton population of Georges Bank. *J. mar. Res.* **6**, 104.
RILEY, G. A. (1955). Review of the oceanography of Long Island Sound. *Pap. mar. Biol. and Oceanogr. Deep-Sea Res.*, **3** (suppl.), 224.
RILEY, G. A. (1956). Oceanography of Long Island Sound, 1952–1954. IX. Production and utilization of organic matter. *Bull. Bingham oceanogr. Coll.* **15**, 324.
RILEY, G. A. & CONOVER, S. A. M. (1956). Oceanography of Long Island Sound, 1952–1954. III. Chemical oceanography. *Bull. Bingham oceanogr. Coll.* **15**, 47.
RILEY, G. A., STOMMEL, H. & BUMPUS, D. F. (1949). Quantitative ecology of the plankton of the western North Atlantic. *Bull. Bingham oceanogr. Coll.* **12**, Art. 3.
RODHE, W. (1948). Environmental requirements of freshwater plankton algae. *Symb. bot. upsaliens.* **10**, 1.
RUSSELL, F. S. (1935). A review of some aspects of zooplankton research. *Rapp. Cons. Explor. Mer*, **95**, 5.
RYTHER, J. H. (1954*a*). The ecology of phytoplankton blooms in Moriches Bay and Great South Bay, Long Island, New York. *Biol. Bull., Woods Hole*, **106**, 198.
RYTHER, J. H. (1954*b*). The ratio of photosynthesis to respiration in marine plankton algae and its effect upon the measurement of productivity. *Deep-Sea Res.* **2**, 134.
RYTHER, J. H. (1956). Photosynthesis in the ocean as a function of light intensity. *Limnol. and Oceanogr.* **1**, 61.
RYTHER, J. H. & YENTSCH, C. S. (1957). The estimation of phytoplankton production in the ocean from chlorophyll and light data. *Limnol. and Oceanogr.* **2**, 281.
SETTE, O. E. (1955). Consideration of midocean fish production as related to oceanic circulatory systems. *J. mar. Res.* **14**, 398.
SEIWELL, H. R. & SEIWELL, G. E. (1938). The sinking of decomposing plankton in sea water and its relationship to oxygen consumption and phosphorus liberation. *Proc. Amer. phil. Soc.* **78**, 465.
SPOEHR, H. A. & MILNER, H. W. (1949). The chemical composition of *Chlorella*; effect of environmental conditions. *Plant Physiol.* **24**, 120.
STEELE, J. H. (1956). Plant production on the Fladen Ground. *J. mar. biol. Ass. U.K.* **35**, 1.
STEELE, J. H. (1957*a*). A comparison of plant production estimates using ^{14}C and phosphate data. *J. mar. biol. Ass. U.K.* **36**, 233.
STEELE, J. H. (1957*b*). Notes on oxygen sampling on the Fladen Ground. *J. mar. biol. Ass. U.K.* **36**, 227.
STEELE, J. H. (1958). Plant production in the northern North Sea. *Mar. Res. Scot.*, 1958, no. 7.
STEEMAN NIELSEN, E. (1952). The use of radioactive carbon (C^{14}) for measuring organic production in the sea. *J. Cons. int. Explor. Mer*, **18**, 117.
STEEMAN NIELSEN, E. (1954). On organic production in the oceans. *J. Cons. int. Explor. Mer*, **19**, 309.
STEEMAN NIELSEN, E. (1955). An effect of antibiotics produced by plankton algae. *Nature, Lond.*, **176**, 553.

STEEMAN NIELSEN, E. & AL KHOLY, A. A. (1956). Use of ^{14}C-technique in measuring photosynthesis of phosphorus or nitrogen deficient algae. *Plant Physiol.* **9**, 144.
STEEMAN NIELSEN, E. & JENSEN, E. A. (1957). Primary oceanic production. *'Galathea' Rep.* **1**, 49.
SVERDRUP, H. U. (1953). On conditions for the vernal blooming of phytoplankton. *J. Cons. int. Explor. Mer*, **18**, 287.
SVERDRUP, H. U., JOHNSON, M. W. & FLEMING, R. H. (1942). *The oceans; their physics, chemistry and general biology*. New York: Prentice-Hall.
TALLING, J. F. (1957). The phytoplankton population as a compound photosynthetic system. *New Phytol.* **56**, 133.
WAKSMAN, S. A. & RENN, C. E. (1936). Decomposition of organic matter in sea water by bacteria. *Biol. Bull., Woods Hole*, **70**, 472.

The Balance between Phytoplankton and Zooplankton in the Sea

By

E. Steemann Nielsen,
Copenhagen

Introduction

There is at present a keen interest in the study of marine plankton, the reason for which is the possibility of using more of the products of the sea as human food. Whether the plankton itself is considered a potential source of human food or whether the increase in food from the sea has to be met by the same organisms as today — primarily fish — it is of importance to understand the production of phytoplankton and the relationship between phytoplankton and zooplankton, the path by which the organic matter produced finally reaches the fish.

Rather different views about the problems connected with marine plankton have been expressed from time to time. Without in any way attempting to review the study of plankton as a whole we may here consider one particular problem. Let us focus our attention on the relationship between phyto- and zooplankton. Before discussions on this subject can begin, however, a number of minor points must be cleared up. If this is not done, there is little sense in discussing the main problem. These questions are:— How can (1) the phytoplankton and (2) the zooplankton be adequately collected and enumerated? (3) Is it possible to obtain by random samples representative estimates of the phyto- and zooplankton?

1. The standing crop of phytoplankton

Very few of the assessments of the standing crop of phytoplankton made in past years can be considered sufficiently reliable. Even normally reliable techniques may in certain circumstances give rather poor results. It is, however, not the aim of the present paper to discuss the different techniques for estimating the standing crop of phytoplankton. This was one of the main topics of the symposium on primary production held at Bergen in September 1957.

In the pioneering stage of plankton studies the fine bolting silk (No. 20 or No. 25) used for the nets was thought to be able to retain practically the whole bulk of plankton algae found in the sea. In 1908, however, LOHMANN showed irrefutably that net-caught samples gave very incomplete pictures of the quantity of phytoplankton present in the sea. A water sample may contain very much phytoplankton while a fine-mesh net, used at the same time, might retain practically nothing.

Silk nets nevertheless continued to be used extensively in quantitative phytoplankton investigations, in spite of several further warnings (cf. STEEMANN NIELSEN, 1938; GESSNER, 1944). The introduction of the measurements of the chlorophyll present in net samples by KREPS and VERJBINSKAYA (1930) and others led to no improvements. The use of nets for obtaining samples from which to estimate the standing crop of phytoplankton has undoubtedly given rise to much confusion in the study of plankton.

2. The standing crop of zooplankton

The zooplankton is a very heterogeneous group. It must be admitted that a single method of sampling cannot produce adequate figures for all kinds of zooplankton. The first step in the investigation must be to determine which method is to be used. A large part of the quantitative work done on zooplankton was undertaken in connexion with investigations on the feeding of fish, above all the herring in European waters. The methods commonly used in zooplankton studies are thus especially adapted for quantitative capture of the organisms in which herring and other fish feed.

A Hensen net (No. 3 silk) thus seems to be excellent for estimating the quantity of adult *Calanus* present in the sea, this copepod being the principal food of the herring, at least in the North Sea. Quantitative determination of zooplankton with a net of No. 3 silk, however, may give quite misleading results when the inter-relationship between phyto- and zooplankton is being studied. An example of this is given below.

During the last war fairly intensive plankton studies were made in the Danish Isefjord in the course of which zooplankton samples were taken by two different techniques used simultaneously (cf. JESPERSEN, 1949). In a particular locality at which the depth of the water was 5 m., 8-L. water samples were collected at all seasons of the year from the surface and from 4 m. and filtered through No. 25 silk, which retains all metazoa. At the same time two vertical hauls with an open net (diameter 30 cm. No. 3 silk) were made from 4 m. to the surface. The counts from both types of plankton catches have been published by JESPERSEN (1949).

Table 1 shows the numbers of holoplankton animals estimated from the two 8-litre water samples (= 16 litres) on 23 days at various times of the year, expressed as percentages of the numbers in the two net samples. The volume of water filtered by the net during the two hauls was in all probability about 150 litres. If all zooplankton organisms had been taken in the same proportion by the two techniques, we should have expected the maximum percentage in Table 1 to be about 12. It is, however, in every instance higher, varying from 16 to 12,600. No. 3 silk is much too coarse to retain all zooplankton organisms. The copepod nauplii, which are usually present in by far the highest numbers, are evidently not retained at all by No. 3 silk.

Another example may be given. In a recent work on the Sargasso Sea, FISH's (1954) estimates of the numbers of zooplankton organisms present are only about 1% of those made by HENTSCHEL (1937) for the corresponding eddy in the South Atlantic. The latter used the technique mentioned below, whereas FISH used a Clarke-Bumpus sampler (No. 2 silk bolting-cloth).

3*

Table 1

The number of holoplanktonic animals in two 8-litre samples filtered through silk No. 25 as percentages of the numbers in net-caught samples (about 150 litres of water filtered through silk No. 3)

The measurements, which were made in three different years, arranged according to seasons.

Month	Day	Year	%	Month	Day	Year	%	Month	Day	Year	%
1	20	1943	42	5	28	1943	115	9	30	1941	23
2	12	1943	300	6	18	1943	410	10	14	1941	48
3	5	1943	39	7	8	1943	12,600	10	28	1941	123
3	26	1943	129	7	21	1942	54	11	19	1941	37
4	16	1943	35	7	30	1943	234	11	27	1942	27
4	23	1942	101	8	20	1943	1,450	12	10	1941	42
5	7	1943	16	8	25	1942	145	12	18	1942	43
5	19	1942	350	9	29	1942	175				

Although one of the smaller plankton animals not retained by No. 3 or No. 2 silk very likely eats considerably less phytoplankton per day than one of the larger specimens caught by the silk, at least some periods exist during which by far the major part of the phytoplankton is eaten by the zooplankton of the size not caught by the coarse silk. As the proportion between number of zooplankton organisms retained by the net and number not retained is highly variable, the relation between the phyto- and the zooplankton can hardly be adequately assessed when gear of such coarse silk as No. 3 or No. 2 is used for collecting the zooplankton.

What appeared to be an adequate method for sampling zooplankton in studies of the phytoplankton-zooplankton inter-relationship was introduced by LOHMANN (1908). Large water samples were drawn by a pump from different depths and filtered through the finest bolting silk. HENTSCHEL (1932, 1933) concentrated the plankton in 4-litre samples by sedimentation before counting. Several modifications of LOHMANN's technique have been put forward, but most of them involve the use of coarser silk, thereby missing one of the most essential parts of LOHMANN's technique. It is easy to understand why most zooplankton specialists have avoided the use of the finest bolting silk. The immense masses of phytoplankton often found — above all, the chains of *Chaetoceras* and other diatoms — may make counting the zooplankton very tiresome and slow.

If only limited time is available, it is preferable to count the relatively few zooplankton organisms present in the mass of phytoplankton obtained in the small water samples filtered through the finest bolting silk — or similar material — instead of counting the numerous zooplankton organisms in large water samples filtered through coarse silk and therefore free of phytoplankton. Water samples of only 4 litres would be considered *a priori* too small to represent the zooplankton in tropical and subtropical parts of the open ocean. This is of course true if the quality of the plankton has to be taken into account but — HENTSCHEL's investigations on the "Meteor" expedition showed that a reliable assessment of the total herbivorous zooplankton might be obtained in that way.

Larger water samples are necessary, however, if a more detailed picture of conditions in open waters is wanted. The 100-litre water-sampler equipped with a net of No. 25 silk — STEEMANN NIELSEN (1935, 1943) — may be mentioned in this connexion. Nets of the finest bolting silk should also be applicable for this purpose if a flow-meter can be used and works satisfactorily in conjunction with No. 25 silk. An ordinary Clarke-Bumpus sampler (CLARKE, 1940), which is made either of No. 2 silk (21 strands per cm.) or of No. 10 silk (43 strands per cm.), cannot be considered satisfactory for sampling all herbivores. Many copepod nauplii are not retained by No. 10 silk.

3. Patchiness of plankton

Somewhat divergent views are held about the possibility of obtaining representative assessments of the phyto- and the zooplankton by random samples. Localities with complicated and variable hydrographic conditions — for example, fronts between different water masses — must of course be excluded from consideration. In coastal areas considerable differences occur between the hydrographic conditions in neighbouring localities. The conditions for plankton production thus also vary.

On the other hand, large hydrographically uniform regions are to be found in the oceans. Investigations carried out in the Atlantic by the two German expeditions — the "Deutschland" in 1911 and the "Meteor" in 1925–27 — showed that different water masses — particularly those in the tropics and the subtropics — could be distinguished by the numbers of plankton organisms present as clearly as by their physical properties (cf. LOHMANN, 1913, 1920; HENTSCHEL, 1932, 1933–1936). This material was specially suitable for estimates of phytoplankton. HENTSCHEL was able to show that both quantitatively and qualitatively the plankton is fairly uniform in the various parts of the ocean. It was of course not fitted for determinations of the qualitative composition of the zooplankton.

LOHMANN's and HENTSCHEL's assessments of the standing crop of phytoplankton in the oceans made by centrifuging the samples and counting the living algae have been corroborated in recent years by measurements of the production of organic matter by the phytoplankton made with the carbon-14 technique (cf. STEEMANN NIELSEN, 1954; STEEMANN NIELSEN and AABYE JENSEN, 1957). The euphotic water masses in the different parts of the oceans may be adequately characterized by their capacity for producing organic matter. In regions with nearly the same hydrographical conditions the rates of production per unit of surface are also nearly the same. The production of phytoplankton in the oceans must in general be regarded as a steady-state process. The conditions in fairly high latitudes are of course influenced by seasonal changes. Seasonal changes may also be found in low latitudes, however, as for instance, in regions in which monsoons occur.

Violent storms may also sometimes give rise to exceptional hydrographic conditions, for instance, in some parts of the tropics in which water relatively rich in nutrients occurs not very far below the surface. After such storms, which may occur only at intervals of years, extraordinarily favourable conditions for the growth of plankton algae may be present for a time; this again affects the production of zooplankton (cf. SETTE, 1955).

These instances, however, do not invalidate the general rule that a particular oceanic region is characterized by a fairly stable rate of primary organic production.

As mentioned above, the concentration of metazoa in a definite oceanic area such as the South Atlantic Eddy was found by HENTSCHEL (1933) to be fairly constant. This does not necessarily mean, however, that the population of metazoa does not change qualitatively. According to FISH (1954) there seems to be a fairly marked biological cycle in the Sargasso Sea. If a coarse net is employed, periods of relatively high (although absolutely low) zooplankton concentrations alternate with periods of low concentrations. As FISH mentions, no nauplii or early stages of the small copepods were caught. This suggests that during the periods when the numbers of animals taken by the coarse net were small, the main bulk of the zooplankton consisted of nauplii and early stages of copepods. It thus seems very likely that the fluctuation found by FISH are explained principally by the alternations of periods with a maximum of adult animals with those in which the zooplankton consists mainly of juvenile animals not caught by the nets employed. It may well be that the rate of grazing is unaffected to a large extent by these fluctuations in the zooplankton.

Patches of certain zooplanktonic species seem to occur fairly often both in coastal water and in oceanic water on the continental shelf. Patches of the copepod *Calanus finmarchicus* are regularly met with in the North Sea (CUSHING, 1955). Such a distribution is easily explained by the complexity of the hydrographical conditions. It is somewhat doubtful, on the other hand, whether real patches of all kinds of zooplankton (not due to diurnal vertical migration) occur in the wide uniform oceanic regions far from the influence of the continental shelf or of divergences. According to the extensive "Dana" investigations published by JESPERSEN (1937) no patchiness appeared as a rule in the volume of zooplankton of all kinds (herbivores and carnivores) taken by stramin nets. The zooplankton at depths down to 300 metres was always fished by night only. This is undoubtedly an important point to be remembered.

4. The possibility of an inverse phytoplankton-zooplankton relationship

Phytoplankton is the food of the herbivorous zooplankton. In the long run therefore a large population of these animals is only possible where the production of phytoplankton is high. No-one seems ever to have doubted this. When we speak about an inverse relation between phytoplankton and zooplankton, we refer only to the momentary state of affairs, not the general relationship lasting over the whole year.

It is *a priori* possible for the sea to assume either more or less stationary conditions in which for instance the zooplankton population matches the production of phytoplankton or variable conditions in which periods of flourishing phytoplankton alternate with periods in which the standing crop of phytoplankton is eaten right down by the herbivorous zooplankton.

In the last 25 years the latter conception has been very prevalent. The sea is regarded as a place "where the animals enjoy alternating periods of overeating and starvation" and where "an equilibrium between standing crop of plants, herbivores and carnivores is continually passing in and out of balance" (HARVEY, 1955, pp. 114 and 118).

On the other hand, HENTSCHEL's investigations in the South Atlantic (see Fig. 1) showed striking similarities between the distributions of phytoplankton and zooplankton populations, although his material was collected in innumerable ocean sections made at all seasons of the year. In the tropics and

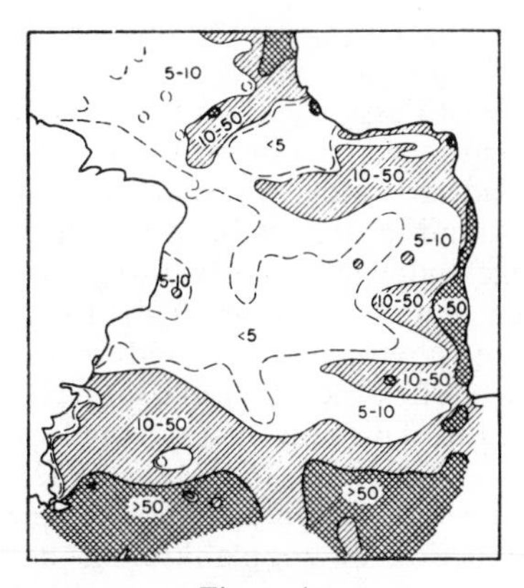

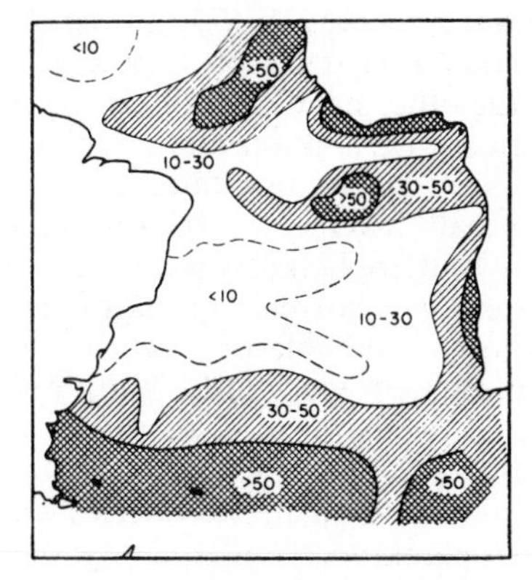

Figure 1 a. Figure 1 b.

(a) Distribution of plankton organisms (numbers in thousands per litre) in the upper 50 m. layer.

(b) Distribution of zooplankton (metazoa) (numbers per 4 litres) in the upper 50 m. layer. After HENTSCHEL, 1933.

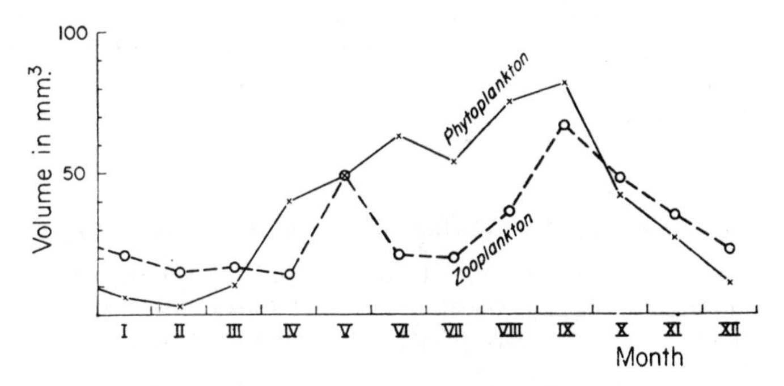

Figure 2. Monthly averages of the volumes of zooplankton and phytoplankton near Kiel. After LOHMANN, 1908.

subtropics the standing crop of both phytoplankton and herbivores must be fairly constant throughout the year.

A direct inter-relationship is also found in coastal waters. Figure 2 (after LOHMANN, 1908) shows the volumes (Rechenvolumen) of phytoplankton and zooplankton present in the Kieler Fjord in the course of a year. It is obvious that there is a direct relation between the quantities of phytoplankton and zooplankton. In the summer months the quantity of phytoplankton exceeds that of zooplankton. On the other hand in the darkest months of the year, when phytoplankton production is greatly reduced, the volume of zooplankton exceeds that of phytoplankton.

No assessments of zooplankton were made by the "Galathea" but the data collected by the expedition clearly show that the standing crop of phytoplankton in tropical and subtropical parts of the ocean remained fairly constant. This requires that the rate of grazing by the zooplankton was fairly even. It may be added that investigations made by the Danish Fishery Investigations with the carbon-14 technique in the northern part of the Atlantic in the course of 4 summers (1954–1957) showed that for a given region the rate of organic production at that season is practically the same each year.

On the other hand, it is of course incorrect to assert that an inverse relation between phytoplankton and zooplankton can never obtain in the sea. Nevertheless such a state of affairs must be considered rather unusual. An example of "normal" inverse interrelation may be found in regions in which the conditions for phytoplankton production change suddenly from bad to good. This may occur in relatively high latitudes when in winter the water is unstratified down to considerable depths. In the spring, phytoplankton production once started proceeds vigorously, but it only starts when the water masses have been stabilized by the heating of the surface layer. This was first pointed out by BRAARUD and KLEM (1931). According to MARSHALL and ORR (1952), egg production in zooplankton species which hibernate is stimulated by the growth of phytoplankton. Consequently the production of zooplankton is delayed compared with that of phytoplankton. A short period with much phytoplankton and little zooplankton may thus occur. As the rate of grazing is low for a time due to the small stock of grazers a large crop of phytoplankton is quickly produced.

The use of net-caught samples for estimating the standing crop of phytoplankton has undoubtedly been the main background for the oft-repeated claims for the existence of an inverse relationship between phytoplankton and zooplankton in the sea. The different theories which try to explain it need not therefore be discussed here.

5. How is the balance between phytoplankton and zooplankton established?

It is of very great interest to try to understand the way in which Nature is able to maintain in a given oceanic region a fairly stable standing crop of plankton algae which regularly produces about the same amount of organic matter daily.

Plankton algae grown under laboratory conditions behave quite differently from those normally seen in nature. In the laboratory, scarcity of nutrients — for example those containing nitrogen and phosphorus — soon effects an absolute deficiency of these substances. The algae reach a state in which they do not grow at all. The respiratory rate is more or less unaffected, but the rate of photosynthesis is much reduced. After a fresh supply of the lacking nutrient has been added to a culture in which growth has stopped entirely, it takes a considerable time — in *Chlorella* more than 24 hours — before growth begins again (AL KHOLY, 1956). The rate of photosynthesis is unaffected for several hours at least (cf. STEEMANN NIELSEN and AL KHOLY, 1956).

If the growth of algae were not controlled by the grazing of animals or by other means, we should expect at least in the oligotrophic, tropical parts of the ocean periods with relatively much phytoplankton alternating with long periods in which practically no algae were present. However, HENTSCHEL's investigations on the "Meteor" expedition showed indisputably that on the contrary the standing crop of algae remains exceedingly steady in regard to both place and time.

In a particular oceanic region and at a particular depth, the curves showing the dependence of photosynthetic rate in the plankton on light intensity are, in principle, all similar. To the 12 series of experiments made with oceanic surface water on the "Galathea" expedition (STEEMANN NIELSEN and AABYE JENSEN, 1957) may be added about 100 series more made with water from all depths of the photic layer obtained during cruises of the "Dana" in the North Atlantic and the Arctic during the summers of 1956 and 1957. In every experiment without exception, the rate of respiration is low in relation to the optimal rate of respiration. This is true both for eutrophic and oligotrophic water. A situation such as that suggested by RYTHER (1954) in which the rate of respiration in oligotrophic regions is about the same as the rate of optimal photosynthesis has never been met with. As shown by STEEMANN NIELSEN and AL KHOLY (1956) the occurrence in nature of such a situation is not very likely.

Accordingly, an equilibrium must normally exist between the growth of algae and their disappearance through grazing and sinking.

It is not at present possible to say with any certainty how this equilibrium is established. If the algae did not disappear at a fairly steady rate, the population of plankton algae would hardly remain constant. STEEMANN NIELSEN and AABYE JENSEN (1957) reckoned that in the oligotrophic, tropical parts of the ocean the daily net production of organic matter in the whole photosynthetic layer is about 25% of the organic matter contained in the algae. It was shown at the same time that almost all the loss of algae is caused by grazing either in the photic zone or — after they have sunk — immediately below this zone. The rate of grazing must thus be reasonably constant in these parts of the ocean.

As static conditions seem to prevail in most parts of the ocean, the size of the zooplankton population must be assumed to depend directly on the extent of the algal production. The detailed mechanism of this undoubtedly rather complicated and intricate relation between zooplankton and phytoplankton production is at present beyond our comprehension. In fact we may just as well say that the size of the phytoplankton crop depends directly on the rate of grazing.

In some way — as yet unknown — the growth of algae in oligotrophic regions must depend on the combined effects of grazing by the zooplankton, of being limited by the lack of some nutrient, and other circumstances. It seems fairly unlikely that the rate of replenishment of nutrients directly determines the rate of primary production, although the idea put forward *inter alia* by KETCHUM (1947) must nevertheless be considered valid: "It is the rate of replenishment, and not the concentration observed at a given time, which determines the fertility of an aquatic environment". The effect of replenishment is of an indirect nature.

The surface water of the Sargasso Sea can be characterized as extremely unproductive. About the lowest productivity observed anywhere in the oceans by the "Galathea" was found here (STEEMANN NIELSEN, 1954). Athough the water is extremely unfertile, it is nevertheless usually possible to detect small amounts of inorganic phosphate and nitrate even at the surface. For the western part of the Sargasso Sea, RILEY (1949, Fig. 6) has published a considerable body of data averaged by two-degree squares of latitude and longitude. East of 65°W. and south of 30°N., the concentration of phosphate-P is about 0·05 μg.-atom per litre, i. e., about the same as in many coastal areas during the height of production. On the "Galathea" cruise similar concentrations of inorganic phosphate were found in the surface water at all Sargasso Sea stations. Measureable concentrations of nitrate are also present in the region, as shown by both THOMSEN (1937) and RILEY (1949).

According to the measurements of primary organic production in the Sargasso Sea, the rate of photosynthesis must be about 1 mg. C per m.[3] per day near the surface (cf. STEEMANN NIELSEN and AABYE JENSEN, 1957, Table 20). If we may assume with REDFIELD (1934) that roughly 1 mg. P is assimilated for every 50 mg. C, the rate of phosphate assimilation may be estimated at about 0·02 mg. P per m.[3] per day. This is about 1 per cent. of the concentration of inorganic phosphate normally present here (0·05 μg.-atom P/l. = 1·5 mg. P/m.[3]).

If it is assumed that conditions in the surface water of the Sargasso Sea are stable, the replenishment of inorganic phosphate may also be reckoned at about 0·02 mg. P per m.[3] per day. As fairly phosphate-rich water does not occur in the Sargasso Sea at depths less than several hundred metres and as the water masses in the eddy show a tendency to descend, the replenishment of inorganic phosphate in the surface water must be due primarily to regeneration within the surface layer itself. The daily phosphorus turnover seems to be extremely slow in relation to the concentration of dissolved inorganic P present at a given time. The turnover of phosphorus in relation to the P-content of the algae is, on the other hand, not slow; it must be of about the same amount as that of carbon in relation to the carbon content of the algae, that is, about 25% per day (STEEMANN NIELSEN and AABYE JENSEN, 1957).

The plankton algae very likely grow more slowly in the unproductive regions of the ocean than in the productive. The difference between the growth-rates is probably not too large, however. Calculations made (STEEMANN NIELSEN and AABYE JENSEN, 1957) indicates that the growth-rates in productive regions are about double the rates in unproductive. RODHE (1948) has arrived at a somewhat similar view for fresh water. He writes (p. 101):— "it is not said that the relative growth, consequently the increase per cell, is always greater in highly productive waters than in those poor in specimens. The results of the previous experiments give ground for the supposition that this should not always be the case".

Plankton algae must thus be able to assimilate phosphorus at a sufficiently high rate from very dilute concentrations of inorganic phosphate, provided these concentrations remain constant, either through a high regeneration rate, as in some eutrophic areas (cf. STEEMANN NIELSEN, 1951) or as the result of a very small standing crop of algae, as in the Sargasso Sea. In some places organic phosphate is perhaps also assimilated, but this can hardly be the case in an oligotrophic oceanic area.

A concentration of phosphate of 1·5 μg. P/l. — such as found in the Sargasso Sea — must be considered adequate for the normal growth of many species of plankton algae. According to RODHE (1948) the growth-rate of the fresh-water diatom *Asterionella formosa* in lake water is constant down to a concentration of 2 μg. P-P_2O_5/l. The flagellates *Dinobryon divergens* and *Uroglena americana* were shown to have still lower phosphorus requirements.

It could of course be argued that the algal production in the Sargasso Sea is limited by some nutrient other than phosphate or nitrate, the availability of the latter being at least as good as that of phosphate. Such an idea cannot be absolutely ruled out; although it seems rather improbable. As already pointed out on p. 184, in laboratory experiments absolute deficiency of a substance has the effect of transforming the algae — at least species like *Chlorella pyrenoidosa* — into a state in which they do not grow at all. It was shown that such conditions do not occur in the relatively unproductive parts of the oceans.

Hence it must be assumed that algal growth is directly controlled by factors other than the deficiency of necessary substances. One of these factors is very likely the rate of grazing by zooplankton. Another is possibly the production of auto-inhibitors by the algae themselves (PRATT, 1942; JØRGENSEN, 1956). The puzzling problem thus is: How is it that Nature arranges matters so that as a rule the grazing of the zooplankton maintains the standing crop of phytoplankton at the size best suited to prevailing growth conditions?

The present extent of our knowledge does not seem sufficient to allow any definite answer to be given. We suspect, however, that external metabolites given off by both algae and zooplankton may have at least some connexion with the establishment of the balance between producers and consumers in the ocean. As LUCAS (1955) so convincingly presented the case, we are now obliged to believe that ecological inter-relationships exist in the sea which are as important as the already familar relationships existing between the organism and its physical environment, and between prey and predator. It may be predicted that an essential part of oceanographic study in the coming years will be work along such lines.

Another possible prime factor in the establishment of a balance between producers and consumers in the ocean is natural selection among the species occurring in the different areas. This concerns, of course, both plants and animals. The most abundant species are almost invariably of a kind best adapted to the particular conditions. In the long run the best results are undoubtedly obtained under steady-state conditions. The organisms which are able to achieve this state will thus oust all competitors.

Summary

A direct relationship between phytoplankton and zooplankton is normal in the sea. The claims for an inverse relationship and the various theories which try to support this conception seem mainly to result from the use of net-caught samples for estimating the standing crop of phytoplankton. The use of nets of coarse silk may at the same time lead to the amount of herbivorous zooplankton being underestimated, expecially in periods when small organisms — for example, nauplii — predominate.

It is shown that as a rule grazing by the zooplankton maintains the standing crop of phytoplankton at the size best suited to prevailing conditions. The result is that with stable hydrographic conditions such as occur mainly in tropical and subtropical regions, the standing crop of phytoplankton and its production is stable. The circumstances which may lead to establishment of conditions of balance are discussed.

References

AL KHOLY, A. A., 1956. "On the assimilation of phosphorus in *Chlorella pyrenoidosa*". Physiol. Plant., **9**: 137.

BRAARUD, T., & KLEM, A., 1931: "Hydrographical and chemical investigations in the coastal waters off Møre and in the Romsdalsfjord". Hvalrådets Skr., **1**.

CLARKE, G., 1940. "Comparative richness of zooplankton in coastal and offshore areas of the Atlantic". Biol. Bull., **78**: 226.

CUSHING, D. H., 1955. "Production and a pelagic fishery". MAFF, Fish. Invest., Ser. 2, **8** (7).

FISH, C. J., 1954. "Observations on the biology of boreo-arctic and subtropical oceanic zooplankton populations". Symposium on Marine and Freshwater Plankton in the Indo-Pacific. Bangkok.
GESSNER, F., 1944. "Der Chlorophyllgehalt der Seen als Ausdruck ihrer Produktivität". Arch. f. Hydrobiol., **40**: 687.
GRØNTVED, J., & STEEMANN NIELSEN, E., 1957. "Investigations on the phytoplankton in sheltered Danish marine localities". Danm. Fisk.- og Havunders., Ser. Plankton, **5** (6):5.
HARVEY, H. W., 1955. "The Chemistry and Fertility of Sea Waters". Univ. Press, Cambridge.
HENTSCHEL, E., 1932. "Die biologischen Methoden und das biologische Beobachtungsmaterial der 'Meteor'-Expedition". Wiss. Ergeb. Deutsch. Atlant. Exped. a. d. Forsch. u. Vermessungsschiff "Meteor" 1925–1927, **10**.
— 1933–1936. "Allgemeine Biologie des Südatlantischen Ozeans". Ibid., **11** (1–2).
JESPERSEN, P., 1937. "Quantitative investigations on the distribution of macroplankton in different oceanic regions". Dana-Rep., **7**.
— 1949. "Investigations on the occurrence and quantity of holoplanktonic animals in the Isefjord 1940–43". Medd. Komm. Danmarks Fisk.- og Havunders., Ser. Plankton, **5** (3).
JØRGENSEN, E. G., 1956. "Growth inhibiting substances formed by algae". Phys. Plant., **9**: 712.
KETCHUM, B. H., 1947. "The biochemical relations between marine organisms and their environment". Ecol. Monogr., **17**: 309.
KREPS, E., & VERJBINSKAYA, N., 1930. "Seasonal changes in the Barents Sea". J. du Cons., **5**: 329.
LOHMANN, H., 1908. "Untersuchungen zur Feststellung des vollständigen Gehaltes des Meeres an Plankton". Komm. Wiss. Unters. d. Deutsch. Meere in Kiel und d. Biol. Anst. Helgoland. Wiss. Meeresunters., N.F., Abt. Kiel., **10**: 131.
— 1920. "Die Bevölkerung des Ozeans mit Plankton nach den Ergebnissen der Zentrifugenfänge während der Ausreise der 'Deutschland' 1911. – Zugleich ein Beitrag zur Biologie des Atlantischen Ozeans". Arch. f. Biontologie, **4**.
LUCAS, C. E., 1955. "External metabolites in the sea". Pap. Mar. Biol. and Oceanogr., Suppl., Deep-Sea Res., **3**.
MARSHALL, S. M., & ORR, A. P., 1952. "On the biology of *Calanus finmarchicus* VII. Factors affecting egg production". J. Mar. Biol. Assoc. U.K., N.S., **30**: 527.
PRATT, R., 1942. "Some properties of the growth-inhibiting substance formed by *Chlorella vulgaris*". American J. Bot., **27**: 52.
REDFIELD, A. C., 1934. "On the proportion of organic derivates in sea water and their relation to the composition of plankton". James Johnstone Memorial Volume, p. 176.
RILEY, G. A., STOMMEL, H., & BUMPUS, D. F., 1949. "Quantitative ecology of the plankton of the western North Atlantic". Bull. Bingh. Oceanogr. Collect., **12** (3).
RODHE, W., 1948. "Environmental requirements of fresh-water plankton algae. Experimental studies in the ecology of phytoplankton". Symbol. Botan. Uppsala, **10**: 1.
RYTHER, J. H., 1954. "The ratio of photosynthesis to respiration in marine plankton algae and its effect upon the measurement of productivity". Deep-Sea Res., **2**: 134.
SETTE, O. E., 1955. "Consideration of mid-ocean fish production as related to oceanic circulatory systems". J. Mar. Res., **14**: 398.
STEEMANN NIELSEN, E., 1935. "Eine Methode zur exakten quantitativen Bestimmung von Zooplankton. Mit allgemeinen Bemerkungen über quantitative Planktonarbeiten". J. du Cons., **10**: 302.
— 1938. "Über die Anwendung von Netzfängen bei quantitativen Phytoplanktonuntersuchungen". Ibid., **13**: 197.
— 1943. "Über das Frühlingsplankton bei Island und den Faröer-Inseln". Medd. Komm. Danmarks Fisk.- og Havunders., Ser. Plankton, **3** (6).
— 1951. "The marine vegetation of the Isefjord. A study on ecology and production". Ibid., Ser. Plankton, **5** (4).
— 1954. "On organic production in the oceans". J. du Cons., **19**: 309.
STEEMANN NIELSEN, E., & AABYE JENSEN, E., 1957. "Primary oceanic production. The autotrophic production of organic matter in the oceans". "Galathea" Rep., **1**: 49.
STEEMANN NIELSEN, E., & AL KHOLY, A. A., 1956. "Use of ^{14}C-technique in measuring photosynthesis of phosphorus or nitrogen deficient algae". Physiol. Plant., **9**: 144.
THOMSEN, H., 1937. "Hydrographical observations made during the 'Dana'-Expedition 1928–30". Dana-Rep., **12**.

The Seasonal Variation in Oceanic Production as a Problem in Population Dynamics

By

D. H. Cushing

Fisheries Laboratory, Lowestoft

Introduction

The quantity of oceanic production is probably dependent in some way on the quantity of available nutrient and so the low quantities of phosphorus in the Tropics are taken to indicate low levels of production. This view of oceanic production is the chemical one which has given minimal and consistent results in such well studied areas as the English Channel and the Atlantic coast of the United States.

The view of production presented in this paper is a biological one based primarily on the conflict of populations. Certain simplifications are made, treating the whole algal community as a prey and the whole herbivore community of zooplankton as a predator. This is justified on the grounds that although many algal species and many species of zooplankton are mixed together, there is no emigration or immigration except on a minute scale and that the algae reproduce in the common sunlight and are probably eaten in almost a random manner in the mixing waters.

There are three sections to this paper, firstly a short description of a model of plankton production[1]), secondly a discussion of production as a predator/prey process, and lastly a description of seasonal differences as functions of this process.

I. A Model of Plankton Production in the Central North Sea

It was assumed that there was only one algal species and that the individuals were of constant size (10,000 μ^3). Specific differences and size differences were neglected because it was thought that they were less marked for the present purpose than differences in reproductive rate. The herbivorous zooplankton

[1]) A fuller description will appear in Cushing, D. H. (in press) "On the nature of production in the sea". Fish. Invest., Lond., Ser. 2, **22** (6).

were placed in two groups, *Calanus* and "*other copepods*": individuals of the former group are about ten times bigger than those of the latter.

For the purposes of the present model, the predator/prey process may be stated formally as follows: —

$$\frac{dP}{dt} = (R-G)\,P \tag{1}$$

$$\frac{dH}{dt} = (R_H-M_H)\,H \tag{2}$$

where P is the number of algae per unit volume or beneath unit surface,
where H is the number of herbivores per unit volume or beneath unit surface,
where R is the daily algal reproductive rate,
where G is the daily rate of grazing mortality on the algae, combined for both herbivore groups,
where R_H is the daily reproductive rate of either group of herbivore,
where M_H is the daily mortality rate of either group of herbivore.

The second equation is operated separately for *Calanus* and for other copepods. The parameters of the two equations are related as follows: —

$$G = f(H) \tag{3}$$

$$R_H = f(P) \tag{4}$$

The operation of the model is a little complicated for three reasons. The first is that the algal reproductive rate is dependent upon the interplay of two physical factors as spring progresses in the sea, the increase of radiant energy and the decrease of vertical turbulence. The second complication is that the relation between the grazing mortality coefficient and the number of herbivores is not a simple one, nor is that between the herbivore reproductive rate and the number of algae. The third point is that as a copepod lives for about 8–12 weeks, there are rapid size changes causing quick changes in reproductive capacity and in grazing capacity; consequently, the herbivore population is most readily dealt with in stages.

In the depth of winter in temperate seas, there is no increase in either the algal or herbivore populations. When winds are high and the sun is low in the sky in the wintertime, the algae are not allowed to remain long enough in the shallow photic layer to do more than respire. Consequently, no reproduction can take place. The numbers of herbivores are very small and the effect of grazing is also very small and so the algal populations remain virtually constant. This level of algal numbers is well below the threshold for herbivore egg production and so the herbivores cannot reproduce and if mortality is low, their populations must also remain constant.

In spring, the total quantity of photosynthesis is proportional to the daily income of radiant energy, whereas the total quantity of respiration is constant. So production can only start when the total photosynthesis in a day exceeds the total respiration. In the sea in winter, the mixed layer is much deeper than the photic layer and so the algae only spend part of the day in the light; production can only start when the mixed layer has become shallow enough to allow the total photosynthesis in the mixed column to exceed the total respiration (SVERDRUP, 1953).

The algal reproductive rate was estimated in the following way. LUND (1950) has published data on the division rates of *Asterionella formosa* Hass, a diatom, in Windermere at 5 m and at 7 m, with contemporaneous measurements of light ashore in kiloluxes. SVERDRUP, JOHNSON, and FLEMING (1942) give factors for converting lux to energy, and the surface energy received to that used in photosynthesis. Thus, it was possible to relate division rates to energy. This relation was used in the central North Sea with JERLOV's (1951) classification of water types; so differences in the spectral composition of light between Windermere and the North Sea were adjusted. The daily income of energy, taken from SVERDRUP *et al.* (1942) corrected for length of day and the sun's altitude and the appropriate energy extinction coefficients, was used to estimate a daily algal division rate.

The depth of the mixed layer was derived from average measures of wind speed (Service Hydrographique, ICES 1950–53) in the central North Sea using EKMAN's formula for the depth of frictional resistance (SVERDRUP *et al.*, 1942); then the depth of the photic layer was expressed as a fraction of the mixed layer. The estimated algal reproductive rates were reduced by this fraction, which expresses the effect of the wind. Thus, the algal reproductive rates were estimated only from weekly measures of sunlight and of wind strength.

The grazing rate of the herbivores was based on the assumption that copepods find their algal food by encounter and that the battery of sensory hairs along the antennules are tactile appendages. The transverse cross section of the animal including all appendages, called the "contact surface", was measured. Some estimates of minimum distance travelled in a day were available from the literature on the vertical migration of plankton. This distance multiplied by the "contact surface" gives a mechanical "volume swept clear" per day (*Sm*). A copepod can only eat one diatom at a time and so when diatoms are dense, this "mechanical" volume swept clear is reduced in proportion to the sizes and numbers of algal cells in the following way: —

$$1 - p = p.v.x.y$$

$$\text{i.e., } S_o = \frac{S_m}{1 + v.x.y}$$

where S_o is the operational "volume swept clear" $= p.\ S_m$,
where x is the time to eat one algal cell in seconds,
where v is S_m sec.,
where y is the number of algae/ml.

Then $G = S_o.H$ in ml swept clear/1000 ml, H being the number of herbivores per litre. It is possible to add the grazing effects of both herbivore populations to obtain a combined daily grazing mortality coefficient for use in the algal equation. Thus, although the two herbivore populations are operated separately, their grazing effect is combined in the algal equation.

The herbivore reproduction rate is based on some experiments of MARSHALL and ORR (1952), showing that egg production of copepods increases with algal numbers; it is not a linear relation, but an asymptotic one, starting at a fairly high threshold value. The mortality of nauplii and younger copepodites was estimated from the detailed studies by NICHOLLS (1933) in Loch Striven.

There is no need to describe the working of the model in further detail. It does fit the few field observations fairly well. The peak quantity of algae in

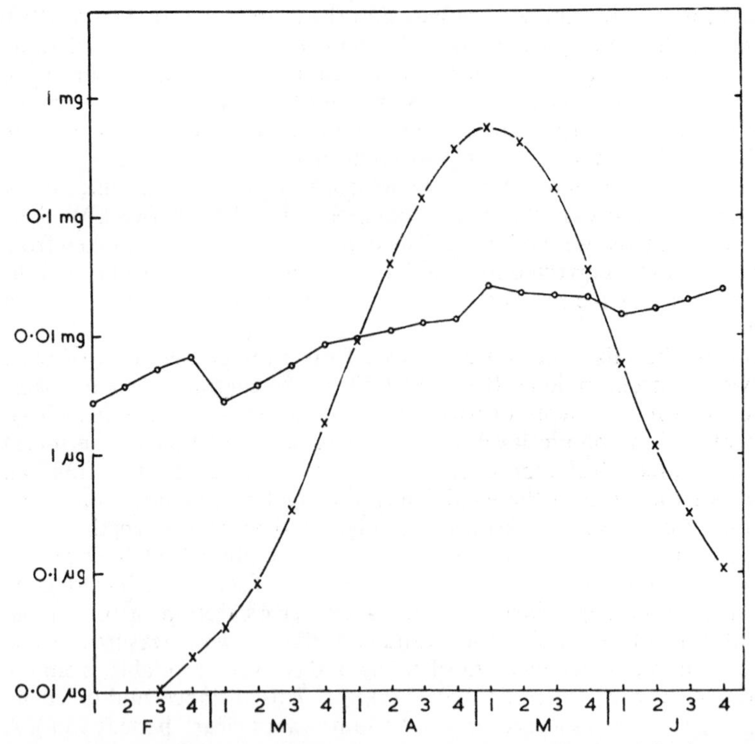

Figure 1. Theoretical algal production (×—×) in wet wt./litre/week in the central North Sea from February to June. The herbivore production (o — o) in wet wt./litre/week for the same period is also shown.

the central North Sea is about 0·5 mm³/l at the end of April or the beginning of May; subsequently, the numbers decrease very rapidly. The numbers of *Calanus* increase from 0·01/l to about 0·4–0·5/l in May and the "other copepods" from 0·1/l to 4–5/l in June. Figure 1 shows that the model runs efficiently within these limits. Because of the large number of variables used, this is not necessarily an adequate test. It becomes more adequate when it is realized that the theoretical values of the reproductive rate were close to those observed at sea in 1949; also that over the season of the spring outburst, the herbivores get enough to eat and the total herbivore production amounts to about one-seventh of the total algal production.

II. The Model as a Predator/Prey Process

As in the classical predator/prey studies (LOTKA, 1925; VOLTERRA, 1928), the parameters of the two equations are interdependent. There are, however, certain differences between this model and the classical ones. The first is that the parameter which drives the whole system, the algal division rate, is dependent upon two physical factors, radiant energy and vertical turbulence, the sun and the wind. The second difference is that the parameters linking the two equations are not linearly related. As algal numbers increase, the total mortality coefficient due to herbivore grazing decreases. So the grazing mortality rate on the algae depends inversely upon the numbers of algae and directly upon the numbers of herbivores. Herbivore egg production is dependent upon algal numbers, and upon herbivore numbers, the algal function being asymptotic in egg production.

There are two further differences associated with the herbivore populations. The first is that there is a threshold in the relation between egg production and algal numbers (at $10^6/\mu^3/l$). Consequently production of herbivores cannot start until eggs are produced at the threshold level of algal numbers. The second difference is that it takes perhaps three weeks for the herbivore egg to develop into a copepodite of significant grazing capacity. The two factors, the threshold to egg production and the development of the copepodite give rise to two delay periods, which are added. This combined factor is analogous to the delayed density dependent effects described for predators by VARLEY (1947).

During this delay period, which may be as much as three weeks, the daily algal division rates rise from 0·1 to 0·5. Consequently the algal numbers increase during the spring outburst by one hundred to one thousand times or more. The control of algal numbers can only take place after the end of the delay period, when the first generation of herbivores increases in size and grazing capacity and so the grazing rate on the algae tends to equal the reproductive rate of the algae. Thus, the algal reproductive rate may be said to drive the productive system and algal populations are controlled by the grazing of the herbivores.

It has been shown by me experimentally that *Temora longicornis* Boeck, a copepod, will eat up to ten times its immediate metabolic needs, if given the chance, in a large vessel (CUSHING, 1958). There are two interesting consequences of this observation. The first is that there is a high rate of nutrient regeneration, which is immediate; and in the sea, I have observed that *Calanus finmarchicus* (Gunn.) in April will excrete into the water much more phosphorus than might be expected from the carbon/phosphorus ratio in copepod flesh. The second consequence is that there is a large production of particulate excreta, which may provide a partial basis for feeding in the less productive season of high summer.

If immediate nutrient regeneration is important then the uptake of nutrients must be balanced to some extent by their regeneration as production proceeds. If uptake were proportional to production and regeneration to the quantity grazed, then there would be a rough balance between the two until about the time of peak algal numbers. The quantity produced is maximal just before that time and the quantity grazed is maximal a little later, because the standing stock is then being rapidly eaten up. So the greatest increment to herbivore material is added just after the time of peak algal numbers; in Windermere, the quantity of silica locked in particulate excreta and living material at this time was one-third of the total silica budget for the year. This large fraction of nutrient lost to the productive system coincides with the peak of algal numbers, because both are functions of the increased grazing capacity, which at this time yields the greatest quantity grazed and also controls the algal numbers. These relations are set out diagrammatically in Figure 2.

If this argument is true, then the decline of nutrients in spring and early summer does not limit the production in the spring outburst, although the consequent lack of nutrients might well restrict it later. The decline in nutrients

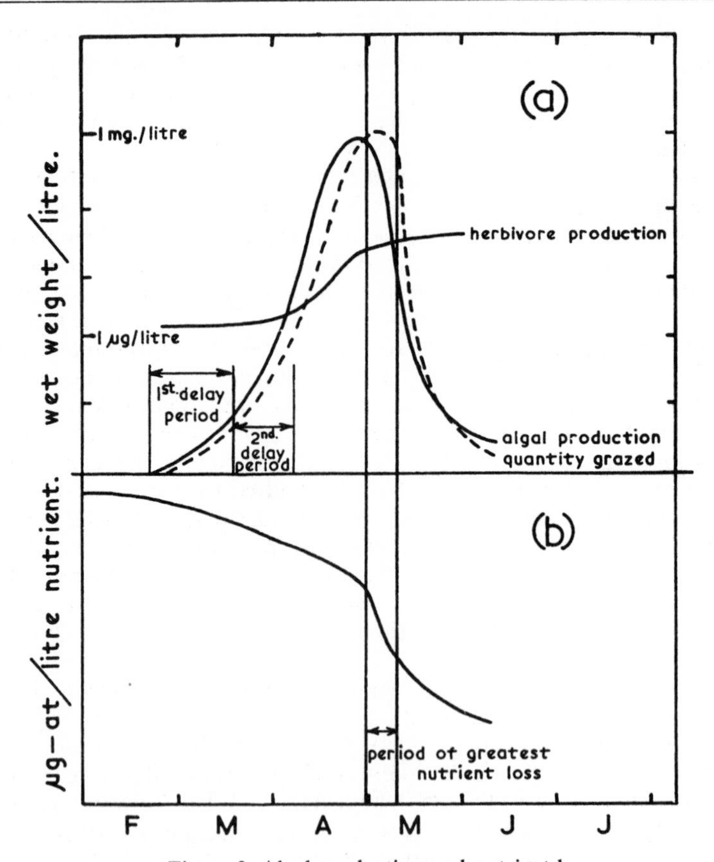

Figure 2. Algal production and nutrient loss.
(a) Weight of algae produced and weight of algae eaten from February to June. Weight of herbivores produced from February to June.
(b) The reduction in the quantity of nutrients from February to June.

might be described as a symptom of production and in the case of phosphorus and nitrogen, it would be exactly related to the increase in the herbivore population rather than to the rapid changes of the transitory phytoplankton.

Summarizing, the spring outburst of production in temperate seas may be visualized as a predator/prey system, which proceeds in an unbalanced manner because of the delay period to the grazing capacity. The consequence of the delay is firstly, the unfettered algal outburst, secondly the excessive herbivore grazing and thirdly, as a consequence, the decline in nutrients.

III. The Seasonal Cycle of Production in Different Regions

In temperate and arctic waters there is no production in winter because the photic layer is only a small proportion of the mixed layer. But, production in the Sargasso Sea and in the Mediterranean continues throughout the winter; in fact, BERNARD (1939) and RILEY, STOMMEL, and BUMPUS (1949) have shown

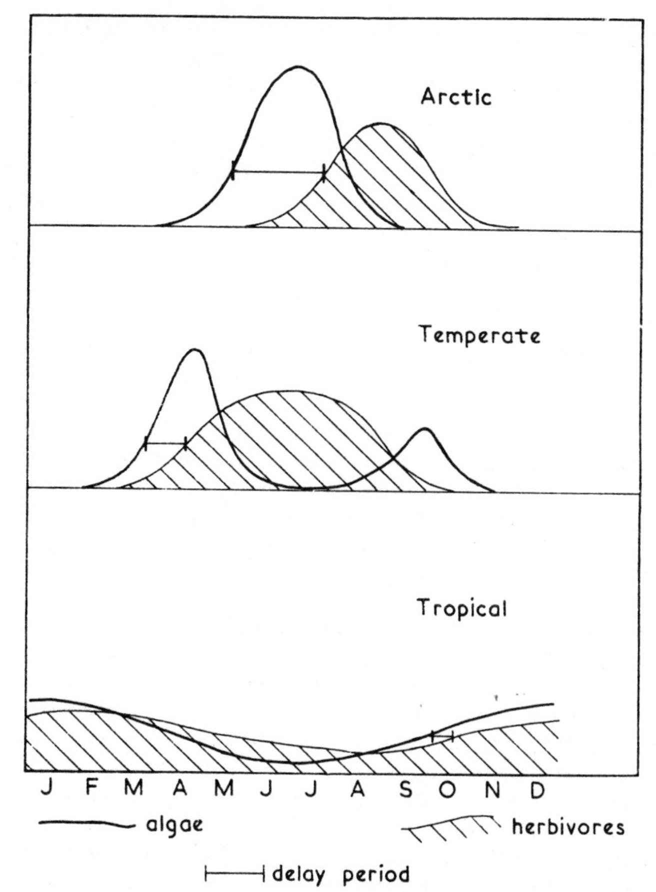

Figure 3. Diagram representing the seasonal amplitudes in algal and herbivore production in different latitudes.

that the highest algal numbers are found there in December. The seasonal range of algal numbers is least in the Tropics where production is continuous and most in high latitudes where production is limited to a short season.

The three main patterns of seasonal production are shown in Figure 3, tropical, temperate, and arctic (or antarctic). In the Arctic the daily radiation increases from zero in March to a maximum in June, with a 24 hour day, and the range in reproductive rate is greatest in the Arctic. The delay period in reaching the threshold in food for egg production is long, because the rate of change of radiation is slow at first; the delay period is further extended by the slow development of herbivores in cold water. Towards the end of the extended delay period, the algal reproductive rate will be higher than elsewhere because the daylight is continuous and very high relative to that at the beginning of the delay period; this factor combined with the long delay periods gives the algal standing stock the greatest chance to be very large and in high latitudes it may not be controlled by grazing. The algae reach maximum numbers at

3*

midsummer, with a single long-lived brood of herbivores following in late summer and in early autumn. The delay period in the Arctic is the longest of all and as a consequence the productive system is the most unbalanced.

In temperate waters, the productive cycle starts earlier. The range in the daily algal reproductive rate will necessarily be less, the herbivores will develop more quickly in the somewhat warmer water (6°–10°C, as opposed to 0°–3°C in the Arctic) and so the delay period will be relatively shorter than that in the Arctic. In fact, it is short enough to allow the algal populations to be certainly controlled by grazing. The herbivore population builds up to midsummer and when it declines again in early autumn, a secondary algal outburst sometimes takes place.

In the Tropics the delay period is small and so the increase in algal reproductive rate during the delay is also small and the algal production takes place early enough in the year for it to be described as a winter outburst. In temperate and arctic waters, the seasonal amplitude in algal numbers may reach a thousand times, but in tropical waters it may be as little as five times. In other words, in the Tropics, production is a continuous or even a steady state process, in which the algal reproductive rate is always more or less balanced by or equal to the grazing rate of the herbivores. There is a narrow seasonal range in reproductive rate, the shortest of delay periods and, in this low but continuous cycle, there is very little seasonal variation in nutrient content.

Considering the whole range of regional variation, the most important single physical factor is the lack of sufficient winter sunlight away from the equator. The indirect effect of this on the predator/prey process is to introduce a delay period, which increases with latitude; also, the longer the delay period is, the more unbalanced do the productive cycles become. On a flat earth with no seasons, oceanic production could well be in a true steady state, in which the algal reproductive rate and the herbivore grazing rate were approximately equal all the time. In the high latitudes of our round earth, the two parameters, however, can be so different that enormous algal populations are produced.

Our present ideas of the productive capacity of the tropical oceans are usually related to the areas of upwelling. For example, in the Benguela Current and in the Peru Current, cold nutrient-rich water is brought up to the surface and, in these regions, dense algal populations are found. The seasonal picture, as described above, is completely changed because the productive season depends upon the season of upwelling. The upwelling water is cold and turbulent and must contain a low plankton population; the algal reproductive rate would necessarily be slow until the turbulence were reduced and it is likely that there would therefore be some sort of delay period in the system and algal production could increase greatly during that delay period. Hence, an unbalanced system would be set up, yielding high algal numbers and, in the Peru Current at least, a guano industry. Thus, it might well be that the areas of high algal numbers were also effects of a delay period as in higher latitudes; then, the presence of large quantities of nutrients in such areas, is, if not quite irrelevant, perhaps merely fortuitous.

In the extremes there are two entirely opposed types of productive cycle, tropical and arctic. The tropical cycle is of low amplitude, is continuous and is characterized by constant low nutrients and low peak algal standing stock; the arctic cycle is unbalanced, discontinuous and is characterized by a nutrient cycle and a very high peak algal standing stock. The predator/prey process described here is only part of the productive cycle and an exact comparison of tropical and arctic production is not yet possible because the higher tropic levels cannot yet be adequately sampled. However, within the primary predator/prey process, it is likely that the tropical cycle is more efficient merely because the longer the delay period the greater the chance of the algae dying in other ways than by grazing. This is perhaps why the diatomaceous ooze is found on the floor of the Southern Ocean. It may be said that the primary predator/prey process comprises the greatest part of the productive cycle; consequently, there is some reason to believe that tropical oceans are less barren than considered hitherto.

Summary

A model of plankton production in temperate seas is described. The algal community is treated as a single prey and the zooplankton community as two predators, *Calanus* and "other copepods". Algal division rates are derived from average measures of radiant energy and wind strength. The coefficient of grazing mortality of algae is derived from an encounter theory of grazing, by means of which *Calanus* "sweeps clear" 1–3 l/day. The herbivore reproductive rate as number of eggs/day is dependent upon the quantity of food available, from a threshold upwards.

Considered only as a predator/prey process, there is a delay period to the grazing capacity of the herbivores. Eggs cannot be produced until a threshold in algal food is reached and then the egg takes perhaps three weeks to develop into a copepodite of significant grazing capacity. During this delay period algal numbers increase by one hundred to one thousand times. This outburst is controlled in the model by herbivore grazing.

To achieve this, grazing is five to ten times in excess of the animals' immediate needs, which may also be shown in the immediate regeneration of phosphorus. It is suggested that the decline of nutrients in the spring takes place just after the algal outburst. Any nutrient observation strikes a balance between uptake and regeneration; the peak of the algal outburst is the peak of production and most nutrient is then being transferred to herbivore flesh. So the gap between uptake and regeneration is then greatest and the nutrient content falls.

It was suggested that seasonal patterns of production differ mainly in the length of the delay period, which increases with latitude. The production in upwelling regions was fitted into the same scheme.

References

BERNARD, F., 1939. "Étude sur les variations de fertilité des eaux méditerranéennes. Climat et nanoplancton à Monaco en 1937–38". J. Cons. int. Explor. Mer, **14**: 228–41.

CUSHING, D. H., 1958. "The effect of grazing in reducing the primary production: a review". Rapp. Cons. Explor. Mer, **144**: 149–54.

CUSHING, D. H., (in press). "On the nature of production in the sea". Fish. Invest., Lond., Ser. 2, **22** (6).

JERLOV, N. G., 1951. "Optical studies of ocean waters". Rep. Swedish Deep-Sea Exped. 1947–48. **3**: 1–59.

LOTKA, A. J., 1925. "Elements of physical biology". 460 pp. Williams and Wilkins Co., Baltimore.

LUND, J. W. G., 1949. "Studies on *Asterionella*. I. The origin and nature of the cells producing seasonal maxima." J. Ecol., **37**: 389–419.
MARSHALL, S. M., & ORR, A. P., 1952. "On the biology of *Calanus finmarchicus*. VII. Factors affecting egg production". J. mar. biol. Ass. U. K., **30**: 527–48.
MARSHALL, S. M., & ORR, A. P., 1955. "On the biology of *Calanus finmarchicus*. VIII. Food uptake, assimilation and excretion in adult and Stage V *Calanus*". J. mar. biol. Ass. U. K., **34**: 495–529.
NICHOLLS, A. G., 1933. "On the biology of *Calanus finmarchicus*. 3. Vertical distribution and diurnal migration in the Clyde sea area". J. mar. biol. Ass. U. K., **19**: 139–64.
PEARSALL, W. H., & ULLYOTT, P., 1934. "Light penetration into fresh water. III. Seasonal variations in the light conditions in Windermere in relation to vegetation". J. exp. Biol., **11**: 89–93.
RILEY, G. A., STOMMEL, H., & BUMPUS, D. F., 1949. "Quantitative ecology of the plankton of the western North Atlantic". Bull. Bingham oceanogr. Coll., **12** (3).
Service Hydrographique de Cons. Perm. int. Explor. Mer. Wind Averages for 10-day periods, Jan.-June, 1950–53. (Charts).
SVERDRUP, H. U., JOHNSON, M. W., & FLEMING, R. H., 1942. "The Oceans". 1087 pp. Prentice-Hall Inc., New York.
SVERDRUP, H. U., 1953. "On conditions for the vernal blooming of phytoplankton". J. Cons. int. Explor. Mer, **18**: 287–95.
VARLEY, G. C., 1947. "The natural control of population balance in the knapweed gall-fly (*Urophora jaceana*)". J. Anim. Ecol., **16**: 139–87.
VOLTERRA, V., 1928. "Variations and fluctuations in the number of individuals in animal species living together". J. Cons. int. Explor. Mer, **3**: 3–51.

Zooplankton Species Groups in the North Pacific

Co-occurrences of species can be used to derive groups whose members react similarly to water-mass types.

E. W. Fager and J. A. McGowan

Oceanic zooplankton are small animals which spend their entire life suspended in water. They are among the most abundant and widespread macroscopic organisms on earth. In general there are more species and more individuals per unit volume in the upper layers of the oceans than there are in the deeper waters. Although the individual species inhabit very large areas of the oceans, most of them have well-defined patterns of distribution and abundance. Many of these patterns show a remarkable similiarity to the patterns of distribution of water masses (*1, 2*). The latter are, however, defined by the temperature-salinity characteristics of the water below the depths that many zooplankton species inhabit. We have very little information to indicate what immediate environmental conditions determine the abundance and distribution patterns of zooplankton in the upper 150 meters.

Heretofore, most taxonomic and distributional studies of zooplankton have been concerned with the distribution of individual species. Large-scale sampling on three oceanographic expeditions in the North Pacific Ocean (Fig. 1) has provided data that make it possible to study groups of species. Study of the data by computer techniques has shown that certain species frequently occur together, and that these species groups characterize particular habitats, and it suggests that the groups are composed of species that have similar reactions to properties of the environment. The group characterizing the Subarctic Water Mass has been examined in detail in an effort to determine which of the usual hydrographic measurements of the physical-chemical properties of the environment are correlated with the overall abundance of the group. The results suggest that, for this group of species, the history of the water may be the most important property and that, from the standpoint of the biologist, the hydrographic properties usually measured on oceanographic expeditions may not be the right ones.

In this analysis only four kinds of zooplankton were studied: Chaetognatha, Euphausiacea, Heteropoda, and Pteropoda. However, they are among the more abundant members of the zooplankton community. The lack of data on the most abundant animals, the Copepoda, is a serious deficiency. Even for the zooplankton treated we have very little of the information that is basic to an understanding of zooplankton community and population dynamics. We need data on the food, feeding methods, and metabolic rates; the frequency of reproduction; the number of young produced; the causes and rates of mortality; and the relation of behavior to survival and selection of habitat. Nevertheless, the preliminary analysis has served to reveal relationships not readily observed in the untreated data and to indicate areas where more intensive work is needed.

Materials and Methods

The three collecting expeditions were POFI-5 (1950), Transpac (1953), and Troll (1955). All were made during the summer months. Expedition POFI-5 (36 stations) covered two parallel north-south lines of stations, one line starting from Hawaii and going south to about 5°S and the other covering the same north-south range about 750 miles west of the first (*3*). Transpac (141 stations) crossed the Pacific from California to Japan on a track generally north of 40°N, came south along the coast of Japan, and returned by a more southern route, approximately between 20°N and 40°N (*4*). Troll [84 stations; zooplankton studied for only 24 stations, most of them south of 20°N (only these stations plotted)] started at Hawaii, went westward approximately between 5°N and 20°N and then, near the Philippines, turned northward toward Japan (*5*). The station positions for the three expeditions are shown in Fig. 1.

At all of the stations oblique net tows were taken to depths of 150 or 200 meters. The nets had a diameter, at the mouth, of 1 meter and were made of silk grit gauze with a mesh size of 0.65 millimeter. The amount of water filtered by each net was recorded by a flowmeter suspended in its mouth. The amounts ranged from 381 to 1943 cubic meters, but most were between 500 and 1200 cubic meters. After the organisms caught in these tows had been identified and enumerated, the numbers of individuals per 1000 cubic meters were calculated. Hydrographic measurements were made at the same stations.

Chaetognaths were identified by Bieri and (one species) Alvariño; euphau-

Dr. Fager is associate professor of biology and chairman of the department of oceanography and Dr. McGowan is assistant professor of oceanography at the Scripps Institution of Oceanography, La Jolla, California.

[144]

Reprinted from Science, May 3, 1963, Vol. 140, No. 3566, pages 453-460

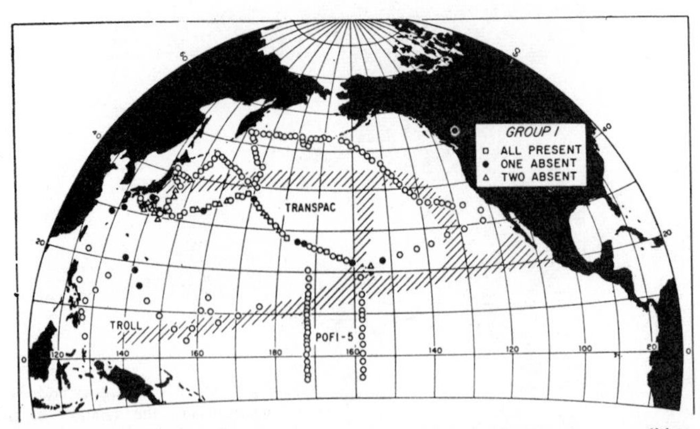

Fig. 1. Distribution of zooplankton group I in the North Pacific. The three expeditions from which material was obtained are indicated by name. The locations of all stations are shown, even those (open circles) where no group I species were found. (Shaded areas) The approximate boundaries of water masses [after R. Bieri (1)].

sids, by Brinton; and heteropods and pteropods, by McGowan (6). All species used in this study satisfy the following criteria: They occurred with high frequency in some of the regions of the North Pacific sampled during the expeditions; their vertical ranges fall within, or overlap to a considerable degree, the depth of sampling; their taxonomic status is clear.

The method which was used to determine recurrent groups of these species has been described (7). The index of affinity between species originally proposed does not follow the hypergeometric distribution exactly (8). It has, therefore, been replaced by the geometric mean of the proportion of joint occurrences, corrected for sample size:

$$[J/(N_A N_B)^{1/2}] - 1/2(N_B)^{1/2}$$

where J is the number of joint occurrences; N_A is the total number of occurrences of species A; N_B is the total number of occurrences of species B; and species are assigned to the letters so that $N_A \leqq N_B$. Pairs of species for which this expression was equal to or greater than 0.50 were considered to show affinity; those for which the values were lower were considered not to show affinity. This breakpoint was chosen because it was felt that species should be found together in somewhat more than "half" their recorded occurrences if they are to be grouped together. The grouping procedure, based on this dichotomy, leads to definition of the largest groups within which all possible pairs of species show affinity. All species within a group are, therefore, rather frequent members of each other's environment. Because of this, the groups appear to be particularly appropriate units within which to examine interspecific relationships.

Calculation of the affinities and the preparation of the incidence matrix on the basis of the dichotomy and the grouping procedure have been programmed for the CDC 1604 computer (9). The operation is at present limited to 150 species; slight modifications should make it possible to increase this number to somewhat over 200 species.

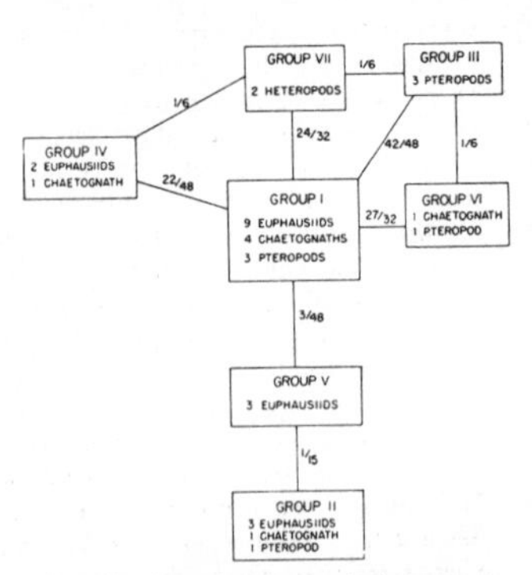

Fig. 2. Composition of zooplankton groups and interrelationships between the groups (see Table 1). Fractions are the ratios of the number of observed species-pair connections between groups to the maximum number of possible connections; for example, there are six possible intergroup species pairs between groups IV and VII but only one of these pairs showed affinity at the "significance" level used.

Species Groupings

Of the 48 species of organisms used in this study (euphausids, 20 species; chaetognaths, 9 species; pteropods, 11 species; heteropods, 8 species), 34 could be arranged in seven groups; five others had affinities with some, but not all, of the members of one or the other of the groups; and nine species had no affinities with any of the others (see Table 1 and Fig. 2). Six of the nine species without affinities were heteropods; the remaining three belonged to the other taxonomic groups, each to a different group.

The first group (I) obtained in the analysis consisted of 16 species (Table 1)—nine euphausids, four chaetognaths, and three pteropods. Three additional species [*Hyalocylix striata* (P), *Limacina trochiformis* (P), and *Euphausia brevis* (E)] had multiple interspecific connections with some but not all of the members of this group (affinities with 14, 12, and 8 of the 16 species in the group, respectively). All of the smaller groups, except II and V, also had many interspecific connections with group I (see Fig. 2).

The second group (II) was an assemblage of five species: three euphausids, one chaetognath, and one pteropod (Table 1). It had connections with only one other group in the analysis; out of 15 possible species pairs, it was connected with group V through a single pair. It was equally sharply separated geographically. Two species [*Thysanoessa inermis* (E) and *Sagitta scrippsae* (C)] had four affinities and one affinity, respectively, with members of this group. The case of *Thysanoessa inermis* is particularly interesting because it could be substituted for *Tessarabrachion oculatus* to form an alternate group of five species. This alternate group, however, did not appear to be as homogeneous in its relation to the environment, to judge by the agreement in relative abundances of the species.

Group V, which was composed of three species of euphausids (Table 1), appeared to form a connecting link between groups I and II, but the connections were limited (3 of 48 possible interspecific connections to group I and 1 of 15 possible connections to group II), and the group seems quite distinct.

Groups III, IV, VI, and VII consisted, respectively, of three pteropods; two euphausids and one chaetognath; one pteropod and one chaetognath; and two heteropods (Table 1). Each of these groups had numerous interspecific connections with group I (42 of a possible 48, 22 of a possible 48, 27 of a possible 32, and 24 of a possible 32, respectively; see Fig. 2). None were connected with any other group by more than one species pair.

Among the 34 species which could be arranged in the seven recurrent groups, species within each genus show considerable morphological similarity, and this implies functional similarity. As the groups were formed on the basis of frequent co-occurrence of the animals, one would, therefore, expect them to be as generically diverse as possible, if functionally similar species tend not to occur together (*10*). In order to test this expectation, a program was written for the CDC -1604 computer which formed all possible arrangements of the 34 species into groups of the same composition (above the level of genus) as the observed groups—all possible sets of nine species of euphausids, four species of chaetognaths, and three species of pteropods for group I, and so on. The sets were drawn from the species lists, only the genera being taken into account. For example, the seven species of *Euphausia* were considered equivalent, the five species of *Nematoscelis* were considered equivalent, and so on. There were 2,227,680 different arrangements. For each arrangement the number of congeneric pairs within each group was calculated. For example, in group I the maximum number of congeneric pairs would be observed if the nine euphausids consisted of the seven species of *Euphausia* and two species of either *Nematoscelis* or *Stylocheiron* or *Thysanoessa* (22 congeneric pairs); the four chaetognaths were all in the genus *Sagitta* (six congeneric pairs); and the three pteropods were all in the genus *Limacina* (three congeneric pairs). The numbers of congeneric pairs were summed over the seven groups. The maximum total of congeneric pairs for the seven groups was 37, the minimum was 10, and the value for the groups as formed from the zooplankton data was 15. Of the more than 2 million different arrangements, 73,044 had a total of 15 or fewer congeneric pairs. If all arrangements are assumed to be equally likely, the probability of the observed value, or a smaller value, is .0328. This constitutes evidence of considerable selection against congeneric pairs within the recurrent groups.

Distribution Patterns of Groups

Certain patterns appear when the stations at which the groups occurred are plotted on a chart of the North Pacific (Figs. 1, 3, and 4). In all cases the group patterns are consistent with the individual species patterns arrived at independently by Bieri, Brinton, and McGowan (*1, 2*), although some of the individual species within the groups have more extensive ranges of occurrence than the groups do.

Table 1. Species composition of recurrent groups. C, chaetognath; E, euphausid; H, heteropod; P, pteropod.

Species
Group I
Euphausia hemigibba (E)
Euphausia mutica (E)
Euphausia recurva (E)
Euphausia tenera (E)
Nematoscelis atlantica (E)
Nematoscelis microps (E)
Nematoscelis tenella (E)
Stylocheiron carinatum (E)
Stylocheiron suhmii (E)
Pterosagitta draco (C)
Sagitta enflata (C)
Sagitta hexaptera (C)
Sagitta pacifica (C)
Creseis virgula (P)
Limacina bulimoides (P)
Limacina inflata (P)
Associated:
Euphausia brevis (E)
Hyalocylix striata (P)
Limacina trochiformis (P)
Group II
Euphausia pacifica (E)
Thysanoessa longipes (E)
Tessarabrachion oculatus (E)
Sagitta elegans (C)
Limacina helicina (P)
Associated:
Thysanoessa inermis (E)
Sagitta scrippsae (C)
Group III
Cavolinia inflexa (P)
Clio pyramidata (P)
Styliola subula (P)
Group IV
Euphausia diomediae (E)
Nematoscelis gracilis (E)
Sagitta robusta (C)
Group V
Euphausia gibboides (E)
Nematoscelis difficilis (E)
Thysanoessa gregaria (E)
Group VI
Limacina lesueuri (P)
Sagitta pseudoserratodentata (C)
Group VII
Atlanta lesueuri (H)
Atlanta turriculata (H)
No affinities
Euphausia paragibba (E)
Sagitta ferox (C)
Cavolinia longirostris (P)
Carinaria japonica (H)
Oxygyrus keraudreni (H)
Protatlanta souleyeti (H)
Pterosoma planum (H)
Pterotrachea hippocampus (H)
Pterotrachea minuta (H)

The distribution of the major water masses of the North Pacific is also shown in Figs. 1, 3, and 4. Comparison of this with distributions of the groups makes it evident that each group is confined to, or occurs much more frequently in, a specific water mass. Group V is an exception, for it mainly occupies an area which is hydrographically transitional between two water masses.

Perhaps the most interesting pattern is that shown by group I (Fig. 1). The 16 individual species comprising this group are widely distributed in the Equatorial and Central water masses of the Pacific. The group, however, had a more restricted distribution, being found largely in the Central Water Mass and in the Kuroshio and its extension. This was true no matter whether only samples from those stations where all members of the group were found were considered or whether samples from stations where all but one or all but two of the species of the group were found were added (see Fig. 1). When samples from the 47 stations where 14, 15, or 16 species of the group were found were considered, the group was found west of the 180-degree meridian much more frequently than expected [the expected ratio of occurrences, based on the west-east ratio of all samples taken between 10° and 40°N, was 27/20; the observed ratio was 39/8; χ^2 (1 degree of freedom) = 11.51, $p < .001$]. This discrepancy was due, at least partially, to an increase in the abundance of its members in the Kuroshio extension, where the waters appear to be "enriched" along the southern boundary of the Subarctic Water Mass.

Group II consists of five species which have been termed "Subarctic" (*1, 2*) because of their approximate limitation in the Pacific to the Subarctic Water Mass. The group was found almost exclusively north of 40°N (Fig. 3); the observed numbers of occurrences east and west of the 180-degree meridian in this region did not differ from the numbers expected on the basis of the west-east ratio for all samples taken in the area north of 40°N. Group II's only areal overlap with other groups occurred at two stations on the southern border of the subarctic region. This group has the most coherent distribution of any of the seven groups, the euphausids, chaetognath, and pteropod in it being frequently part of each other's biological environment over a very large geo-

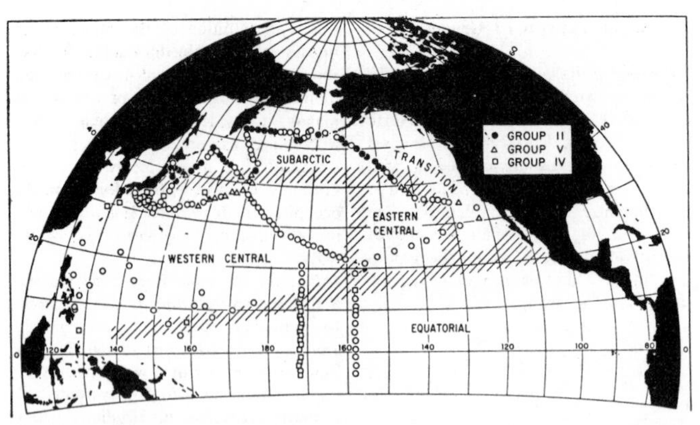

Fig. 3. Distribution of zooplankton groups II, IV, and V in the North Pacific. The water masses are indicated by name. The locations of all stations are shown, even those (open circles) where no group II, IV, or V species were found. (Shaded areas) The approximate boundaries of water masses [after R. Bieri (1)].

graphical range of waters with considerable hydrographic similarity.

The three euphausids comprising group V were found in an area that is hydrographically transitional between the Subarctic Water Mass to the north and the Central Water Mass to the south (Fig. 3). As would be expected, its areal overlap was greatest with groups I and II, although still quite limited. This, in conjunction with the weak, but essentially equal, interspecific connections with these groups marks group V as a transition-zone group. Several species in the other taxa are also thought to be transition-zone species. However, they occurred too infrequently in this series of samples to appear in the groupings. There was some indication that group V occurred west of the 180-degree meridian more frequently than expected, but the observed distribution of occurrences differed from the expected distribution only at the 25-percent level.

Group IV was found only in the Equatorial Water Mass (mostly south of 10°N) and in the Kuroshio (Fig. 3). Most of its areal overlap with group I was in the latter region. It may have been somewhat more frequent east of the 180-degree meridian, but the evidence is not very strong ($p \sim .30$, on the basis of all samples taken south of 10°N).

Over 50 percent of the occurrences of groups III, VI, and VII were in samples which also contained group I; the overlap among the three smaller groups was noticeably less (21 to 39 percent). On the basis of the overlap with group I and the multiple interspecific connections with it (see Fig. 2), they might be considered subgroups. However, groups III and VI are formed of species which are known to be limited to the Central Water Mass. Keeping these groups separate serves to emphasize their status as additions which alter the expression of the main group in certain localities. By contrast with findings for group I, the west-east distribution (in relation to the 180-degree meridian) of samples containing group III (23 samples) or group VI (31 samples) was found to be almost exactly that expected (expected west-east ratios, 13.2/9.8 and 17.8/13.2 for groups III and VI, respectively; observed ratios 13/10 and 18/13). Apparently these species of the Central Water Mass differ from the species of group I in their lack of response to the "enrichment" near the boundary between the Central and the Subarctic water masses. The distribution of group VII, on the other hand, parallels that of group I so closely [significantly more frequent west of the 180-degree meridian ($p < .01$)] that it is probably best to consider it a subgroup.

Abundances of Groups

The initial groupings were based entirely on the presence or absence of species and did not take account of the wide ranges of abundance of the various species. If it is assumed that species are most abundant in areas that they find most suitable as habitats, and least abundant in those areas within their range that are least suitable, then the relative abundances between samples can be used to determine whether each group as a whole shows agreement among its component species as to the "best" and "worst" habitats (7). For this purpose, the abundances of each species were ranked over the set of samples within which the group occurred, the ranks for each sample were then summed, and these sums of ranks were used as a basis for calculating concordance (strongly positive correlation of species relative abundances) within the group.

The 16 species of group I showed very significant concordance ($p < .005$, on the basis of data for the 47 stations at which 14, 15, or 16 of the species occurred together). The "best" habitats for these species were clustered in the core of the Kuroshio and in its extension eastward; the "worst" habitats were near the northern boundaries of the group's area of distribution. In this case the "best" habitats for the group were in its region of most frequent occurrence. As the individual species comprising the group are much more widely distributed in the Central and Equatorial water masses than the group seems to be, one might conclude that the apparent general abundance and high frequency of occurrence of the group in the area of mixing is an artifact resulting from the grouping procedure or from limitations in the station pattern or from our inability to collect species where they are uncommon. It seems unlikely, however, that there would be such very strong agreement in relative abundances if the group did not have some biological reality, at least in the sense of similarity of reactions of the species to environmental conditions.

The five species of group II also showed very significant concordance ($p < .005$, on the basis of data for 34 stations at which all five species were found). For the species of group II, the "best" habitats were in the eastern sector of the Subarctic Water Mass, just north of the southern limit of the group range; the "worst" habitats were on the southern edge of the group range, both immediately adjacent to the "best" areas and off Japan. The west-east distribution of "best" habitats was strongly biased toward the east, although, as already noted, the group

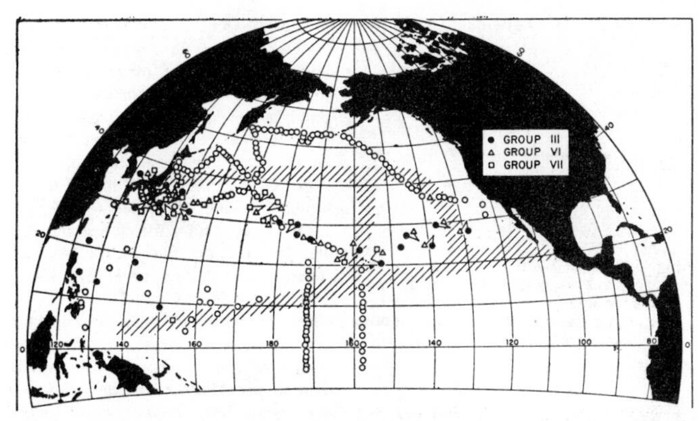

Fig. 4. Distribution of zooplankton groups III, VI, and VII in the North Pacific. The locations of all stations are shown, even those (open circles) where no group III, VI, or VII species were found. (Shaded areas) The approximate boundaries of water masses [after R. Bieri (1)].

showed no bias in frequency of occurrence east and west of the 180-degree meridian.

For the three species of group V there was a certain amount of agreement on the "best" and the "worst" habitats ($p \sim .15$, on the basis of data for 19 stations at which all three species occurred), but there was no consistent pattern of distribution of "best" and "worst." In fact, of the four "best" and the four "worst" samples, in two cases a "best" and a "worst" sample were immediately adjacent to each other.

The three species in group IV also showed some degree of concordance ($p < .10$, on the basis of data for 26 stations at which all three species were found). The distribution of samples taken in the Equatorial Water Mass was such that no certain idea of the west-east distribution of "best" and "worst" samples can be obtained. There was, however, a clear north-south difference; the "worst" samples occurred along the northern boundary of the area occupied by the group, and the "best" samples were taken in the southern part of the area covered by the expeditions. This is what would be expected of a group of Equatorial Water Mass species.

Groups III and VI did not show significant agreement on what were the "best" and "worst" habitats. The three species of pteropods comprising group III showed some positive correlation of abundances ($p = .35$, on the basis of data for 23 stations at which all three species occurred); the abundances of the two species that made up group VI (one pteropod, one chaetognath) were somewhat negatively correlated ($p = .40$, on the basis of data for 31 stations at which both species were found). In contrast to this, the abundances of the two species of heteropods which formed group VII were very strongly positively correlated ($p < .001$, on the basis of data for 30 stations at which both species occurred); the "best" samples were all located in the region of the Kuroshio off Japan and in its eastward extension, and the "worst" samples were in the southern and eastern parts of the area occupied by the group. The distribution of group VII thus agrees with that of group I not only in general extent and in regions of greatest frequency but also in detail in regard to the location of the "best" habitats.

Group II and Its Environment

Because group II was a particularly well defined, coherent entity in terms of its very few interspecific connections with other groups and its geographic distribution, a more detailed analysis of the group was attempted. Its characteristics, as developed so far, are briefly as follows. It consists of three euphausids, one chaetognath, and one pteropod. It occurs generally across the North Pacific, north of 40°N, being largely confined to the Subarctic Water Mass. There is strong evidence of agreement among the species in regard to "best" and "worst" habitats; in terms of the relative abundance of the group as a whole, the "best" localities were east of the 180-degree meridian and near the southern border of the group range, the "worst" were on the southern border.

Table 2 gives some statistics for the individual species. The estimates of frequency, abundance, average rank, dominance, and dispersion are based on the data from samples taken at 62 stations considered, from their hydrographic properties, to be in the Subarctic Water Mass. The two relative measures, average rank and dominance, reflect relations among the five species only and do not take into account the other organisms in the samples. The information under the column headings "Fidelity" and "Vitality" is drawn from general knowledge of the species.

Tessarabrachion oculatus, which has the most restricted general distribution, was found in only 40 percent of the shallow (0 to 150 m) samples; however, it was present in the deeper tows at a number of additional stations. The other four species of Table 2 were found in about 90 percent or more of the samples. *Tessarabrachion oculatus* was also the least abundant species; the median number of individuals per 1000 cubic meters in those samples in which the species occurred was 8. The median abundances of *Thysanoessa longipes*, *Euphausia pacifica*, and *Limacina helicina* were about equal, at about 370 individuals per 1000 cubic meters. The first two of these species were also similar in average rank, and both were definitely above *L. helicina* in this characteristic. Of the five species, *Sagitta elegans* had the highest frequency, abundance, and average rank. This species is a predator; the other species in this group are not. Its relative position emphasizes the limited nature of the data so far available. Clearly, not all of its prey are included.

The measure of dominance used indicates with what frequency a species was the most abundant, or one of the more abundant, of the five species. In 52 of the 62 samples, one species contributed 50 percent or more of the individuals of the five species; in the remaining ten samples, two species were needed to make up 50 percent or more of the individuals. Again, *Sagitta elegans* had the highest rating, but *Euphausia pacifica* was not far below it; *Thysanoessa longipes* and *Limacina helicina* were definitely secondary species, while *Tessarabrachion oculatus* was never dominant. The use of numbers of individuals instead of weight could obscure the real relationship. A consideration of the size range of individuals of the five species in these samples, however, suggests that this is not a problem. The individuals of *Limacina helicina* are, on the average, smaller than individuals of the other four species, which are all about the same size. Changing numbers to weight would, therefore, have little effect except to move *L. helicina* somewhat further toward *Tessarabrachion oculatus.*

The changes in order of the species, particularly the three intermediate species, that occur as the statistic examined is changed indicate that no one measure is likely to be completely satisfactory as a criterion of the position of a species in a community.

For all the species the ratio of variance to mean is above the value (1.0) expected for a random (Poisson) distribution by 2 to 4 orders of magnitude. This indicates that all the species have highly clumped distributions, even within this geographic area where conditions are, presumably, everywhere within their tolerance ranges. Such inflated variances might arise from the presence of a few excessively large or small values, or it might be a general property of the distributions. An indication of which of these factors is the cause can be obtained from the ratio of the calculated variance to a percentile estimate of the variance, $(P_{84} - P_{16})^2/4$, based on the value below which 16 percent of the values fall (P_{16}) and that above which 16 percent of the values fall (P_{84}). Extreme values in the tails of the distribution are thus not included. If the ratio is not far from 1.0, these distributions, although flat-topped, are probably more or less symmetrical; if the ratio is large, a few outsize values have contributed to inflation of the variance.

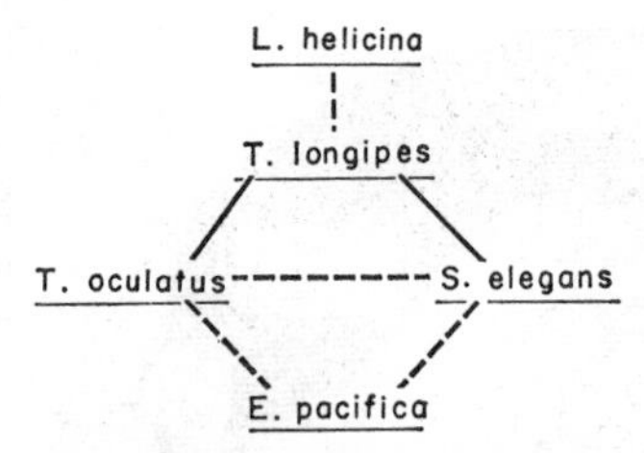

Fig. 5. Interrelationships among members of zooplankton group II. (Solid lines) Strong positive correlation of abundances ($p < .005$); (dashed lines) weaker positive correlation of abundances ($.03 < p < .07$).

For *Thysanoessa longipes, Euphausia pacifica,* and *Sagitta elegans* the ratios were 3.5, 4.4, and 1.2, respectively; for *Tessarabrachion oculatus* and *Limacina helicina,* the ratios were 329 and 406. The first three species seem to have, generally, non-Poisson distributions; for the latter two species the main cause of the large values of the variance appears to be the presence of a relatively few samples containing unusually large numbers of individuals.

Only one of the species, *Tessarabrachion oculatus,* occurs almost wholly in the Subarctic Water Mass. The others can be listed in a series, with more and more extensive ranges outside of this region: *Thysanoessa longipes, Euphausia pacifica, Sagitta elegans, Limacina helicina.* The extensions are both south along the west coast of North America in the California Current, which is made up partially from subarctic water, and north across the Arctic Ocean into the North Atlantic. Therefore, although these five species can be considered members of a subarctic community within the area studied, it is evident that in other areas they would be members of communities composed of other species. However, the geographic region within which they occur together is very extensive, and it is important to understand their interrelations within it.

Because almost all growth stages of the species were found in the tows, it seems reasonable to assume that they breed successfully in this region of the North Pacific. Little can be said about seasonal changes in abundance, for the samples available were taken only during the summer season. However, Beklemishev (*11*) has shown that *Limacina helicina* is present during the winter in this area.

The area within which group II occurred may be divided into sectors on the basis of the dynamic topography (*12*); such a division gives a system of gyres and currents. For the purpose of this article, these may be designated the Oyashio-West Bering Sea gyre; the center portion of the North Pacific-Gulf of Alaska gyre; the

Table 2. Statistics for the five species of zooplankton that comprise the Sub-Arctic group (II): *Tessarabrachion oculatus, Thysanoessa longipes, Euphausia pacifica, Sagitta elegans,* and *Limacina helicina.*

Species	Frequency *	Abundance †	Average rank ‡	Dominance §	Dispersion (aggregated) ‖	Fidelity ¶	Vitality **
T. oculatus	25/62	2–744 (8)	1.24	0/62	430	Aleutians S to 40°N in Pacific	All stages present
Th. longipes	59/62	4–9,784 (413)	3.35	13/62	2,275	Arctic Ocean S to 40°N in Pacific	All stages present
E. pacifica	55/62	4–22,278 (345)	3.36	23/62	6,950	60°N in Bering Sea, S to 25°N in Calif. Current to 35°N in Central Pacific	All stages present
S. elegans	60/62	17–7,900 (1,405)	4.15	29/62	1,830	N. Atlantic, Arctic Ocean, S to 40°N in Pacific	All stages present
L. helicina	58/62	20–75,770 (360)	2.90	7/62	52,400	N. Atlantic, Arctic Ocean, S to 30°N in Calif. Current to 35°N in Central Pacific	All stages present

* Proportion of samples in which the species was found (total number of samples, 62). *T. oculatus* occurred in deeper tows at nine additional stations. † Range and median (in parentheses) of numbers of individuals per 1000 cubic meters in samples in which species was found. ‡ Species were ranked within each sample on the basis of numbers of individuals. Ranks for each species were averaged over the 62 samples (1, least abundant; 5, most abundant). § Proportion of samples in which the species was among those making up 50 percent of the individuals; summation in each sample was begun with the most abundant species. ‖ The ratio of variance to mean; the expected value for a random (Poisson) distribution is 1.0. ¶ Degree of restriction to the Subarctic Water Mass. ** Proportion of life lived in the Subarctic Water Mass.

periphery of the North Pacific-Gulf of Alaska gyre; and the North Pacific Drift. The expected frequencies of occurrence in these sectors were calculated on the basis of frequencies observed in sampling the sectors. The observed distribution of frequencies of occurrence of the group did not differ significantly from expectation. This finding is in agreement with the earlier calculation which suggested that there was no west-east trend in frequencies of occurrence. However, with respect to the relative abundance of the group as a whole, the abundance in samples taken in the sector designated "periphery of the North Pacific-Gulf of Alaska gyre" was much greater than expected and the abundance in the sectors designated "North Pacific Drift" and "Oyashio-West Bering Sea gyre" was much less than expected. [According to the Kruskal-Wallas one-way analysis of variance by ranks, $p < .01$ (*13*).]

It has already been pointed out that the five species comprising the group showed very significant concordance in their relative abundances in the 34 samples in which all occurred. The rank correlations between relative abundances for all possible pairs of the five species (ten pairs) were examined in these samples. Even though this procedure raises the question of multiple tests of interrelated data, the correlation coefficients can still be used as qualitative indicators of the direction and strength of relationships. All the correlations were positive. Those between *Thysanoessa longipes* and *Tessarabrachion oculatus* and between *Thysanoessa longipes* and *Sagitta elegans* were very significant ($p < .005$). Four others were significant at about the 5-percent level: *Sagitta elegans* and *Tessarabrachion oculatus, Limacina helicina* and *Thysanoessa longipes, Sagitta elegans* and *Euphausia pacifica*, and *E. pacifica* and *Tessarabrachion oculatus*. None of the other four pairs showed significant correlation. These relations are shown in Fig. 5. They suggest that the three species *Thysanoessa longipes, Tessarabrachion oculatus*, and *Sagitta elegans* may be alike in their response to environmental conditions. Further evidence for this is the fact that it is just these three species which have distributions with a definite southern limit at 40°N in the Pacific; the distributions of the other two species, *Limacina helicina* and *Euphausia pacifica*, extend appreciably further south along the west coast of North America (see Table 2).

The sums of the ranks of abundances of the individual species in the 34 samples, used in calculating the concordance, were also used as the dependent variable in a multiple regression analysis. Hydrographic and other data taken at the same 34 stations were treated as the ten "independent" variables; it is evident from the list given below that some of these "independent" variables are more or less correlated.

1) Difference in shape of the temperature-salinity (T-S) diagram from that of the "best" station. A planimeter was used to determine the area enclosed by the T-S diagram of Transpac station 18 (the "best" station), the T-S diagram of the station being considered, and two straight lines, one connecting the bottom points of the two diagrams, the other connecting their top points. The T-S diagrams were plotted from depths of 50 to 1000 meters (*14*).

2) Depth of thermocline, estimated from bathythermograph cards.

3) Time of day of sampling, measured in hours from the nearest midnight on the assumption that effects due to vertical migration would be symmetrically distributed around midnight.

4) Depth at which the O_2 concentration fell to 2 milliliters per liter, estimated by linear interpolation between adjacent measured values.

5) Temperature at 10 meters, estimated by linear interpolation between adjacent measured values.

6) Temperature at 100 meters, estimated by linear interpolation between adjacent measured values.

7) Salinity at 10 meters, estimated by linear interpolation between adjacent measured values.

8) Salinity at 100 meters, estimated by linear interpolation between adjacent measured values.

9) Latitude, measured in degrees, with decimal equivalents of minutes and seconds.

10) Longitude, measured in degrees, with decimal equivalents of minutes and seconds; longitudes for stations west of 180° were expressed in terms of 180° + number of degrees west of 180°; for example, 159°E = 180° + 21° = 201°.

The depths for temperature and salinity were chosen to represent conditions near the shallowest and the deepest levels through which the net tows were taken.

The overall regression was significant ($p < 0.05$; standard error of estimate, 24.365). It accounted for about 56 percent of the variability of the dependent variable. Only two of the individual "independent" variables had significant regression coefficients; greater relative abundances were associated with T-S diagrams having shapes more like that of the "best" station ($p < .01$) and with lower salinities at 10 meters ($p < .05$). Both of these associations appear to indicate a property, or properties, related to the history and quality of the water; probably neither change in the shape of the T-S curve nor change in salinity is, of itself, the cause of the observed changes in relative abundance of the group.

This is particularly true of the salinity at 10 meters, for it seems unlikely that the observed small differences in salinity could have such an effect on the abundances of the species. More probably, the lower salinities indicate water which came from a region with higher precipitation than evaporation—that is, from the Oyashio or the Gulf of Alaska. Although the other regression coefficients were not significant, they are interesting as indicators of possible trends. Greater relative abundances were associated with greater depth of thermocline; with times farthest from midnight; with greater depth at which the oxygen content was reduced to 2 milliliters per liter; with higher temperature at 100 meters and lower temperature at 10 meters; with lower salinity at 100 meters; and with the more southern and western locations. Many of these associations suggest that more thorough vertical mixing creates more favorable conditions for these organisms.

Discussion

It is evident that multispecies groups occur in the zooplankton and that these groups can be identified by the procedure used in our study. If samples are taken in such a way that each contains animals from a single environment, the groups can be examined for evidence of interspecific relations, for all species within a group will be frequent members of each other's environment. The details of group composition may change with more extensive sampling in the North Pacific, but it seems likely that the changes will represent finer resolution and not fundamental alterations. For example, some of the species placed in group I seem to have stronger connections with the Central Water Mass, while others have stronger connections with the Equatorial Water Mass. With further sampling, therefore, group I might be split into Central and Equatorial groups.

The difficulties introduced by the sampling pattern become evident when the patterns of distribution of the samples representative of each group are considered in terms of frequency and abundance. Again, group I provides an example. The samples available indicate that the group was both more frequent and more abundant west of the 180-degree meridian, in particular off the coast of Japan. It is, however, known that many of the individual species in the group are more abundant toward the northern edge of their range and less abundant in the center. The southern track of the Transpac expedition started near the northern edge of the range of group I, off the coast of Japan, and then went south and east to Hawaii. The observed greater frequency of capture of the complete group and the greater relative abundance of the group at the stations off the coast of Japan may, therefore, represent the effects of the pattern of sampling, which seems to have made a north-south gradient in abundance appear to be an east-west gradient.

This possibility serves to emphasize the need for supplementary information on the individual species, for caution in interpretation, and for more extensive sampling, but it should not be allowed to overshadow the usefulness of the grouping procedure as a means of summarizing, of picking out groups of species which share a habitat, and of getting a preliminary idea of the geographic distribution patterns of the groups.

The strong evidence for group selection of water masses and the frequent strongly positive correlation of species relative abundances within groups suggest that the groups are composed of species with similar reactions to properties of the environment. The regression analysis of the relative abundances of group II was disappointing in that it indicated that many of the usually measured properties of the water—temperature, thermocline depth, oxygen content, and so on—were not closely related to the differences in abundance. Instead, the results suggest either that the organisms are reacting to a complex of factors, including the history of the water, or that the usual hydrographic procedures are not measuring the right things from the standpoint of the biologist. It will be important to examine the boundaries between groups in greater detail, to attempt to correlate other properties of the water—physical, chemical and biological—with abundances, and to supplement these field observations with laboratory studies of the effect of changes in various environmental parameters on the animals and their functions. Thus, we may get an understanding of the causes of the evident, and often sharply defined, patterns of zooplankton distribution and abundance (*15*).

References and Notes

1. R. Bieri, *Limnol. Oceanog.* **4**, 1 (1959).
2. E. Brinton, *Bull. Scripps Inst. Oceanog. Univ. Calif.* **8**, 51 (1962); J. A. McGowan, thesis, University of California, San Diego (1960).
3. J. E. King and J. Demond, *U.S. Fish Wildlife Serv. Fishery Bull.* **82**, 111 (1953).
4. J. A. McGowan, *Deep-Sea Res.* **6**, 125 (1960).
5. J. H. Harley, "Operation Troll, Joint Preliminary Report" (U.S. Atomic Energy Commission and Office of Naval Research, Washington, D.C., 1956).

6. We thank Drs. Alvariño, Bieri, and Brinton, who allowed us to use some of their data, and Dr. M. W. Johnson, under whose general guidance the zooplankton program at Scripps Institution of Oceanography was initiated and developed.

7. E. W. Fager, *Ecology* **38**, 586 (1957).
8. W. H. Kruskal, personal communication.
9. The computer program was written by Mrs. R. Mitchell.
10. G. Hardin, *Science* **131**, 1292 (1960).
11. K. V. Beklemishev, *Tr. Inst. Okeanol. Akad. Nauk SSSR* **45**, 142 (1961).
12. N. W. Rakestraw *et al.*, Eds., *Oceanic Observations of the Pacific, 1955: The NORPAC Data*, the *NORPAC Atlas* (Univ. of California Press, Berkeley, 1960), plate 9.
13. M. W. Tate and R. C. Clelland, *Nonparametric and Shortcut Statistics* (Interstate, Danville, Ill., 1957), pp. 109–111.
14. T-S diagrams are plots of temperatures against the corresponding salinities of subsurface water in a given area. Water masses are defined by the shape and position of the plot [see H. U. Sverdrup, M. W. Johnson, R. H. Fleming, *The Oceans* (Prentice-Hall, Englewood Cliffs, New Jersey, 1942), pp. 141–146]. Because of local short-term variations of temperature and salinity in the upper layers, the use of T-S diagrams to identify water masses is generally considered valid only for depths greater than those we used. We have, however, set our upper limit at 50 meters so as to include at least part of the vertical range of our organisms while avoiding some of the extreme local variations in the mixed layer (mean depth of mixed layer, 24 m). The lower limit was set at 1000 meters to include some index of the "conservative" properties of the water masses that are not affected by the factors responsible for the local, shallow variations.

15. This work was partially supported by the National Science Foundation (grants G-7141 and G-19417) and by the Marine Life Research Program, the Scripps Institution's component of the California Cooperative Oceanic Fisheries Investigations, a project sponsored by the Marine Research Committee of the State of California.

[151]

Benthic Ecology

Reprinted from
Netherlands Journal of Sea Research
3, *2, 1966, p. 267–293*

SOME FACTORS INFLUENCING THE RECRUITMENT AND ESTABLISHMENT OF MARINE BENTHIC COMMUNITIES

by

GUNNAR THORSON
(*Marine Biological Laboratory, Helsingør, Denmark*)

In a series of papers (1955, 1956, 1957) the present author called the attention to the fact, that the macrofauna communities of marine sediment bottoms (or "level" bottoms) originally described by C. G. JOH. PETERSEN (1913, 1918) seem, roughly, to have "parallels" in many areas of our globe, irrespective of latitude. This would mean—again roughly—that the same type of sediment substratum at about the same depth, whether in cold, temperate or warmer regions, would be inhabited by a series of macrofauna-communities, in which the quantitatively predominating animals will belong to the same *genera* but to different *species*.

So, already in 1957 (pp. 505–507) I could refer to an arctic *Macoma calcarea* comm., a boreal *M. baltica* comm., a North Pacific *M. nasuta—M. secta*-comm., and a boreal-Pacific *M. incongrua* comm., all preferably associated with mixed bottoms in the intertidal zone or in quite shallow water. To these may now be added a *Macoma mitchelli* comm. from the Texas coast (REID *et al.*, 1956), although in this case quantitative bottom samples are not available.

Similar references (1957, pp. 507–508) could be given to a boreo-lusitanian *Tellina tenuis—T. fabula* comm., a Mediterranean *T. distorta—T. donacina* comm., a North-West-Atlantic *T. tenera* comm. and a New Zealand *T. lilacina* comm., all associated with pure sandy bottoms in the intertidal or in quite shallow water. To these we can add today a *T. buttoni* comm. off Southern California (BARNARD, 1963; JONES and BARNARD, 1963) and a *T.* (= *Strigella*) *trotteriana* comm. in the intertidal of Madagascar (PICHON, 1962).

The review (1957, pp. 508–510) also comprised an Arctic *Venus fluctuosa* comm., a boreal *V. gallina* (= *striatula*) comm., and a Mediterranean *V. verrucosa* comm., all associated with sandy bottoms at roughly 10–40 m depth, to which may be added a deeper living boreal *V. fasciatum* comm. associated with shelly sand. Today we further know a *Venus declivis* comm. from shelly sand off Guinea and Senegal (LONG-

HURST, 1958), a *V.* (= *Mercenaria*) *campechiensis* comm. from Florida (MCNULTY *et al.*, 1962), a *V. casina* comm. from the Channell off Roscoff (CABIOCH, 1961), and a *V.* (= *Austrovenus*) *stutchburgi* comm. from New Zealand (RALPH and YALDWYNN, 1956).

In 1957 (p. 510) we knew a temperate-European *Syndosmya* (= *Abra*) *alba* comm. from mixed and soft bottoms from 5–70 m depth. Today we can add a *S. ovata* comm. from the Asov Sea (now also introduced in the Caspian Sea) (SAENKOVA, 1956), and a *S. longicallis* comm. and a *S. profundorum* comm. in the bathyal and abyssal zones of the Mediterranean respectively, thus both associated with muddy bottoms, but living at greater depth than the *S. alba* and *S. ovata* communities (PÉRÈS and PICARD, 1956, 1960).

Similarly, in 1957 (pp. 510–513) we had records for a boreal-Mediterranean *Amphiura filiformis—A. chiajei* comm., a Black Sea *A. florifera* comm., a Japanese *A. aestuarii* comm., and a New Zealand *A. rosea* comm., all associated with muddy bottoms at 10 to about 100 m depth. We now further know a Black Sea *A. stepanovi* comm. (BACESCO, 1963; BACESCO and MARGINEANU, 1958), and a New Zealand *A. norae* comm. (HURLEY, 1964).

Closely similar to the *Amphiura* communities and also living on muddy bottoms in somewhat deeper water are the *Amphiodia-Amphioplus* communities (1957, pp. 513–514). We so far knew an *Amphiodia cratoderma* comm. from the Japan Sea, an *Amphiodia occidentale* comm. from the Vancouver–region, an *Amphiodia sp.* comm. from off Chile, and an *Amphioplus macraspis* comm. from the Japan Sea. Today we can add an *Amphiodia urtica* comm. from Southern California (HARTMAN, 1960; BARNARD and ZIEZENHENNE, 1961), an *Amphioplus congensis* comm. from Guinea and Senegal (LONGHURST, 1958) and—with still too few details to specify them—2 new *Amphioplus-communities* from off Florida (MCNULTY *et al.*, 1962; THOMAS, 1962).

So, for these 6 groups of "parallel" communities our knowledge from 1957 to 1965 has been increased as follows:

Type of community	*number of "parallel" communities known*	
	in the year 1957	*in the year 1965*
Macoma communities	4	5
Tellina communities	4	6
Venus communities	4	8
Syndosmya (= *Abra*) comm.	1	4
Amphiura communities	4	6
Amphiodia-Amphioplus comm.	4	8
Total	21=100%	37=177%

which shows, that the number of known "parallels" is steadily increasing as new bottom areas are studied from a quantitative point of view.

HARTMAN (California, 1955), KNOX (New Zealand, 1961) and DAY (South Africa, 1963), however, instead of finding "parallel" communities with a few predominating species, described sediment bottoms inhabited by a large number of species occurring only in a few specimens each. My personal studies (420 grab hauls) in January-March 1966 off the west coast of Thailand gave a similar result with no "parallels" at all. Mr. A. Gallardo, M.sc., found the same tendency for about 500 grab hauls taken off South Vietnam (verbal information). So, for large tropical and parts of the warm temperate bottom areas the concept of "parallel" bottom communities must be radically revised, and I hope elsewhere to give a new evaluation of the whole problem.

All communities described so far have, however, been based on samples of the *macro*benthos only, i.e. on animals large enough to be retained by a sieve with mesh-openings of 2 mm. This first of all means that the *micro*benthos: the flora and fauna of bacteria, diatoms, flagellates, amoeba and ciliates, has been fully disregarded. This so far seems regrettable, but imperative, since a study of this fraction of the benthos claims quite other methods and a time-consuming and comprehensive teamwork.

Using 2 mm-meshes for our sieves, however, also means that the *meio*benthos (MARE, 1942), i.e. the fauna of intermediate size comprising foraminiferans, gastrotrichs, kinorhynchs, turbellarians, small oligochaetes, small polychaetes, nematodes, ostracods, copepods and cumaceans, is missing in our calculations. Since now this fraction of the benthos is known by far to outnumber the macrobenthos as concerns the quantity of specimens per square unit, since the metabolic rate of these small animals may roughly be estimated as 5 times higher per weight unit than that of the macrobenthos (ZEUTHEN, 1947), and since at least some meiobenthos-groups are known to develop more generations in a shorter time than the macrobenthos, it seems not at all unlikely that the annual production of the meiobenthos will surpass that of the macrobenthos living in the same bottom area. If this is so, the value of the macrobenthic communities as indicators of *productivity* may be severely questioned.

Although it is not the main item for the present paper it, therefore, may seem reasonable to discuss in a few rough outlines this provoking aspect, before we take up the questions on the recruitment and establishment of the macrobenthos communities.

If we compare the number of specimens per square meter of sediment bottom from four geographical areas from which we know that the sorting and counting of the quantitative samples has been undertaken with special care, we find the following number of specimens to be retained by a 2 mm mesh sieve:

Faxa Bay, Iceland, 33–154 m, average for 5 communities (EINARSSON, 1941) . . .	260 specimens/sq.m.
Öresund, Denmark, 16–22 m, average for 2 communities (unpubl. counting by Mr. Willy Nicolaisen, M.sc.)	1684 specimens/sq.m.
English Channel, off Plymouth, 15–70 m, all types of sediment bottoms from 18 stations (HOLME, 1953)	171 specimens/sq.m.
Laguna Veneta, Adriatic Sea, 7–21 m, average for 3 communities (VATOVA, 1940)	114 specimens/sq.m.
Average for all four areas	557 specimens/sq.m.

So, the samples from Øresund, comprising a sand-bottom *Venus* comm. from 16–19 m depth and a mud-bottom *Amphiura* comm. from 22 m depth, show by far the highest figures for macrobenthos specimens per sq.m. This must at least partially be ascribed to the extraordinarily great care shown when sampling and counting this material.

The paper of WIGLEY and MCINTYRE (1964) from Martha's Vineyard, U.S. Atlantic coast, gives us an impression of the importance of the size of the meshes in the sieves when separating macro- and meiobenthos. For samples comprising all types of sediment bottoms from 40–570 m depth sifted through 1 mm-mesh openings an average of 2490 specimens was retained per sq. m., i.e. a much higher figure than for any of the samples mentioned above and sifted through 2 mm-mesh openings. This is in good agreement with HOLME's finding (1953, p. 21, table VIII) that the ratio of bottom animals retained by a 2.2- and a 1.2.-mesh opening respectively will roughly be 30 to 70, or well over double the number of specimens in the finer screen.

In contrast to the numbers for macrobenthic animals per sq. m. given above, the meiobenthos of a similar square area will roughly comprise an average of 100.000 to 500.000 or more specimens. So, for example, when the harpacticids predominate, their number per sq. m. may average 110.000 specimens (Danish Waddensea, SMIDT, 1951), 120.000–198.000 (Banyuls, BOUGIS, 1946, 1950), 173.000 (Danish Fjords, BREGNBALLE, 1961), or 486.000 (Whitstable, England, PERKINS, 1958). When the nematodes predominate, their number per sq. m. may average 225.000 specimens (Danish Waddensea, SMIDT, 1951), 251.000 (Clyde Sea, MOORE, 1931), 293.000 (Arcachon, BOISSEAU and RENAUD, 1955), 486.000 (Buzzards Bay, WIESER, 1960a), 435.000–640.000 (Banyuls, BOUGIS, 1946, 1950), 790.000 (Danish Fjords, BREGNBALLE, 1961) or even 1.750.000 specimens (Woods Hole, WIESER and KANWISHER, 1961), 3.639.000 (Bristol Channel, REES, 1940) and 5.200.000 (Whitstable, England, PERKINS, 1958).—Also such meiobenthic groups like small polychaetes, oligochaetes, kinorhynchs, tardigrades and ostracods may in suitable places average from 10.000 to more than 100.000 specimens per sq.m. (see also WIESER, 1960b).

WIESER (1960a) and SANDERS (1960) were the first who on a larger scale examined the number-to-weight-ratio of macro- as well as meiobenthos of the same localities, using for their test soft bottom communities at 19 m depth from Buzzards Bay. SANDERS (1960, p. 152) found about 930.000 meiofaunal animals per sq. m. in his samples. Although the number of meiofauna species thus was about 100 times larger than that of the macrofauna, the weight of the meiofauna will only amount to 3% of that of the macrofauna. It is difficult from these papers to see what size-limits the authors have used to separate meio- and macrofauna, but since WIESER (1960a, p. 121) refers to MARE's (1942) definition of meiofauna, it seems reasonable to assume that their figures will correspond fairly well to such animal sizes which will pass a 2 mm-mesh screen.

MCINTYRE (1961) found on mud bottoms in the North Sea, at 140 m depth, that the weight per square unit of animals passing a 1.3 mm-mesh screen was 28% of that of the "macrofauna" retained by the screen, i.e. a surprisingly high figure. From Loch Nevis in Scotland on a mud-bottom, 90–110 m depth, and using the same technique he found however, the weight of the meiofauna to make 6.6% only of that of the "macrofauna". Finally, WIGLEY and MCINTYRE (1964) for 10 stations between 40 and 567 m depth in Martha's Vineyard, U.S. Atlantic coast, by using a 1 mm-mesh opening found the weight of the meiobenthos to be only 4.2% of that of the "macrofauna". If the authors had used a 2 mm-mesh screen, this figure for the meiofauna-weight probably would have been about double as large (see HOLME, 1953, p. 21, table VIII), but even then the meiofauna weight would probably be less than 10% of that of the macrofauna of the same area.

A review of the recent quantitative literature on meiobenthos clearly shows that the number of specimens per square unit will usually reach its peak in quite shallow water, preferably in mud- and sand-flats of the tidalzone, and WIESER (1960b, Abb. 1) in his compilation shows how the number of meiofauna specimens per sq. m. decreases with increasing depths, thus showing just the same tendencies as are known for the macrofauna.

Although the weight of the meiobenthos thus will only rarely surpass 10% of the weight of the macrobenthos from the same sediment bottom area, we have to include in our calculations that some meiofauna-groups will reproduce at much shorter intervals, i.e. raise more generations in a given time and have a shorter life-span, than is the rule among most macrobenthic animals which will, in most cases, have one well defined and well limited season of reproduction each year.

Thus, in the *harpacticoid copepod*, *Tisbe furcata*, the minimum time between two generations will usually be 19–24 days and the life-span of the individual specimens will on an average hardly surpass 40–50 days (JOHNSON and OLSON, 1948). For another species, *Tigriopus fulvus*, the time from hatching to the adult stage may take roughly 1–2 months, and the life-span of a female may roughly be 1–2 1/2 months, during which at least 4 batches of eggs can be produced (FRASER, 1936).—The *nematodes* make another prolific group. Thus, NIELSEN (1949) found for terrestrial nematodes a life-span of about 1 month, and according to kind verbal information from Dr. S. Gerlach and Mr. Bent Muus, M.sc., this also seems to hold good for some of the smaller marine benthic species. For large species of nematodes, like *Enoplus communis*, WIESER and KANWISHER (1960) have shown that they have an annual life cycle, thus reproducing only once per year.—Similar tendencies seem to occur in the *oligochaeta*. For terrestrial soil *Enchytraeidae*, Dr. Bent Christensen has kindly informed me, that the smaller species will usually have 2–3 generations per year, while the larger species seem to be annual with only one yearly season of reproduction. For the Baltic and the North Sea, VON BÜLOW (1957, p. 113) has shown that, while many species of marine oligochaeta are found to be sexually mature throughout the year, still more species will have only one restricted spawning season per year, which will naturally limit the number of generations produced. Finally, the marine oligochaete, *Tubifex costatus*, seems to need not less than two full years to attain maturity (BRINKHURST, 1964).

For the *ostracods*, ELOFSON (1941, pp. 396–87) from the Swedish W. Coast mentions 30–40 days' life cycles for 3 species, 60 days' life cycles for two species, 10–12 months' life cycles for two species, and a life-span of two-three years for four species. Mr. Bent Theisen, M.sc., has kindly informed me, that in quite shallow water off Nivå, Denmark, most species of ostracods will have 3 generations per year, while one of the species examined has 1 annual generation only.

Even for the *foraminiferans*, the predominating group of unicellulars among the meiobenthos, some data on the life-span are available. Thus MYERS (1942, p. 340) for the common sediment bottom species, *Elphidium crispum*, estimated its sublittoral life-span to be "... one or two years, and the completion of the life cycle requires two, three or even four years". He adds, however (p. 332) that "The life-span of *E. crispus* is perhaps longer than that of most foraminifera living in the same life zones", informing us that *Discorbis globularis* grows up to the 19-chamber stage in 52 days, and that *Patellina corrugata* matures in 21 days.

Before we draw our conclusion we must, however, remember, that the meiobenthos besides the groups mentioned above, i.e. the *true* meiobenthos which will never grow larger than 2 mm, also comprises huge numbers of young, newly settled specimens of the macrobenthos. While the true meiobenthos will usually predominate as to *number* of animals, the young bottom stages of the macrobenthos will often predominate as to the *weight*. So, in a comprehensive series of quantitative samples (of 1/50 sq. m. each) from 16–19 m sandy bottoms and 27 m muddy bottoms taken at 2–3 weeks' intervals during 3 years under the supervision of Mrs. Kirsten Muus, M.sc., in the northern part of the Öresund, Denmark, our general impression on an all-year basis is, that the young just settled stages of the macrobenthos will on an average make about 50% of the weight of the whole meiobenthos sample, and probably more.

Realizing now 1) that the total weight of the meiobenthos will only rarely surpass 10% of the weight of the macrobenthos, 2) that in many cases half the weight of the meiobenthos will be made by the just settled young of the macrobenthos which will usually grow at such a speed that they will soon grow out of the meiobenthos-size to be registered as true macrobenthos, and 3) that the true meiobenthos, although producing a somewhat larger number of generations per time unit than the macrobenthos, is by far not so prolific as has been supposed, the conclusion must be, that even the meiofauna weight multiplied by the average number of generations of meiofauna-animals per year (probably about 3 generations) will usually be far from coming up to the average weight of the standing crop of the macrobenthos inhabiting the same area, even if we estimate an average "standing crop" of a macrofauna to comprise, roughly, 2 years' production (see p. 287).

So, most of the macrobenthic communities seem roughly to be representative for the production on the bottom, where they live. They will accordingly be valid as "indicator-communities" and therefore, be worth while to study in detail.

In the Arctic-Atlantic Sea areas, most species of benthic invertebrates will have a non-pelagic mode of reproduction. Already in cold-temperate coastal areas will, however, species reproducing by planktotrophic pelagic larvae be dominant to lead more and more as we approach the warm temperate and tropical shelf-areas. So, in the boreo-atlantic region, planktotrophic pelagic larvae are found in 55–65% of all benthic species of invertebrates, and on the tropical shelf up to 80–85% of all species will reproduce by this larval type (THORSON, 1950, pp. 11–12). During their long pelagic life, larvae of this type may be transported over wide distances and they may, like parachutists, settle in new areas and found new populations there. The ability of these larvae to settle down and occupy any area of the bottom which might be available to receive them, makes it understandable, that especially such macrobenthic species which are quantitatively predominant on the bottom will often originate from this very type of larvae. To understand how a sediment bottom community is established, therefore, first of all will mean to examine the conditions which the planktotrophic pelagic larvae will have to meet during their development and settling.

During a roughly 2–4 weeks' stay in the free water masses (THORSON, 1961), these larvae will, however, meet so many barriers, that only a highly reduced and selected set of larval species will finally get into contact with that type of bottom substratum on which they are going to spend their adult life.

The first selection in the plankton is dependant on the availability of food and on the temperature of the water during the pelagic life of the larvae. By introducing water masses with similar stocks of larvae into heated (18–20°C) or non-heated (a few centigrades lower) concrete tanks in the Danish Limfjord, SPÄRCK, (1926, pp. 278–279) could show how deeply the temperature influenced the larvae. In the heated tanks, the bottom was soon fully covered by settling polychaetes (*Nereis*, *Polydora*), and nothing else. In the non- heated tanks a much larger number of species (besides the polychaetes also *Ascidiella*, *Mytilus*, *Cardium* and *Mya*) had passed metamorphosis and settled. The same may take place in the sea also. A small change in the temperature or the food conditions from year to year may fully change the chances for the individual species of larvae to carry through their planktonic development. Thus, the larval stock surviving till metamorphosis will *not* have the same species-composition each year, and it is from this surviving stock of larval species that the benthic communities receive their most important annual "blood transfusion".

The second selection is especially associated with coastal areas with a brackish water mass, or with estuaries, river outlets etc. Such areas will often have a stratification of the water masses with a brackish surface layer and a more saline layer at the bottom separated by an often very distinct discontinuity-layer. In such areas, a larval species which when approaching metamorphosis is photopositive (and/or geonegative) and prefers a low salinity will stick to the surface layer, thus never getting into contact with e.g. a muddy bottom in deep water. Similarly, a larval species which when approaching metamorphosis is photonegative (and/or geopositive) and prefers a high salinity will stick to the deeper water layer and will usually never get into contact with e.g. a coarse sandy bottom in shallow water. It seems evident that such a stratification of the larvae will furthermore restrict the number of species, from which each community can be recruited. (For exceptions, see BANSE, 1964, p. 110).

The third selection takes place when the larvae have to "decide" on which type of substratum they are going to settle. In a series of fine experiments summarized in a paper from 1952, DOUGLAS WILSON has shown, that larval polychaetes ready to metamorphose are often able to postpone metamorphosis until they find a substratum suitable for adult life. Meanwhile they may continue their pelagic life swimming and drifting over the bottom for several days or even weeks, and testing the substratum at intervals. Similar observations have been done for echinoderms (MORTENSEN, 1921, pp. 105–106, 1938), for other polychaetes (SMIDT, 1951) and for prosobranchs (SCHELTEMA, 1961). WILSON's experiments on polychaete-larvae, preferably those on *Ophelia*, were undertaken in petri-dishes, where the larvae by swimming 1 or 2 centimeters only had a chance to discriminate between heaps of sand which might be more or less attractive, neutral or more or less repellent to them as a future substratum. In nature, the larvae will *not* get a similar opportunity to compare a series of substrata by swimming a short distance only. WILSON (1953a, p. 417 and 421, 1953b, p. 216, 1954, p. 366), however, also found that larvae which had postponed metamorphosis for some time were less critical towards the type of substratum than younger ones. Far from questioning WILSON's main-thesis: That the larvae may discriminate between attractive and non-attractive substata, a fact shown so convincingly that it can be accepted as a "biological rule", we have, however, to find out what will happen *in nature*, when a larval swarm ready to metamorphose and drifting along the bottom will for the first time meet a substratum which they might "accept", although it is far from ideal for their settling. The larvae cannot know, that if they continued to drift over the bottom for perhaps 10–20 kilometers more, they might meet a much more attractive substratum. It seems reasonable to assume, that such larvae, at least if they have already postponed their metamorphosis for some time and are in their less critical phase, will accept and accordingly settle in a bottom substratum much less attractive than the one they would have preferred, had they been given a "free choice".

The consequence of this must be, that the distributional pattern of larvae on natural bottom substrata must be much less delicate, i.e. much more coarse, than in the experiments undertaken in the laboratory.

Actually, a survey of the Helsingør-laboratory under the supervision of Mrs. Kirsten Muus, M. sc. (see p. 286) where quantitative samples of 1/50 sq.m. were regularly taken and counted at 2–3 week's interval for a 3-year period on a 16–19 m sand-bottom *Venus*-community and a 27 m mud-bottom *Amphiura*-community, has clearly given the proof that these conclusions are justified. *Amphiura filiformis* (and partially *A. chiajei*) are quantitative dominants in the mud-bottoms at 27 m, but they never seem to occur as adults or even as half-grown specimens in the sand-bottoms of 16–19 m depth where the *Venus*-community prevails. Nevertheless, it was found that at the time when the main settling of the young *Amphiuras* takes place, not less than 3.500 young of *Amphiura* (probably mostly *A. filiformis*) will settle per sq. m. of a sand-bottom with a *Venus-community* which must be very little attractive if not directly repellent to them. The fact that the peak of settlement occurs two weeks earlier on the mud bottom than on the sand-bottom, furthermore supports the suspicion, that the sand-bottom has been populated by the "less critical" larvae, which have had to postpone their metamorphosis for some time. Furthermore, it is evident from the countings, that the young *Amphiura's* on the sand-bottom are nearly exterminated already two weeks later, while the young *Amphiuras* on the mud-bottom in spite of a heavy wastage will be numerous enough to renew the local population. Finally, it should be stressed, that the settlement of young *Amphiura's* is twice as heavy on the mud-bottoms where the species usually lives as on the less attractive sand-bottom. This is just one example of a series which will be published and discussed by Mrs. Kirsten Muus in detail during the coming years. (See also *Muus*, 1966).

Nevertheless, it is quite obvious that a discrimination of the substratum takes place when the larvae settle on the bottom. It, therefore, will be a limited and very selected set of young species which will settle on each type of substratum to start their mutual "survival-of-the-fittest-fight" there.

This "fight" may be carried out on several fronts. It may comprise the competition for space and for food between different species or between young and older individuals of the same species (THORSON, 1957, p. 483), and it may be the fight between predators and their prey. When focusing upon this latter point, we usually think in the terms of *macro*-predators like sea-stars, brittle-stars, predatory gastropods, crabs, etc., but besides this we must remember that we also have various predators on the *meiobenthos*-level, and that many of the young of the macrobenthic animals will start their life on the bottom as members of the meiobenthos.

Table I shows the body length or largest diameter just after settling of 186 species belonging to different groups of benthic invertebrates. This table is based: for the *polychaetes* on HANNERZ (1956), OKUDA (1950), THORSON (1946), and WILSON (1932, 1933, 1936, 1948), for the *echinoderms* on GEMMILL (1914), Kirsten Muus (unpubl.), HÖRSTADIUS (1939), MORTENSEN (1921, 1931, 1938), Nina Rehfeldt (unpubl.), TATTERSALL and SHEPPARD (1934), THEEL (1898), THORSON (1946) and URSIN (1960), for the *lamellibrachs* on LOOSANOFF and DAVIS (1963), SULLIVAN (1948), STAFFORD (1912), JØRGENSEN (1946) and YOSHIDA (1960), for the *prosobranchs* on INO (1953), MURAYAMA (1935) and THORSON (1946), for the *tectibranchs* on THORSON (1946), and for the *decapod crustaceans* on HART (1960),

TABLE I

The length of different groups of benthic invertebrates just after settling.

Systematic Group	*Number of species used for the calculation*	*Length of the young bottom stage at settling*			
		Meiofauna: less than 2 mm		*Macrofauna: 2 mm or more*	
		number	%	*number*	%
Polychaeta	41	29	70.7	12	29.3
Echinodermata	24	23	95.9	1	4.1
Lamellibranchia	23	23	100.0	0	0.0
Prosobranchia	30	28	93.3	2	6.7
Tectibranchia	8	8	100.0	0	0.0
Decapoda	60	11	18.3	49	71.7
Total	186	122	65.6%	64	34.4%

KNUDSEN (1958, 1959 a–b, 1960), LEBOUR (1928, 1930, 1931 a–b, 1932, 1940 a–b) and PIKE and WILLIAMSON (1959, 1960, 1961, 1964).

From this table, which by no means claims to give a complete list, but only a test-sample of species selected at random, it is seen that 96–97% of all echinoderms and molluscs will start their life on the bottom as members of the smaller meiofauna. Among the polychaetes, roughly 70% will start their bottom-life on the meiofauna-level, and those large enough to be referred to the macrofauna already at settling nearly all belong to the *Spionidae* or related groups.—Among the decapod crustaceans only 18%, making a fraction of the group *Reptantia*, are at settling small enough to be counted among the meiofauna, while 52% of all *Reptantia* plus all the species of *Natantia* examined will start their life on the bottom as members of the macrofauna.

Taken as a whole, 2/3 of all the species considered in Table I thus start their life as members of the meiofauna. It must, however, be remembered that for most of them the upgrowth and volume increase will be so quick, that their "membership" of the meiofauna will usually be very short, in most cases significantly less than one month. But inside this short interval they will often have to pay a heavy tax to their enemies, and among them also to the predaceous element of the meiofauna. A few examples will show this:

Mrs. Grete Møller Christensen, M.sc., working at the Helsingør-laboratory, has kindly informed me, that one 5–6 mm long turbellarian, *Discocelides longi* kept in her culture dishes at 17–18°C, during two experimental periods of 7 and 10 days each consumed an average of 3 and 2.6 spat respectively of the 1–2 mm long just settled lamellibranch *Spisula elliptica* per day. At two occasions the turbellarian was observed to swallow, digest and eject the empty shells of a *Spisula* inside a period of 8 and 20 minutes respectively. We do not know at present, how frequent just this species of turbellarian is on our natural bottoms, but since turbellarians all seem to be predators (GLYNN, 1965; JENNINGS, 1962; MAMKAEV and SERAVIN, 1963; WOELKE, 1956), and since in other sediment bottoms they may occur in numbers per sq. m. averaging 200–1.800 spms. (Öresund, KROGH and SPÄRCK, 1936), 2.800 spms. (Buzzards Bay, WIESER, 1960), 4.900 spms. (Plymouth Sound, MARE, 1942) and 9.500 spms. (Arcachon, RENAUD-DEBYSER, 1958), the perspective of such a predation seems to be a serious one. Mrs. Grete Møller Christensen in her cultures of *Spisula* spat also observed that the very youngest spat (i.e. less than 1 mm) was attacked and in large quantities eaten by a nematode (so far un-identified), whereas this nematode could apparently do no harm to the larger spat, i.e. 1–2 mm or older. Finally, she has observed in the same cultures, that also harpacticoid copepods may attack, kill and eat the tiny lamellibranch spat. Realizing how numerous such copepods are in many meiofaunas, again here we have to focus a serious menace for the settling young of the macrofauna.

One more example may illustrate the predatory activity of such organisms, which hitherto we have more or less disregarded as enemies in the food chains of sediment bottoms. The large foraminiferan, *Astrorhiza limicola*, occurring in a number of roughly 50 spms. per sq. m. at the sand-bottoms at 30–70 m depths off the Northumberland coast have been studied by BUCHANAN and HEDLEY (1960), and proved to be a voracious predator. Its pseudopodia, which may extend to distances of more than 6–7 cm from the central protoplasmic mass of each individual are very adhesive and can not only tenaciously hold a prey but also digest it. In this way, *Astrorhiza* was seen to capture and digest copepods, nematodes, amphipods and even full grown cumaceans, 2–3 cm long caprellids, and recently metamorphosed specimens of *Echinocardium flavescens*. If we use the figure given above of roughly 50 specimens of *Astrorhiza* per sq. m. in the sandy-bottoms where it predominates and take the "radius" of its pseudopodia as roughly 6 cm, this will mean that the pseudopodia of all *Astrorhizas* inside 1 sq. m. of bottom may together be active over an area of about 5.000 sq. m. or half the total bottom area. This area, accordingly, will be a "danger-zone" for many other animals, among them the newly settled young of the macrofauna.

The next group of potential predators to be considered and which may be especially fatal for quite young specimens of the macrofauna, are the so-called "harmless detritus-feeders". Among them we often find species, which are designated as "non-selective detritus-feeders", a term which means that they transport large quantities of un-sorted detritus through their digestive tract. In nearly all cases examined, this detritus is collected by the feeding animals from the very surface layer of the bottom, i.e. from the layer where the newly settled young of the macrofauna will establish themselves after having metamorphosed and left the plankton. It thus seems reasonable to assume, that this un-sorted detritus will contain at least in certain seasons large quantities of such newly settled young which, therefore, automatically will pass the digestive tract of these non-selective detritus-feeders, and that these tiny and often very sensitive young bottom-stages might be killed or even digested during this passage.

It seems reasonable to give examples on these impressive quantities of bottom material, which may pass the digestive tract of non-selective detritus-feeders of widely varying systematic groups per animal and per unit of time.

Table II gives the rate of consumption for 3 different species of *polychaetes* and for one *enteropneust.* It has also been calculated how much bottom material will pass the digestive tract for each sq. m. of bottom in such areas where the animals predominate.

TABLE II

Rate of consumption of unsorted bottom material in vermes

Species	*Locality*	*Weight of bottom material consumed*		*Authors*
		per animal/year	*per sq. m/year*	
Thoracophelia mucronata	California	84 g wet weight	850–2520 kilo	Fox *et al.*, 1948 Fox, 1950 Dales, 1952
Melinna palmata	Black See	103 g wet weight	156–316 kilo	Dragoli, 1961, 1962
Clymenella torquata	New England	0.15–0.37 g wet weight	26–244 g	Mangum, 1964
Balanoglossus gigas	Brazil	73–91 kilo wet weight	73–91 kilo	Burdon-Jones, 1962

For *Thoracophelia*, it might be added that each animal will empty and refill its digestive tract each 15 minutes (fox et al., 1948) i.e. so quickly, that many tiny animals may probably survive the passage. For *Balanoglossus* it should be remembered that this very large species, 1–1 1/2 m long, will roughly be found in 1 specimen per square meter in the zone, in which it is most frequent.

But also among the *echinoderms* we may find examples of a transport of very large quantities of detritus through the digestive tract. *Echinocardium cordatum*, the nearly cosmopolitan irregular sea-urchin, showed an average of 3 g of detritus in the digestive tract of each of 42 adult speci-

TABLE III

Rate of consumption of unsorted bottom material by some tropical shallow water *Holothurians*

Species	*Locality*	*Weight of bottom material consumed*		*Author*
		per animal/year	*per sq. m/year*	
Holothuria atra	Marshall Isl.	70 kilo dry weight	70 kilo dry weight	Bonham and Held, 1963
	Palao Isl.	31 kilo wet weight	16 kilo wet weight	Yamanouti, 1939
Holothuria vitreus	Palao Isl.	27 kilo wet weight		Yamanouti, 1939
Holothuria edulis	Palao Isl.	22 kilo wet weight		Yamanouti, 1939
Holothuria flavomaculata	Palao Isl.	9 kilo wet weight		Yamanouti, 1939
Holothuria scabra	Palao Isl.	82 kilo wet weight		Yamanouti, 1939
Holothuria bivittata	Palao Isl.	45 kilo wet weight		Yamanouti, 1939
Holothuria floridana	Samoa Isl.	30 kilo dry weight		Mayor, 1924
Stichopus moebii	Bermuda	42 kilo dry weight	6–7 kilo dry weight	Crozier, 1918
Stichopus variegatus	Palao Isl.	18 kilo wet weight		Yamanouti, 1939

Total for 9 species of Holothurians: 9 to 82 kilo wet weight and 30 to 70 kilo dry weight of bottom material passing the digestive tract of one animal per year.

mens examined (BLEGVAD, 1914, appendix p. 29). Observations have shown that this species will empty and refill its digestive tract at least once in 24 hours, probably much more often. So, each *Echinocardium* will swallow a minimum of 90 g of unsorted detritus per month. PETERSEN (1913, appendix pp. 10–12) found for 135 quantitative samples covering 15.5 m^2 of a sandy bottom *Venus* community from random areas of Danish waters an average of 22.4 adult *Echinocardium* per sq. m., which again means that the *Echinocardium*-population of each sq. m. will swallow a minimum of about 2.000 g bottom material per month.

In the digestive tract of 79 adult *Echinocardium* from scattered localities in Danish Seas, BLEGVAD (1914, appendix pp. 29–30) found a total of 29 small lamellibranchs (25 of them being *Spisula subtruncata*) of about 1 mm length. With one emptying and refilling of the digestive tract per day this means a swallowing of 0.37 lamellibranch per *Echinocardium* per day or 2.5 small lamellibranchs per week. Estimating the season during which these quickly growing small *Spisula's* are present at the bottom at about 4 weeks and using the figures given above, we find that the adult population of *Echinocardium* of one sq. m. may swallow and probably kill $2.5 \times 4 \times 22.4 = 224$ young lamellibranchs inside this period, and probably significantly more.

In the *holothurians*, however, the passage of unsorted bottom-material through the digestive tract may reach much higher figures. Table III, p. 280, shows the enormous quantities ingested by tropical shallow water species.

Here again, we know that in some cases the passage may be so quick that even small cladocerans, copepods and sipunculids and "nearly half the total number of ostracods" may survive the transport (BERTRAM, 1936, for *Holothuria impatiens*).

Another group, in which such a transport of unsorted bottom material may be most dangerous to the young of the macrofauna, is the *decapod crustaceans*.

Table IV shows some calculations for the sand-sifting activity of our common hermit crab, *Pagurus bernhardus*, and for the Pacific small "lobster", *Callianassa californica*.

For the hermit crab the calculation is extraordinarily modest, since we have only considered 50% of the material stirred up by the cheliped as passing its mouth parts and only the warm part of the year has been regarded, because our experiments were made only during that season. Again we will, however, see the impressive amount of bottom material with its content of newly settled young, which in these cases will pass the veritable crushing-mill of the mouth parts before being finally swallowed.

Another group, the feeding ecology of which must be re-evaluated, is that comprising the *omnivorous* or *polyphagous* animals. It seems as if

TABLE IV

Rate of consumption of unsorted bottom material by one boreal and one warm temperate species of Decapoda.

Pagurus bernhardus:

Medium sized specimen, weight without shell about 2.8 g. Experiment, Elsinore Aquaria, August 1965 at 17.5 °C. Very rough calculation. Unpublished.

Average sand-shoveling activity of left cheliped: 30.6 times/minute
Average quantity of bottom material carried towards
the mouth: 0.83 ml/minute = 50 ml/hour = 1200 ml/24 hours
Only about 50% of this seems to be sifted by mouth parts = 600 ml/24 hours
For the warm half-year this will make . . . . 108 liter/6 months

Callianassa californiensis:

Adult animals, 5–10 cm long, sifting sand. Experiment from California. (MacGinitie, 1934, 1935).

Average volume of sand sifted through the mouth-parts of one animal:
20 to 50 cm^3/24 hours = ... 7.3 to 18.3 liter/year
and there are many specimens per sq.m.

many species, probably the majority, inside this group if given a choice will prefer animal food to anything else. Thus, the sea-urchin, *Echinus esculentus* may feed on a pure diet of *Laminaria*, but seems to prefer laminarians covered by dense crusts of living bryozoans (*Membranipora*) in contrast to clean algae, and when living intertidally "feeds very largely on barnacles" (ELMHIRST, 1922). A similar preference for algae covered by *Membranipora* as compared to "pure" algae was found for North West Pacific specimens of *Strongylocentrotus droebachiensis* (SEVILLA, 1961).—The brittle star, *Ophiura texturata*, can be found with any diet from pure mud and detritus to fairly large lamellibranchs in its stomach (BLEGVAD, 1914, pp. 25–26), simply depending on whether lamellibranchs have been available on the bottom in the locality and season, when the ophiurans were examined. The crab, *Cancer magister*, is known to be omnivorous, but prefers living lamellibranchs to any other food (BUTLER, 1954).—The hermit crab, *Pagurus bernhardus*, "obtains its food largely by scooping up loose material from the sea-bottom" (ORTON, 1927, p. 911), but has also been found to take lamellibranchs, echinoderms, crustaceans and polychaetes (HUNT, 1925).

The fact that the stomach of such "omnivorous" animals will in most cases be filled up with a mixture of vegetable matters, detritus, and animal food may reduce their significance as 'true" predators, since their digestive tract will often be so distended that they may be unable to swallow more, should a prey suddenly turn up. On the other hand, their varied diet may allow them to survive even during seasons when no food animals are present, while a "specialized" predator simply must starve to death if the food animals on which it is fully dependent disappear. In the German Wadden Sea, in the year 1928, a heavy settling of the lamellibranchs *Spisula subtruncata* (prey-animal), the prosobranch *Natica alderi* (specialized predator, only feeding on lamellibranchs), and the brittle star *Ophiura texturata* (omnivorous, but prefering animal food) took place simultaneously. Just after settling, the *Natica's* as well as the *Ophiura's* preyed voraciously on the *Spisula's*, which for that reason soon were wiped out. As soon as the lamellibranchs had disappeared, the *Natica's* started starving and soon died out, while the *Ophiura's* simply changed over to another diet and survived. Being omnivorous the ophiurans could "run idle" on detritus, thus being ready to attack any new spat-fall of prey animals which might later on settle in the area (HAGMEIER, 1930, p. 142 and p. 151). In such a case it may really be questioned, which feeding type will be the most destructive for the newly settled young of the macrofauna: the "specialized predator" or the "omnivorous animal" preferring living prey to anything else.

Table V shows an experiment from the Helsingør-aquaria, Aug. 1965, to throw light on the importance of detritus feeding and predation respectively as the preferred way of feeding in the omnivorous *Pagurus bernhardus*.

It will be seen that a single medium-sized hermit crab in 24 hours "trenched" and sifted a sand-bottom area of 2 sq. dm. and 1 cm deep so thoroughly, that about 85% of the 60 young *Spisula*-spat which were installed and naturally dug down in the sand, were found and eaten. In this case it is obvious, that even "... when scooping up loose material from the sea-bottom", i.e. when feeding in what is regarded as a non-predatory way, the hermit may destroy huge numbers of spat which will be crushed when passing its mouth parts.

When, finally, we return to the "*true*" *predators* on the bottom, there is no much point in summarizing here such data on predation by shore-crabs, drilling-snails, and sea-stars of the *Asterias*-type, the destructive effects of which are generally known. It seems more important to call the attention to a few facts about some well-known "true" predators, which may force us to revise our aspects.

If we examine a random population of the prosobranch, *Nassarius reticulatus*, taken off Helsingør and preserved immediately after the capture, roughly 80% of the specimens will have empty or nearly empty stomachs. The interpretation of this observation may be 1) that like most other predators they have a very rapid digestion, or 2) that they are actually starving and the average population, thus, is undernourish-

TABLE V

Predation by medium sized spms. of *Pagurus bernhardus* L. (average weight without shell 2.8 g; average length of carapax 14.7 mm) on young spat of *Spisula subtruncata* (0.5–3.15 mm). Each hermit crab was for 24 hours by 17.5 °C exposed to 60 young *Spisula's*, dug down in a 2 dm^2 sand flat, 1 cm deep, Elsinore, August 1965.

Aquarium no.	*No. of* Spisula's *by start of exper.*	*No. of* Spisula's *left*	*No. of* Spisula's *disappeared (in aquaria no. 1–10 prob. eaten)*
1	60	1	59
2	60	9	51
3	60	1	59
4	60	2	58
5	60	2	58
(6)	(60)	(13)	(47)
7	60	4	56
8	60	0	60
9	60	1	59
10	60	3	57
	540	23	517, or 95%
11*	60	54	6
12*	60	54	6
13*	60	54	6
	180	162	18, or 10%

So 85% of the *Spisula's* had been found and eaten in 24 hours.

The hermit crab in aquarium no. 6 moulted during the experiment, which is therefore disregarded.

* control aquarium without hermit crab.

ed, or 3) that it is caused by a combination of both. Now, VINOGRADOVA (1950, in Russian, here kindly summarized for me in English by Dr. S. Mileikovsky) for *Nassarius reticulatus* var. *pontica* made a series of most interesting experiments at Sevastopol in the Black Sea. She compared the number of egg-capsules and of eggs produced per time unit in females of this species 1) in their natural surroundings, 2) in aquaria where they had access to any amount of food they might want to eat, and 3) in aquaria with the same optimal food-conditions plus an addition of vitamin B^1, C and D. The result was clear: By far the largest number of egg-capsules and eggs were spawned by the females feeding under optimal conditions plus vitamins. Next come the females which were optimally fed, but without vitamins, and far behind came the females from natural surroundings with a much smaller rate of reproduction. So, the populations in nature seem to be permanently undernourished. This, of course, will mean that they are always hungry and ready to attack and eat a prey any time it turns up, but it also means that any effort to reduce the number of such predators on the bottom, for instance by fishing as many of them up as possible, will only result in a more prolific ratio of reproduction for the surviving specimens, which now have got less competition for the food. So, next year, the larval settlement may well be so much larger, that the spatfall on the bottom will surpass the density of the population, before we tried to reduce it.

Another example giving new aspects on predation was presented by CHRISTENSEN (1962) during his experiments in Helsingør with *Astropecten irregularis*. This voracious sea-star, inhabiting a sandy-bottom *Venus* community in northern seas, and which may contain more than 400 lamellibranch-spat in its stomach at the same time, will in our area feed on *Spisula subtruncata* (a quickly growing, short lived species with a high metabolic rate) as well as on *Venus gallina* (a slowly growing species with a long life span and a low metabolic rate). Since now each specimen of *Venus* with a life span roughly 4 times longer than that of *Spisula* may also be in danger of being eaten by *Astropecten* for a 4 times longer period than this, it may be asked why *Venus* has not been exterminated by the predation of *Astropecten* already long ago. This has, however, *not* happened, and CHRISTENSEN (l.c.) explains why. Firstly, the water pumping rate of the oxygen-demanding *Spisula* will create a water current much more active than that of *Venus*, and this current helps *Astropecten* to scent and find a *Spisula* much more often than a *Venus*. This high oxygen consumption will furthermore force a *Spisula* to open its valves even in the stomach of *Astropecten* where it will, therefore, rapidly be killed and digested, while a *Venus* in the same situation can stand an oxigen deficit, i.e. it will firmly close its solid valves and may be spitted out of the stomach of the sea-star up to 18 days later—and fully alive. This experiment also shows in a nearly shocking way, that if we evaluate the rate of feeding of a predator of this type simply on the proportions of the prey animals found in the stomachs of preserved specimens, we may get absolutely misleading results.

A group of predators hitherto disregarded in all calculations from the sediment-bottoms, is the actinians, in this case the Edwardsiidae and the Cerianthidae. When a grab-haul is sorted and sifted, these animals will usually be overlooked, since when disturbed they contract vigorously and reduce their body-volume very much. This probably is the reason why, in PETERSENS (1923, appendix) original grab-series making the basis for his *Venus gallina* community, actinians are hardly mentioned. The extraordinarily careful counting by Mr. Willy Nicolaisen, M.sc., on the *Venus*-community North of Helsingør has, however, shown a denisty of 260 Edwardsiidae and 65 *Cerianthus* per sq. m. in the areas where they are most common. The Edwardsiidae often sit so deeply that their slender tentacles are distended immediately over the bottom-substratum. If roughly, we calculate the tip of such a tentacle to be 5–6 mm long taken from the center of the mouth of the actinian, this will mean that the tentacles of each of the 260 specimens of Edwardsiidae will cover about 1 sq. cm., or a total of 2–3% of each sq. m. of bottom. In such areas, the nematocyst-loaden tentacles of the *Edwardsia's* may be extremely dangerous also to metamorphosing larvae of the macrobenthos drifting over the bottom ready to settle.

The aim of this review has been to call the attention to the following two facts: *Firstly*, that there are many barriers which the planktonic larvae of the macrobenthos have to meet, before they finally settle on the bottom, and that each type of bottom substratum will, accordingly, be recruited by a very limited and selected set of larval species. *Secondly*, that this selected set of species also after metamorphosis and settling must pass a new series of barriers on the meio-fauna- as well as the macrofauna-level, and pass a heavy "survival-of-the-fittest-fight", in which not only the traditional predators (e.g. crabs, gastropods, sea-stars), but also animals from many groups hitherto regarded as fully or at least relatively harmless, take their heavy tax.

Actually, so many different types of true and even potential predators are ready to attack the macrobenthos at all levels of its development and upgrowth, that it may seem surprising that, in spite of all, a number of macrobenthic animals large enough to secure the next generation can survive to full adult size and maturity.

In the survey from the sandy-bottom *Venus*-community in the Øresund at 18 m depths (see p. 276), Mrs. Kirsten Muus, M. sc. has counted the number per sq. m. of all newly settled young of the macrobenthos, based on series of samples taken at 2–3 weeks' intervals throughout the year 1964. If from these figures and for each species we only consider the average number per sq. m. for the month during which it has its *maximal* occurrence, the addition of these figures for all species concerned will make a total of roughly 30.000 specimens per sq. m. But then we have only got figures for the maximal occurrence of the species, and already one or two months later it will be a new set of young we find on the bottom. It, therefore, is a *very* modest estimation if we assume that during all other months of the year each species has at least been represented on the bottom with a number of young equal to that found during the month of maximal occurrence. In this case, 60.000 young of the macrobenthos per year and sq. m. will be the *minimum* recruitment on this type of bottom. Mr. Nicolaisen, M.sc., in his countings, (see p. 270) from the same locality found an average number of 1.684 specimens per sq. m. to be retained by the 2 mm-meshes of a sieve. So, this figure indicates the average number of specimens of true macrobenthos in the community. Knowing fairly well the conditions for upgrowth in a *Venus* community, it seems reasonable to assume, that this "standing crop" of macrobenthos will represent roughly an average of 2 years' production (see p. 273). If so, it is also based on 2 years' spat-fall, which again means that to recruit a standing crop of this number, 2 × 60.000 = 120.000 newly settled young, or probably more, seem to be needed. This will mean that only about 1.4% of the young settling on the bottom will grow up to the size retained by a 2 mm-mesh screen, and probably less than that.

It still can be taken for granted that the most heavy wastage per time unit during the development of a species of macrofauna-invertebrate from a sediment bottom will take place during the pelagic life (THORSON, 1950), but it will be seen, that also the loss after settling is impressive and in good accordance with the large number of enemies, which have to share the prey.

REFERENCES

BACESCO, M., 1963: Contribution à la biocoenologie de la Mer Noire. L'étage périazoique et le facies dreissenifere. Leurs characteristiques. Rapp. Proc. Verb. Réun., **17,** 107–122.

BACESCO, M. and C. MARGINEANU, 1958: Elements méditerranéennes nouveaux dans la faune de la Mer Noire rencontrés dans les eaux de Roumelic (Nord-Ouest-Bosphore). Arch. Oceanogr. Limnol., **11,** Suppl., 63–74.

BANSE, K., 1964: On the vertical distribution of Zooplankton in the sea. Progr. in Oceanogr., **2,** 53–125.

BARNARD, J. L., 1963: Relationship of benthic Amphipoda to invertebrate communities of inshore sublittoral sands of Southern California. Pac. Nat., **3** (15), 439–468.

BARNARD, J. L. and F. C. ZIESENHENNE, 1961: Ophiuroid communities of Southern Californian coastal bottom. Ibid., **2,** 131–152.

BERTRAM, G. C. L., 1937: Some aspects of the breakdown of coral at Ghardaqa, Red Sea. Proc. Zool. Soc. Lond., 1936, 1011–1026.

BLEGVAD, H., 1914: Food and conditions of nourishment among the communities of invertebrate animals found on or in the sea bottom in Danish waters. Rep. Danish Biol. Stat., **22,** 41–78.

BOISSEAU, J. P. and J. RENAUD, 1955: Répartition de la faune interstitielle dans un segment de plage sablo-vaseuse du bassin d'Arcachon. C. R. Acad. Sci., **241,** 123–125.

BONHAM, K. and E. E. HELD, 1963: Ecological observations on the sea cucumbers *Holothuria atra* and *H. leucospilota* at Rongelap Atoll, Marshall Islands. Pac. Sci., **17,** 302–314.

Bougis, P., 1946: Analyse quantitative de la micro-faune d'une vase marine a Banyuls. C. R. Acad. Sci., **222,** 1122–1124.

Bougis, P., 1950: Méthode pour l'étude quantitative de la microfaune des fonds marins. Vie et Milieu, **1,** 1–15.

Bregnballe, F., 1961: Plaice and Flounder as consumers of the microscopic bottom fauna. Medd. Danm. Fisk.-og Havunders., N.S., **3,** 133–182.

Brinkhurst, R. O., 1964: Observations on the biology of the marine oligochaete *Tubifex costatus*. J. Mar. Biol. Assoc. U.K., **44,** 11–16.

Buchanan, J. B. and R. H. Hedley, 1960: A contribution to the biology of *Astrorhiza limicola* (Foraminifera). J. Mar. Biol. Assoc. U.K., **39,** 549–560.

Burdon-Jones, C., 1962: The feeding mechanism of *Balanoglossus gigas*. Bol. Fac. Fil. Cien.-Letr. Univ. S. Paulo, 261, Zoologica **24,** 255–280.

Butler, T. H., 1954: Food of the commercial crab in the Queen Charlotte Islands region. Fish. Res. Bd. Canad., Pac. Progr. Repts., **99,** 3–5.

Bülow, T. v., 1957: Systematisch-autökologische Studien an eulitoralen Oligochaeten der Kimbrischen Halbinsel. Kiel. Meeresforsch., **13,** 69–116.

Cabioch, L., 1961: Etude de la repartition des peuplements benthiques au large de Roscoff. Cah. Biol. Mar., **2,** 1–40.

Christensen, A. M., 1962: Some aspects of prey-predator relationships in marine level-bottom animal communities. First Nat. Coastal and Shallow Water Research Conference (Abstract), 69–70.

Crozier, W. J., 1918: The amount of bottom material ingested by holothurians (*Stichopus*). J. Exp. Zool., **26,** 377–392.

Dales, R. Ph., 1952: The larval development and ecology of *Thoracophelia mucronata* (Treadwell). Biol. Bull. Woods Hole, **102,** 232–242.

Day, J. H., 1963: The complexity of the biotic environment. System. Assoc. Publ. no. 5, Speciation of the sea, 31–49.

Dimon, A. C., 1908: The mud snail: *Nassa obsoleta*. Cold Spring Harbor Monographs **5,** 1–48.

Dragoli, A. L., 1961: Peculiar feeding habits in the Black Sea polychaete *Melinna palmata* Grube. Dokl. Akad. Nauk SSSR, **138,** 534–535.

Dragoli, A. L., 1962: On the ecology of a Black Sea polychaete *Melinna palmata* Grube. Voprosy ekologii, Vysshaya Shkola, Moscow, **5,** 55–57.

Einarsson, H., 1941: Survey of the benthonic animal communities of Faxa Bay (Iceland). Medd. Danm. Fisk.-og Havunders., Ser. Fiskeri, **11,** 1–46.

Elmhirst, R., 1922: Habits of *Echinus esculentus*. Nature, **110,** 667.

Elofson, O., 1941: Zur Kenntnis der marinen Ostracoden Schwedens. Zool Bidr., Uppsala, **19,** 217–534.

Fox, D. L., 1950: Comparative metabolism of organic detritus by inshore animals. Ecology, **31,** 100–108.

Fox, D. L., C. C. Sheldon and B. H. McConnaughey, 1948: A biochemical study of the marine annelid worm, *Thoracophelia mucronata*. Sears Found., J. Mar. Res., **7,** 567–585.

Fraser, J. H., 1936: The occurrence, ecology and life history of *Tigriopus fulvus* (Fischer). J. Mar. Biol. Assoc. U.K., **20,** 523–536.

Gemmill, J. F., 1914: The development and certain points in the adult structure of the starfish *Asterias rubens* L. Philos. Trans. R. Soc., Ser. B, **205,** 217–294.

Glynn, W., 1965: Community composition, structure, and interrelationship in the marine intertidal *Endocladia muricata—Balanus glandula* association in Monterey Bay, California. Beaufortia, **12,** 1–198.

Hagmeier, A., 1930: Eine Fluktuation von *Mactra* (*Spisula*) *subtruncata* da Costa an der ostfriesischen Küste. Ber. Deutsch. wiss. Komm. Meeresf., N.F., **5,** 126–155.

HANNERZ, L., 1956: Larval development of the polychaete families Spionidae Sars, Disomidae Mesnil, and Poecilochaetidae n. fam. in the Gullmar Fjörd (Sweden). Zool. Bidr. Uppsala, **31,** 1–204.

HART, J. F. L., 1960: The larval development of British Columbia Brachyura. II. Majidae, subfamily Oregoninae. Canad. J. Zool., **38,** 539–546.

HARTMAN, O., 1955: Quantitative survey of the benthos of the San Pedro Basin, Southern California. Allan Hancock Pacific Exped., **19,** 1–187.

HARTMAN, O., 1960: The benthonic fauna of Southern California in shallow depths and possible effects of wastes on the marine biota. Waste Disposal in Marine Environments, 57–81.

HOLME, W. A., 1953: The biomass of the bottom fauna in the English Channel off Plymouth. J. Mar. Biol. Assoc. U.K., **32,** 1–49.

HUNT, O. D., 1925: The food of the bottom fauna of the Plymouth fishing grounds. Ibid., **13,** 560–598.

HURLEY, D. E., 1964: Benthic ecology of Milford Sound. New Zeal. Dept. Sci. indust. res. Bull., **157,** 78–89.

HÖRSTADIUS, S., 1939: Über die Entwicklung von *Astropecten aranciacus* L. Publ. St. Zool. Napoli, **17,** 221–312.

INO, T., 1952: Biological studies on the propagation of Japanese Abalone (genus *Haliotis*). Bull. Tokai Reg. Fish. Res. Lab., **5,** 1–102.

JENNINGS, J. B., 1962: Further studies on feeding and digestion in triclad Turbellaria. Biol. Bull. Woods Hole, **123,** 571–581.

JOHNSON, M. W. and J. B. OLSON, 1948: The life history and biology of a marine harpacticoid copepod, *Tisbe furcata* (Baird). Ibid., **95,** 320–332.

JONES, G. F. and J. L. BARNARD, 1963: The distribution and abundance of the inarticulate brachiopod *Glottidia albida* (Hinds) on the mainland shelf of Southern California. Pac. Nat., **4,** 27–52.

JÖRGENSEN, C. BARKER, 1946: *Lamellibranchia,* in: Medd. Danm. Fisk.-og Havunders., **4,** Ser. Plankton, 277–311.

NIELSEN, C. OVERGAARD, 1949: Studies on the soil microfauna 2. The soil inhabiting nematodes. Nat. Jutl., 1–131.

KNOX, G. A., 1961: The study of marine bottom communities. Proc. R. Soc. N. Z., **89,** 167–182.

KNUDSEN, J. W., 1955: Life cycle studies on the Brachyura of Western North America, I. General culture methods and the life cycle of *Lophopanopeus leucomanus leucomanus* (Lockington). Bull. South. Calif. Acad. Sci., **57,** 51–59.

KNUDSEN, J. W., 1959a: Life cycle studies of the Brachyura of Western North America, II. The life cycle of *Lophopanopeus bellus diegensis* Rathbun. Ibid., **58,** 57–64.

KNUDSEN, J. W., 1959b: Life cycle of the Brachyura of Western North America, III. The life cycle of *Paraxanthias taylori* (Stimpson). Ibid., **58,** 138–145.

KNUDSEN, J. W., 1960: Life cycle of the Brachyura of Western North America, IV. The life cycle of *Cycloxanthops novemdentatus* (Stimpson). Ibid., **59,** 1–8.

KROGH, A. and R. SPÄRCK, 1936: On a new bottom-sampler for investigation of the microfauna of the sea bottom with remarks on the quantity and significance of the benthonic microfauna. K. Dansk. Vidensk. Selsk. Biol. Medd., **23,** 862–879.

LEBOUR, M. V., 1928: The Larval stages of the Plymouth Brachyura. Proc. Zool. Soc. Lond., 1928, 473–560.

LEBOUR, M. V., 1930: The larvae of the Plymouth Galatheidae. I. *Munida bamffica, Galathea strigosa,* and *Galathea dispersa.* J. Mar. Biol. Assoc. U.K. **17,** 175–181.

LEBOUR, M. V., 1931a: The larvae of the Plymouth Caridea. I. The larvae of the Crangonidae. II. The larvae of the Hippolytidae. Proc. Zool. Soc. Lond., 1931, 1–9.

LEBOUR, M. V., 1931b: The larvae of the Plymouth Galatheidae. II. *Galathea squamifera* and *Galathea intermedia*. J. Mar. Biol. Assoc. U.K., **17**, 385–390.

LEBOUR, M. V., 1932: The larvae of Plymouth Caridea. II. The larval stages of *Spirontocaris cranchii* (Leach). Proc. Zool. Soc. Lond., 1932, 131–137.

LEBOUR, M. V., 1940a: The larvae of the Pandalidae. J. Mar. Biol. Assoc. U.K., **24**, 239–252.

LEBOUR, M. V., 1940b: The larvae of the British species of *Spirontocaris* and their relation to *Thor* (Crustacea Decapoda). Ibid., **24**, 505–514.

LONGHURST, A. R., 1958: An ecological survey of the West African marine benthos. Fish. Publ. No. 11, Colonial Office, Lond., 102 pp.

LOOSANOFF, V. L. and H. C. DAVIS, 1963: Rearing of bivalve molluscs. in: Russell, F.S.: Advances in Marine Biology, 1: 1–136.

MACGINITIE, G. E., 1934: The natural history of *Callianassa californiensis* Dana. Amer. Midl. Natur., **15**, 166–177.

MACGINITIE, G. E., 1935: Ecological aspects of a California marine estuary. Ibid., **16**, 629–765.

MCINTYRE, A. D., 1961: Quantitative differences in the fauna of boreal mud associations. J. Mar. Biol. Assoc. U.K. **41**, 599–616.

MCNULTY, J. K., R. C. WORK and H. B. MOORE, 1962: Level sea bottom communities in Biscayne Bay and neighbouring areas. Bull. Mar. Sci. Gulf Caribb., **12**, 204–233.

MAMKAEV, Y. V. and L. W. SERAVIN, 1963: Feeding habits of the acoelous turbellarian *Convoluta convoluta* (Abildgaard). Zool. J., Moscow, **42**, 197–205.

MANGUM, C. P., 1964: Activity patterns in metabolism and ecology of polychaetes. Comp. Biochem. and Physiol., **11**, 239–256.

MARE, M. F., 1942: A study of a marine benthic community with special reference to the microorganisms. J. Mar. Biol. Assoc. U.K., **25**, 517–554.

MAYOR, A. G., 1924: Causes which produce stable conditions in the depth of the floors of Pacific fringing reef-flats. Pap. Dep. Mar. Biol. Carnegie Inst., Wash., **19**, 27–36.

MOORE, H. B., 1931: The muds of the Clyde Sea area, III. Chemical and physical conditions, rate and nature of sedimentation, and fauna. J. Mar. Biol. Assoc. U.K., **17**, 325–358.

MORTENSEN, TH., 1921: Studies on the development and larval forms of echinoderms. 266 pp.

MORTENSEN, TH., 1931: Contributions to the study of the development and larval forms of echinoderms I-II. K. Dansk. Vidensk. Selsk. Skr., Naturv. and Math. Raekke 9, **4**, 1–39.

MORTENSEN, TH., 1938: Contributions to the study of the development and larval forms of echinoderms. IV. Ibid., **7**, 1–59.

MURAYAMA, S., 1935: On the development of the Japanese Abalone, *Haliotis gigantea*. J. Coll. Agric., **13**, 227–233.

MUUS, K., 1966: A quantitative 3-year survey on the meiofauna of known macrofauna communities in the Öresund. Veröff. Inst. Meeresf. Bremerhaven, Sonderbind II, 289–292.

MYERS, E. H., 1942: A quantitative study of the productivity of Foraminifera in the sea. Proc. Amer. Philos. Soc., **85**, 325–341.

OKUDA, S., 1950: Studies on the development of Annelida Polychaeta I. J. Fac. Sci. Hokkaido Imp. Univ., Ser. 6, **9**, 116–219.

ORTON, J. H., 1927: On the mode of feeding of the hermit crab, *Eupagurus bernhardus* and some other Decapoda. J. Mar. Biol. Assoc. U.K., **14**, 909–921.

PÉRÈS, J. M. and J. Picard, 1956: Nouvelles observations biologiques effectuées avec la Bathyscaphe F.N.R.S. III, et considérations sur le système aphotique de la Mediterranée. Bull. Inst. Océanogr. Monaco, no. 1075, 1–10.

PÉRÈS, J. M. and J. Picard, 1960: Considérations sur l'étagement du formations benthiques. Rec. Trav. Stat. Mar. d'Endoume, **33,** 11–16.

PERKINS, E. J., 1958: The food relationship of the microbenthos with particular reference to that found at Whitstable, Kent. Ann. Mag. Nat. Hist., Ser. 13, **1,** 64–77.

PETERSEN, C. G. Joh., 1913: Valuation of the Sea II. The animal communities of the sea-bottom and their importance for marine zoogeography. Rep. Danish Biol. Stat., **21,** 1–44.

PETERSEN, C. G. JOH., 1918: The sea-bottom and its production of fish-food. A survey of the work done in connection with the valuation of the Danish waters from 1883–1917. Ibid., **25,** 1–62.

PICHON, M., 1962: Note préliminaire sur la répartition et le peuplement des sables fins et des sables vaseux non-fixés de la zone intertidale dans la région de Tuléar. Rec. Trav. Sta. Mar. d'Endoume, fasc. hors sér., suppl. 1: 220–235.

PIKE, R. B. and D. G. WILLIAMSON, 1959: Crustacea Decapoda: Larvae. XI. Paguridea, Coenobitidea, Dromiidea, and Homolidea. Zooplankton sheet 81, Cons. Intern. l'Expl. Mer, 1, 1958.

PIKE, R. B. and D. G. WILLIAMSON, 1960: Larvae of Decapoda Crustacea of the families Diogenidae and Paguridae from the Bay of Naples. Pubbl. Staz. zool. Napoli, **31,** 493–552.

PIKE, R. B. and D. G. WILLIAMSON, 1961: The larvae of *Spirontocaris* and related genera (Decapoda, Hippolytidae). Crustaceana, **2,** 187–208.

PIKE, R. B. and D. G. WILLIAMSON, 1964: The larvae of some species of Pandalidae (Decapoda). Ibid., **6,** 265–284.

RALPH, P. and J. C. YALDWYN, 1956: Sea floor animals from Otago Harbour. Tuatara, **6,** 57–85.

REES, C. B., 1940: A preliminary study of the ecology of a mud-flat. J. Mar. Biol. Assoc. U.K., N.S., **24,** 185–199.

REID, G. K., A. INGLIS and H. D. HOESE, 1956: Summerfoods of some fish species in East Bay, Texas. Southwest. Naturalist, **1,** 100–104.

RENAUD-DEBYSER, J., 1958: Contribution à l'étude de la faune interstitielle du Bassin d'Arcachon. 15th Intern. Congr. Zool., Sect. 4, paper 7.

SAENKOVA, A. K., 1956: New components of the fauna of the Caspian Sea. Zool. Zhurnal, **35,** 678–679.

SANDERS, H., 1960: Benthic studies in Buzzards Bay III. The structure of the soft-bottom community. Limnol. and Oceanogr., **5,** 138–153.

SCHELTEMA, R. S., 1961: Metamorphosis of the veliger larvae of *Nassarius obsoletus* (Gastropoda) in response to bottom sediment. Biol. Bull. Woods Hole, **120,** 92–109.

SEVILLA, J. Z., 1961: On the food preferences of *Strongylocentrotus droebachiensis*. Unpubl. Student Rep., Spec. Project (Zool. 533), Friday Harbor Lab., Univ. of Washington, 14 pp.

SMIDT, E., 1951: Animal production in the Danish Waddensea. Medd. Danm. Fisk.- og Havunders., Ser. Fiskeri, **11,** 1–151.

SPÄRCK, R., 1926: On the food problem in relation to marine zoogeography. Physiol. papers dedicated to prof. August Krogh, 268–283.

STAFFORD, J., 1912: On the recognition of bivalve larvae in plankton collections. Contr. Canad. Biol., 1906–1910, 221–242.

SULLIVAN, C. M., 1948: Bivalve larvae of Malpeque Bay. P.E.G. Fish. Res. Bd. Canad., **77,** 1–36.

TATTERSALL, W. M. and E. M. SHEPPARD, 1934: Observations of the Bipinnarias of the Asteroid genus *Luidia*. James Johnstone Mem. Vol., Lancashire Sea-Fish. Lab., 34–61.

THEEL, HJ., 1892: On the development of *Echinocyamus pusillus* (O. F. Müller). Nova Acta Regiae Sox. Sci., **15,** 1–57.

THOMAS, L. P., 1962: The shallow water amphiurid brittle stars (Echinodermata, Ophiuroidea) of Florida. Bull. Mar. Sci. Gulf Caribb., **12,** 623–694.

THORSON, G., 1946: Reproduction and Larval development of Danish marine bottom invertebrates. Medd. Danm. Fisk.-og Havunders., Ser. Plankton, **4,** 1–523.

THORSON, G., 1950: Reproductive and larval ecology of marine bottom invertebrates. Biol. Rev., **25,** 1–45.

THORSON, G., 1955: Modern aspects of marine level-bottom animal communities. J. Mar. Res., **14,** 387–397.

THORSON, G., 1956: Marine level-bottom communities of recent seas, their temperature adaptation and their "balance" between predators and food animals. Trans. New York Acad. Sci., Ser. 2, **18,** 693–700.

THORSON, G., 1957: Bottom communities (Sublittoral or shallow shelf). Chapter 17 in: Geol. Soc. Amer. Memoir 67, **1,** 461–534.

THORSON, G., 1961: Length of pelagic larval life in marine bottom invertebrates as related to larval transport by ocean currents. Oceanography. Amer. Assoc. Adv. Sci. Publ. No. 67, 455–474.

URSIN, E., 1960: A quantitative investigation of the echinoderm fauna of the Central North Sea. Medd. Danm. Fisk.-og Havunders., N.S., **2,** 1–204.

VATOVA, A., 1940: Le zoocenosi della Laguna Veneta. Thalassia, **3,** 1–28.

VINOGRADOVA, Z. A., 1950: Material about the biology of the Black Sea Molluscs. Trudi Karadag Biol. Stat. Acad. Sci. Ukrainskaya SSR, **9,** 100–158. (In Russian).

WIESER, W., 1960a: Benthic studies in Buzzards Bay II. The meiofauna. Limnol. and Oceanogr., **5,** 121–137.

WIESER, W., 1960b: Populationsdichte und Vertikalverteilung der Meiofauna mariner Böden. Int. Rev. ges. Hydrobiol., **45,** 487–492.

WIESER, W. and J. KANWISHER, 1960: Growth and metabolism in a marine nematode, *Enoplus communis* Bastian. Z. Vergl. Physiol., **43,** 29–36.

WIESER, W. and K. KANWISHER, 1961: Ecological and physiological studies on marine nematodes from a small salt marsh near Woods Hole, Mass. Limnol. and Oceanogr., **6,** 262–

WIGLEY, R. L. and A. D. MCINTYRE, 1964: Some quantitative comparisons of offshore meiobenthos and macrobenthos south of Martha's Vineyard. Ibid., **9,** 485–493.

WILSON, D. P., 1932: The development of *Nereis pelagica* L. J. Mar. Biol. Assoc. U.K., **18,** 203–217.

WILSON, D. P., 1933: The larval stages of *Notomastus latericeus* Sars. Ibid. **18,** 511–518.

WILSON, D. P., 1936: The development of *Andouinia tentaculata* (Mtg.). Ibid. **20,** 567–579.

WILSON, D. P., 1948: The larval development of *Ophelia bicornis* Savigny. Ibid. **27,** 540–553.

WILSON, D. P., 1952: The influence of the nature of the substratum on the metamorphosis of the larvae of marine animals, especially the larvae of *Ophelia bicornis* Savigny. Ann. Inst. Océanogr., **27,** 49–156.

WILSON, D. P., 1953a: The settlement of *Ophelia bicornis* Savigny larvae. The 1951 experiments. J. Mar. Biol. Assoc. U.K., N.S., **31,** 413–438.

WILSON, D. P., 1953b: The settlement of *Ophelia bicornis* Savigny larvae. The 1952 experiments. Ibid., **32,** 209–233.

WILSON, D. P., 1954: The attractive factor in the settlement of *Ophelia bicornis* Savigny. Ibid., **33,** 361–380.

WOELKE, C. E., 1956: The flatworm *Pseudostylochus ostreophagus* Hyman, a predator of oysters. Proc. Natl. Shellfish Assoc. for 1956. **47,** 62–67.

YAMANOUTI, T., 1939: Ecological and physiological studies on the holothurians in the Coral Reef of Palao Islands. Palao Trop. Biol. Stat. Studies, **4,** 603–635.

YOSHIDA, H., 1960: On the early life-history of *Tapes variegata* Sowerby. J. Shimonoseki Coll. Fish., **10,** 115–118.

ZEUTHEN, E., 1947: Body size and metabolic rate in the animal kingdom. C. R. Lab. Carlsberg, ser.-chem., **26,** 20–165.

SOME QUANTITATIVE COMPARISONS OF OFFSHORE MEIOBENTHOS AND MACROBENTHOS SOUTH OF MARTHA'S VINEYARD

Roland L. Wigley
Bureau of Commercial Fisheries, Woods Hole, Massachusetts

and

A. D. McIntyre
Marine Laboratory, Aberdeen, Scotland

ABSTRACT

Quantitative samples were taken from the offshore area south of Martha's Vineyard, Massachusetts, in June 1962, to compare the macrobenthos with the meiobenthos. Macrobenthos infauna ranged from 700 to 5,500 individuals per m^2 and (excluding individuals larger than 5 g) from 21 to 82 g/m^2. The principal taxonomic groups were Crustacea, Mollusca, Echinodermata, and Polychaeta. Meiobenthos ranged from 117,000 to 988,000/m^2 in number of individuals and from 0.6 to 4.6 g/m^2 in weight. Major taxonomic groups were Nematoda, by far the most numerous component, Polychaeta, Kinorhyncha, and Foraminifera. Meiobenthos populations in offshore areas are shown to be at least as rich as those previously known from shallow coastal waters. Numbers and weight of both macrobenthos and meiobenthos were lowest in deep water and highest at shallow or mid-depths. Meiobenthos was most plentiful on sandy sediments, and, although this applied to numbers of macrobenthos, the weight of the latter was generally highest on fine or moderately fine sediments. The ratio of macro- to meiobenthos over all samples was numerically 1 : 70 and by weight 24 : 1. These ratios were remarkably consistent from station to station in spite of variations in faunal components and the environment.

INTRODUCTION

Quantitative comparisons of subtidal benthic meiofauna with macrofauna have been made in only five investigations. Krogh and Spärck (1936) studied material from the vicinity of Copenhagen, Denmark; Mare's (1942) material was from the English Channel off Plymouth; Sanders (1958) and Wieser (1960) studied the fauna of Buzzards Bay, Massachusetts; and McIntyre (1961) investigated two areas off Scotland.

A series of samples collected in June 1962 from the research vessel *Delaware* made it possible to compare relative quantities of meiobenthos and macrobenthos in various substrata and at different depths off southern New England. There appear to be no reports in the literature giving quantitative data on either macrobenthos or meiobenthos populations for this area. Although our observations are limited to relatively few samples, they extend the range of our knowledge considerably and are the first measurements of this kind from offshore waters in the western Atlantic Ocean.

We wish to thank Roger B. Theroux and Timothy W. Robbins for assistance at sea and in the laboratory.

MATERIALS AND METHODS

Data presented here are based on samples from *Delaware* Cruise 62-7, representing 10 stations (46–55) spaced approximately at 10-mile (18-km) intervals along longitude 71°00′ W south of Martha's Vineyard. Station positions, depths, and bottom temperatures are listed in Table 1 and illustrated in Fig. 1. Water temperatures were measured with a bathythermograph attached to the grab, except at Stations 54 and 55. At these two deepwater stations, sediment temperature was measured, using a standard bulb thermometer. Water depths ranged from 40 m near shore to 567 m on the continental slope.

Bottom water temperatures ranged from 5.0C (41F) in deep water to 10.8C (51.5F) near the periphery of the continental shelf.

TABLE 1. *Station location and physical features. All stations located on longitude 71°00′ W*

Station	Depth (m)	Bottom water temperature (C)			N lat
		June 1962	Approximate annual range*		
			Minimum	Maximum	
46	40	9.7	3	14	41°10′
47	51	6.4	3	14	41°00′
48	58	6.1	4	14	40°50′
49	69	6.3	4	12	40°40′
50	84	6.3	6	11	40°30′
51	99	9.4	8	11	40°21′
52	146	10.8	10	12	40°10′
53	179	10.8	11	11	40°06′
54	366	6.6†	5‡	7‡	39°59′
55	567	5.0†	4‡	5‡	39°56′

* Based on data from Bigelow (1933).
† Bottom sediment temperature.
‡ Based on data from Townsend (1901).

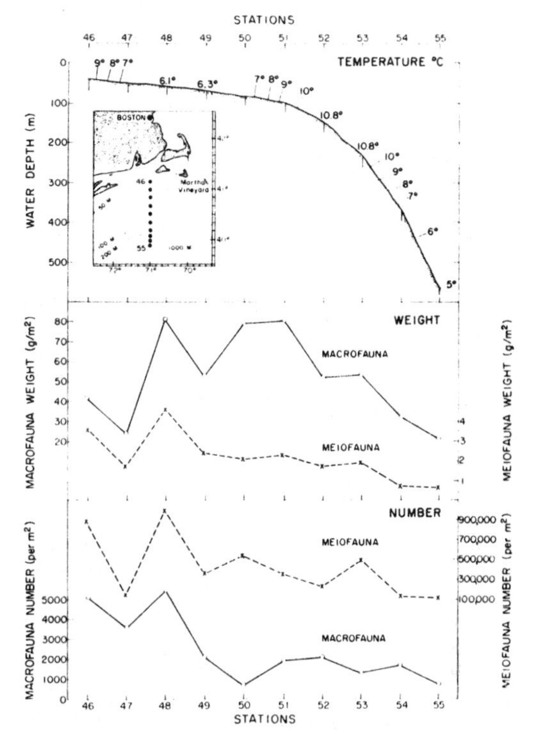

FIG. 1. Water depth-temperature profile together with graphs of macrofauna and meiofauna weights and numbers, plus station location chart.

In addition to the June temperatures, the temperature extremes during a 1-year period are pertinent and are included in Table 1, but only approximate values are available. They ranged from 3 to 14C at shallow-water stations, 10 to 12C near the continental shelf edge, and 4 to 7C at deepwater stations. The temperature regime of bottom waters south of Martha's Vineyard is complicated by a moderately cold water mass midway out on the shelf and a warm water mass near the shelf edge. Occasional shifts in location of these masses result in wide temperature fluctuations at any given location over a time span of several years or more. Minimum annual bottom water temperatures on the shelf generally occur in February–March, and maximum temperatures occur in September–November; June temperatures are intermediate between the two extremes (Townsend 1901; Bigelow 1933; data in the hydrographic files at the Woods Hole Oceanographic Institution).

Bottom sediments varied from predominantly gravels and sands inshore to silt clays at offshore stations. Particle size analysis of the deposit for each station is listed in Table 2.

At each station, two hauls were made with a 0.1 m^2 Smith-McIntyre grab, and the materials for macrobenthos studies were washed through a 1-mm screen. From one grab haul at each station, a meiobenthos core was collected in the following way. The grab was carefully kept in an upright position until the water had drained off; then a top plate on the grab was removed, exposing the surface of the deposit. After ensuring that this surface was even and undisturbed, a core sample was taken using a plastic tube of 34.8 mm inside diameter, giving a surface area of 9.50 cm^2. The top 4 cm of the core was then pushed out from below with a plunger, and preserved in 5% formalin. A possible source of error in this method of collecting meiobenthos is that a downrush of water when the grab strikes bottom might disturb the surface layer before the grab bites. This was not considered a serious objection, since the type of grab used has bucket tops made of fine-mesh screen instead of solid plates, which is thought to reduce the wash considerably.

TABLE 2. *Sediment grain size composition, expressed as percentage weight*

Station	Grain size classes (mm)							Mean grain size (mm)	Sorting coefficient*
	> 2.00	2.000–1.000	1.000–0.500	0.500–0.250	0.250–0.125	0.125–0.062	< 0.062		
46	0.10	0.40	34.20	55.72	9.26	0.32	0	0.58	0.50
47	12.12	49.27	25.06	11.47	1.63	0.14	0.31	1.47	0.90
48	0	0.59	4.57	31.91	46.17	2.31	14.45	0.25	0.80
49	0.65	2.06	5.02	14.57	17.90	9.27	50.53	0.12	1.92
50	0	0	1.24	1.42	1.55	6.25	89.54	0.04	0.87
51	1.31	0.70	10.98	19.22	10.46	3.31	54.02	0.13	2.38
52	7.30	2.89	8.58	37.89	13.05	2.16	28.13	0.14	2.56
53	0.58	2.14	4.78	48.45	16.42	2.14	25.49	0.13	1.98
54	0	0	1.16	3.34	18.52	9.43	67.55	0.08	1.47
55	1.55	1.91	5.57	20.24	26.69	16.68	27.36	0.16	1.65

* Phi deviation measure.

In the laboratory, the macrobenthos collections were sorted and the animals identified, counted, and weighed in the usual way. External moisture was removed from the specimens by means of blotting paper. Mollusk shells and Foraminifera tests of living specimens are included in the weights. The procedure for processing meiobenthos samples was as follows. The core was first washed through a 1-mm screen and all animals retained were recorded and returned to the appropriate macrobenthos sample. The filtrate was further washed through a 74-μ screen, and the residue on this screen was placed overnight in an aqueous solution of rose bengal for staining. It was later examined in 70% alcohol, and all stained animals were picked out by direct search under a binocular microscope. The filtrate of the 74-μ screen was not examined, since it was known from unpublished studies of North Sea material that this fraction was likely to contain insignificant faunal weights. Wet weights of nematodes were calculated from measurements of length and width, assuming a specific gravity of 1.13 (*see* Wieser 1960). Approximations for other groups were obtained from weighings of large numbers of individuals.

RESULTS

Macrobenthos

All living organisms retained by the 1-mm mesh sieving screen were considered macrobenthos forms. Components of this group ranged from small organisms, such as nematodes and Forminifera that weighed less than 0.001 g each, to large lamellibranchs that individually weighed more than 100 g. Numbers of individuals and wet weights of the macrobenthos at each station are listed in Table 3. The values given are means for a pair of samples from each station. For purposes of comparison with previous studies, the macrobenthos data have been separated into two major categories, 1) infauna and 2) epifauna. The infauna has been subdivided into two size groups: Group A—specimens weighing less than 5 g each; Group B—specimens weighing more than 5 g each.

Three major macrofaunal groups are evident from this sampling, and their distribution corresponds to biotopic and geographic features. Beginning inshore and progressing seaward, the groups are as follows: 1) an inner continental shelf group characterized by amphipods and polychaetes, 2) an outer continental shelf group characterized by echinoderms and mollusks, and 3) a continental slope group characterized by a general paucity of fauna but dominated by polychaetes and including small numbers of certain miscellaneous groups.

Macrofauna in the inner continental shelf group occurred at Stations 46–48. Depths at these stations ranged from 40 to 58 m; bottom temperatures ranged from 6.1 to 9.7C; and the sediments were principally sands or fine gravels. Burrowing amphi-

TABLE 3. *Macrobenthos wet weight (g/m²) and number of individuals (per m²) at each station*

	Stations									
	46	47	48	49	50	51	52	53	54	55
Infauna										
Group A (Individuals weighing less than 5 g each)										
Crustacea										
Number	4,235	2,990	3,465	790	20	55	5	25	20	5
Weight	22.26	12.35	33.07	3.60	0.05	1.50	0.02	0.08	0.10	0.02
Mollusca										
Number	45	0	155	165	210	525	320	260	90	25
Weight	1.45	0	1.18	10.98	2.98	9.33	13.32	6.35	0.88	2.15
Echinodermata										
Number	15	0	15	15	105	740	455	210	30	10
Weight	0.05	0	0.08	0.02	25.85	33.15	8.05	36.90	2.02	10.02
Polychaeta										
Number	600	440	1,735	1,105	325	270	955	325	795	425
Weight	11.15	10.05	46.62	37.98	47.55	6.37	24.70	8.08	12.90	7.95
Miscellaneous										
Number	245	155	125	70	40	330	350	460	760	265
Weight	6.70	1.90	0.60	0.55	2.27	30.57	5.37	1.91	16.08	0.95
TOTAL										
Number	5,140	3,585	5,495	2,145	700	1,920	2,085	1,280	1,695	730
Weight	41.61	24.30	81.55	53.13	78.70	80.92	51.46	53.32	31.98	21.09
Group B (Individuals weighing more than 5 g each)										
Number	—	—	15	—	15	5	—	5	—	—
Weight	—	—	1,274.10	—	155.55	46.75	—	23.45	—	—
Epifauna										
Number	—	—	5	—	25	5	55	5	—	—
Weight	—	—	0.03	—	32.45	0.02	7.20	0.10	—	—
GRAND TOTAL										
Number	5,140	3,585	5,515	2,145	740	1,930	2,140	1,290	1,695	730
Weight	41.61	24.30	1,355.68	53.13	266.70	127.69	58.66	76.87	31.98	21.09

pods were particularly abundant, averaging 3,000 to 4,000/m² in number and from 10 to 30 g/m² in weight. Some arenaceous Foraminifera and polychaetes in the families Scalibregmidae and Sabellidae were less abundant, but they occurred only at these stations.

Faunal composition at Station 49 appears to be transitional between the inner and outer continental shelf groups. Amphipods at this station constituted a significant number, but they were much less abundant than that expected in a typical inner shelf association. The occurrence of large quantities of mollusks is characteristic of the outer shelf group. Polychaetes, particularly the scoleciform types, were dominant in both weight and number.

Outer continental shelf fauna occurred at Stations 50 through 53, where the depths ranged from 84 to 179 m and bottom temperatures ranged from 6.3 to 10.8C. Sediments were of two types: at Stations 50 and 51, silt clay predominated, while at Stations 52 and 53, medium sand predominated. Mollusks were relatively numerous on this portion of the shelf, ranging in number from 200 to 500/m²; lamellibranchs were the principal group. Echinoderms, consisting mostly of brittle stars plus a few holothurians, averaged 100 to 700/m² and were a characteristic component of the outer shelf macrofauna.

Continental slope fauna was represented at Stations 54 and 55. These are the two deepest stations, 366 and 567 m, respectively. Bottom temperatures ranged from 5.0 to 6.6C, with very little seasonal variation, and the sediments were composed primarily of silt clay and fine sand. Macrofauna composition at these localities was not particularly characteristic, except for the occurrence of tanaids, harpacticoid copepods, and pogonophores. Polychaetes,

TABLE 4. *Meiobenthos wet weight (mg/10 cm²) and number of individuals (per 10 cm²) at each station*

Faunal group	Stations									
	46	47	48	49	50	51	52	53	54	55
Number/10 cm²										
Nematoda	685	50	924	328	507	302	202	438	117	110
Kinorhyncha	19	2	15	–	2	–	–	–	–	–
Foraminifera	4	4	14	20	7	40	25	32	6	4
Copepoda	122	45	15	2	–	–	4	15	–	1
Ostracoda	2	1	7	–	–	1	1	1	–	–
Gastrotricha	7	5	–	–	–	–	–	–	–	–
Halicaridae	–	1	1	1	–	–	–	–	–	–
Nauplii	1	9	2	–	–	–	–	–	–	–
Polychaeta	30	10	10	6	19	5	3	4	4	1
Amphipoda	3	–	–	4	–	–	–	–	–	–
Isopoda	–	–	–	–	–	–	–	2	–	–
Cumacea	1	–	–	–	–	1	–	–	–	–
Lamellibranchia	1	–	–	–	1	2	–	–	–	1
Echinodermata	–	–	–	1	1	–	–	–	–	–
Total	875	127	988	362	537	351	235	492	127	117
Per cent Nematoda	78	39	94	91	94	86	86	89	92	94
Weight mg/10 cm² (values are equivalent to g/m²)										
Nematoda	1.016	0.155	2.052	1.066	0.923	0.570	0.349	1.004	0.243	0.359
Kinorhyncha	0.947	0.010	0.074	–	0.032	–	–	–	–	–
Foraminifera	0.046	0.038	0.163	0.242	0.063	0.526	0.244	0.389	0.053	0.038
Polychaeta	0.958	1.410	2.000	0.810	0.989	1.105	1.116	0.442	0.453	0.105
All other groups	0.632	0.074	0.268	0.289	0.105	0.132	0.037	0.060	—	0.105
Total	3.599	1.687	4.557	2.407	2.112	2.333	1.746	1.895	0.749	0.607

400 to 800/m², were the most abundant group, and the nereidiform and spioniform types were particularly common. Foraminifera were also quite abundant (200 to 500/m²), but because of their small size their weight contribution was small (average 1.3 g/m²). Pogonophora are generally rare throughout the world, so it was interesting to find representatives of this group in all samples from the continental slope. Their average density was 28/m².

Although there was considerable variation in weight and number from one station to another, quantitative relationships of the macrofauna with both water depth and sediment type can be detected. Numbers of individuals were highest, up to about 5,000/m², in shallow water and in coarser deposits, dropping to about one-fifth of this value in deeper water and in finer deposits. Weight values showed a slightly different trend, being highest at intermediate depths and in the finer sediments.

Meiobenthos

All animals passing through a 1-mm mesh but retained on a 74-μ mesh are here regarded as meiobenthos. Numbers and weights of these animals from each station (Table 4) indicate that there are three meiobenthos zones or groupings, geographically similar to those described for macrobenthos.

The inner shelf group is characterized by a varied and abundant meiobenthos, 11 major taxonomic groups being represented. The number of individuals averaged 662/10 cm², with a corresponding wet weight of 3.28 mg.

In the outer shelf group, including Station 49, the fauna is both less varied and, except for calcareous Foraminifera, less numerous, with an average number of 399 individuals/10 cm² and a weight of 2.10 mg. Within this group, the largest population occurred at Station 50, which had the highest silt clay content in the deposit.

At the slope stations, the fauna was relatively sparse, with nematodes present in only small numbers and the other groups much reduced or absent. The average numbers and weights per 10 cm^2 were 122 and 0.68 mg, respectively.

Gastrotrichs, kinorhynchs, and, to a lesser extent, copepods and ostracods tended to be confined to shallower water, while Foraminifera were most abundant at intermediate depths. Nematoda comprised the major group, being present in all samples, and they were by far the most numerous animals. They were least abundant on a substratum of gravel, where they made up only 39% of the fauna, but at all other stations they composed between 78% and 94%, numerically, of the total fauna.

A few of the animals listed in Table 4 are juveniles whose adults would have been retained on the 1-mm screen. These include the lamellibranchs and echinoderms, as well as some of the polychaetes and crustaceans (except copepods). The greater part, however, is made up of animal groups most of whose species would pass through a 1-mm screen (or even through a 0.5-mm screen) at any stage of their life history.

Since the 74-μ screen was the smallest used, any animals passing through this were lost. Unpublished work regarding the North Sea suggests that material lost in this way would consist largely of nematodes (mainly immature stages) and that, while their numbers may be significant, the weights would be small. Another source of error is that only the top 4 cm was examined. This fraction of the core contains the greater part of the fauna, but small numbers of animals, mainly nematodes, must have been lost. For these reasons, the values given in Table 4 must be regarded as minimum estimates.

DISCUSSION

Benthic fauna in the area studied was quantitatively moderate to rich, compared with nearby inshore areas and north European localities. Although considerable variation was evident, both macro- and meiobenthos exhibited similar station-to-station abundance trends. Numbers of individuals in the fauna in both categories were highest in coarse-grained sediments in shallow water and considerably lower in fine-grained sediments in deep water. Also, biomass in both categories was moderately high in shallow water, highest at intermediate depths, and lowest in deep water.

Results from two stations, 47 and 48, were strikingly different from all the rest. The first carried a comparatively sparse fauna, the other was unusually rich. At Station 47, where sediments consisted chiefly of fine gravels and coarse to medium sands, we found moderately low quantities of both macro- and meiobenthos, with numbers of individuals averaging 3,600 and 127,000/m^2, respectively, and corresponding weights of 24 and 1.7 g/m^2. Nematodes in particular were sparse in the meiobenthos, although they were relatively numerous in the macrobenthos. Not a single mollusk or echinoderm was present. Unsuitable substratum combined with unfavorable biological or hydrographic conditions may account for this limited taxonomic representation and rather low faunal density. Conversely, in medium and fine sand sediments of adjacent samples, Station 48, there was an exceptionally abundant benthic fauna. Average number and weight of the small (Group A) macrobenthic infauna were 5,500/m^2 and 82 g/m^2. In addition to this were its large infaunal organisms composed exclusively of the bivalve mollusk *Arctica islandica* Linnaeus; they averaged 1,274 g/m^2. Thus, the total of all macrobenthos as this locality amounted to 1,356 g/m^2. A pertinent factor that may in part account for this unusually rich benthic fauna is the abundant zooplankton in the overlying waters (Bigelow and Sears 1939).

Table 5 gives summary data for comparing quantities of macro- and meiofauna at the 10 stations south of Martha's Vineyard. Weights of the former ranged from 21 to 82 g/m^2, and of the latter from 0.6 to 4.6 g/m^2. All stations considered, the macrobenthos weight was 24 times that of meiobenthos, the ratio ranging from 12 : 1 to 43 : 1 (Fig. 2). There is no obvious connection between this ratio and the sediment type, but the general trend was for the ratio to increase with water depth due to the fact that meiobenthos weights de-

TABLE 5. *Density of macro- and meiobenthos expressed as numbers of individuals and weights per m²*

Station	Number of individuals			Weight		
	Macro-benthos*	Meio-benthos	Ratio	Macro-benthos*	Meio-benthos	Ratio
46	5,140	875,000	1:170	41.61	3.60	12:1
47	3,585	127,000	1: 35	24.30	1.69	14:1
48	5,495	988,000	1:180	81.55	4.56	18:1
49	2,145	362,000	1:170	53.13	2.41	22:1
50	700	537,000	1:770	78.70	2.11	37:1
51	1,920	351,000	1:185	80.92	2.33	35:1
52	2,085	235,000	1:110	51.46	1.75	29:1
53	1,280	492,000	1:380	53.32	1.90	28:1
54	1,695	127,000	1: 75	31.98	0.75	43:1
55	730	117,000	1:160	21.09	0.61	35:1
Average	2,477	421,100	1:170	51.81	2.17	24:1

* Group A infauna. For explanation, *see* p. 487.

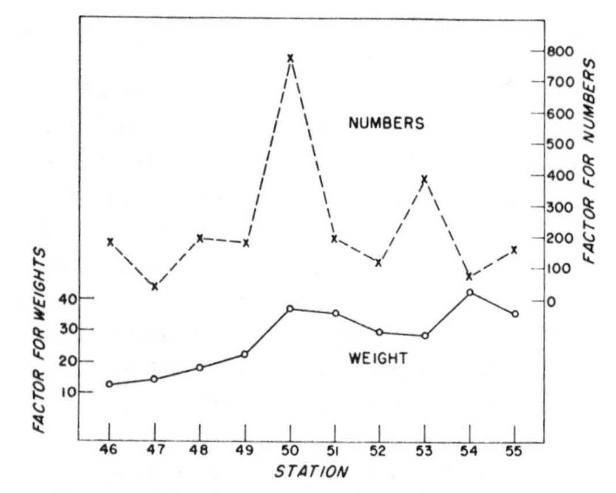

FIG. 2. Factors by which meiobenthos exceeds macrobenthos numbers (top graph) and macrobenthos exceeds meiobenthos weights (lower graph) at each station.

creased relatively more than the macrobenthos as the water deepened. Numbers of individuals had a much wider range, varying in the macrobenthos from 700 to 5,500/m² and in the meiobenthos from 117,000 to 988,000/m². Thus, the latter exceeded the former by factors ranging from 35 to 770, and, considering all stations, the factor averaged 170. The striking feature is the high degree of uniformity (Fig. 2). At seven of the stations, the factor was remarkably close to the average, and only at Station 50 was there an extreme deviation. The high ratio here was due to low numbers of macrobenthos rather than to exceptionally high meiobenthos.

Since workers in different areas have employed various methods in processing macrobenthos and meiobenthos, particularly in regard to sieving screen mesh sizes, precise comparisons between our results and others are precluded. However, it is appropriate briefly to review two main aspects: 1) nematode density, and 2) relative quantities of macrobenthos versus meiobenthos, reported for other areas to provide a basis for generalized comparisons with our findings.

One common feature of all reports on subtidal meiobenthos is the high proportion

TABLE 6. *Numbers of nematodes per 10 cm² recorded by various workers, listed according to water depth and sediment type*

Water depth (m)	Number of nematodes (per 10 cm²)	Sediment type	Location	Authority
18	798	Mud	Massachusetts	Wieser (1960)
18–30	364	Sand	Massachusetts	Wieser (1960)
24	252	Mud	Scotland	Moore (1931)
30	474	Sand	Mediterranean Sea	Bougis (1950)
45	<90	Mud	English Channel	Mare (1942)
90–110	86	Mud	Scotland	McIntyre (1961)
116	157	Mud	Scotland	Moore (1931)
140	62	Mud	Scotland	McIntyre (1961)
40–58	805	Sand	Off southern New England	Present work
51	50	Sandy gravel	Off southern New England	Present work
69–99	379	Silt clay	Off southern New England	Present work
146–179	320	Medium sand	Off southern New England	Present work
366–567	114	Fine sand, silt clay	Off southern New England	Present work

of nematodes in the samples. Data for this are given in Table 6. Certain workers have used screens of 100 μ or coarser (Moore 1931; Mare 1942; McIntyre 1961), so their figures are certainly underestimates, but Bougis (1950) corrected his data for loss of small nematodes, and Wieser (1960) gives total counts. The present work gives values for a greater variety of sediments and depths than were previously available, and, even though they must be regarded as minimum estimates, they do show that at moderate depths offshore, the populations are as numerous as in inshore areas; that in water depths of 100–200 m, high populations are still found; and that, even on the continental slope, significant numbers of nematodes occur—indeed, there were higher populations there than on sandy gravel sediments in shallow water.

Only a few previous investigations have been made that quantitatively compare the subtidal macrobenthos with the meiobenthos. Reliable data on this subject are available for four areas: 1) in the English Channel, the macrobenthos : meiobenthos ratio was ca. 90 : 1 for weights and ca. 1 : 60 for numbers of individuals (Mare 1942); 2) in Buzzards Bay, Massachusetts, the ratio was ca. 33 : 1 for weights (H. L. Sanders, personal communication), and the ratio for numbers of individuals ranged from ca. 1 : 30 to 1 : 140 (Wieser 1960); 3) on the Fladen ground in the central North Sea, the ratio was ca. 40 : 1 for weights and ca. 1 : 40 for numbers of individuals (McIntyre 1961); and 4) in Loch Nevis, western Scotland, the ratio was ca. 15 : 1 for weights and ca. 1 : 100 for numbers of individuals (McIntyre 1961). Despite the varied taxonomic composition of the fauna from one locality to another and the different processing methods used, the macrobenthos : meiobenthos ratios were unusually consistent from one area to another.

Macrobenthos : meiobenthos ratios for samples south of Martha's Vineyard agree well with ratios from the localities listed above. Ratios for numbers of individuals in the offshore Martha's Vineyard samples range from 1 : 35 to 1 : 770, compared with 1 : 30 to 1 : 140 for the other areas. Weight ratios for the offshore Martha's Vineyard samples range from 12 : 1 to 43 : 1, compared with 15 : 1 to 90 : 1 reported for other localities.

CONCLUSIONS

The benthic macrofauna and meiofauna on the continental shelf and upper continental slope south of Martha's Vineyard were moderate to rich, compared with other regions.

Numbers of individuals and weights comprising the macro- and meiobenthos were lowest in deep water and highest at shallow and mid-depths.

A major portion of the macrofauna was composed of four taxonomic groups. Crustacea were common only in inshore, shallow waters; Mollusca and Echinodermata were prevalent at moderate depths along the outer portion of the continental shelf; Polychaeta were rather abundant at all stations.

A small number of large (greater than 5 g) organisms radically increased the total biomass values at a few stations.

Meiofauna formed a moderate-to-small share of the benthic biomass.

Numerically, nematodes were the principal meiofauna component at all depths sampled, and they, together with polychaetes, kinorhynchs, and Foraminifera, were the main meiofauna groups in terms of weight.

The relative quantity of macrobenthos compared to meiobenthos was quite similar from station to station in the offshore Martha's Vineyard region, and as a group these proportions agree rather closely with similar ratios based on samples from European and inshore New England waters. Rather good consistency was evident in the macrobenthos : meiobenthos ratios between different areas and between samples within a given area, even though in many cases the taxonomic composition and biotopic features differed markedly.

REFERENCES

BIGELOW, H. B. 1933. Studies of the waters on the continental shelf, Cape Cod to Chesapeake Bay. I. The cycle of temperature. Papers Phys. Oceanog. Meteorol., **2**(4). 135 p.

———, AND M. SEARS. 1939. Studies of the waters of the continental shelf, Cape Cod to Chesapeake Bay. III. A volumetric study of the zooplankton. Mem. Museum Comp. Zool., **54**: 183–378.

BOUGIS, P. 1950. Méthode pour l'étude quantitative de la microfaune des fonds marins (Meiobenthos). Vie Milieu, **1**: 23–38.

KROGH, A., AND R. SPÄRCK. 1936. On a new bottom-sampler for investigation of the micro fauna of the sea bottom with remarks on the quantity and significance of the benthonic micro fauna. Kgl. Danske Videnskab. Selskab, Biol. Medd., **13**(4): 3–12.

MCINTYRE, A. D. 1961. Quantitative differences in the fauna of boreal mud associations. J. Marine Biol. Assoc. U.K., **41**: 599–616.

MARE, M. 1942. A study of a marine benthic community with special reference to the micro-organisms. J. Marine Biol. Assoc. U.K., **25**: 517–554.

MOORE, H. B. 1931. The muds of the Clyde sea area. J. Marine Biol. Assoc. U.K., **17**: 325–358.

SANDERS, H. L. 1958. Benthic studies in Buzzards Bay. I. Animal-sediment relationships. Limnol. Oceanog., **3**: 245–258.

TOWNSEND, C. H. 1901. Dredging and other records of the United States Fish Commission steamer *Albatross*, with bibliography relative to the work of the vessel. Rept. U.S. Comm. Fish and Fisheries for 1900, p. 387–562.

WIESER, W. 1960. Benthic studies in Buzzards Bay. II. The meiofauna. Limnol. Oceanog., **5**: 121–137.

BENTHIC STUDIES IN BUZZARDS BAY
III. THE STRUCTURE OF THE SOFT-BOTTOM COMMUNITY[1]

Howard L. Sanders

Woods Hole Oceanographic Institution, Woods Hole, Mass.

ABSTRACT

A series of 24 samples was taken at periodic intervals at a single locality in Buzzards Bay from February 1956 to January 1958. With the exception of 4 samples which departed markedly from the others in regard to sediment composition, very high indices of faunal affinity were obtained within the series. The silt-clay fraction of the sediment in the 20 samples with high faunal indices varied from 78–91%, and when the contained species were ranked by abundance, the commoner species achieved approximately the same rank from sample to sample. Thus, 8 or 9 species are a constant part of the biological environment of this association, and there is an approximately constant numerical proportionality among them. Seventy-nine species were obtained from the 20 samples of typical sediment composition. However, 11 species comprised almost 95% of the total fauna by number while the lamellibranch *Nucula proxima* (58.86%) and the polychaete *Nephthys incisa* (17.21%) were commonest. Ten species make up 95% of the total biomass. Thus the community, both numerically and by biomass, is characterized by pronounced dominance of few species. The niches of the 9 most common species, representing 93.5% of the community by number, were categorized by method of feeding, kind of food and spatial distribution. Considering only these few factors, it was possible to demonstrate certain differences in niche preference. The relationship of animal numbers to animal weight revealed the expected pyramid of numbers with very few organisms in the largest weight categories and numerous individuals in the smaller weight groups. However, when the biomass within each weight category was determined, the expected inverted pyramid did not result. Numbers are considered more valid than biomass in quantitative benthic studies, since only 0.15% of the fauna constitute 55.17% of the entire weight. The presence or absence of a rare, randomly distributed large animal may alter the biomass of an individual sample by two orders of magnitude.

INTRODUCTION

This paper, the third in a series dealing with the benthic fauna of Buzzards Bay, Massachusetts, is concerned primarily with the community of animals present in the finer sediments. The term "community" is used in this study to mean a group of species that show a high degree of association by tending to reoccur together.

I am indebted to Dr. Wolfgang Wieser for numerous discussions regarding this study and for making available to me some of his unpublished data concerning the meiofauna.

METHODS

From the initial survey (Sanders 1958), Buzzards Bay station R (41° 30′ N 70° 31′ W) was selected as being characteristic of the *Nephthys incisa-Nucula proxima* community, the fauna typically present in the softer sediments. While most of the sediment at this locality is composed largely of silts and clays, subsequent samplings did reveal patchiness. At approximately monthly intervals from February 1, 1956 to March 3, 1957, paired quantitative samples were taken at this station, which is situated in about 19 m of water, using a modification of the Forster anchor dredge. For the following 10-month period (April 1957 to January 1958) single samples were collected at appreciably longer intervals of time.

A small aliquot of the sediment from each sample was retained and analysed for particle-size composition. The sands were divided into various size categories using a graded series of sieves while the silts and clay fractions were determined by the pipetting method (Soil Survey Staff 1951). The results of the mechanical analyses are given in Table 1.

[1] Contribution No. 1033 from the Woods Hole Oceanographic Institution. This research was supported by National Science Foundation Grant G–4812.

TABLE 1. *The particle size analysis by per cent weight of the 24 samples from station R in Buzzards Bay*

Sample	Date	Gravel (8.0–4.0 mm)	Gravel (3.9–2.0 mm)	Sand (1.9–1.0 mm)	Sand (0.9–0.5 mm)	Sand (0.49–0.25 mm)	Sand (0.24–0.125 mm)	Sand (0.124–0.062 mm)	Silt (61–31μ)	Silt (30–16μ)	Silt (15.8 μ)	Silt (7–4 μ)	Silt (3–2 μ)	Clay (<2μ)	0/0<50μ
1RA	2/1/56		0.6	0.2	0.1	0.3	1.4	3.3	24.4	21.1	10.1	12.0	8.1	18.4	85.8
1RB	2/1/56		0.6	0.4	0.8	1.3	2.1	3.2	26.7	16.0	13.4	11.4	7.4	16.7	82.5
2RA	3/13/56			0.1	0.5	2.0	3.2	3.8	34.9	25.4	5.3	7.4	0.4	17.0	78.5
2RB	3/13/56			0.1	0.5	1.4	1.7	3.3	29.3	17.6	11.7	11.7	11.7	11.0	83.0
3RA	4/27/56			0.2	1.3	3.4	5.3	4.6	23.7	12.5	13.8	8.1	10.6	6.5	67.1
4RA	6/6/56				0.6	1.5	1.6	2.6	24.7	15.9	14.8	13.2	5.4	19.7	85.3
4RB	6/6/56		1.2	1.4	5.7	13.6	10.1	3.8	18.5	12.1	7.8	6.3	7.8	11.7	57.9
6RA	7/12/56	1.8	5.5	5.7	15.8	26.2	15.7	0.4	8.9	3.9	3.6	3.9	3.3	5.3	25.9
6RB	7/12/56				0.1	0.4	1.1	1.8	18.4	15.4	7.7	19.9	14.1	21.1	90.3
7RA	8/14/56			0.1	0.1	0.3	1.0	2.9	32.1	17.5	12.7	12.2	11.6	9.5	84.7
7RB	8/14/56			0.1	0.2	0.5	0.8	2.2	28.0	15.9	23.3	10.1	9.0	9.9	86.7
8RA	9/6/56			0.1	0.3	0.5	1.2	2.7	31.8	14.7	13.9	13.1	7.3	14.4	84.4
8RB	9/6/56		0.6	0.2	0.4	1.0	3.4	3.8	29.1	16.4	12.9	10.7	7.6	13.9	80.7
9RA	10/4/56		0.2	0.9	5.4	16.8	16.1	6.4	15.1	8.7	7.4	4.8	5.2	13.0	49.1
9RB	10/4/56		0.2	1.3	0.4	3.8	4.1	4.0	20.1	15.9	11.8	9.4	8.1	20.9	79.4
10RA	11/5/56			0.2	0.3	0.4	1.0	2.2	23.6	16.4	14.6	15.5	4.6	21.2	87.9
10RB	11/5/56			0.2	0.2	0.5	1.1	2.3	25.8	16.5	15.5	12.7	9.6	15.6	86.9
11RA	12/19/56		1.3	0.2	0.1	0.2	1.3	4.7	29.6	10.6	17.3	11.8	8.4	14.5	82.1
11RB	12/19/56			0.1	0.1	0.2	1.7	5.5	17.5	17.1	14.5	12.8	9.8	20.7	86.5
12RA	3/6/57			0.1	0.2	0.5	1.2	2.2	26.9	17.4	12.8	13.1	7.1	18.5	86.7
12RB	3/6/57		0.3	0.1	0.1	0.4	1.1	2.7	21.5	19.4	4.5	21.8	8.3	19.8	88.0
14R	6/20/57	0.1	0.2	0.4	1.0	2.2	3.7	5.3	25.4	14.9	12.7	9.7	0.2	24.2	78.4
15R	9/12/57			0.1	0.1	0.6	1.7	3.5	27.2	17.9	11.6	7.6	9.6	20.1	84.8
17R	1/30/58		0.3	0.2	0.4	1.2	1.8	3.0	25.0	14.7	12.2	11.0	7.9	22.3	84.6

The volume of the remainder of the sample was measured and the contents were passed through a screen with 0.2 mm mesh openings. The writer has demonstrated in previous studies using a series of graded screens, that very large fractions of the smaller sized species and most of the younger stages of the larger forms are not retained by the larger meshed screens. Obviously the loss of such faunal components would seriously compromise the validity of the results, and the use of a very fine seive is essential to this study.

The sediment

The predominant sediment components at this station are the silts and clays (see Table 1). Hereafter the term "% silt-clay" is used to mean the part of the sediment that is less than 50μ. This fraction is listed in the column at the extreme right in Table 1. Such a procedure was adopted in order to make the present investigation comparable with earlier studies (Sanders 1956, 1958). In most cases this silt-clay fraction makes up from 78 to 91% of the samples. If an undisturbed core is taken, the sediment can be clearly divided into two parts; a lower consolidated gray silty zone and a brown unconsolidated flocculent layer which comprises the top two or three centimeters of the substratum. The upper layer, as observed from core samples, is readily moved even by the slightest disturbance. Apparently this flocculent layer sloshes continuously back and forth over the consolidated sediment. The layer is appreciably thicker in such an environment than on coarser sediments and its accumulation here is most probably the result of the much reduced bottom currents characteristic of areas of deposition.

It is reasonable to suppose that the organic content of the flocculent layer should be appreciably higher than that of the underlying sediment since the "rain" of detritus from the water column as well as broken fragments of seaweed are being incorporated continuously into it. When samples from the two zones were placed in an oven at 500°C for 24 hours, 7.4% of the flocculent sample was burned as compared to 6.0% for the consolidated sediment sample. It would seem then that the amount of organic matter present in the two layers is not appreciably different and therefore the food resources for the deposit-feeders are approximately as abundant in one zone as in the other.

However, there may be a qualitative difference in the organic matter of the two zones. Some very preliminary measurements of free small chain sugars in the sediment seem highly suggestive. Detectable quantities of these substances were found only in the uppermost two centimeters. Such sugars should serve as a ready food source for the deposit-feeders. The organic matter in the consolidated sediments, being separated from the sources of detritus both in time and space, may be refractory or "fossil" in nature and thus less utilizable as a food source.

The distribution of the animals in the sediment column further supports such a conjecture, although anaerobic conditions deeper in the sediment may largely determine the pattern. A series of cores was split approximately at the interface between the two sediment zones. It should be emphasized that there was considerable contamination of one zone by the other. Yet nearly 75% of the animals were present in the upper zone and the remaining animals were limited to the top 2 cm of the consolidated layer. In another series of cores, none of the approximately one hundred specimens were found below a depth of 6 cm.

Faunal homogeneity

The numerical distribution of the fauna in the 24 samples comprising this series is shown in Table 2.

It is obviously necessary in a community study to demonstrate that the species components of the included samples show faunal similarity. To measure such affinities a matrix method, the "trellis diagram," was adopted. The percentage composition of the various species in each sample is determined. The samples are then arranged linearly at right angles along the ordinate and abcissa and all possible pairs are compared regarding their faunal content. The resultant value, the "index of affinity," is a measure of the percentage of the fauna common to a pair of samples and it is obtained by summing the smaller percentages of those species present in both samples. The table is then rearranged so that the samples with the highest values are brought into closer proximity. In this way the samples that are most ecologically alike are grouped together. The resulting diagram is shown in Figure 1.

It is immediately evident from this figure that samples 9RA, 6RA, 4RB and 3RA have low indices of affinity; that is, these 4 samples have a markedly different faunal composition from the other 20. Inspection of Table 1 suggests a possible reason for this difference. These same 4 samples agree in having much lower percentages of silt-clay. Whereas the sediments of the 20 samples with high indices of affinity range from 78.4 to 90.3% silt-clay, with a mean of 84.3%, the silt-clay percentages of the 4 samples in question are as follows: 3RA, 67.1; 4RB, 57.9; 9RA, 49.1; 6RA, 25.9. Another sample, 3RB, was omitted from consideration when it became evident that a part of sample 3PB from the sand bottom community series was accidentally included with it.

An average index of affinity value for all possible sample pairs (276) is 56.6. If the four aberrant samples are excluded, the value of the remaining paired combinations (190) becomes 69.3. In order to determine the significance of this number, a comparison with other values, either given in the literature or calculated from other studies, is summarized in Table 3.

From Wieser's (1960) investigation of the meiofauna of Buzzards Bay, the average correlation coefficient is given for all samples and for the four samples collected at station R with a homogenous sediment composition. Correlation values were calculated for the macrofaunal survey of Buzzards Bay (Sanders 1958) and indices are given for all stations included in the survey, those stations carrying a fauna largely characteristic of the *Ampelisca* spp. com-

TABLE 2. *Showing the actual number of various species in each of the R station samples*

Species	1RA	1RB	2RA	2RB	3RA	3RB	4RA	4RB	6RA	6RB	7RA	7RB	8RA	8RB	9RA	9RB	10RA	10RB	11RA	11RB	12RA	12RB	14R	15R	17R
Polychaeta:																									
Nephthys incisa	143	53	13	45	13	77	36	17	9	325	171	550	223	23	64	45	187	154	19	105	242	129	129	36	61
Ninoe nigripes	17	8	11	25	5	77	12	32	60	22	23	94	70	5	108	12	84	47	2	19	76	60	34	13	37
Lumbrinereis tenuis	93	24	2	25	3	15	2	10	15	15	16	32	8	—	4	3	10	18	2	10	11	5	2	4	8
Lumbrinereis fragilis						4		1														1			2
Arabella iricolor		1				5	1	2	5	1			2		1	1	1	2		1	4	3		1	1
Drilonereis longa		1		1		6			1												1	1			
Diopatra cuprea						1			2			2			3						1				
Harmothoe extenuata										1															
Pholoe minuta	1					6		1							3	3									
Syllis cornuta		1				7		1					1		2			1							
Phyllodoce arenae																									1
Paranaitis speciosa								1																	
Nereis grayei									1															1	
Glycera americana						4		1	1				1				2				5	17		1	2
Dorvillea caeca	57				22	14				1		1	9				3	4		14					
Scolopos acuta						1																			
Scalibregma inflatum						6															2		1		
Aricidea jeffreysii	12	4	4	5		36		2					1		2		1	1			10	2		2	1
Paraonis gracilis						1	1	1		1	1	4	1	2		2	2	4	1	1	1		1		1
Tharyx sp.	57		4	15	1	14	1	2	2		1	2			1	3	2			3	2	1	1	2	34
? *Spio filicornis*	63	1	4	5	1	7	1			2		2	5			2	7			3	1		1	3	
Capitellid	14	2	1			4			1	2						1	2					2		2	2
Spiochaetopterus oculatus																				1	1	2	2	1	1
Flabelligera affinis								1	4	1					1					1					1
Maldanopsis elongata	1		1							1	1		2	1	4	3	4	1			3	3			1
Praxillella praetermissa					1																				
Melinna cristata	2	4	1	1		1	1			1		1	1	1	3		3	2							
Ampharete arctica						4		1		20		2			1										1
? *Nicolea* sp.					1																				
Asabellides oculata																						1			
Polycirrus sp.				2		10				1															
Pectinaria gouldii	5	1		2		1		1				1			1		1	1							
Exogone sp.									1			1													
Oligochaeta:						1											2				1			5	

Turbellaria:																									
Anaperus sp.																					2				5
Flatworm b.															1			1							1
Nemertinea:																									
Tubolonus pellucidus		2	1	2		1	2									1				1				2	8
Micrura leidyi	6	1					1				1													1	1
Sipunculoidea:																									
Phascolon strombi				5				1															1		
Phascolosoma verrilli	1				1				4	1															
Enteropneusta:																									
Dolichoglossus Kowalevskii	1	3	2		1	2																			1
Tunicata:																									
Bostrichobranchus pilularis				5		6										1	3	2	2	6	3	1	14		
Anthozoa:																									
Cerianthus americana	4	1							1							2						1		1	1
Edwardsia sp.								1				1	1			1						1		1	1
Mollusca:																									
Nucula proxima	1940	1180	82	422	9	274	116	1	3	256	241	1170	743	40	3	184	504	290	8	105	560	280	592	551	295
Nucula delphinodonta	4	4		1			1	1	1	10		9	1		2		2					2	1	1	1
Yoldia limatula		17	1	5		2	1			4	1	6		1				1					1	1	
Macoma tenta		8	1	3		2		1		1						1	1	2					2		2
Callocardia morrhuana	73	49	3	43	4	19	5	21	2	31	29	128	8	2	1	3	18	12	2	10	36	19	22	1	11
Mulinia lateralis	3	2		3		1										1			1						
Astarte subaequilatera	2										1	3						1				1	1		1
Astarte undata								2								1									
Arca pextata	2					1						2					1				1				
Cerastoderma pinnulatum	1		1																		1	1			
Pandora gouldiana																					1				1
Commensal clam								5															3		
Cylichna orzya	138	117	17	63	2	36	2			29	14	35	5		1	5	28	29	1	2	38	25	13	16	71
Retusa caniculata	17	11	4	13		5	2			12	2	11	1		1	1	6	10	1	1	10	9	4	1	4
Acteon punctostriatus	3	2		3		1									1			1							
Turbonilla a	13	34	1	10	1	11	3	4		3		2	4			8	17	12	1	7	7	8	6	4	5
Turbonilla b	1								4						26						1	1			
Crepidula plana	1		1																		1	1			
Anachis avara															1										

TABLE 2. *Continued*

Species	1RA	1RB	2RA	2RB	3RA	3RB	4RA	4RB	6RA	6RB	7RA	7RB	8RA	8RB	9RA	9RB	10RA	10RB	11RA	11RB	12RA	12RB	14R	15R	17R
Mitrella lunata		1								3					3										1
Nassarius triviattatus	1	3		7					4	2	5	11	3	1	3		1			1		2	3	5	
Eupleura caudata			1	2				1		1							1					1		1	
Bela turricola	2	2		1								1					1					1			
Natica pusilla				1				1										3							
Polynices triseriata		1		1		1							1								1				
Gastropod a												1													
Crustacea:																									
Hutchinsoniella macracantha	2			2	23	3	—75—				15	1	—68—		—25—		—58—		1	28	—68—			10	2
Ampelisca spinipes	3	5		1		168	1	17	2		2	9			32	4	13	10	1	2			1	1	1
Ampelisca macrocephala						36										1				1					
Byblis serrata						9		2																	
Leptocheirus pinguis								34	84						3						1		4		
Unciola irrorata	18	1			9			7	1						17	2		1		2	15	1	1		17
Corophium acherusicum	1								1			1				1	1		1		1	2		1	
Corophium insidiosum									1										1						
Corophium tuberculatum															2										
Siphonoecetes smithianus						1																			
Lysianopsis alba	1																								
Stenothoe cypris	1																								
Batea secunda					2													1							
Hyperia galba	1																								
Aeginella longicornis									1																1
Amphipod A								2	1								1								
Edotea triloba	1							1			1	1				1	7	2		3	1	4	2		14
Oxyurostylis smithi	1	2		1		1						1											1	3	
Diastylis sp.	1	1		1		1										1		1					2	1	
Eudorella emarginata	1																								
Petalosarsia declivus	1																								
Pagurus annulipes									1			1													
Libinia sp.	1																								
Pinnixia cylindrica									1															1	
Pycnogonida:																									
Anoplodactylus lentus												1									1				1
Ophiuroidea:																				1					

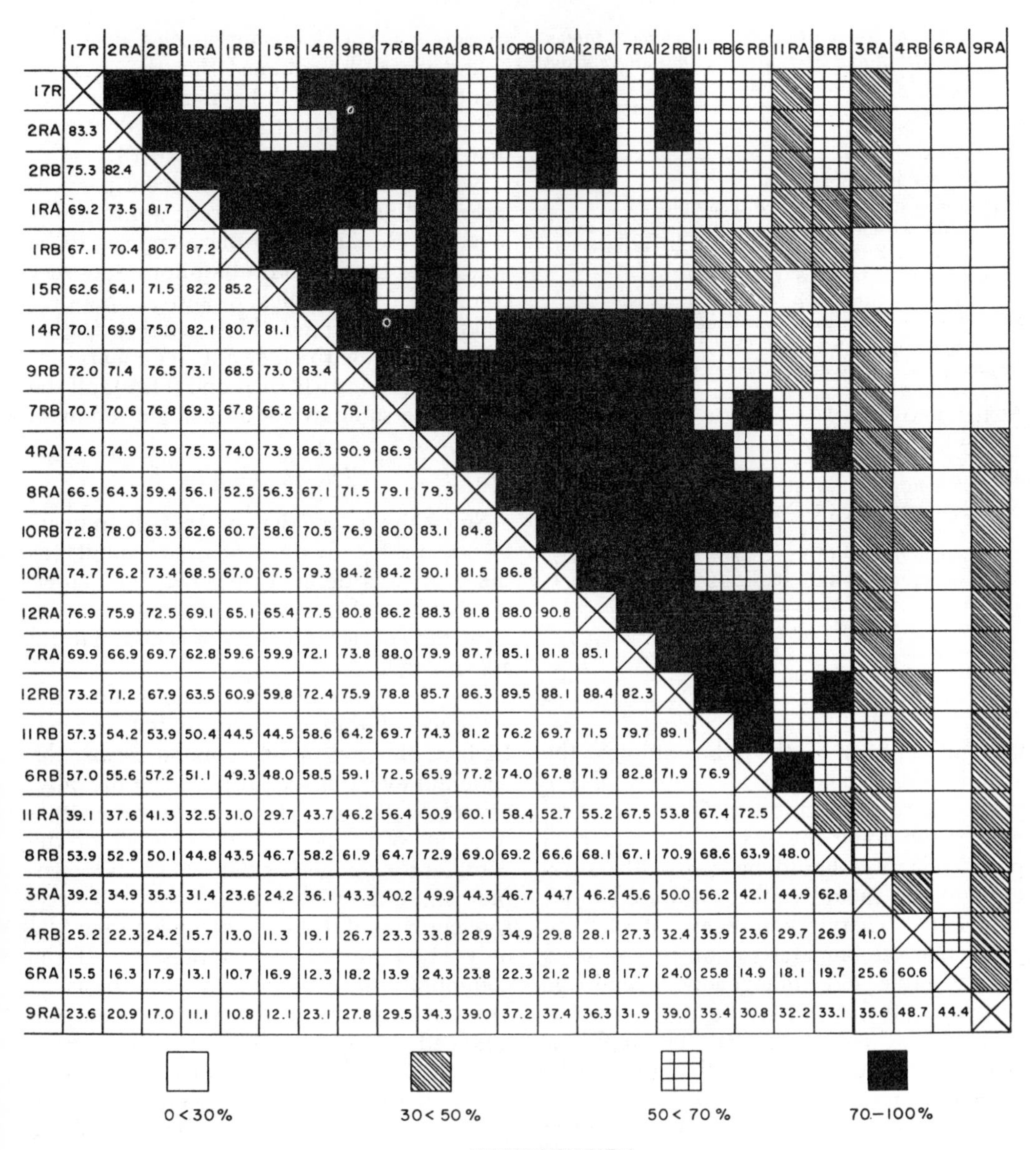

	17R	2RA	2RB	1RA	1RB	15R	14R	9RB	7RB	4RA	8RA	10RB	10RA	12RA	7RA	12RB	11RB	6RB	11RA	8RB	3RA	4RB	6RA	9RA
17R																								
2RA	83.3																							
2RB	75.3	82.4																						
1RA	69.2	73.5	81.7																					
1RB	67.1	70.4	80.7	87.2																				
15R	62.6	64.1	71.5	82.2	85.2																			
14R	70.1	69.9	75.0	82.1	80.7	81.1																		
9RB	72.0	71.4	76.5	73.1	68.5	73.0	83.4																	
7RB	70.7	70.6	76.8	69.3	67.8	66.2	81.2	79.1																
4RA	74.6	74.9	75.9	75.3	74.0	73.9	86.3	90.9	86.9															
8RA	66.5	64.3	59.4	56.1	52.5	56.3	67.1	71.5	79.1	79.3														
10RB	72.8	78.0	63.3	62.6	60.7	58.6	70.5	76.9	80.0	83.1	84.8													
10RA	74.7	76.2	73.4	68.5	67.0	67.5	79.3	84.2	84.2	90.1	81.5	86.8												
12RA	76.9	75.9	72.5	69.1	65.1	65.4	77.5	80.8	86.2	88.3	81.8	88.0	90.8											
7RA	69.9	66.9	69.7	62.8	59.6	59.9	72.1	73.8	88.0	79.9	87.7	85.1	81.8	85.1										
12RB	73.2	71.2	67.9	63.5	60.9	59.8	72.4	75.9	78.8	85.7	86.3	89.5	88.1	88.4	82.3									
11RB	57.3	54.2	53.9	50.4	44.5	44.5	58.6	64.2	69.7	74.3	81.2	76.2	69.7	71.5	79.7	89.1								
6RB	57.0	55.6	57.2	51.1	49.3	48.0	58.5	59.1	72.5	65.9	77.2	74.0	67.8	71.9	82.8	71.9	76.9							
11RA	39.1	37.6	41.3	32.5	31.0	29.7	43.7	46.2	56.4	50.9	60.1	58.4	52.7	55.2	67.5	53.8	67.4	72.5						
8RB	53.9	52.9	50.1	44.8	43.5	46.7	58.2	61.9	64.7	72.9	69.0	69.2	66.6	68.1	67.1	70.9	68.6	63.9	48.0					
3RA	39.2	34.9	35.3	31.4	23.6	24.2	36.1	43.3	40.2	49.9	44.3	46.7	44.7	46.2	45.6	50.0	56.2	42.1	44.9	62.8				
4RB	25.2	22.3	24.2	15.7	13.0	11.3	19.1	26.7	23.3	33.8	28.9	34.9	29.8	28.1	27.3	32.4	35.9	23.6	29.7	26.9	41.0			
6RA	15.5	16.3	17.9	13.1	10.7	16.9	12.3	18.2	13.9	24.3	23.8	22.3	21.2	18.8	17.7	24.0	25.8	14.9	18.1	19.7	25.6	60.6		
9RA	23.6	20.9	17.0	11.1	10.8	12.1	23.1	27.8	29.5	34.3	39.0	37.2	37.4	36.3	31.9	39.0	35.4	30.8	32.2	33.1	35.6	48.7	44.4	

FIG. 1. A comparison of the average faunal index of affinity at station R with indices from other studies.

munity, and the stations predominantly with faunal elements present in the *Nephthys incisa-Nucula proxima* community. Renkonnen (1944) used the trellis diagram for a study of beetles inhabiting the supralittoral zone of a lake in southwest Finland. From his analysis he recognized 5 communities; the average correlation values within each of his communities are also given in Table 3. Finally Macfadyen (1954) used the same technique, together with other methods of analysis, in his investigation of the invertebrate fauna of Jan Mayen Island (East Greenland). Measuring the affinites among the Collembola species, he found two associations and the average correlation value within each association is included.

Table 3 reveals that the present study gives by far the highest index of affinity. This value was obtained despite the fact

TABLE 3. *A comparison of the average faunal index of affinity at station R with indices from other studies*

	Index of Affinity
Macrofauna at station R (all samples)	56.6
Macrofauna at station R less 5 samples of aberrant sediment composition	69.3
Meiofauna of Buzzards Bay (Wieser 1960)	27.2
Meiofauna of R samples with similar sediment composition (*op. cit.*)	55.7
Macrofauna of Buzzards Bay (Sanders 1958)	21.4
Ampelisca spp. community (Sanders 1958)	31.2
Nephthys incisa–Nucula proxima community (Sanders 1958)	37.2
Beetle fauna from lake shore in SW Finland (Renkonnen 1944)	
Total of 40 stations	24.3
Community 1 (1 station)	—
Community 2 (4 stations)	54.2
Community 3 (23 stations)	33.5
Community 4 (15 stations)	34.2
Community 5 (16 stations)	36.9
Collembola of Jan Mayen Island (Macfadyen 1954)	
Association 1 (11 species)	24.7
Association 2 (9 species)	24.6

that, in contrast to the other studies, the samples were taken at all seasons of the year. Because recruitment is not synchronous in the dominant species, the numerical proportionality should be constantly altering and one should expect a lowered index of affinity. Therefore, it seems valid to conclude that the fauna at station R represents a very stable and homogenous assemblage except in the aberrant sediments. However, it should be mentioned that station R was selected because it was thought to be faunally homogeneous.

There are certain weaknesses inherent in this method of analysis. No discrimination is made against ubiquitous species (see Macfayden 1954). Furthermore, the method is not quantitative and unless care is taken gross errors may be introduced. For example, species A at a station X carrying a sparse fauna may comprise 80% of the population while at another station, Y, with a more numerous fauna the same species may constitute only 2% of the population. It is conceivable, however, that species A may be numerically more common at Station Y. Yet by the method of analysis used, species A is the dominant species at station X and it will be assumed that the environmental conditions will be more favorable at a station (X) where the species is dominant than at a station where it is not (Y).

Neither of these criticisms seem particularly pertinent to the present study, however. In Buzzards Bay there appears to be no species that is both ubiquitous and very common (see Tables 3 and 4, Sanders 1958). Concerning the second point, if the samples with aberrant sediments are omitted (3RA, 4RB, 6RA and 9 RA), all the remaining samples are of a single sediment type.

The community

Since the association of animals appears to possess the features of a community as the term was defined earlier in this paper, it would be of interest to attempt to determine the species structure of such a natural assemblage.

The data presented in Table 4 show the evidence of dominance by a few species. In fact, the two most common forms, the lamellibranch *Nucula proxima* and the polychaete *Nephthys incisa* make up respectively about 59% and 17% by number of the community. Together the two species comprise more than 76% of the fauna. Ninety-five per cent of the assemblage is formed by only 11 of the 79 species present. If the community is evaluated on the basis of biomass rather than number the same pronounced dominance by a few species is evident. Only 13 species make up 95% of the biomass. The per cent composition by weight of these species are shown in Table 4. Only some of the numerically more abundant species contribute significantly to the biomass (*Nephthys, Nucula, Callocardia* and *Ninoe*) while certain numerically rare forms such as the tunicate *Bostrichobranchus pilularis,* the nemertean *Micrura leidyi,* and the polychaetes *Arabella iricolor* and *Lumbrinereis fragilis,* because of their large individual size, represent major biomass components of the fauna.

Such a pattern could result from the bias of a few unusually large samples. To test such a possibility, each sample was ranked

TABLE 4. *The structure of the* Nephthys incisa–Nucula proxima *community at station R. The species present in the 20 samples characterized by "typical" sediment composition (78–91% silt-clay) are listed according to their numerical abundance. The totals given (column 2) are the sum of the numbers of individuals of the given species in the 20 samples. Columns 3, 4, 5, 6 and 7 list respectively the cumulative per cent by number, the per cent composition by number, the dry weights, the per cent composition by weight and feeding type.*

Rank by no.	Species	No.	Cumulative % by no.	% of fauna by no.	Dry wt. (g.)	% by wt.	Rank by wt.	Feeding Type[1]
1	*Nucula proxima*	9195	58.86	58.86	2.9694	13.98	4	SDF
2	*Nephthys incisa*	2689	76.07	17.21	6.1489	28.95	1	NSDF
3	*Ninoe nigripes*	671	80.37	4.30	0.3791	1.78	8	NSDF
4	*Cylichna orzya*	647	84.51	4.14	0.0782			?
5	*Callocardia morrhuana*	525	87.87	3.36	3.1803	14.97	3	SF
6	*Hutchinsoniella macracantha*	307	89.84	1.97	0.0583			NSDF
7	*Lumbrinereis tenuis*	286	91.67	1.83	0.0620			NSDF
8	*Turbonilla* a	145	92.60	0.93	0.0053			?
9	?*Spio filicornis*	135	93.46	0.86	0.0186			SDF
10	*Retusa caniculata*	120	94.23	0.77	0.0440			?C
11	*Dorvillea caeca*	103	94.89	0.66	0.0099			?C
12	*Tharyx* sp.	89	95.46	0.57	0.0100			SDF
13	*Unciola irrorata*	58	95.83	0.37	0.0286			SDF
14	*Ampelisca spinipes*	54	96.18	0.35	0.0324			SF
15	*Nassarius triviattatus*	45	96.47	0.29	0.1726	0.81	9	NSDF
16	*Bostrichobranchus pilularis*	40	96.73	0.26	4.9878	23.49	2	SF
17	*Yoldia limatula*	39	96.98	0.25	0.1084	0.51	13	SDF
18	*Nucula delphinodonta*	37	97.22	0.24	0.0060			SDF
19	*Edotea triloba*	37	97.46	0.24	0.0381			S
20	*Paraonis gracilis*	31	97.66	0.20	0.0032			SDF
21	*Melinna cristata*	25	97.82	0.16	0.0461			SDF
22	Capitellid	24	97.97	0.15	0.0007			NSDF
23	*Ampharete arctica*	24	98.12	0.15	0.0154			SDF
24	*Macoma tenta*	22	98.26	0:14	0.0094			SDF
25	*Arabella iricolor*	19	98.36	0.12	0.8000	3.77	6	?C
26	*Tubolonus pellucidus*	19	98.50	0.12	0.0019			?C
27	*Maldanopsis elongata*	17	98.61	0.11	0.0488			NSDF
28	*Mulinia lateralis*	12	98.69	0.08	0.0059			SF
29	*Pectinaria gouldii*	11	98.76	0.07	0.0185			NSDF
30	*Micrura leidyi*	11	98.83	0.07	1.0263	4.83	5	C
31	*Oxyurostylis smithi*	9	98.89	0.06	0.0071			
32	*Cerianthus americana*	9	98.95	0.06	0.1578	0.74	10	
33	*Corophium acherusicum*	9	99.01	0.06	0.0049			
34	*Diastylis* sp.	8	99.06	0.05	0.0063			
35	*Acteon punctostriatus*	8	99.11	0.05	0.0004			
36	*Dolichoglossus kowalevskii*	8	99.16	0.05	0.0124			
37	*Spiochaetopterus oculatus*	8	99.21	0.05	0.0063			
38	*Bela turricola*	8	99.26	0.05	0.0053			
39	*Astarte subaequilatera*	8	99.31	0.05	0.0029			
40	*Eupleura caudata*	7	99.35	0.04	0.0069			
41	*Leptocheirus pinguis*	6	99.39	0.04	0.0213			
42	*Phascolon strombi*	6	99.43	0.04	0.0352			
43	*Anaperus* sp.	6	99.47	0.04	0.0004			
44	Oligochaeta	6	99.51	0.04	0.0002			
45	*Arca pextata*	6	99.55	0.04	0.0027			
46	*Mitrella lunata*	5	99.58	0.03	0.0033			
47	*Edwardsia* sp.	5	99.61	0.03	0.0500			
48	*Diopatra cuprea*	4	99.64	0.03	0.0468			
49	*Flabelligera affinis*	4	99.67	0.03	0.1403	0.66	11	
50	*Natica pusilla*	4	99.70	0.03	0.0017			
51	*Cerastoderma pinnulatum*	4	99.73	0.03	0.1382	0.65	12	
52	*Lumbrinereis fragilis*	3	99.75	0.02	0.3954	1.86	7	NSDF
53	*Glycera americana*	3	99.77	0.02	0.0480			
54	*Pholoe minuta*	3	99.79	0.02	0.0012			

TABLE 4. *Continued*

Rank by no.	Species	No.	Cumulative % by no.	% of fauna by no.	Dry wt. (g.)	% by wt.	Rank by wt.	Feeding type[1]
55	Flatworm b	3	99.81	0.02	0.0008			
56	*Anoplodactylus lentus*	3	99.83	0.02	0.0008			
57	*Polynices triseriata*	3	99.85	0.02	0.0047			
58	*Syllis cornuta*	2	99.86	0.01	0.0009			
59	*Polycirrus* sp.	2	99.87	0.01	0.0033			
60	*Phascolosoma verrilli*	2	99.88	0.01	0.0007			
61	*Ampelisca macrocephala*	2	99.89	0.01	0.0160			
62	*Turbonilla* b	2	99.90	0.01	0.0001			
63	*Pandora gouldiana*	2	99.91	0.01	0.0062			
64	*Crepidula plana*	2	99.92	0.01	0.0021			
65	*Batea secunda*	1						
66	*Astarte undata*	1	99.93	0.01				
67	*Nereis grayei*	1						
68	*Harmothoe extenuata*	1	99.94	0.01				
69	*Phyllodoce arenae*	1						
70	?*Nicolea* sp.	1	99.95	0.01				
71	*Asabellides oculata*	1						
72	*Corophium insidiosum*	1	99.96	0.01				
73	*Eudorella emarginata*	1			0.0150			
74	*Petalosarsia emarginata*	1	99.97	0.01				
75	*Pagurus annulipes*	1						
76	*Libinia* sp.	1	99.98	0.01				
77	*Pinnixia cylindrica*	1						
78	Ophiuroid	1	99.99	0.01				
79	Gastropod a	1						
	Total	15,622			21.2374			
	Number/m^2	8,985						
	Biomass/m^2				12.1858			

[1] SDF = Selective deposit-feeder
NSDF = Non-selective deposit-feeder
SF = Suspension-feeder
C = Carnivore
S = Scavenger

separately (Fager 1957), thus giving equal weight to each of the samples. Table 5 lists the frequency of a given species appearing as one of the 10 most abundant species (maximum frequency = 20); and the quantitative importance of the species. This latter factor is determined by ranking the species from 1 to 10 by abundance within each sample. A rank of 1 is given a value of 10 points; a rank of 2 equals 9 points; 3 is equivalent to 8 points,....., and a rank of 10 equals one point. Thus if a species is ranked first in all 20 samples, it will have 200 points, the highest possible score.

The results in Table 5 verify the findings given in Table 4. *Nucula proxima,* with almost the maximum possible points, is the most abundant species in 18 samples and is second most common in the remaining two giving a score of 198. *Nephthys incisa,* ranking first in two, second in 13 and third in 5 samples has the very high score of 177. Other species with high scores are the polychaete, *Ninoe nigripes* (132); the gastropod, *Cylichna orzya* (177); the lamellibranch, *Callocardia morrhuana* (109); the crustacean, *Hutchinsoniella macracantha* (83); and the polychaete, *Lumbrinereis tenuis* (62). Each of the above species is present in at least 80% and is one of the 10 most common species in 65–100% of the samples. Two other gastropods, *Turbonilla* a and *Retusa caniculata,* with somewhat lower scores (38 and 30) are present in 18 and 19 of the samples respectively. The analysis demonstrates, therefore, that these 9 species represent a very significant and constant part of the biological environment of this community. It follows that some knowledge of the niches of these species

TABLE 5. *Faunal frequency evaluation of the 20 samples with homogeneous sediment composition*

Species	1	2	3	4	5	6	7	8	9	10	Frequency	Frequency as one of ten most common species	Biological index value
Nucula proxima	18	2									20	20	198
Nephthys incisa	2	13	5								20	20	177
Ninoe nigripes			6	9	3				1	1	20	20	132
Cylichna orzya		4	2	4	2	2	3	1			19	18	117
Callocardia morrhuana			4	3	4	4	1	2	1		20	19	109
Hutchinsoniella macracantha		1	3	2	4	1	1	1			17	13	83
Lumbrinereis tenuis				1	1	5	2	3	3	1	19	16	62
Turbonilla a				1	1	1	2	2	2	2	18	11	38
Retusa caniculata							1	6	3	2	19	12	30
?*Spio filicornis*					1	2			1	2	16	6	20
Dorvillea capeca					2		2				11	4	20
Nassarius triviattatus						1	1	2		2	13	6	17
Ampelisca spinipes							1	1	4		14	6	15
Bostrichobranchus pilularis					1	1			1		10	3	13
Paraonis gracilis					1	1				1	15	3	12
Tharyx sp.							2	1		1	12	4	12
Unciola irrorata						1	1		1		9	3	11
Yoldia limatula							1	1		1	11	3	8
Edotea triloba							1			1	11	2	5
Ampharete arctica						1					2	1	5

would help in understanding the community structure.

The *Nephthys incisa-Nucula proxima* community can be considered to belong to the category defined by Elton (1946) as "animal communities in which species live on comparatively few basic sources of food." Besides aquatic benthic deposit-and suspension-feeders, he includes log-dwelling herbivores, soil animals, zooplankton, drift-line animals and blood-sucking ectoparasites in this category. In the *Nephthys incisa-Nucula proxima* community a minimum of 87.5% of the animals by number are deposit-feeders, obtaining their food from detritus either on or in the sediment. The other primary feeding type, the suspension-feeder, is present in much smaller numbers (4.3%) as might be expected from the nature of the sediment (see Sanders 1958).

Nucula proxima, the overwhelming dominant member of the community, normally lies just below the surface of the sediment feeding on the sediment immediately beneath it by long appendages derived from the palp. The inner surfaces of these structures are grooved and lined with ciliated epithelium (Drew 1899; Yonge 1939). Only the fine particles are moved along the groove to the palps where they are passed again by cilia to the mouth. *Nucula* then is a selective deposit-feeder.

The genus to which the other very abundant species, *Nephthys incisa*, belongs is considered to be carnivorous (Blegvad 1914; Smith 1932; Mare 1942; Holme 1949; Savilov 1957; Longhurst 1958). Yet in an earlier study in Long Island Sound (Sanders 1956) the writer has shown that the gut contents of those individuals examined were filled with sediments and there was no evidence of animal remains, indicating that the animal was primarily a deposit-feeder. Turpaeva (1957) also regards members of this genus as deposit-feeders.

These conclusions were further verified in the present study. Generally the gut contents were of two sorts. They either consisted of fine silt particles indicating the animal was ingesting the sediment from the lower gray consolidated zone or of large detrital particles, a few sand grains, intact diatoms including such benthic forms as *Gyrosigma* and various naviculoids as well as empty diatom frustules of both pelagic and benthic forms. Apparenly this actively burrowing form swallows the medium through which it moves indiscriminately ingesting sediment from both zones of the habitat. This does not imply that *Nephthys incisa* cannot and would not act as a carnivore under certain circumstances. However in the community under discussion it represents such a significant fraction of the fauna both in regard to number and biomass (see Table 4) that it would not be feasible for the animal to be primarily predaceous and, indeed, the contents of the gut would certainly not support any such idea.

Third in abundance in this association is another polychaete, *Ninoe nigripes,* belonging to the family Lumbrinereidae. Being an errant polychaete with well developed jaws, and because other lumbrinereids are considered to be predators, this form too should be carnivorous. However, as in the case of *Nephthys,* stomach analysis revealed that the species is a non-selective deposit-feeder but unlike *Nephthys* the gray silty sediments were absent from the gut. The evidence therefore indicates that *Ninoe* feeds almost exclusively from the brown flocculent zone since the gut contents usually reveal large pieces of detritus, diatom frustules, benthic diatoms and few sand grains. Furthermore, *Ninoe* differs from *Nephthys* in being sedentary. It constructs a simple slime tube and the slow movement of the flocculent layer replenishes the source of its food.

The minute opistobranch gastropod, *Cylichna orzya,* is numerically the fourth most common species. In contrast to the previously mentioned forms, this animal is found only at the surface of the sediment. Because of its small size, the gut contents could not be satisfactorily analyzed.

The lamellibranch, *Callocardia morrhuana,* fifth in rank by number and third in rank by weight, is a representative of the other major primary consumer group, the suspension-feeders. It draws in its food by means of the inhalant siphon from the suspended matter in the overlying water column.

The sixth ranking species, surprisingly, is the remarkably primitive and supposedly very rare crustacean, *Hutchinsoniella macracantha.* The limb movements of this animal are metachronal and, as the space between two consecutive appendages enlarge, a suction chamber is formed from the interlimb space. The deposited material that is temporarily stirred up by the action of the large terminal spines on the inner ramus of the limb is sucked into this space. Subsequently, this material is squeezed into the mid-ventral food groove, where it is passed forward by spine-like processes at the base of the thoracic and maxillary limbs. The material in the food groove, gut, and in the fecal pellets was shown to be composed of detritus, diatom frustules, and sand grains. It is therefore a non-selective deposit-feeder eating in the flocculent non-consolidated sediment. A detailed study of the functional morphology of *Hutchinsoniella* is at present being undertaken by the writer.

The 6 species just discussed represent almost 90% of the entire macrofaunal assemblage. As they differ in spatial distribution, in their diet and methods of obtaining it, the niches of these animals are clearly different.

Another polychaete, *Lumbrinereis tenuis,* is ranked seventh in abundance in this community. Members of the genus *Lumbrinereis* (=*Lumbriconereis*) are classified as carnivores (Smith 1932, Savilov 1957, Longhurst 1958). The few gut contents of the species examined contained detritus. Like *Ninoe* the animal is sedentary and ensheathes itself in a slime tube. Furthermore, it feeds in the same manner and on the same sort of food as *Ninoe.* The few factors evaluated show no difference in the niches of these two lumbrinereid polychaetes.

The small pyramidellid gastropod designated as *Turbonilla* a, the eighth most common species, presents an enigma. All pyramidellids are ectoparasites (Fretter and Graham 1949); each species has a specific host, a sessile organism that protrudes part of its body to feed. All known hosts are either polychaetes and coelenterates that feed by tentacles or molluscs which are with a single exception epifaunal. In the former cases the pyramidellid pierces the tentacles or body wall with a stylet at the tip of a long eversible proboscis and sucks in the blood or body fluids; in the latter situation it is the mantle that is pierced. There is, however, one report where the siphon of a lamellibranch was parasitized (Medcof 1948). As indicated in an earlier paper (Sanders 1958) it is difficult to accept this specific pyramidellid as being an ectoparasite since in slightly coarser sediments (50–70% silt-clay) it may comprise 25–50% of the fauna. Furthermore, there seem to be no suitable hosts among the species living in the same habitat, when abundance, size and mode of life are considered. The lamellibranch *Callocardia* might provide a possible exception; however the great majority of these clams are of the same small size as their possible parasite, a highly unlikely relationship. It is tentatively suggested that, in this environment, *Turbonilla* a is a deposit-feeder, using its buccal pump to draw in the flocculent superficial sediment.

The small spionid polychaete, ?*Spio filicornis,* the ninth most abundant member of the community, utilizes still another mode of feeding. Two large tentacles on the head protrude above the substratum and move over the surface of the surrounding sediment. A ciliated groove situated along the length of the tentacle draws in and carries fine superficial particles to the mouth. The animal therefore is a selective deposit-feeder exploiting only the finer deposited material at the very surface of the sediment.

The remaining 6.5% of the fauna consists of 70 species, none of which make an appreciable numerical contribution to the community. Some members of this group use the same methods of feeding as the common species. However, the significant finding in this study is that definite niche differences can be demonstrated for those few species that make up the vast majority of the fauna.

This study also demonstrates the danger of assuming a carnivorous feeding habit, particularly in the case of polychaetes, from morphology or by feeding experiments in the laboratory. Because of this, two other predatory polychaetes, *Arabella iricolor* and *Lumbrinereis fragilis* were examined to provide data on their food. While both species are of little importance numerically, they are relatively large animals and rank fifth and seventh respectively in regard to biomass (see Table 4). The gut contents of the two specimens of *Arabella* were empty while the single representative of *Lumbrinereis fragilis* contained detritus, numerous centric diatoms and sand grains.

Evidence has now been presented which indicates that biological interaction, or perhaps a better expression would be biological accommodation, plays an important part in the structure of the animal assemblage at station R. It should not be inferred that such a community therefore represents a sharply circumscribed group of species, each with identical distribution patterns. On the contrary, frequently both the sediment ranges of the species and the sediment composition where the maximal numbers are found vary significantly (see Sanders 1958). In fact, the species components of the benthic fauna in Buzzards Bay constitute a continuum varying with the gradual change in sediment composition. Implicit in such distribution patterns is the effect of physical factors. However, as demonstrated in the present study, at a given point (station R) on the sediment spectrum the species structure of the association is similar even when the sampling is done at various times of the year. The interpretation made is that accommodations among the species, biological effects, play a major role in determining the structure of a community. Thus the faunal proportions will vary at different locations on the sediment spectrum in Buzzards Bay and at any point on the spectrum the community structure will be determined by both physical and biological factors. Obviously the gradients of physical factors other than sediment composition play a role. For example, the lamellibranch, *Macoma tenta,* appears to be much more abundant in the same type of sediment in slightly shallower environments. Because of the greater abundance of *Macoma,* it would be interesting to observe what modifications have occurred in the structure of the community at such a location in contrast to that found at station R.

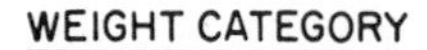

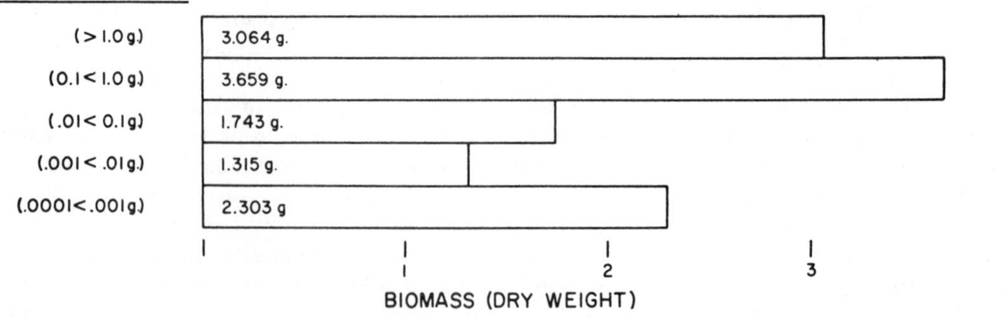

FIG. 2. The relationship of numbers (left) to weight (right) for the benthic community at station R.

For further comments on the subject of biological versus physical regulation see Jones (1950), Hedgpeth (1954) and Longhurst (1958).

Animal numbers and animal weight

Another aspect of the structure of an animal community is the relationship of numbers to weight. It is a common assumption that the smaller the animals the more abundant they will be. Therefore, if the individuals are grouped by number into size or weight categories, a pyramid can be constructed with the smallest size or weight category forming the base and the largest category the apex of the pyramid. Conversely, if the total weight rather than the number is calculated for each of the categories, an inverted pyramid results, with the largest biomass present in the largest size category and the smallest biomass in the smallest size group. However, there is little field evidence in support of this.

The present data lend themselves to the comparison of numbers to weight (Fig. 2). Only the data from the 20 samples with high indices of faunal affinity (*i.e.*, those with similar sediment composition) were used. The intervals chosen were logarithmic rather than arithmetic, thus each category is an order of magnitude larger than the preceding category. Those animals weighing less than 0.0001 g are only partially sampled by a screen with 0.2 mm mesh and therefore those smaller than 0.0001 g are not included in Figure 2. Thus the analysis has validity only for those organisms weighing 0.0001 g or more. Within the range from 0.0001 to more than 1.0 g, there is a constant decrease in numbers from one interval to another (see Fig. 2) which is in agreement with the accepted concept. The average increase in number of individuals is $\times$ 8.5 as one moves from one category to that containing the next smallest animals. A single category, the interval 0.0001–0.001 g, includes 69.4% of the fauna by number.

When the same treatment is applied to total weight or biomass within each of the categories, no trend is evident (see Fig. 3) and the expected inverted pyramid does not materialize. Despite the fact that this situation differs from the theoretical, one should be cautious about generalizing from the results. Obviously the biomass histogram describes the structure of this particular community and to a considerable degree, the form of the histogram reflects the nature of the dominant species. The larger than expected biomass value in the 0.0001 $<$0.001 g interval is undoubtedly due to concentrations of adult *Nucula* and juvenile *Nephthys*. Possibly different kinds of animal communities can be described by characteristic biomass histograms and it may prove feasible to use such histograms as another tool in describing communities.

The present analysis brings out the uncertainty of our knowledge of the feeding habits of many benthic invertebrates and, therefore, a precise definition of number and weight relationships from the trophic aspect cannot be attempted at this time. It

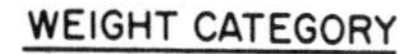

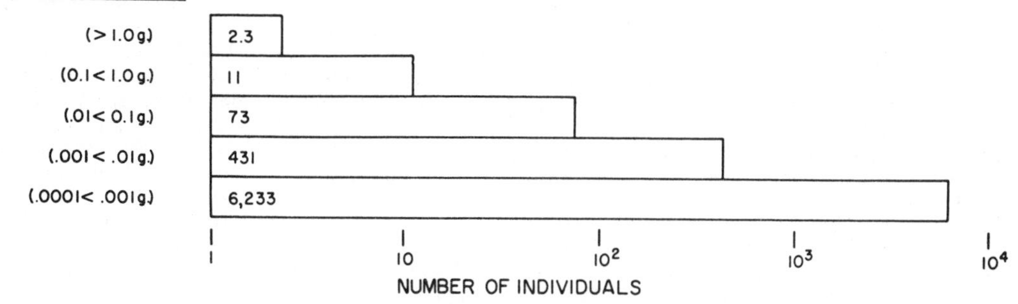

FIG. 3. The relationship of weight categories (left) to biomass (right) for the benthic community at station R.

can be stated, however, that true carnivores make up, at most, about 8% of the fauna by number. The actual percentage is probably much lower.

Finally, some mention should be made of the meiofauna, such as nematodes, kinorhynchs, ostracods and benthic copepods and the contribution it makes to the histograms of numbers and weights at station R. The 4 samples from R station that showed the typical sediment composition in Wieser's study (1960) are considered.

There are about 930,000 meiofaunal animals per m^2, which indicates that they are about a hundred times more numerous than the macrofauna. However, their biomass of 0.3662 g/m^2 is only 1/33 the weight of the macrofauna. The most overwhelmingly abundant group is the nematodes, which comprise 97.66% of the meiofauna by number.

Because it is very difficult to obtain individual weights for such small animals, average weights were used. Therefore the weight ranges within each group cannot be determined and, as a result, it is impossible to assign the proper fraction of each group to its correct weight category.

Wieser found that the average weight of a nematode is 0.00028μg, a kinorhynch, 0.003μg, a benthic copepod, 0.0017μg, and an ostracod, 0.018μg. Thus there is little overlap between the macro- and the meiofauna. Obviously, since the meiofauna consists overwhelmingly of nematodes and the average weight of these animals is appreciably less than the other groups, much of the meiofaunal biomass must be present in the smaller weight categories. Therefore, like the macrofauna, it would seem highly unlikely that a biomass pyramid of the meiofauna at station R would approximate the theoretical inverted form.

In the light of what has been shown regarding animal weights and numbers, it may be convenient here to consider the relative validity of numbers and weights in quantitative faunal studies. Generally speaking, the larger animals are less abundant than smaller animals. Only 13.3 specimens in the largest two categories (see Fig. 2), representing less than 0.15% of the fauna by number, constitute 55.17% of the entire biomass. Since the latter figure represents the average percentage of 20 samples, the effect of these animals on the biomass of individual samples would be much more extreme. It would seem that the presence or absence of the rare, randomly distributed large animals effectively determines the biomass of a sample (see also Sanders 1956). Unless this component is separately assayed, comparative biomass values become relatively meaningless. Because of this difficulty, the writer feels that animal numbers are the more valid measurement providing, of course, that a standard screen with small enough apertures is used so that an adequate number of animals is captured.

REFERENCES

BLEGVAD, H. 1914. Food and conditions of nourishment among the communities of invertebrate animals found on or in the sea bottom in Danish waters. Rep. Danish biol. Sta., **22:** 41–78.

DREW, G. A. 1899. Some observations on the habits, anatomy and embryology of members of the Prosobranchia. Anat. Anz., **15:** 493–519.

ELTON, C. 1946. Competition and the structure of ecological communities. J. Anim. Ecol., **15:** 54–68.

FAGER, E. W. 1957. Determination and analysis of recurrent groups. Ecol., **38:** 586–595.

FRETTER, V., AND A. GRAHAM. 1949. The structure and mode of life of the Pyramidellidae, parasitic opistobranchs. J. Mar. Biol. Ass. U. K., **28:** 493–532.

HEDGPETH, J. W. 1954. Bottom communities of the Gulf of Mexico. U. S. Fish and Wildlife Service, Fishery Bull., 89, **55:** 203–214.

HOLME, N. A. 1949. The fauna of the sand and mud banks near the mouth of the Exe. J. Mar. Biol. Ass. U. K., **28:** 189–237.

JONES, N. S. 1950. Marine bottom communities. Biol. Rev., **25:** 283–313.

LONGHURST, A. R. 1958. An ecological survey of the west African marine benthos. Colonial Office. Fish. Publ., **11:** 1–102.

MACFADYEN, A. 1954. The invertebrate fauna of Jan Mayen Island (East Greenland). J. Anim. Ecol., **23:** 261–297.

MARE, M. F. 1942. A study of a marine benthic community with special reference to the microorganisms. J. Mar. Biol. Ass. U. K., **25:** 517–554.

MEDCOF, J. C. 1948. A snail commensal with the soft-shell clam. J. Fish. Res. Bd. Can., **7:** 219–220.

RENKONNEN, O. 1944. Die Carabiden- und Staphylinidenbestaende eines Seeufers in SW-Finnland. Ann. Entom. Fenn., **10:** 33–104.

SANDERS, H. L. 1956. Oceanography of Long Island Sound, 1952–1954. X. Biology of marine bottom communities. Bull. Bingham Oceanogr. Coll., **15:** 345–414.

———. 1958. Benthic studies in Buzzards Bay. I. Animal-sediment relationships. Limn. & Oceanogr. **3:** 245–258.

SAVILOV, A. I. 1957. Biologic aspect of the bottom-fauna of the northern Okhotsk Sea. Trudy Inst. Okeanol., **20:** 88–170. (*In Russian.*)

SMITH, J. E. 1932. The shell gravel deposits and infauna of the Eddystone grounds. J. Mar. Biol. Ass. U. K., **18:** 243–278.

SOIL SURVEY STAFF. 1951. Soil Survey Manual, U. S. Dept. Agr. Handbook No. **18:** 1–503.

TURPAEVA, E. P. 1957. Food relationships between the dominant species of marine bottom biocenoses. Trudy Inst. Okeanol., **20:** 171–185. (*In Russian.*)

WIESER, W. 1960. Benthic studies in Buzzards Bay. II. The meiofauna. Limn. & Oceanogr., **5:** 121–137.

YONGE, C. M. 1939. The protobranchiate Mollusca; a functional interpretation of their structure and evolution. Phil. Trans. B, **230:** 79–147.

ERRATA: The following corrections are required to Vol. V No. 2 of Limnology Oceanography.

p. 144 The legend of Figure 1 should read—"Trellis diagram showing the degree of faunal simil among the 24 samples from Station R.

pp. 151–152 The histograms in Figures 2 and 3 should be interchanged with their legends remai on the pages as now printed.

Deep-Sea Ecology

Deep-Sea Research, 1965, Vol. 12, pp. 845 to 867. Pergamon Press Ltd. Printed in Great Britain.

An introduction to the study of deep-sea benthic faunal assemblages along the Gay Head–Bermuda transect*

H. L. SANDERS, R. R. HESSLER and G. R. HAMPSON

(*Received* 2 *August* 1965)

Abstract—This paper is the introduction to a long-term, detailed study of open ocean, benthic assemblages. Attention will be paid to taxonomy, community structure, physiology, recruitment, patchiness, modes of feeding, zoogeography, and many other related topics. The major area of study is a transect between Massachusetts, U.S.A., and Bermuda, in the North Atlantic Ocean. Quantitative samples were taken with an anchor dredge and were washed through a 0·42-millimeter-aperture screen. The sediments along the transect range from fine sands of terrigenous origin on the upper continental slope, to silty clays and clays composed primarily of pelagically derived calcium carbonate in the Sargasso Sea. With increasing depth, temperature at the stations not only decreases, but shows more restricted fluctuation.

Analysis of results shows much greater faunal densities than those reported from previous studies; this probably results from our using a smaller screen aperture. Each region of the transect supports a characteristic number of animals per square meter in a general trend of decreasing density with increasing depth and distance from the continent : outer continental shelf, 6,000–13,000; upper continental slope, 6,000–23,000; lower continental slope, 1,500–3,000; abyssal rise, 500–1,200; abyss under the Gulf Stream, 150–270; abyss in the Sargasso Sea, 31-130; lower Bermuda Slope, 140–300; upper Bermuda Slope, 500–850.

From all but three stations, Polychaeta, Crustacea, Pelecypoda and Sipunculoidea combined comprised 85–100% of the fauna, with the former group being particularly important. Pogonophorans, ophiuroids, anemones, solenogastres, gastrods, and scaphopods were also important, but less abundant. The dominance of polychaetes tended to decrease with distance from the continent, while the crustaceans correspondingly increased in importance. Attempts to relate the density of animals to the amount of organics in the sediment provided no clear correlation. It is suggested that this results from the fact that the analytical techniques do not differentiate between labile and more refractory organic compounds.

DEEP-SEA benthic investigations have been almost exclusively surveys of a wide-ranging nature, covering vast areas of the ocean floor. By necessity, the numbers of samples collected in any specific region have been sparse. Such studies give an approximation of the kinds of animals and quantity of life present in different regions and depths of the world ocean.

The objectives of our programme are to study open ocean benthic communities in detail and, therefore, we have confined ourselves to a restricted region of the North Atlantic, a transect between southern New England and the Bermuda islands (Fig. 1). Besides the stations comprising the present study there are a number of additional stations included on this chart which have never been formally analyzed. Conveniently located between two bases of operation, the Woods Hole Oceanographic

*Contribution No. 1666 from the Woods Hole Oceanographic Institution. This research was supported by National Science Grants G-4812, G-15638, GB-563 and Office of Naval Research Contract Nonr-1135 (02).

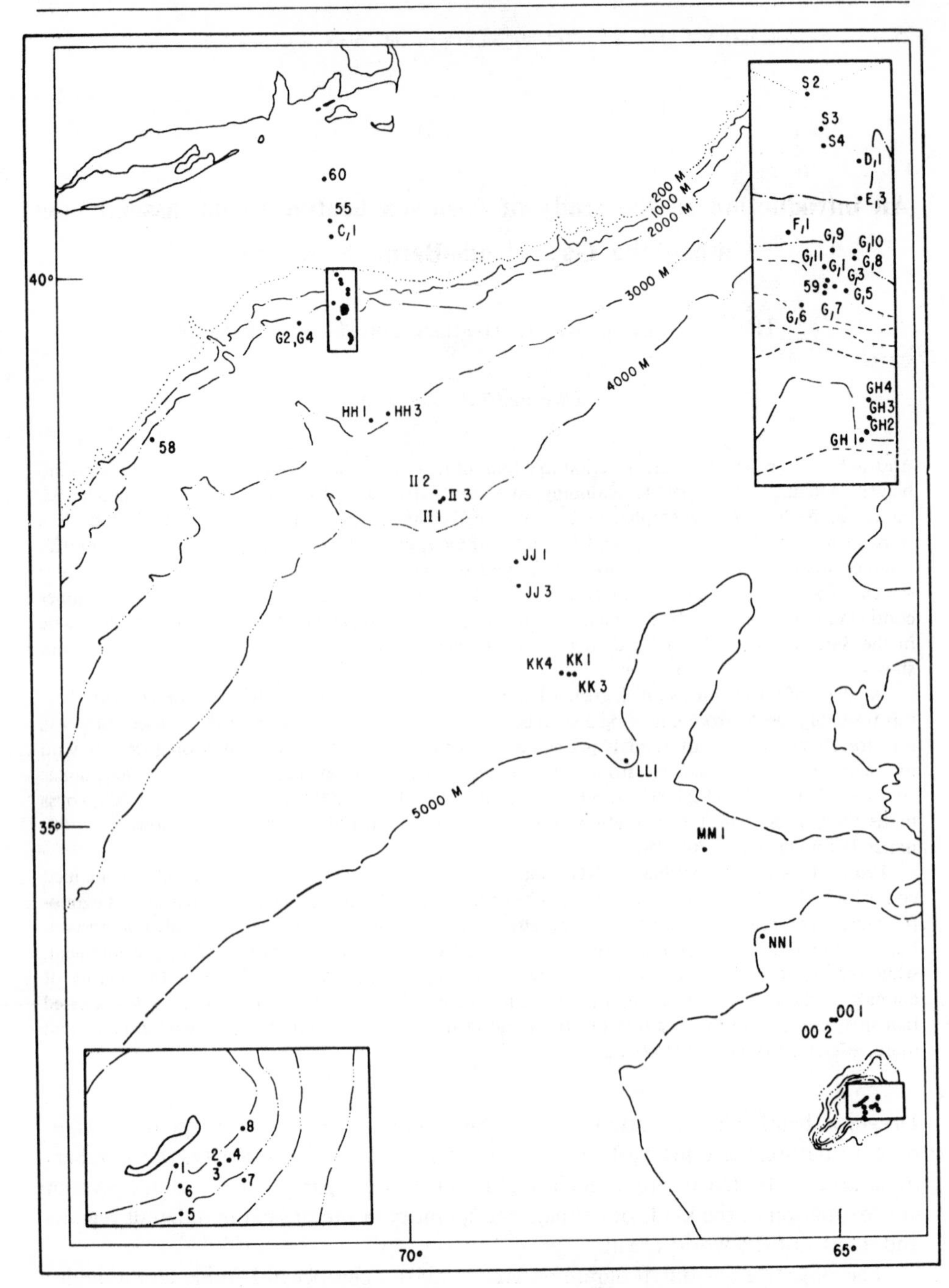

Fig. 1. Map showing the location of benthic transect stations. The fauna from a number of these stations have not as yet been analyzed.

Institution and the Bermuda Biological Laboratory, the study area crosses a diversity of benthic environments which differ in depth, temperature, and sediment composition : the continental shelf, continental slope, abyssal rise, abyssal plain, and the Bermuda Slope. Close proximity to the region of study allows us to make frequent trips to various portions of the transect. Another advantage in selecting this particular area is that more is probably known about the physical, chemical and biological aspects of the water column than any other comparable deep-sea region (RICHARDS, 1958; KETCHUM, CORWIN and VACCARA, 1958; KETCHUM, RYTHER, YENTSCH and CORWIN, 1958; MCGILL, CORWIN and KETCHUM, 1964; VACCARO, 1963; RYTHER and MENZEL, 1960; MENZEL and RYTHER, 1960; HULBURT, RYTHER and GUILLARD, 1960; HULBURT, 1962; HULBURT and RODMAN, 1963; RILEY, STOMMEL and BUMPUS, 1949; RILEY, 1951, 1957; GRICE and HART, 1962).

We expect this investigation to be a long-term, continuing programme concerned with many diverse phases of deep-sea benthic biology such as structure of animal communities, physiology, periodicity and mode of reproduction, variations of patchiness in faunal distribution, modes of feeding, taxonomy of major animal groups, zoogeography, faunal boundaries and reasons for faunal discontinuities. Certain of these aspects will be the research efforts of others. The present paper is an introduction to our programme, but deals as well with a quantitative evaluation of the transect.

Station names

The variety of station names used on our transect is admittedly confusing and results from having altered the numbering system too many times. Initially we attempted to use the station numbers of the Atomic Energy Commission hydrographic cruises (stations HH-00) whose positions we duplicated as our basic sampling scheme. Different letters were used on the continental and Bermuda slopes where our stations and the A.E.C. stations did not coincide. With time, additional stations were added to the transect and it became obvious that the numbering system had become too cumbersome. Starting with *Atlantis I* cruise 298 in the fall of 1963 we commenced a consecutive numbering system beginning with the number 55, that having been approximately the 55th station taken.

The stations in this programme of study of open ocean benthic communities will be known as W.H.O.I. Benthic Stations (WHOI B.S.) to avoid confusion with station numbers from any other programmes.

MECHANICS OF SAMPLING

Dredging

Bottom samples were collected by means of a modified anchor dredge (Fig. 2.) This device is designed so that it will dig to a given depth and therefore allow quantitative sampling. The area of the sample obtained is its volume (which is easily measured on board ship) divided by the depth to which the dredge is known to bite.

The use of this type of dredge was dictated by a desire to obtain large samples which would be quantitative for the full depth of the sediment sampled. Such a dredge avoids the disadvantage of grab techniques where the full area is realized only in the shallower part of the sample.

The dredge consists of two main parts, a metal frame with planing surface and dredge mouth, and a cloth bag to retain the collected sediment (Fig. 2b).

The metal frame is a flat, rectangular box, open at both ends. Rigidly attached to the box at the forward open end is the planing surface which subdivides the forward opening into two equal, horizontal areas. This is a large metal plate on whose front end is shackled the towing wire. By virtue of its being towed from its front end and having the rest of the dredge behind it, the planning surface rides on the surface of the sediment and prevents the dredge from digging deeper than the level of the planing surface.

Because the forward opening of the box is subdivided into two equal mouths, it is possible for the dredge to land on the bottom on either side and still operate properly. Each mouth has a narrow biting plate (the biting edge) on its lower margin. This plate angles downward and has a sharp forward edge. The biting plate causes the dredge to dig into the sediment as far as possible, that is, until the planing surface prevents further penetration. Thus the dredge is designed to bite to a uniform depth. This depth is eleven centimeters, which is deep enough to capture all but the largest of burrowing organisms.

A cloth collecting bag is clamped over the rear opening of the box. This bag is essentially a long tube whose cod end is tied shut, allowing it to retain the sample. Most of the bag is made of canvas, but the cod end is heavy burlap which is more pliant and therefore easier to tie off. The cloth bag is given extra support by being surrounded by a one-inch mesh nylon net.

A safety wire is employed to insure recovery of the dredge in the event of its becoming hung up on some bottom obstruction. This wire runs from an attachment inside the box, back into the collecting bag, out through the cod end, and then forward on the outside of the dredge, finally attaching to the towing wire in front of a weak link that is intercalated between the towing wire and the planing surface. If the dredge hangs up, the weak link breaks. The dredge may then be broken free of the obstruction by lifting it backward by means of the safety wire. The total weight of the empty dredge is about 500 pounds.

By virtue of the above design, the dredge in theory operates as follows : the dredge is lowered to the bottom and is given enough extra wire to allow a reasonable catenary arc (1000–1500 extra meters in 3000–5000 meters depth). It is then towed as slowly as possible. Almost immediately it bites to the depth of the planing surface and remains at the depth for the entire period of dredging. Sediment taken in the mouth moves back through the box and accumulates in the collecting bag (Fig. 2c). On deck, the collected bottom sample is emptied into containers by untying the cod-end opening of the bag.

While the operation of the dredge is sufficiently clear-cut from a theoretical point of view, we felt it was necessary to make actual tests to see if this was indeed the way the dredge behaved. These tests were conducted in the clear, shallow water of Bermuda, with divers observing and photographing the operation.

An attempt was made to dredge in sediments approximating those typically found along our transect, that is, silt-clays without a growth of surface flora. At times a minor percentage of gravel was also present, but this was not felt to be a biasing influence. Tests were also made on a hard, well-sorted, medium-fine sand and a poorly sorted coarse sand bottom covered with a growth of *Thalassia*. The depth of these bottoms varied from 40 to 50 feet, but the coarse sand bottom was at only 20 feet.

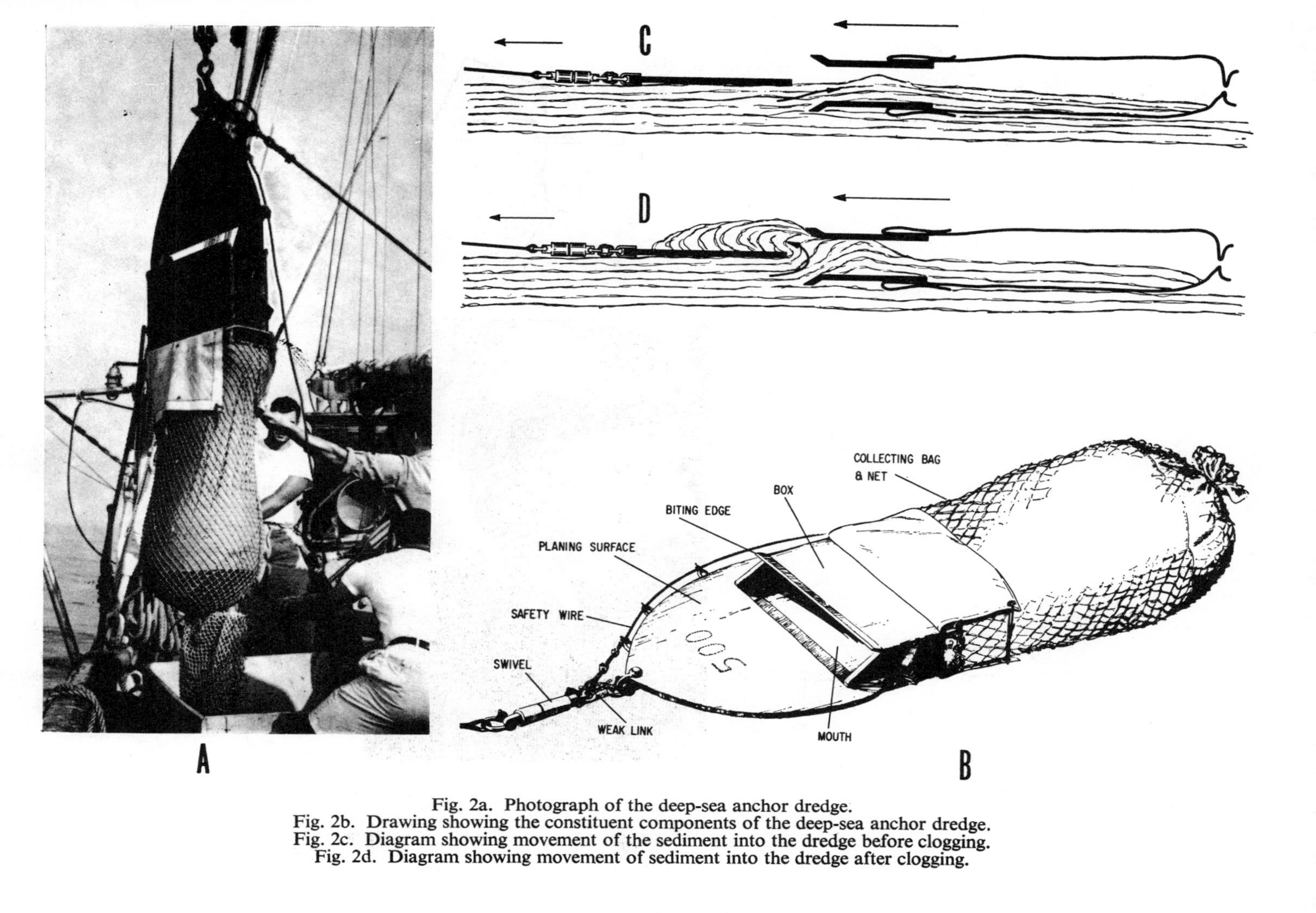

Fig. 2a. Photograph of the deep-sea anchor dredge.
Fig. 2b. Drawing showing the constituent components of the deep-sea anchor dredge.
Fig. 2c. Diagram showing movement of the sediment into the dredge before clogging.
Fig. 2d. Diagram showing movement of sediment into the dredge after clogging.

Fig. 3. Washing and screening the sample.

The dredge was towed with 300 feet of three-eighths inch wire at as slow a speed as possible (about one knot). On occasion the towing speed was higher, and once a very fast tow was attempted.

In general the dredge behaved as postulated. At the biting edge it dug to the level of the planing surface in a short distance and then stayed at that depth for the duration of the tow. The front end of the planing surface remained one to three inches off the bottom because of the upward vector component of the pull from the towing wire.

Sediment went into the box through the lower mouth and then into the bag, apparently in an undisturbed sheet. However, after a short distance the sediment clogged in the box, preventing any further collection. The dredge continued to operate at the same depth, but now the sediment which was still being taken in the lower mouth flowed upward and poured out of the upper mouth onto the planing surface (Figs. 2c, d). It is likely that this clogging is the result of frictional resistance of the sediment becoming so great against the box and bag wall that the sediment sheet buckles rather than continuing to move back into the dredge. This in turn increases the friction even more, causing the sediment to pile up in the front of the dredge until the dredge box is clogged from top to bottom all the way forward to the mouth. To show that the clogging was not due to trapped water the collecting bag was left untied, allowing the water to freely escape; the dredge still clogged. The sediment which accumulated on the planing surface eventually fell off to the sides.

Thus it appeared in most of the tests that the dredge took quantitatively valid samples until it clogged, and than rejected any further sediment. Nevertheless, this testing programme suggested a number of biases which possibly could be occurring at one time or another.

When the dredge first digs in, it samples less than the full depth of sediment. However, the distance over which this occurs is short. If the towing speed is increased, the tip of the planing surface lifts higher off the bottom, which prevents the dredge from biting to full depth. Assuming most of the fauna lives in the top few centimeters of sediment, both processes would give misleadingly high numbers. When towed at too high a speed, the mouth of the dredge lifts completely off the bottom, preventing any sample from being taken.

It seems unlikely that in our deep-water samples this bias is of any importance. In such depths a heavy one-half inch trawl wire is employed, and 1000 to 1500 m of extra wire is let out to give sufficient scope. Probably when the ship is towing, the weight of the wire is so great that even before the wire just preceding the dredge is lifted off the bottom, the dredge is pulled forward. This means that all of the force acting on the dredge is horizontal, no vertical component being present.

In the Bermuda tests the dredge was found to bite to less than full depth in sediments coarser and harder than silt-clays. This was seen to occur on a hard, well-sorted, medium-fine sand substrate and on a poorly sorted, coarse sand, covered with *Thalassia*. On one occasion the dredge bit to only a little more than half the full depth in a silt-clay (although it had a gravelly surface); no explanation has been found for this. Other than this single exception, the dredge dug to full depth in silt-clays, and since deep-sea sediments are typically of such a composition, it is probable that the sediment usually offers no impediment to full biting.

On an irregular bottom the dredge did not rise and fall with the irregularities, but dug at a mean depth, sometimes deeper than full depth, sometimes shallower.

The process of rejection of sediment after clogging may introduce a bias. If the dredge were not completely clogged, but slowly accepted a small portion of the sediment which came into the lower mouth, then it is likely the dredge would preferentially be selecting the less populated, deeper portion of the sediment and rejecting the rich surface layers. We believe this bias, which would result in samples with a lowered faunal density, is probably not important.

Finally, the sample may be biased by winnowing as the dredge is brought through the water column to the surface. That this does not occur is indicated by the fact that when the dredge breaks the surface of the water even after passing through as much as five kilometers of water at a rate of 100 meters a minute, its mouth is very often still packed with the mud that clogged it during dredging. This clogging is so common that it is the most likely explanation for the great variation in the size of our samples. The different sediments would surely tend to clog the dredge at different rates.

Thus, while different biases may play a part during the dredging operation, we feel that in general our samples are relatively valid. The internal consistency of the density of animals between our samples from the same type of habitat (not counting the possibility of an unthought-of systematic error) also supports this conclusion.

Although no direct comparison of our dredge with other quantitative techniques will be attempted here, it should be emphasized that the standard, classically employed grab techniques also fall heir to many biases, such as washing caused by leakage, oblique penetration of the substrate, blowing away of surface organisms by the advance current of water, sediment being forced out of the grab as the jaws close, and incomplete or variable penetration of the substrate (THORSON, 1957). Most of the published evaluations of a grab's efficiency are based on relative success in obtaining a larger volume of sediment or larger number of animals (JOHANSEN, 1927; HAGMEIER, 1930; THAMDRUP, 1938; HOLME, 1949, 1953; URSIN, 1954; see also the summaries in HOLME, 1964 and LONGHURST, 1964). Such tests are usually invalid because they are merely comparisons with some other device. More meaningful are tests involving the use of some absolute standard, such as SMITH and MCINTYRE's (1954) comparison of the amount of sediment taken by their grab to its maximum volumetric capacity, MENZIES, SMITH and EMERY's (1963) comparison of the number of surface animals taken by the Campbell grab to the number of animals photographed just before the sample was taken or LIE and PAMATMAT's (in press) comparison of paired grab and hand dug samples.

In none of these reports, other than that of LIE and PAMATMAT, were underwater observations mentioned as being employed, although this technique appears to reveal biases otherwise difficult to assess (ROLAND WIGLEY, personal communication). We have come to feel that at least some direct, underwater observation of quantitative sampling equipment is essential for proper evaluation of results. Had we not actually seen the rejection process in our dredge, we never would have known of its existence or appreciated its importance.

Washing and sorting

After the sample is removed from the dredge, its volume is measured, and then the sample is washed. This is accomplished by putting the sample in a large garbage can which has a spout near the top, much like a coffee pot. A large diameter water hose is shoved down into the sediment, and a large volume of water running at a low velocity is pumped through the sediment (Fig. 3). The resulting suspension of animals and fine grained sediment pours out of the spout and then through a 0·42 mm mesh screen. The animals are retained on the screen. Large animals are immediately picked out and preserved.

At the end of the washing process there are three fractions : the picked animals, the screen fraction which contains most of the fauna, and a coarse fraction remaining in the can. The latter consists of coarser sediments, tests of pelagic Foraminifera and Gastropoda, as well as benthic Foraminifera, Pelecypoda, Ostracoda, and larger Echinodermata, Polychaeta, etc. The three samples are preserved separately in formalin for two or three days, a time sufficient for fixation. They are then transferred to alcohol for storage. (Earlier in our programme the samples were stored in formalin until they were sorted. This tended to dissolve the calcium carbonate in the shells of the Mollusca and in the integument of Echinodermata and Crustacea). Later in the laboratory, these samples are carefully sorted.

The above technique is time consuming, but it is also extremely gentle, and in general the animals are well preserved and relatively undamaged.

Often the animals are still alive after washing, despite the fact that they have been brought to the surface from considerable depths. It has been usually thought that the tremendous pressure changes invariably kill deep-water organisms. Our findings suggest, however, that pressure may not be too critical in regard to benthic invertebrates, but rather the pronounced temperature change may be lethal. By bringing up the animals in the intact sediment which maintains its original temperature for a long period of time, the organisms are effectively insulated from temperature change except for that period when the sample is being washed on deck.

Because a fine screen is employed, very small animals are retained, but in the deep-sea benthos, animals normally considered to be macrofaunal (Polychaeta, Isopoda, Tanaidacea, etc.) are most commonly extremely small. As a result, the measured faunal density at our stations is much higher than that of stations taken in some other programmes where coarser screens were used.

Temperature

From unpublished temperature measurements collected during the course of a number of A.E.C. cruises from this institution, we have constructed temperature-depth curves for the lower part of the water column for many of our stations along the transect. In Figs. 4a, b, c, d and e the water depth at each locality is designated by the position of the station letter or number on the curve. (Since the positions of the A.E.C. stations do not agree precisely with our own and because of the irregularities of the bottom, the deepest recorded temperature readings are not necessarily coincident with the given benthic depth). These figures, together with the information provided in Fig. 1 and Table 1, clearly show that the shallower the water depth, the higher is the mean temperature and the larger is the temperature range. (The temperature values for stations S3, D and GH are extrapolated from data of neighbouring stations at equivalent depths.) At depths greater than 700 m, the variation in temperature is markedly reduced. (Fig. 4b). At 800 m on the Continental Slope (Table 1 and Fig. 4b) and 1700 m on the Bermuda slope (Table 1 and Fig. 4e), the

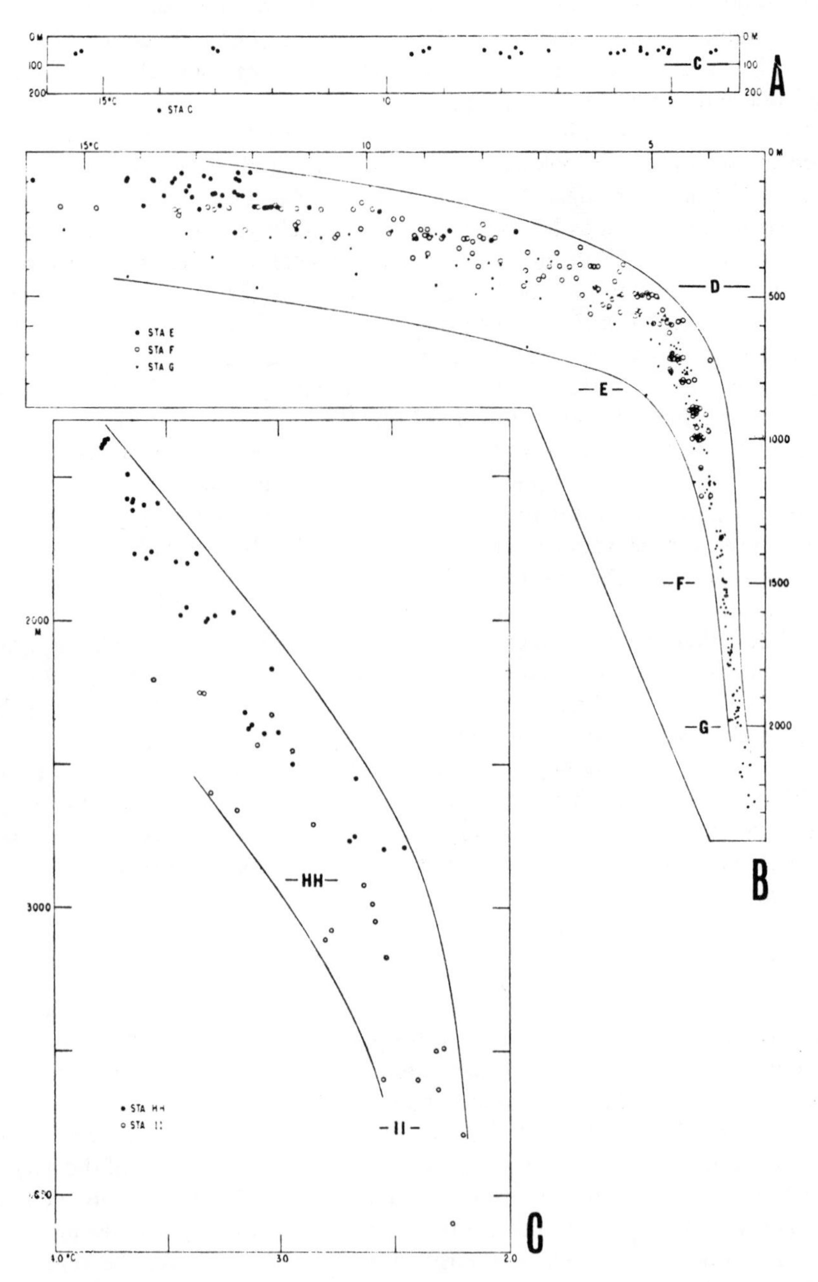

Fig. 4. Temperature-depth graphs for a number of stations of the transect. A. Continental shelf station C. B. Continental slope Stas. D, E, F and G. C. Abyssal rise stations HH and II.

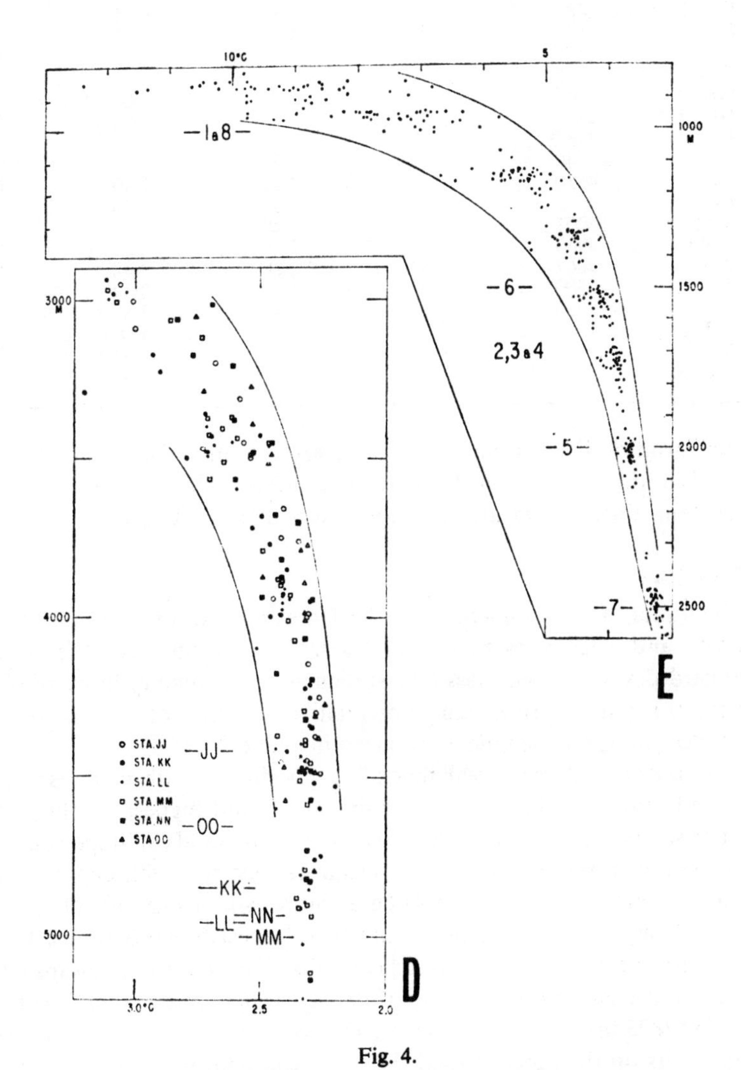

Fig. 4.

D. Abyssal plain Stas. JJ, KK, LL, MM, NN and OO. E. Bermuda slope Stas. 1 to 8.

Table 1. *Station temperatures* (°C).

Station	*Depth* (m)	*Mean temp.*	*Minimum*	*Maximum*	*Range*
R	20		− 1·0	23·0	24
C	97		5·0	15·5	10·5
Sl. 3	300	7·91	7·3	12·4	5·1
D	487	5·23	4·87	6·25	1·38
E	824	4·41	4·20	4·60	0·40
F	1500	3·73	3·67	3·85	0·18
G	2086	3·47	3·34	3·65	0·31
GH	2500	2·91	2·67	3·07	0·40
HH	2873	2·59	2·45	2·70	0·25
II	3752	2·33	2·26	2·41	0·15
JJ	4540	2·29	2·25	2·32	0·07
KK	4850	2·27	2·20	2·30	0·10
LL	4977	2·32	2·30	2·34	0·04
MM	5001	2·30	2·29	2·35	0·06
NN	4950	2·30	2·29	2·32	0·03
OO	4667	2·33	2·28	2·42	0·14
Ber. 7	2500	3·26	3·08	3·39	0·31
Ber. 5	2000	3·67	3·48	3·86	0·38
Ber. 2, 3 & 4	1700	3·91	3·72	4·20	0·48
Ber. 6	1500	4·17	3·85	4·53	0·68
Ber. 1 & 8	1000	7·37	5·7	9·0	3·3

temperature variation is less than 0·5°C. At greater depths, the amplitude of change gradually and continuously diminishes until at our deepest stations (LL, MM and MN), at about 5000 m, the temperature variation is less than 0·1°C (Table 1 and Fig. 4d).

Sediment analyses

A small aliquot of the sediment was collected at most of our stations for particle-size, organics, and calcium carbonate analyses. These results, together with mode, mean, standard deviation, skewness (measurement of asymmetry in curve of particle size distribution), kurtosis (peakedness or relative influence of extreme deviation in curve), and designation of sediment are given in Table 2.

The mechanical analyses reveal that all our sediments are composed primarily of fine-grained particles. The coarsest sediments, found highest on the continental slope, are fine sands at Slope Sta. 2 (200 m) and very fine sands at Slope Sta. 3 (300 m). Thereafter there is a gradual decrease in mean particle size with depth so that silty sands occur somewhat lower on the slope at Slope Sta. 4 (400 m), D # 1 (487 m), and E # 3 (824 m), while at the remaining two deepest Slope Stations, F (1500 m) and G (2086 m), the sediments become clayey silts. This tendency toward decreasing grain-size continues onto the abyssal rise with clayey silts present at GH # 1 (2500 m) and HH # 3 (2873 m). Farther along the transect, the sediments are, in sequential order, clayey silts on the outer abyssal rise and neighbouring stations on the abyssal plain (II # 2, 3752 m; JJ # 3, 4540 m; LL # 1, 4977 m), silty clays farther out on the abyssal plain (MM # 1, 5001 m), and clays on the abyssal plain near Bermuda, (NN # 1, 4950 m; and 00 # 2, 4667 m).

A similar trend can be demonstrated with the calcium carbonate content of the sediment. On the upper half of the continental slope these values are low, less than 10%, with no discernible gradient with depth. Benthic molluscs are important contributors of this compound at the shallower stations. At greater depths, the calcium carbonate content increases at a continuous rate : 14% at Sta. F # 1 (1500 m),

20% at G # 1 (2086 m), 26% at GH # 1 (2500 m), 33% at HH # 3 (2873 m). From Stas. II # 2 (3752 m) through LL # 1 (4977 m), the calcium carbonate content remains relatively constant, varying from 28 to less than 33%. Thereafter the calcium carbonate of the sediment again increases. At MM # 1 (5001 m), next station along the transect, the amount rises to almost 40%, at NN # 1 (4950 m), more than 50%, and at 00 # 2 (4667 m), our most distant abyssal station in the Sargasso Sea, the value is almost 66%. Our three calcium carbonate measurements from the Bermuda Slope (Bermuda Stas. 7 at 2500 m, 5 at 200 m and 6 at 1500 m are 96, 97·3, and 99·1%, respectively.

In contrast to the upper continental slope stations, the calcium carbonate elsewhere in the transect is almost entirely of pelagic origin. Throughout the transect from Station HH # 3 (2873 m), on the abyssal rise, through the Sargasso Sea section of the transect and up the Bermuda Slope, the sediments can be classified as *Globigerina* oozes since the sediments contain more than 30% calcium carbonate derived from pelagic Foraminifera. On the Bermuda Slope, shells of pteropod molluscs are conspicuous elements of the sediments and, therefore, such sediments might be designated as mixed *Globigerina* and pteropod oozes.

There is no consistent pattern or gradient along the transect in the distribution of either organic carbon or Kjeldahl nitrogen (see Fig. 8). However, almost an order of magnitude difference exists between maximum organic carbon value of 1·24% (Sta. C # 1) and minimum value of 0·15% (Bermuda Stas. 5 and 6). Lower organic carbon concentrations are found on the Bermuda Slope and the Sargasso Sea abyss than on the abyssal rise and continental slope.

Organic carbon and Kjeldahl nitrogen occur in ratios ranging from 10·3 : 1 to 5·6 : 1 with an average of 7·4 : 1. This small amplitude of variation is not surprising because in organic material carbon and nitrogen are present in a relatively constant ratio. The ratios of organic carbon to organic nitrogen in the present study are on an average lower than the 10 : 1 ratio of TRASK (1939) and the average ratio of 8·8 : 1 found by MIRANOV and BORDOVSKY (1959) for the bottom sediments of the Bering Sea.

Quantitative calculation

In previous quantitative investigations of the benthos in bathyal and abyssal depths, wet weight measurements have been used. The obvious advantage of such a procedure is that a quantitative determination can be made without destroying the specimens for systematic or other purposes. However, this technique has a drawback which in our opinion seriously compromises the results. The bulk of the weight in a sample frequently may be accounted for by the inorganic calcium carbonate present in the molluscs, crustaceans, and echinoderms and it is this, rather than organic biomass, which is measured. A more valid method is the use of decalcified dry weight (SANDERS, 1956).

Another difficulty with biomass measurements generally, is that they are by the usual quantitative sampling methods (in which the area sampled is small) often a measurement of the presence or absence of large, rare, and sporadically distributed animals. If one or more such animals happen to be captured, then the biomass may be increased 2–50 times (SANDERS, 1956, pp. 357–359). Because of this inherent difficulty, the meaning of comparisons of samples within a given area and between

Table 2. A partial analysis of calcium carbonate, organic carbon, organic nitrogen, and particl in percent weight of the sediments for the transect stations included in the present study.

<table>
<tr><th></th><th></th><th colspan="4">SAND</th></tr>
<tr><th>Station</th><th>Depth (m)</th><th>Phi-1 (1·000–0·500 mm)</th><th>Phi-2 (0·500–0·250 mm)</th><th>Phi-3 (0·250–0·125 mm)</th><th>Phi-4 (0·125–0·062 mm)</th></tr>
<tr><td>C # 1</td><td>97</td><td>---</td><td>1·0</td><td>1·0</td><td>15·0</td></tr>
<tr><td>Sl. 2</td><td>200</td><td>---</td><td>8·0</td><td>54·0</td><td>24·0</td></tr>
<tr><td>Sl. 3</td><td>300</td><td>2·0</td><td>11·0</td><td>18·0</td><td>45·0</td></tr>
<tr><td>Sl. 4</td><td>400</td><td>---</td><td>5·0</td><td>12·0</td><td>42·0</td></tr>
<tr><td>D # 1</td><td>487</td><td>---</td><td>1·0</td><td>12·0</td><td>52·0</td></tr>
<tr><td>E # 3</td><td>824</td><td>---</td><td>3·0</td><td>12·0</td><td>33·0</td></tr>
<tr><td>F # 1</td><td>1500</td><td>---</td><td>---</td><td>---</td><td>3·0</td></tr>
<tr><td>G # 1</td><td>2086</td><td>0·6</td><td>4·2</td><td>6·4</td><td>8·7</td></tr>
<tr><td>GH</td><td>2500</td><td>---</td><td>---</td><td>---</td><td>1·0</td></tr>
<tr><td>HH # 3</td><td>2873</td><td>---</td><td>---</td><td>---</td><td>0·3</td></tr>
<tr><td>II</td><td>3752</td><td>---</td><td>---</td><td>---</td><td>1·3</td></tr>
<tr><td>JJ # 3</td><td>4540</td><td>---</td><td>---</td><td>0·5</td><td>2·1</td></tr>
<tr><td>LL # 1</td><td>4977</td><td colspan="4">---------------- 0·9 ----------------</td></tr>
<tr><td>MM # 1</td><td>5001</td><td colspan="4">---------------- 0·2 ----------------</td></tr>
<tr><td>NN # 1</td><td>4950</td><td colspan="4">---------------- 1·7 ----------------</td></tr>
<tr><td>OO # 2</td><td>4667</td><td colspan="4">---------------- 0·5 ----------------</td></tr>
<tr><td>Ber. # 7</td><td>2500</td><td></td><td></td><td></td><td></td></tr>
<tr><td>Ber. # 5</td><td>2000</td><td></td><td></td><td></td><td></td></tr>
<tr><td>Ber. # 6</td><td>1500</td><td></td><td></td><td></td><td></td></tr>
</table>

	SILT			CLAY		
Phi-5 (0·062–0·031 mm)	Phi-6 (0·031–0·016 mm)	Phi-7 (0·016–0·008 mm)	Phi-8 (0·008–0·004 mm)	Phi-9 (0·004–0·002 mm)	Phi-10 (0·002–0·001 mm)	Phi-11 to Phi-16 (0·001–0·000016 mm)
37·4	14·0	11·4	6·6	4·5	4·4	4·7
7·0	1·8	1·0	1·1	0·8	0·7	1·9
16·3	2·4	1·6	1·7	0·9	1·0	0·4
23·0	5·5	2·1	3·8	2·3	1·5	1·6
23·2	2·1	1·5	2·2	2·0	1·2	2·0
32·5	5·0	3·1	3·8	3·0	1·9	2·5
22·6	15·3	15·7	16·0	11·2	6·7	9·5
14·8	8·8	9·9	11·6	12·2	9·8	13·0
8·2	8·5	13·2	15·1	14·9	15·0	24·1
4·4	7·1	11·8	18·1	23·7	14·9	19·7
10·6	13·6	13·2	18·3	16·1	12·1	14·8
5·5	18·7	13·6	11·5	10·8	9·0	18·3
7·1	14·5	20·2	17·9	12·5	9·0	7·91
3·3	8·2	14·1	15·7	17·6	12·3	28·6
1·6	4·5	6·5	8·9	14·7	12·9	49·2
1·2	3·9	5·1	9·7	18·4	17·2	44·0

Table 2 (Continued)

A partial analysis of calcium carbonate, organic carbon, organic nitrogen, and particle-size in percent weight of the sediments for the transect stations included in the present study.

Station	*Depth* (m)	*Mode* (*Phi*)	*Median* (*Phi*)	*Mean* (*Phi*)	*Standard Deviation*
C # 1	97	4·50–3·74	4·91	5·56	1·99
Sl. 2	200	2·60- 5·40	2·79	3·19	1·62
Sl. 3	300	3·50–4·50	3·42	3·50	1·51
Sl. 4	400	3·60–4·21	3·80	4·16	1·76
D # 1	487	3·50–5·25	3·71	4·08	1·62
E # 3	824	4·00- 3·58	4·05	4·43	1·88
F # 1	1500	4·6 –2·29 7·2 –1·63	6·57	6·81	2·09
G # 1	2086	4·5 –1·23 8·3 - 1·23	6·66	6·67	2·94
GH	2500	7·7 –1·52 9·2 –1·53	8·27	8·32	2·36
HH # 3	2873	8·4 –2·37	8·34	8·33	1·91
II	3752	5·6 –1·37 7·6 –1·84	7·62	7·67	2·13
JJ # 3	4540	5·4 –1·88	6·97	7·48	2·59
LL # 1	4977	6·6 –2·03	7·39	7·75	2·17
MM # 1	5001	8·4 –1·77	8·47	8·62	2·21
NN # 1	4950	8·6 –1·48	9·92	9·89	2·68
OO # 2	4667	8·7 –1·88	9·65	9·69	2·21
Ber. # 7	2500				
Ber. # 5	2000				
Ber. # 6	1500				

Skewness	*Kurtosis*	*Designation*	*% Calcium Carbonate*	*% Organic Carbon*	*% Organic Nitrogen*
0·49	0·37	Sandy Silt	3·17	1·24	0·1757
1·38	8·86	Find Sand	9·66	0·37	0·041
0·72	4·48	Very Find Sand	5·74	0·33	0·032
0·79	2·98	Silty Sand	6·16	0·60	0·076
1·15	5·78	Silty Sand	5·29	0·42	0·055
0·75	2·32	Silty Sand	6·95	0·52	0·078
0·25	– 0·60	Clayey Silt	14·78	1·03	0·140
0·07	– 0·76	Clayey Silt	19·95	0·59	0·082
0·08	– 0·59	Silty Clay	26·37	0·80	0·122
– 0·01	– 0·32	Silty Clay	32·87	0·56	0·091
0·11	– 0·60	Clayey Silt	33·15	0·44	0·079
0·30	– 0·47	Clayey Silt	29·71	0·34	0·048
0·24	– 0·50	Clayey Silt	27·95	0·23	0·013
0·10	– 0·57	Silty Clay	39·58	0·28	
– 0·03	– 0·34	Clay	50·77	0·31	
– 0·03	– 0·16	Clay	65·77	0·24	
			96·03	0·22	
			97·35	0·15	
			99·16	0·15	

Table 5. The numbers and percent composition (in brackets) of the major taxa from the transect stations.

Station	C # 1 No.		Sl.2 No.		Sl. 3 No.		Sl. 4 No.	
ANNELIDA :								
Non-Polychaeta :	48				1		1	
Polychaeta :	1260	(40·88)	4982	(77·18)	8007	(67·25)	3494	(78·71)
ARTHROPODA :								
Crustacea :	166	(5·39)	331	(5·13)	415	(3·49)	330	(7·43)
Isopoda	*12*		*27*		*45*		*23*	
Amphipoda	*142*		*274*		*294*		*270*	
Tanaidacea			*26*		*67*		*35*	
Cumacea	*12*		*4*		*5*		*2*	
Leptostraca								
Cephalocarida					*4*			
Pycnogonida :			3				2	
MOLLUSCA :	669	(21·71)	794	(12·30)	1847	(15·51)	458	(10·32)
Pelecypoda :	*601*	*(19·50)*	*738*	*(11·43)*	*1709*	*(14·35)*	*389*	*(8·76)*
Gastropoda :	*68*		*12*		*16*		*9*	
Scaphopoda :			*31*		*41*		*21*	
Aplachophora :			*13*		*81*		*39*	
SIPUNCULOIDEA :	3	(0·10)	223	(3·45)	1516	(12·73)	80	(1·80)
ECHIURIDA :								
PRIAPULOIDEA :								
NEMERTINEA :	18		13		18		21	
POGONOPHORA :			9		14		11	
ECHINODERMATA :	721		92		65		17	
Ophiuroidea :	*699*		*32*		*25*		*17*	
Asteroidea :			*59*		*32*			
Echinoidea :					*8*			
Crinoidea :								
Holothurioidea :	*22*							
COELENTERATA :	168		4		12		25	
PORIFERA :								
TUNICATA :								
BRYOZOA :	8		2		6		many	
INCERTA SEDIS :	21		2		6			
TOTAL	3082		6455		11907		4439	
No./lm²	5314		12910		21263		6081	

D # 1 *No.*	E #3 *No.*	F # 1 *No.*	G # 1 *No.*	GH # 1 *No.*	GH # 4 *No.*
	53	15	89	5	1
2924 (57·17)	2201 (73·17)	698 (70·01)	737 (65·80)	265 (72·60)	251 (83·95)
432 (8·45)	330 (10·97)	141 (14·14)	132 (11·79)	51 (13·97)	15 (5·02)
37	*44*	*63*	*35*	*19*	*6*
299	*134*	*39*	*64*	*9*	*2*
91	*1417*	*37*	*30*	*22*	*7*
4	*5*	*2*	*3*	*1*	
1					
1					
1409 (28·05)	228 (7·58)	55 (5·52)	91 (8·13)	13 (3·56)	10 (3·34)
1218 (23·81)	*218 (7·25)*	*54 (5·42)*	*70 (6·25)*	*11 (3·01)*	*9 (3·01)*
44	*10*	*1*	*18*	*2*	
69			*3*		
78					*1*
29 (0·57)	14 (0·47)	23 (2·31)	13 (1·16)	7 (1·92)	10 (3·34)
		1			
		4	26		
27	14	26	8	9	
32	20	17			
19	5	546	16	6	6
18	*3*	*3*	*11*	*3*	*4*
		2			
			4	*3*	*1*
	1		*1*		
1	*1*				*1*
48	61	6	8		1
	1	1		7	
				1	
107	8	1		1	
77	73				5
5115	3008	997	1120	365	299
8669	2978	1719	2154	521	467

Table 5. The numbers and percent composition (in brackets) of the major taxa from the transect stations.

Station	HH # 3 No.	II # 2 No.	JJ # 1 No.	JJ # 3 No.	KK # 1 No.	LL # No.
ANNELIDA :						
Non-Polychaeta :	56	3		1	1	
Polychaeta :	358 (56·29)	144 (36·83)	137 (51·89)	48 (47·52)	66 (58·44)	32 (4
ARTHROPODA :						
Crustacea :	136 (21·38)	32 (8·18)	84 (31·82)	31 (30·69)	31 (27·43)	33 (4
Isopoda	*45*	*12*	*43*	*7*	*5*	*12*
Amphipoda	*14*	*3*	*2*	*2*		*5*
Tanaidacea	*76*	*17*	*39*	*22*	*25*	*16*
Cumacea	*1*				*1*	
Leptostraca						
Cephalocarida						
Pycnogonida :						
MOLLUSCA :	6 (0·94)	12 (3·07)	14 (5·30)	11 (10·89)	1 (0·88)	1 (1
Pelecypoda :	*6 (0·94)*	*12 (3·07)*	*11 (4·17)*	*8 (7·92)*	*1 (0·88)*	*1 (1*
Gastropoda :			*2*	*2*		
Scaphopoda :				*1*		
Aplachophora :			*1*			
SIPUNCULOIDEA :	41 (6·45)	11 (2·81)	15 (5·68)	8 (7·92)	6 (5·31)	
ECHIURIDA :		1				
PRIAPULOIDEA :			2			
NEMERTINEA :		1	2			
POGONOPHORA :	38	179 ?				
ECHINODERMATA :	1		3	1	2	
Ophiuroidea :			*2*	*1*	*2*	
Asteroidea :						
Echinoidea :	*1*		*1*			
Crinoidea :						
Holothurioidea :						
COELENTERATA :						
PORIFERA :		5		1		
TUNICATA :						
BRYOZOA :						
INCERTA SEDIS :		3	7		6	
TOTAL	636	391	264	101	113	67
No./lm^2	748	1003	264	158	92	55

MM # 1 *No.*	NN # 1 *No.*	OO # 2 *No.*	Ber. 7 *No.*	Ber. 5 *No.*	Ber. 6 *No.*	Ber. 8 *No.*
14 (51·85)	30 (58·22)	38 (65·52)	34 (37·36)	59 (66·29)	90 (43·27)	112 (34·15)
9 (23·83)	19 (37·25)	11 (18·97)	37 (40·66)	14 (15·73)	89 (42·79)	121 (36·89)
3	*7*	*3*	*7*	*8*	*22*	*16*
	1		*5*		*3*	*8*
6	*11*	*8*	*25*	*6*	*64*	*97*
2 (7·41)	1 (1·96)	5 (8·62)	10 (10·99)	1 (1·12)	13 (6·25)	18 (5·48)
2 (7·41)	*1 (1·96)*	*4 (6·90)*	*10 (10·99)*	*1 (1·12)*	*11 (5·29)*	*3 (0·91)*
					2	
						15
		1				
	1 (1·96)		3 (3·30)	12 (13·48)	8 (3·85)	9 (2·74)
					1	2
						57
				2	1	2
						2
				2	*1*	
		3				
			5	1	4	2
			2			
2		1				4
27	51	58	91	89	208	328
33	38	126	120	189	178	729

different regions become highly questionable. It is for this reason that we have chosen to use numbers rather than weight in the present investigation.

Faunal densities and composition

Despite the fact that benthic animals have been collected from the deep-sea for about a century, the benthic fauna was not evaluated quantitatively until 1950 when the *Galathea* Expedition obtained 13 samples off the West Coast of Africa in depths from 800 to 3660 m. A 0·2 m^2 Petersen grab was used and the number of animals found ranged from 5 to 85/m^2 with a mean value of 27 (SPÄRCK, 1951). SPÄRCK concluded that on the average there are about ten animals/m^2 on the deep-sea floor. Seven samples were also taken from hadal trenches in the Indo-Pacific region and these yielded from 0 to 60 animals/m^2 (SPÄRCK, 1956).

More recently, extensive deep-sea benthic investigations have been carried out by Russian biologists, particularly in the Pacific, Antarctic, and Indian oceans (BELYAEV 1959, 1960, 1960a; BELYAEV and SOKOLOVA, 1960; BELYAEV and VINOGRADOVA, 1961, 1961a; BELYAEV, VINOGRADOVA and FILATOVA, 1960; BIRSTEIN, 1959; BIRSTEIN and SOKOLOVA, 1960; CHUKHCHIN, 1963; FILATOVA, 1959, 1960; FILATOVA and BARSANOVA, 1964; FILATOVA and LEVENSTEIN, 1961; KUZNETSOV, 1960, 1961, 1963, 1964; SOKOLOVA and PASTERNAK, 1962; VINOGRADOVA, 1962; ZENKEVITCH, 1958; ZENKEVITCH and FILOTOVA, 1958, 1960; ZENKEVITCH, BARSANOVA and BELYAEV, 1960; ZENKEVITCH, BELYAEV, BIRSTEIN and FILATOVA, 1959). A general English language review of these studies is given by VINOGRADOVA (1962). Only certain of their quantitative investigations are expressed in terms of animal numbers, biomass being the major quantitative criterion.

In the paper by FILATOVA and LEVENSTEIN (1961) on the benthos of the northeast Pacific area, from 5 to 90 animals/m^2 were found, with an average of 26, at the 18 stations listed, (excluding Foraminifera, nematodes, and ostracods to make these results comparable to our own). Most of the stations were from depths greater than 4000 m. The samples were obtained with an Okean-50 grab, and a No. 140 (0·5 mm) mesh-sized net was used in the sieving. These are standard procedures in the Russian deep-water quantitative studies.

In a study from the eastern Mediterranean Sea (CHUKHCHIN, 1963), five samples collected from the depth interval of 100–200 m gave values of 16–764 animals/m^2 with a mean number of 290, ten samples from the 200 to 1000 m interval ranged from 0 to 104 animals/m^2 with a mean numerical value of slightly less than 21, and nine samples from the 1000 to 3000 m interval yielded only 0 to 4 animals/m^2 with a mean of less than two organisms/m^2. Instead of the usual 0·5 mm mesh-sized screens, 2 mm and 1 mm mesh-sizes were used in this study.

Three samples taken in the Java Trench (BELAEV and VINOGRADOVA, 1961) at depths between 6000 and 7000 m yielded 3 (0·2 m^2 Petersen grab), 5 (0·25 m^2 Okean), and 10 (0·25m^2 Okean) organisms or 15, 20, and 40 animals/m^2. We were able to extract from the data given for seven of a larger number of samples taken at slope depths of from 100 to 1000 m in the western part of the Bering Sea (FILATOVA and BARSANOVA, 1964) the number of benthic animals per m^2. These varied from 68 to 4696 organisms per m^2 with a mean value of 895. KUZNETSOV (1964) gave minimal values for the same region that ranged from 152 to 1265/m^2 for six samples collected at depths of 100 to 1000 meters, with a mean of 521. A single sample collected at

1805 m yielded more than 132 animals/m², while four stations from the 2000 to 3000 m interval gave values of from 22 to more than 359 animals/m² or an average value of more than 118 individuals.

A major paper by KUZNETSOV (1963) presents the extensive investigations carried out on the shelf and slope in the southern Kamchatka and northern Kurile Islands region during 1949 to 1955. The average numbers of benthic organisms obtained at different depth intervals for a very large number of samples were 102/m² for the 0–50 m depth range, 94 for the 50–100 m interval, 111 for the 100–200 m range, 245 for the 200–500 m interval, 284 for the depths of 500–1000 m, and 26 for the 1000–2000 m depth range. The numbers for a specific station may be very much greater than these averages.

Quantitative benthic studies have also been undertaken in the deep (627-2107 m) enclosed basins off southern California (HARTMAN and BARNARD, 1958). The sampling gear employed were an orange-peel grab and a Campbell grab. Screens with mesh apertures of about 1·0 mm were used in the washing. The faunal densities of these somewhat anaerobic basins ranged from 12 to 117 animals/m².

Finally, WIGLEY and MCINTYRE (1964) obtained bottom samples from a line of ten stations running south of Martha's Vineyard, off southern New England. The five deeper stations (99–567 m) were at sublittoral to shallow bathyal depths in the immediate region of our transect. In sequential order with depth, the number of

Table 3. Depth, latitude, longitude, number of animals collected, and number of animals per square meter for the transect stations included in the present study.

Station	*Depth* (m)	*Latitude*	*Longitude*	*No. animals in sample*	*No. animals*/m²
55	75	40° 27·2′N	70° 47·5′W	3791	13073
C # 1	97	40° 20·5′N	70° 47′W	3082	5314
Sl. 2	200	40° 01·8′	70° 42′	6455	12910
Sl. 3	300	39° 58·4′	70° 40·3′	11907	21263
Sl. 4	400	39° 56·5′	70° 39·9′	4439	6081
D # 1	487	39° 54·5′	70° 35′	5115	8669
E # 3	823	39° 50·5′	70° 35′	3008	2979
F # 1	1500	39° 47′	70° 45′	997	1719
G # 1	2086	39° 42′	70° 39′	1120	2154
GH # 1	2500	39° 25·5′	70° 35′	365	521
GH # 4	2469	39° 29′	70° 34′	299	467
HH # 3	2870	38° 47′	70° 08′	636	748
II # 1	3742	37° 59′	69° 32′	——*	——*
II # 2	3752	38° 05′	69° 36′	391	1003
JJ # 1	4436	37° 27′	68° 41′	264	264
JJ # 3	4540	37° 13·1′	68° 39·6′	101	158
KK # 1	4850	36° 23·5′	68° 04·5′	113	92
LL # 1	4977	35° 35′	67° 25′	67	55
MM # 1	5001	34° 45′	66° 30′	27	33
NN # 1	4950	33° 56·5′	65° 50·7′	51	38
OO # 2	4667	33° 07′	65° 02·2′	58	126
Ber. 7	2500	32° 15′	64° 32·6′	91	120
Ber. 5	2000	32° 11·4′	64° 41·6′	89	189
Ber. 4	1700	32° 17′	64° 35′	217	271
Ber. 3	1700	32° 16·6′	64° 36·3′	126	274
Ber. 2	1700	32° 16·6′	64° 36·3′	189	215
Ber. 6	1500	32° 14·3′	64° 42′	208	178
Ber. 8	1000	32° 21·3′	64° 33′	326	729
Ber. 1	1000	32° 16·5′	64° 42·5′	243	528

*Sample excluded from quantitative analysis because of small size.

macrofaunal animals/m² found at these five stations were 1930 (99 m), 2140 (146 m), 1290 (179 m), 1695 (366 m), and 730 (567 m). The samples were collected with a 0·1 m² Smith-McIntyre grab and washed through a screen with 1·0 mm mesh apertures.

Faunal densities in present study

Samples from the Gay Head-Bermuda transect have yielded much greater densities of animals (see Table 3) than the above studies. Each region along the transect seems to support its own characteristic number of animals : outer continental shelf (stations 55 and C # 1), 6000–13000/m²; upper continental slope (Slope Stas. 2, 3, 4 and Sta. D # 1), 6000–23000; lower continental slope (stations E # 3, F # 1, and G # 1), 1500–3000; abyssal rise (GH # 1, GH # 4, HH # 3 and II # 2), 500–1200; abyss under the Gulf Stream (JJ # 1 and JJ # 3), 150–270; abyss in the Sargasso Sea (KK # 1, LL # 1, MM # 1, NN # 1 and OO # 2), 30–130; lower Bermuda Slope (Stas. 7, 5, 2, 3, 4 and 6), 120–300; upper Bermuda Slope (Stas. 1 and 8), 500–750. Nematodes, Foraminifera, ostracods, and harpacticoid copepods were collected but are not included in the numerical results because these faunal components are very inadequately sampled by our methods.

Because the numbers for our quantitative studies are appreciably greater than those found in other investigations of the deep-sea benthos, conceivably the density of animals present in our region of study is abnormally high. To test this possibility samples were collected in April, 1963, at comparable depths off the coast of South America. The numerical results (Table 4) are of the same general magnitude as those from our stations in the North Atlantic.

Table 4. Depth, latitude, longitude, and number of animals per square meter from four stations off the northeast coast of South America.

CHAIN 35 station	*Depth* (m)	*Latitude*	*Longitude*	*Sample size* (m²)	*No. of animals*	*No. per* m²
33	535	07° 53·5′N	54° 33·3′W	0·48	2612	5427
12	790	07° 09′S	34° 25·5′W	0·68	1284	1883
34	1500	08° 46′N	53° 46′W	0·73	377	519
19	4525	07° 33′N	45° 02·5′W	0·38	21	55

It appears unlikely that the gross differences between our findings and those of other studies can be attributed to the sampling gear. These differences probably stem from the fact that we use fine-meshed screens (0·42 mm apertures) which collect a larger fraction of the small macrofauna, the numerically overwhelming size group of the benthos. In general, the results from our limited number of stations are in agreement with Russian investigations (see Vinogradova, 1962) which showed that usually there is a decrease in the abundance of animal life with depth and distance from major land masses.

Because we have appreciable numbers of specimens which were collected quantitatively, we can make certain observations about the composition of the open ocean infauna. All our samples were characterized by diverse groups of invertebrates represented by numerous species. Three groups, the Polychaeta, Crustacea, and Pelecypoda, were present in every sample while a fourth, the Sipunculoidea, was absent from only three samples. These four faunal elements composed 85–100%

of the fauna by number at all stations except C, II # 2, and Bermuda 8 (see Table 5). At the latter two stations, concentrations of Pogonophora were responsible for the reduced percentages; at Sta. C on the continental shelf, Ophiuroidea and anemones were the causal factors.

In addition to the animals listed above, solenogastres, gastropods, and scaphopods were frequent constituents of our anchor dredge samples. Some of the less common groups were the priapulids (SANDERS and HESSLER, 1962), echiuroids, asteroids, and hexactinellid sponges.

The single most abundant group, the Polychaeta, was always represented by numerous species. For the detailed analysis of the polychaetes of our transect, see HARTMAN (1965). They formed 34 to 84% by numbers of all samples. Each portion of the transect supported its own number of animals (Table 5 and Fig. 5) : the continental shelf, about 2000/m² or 41% of all animals by number; the upper continental slope, from about 4700–14,000/m² or 57–79%; the lower continental slope, from 1400–2200/m² or 65–74%; the abyssal rise, about 370–420/m² or 37% (at II # 2, the per cent composition may be artificially low because many of the very abundant opaque black pogonophoran tubes counted in the analysis might actually be empty) to 84%; the Gulf Stream region, 75–140/m² or about 50%; the abyss of the Sargasso Sea, 15–85/m² or 47–66%; the lower Bermuda Slope, 45–125/m² or 37–67%; the upper Bermuda Slope, about 110/m² or 34%.

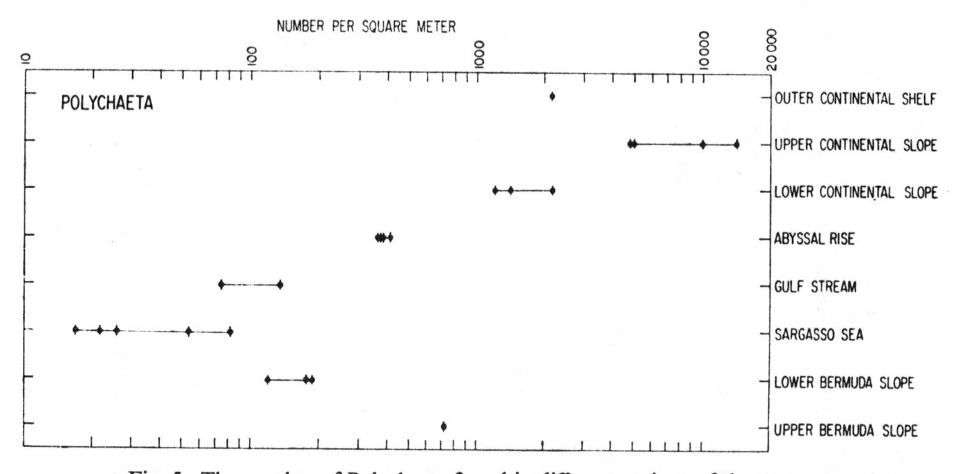

Fig. 5. The number of Polychaeta found in different regions of the transect.

The 266 polychaete species named by HARTMAN belong to 50 families. (A few samples from the Atlantic Ridge not obtained from our present investigations were included in the paper). The best represented families are the Maldanidae with 24 species, the Paranoidae with 15, Syllidae, Spionidae, and Ampharetidae with 13, and the Phyllodocidae with 10. HARTMAN points out that her results differ from the world-wide findings where the best represented families are the Polynoidae, Syllidae, Nereidae, and Serpulidae (HARTMAN, 1959).

The second most abundant group, the Crustacea, represented 3·5 to almost 50% of the fauna in the samples (Table 5) and like the polychaetes, were always represented by numerous species. Each region of the transect supported a characteristic number

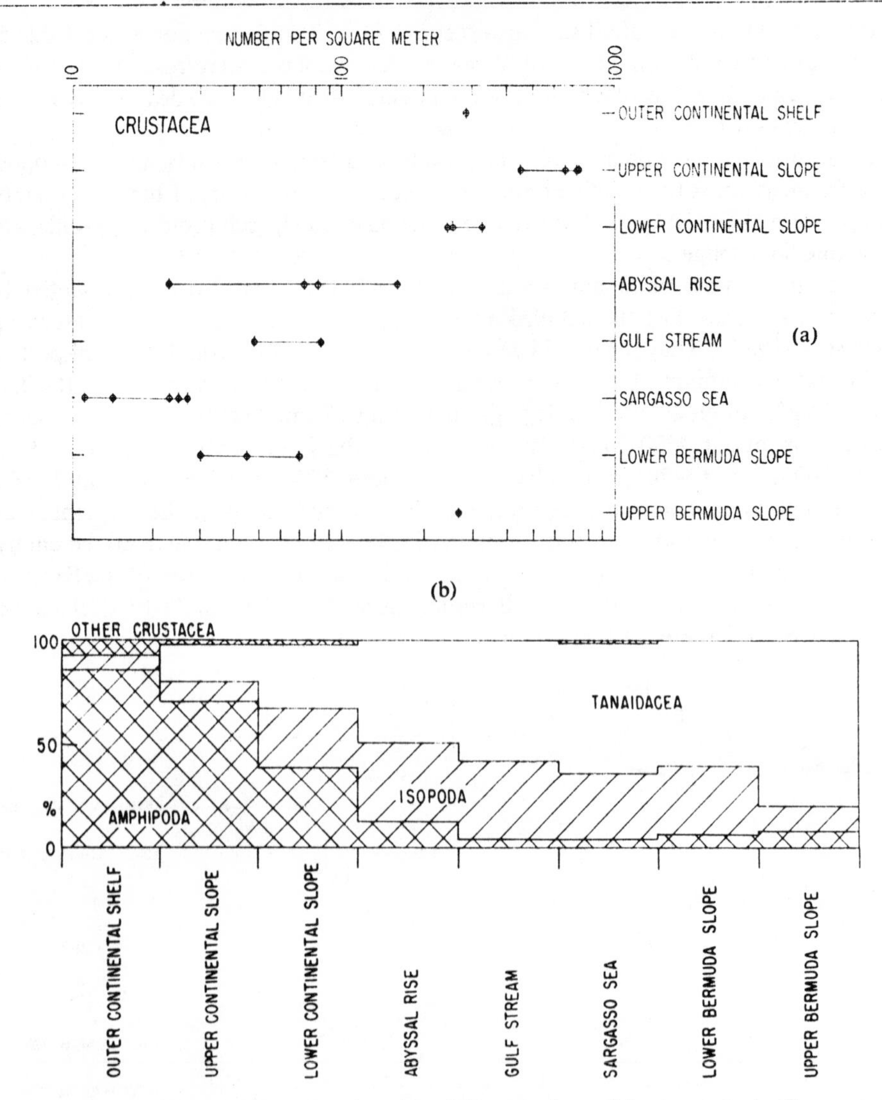

Fig. 6. a. The number of Crustacea found in different regions of the transect. b. The percent composition of the quantitatively important major crustacean taxa in the different regions of the transect.

of Crustacea (Table 5 and Fig. 6a) : the outer continental shelf, almost 300/m² or about 5·4% of the total fauna; the upper continental slope, 450–750/m² or 3·5 to 8·5%; the lower continental slope, 240–330/m² or 11–14%; the abyssal rise, 23–330/m² or 5–21·4%; the Gulf Stream region, 50 to 85/m² or 30·7 to 31·8%; the abyss of the Sargasso Sea, 11 to 27/m² or 19 to 50%; the lower Bermuda Slope, 30 to 76/m² or 15 to 43%; the upper Bermuda Slope, about 270/m² or 37%.

Within the Crustacea, three orders, the Amphipoda, Isopoda, and Tanaidacea formed 97% of the macrofauna encountered. (The ostracods and harpacticoid copepods were neglected quantitatively since they are too difficult to sample adequately with our techniques). Other groups, such as the Cumacea, Leptostraca (HESSLER and SANDERS, in press), and the Cephalocarida (HESSLER and SANDERS, 1964) were only occasionally found in these infaunal samples.

The Amphipoda were the overwhelmingly dominant group at the shallower stations at the northern end of the transect where they comprise 85% of the crustacean fauna on the outer continental shelf and 70% on the upper continental slope. Thereafter, the amphipods formed a progressively smaller fraction of the crustacean fauna along the transect; 39% on the lower continental slope, 12·5% on the abyssal rise, between 4 and 4·5% in both the Gulf Stream and Sargasso Sea areas, and 6 to 7% on the Bermuda Slope (Fig. 6b).

The Tanaidacea showed just the opposite distribution pattern to that of the Amphipoda. They were absent from the sample on the outer continental shelf and formed 18·5% of the crustacean fauna in the upper continental slope samples. Thence their percent contribution continuously increased, 31% on the lower continental slope, 50% on the abyssal rise, about 59% in the Gulf Stream region, 64% in the Sargasso Sea, 61% on the lower Bermuda Slope, and 80% on the upper Bermuda Slope (Fig. 6b).

The Isopoda, too, were relatively less common at the outer continental shelf and upper continental slope stations where they represented only 7·3 and 9·7% of the crustacean fauna. On the lower continental slope they formed 28%. On the abyssal rise, Gulf Stream, Sargasso Sea, and lower Bermuda Slope region their percent contribution remained rather constant, varying only from 31 to 37%. On the upper Bermuda slope they constituted 13·4% of the Crustacea (Fig. 6b). Note in Table 5, that as with the Crustacea as a whole, of the groups Amphipoda, Isopoda, and Tanaidacea, the various distinctive regions along the transect each supports a characteristic number of individuals.

The Pelecypoda, the remaining major group present in all of our samples, comprised from less than one percent to almost 24% of the total fauna. As with the polychaetes and crustaceans, bivalves are represented in each region by a characteristic number; the outer continental shelf, 825/m^2 or 19·5%, the upper continental slope, 500–2000/m^2 or 8·75 to almost 24%; the lower continental slope, almost 100–220/m^2 or 5·4–7·2%; the abyssal rise, 7–31/m^2 or less than one to slightly more than 3%; the Gulf Stream region, 12/m^2 or 6%, the Sargasso Sea, one to less than 10/m^2 or less than one to 7·4%; the lower Bermuda Slope, 2 to 13/m^2 or 1–11%; and the upper Bermuda Slope, 7/m^2 or less than one percent of the total fauna (Table 5 and Fig. 7A).

Two of the pelecypod orders numerically dominated. They are the Eulamellibranchiata and the Protobranchiata which together form 93% of the Pelecypoda. On the outer continental shelf and upper continental slope, the eulamellibranchs are the dominant element forming 77 and 95% of the bivalves. Thereafter, their percent composition diminished continuously with distance from the continent; less than 40% on the lower continental slope; 29% on the abyssal rise; 27% in the region of the Gulf Stream; 21% in the Sargasso Sea; and absent from the samples collected on the Bermuda Slope (Table 5 and Fig. 7b). The protobranchs showed the contrasting distribution pattern. They formed 13 and 3·3% of the pelecypod fauna on the outer continental shelf and upper continental slope, 55% on the lower continental slope, 71% on the abyssal rise, 73% in the region of the Gulf Stream, 75% in the Sargasso Sea, 61% on the lower Bermuda Slope, and 100% on the upper Bermuda Slope.

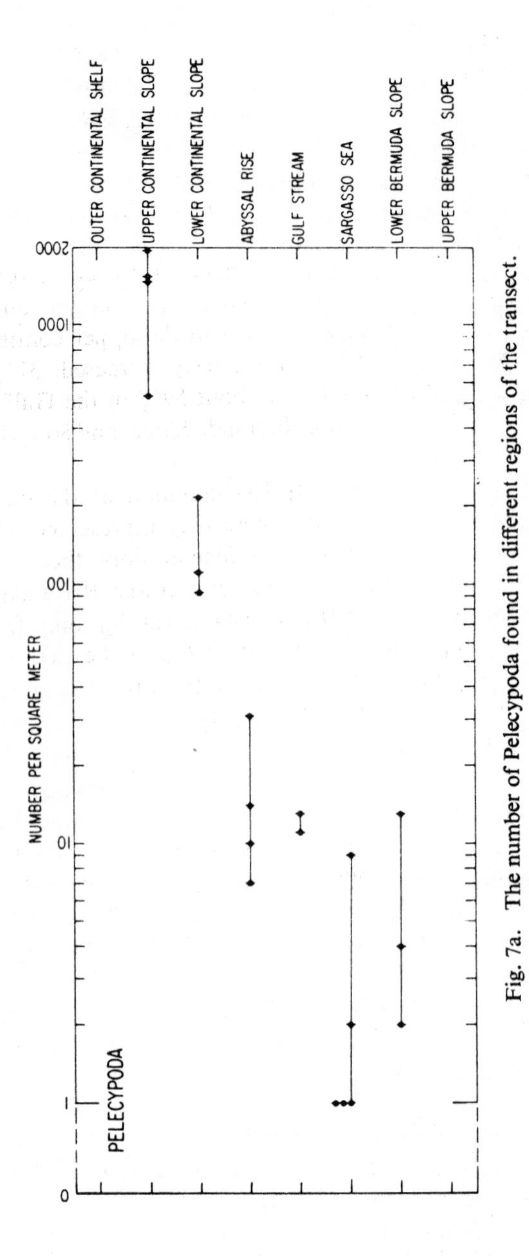

Fig. 7a. The number of Pelecypoda found in different regions of the transect.

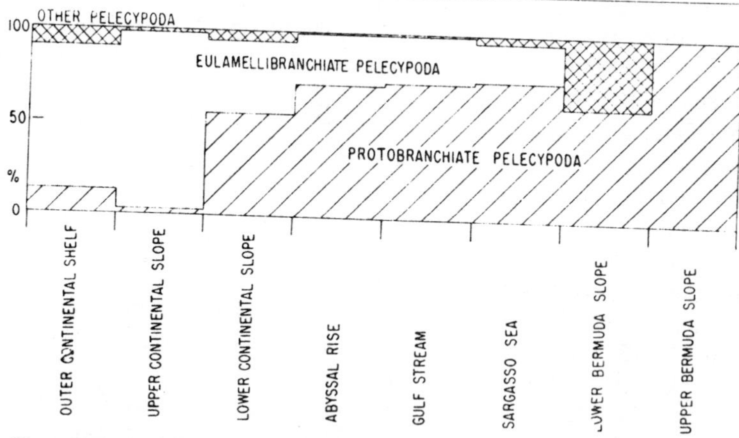

Fig. 7b. The percentage composition of the quantitatively important major pelecypod taxa in the different regions of the transect.

Organic content and animal density

The numerically reduced populations in the Sargasso Sea, Gulf Stream area, and to a lesser extent, on the abyssal rise, almost certainly must be related to the amount or the rate of influx of available food. The vast majority of the animals in all our samples are deposit-feeders and, therefore, the amount of organic matter present in the sediment (as determined through organic carbon and Kjeldahl nitrogen measurements) should give a reasonable approximation of the amount of potential food. Yet when the animal members per square meter are plotted against the organic carbon and Kjeldahl nitrogen values for the various stations, no consistent pattern emerges (Fig. 8).

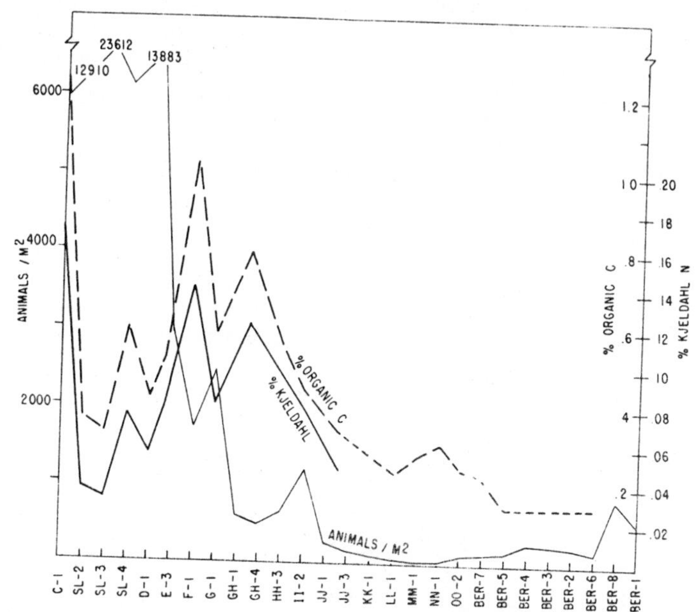

Fig. 8. Animal numbers per square meter, and percent organic carbon and Kjeldahl nitrogen of the sediments found at the various stations of the transect.

The maximum organic carbon value is only eight times that of the minimum. The maximum value for animal numbers exceeds by 600 times the minimum value. Further, there is no discernible overall trend, between carbon or nitrogen values and animal numbers. In fact, some of the lower values for organic content occur on the upper Continental Slope where largest animal numbers are found. (The use of biomass rather than numbers would not basically alter the picture since, if anything, benthic animals, on the average, appear to be slightly smaller at greater depths when compared to the slope regions.)

Organic carbon and Kjeldahl nitrogen measurements do not tell us what fraction of the organic matter is refractory, and what fraction is non-refractory and thus available to the benthic deposit-feeders as food. Therefore, one should not expect to find any simple ratio between organic carbon or nitrogen values and animal numbers. Possibly, in the low depositional environment of the Sargasso Sea, the amount of organic matter is the result of accumulation over an extended period of time and, therefore, may be made up almost entirely of refractory elements. The continental slope, on the other hand, has a higher deposition rate, and a larger percentage of the organic matter in the upper ten centimeters of the sediment (where most of the infauna occur) may be more labile. Through a combination of significant animal activity and a rapid cycling of the labile component brought about by the high animal density, and a greater rate of deposition, the older, more refractory organic matter would be buried at a more rapid rate.

This interpretation finds support from the comparative experiments of WAKSMAN (1933) using sediments obtained from the shallow inshore regions of Buzzards Bay and Cape Cod Bay, Massachusetts, and abyssal sediments, some of which were obtained from localities near our transect (in area of Stas. NN # 1 and OO # 2 and near Bermuda Sta. 7). He found that the organic matter in the samples from the deep-sea oxidized much less readily than that in the near shore sediments indicating that the organic matter in the deep-water sediments is more refractory.

What is needed is a more detailed study of the chemical composition of organic constituents, which would allow determination of what percentage of the organic compound is available to the animals. Analysis of such data may then give closer agreement between amount of labile organics and density of animals

Acknowledgements—We are indebted to Mr. DAVID OWEN who participated in the testing of the dredge and who took excellent bottom photographs which has allowed us to compare and contrast our samples with others taken at the same time and place (OWEN, SANDERS and HESSLER, in press). We also thank Drs. K. O. EMERY, J. HÜLSEMANN and J. S. SCHLEE of this Institution and Mr. THEODORE LODER of Lehigh University for the particle-size and organic analyses of our samples. Dr. RUDOLF S. SCHELTEMA kindly made the data from a number of his benthic samples available to us.

REFERENCES

BELYAEV G. M. (1959) Some regularities in the quantitative distribution of bottom fauna in the Southern Ocean. *First Intern. Oceanogr. Congr. Preprints, Amer. Assoc. Adv. Sci.*, 367–369.

BELYAEV G. M. (1960) Some regularities in the quantitative distribution of the bottom fauna in the western Pacific. (In Russian). *Trudy Inst. Okeanol., Akad. Nauk, SSSR*, **41**: 98–105.

BELYAEV G. M. (1960a) Quantitative distribution of benthos in the Tasmanian Sea and in the Antarctic waters south of New Zealand. (In Russian). *Dokl., Akad. Nauk, SSSR*, **130** (4): 875–878.

BELYAEV G. M. and SOKOLOVA M. N. (1960) Investigation of the bottom fauna of the Mariana Trench. (In Russian). *Trudy Inst. Okeanol., Akad. Nauk, SSSR*, **41**: 123–127.

BELYAEV G. M. and VINOGRADOVA N. G. (1961) An investigation of the Java Trench deep-sea bottom fauna. (In Russian). *Okeanologiya, Akad. Nauk, SSSR*, **1**: 125–132.

BELYAEV G. M. and VINOGRADOVA N. G. (1961a) The quantitative distribution of bottom fauna in the northern part of the Indian Ocean. (In Russian), *Dokl., Akad. Nauk, SSSR*, **138** (5) : 1191–1194.

BELYAEV G. M., VINOGRADOVA N. G. and FILATOVA Z. A. (1960) Investigations of the bottom fauna in the deep-water trenches of the southern Pacific. (In Russian). *Trudy Inst. Okeanol., Akad. Nauk, SSSR*, **41**: 106–122.

BIRSTEIN J. A. (1959) The ultra-abyssal fauna of the Pacific Ocean. *First Intern. Oceanogr. Congr. Preprints, Amer. Assoc. Adv. Sci.* 370–372.

BIRSTEIN J. A., and SOKOLOVA M. N. (1960) Bottom fauna of the Bougainville Trench. (In Russian). *Trudy Inst. Okeanol., Akad, Nauk SSSR*, **41**: 128–131.

CHUKHCHIN V. D. (1963) Quantitative distribution of benthos in the eastern part of the Mediterranean Sea. (In Russian). *Trudy Sevastopol Biol. Sta.*, **16**: 215–233.

FILATOVA Z. A. (1959) Deep-sea bottom fauna communities (complexes) of the northern Pacific. *First Intern. Oceanogr. Congr. Preprints, Amer. Assoc. Adv. Sci.*, 372–374.

FILATOVA Z. A. (1960) On the quantitative distribution of the bottom fauna in the central Pacific. (In Russian). *Trudy Inst. Okeanol., Akad. Nauk, SSSR*, **41**: 85–105.

FILATOVA Z. A. and BARSANOVA N. (1964) The communities of bottom fauna of the western part of the Bering Sea. (In Russian; English summary). *Trudy Inst. Okeanol., Akad. Nauk, SSSR*, **69**: 6–97.

FILATOVA Z. A. and LEVENSTEIN R. J. (1961) Quantitative distribution of the deep-sea bottom fauna in the north-eastern Pacific. (In Russian; English summary). *Trudy Inst. Okeanol., Akad. Nauk, SSSR*, **45**: 190–213.

GRICE G. D. and HART A. D. (1962) The abundance, seasonal occurrence and distribution of the epizooplankton between New York and Bermuda. *Ecol. Monogr.*, **32**: 287–309.

HAGMEIER A. (1930) Eine Fluctuation von *Mactra* (*Spisula*) *subtruncata* da Costa an der ostfriesischen Küste. *Ber. deutsch. wiss. Komm. Meeresforsch.*, **5**: 126–155.

HARTMAN O. (1959) Catalogue of the polychaetous annelids of the world. Pt. 1. *Allan Hancock Foundation Publ., Occas. Pap.*, **23**: 1–353.

HARTMAN O. (1965) Deep-water benthic polychaetous annelids off New England to Bermuda and other north Atlantic areas. *Allan Hancock Foundation Publ., Occas. Pap.*, **28**: 1–378.

HARTMAN O. and BARNARD J. L. (1958) The benthic fauna of the deep basins off southern California. *Allan Hancock Pacific Exped.*, **22** (1): 1–67.

HESSLER R. R. and SANDERS H. L. (1964) The discovery of Cephalocarida at a depth of 300 meters. *Crustaceana*, **7** (1): 77–78.

HESSLER R. R. and SANDERS H. L. (in press). Bathyal Leptostraca from the continental Slope off northeastern United States. *Crustaceana*.

HOLME N. A. (1949) A new bottom-sampler. *J. mar. biol. Ass. U.K.*, **28**: 322–332.

HOLME N. A. (1953) The biomass of the bottom fauna in the English Channel off Plymouth. *J. mar. biol. Ass. U.K.*, **32**: 1–49.

HOLME N. A. (1964) Methods of sampling the benthos. *Advances in Marine Biology*. F. S. RUSSELL, Editor, Academic Press, **2**: 171–260.

HULBURT E. M. (1962) A note on the horizontal distribution of phytoplankton in the open ocean. *Deep-Sea Res.*, **8**: 72–74.

HULBURT E. M. and RODMAN J. (1963) Distribution of phytoplankton species with respect to salinity between the coast of southern New England and Bermuda. *Limnol. Oceanogr.*, **8** (2): 263–269.

HULBURT E. M., RYTHER J. H. and GUILLARD R. R. L. (1960) The phytoplankton of the Sargasso Sea off Bermuda. *J. Cons. Intern. Explor. Mer.*, **25** (2): 115–128.

JOHANSEN A. C. (1927) Preliminary experiments with Knudsen's bottom sampler for hard bottom. *Medd. Komm. Havunders., ser. Fisk.*, **8** (4): 1–6.

KETCHUM B. H., CORWIN N. and VACCARO R. F. (1958) The annual cycle of phosphorus and nitrogen in New England coastal waters. *J. mar. Res.*, **17**: 282–301.

KETCHUM B. H., RYTHER J. H., YENTSCH C. S. and CORWIN N. (1958) Productivity in relation to nutrients. *Rapp. Proc. Verb. Cons. Intern. Explor. Mer.*, **144**: 132–140.

KUZNETSOV A. P. (1960) Data concerning quantitative distribution of bottom fauna of the bed of the Atlantic. (In Russian). *Dokl. Akad. Nauk, SSSR*, **130** (6): 1345–1348.

KUZNETZOV A. P. (1961) Data on the quantitative distribution of the bottom fauna in the Kamchatka Bay. (In Russian). *Trudy Inst. Okeanol., Akad. Nauk, SSSR*, **46**: 103–123.

Kuznetsov A. P. (1963) Benthic invertebrate fauna of the Kamchatka waters of the Pacific Ocean and the northern Kurile Island. (In Russian; English summary). *Akad. Nauk, SSSR, Inst. Okeanol.*: 1–271.

Kuznetsov A. P. (1964) Distribution of the sea bottom fauna in the western part of the Bering Sea and trophic zonation. (In Russian). *Trudy Inst. Okeanol., Akad. Nauk, SSSR*, **69**: 98–177.

Longhurst A. R. (1964) A review of the present situation in benthic synecology. *Bull. Inst. océanogr. Monaco*, **63** (1317): 1–54.

McGill D. A., Corwin N. and Ketchum B. H. (1964) Organic phosphorus in the deep water of the western North Atlantic. *Limnol. Oceanogr.*, **9** (1): 27–34.

Menzel D. W. and Ryther J. H. (1960) The annual cycle of primary production in the Sargasso Sea off Bermuda. *Deep-Sea Res.*, **6**: 351–367.

Menzies R. J., Smith L. and Emery K. O. (1963) A combined underwater camera and bottom grab : a new tool for investigation of deep-sea benthos. *Int. Rev. ges. Hydrobiol.*, **48**: 529–545.

Mironov S. I. and Bordovsky O. K. (1959) Organic matter in bottom sediments of the Bering Sea. *Intern. Oceanogr. Congr. Preprints, Amer. Assoc. Adv. Sci.* : 970–972.

Owen D. M., Sanders H. L. and Hessler R. R. (In press). Bottom photography as a tool for estimating benthic populations.

Richards F. A. (1958) Dissolved silicate and related properties of some western north Atlantic and Caribbean waters. *J. mar. Res.*, **17**: 449–465.

Riley G. A. (1951) Oxygen, phosphate, and nitrate in the Atlantic Ocean. *Bull. Bingham Oceanogr. Coll.*, **13**: 1–126.

Riley G. A. (1957) Phytoplankton of the north central Sargasso Sea, 1950-52. *Limnol. Oceanogr.*, **2** (3): 252–270.

Riley G. A., Stommel H. and Bumpus D. F. (1949) Quantitative ecology of the plankton of the western North Atlantic. *Bull. Bingham Oceanogr. Coll.*, **12**: 1–169.

Ryther J. H. and Menzel D. W. (1960) The seasonal and geographical range of primary production in the western Sargasso Sea. *Deep-Sea Res.*, **6**: 235–238.

Sanders H. L. (1956) Oceanography of Long Island Sound, 1952–1954. X. The biology of marine bottom communities. *Bull. Bingham Oceanogr. Coll.*, **15**, 345–414.

Sanders H. L. and Hessler R. R. (1962) *Priapulus atlantisi* and *Priapulus profundus*. Two new species of priapulids from bathyal and abyssal depths of the north Atlantic. *Deep-Sea Res.*, **9** (1/2): 125–130.

Smith W. and McIntyre A. D. (1954) A spring-loaded bottom-sampler. *J. mar. biol. Ass. U.K.*, **33**: 257–264.

Sokolova M. N. and Pasternak E. A. (1962) The quantitative distribution of the bottom fauna in the northern part of the Arabian Sea and Bengal Bay. (In Russian). *Dokl. Akad. Nauk, SSSR*, **144** (3): 645–648.

Spärck R. (1951) Density of bottom animals on the ocean floor. *Nature, Lond.* **168** (4264): 112–113.

Spärck R. (1956) The density of animals on the ocean floor. In: *The Galathea Deep-Sea Expedition 1950-1952*, 196–201.

Thamdrup H. M. (1938) Der van Veen-Bodengreifer. Vergleichsversuche über die Leistings-fähigkeit des van Veen- und des Petersen-Bodengreifers. *J. du Conseil.*, **13**: 206–212.

Trask P. D. (1939) Organic content of recent sediments. In: P. D. Trask, Editor, *Recent Marine Sediments, Am. Ass. Petrol. Geol., Tulsa* pp. 428–453.

Ursin E. (1954) Efficiency of marine bottom samplers of the van Veen and Petersen types. *Medd. Komm. Danm. Fisk., Havunders.*, **1** (7): 1–7.

Vaccaro R. F. (1963) Available nitrogen and phosphorus and the biochemical cycle in the Atlantic off New England. *J. mar. Res.*, **21** (3): 284–301.

Vinogradova N. G. (1962) Some problems of the study of deep-sea bottom fauna. *J. oceanogr. Soc. Japan, 20th Anniversary Vol.*, 724–741.

Waksman S. A. (1933) On the distribution of organic matter in the sea bottom and chemical nature and origin of marine sediments. *Soil Sci.*, **36**: 125–147.

Wigley R. L. and McIntyre A. D. (1964) Some quantitative comparisons of offshore meiobenthos and macrobenthos south of Martha's Vineyard. *Limnol. Oceanogr.*, **9** (4): 485–493.

Zenkevitch L. A. (1958) Certain zoological problems connected with the study of the abyssal and ultra-abyssal zones on the ocean. *XVth Intern. Congr. Zool. Sect.* III: 215–218.

ZENKEVITCH L. A., BARSANOVA N. G. and BELYAEV G. M. (1960) Quantiative distribution of bottom fauna in the abyssal area of the World Ocean. (In Russian). *Dokl. Akad. Nauk. SSSR*, 130 (1): 183–186.

ZENKEVITCH L. A., BELYAEV G. M., BIRSTEIN J. A. and FILATOVA Z. A. (1959) The quatitative and quantitative characteristic of the deep-sea bottom fauna. (In Russian). *Itogi Nauku, Dostijenija Okeanologii*, **1**: 106–147.

ZENKEVITCH L. A. and FILATOVA Z. A. (1960) Quantitative distribution of bottom fauna in the northern part of the Pacific at a depth below 2000 m. (In Russian). *Dokl. Akad. Nauk, SSSR*, **133** (2): 451–453.

Deep-Sea Research, 1956, Vol. 4, pp. 54 to 64. Pergamon Press Ltd., London.

Studies of the deep water fauna and related problems

L. A. ZENKEVICH and J. A. BIRSTEIN

Abstract—1. Investigations on the Kuril Trench, carried out in 1953 by the Soviet expedition on the Research Ship *Vitjaz* allowed several general principles to be laid down in the change of the bottom and the pelagic faunas with increasing depth.

2. The biomass of benthos and plankton in the ultra-abyssal zone (> 6,000 m) is 1,000 times less than in the surface zone.

3. With the increase of depth the variety of species of benthos decreases ; its diversity increases at first and then decreases, for the most part within the ultra-abyssal zone.

4. The colouring change of bottom and pelagic animals also takes place differently according to depth. Many pelagic animals are pigmented up to depths of 6-7,000 m. Only the plankton of the ultra-abyssal zone is totally devoid of pigment (except the amphipad *Andaniexis subabyssi* Birst. et M. Vin.). The bottom animals become pigmentless, at depths greater than about 2,000 m. Possibly this dissimilarity is due to vertical migrations of broad range peculiar to pelagic animals.

5. The relative importance of predators of plankton increases with depth. This contrasts with the bottom animals, for which there is an increase in relative importance of detritus eaters and a lesser importance of predators at increasing depths.

6. One of the many sources of food supply for the abyssal and ultra-abyssal fauna is provided by the transportation of organic matter from the zone of photosynthesis down to the great depths by means of vertical migration of zooplankton, performed either within each fixed vertical zone, or having a still broader range. During their vertical migrations the upper layers of deep-sea animals become the prey of the lower dwellers and thus the organic matter produced in the zone of photosynthesis is brought to the greatest depths of the ocean.

7. The body-size comparison of species of the same genus inhabiting different depths shows a tendancy to greater body-size with greater depth. This is true for both bottom and pelagic animals. It cannot be explained by temperature changes.

8. The deep-sea fauna comprises a great quantity of ancient and primative forms.

9. The zoogeographic regions of the ocean ought to be determined from separate vertical zones. In addition, the determination of zoogeographic regions of the abyssal zone will show not only the connections and the hydrological regime of the contemporary seas and oceans but also those of the seas and oceans of the remote past.

10. The ultra-abyssal faunas of each trench must comprise endemic species or even units of a higher taxonomic grade. A set of these species was discovered in the extreme depths of the Kuril Trench.

INVESTIGATION of faunas inhabiting the great depths of the ocean only started in the last half of the 19th century and until recently has been confined to depths of less than about 6 km. In the last seven years, however, new and extensive material has been obtained on the deep-water fauna of all oceanic depths down to the greatest, of more than 10 km. These collections were made mainly by two expeditions: the Soviet expedition on the *Vitjaz* (started in 1949) (Fig. 1)* and the Danish expedition on the *Galathea* (1951-1952). The *Galathea* has explored several deep-sea depressions in different parts of the ocean, but only in respect to their bottom fauna and bacteria.** The work of the *Vitjaz*, though confined to the north-western Pacific and bordering seas, embraces a complex of oceanographic research, including biological investiga-

*Fig. omitted

** *Galatheas Jordomsejling* 1950-52, 1953. Under redaction of BRUUN, GREVE, MIELCHE and SPÄRCK.

tions. A detailed study of the Kuril Trench involving its whole water mass as well as its floor down to the greatest depths, has revealed some important peculiarities in the vertical and horizontal distribution of life through the whole immensity of its 10 km deeps.

Table 1. Vertical faunistic zones of the ocean.

Zones	*Subzones*	*Depth (m)*	*Percentage of the whole surface area of the World Ocean*
Surface	littoral	over 0	0·004
	sublittoral	0–200	7·6
Transitional		200–600	3·5
Abyssal	upper abyssal	600–2,000	5·0
	lower abyssal	2,000–6,000	82·7
Ultra-abyssal†		6,000–10,862	1·2

To begin with, the north-western Pacific is characterised by what appears to be a rather rich deep-water fauna. Nineteen species of pogonophores have been recorded, ten species of echiurids, some scores of amphipod species, echinoderms, especially holothurians and many others. Future investigations will show whether these findings reflect a real quantitative diversity of the deep-water fauna of the Kuril Trench, or whether they are merely the result of the very thorough investigations carried out by the *Vitjaz*.

Table 2. Changes of biomass of zooplankton with increasing depth (after VINOGRADOV, *1954).*

Horizon (m)	0–50	50–100	100–200	200–500	500–1,000	1,000–2,000	2,000–4,000	4,000–6,000	6,000–8,000
Biomass (mg/m^3)	497·6	320·3	246·6	228·0	59·3	21·8	9·3	2·64	0·48

Table 3. Changes of the biomass of benthos with increasing depth (after ZENKEVITCH, BIRSTEIN *and* BELJAEV, *1955).*

Depth (m)	950–4,070	5,070–7,230	8,330–9,950
Biomass (g/m^2)	6·94	1·22	0·32

The material obtained permits a scheme to be drawn of the vertical zoning in the distribution of oceanic faunas. The scheme, originally suggested for the plankton

† For the deep zone over 6,000-7,000 m and the fauna populating it, ZENKEVITCH put forward in 1953 terms such as: super-oceanic depths and superdeepsea fauna. BIRSTEIN (1954) preferred terms: ultra-abyssal zone and ultra-abyssal fauna. BRUUN (1956) proposed the term: hadal.

organisms (BIRSTEIN, VINOGRADOV and TCHINDONOVA, 1954) was found to apply also to benthos and nekton.

Thus the living space occupied by the deep-water fauna is far greater than all the rest of the biosphere and extends over two-thirds of the whole Earth's surface.

Considering the changes occurring in the benthic and pelagic fauna with increasing depth, we discover some striking features of similarity and dissimilarity. First of all we notice the similar characteristics of decrease of biomass with increasing depth. The following indices, illustrating this phenomenon, were obtained for zooplankton and benthos of the Kuril Trench (Tables 2 and 3).

Considering that on the continental shelf of the north-western Pacific the benthic biomass is expressed by a value of some hundred g/m² we must admit that the biomass of plankton and benthos undergo a similar decrease when reaching the ultra abyssal zone, i.e. they diminish by about 1,000 times.

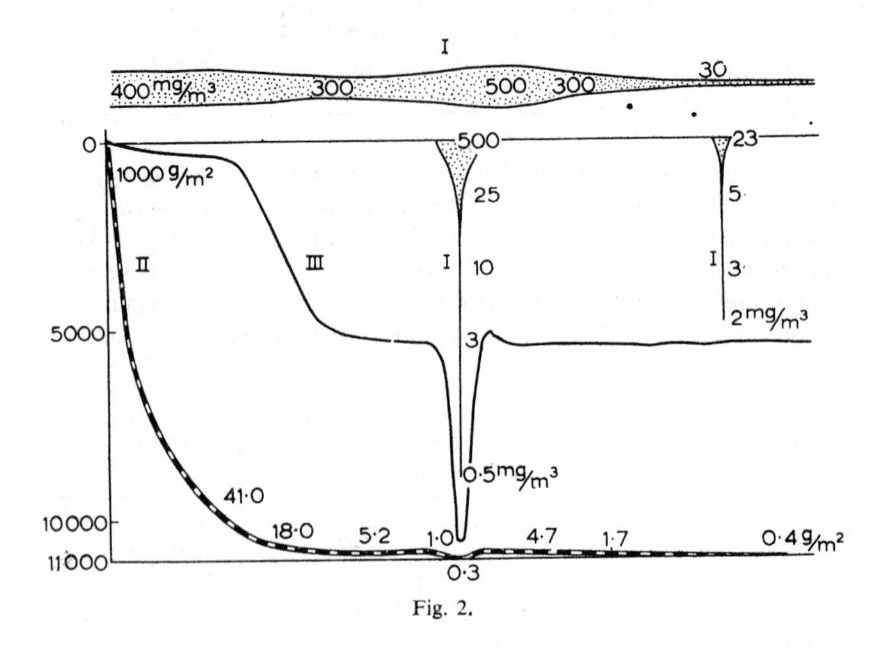

Fig. 2.

The foregoing remarks are well illustrated by Fig. 2 showing a section through the Kuril Trench and the adjacent part of the Pacific (III), with the characteristic biomasses of plankton (I) and benthos (II)* of this region. The biomass of surface plankton is expressed by a value of 300-500 mg/m³ over a considerable area, but it increases abruptly in the region of the junction of warm and cold waters and drops to some few mg/m³ in the zone influenced by the warm Kuroshio waters. The decrease related to depth is still more abrupt. The biomass of benthos also decreases with depth, but, in this case, contrary to that of the plankton there is also a regular decrease with distance from land, observed even on stations taken at identical depths. This is determined by a lesser supply of food, since the nutrient material is produced in the coastal zone and brought by outflow from land.

*The scale of the ordinate bears no relation to the benthos biomass curve (II), but correlates with the bottom relief mass curve (III) only.

If we consider the variations related to depth, of some qualitative indices of the fauna, we find very striking differences between benthos and plankton, and, first of all in the number of species (Fig. 3). It has been demonstrated for several systematic groups of bottom animals, that their diversity decreases with depth.

If we take the total number of species in each systematic group as 100 per cent, this decrease will be expressed by the following indices (Table 4).

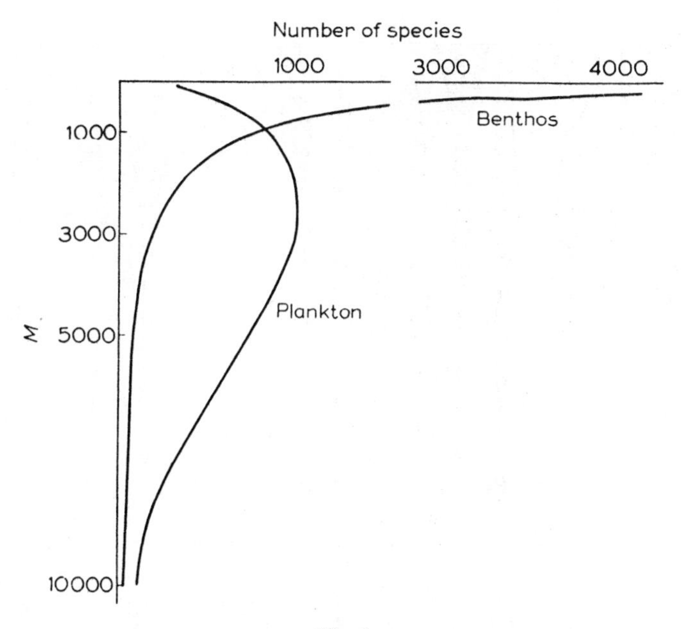

Fig. 3.

*Table 4. Relative specific variety of some groups of bottom animals (percentage of total number of species).**

Depth (m)	*Over* 1,000	*Over* 2,000	*Over* 3,000	*Over* 4,000	*Over* 5,000
Hydroidea	?	?	0·4	0·4	0·2
Polychaeta	?	3·7	2·6	1·7	0·6
Peracarida	?	1·3	?	?	?
Pantopoda	?	9·7	4·5	1·5	?
Bryozoa	?	?	1·2	0·4	0·1
Crinoidea	21·6	7·7	2·9	1·8	0·5
Echinoidea	18·7	?	5·4	1·6	0·5
Ophiuroidea	?	?	5·2	2·2	0·4
Asteroidea	33·3	?	9·3	3·7	1·2
Holothurioidea	22·7	?	12·3	8·4	2·8

These averages, obtained for all the oceans of the world are corroborated by the material collected from the Kuril Trench. An analogous decrease was observed ranging from 100 species at a depth of 1,000 m to 6 species at 9,700-9,950 m. This impoverishment of the bottom fauna of the Kuril Trench is illustrated by Fig. 4.

* After different data, published by XIV Intern. Zool. Congress.

A different regularity was established for the systematic groups of plankton animals investigated in this respect. Their diversity increases with depth, with a maximum in the abyssal zone. This was well demonstrated on copepods by BRODSKY (1952) (Table 5).

Similar data were obtained by BIRSTEIN and VINOGRADOV (1955) for pelagic gammarids (Table 6).

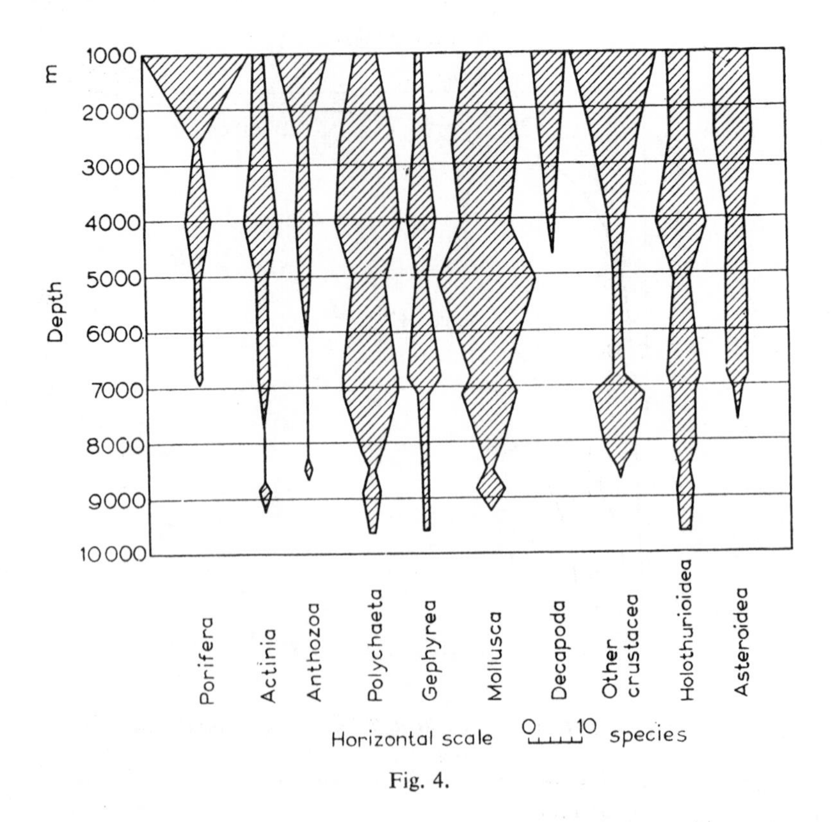

Fig. 4.

Table 5. Number of species of calanids per 1 m^3 at different levels of the Kuril Trench (after BRODSKY, 1952).

Horizon (m)	0–25	25–50	50–100	100–200	200–500	500–1,000	1,000–4,000
Number of species	7	7	9	10	28	30	87

An increase in specific variety related to depth was observed also in chaetognaths, pelagic coelenterates, mysids, prawns and some others. However on the extreme limits of oceanic depths, in the ultra-abyssal zone, the specific variety of plankton shows an abrupt decrease, similar to that of the bottom fauna (Fig. 3).

Another very conspicuous dissimilarity observed in the reaction of benthos and plankton animals to increasing depth is their different colouring.

Strongly pigmented bottom organisms are prevalent at depths lesser than 1,000 m. But in the abyssal zone, and especially in the lower abyssal subzone, white and yellowish pigmentless organisms become predominant. On the contrary the plankton of the abyssal zone is characterised by a deep red colouring; and only representatives of the ultra-abyssal plankton (excepting the amphipod *Andaniexis subabyssi*) are all devoid of pigment.

The intense dark red colouring of abyssal plankton can be considered as a strong argument in favour of the existence of a powerful system of vertical migrations performed by these planktonic organisms. We know that traces of perceptible light penetrate in the water no deeper than for 1·5 km nevertheless red planktonic animals inhabit depths of 6-7 km. It seems quite possible that they ascend periodically to a zone where they become exposed to the action of light and are using the protective properties of their pigment. On the other hand, the unpigmented plankton of the ultra-abyssal zone is evidently fully isolated from the illuminated water-layers. Abyssal bottom animals do not perform vertical migrations nor do they develop protective pigments.

Of special interest are dissimilarities related to depth, observed in the feeding habits of bottom- and plankton animals.

Table 6. *Number of species of pelagic gammarids in different vertical zones of the Kuril Trench* (*after* BIRSTEIN *and* VINOGRADOV, 1955).

Zone and subzone	*Number of species*
Surface	1
Transitional	1
Upper-abyssal	13
Lower-abyssal	12
Ultra-abyssal	6

Prior to the investigations carried out by the *Vitjaz* almost no attention was given to the feeding of deep-sea animals. Some scattered information is found in the literature on analyses of the gut contents of a leptostracan crustacean *Nebaliopsis typica* (Rowett 1943), three species of deep-water crabs (DOFLEIN, 1904) and soft sea-urchins Echinothuriidae (MORTENSEN, 1938); the gut of the latter contained, among other materials, remains of land and coastal vegetation.

SOKOLOVA (in press) has investigated in detail the content of the intestines of bottom animals; it was found that with increasing depth bottom-detritus-eaters gradually acquire a higher relative importance in the bottom fauna, whereas predators and suspension-feeders become less important. At the extreme depths investigated (9,800 m) all suspension-feeders disappear, as well as almost all carnivorous benthos-eaters, with the exception of polychaetes of the genus *Maccellicephala*; of the bottom-detritus-eaters only roughly sorting holothurians of the Elasipoda group and echiuroids are left.

As regards the plankton fauna, the investigations of TCHINDONOVA and VINOGRADOV (in press) have shown that in this case, on the contrary, the relative importance of predators greatly increases with increasing depth. It is only in the surface zone, at night, that pelagic animals utilise the abundant phytoplankton.

When these herbivorous forms return to greater depths they become the prey of numerous deep-water predators. Chaetognaths, pelagic prawns, amphipods and mysids, jellyfish and ctenophores feed on the plant-eaters, and furthermore devour one another. Among the many predators are forms able to use various kinds of food. Thus the mysid *Gnathophausia gigas* and the euphausiid *Bentheuphausia amblyops* attack live animals, eat their excrements and obtain food by means of filtration. On the other hand the leptostracan crustacean *Nebaliopsis typica*, which is provided with a pouch-like diverticulum of the middle-gut feeds exclusively on eggs of deep-sea animals (ROWETT, 1943) and is able to store a considerable supply of food in this intestinal pouch.

As was to be expected, all the deep-sea fishes investigated were found to be predators. Some of them are very selective in their diet. Thus, according to BIRSTEIN and VINOGRADOV (1955a) *Acrolepis* feeds mainly on luminescent animals.

Closely related to the study of the feeding of deep-sea animals is another extremely important, though little studied problem – that of the dynamics of organic matter at great oceanic depths. In approaching this problem we are first of all confronted with the question of the main sources of food supply.

KROGH (1934) suggested two possible sources of food for the abyssal fauna: (1) excrements and corpses of surface animals and (2) dissolved organic matter resulting from the biological activity and decay of surface animals, assimilated by hypothetical deep-water organisms, as yet unknown, which, in their turn, serve as food to deep-water animals. The first source seems improbable, taking into account the considerable rate of mineralisation of dying-off organisms and of their excrements, and the time they need to descend to the immense oceanic depths. The second assumption, not based on any facts as the author himself admits, remains a speculative hypothesis. Recently many investigators (EKMAN, 1953; YONGE, 1953; FAGE, 1953; ZOBELL, 1953) have emphasized the paramount importance of bacteria in the nutrition of deep-sea animals. According to ZOBELL (1953) heterotrophic bacteria are using dissolved and colloidally dispersed organic matter rather than the formed fragments of the inhabitants, of upper water layers. VINOGRADOV (1953, 1955) approaches the question from another standpoint – that of the probable importance of vertical migrations in transporting organic matter produced in the zone of photosynthesis down to the great oceanic depths.* The pelagic crustaceans feed at night at the surface and descend in the daytime to depths of 500 or even 800-1,000 m where they become the prey of the deep-water predators, the latter themselves being unable to ascend to the upper layers.

It may be assumed that when these predators return to their greater depths, they become in their turn the prey of the lower abyssal dwellers. Thus a ladder of migration is formed, by which organic matter produced in the upper layers is brought to the greatest depths of the ocean.

Notwithstanding all these data on the nutrition of deep-sea animals and its sources, the problem of the dynamics of organic matter is far from being solved. We cannot even say with certainty whether deep-sea animals live under conditions of acute food shortage, or whether they are supplied with food in sufficient quantities. In any case some of their food habits and processes of food digestion are very peculiar.

* The significance of vertical migrations in transference of food resources to deep depths is pointed out also by MARSHALL (1954) and GRAY (1956).

Most striking in this respect are the data recently obtained by IVANOV (1955) on the total absence of an intestine in representatives of a most remarkable class of deep-sea animals, the Pogonophora. IVANOV has discovered on the cephalic tentacles of these animals, giant-cell pinnulae which are penetrated by blood vessels; he believes that the pogonophores digest their food externally by means of glandular cells set at the base of the pinnulae. The digested food is passed through the pinnulae into the circulatory system, and distributed through the whole body.

The question of the food supply of abyssal fauna is of great interest in relation to another remarkable peculiarity of deep-sea animals. A comparison of related species of the same genus, but inhabiting different depths, usually discloses a definite tendency to greater body-size with greater depth. This is observed both in bottom and in plankton animals, and seemingly does not depend on the character of food nor on the manner of feeding. This increase in body size can be illustrated by the example of *Amblyops* – one of the eurybathic genera of mysids: the largest of its species was collected by the *Vitjaz* from the greatest depth of the Kurile-Kamchatka trench (Table 7).

Table 7. Depth of habitat and body-size of different species of the genus Amblyops.

Species	*Depth* (*m*)	*Length* (*mm*)
A. kempi Tattersall	700–1,463	16
A. tenuicauda Tattersall	820–1,400	17
A. abbreviata M. Sars	366–1,372	18
A. ohlini Tattersall	1,940–1,980	25
A. crozeti G. O. Sars	2,930	30
A. magna n. sp.	7,800	38

This increase in size with depth is even more strikingly expressed in species of the deep-water isopod *Haplomesus*. Two new species of this genus were discovered by the *Vitjaz* in the same trench.

Up to now six species of this genus had been known: five from the North Atlantic and one from the Polar Basin. The length of the Atlantic species found in depths ranging from 1,350 to 3,276 m varies from 2 to 4·5 mm and the arctic species taken at a depth of 698 m, has a length of 4·5 mm, in comparison the first of the two new species of *Haplomesus* discovered by the *Vitjaz*, from a depth of 8,000-8,200 m measured 14 mm, and the second, from 4,860 m – 7·5 mm.

An analogous increase in size is recorded for cirripeds of the genus *Scalpellum* and some other forms. Thus, for instance, the mysid *Eucopia australis*, characteristic of the lower abyssal subzone is considerably larger than *E. grimaldii* which lives in the upper abyssal subzone. It is interesting that the same tendency to an increase in body-size with transition to life in greater depths is manifest in the ontogeny of some deep-sea animals. Large mature, or nearly mature individuals of many pelagic animals, such as the mysid *Gnathophausai gigas*, the euphausiid *Bentheuphausia amblyops*, the leptostracan *Nebaliopsis typica* – inhabit greater depths than their smaller young stages.

Finally, attention has already been drawn to the giantism of the deep-water representatives of certain systematic groups. For example, there is the foraminiferan

Bathysiphon, common at great depths of the north-western Pacific, whose tubular shell reaches a length of 15 cm.

From this evidence, the question arises of the cause of this increase in size of many deep-sea animals with increasing depth of habitat.

The factor of temperature is obviously of no great importance, as no significant changes of temperature occur between depths of 2-5 km and 6-8 km. Neither can differences in food habits and quality of food affect the body-size of deep-sea organisms, considering that deep-water giantism is observed in forms having different food habits and using various kinds of food. The only conclusion possible is that there are some peculiarities in the metabolism of many deep-water organisms which determine this charactistic phenomenon of giantism.

There is yet another aspect to be considered of the marked changes undergone by deep-water fauna with increasing depth of habitat – namely the relationships of species inhabitating different depths. In the process of systematic analysis of the collections of the *Vitjaz* it becomes increasingly evident that the most ancient and primitive representatives of different groups of animals tend to concentrate in greater depths. Thus among isopods representatives of the most primitive suborder Asellota prevail; among mysids representatives of the most primitive family Lophogastridae and primitive tribe Erythropini; among decapods representatives of the almost extinct ancient family Eryonidae, the primitive family Oplophoridae, known as fossil from the Lower Cretaceous, occur; the stalked sea lilies, preserved from an even more remote past; the family Bonneliidae among the Echiuroidea; the family Aphroditidae among Polychaeta, etc.

Along with these ancient and primitive forms the abyssal fauna also comprise species with a younger phylogeny: such as some ophiuroids, starfishes and others.

Many investigators, especially SVEN EKMAN (EKMAN, 1953) have noticed this conservation of ancient groups and forms in the abyssal depths. This fact is of a great interest not only in itself but also in its bearing on the zoogeography of the ocean. Considering the predominance of ancient forms at greater oceanic depths we may surmise that a zoogeography of oceanic depths would differ from a zoogeography of the sublittoral region, and would reflect not so much the recent relations and hydrographical features of seas and oceans, as the corresponding characteristics of very ancient seas, where the primitive groups were formed, which at present inhabit the great oceanic depths.

There are facts showing evidence that deep-sea faunas on both sides of Central America differ in their composition, whereas the shallow-water faunas of the same parts of the ocean are very much alike. This may be explained by the fact that during the Mesozoic and Palaeozoic era, the shallow-water region existing in place of the isthmus of Panama, prevented an exchange of deep-water faunas between the Atlantic and Pacific Oceans. The dissimilar features of these two faunas seem to be of a more ancient origin than the similar features of shallow-water faunas.

It was also found that the waters beneath the mean depth of the Pacific, are inhabited by some scores of families, genera and species of benthic and planktonic animals, many of which never occur at lesser depths. A considerable part of these species proved to be new: furthermore some unknown taxonomic units of a higher rank were discovered. Thus there were several species of pogonophores, one new family of pelagic amphipods (Vitjazianidae), two ultra-abyssal genera of the family

Bonneliidae (*Jakobia*, *Vitjazema* and *partim Alomasoma*) and others. It is possible that this fauna, isolated systematically from the neighbouring fauna of the lower abyssal subzone, was formed after the appearance of the greatest depths of the Kuril Trench, by a descent of the inhabitants of the lower abyssal to the extreme depths. It may be assumed that analogous ultra-abyssal faunas in other oceanic depressions were formed in a similar way. If this assumption proves to be true, the degree of isolation of the ultra-abyssal fauna in a depression from the fauna in the lower abyssal depths, can be used as an index of the chronology of its extreme depths. According to the geological data available the known oceanic depressions are of diverse ages, so that we may expect each of them to harbour a rather distinct ultra-abyssal fauna.

Due to the zoogeographical heterogeneity of the fauna at different depths the oceans of the world have to be divided into zoogeographical regions not as a whole but by separate vertical zones. This was made manifest during the systematic analysis of some animals from the Kuril Trench. Thus, pelagic gammarids from this trench could be divided, according to the character of their habitat, into five groups : Pan-oceanic, Atlanto-pacific, Arctic, North-pacific and Endemic species. It was found that the ratios of these groups change abruptly from one vertical zone to another (Table 8).

Table 8. *Ratio of zoogeographic groups of pelagic gammarids in different vertical zones of Kuril Trench* (*percentage of species*) (*after* BIRSTEIN *and* VINOGRADOV, 1955).

	Pan-Oceanic	*Atlanto-Pacific*	*Arctic*	*North Pacific*	*Endemic*
Upper abyssal subzone	46·1	38·5	7·7	7·7	–
Lower abyssal subzone	8·3	33·3	–	8·3	50
Ultra-abyssal zone	–	–	–	–	100

There are reasons to assume that the connections between faunas of the Pacific and the Atlantic oceans, established for the upper and lower abyssal subzones, have been produced through different channels: mainly through the Polar Basin for the upper abyssal subzone, and through the Antarctic and Indian ocean – for the lower abyssal subzone.

Institute of Oceanology,
Academy of Sciences of the U.S.S.R.

REFERENCES

BIRSTEIN J. A. and VINOGRADOV M. E. (1955) *Trans. Inst. Ocean.* USSR, vol. XII.
BIRSTEIN J. A. and VINOGRADOV M. E. (1955a) *Zool. J.*, vol. XXXIV, No. 4.
BIRSTEIN J. A., VINOGRADOV M. E. and TCHINDONOVA J. G. (1954) *C. R. Acad. Sci.* USSR, vol. XVC, No. 2.
BRODSKY K. A. (1952) *Invest. Far. East Seas*, USSR, vol. III.
BRUUN A. F., (1956) *Nature*, **177**.
DOFLEIN F. (1904) *Wiss. Erg. deutsch. Tifsee-Exp. Valdivia*, Bd. VI.
EKMAN SV. (1953) *Zoogeography of the Sea.*
EKMAN SV. (1953) XIV *Zool. Congr.*, Copenhagen.

GRAY M. (1956) *Fieldiana : Zoology*, vol. 36, No. 2.
IVANOV A. V. (1955) *C. R. Acad. Sci.*, USSR, vol. 100, No. 1.
IVANOV A. V. (1955) *ibid.*, No. 2.
KROGH A. (1934) *Ecol. Monographs*, No. 4.
MADSEN, E. J. (1953), *XIV Zool. Congr.*, Copenhagen.
MARSHALL N. B. (1954) Aspects of deep-sea biology, London.
MORTENSEN TH. (1938) *Annot. Zool., Japon.*, vol. 17.
ROWETT H. C. Q. (1943) *Discovery Rep.*, XXIII.
VINOGRADOV M. E. (1953) *Priroda*, No. 6.
VINOGRADOV M. E. (1954) *C. R. Acad. Sci.*, USSR, vol. XCVI, No. 3.
VINOGRADOV M. E. (1955) *Trans. Inst. Ocean.*, USSR, vol. XIII.
YONGE C. M. (1953) *XIV Intern. Zool. Congr.*, Copenhagen.
ZENKEVICH L. A. (1952) *Priroda*, No. 2.
ZENKEVICH L. A. (1953) *XIV Intern. Zool. Congr.*, Copenhagen.
ZENKEVICH L. A. (1954) *Union international des sciences biologiques*, ser. B., No. 16.
ZENKEVICH L. A., BIRSTEIN J. A. and BELJAEV G. M. (1954) *Priroda*, No. 2 (German translation in " Urania," Jahrg. 18, H.1, 1955).
ZENKEVICH L. A. BIRSTEIN J. A., and BELJAEV G. M. (1955) *Trans. Inst.* USSR, vol. XII.
ZOBELL CL. E. (1953) *XIV Intern. Zool. Congr.*, Copenhagen.

Deep-Sea Research, 1959, Vol. 5, pp. 205 to 208. Pergamon Press Ltd., London. Printed in Great Britain

The zoogeographical distribution of the deep-water bottom fauna in the abyssal zone of the ocean

N. G. VINOGRADOVA

(*Received* 14 *April* 1958)

Abstract—Analysis of the geographical distribution of certain groups of benthic invertebrates (1031 species) in depths greater than 2000 m showed that 84 per cent of the species are confined to a single ocean and only 4 per cent occur in the Pacific, Atlantic and Indian Oceans. Each ocean has its own characteristic fauna. In the Pacific and Atlantic Oceans, 75 per cent of the deep-sea species are endemic. In the Indian Ocean, the percentage is somewhat lower. However, the genera and families of deep-sea animals are generally widely distributed. The specificity (endemism) of the fauna increases with increasing depth of habitat.

Three zoogeographical regions can be distinguished in the abyssal zone of the ocean : the Pacific-North Indian, the Atlantic and the Antarctic. These may be divided into sub-regions and provinces. One of the chief factors affecting the distribution of deep-sea bottom fauna seems to be the configuration of the ocean floor, especially the macrorelief. Consequently, the boundaries between zoogeographical regions follow underwater ridges and other elevations.

The species of deep-sea bottom invertebrates with a wide vertical range generally have an extensive horizontal range as well. Almost all cosmopolitan deep-sea forms are eurybathic. The stenobathic species, on the contrary, occur only in a restricted area.

THE literature on investigations of the ocean depths by both Russian and foreign expeditions since the middle of the last century, as well as analyses of the deep-water fauna collected by the " *Vitjaz* " in the Far Eastern seas of the Soviet Union and the north-west part of the Pacific Ocean, make it possible to obtain some information on the distribution of 1031 species of deep-water Porifera, Coelenterata, Cirripedia, Isopoda, Decapoda, Pantopoda, Echinodermata (with the exception of Ophiuroidea), and Pogonophora, found at depths of over 2000 m. Of these species, 84 per cent are confined to one ocean, about 15 per cent occur in two oceans and only 4 per cent in all three. The genera and families on the other hand are, for the most part common to all three oceans.

Of the species common to more than one ocean, small numbers (2·1 per cent) are common to the Atlantic and Indian oceans and twice as many to the Indian and Pacific oceans. These figures contradict MADSEN's (1953) theory that the deep-water fauna of the Indian and Atlantic oceans are the more closely related.

In the Pacific and Atlantic oceans, 73·2 per cent and 76 per cent of the deep-water species respectively are endemic. The large percentage of deep-water species found only in the Atlantic indicates that the fauna there is no less distinctive than that in the Pacific. The Indian Ocean, on the other hand, has the smallest proportion of endemic deep-water species.

Another feature peculiar to the Indian Ocean fauna is that it is not uniform throughout the ocean. Only 2·4 per cent of the deep-water species are found both in the northern (north of the line between the Cape of Good Hope and Cape Naturalist) and southern parts. A very large number of species in the Indian Ocean also occur in the Pacific Ocean (47·7 per cent and particularly in its northern part). In the southern Indian Ocean, there is an increasing number of deep-water species typical of the Atlantic, but most of them are Antarctic species found in the other oceans, but only south of 40°S. The distribution of the deep-sea fauna of the Indian Ocean agrees well with the paleogeographic reconstruction of the seas of the late Mesozoic especially with the contours of the Tethys. Accordingly, it reflects the influence of past geologic eras.

Table 1. The number of deep-water species common to the different oceans with relation to the depth at which they occur

Depths at which species occur	*Species common to Antarctic and all other oceans in percentage of Antarctic species (in round numbers)*			*Species common to West and East Atlantic in percentage of total No. of species in given area*	*Species common to western and eastern parts of the North Pacific in percentage of total No. of species in given area*
	Atlantic	*Indian*	*Pacific*		
Deep-sea species rising to depths of less than 2000 m	70	60	27	49	49
Species from depths below 2000 m	15	40	4·3	12·9	12·6
Species from depths below 3000 m	6	10	2·5	7·2	15·6
Species from depths below 4000 m	0	0	0	0	1·2

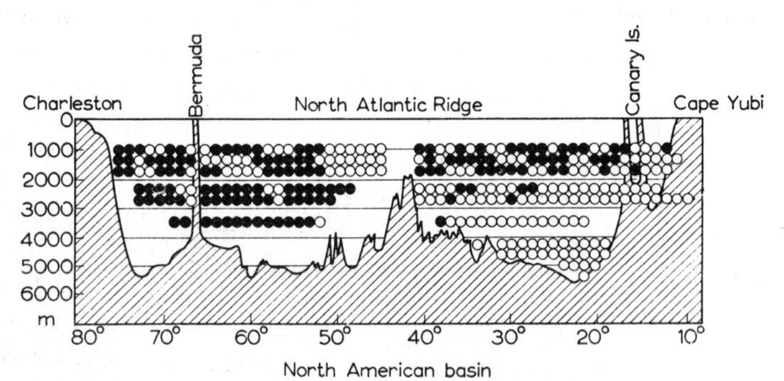

Fig. 2. Vertical cross-section of the North Atlantic Ocean to show the distribution of species of bottom invertebrates of the North American and North African basins, at various depths. The black circles represent the species of the North American basin, the open circles the species of the North African basin.

From the geographical distribution of the abyssal fauna and bottom invertebrates, we have divided the ocean into the following zoogeographical areas (Fig. 1). This division of the Abyssal Zone refers chiefly to the upper-abyssal sub-zone (3000-4500m). The boundaries separating the faunas at greater depths might be drawn rather differently due to the greater degree of isolation of the organisms living there. Thus, with increasing depth the faunas of the various areas become increasingly restricted (Table 1) or, in other words, the area of occurrence of the deep-water species tends

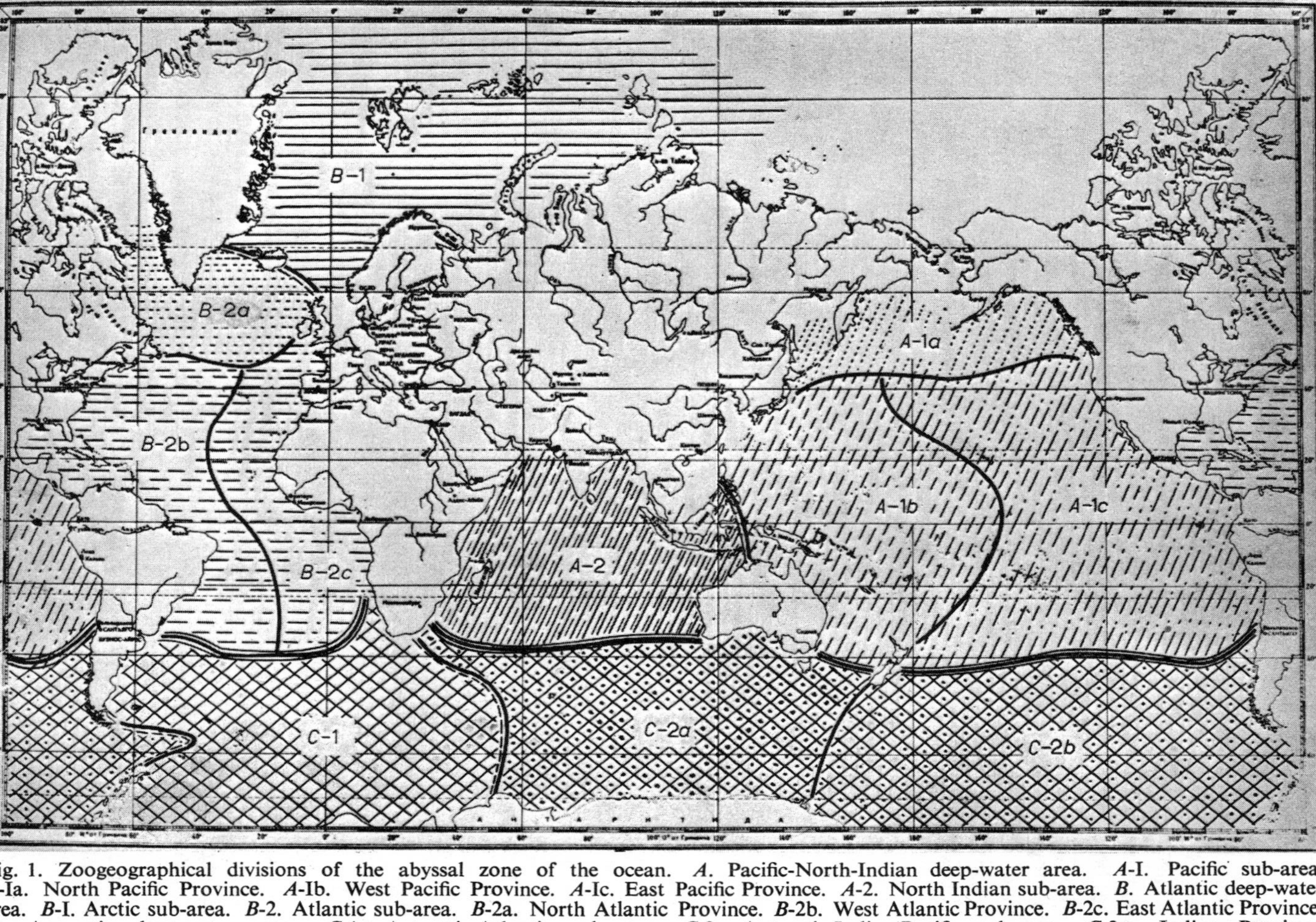

Fig. 1. Zoogeographical divisions of the abyssal zone of the ocean. *A*. Pacific-North-Indian deep-water area. *A*-I. Pacific sub-area. *A*-Ia. North Pacific Province. *A*-Ib. West Pacific Province. *A*-Ic. East Pacific Province. *A*-2. North Indian sub-area. *B*. Atlantic deep-water area. *B*-I. Arctic sub-area. *B*-2. Atlantic sub-area. *B*-2a. North Atlantic Province. *B*-2b. West Atlantic Province. *B*-2c. East Atlantic Province. *C*. Antarctic deep-water area. *C*-1. Antarctic-Atlantic sub-area. *C*-2. Antarctic-Indian-Pacific sub-area. *C*-2a. Indian Province. *C*-2b. Pacific Province.

to decrease with depth. This agrees with the hypothesis for the endemism of the fauna in the ocean trenches at depths greater than 6000 m (ZENKEVICH, *et al.*, 1955 ; BRUUN, 1956).

The distribution of deep-water species is apparently restricted to certain areas by the relief of the bottom, especially by the features of the macrorelief (Fig. 2). Hence, the boundaries of the zoogeographical areas for the deep-water bottom fauna correspond with submarine mountain chains and elevations. The great majority of cosmopolitan deep-water species, which have a comparatively wide vertical depth range are eurybathic. In contrast, stenobathic species, with a limited vertical distribution, occur only in restricted areas. About 30 per cent species of the groups investigated were eurybathic and about 70 per cent stenobathic species (Fig. 3).

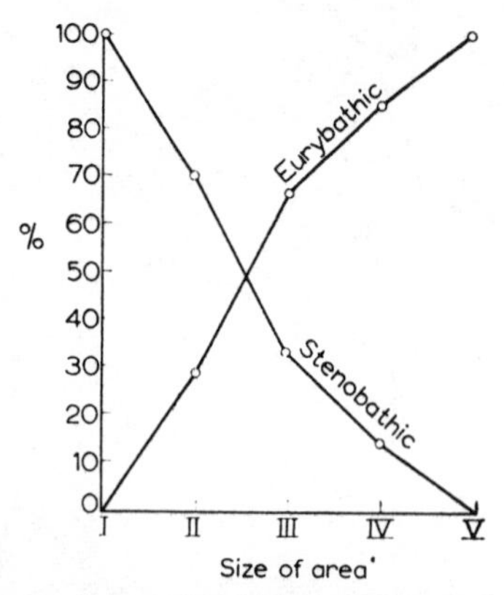

Fig. 3. Graph showing the relation between the extent of the areal distribution and the eurybathic characteristics of the species. The Roman numerals are arbitrary divisions to indicate the extent of these ranges.
I. Range limited to a very restricted portion of an ocean.
II. Range extends over half an ocean.
III. Range confined within the bounds of one ocean.
IV. Range extends throughout two oceans.
V. Range extends throughout all three oceans – or species with a pan-oceanic distribution.

Undoubtedly, there are other factors which also influence the distribution of the deep-water fauna – such abiotic factors as temperature, the nature of the bottom and pressure, and such biotic factors as the presence or absence of pelagic larvae, feeding habits, the distribution of food, etc.

In conclusion, although our knowledge of the causes of the geographical distribution of the deep-water bottom fauna is far from complete, we have a fairly clear idea of the general characteristics of the zoogeographical divisions of the upper abyssal zone of approximately 2000-4500 m, depths for which the greatest amount of data exist.

Institute of Oceanology
U.S.S.R. Academy of Sciences
Moscow, U.S.S.R.

REFERENCES

BRUUN A. F. (1956) The abyssal fauna: its ecology, distribution and origin. *Nature, Lond.* **177** (4520), 1105-1108.

MADSEN F. I. (1953) Some general remarks on the distribution of the echinoderm fauna of the deep sea. *XIV Internat. Zool. Congr., Copenhagen.*
ZENKEVITCH L. A., BIRSTEIN Y. A. and BELYAEV G. M. (1955) The bottom fauna of the Kurile Kamchatka-Trench. (In Russian). *Trudy Inst. Okeanol., Akad. Nauk U.S.S.R.* **12** 345-380.

Note : In studying the deep-sea fauna in depths greater than 2000 metres, the author has used data collected on Soviet and foreign expeditions culled from a search through complete series of numerous periodicals and expedition reports. More thán 622 entries have been made in a card index and hence are too numerous to list here. For the tracks of the major expeditions exploring the oceans, see : G. SCHOTT, " Geographie des Indischen und Stillen Ozeans " (1935) and " Geographie des Atlantischen Ozeans " (1942), published by C. Boysen, Hamburg. Morskoi atlas Vol II, 1953. Isdanie Glavnogo Staba Voenno-Morskikh sil U.S.S.R.

Intertidal Ecology

Reprinted from Ecology, Vol. 27, No. 4, October, 1946
Printed in U. S. A.

CRITICAL TIDE FACTORS THAT ARE CORRELATED WITH THE VERTICAL DISTRIBUTION OF MARINE ALGAE AND OTHER ORGANISMS ALONG THE PACIFIC COAST

Maxwell S. Doty
Department of Botany, Northwestern University

Introduction

In Oregon and California as elsewhere along the Pacific Coast of North America there is a pronounced difference between the levels to which two successive low tides fall. The same is true of the maximum levels of two successive high tides. This results in the establishment of at least six or seven zones in the intertidal vegetational cover. An observer in the intertidal area during a low tide sees these zones occurring one below another as his view shifts from the highest tide levels to the lowest tide levels. While individually they may be from only a few inches to a few feet in height the zones stretch along the entire tidally affected coast.

In the northernmost parts of the hemisphere these zones are dominated by the larger brown algae. As the lower latitudes are reached the brown algae become less and less conspicuous and the smaller red algae more conspicuous. In fact, the whole flora changes, and there might be none of the species or even genera present in a zone in moderate latitudes that characterized the same zone in high latitudes. Or owing to local conditions, within a few feet the species occupying a zone may change completely with none or few of the species present in one community lapping over into adjacent communities. The different communities are identified by one or a few dominant conspicuous species. None of the dominant algal species occurs regularly above *and* below a break between the major zones.

While other authors have correlated the vertical distribution of marine algae with tide levels, the actual phenomena, or tide factors, accounting for the sudden breaks in the flora have not been pointed out. On the following pages there are given what are thought to be some of the fundamental tide factors which will account for the sudden breaks in the vegetational cover and the consequent zonation. No attempt is made to discuss factors of algal ecology other than those concerned with the breaks between the aforementioned zones. Studies of this phase of ecology have not been made previously along the Pacific Coast of the United States, and until Smith ('44) published his Marine algae of the Monterey Peninsula no one had attempted to record algal vertical distribution for even one local area.

Review of Literature

Kemp ('62) studied the algae at Casco Bay, Maine, and named four of the six zones found there according to their dominating algal genus. In this country the only other attempt really to define vertical zonation among the algae is to be found in the works of Johnson and Skutch ('28a, '28b), who in the latter paper reviewed the literature through the middle twenties. They distinguished three major zones based on position in the intertidal zone. The late W. A. Setchell proposed exposure to wave action and sunlight as the principle factors controlling vertical distribution. However, neither Setchell nor Setchell and Gardner together (e.g. '25) were at all consistent or definite in their mention of the vertical limits through which a species might be found. Coleman ('33) at Plymouth, England, was probably the first to stress and actually to use tide levels in correlation with the distribution of marine algae. Chapman ('43) has refined and extended but added little of fundamental nature to Coleman's work.

Chapman summarized well the state of knowledge in the British Isles to date. Repeated observations of dominance and vertical distribution from year to year have been recorded by Delf ('43). From South Africa a long series of littoral-ecology papers touching the present topic has appeared chiefly under the authorships of W. E. Isaac ('37), G. J. Broekhuysen ('40), and T. A. Stephenson ('44, see for bibliography of the long series). Isaac, in particular, figured and discussed the zonation of marine algae in South Africa. Broekhuysen has concerned himself with such effects as may be produced by temperature, salinity, and desiccation. These studies all seem to show that the vertical distribution of marine algae in other parts of the world follows the same fundamental pattern controlled by the same factors as found along the Pacific Coast of the United States.

Johnson and Skutch pointed out that since an alga may not be apparent at all seasons or at all stations the names for zones should not be formed from the name of a dominating alga. Often the vertical dimensions of zones or the vertical distribution of individual species have been recorded as the distance above or below some datum point, and this seems to be the most valuable method if for use only at the originating locality. However, the height of these zones and the breaks between them in respect to mean lower low water, or their individual breadths, are in turn dependent upon the mean range of tides which varies from place to place along the coast. The zones should not, therefore, in studies over a broad geographic area, be named according to such measurements from a datum point. The zone names commonly used, such as "littoral" or "lower-littoral," have become ambiguous by reason of ill and dissimilar definition, and there are not enough of these names acceptable for the number of zones present. Further, these names do not indicate the physical limits or physical nature of the zones.

Tidal Phenomena of the Investigated Region

In quest of a means of describing zonation so that records from one station would be directly comparable with those from another, a study of the Pacific tides was made. These semi-diurnal tides provided the only factor whose vertical variation was consistent over the whole range of coastline within the limits of this study and which was correlated with the zonation observed. The U. S. Coast and Geodetic Survey obligingly provided the series of curves (marigram) made on their tide-predicting machine graphing the tides predicted for 1945 for the entrance to San Francisco Bay. Throughout the following discussion, unless specific reference is made, all references to tide levels and tidal phenomena are to those predicted for this station during 1945. This station was chosen for more data are available on this station than for any other along the Pacific Coast, and because the tides here have the same characteristics as the tides at the studied field stations. The initial clue to the possible nature of the tide factors was gleaned from lectures on marine ecology by Professor Rolf Bolin of Stanford University. Bruce J. Roberts began a study of this sort under Professor Bolin's direction in 1941 at the Hopkins Marine Station, but the work never reached the publication stage.

A study of the 1945 marigram for San Francisco revealed a number of tidal phenomena that might be critical agents in governing the vertical distribution of marine organisms. Figure 1A reproduces a short section of the marigram for the period June 8–30. During this time of year, and again during the winter solstice, the greatest differences between high and low water, the greatest differences between successive high or successive low waters, and the greatest total ranges of the tides occur. The minima of these features occur during the vernal and autumnal equinoxes. The figure illustrates the progression from spring tides to neap

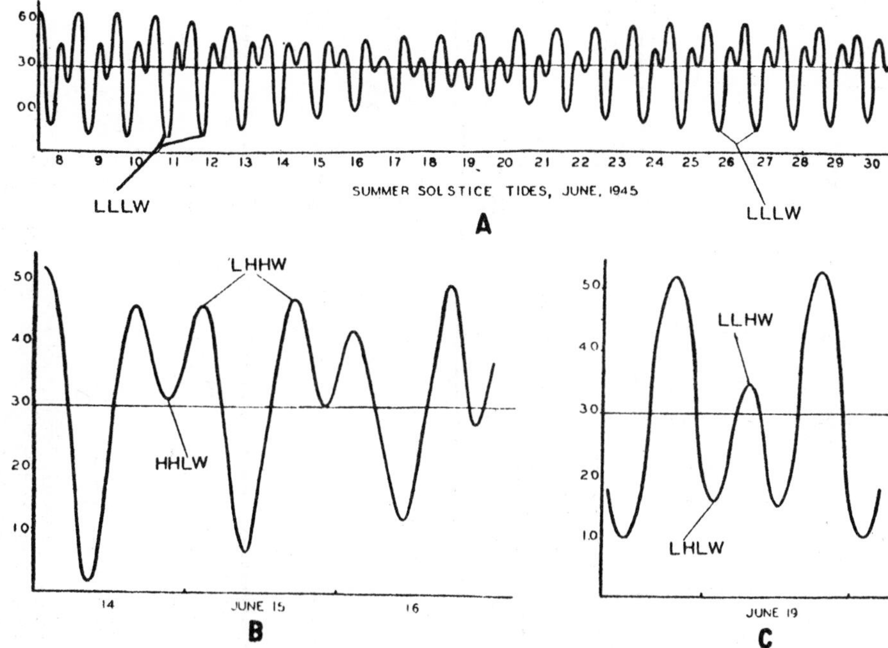

FIG. 1. A, section of June, 1945, marigram for San Francisco with days of month plotted along horizontal axis and tide heights in feet above or below mean lower low water (MLLW) plotted on vertical axis, illustrating in particular the lowest lower low water factor (LLLW) and regular gradual progression from spring to neap to spring, etc., tide series. June 8th through the 15th and 21st through 29th are spring-tide periods separated by a neap-tide period.
B, enlargement of a section of A to show more clearly the critical factors of lowest higher high water level (LHHW) and of the highest higher low water (HHLW).
C, enlargement of a section of B to show more clearly the critical factors of lowest lower high water level (LLHW) and lowest higher low water (LHLW).

tides that occurs during a cycle that is completed approximately every two weeks. The maximum variations from mean tide level occur during spring-tide series and the mimina during the neap-tide series. This is true for all except the higher low tides, whose maximum variations from mean tide level occur during neap-tide series. Figure 3 summarizes the levels of tidal phenomena[1] that may be recognized during such a tide series. Figure 5 illustrates the normal range of variation in these levels during the year. Figures and discussion to follow point out the correlation between some of these phenomena and the limits of vertical range observed among the algal species.

[1] Hydrographic terms and abbreviations used in this paper are those used by the U. S. Coast and Geodetic Survey on hydrographic charts of the region and defined in special publications 135, *Tidal Datum Planes*, and 228, *Tide and Current Glossary*, of the U. S. Department of Commerce, U. S. Coast and Geodetic Survey, or are obvious derivatives from them.

METHODS AND RESULTS

Field procedure

Several benchmarks (Hewatt, '37) have been established on Mussel Point, Monterey County, California. Observations near these benchmarks revealed a break or line of demarcation in the algal flora marking the limits of upward extension of vertical range of some species and the lower limits of vertical range of other species. The elevation of this most distinct and easily recognizable level is found to be the mean lower low water (MLLW) level. By measurements from benchmarks and from observations of the

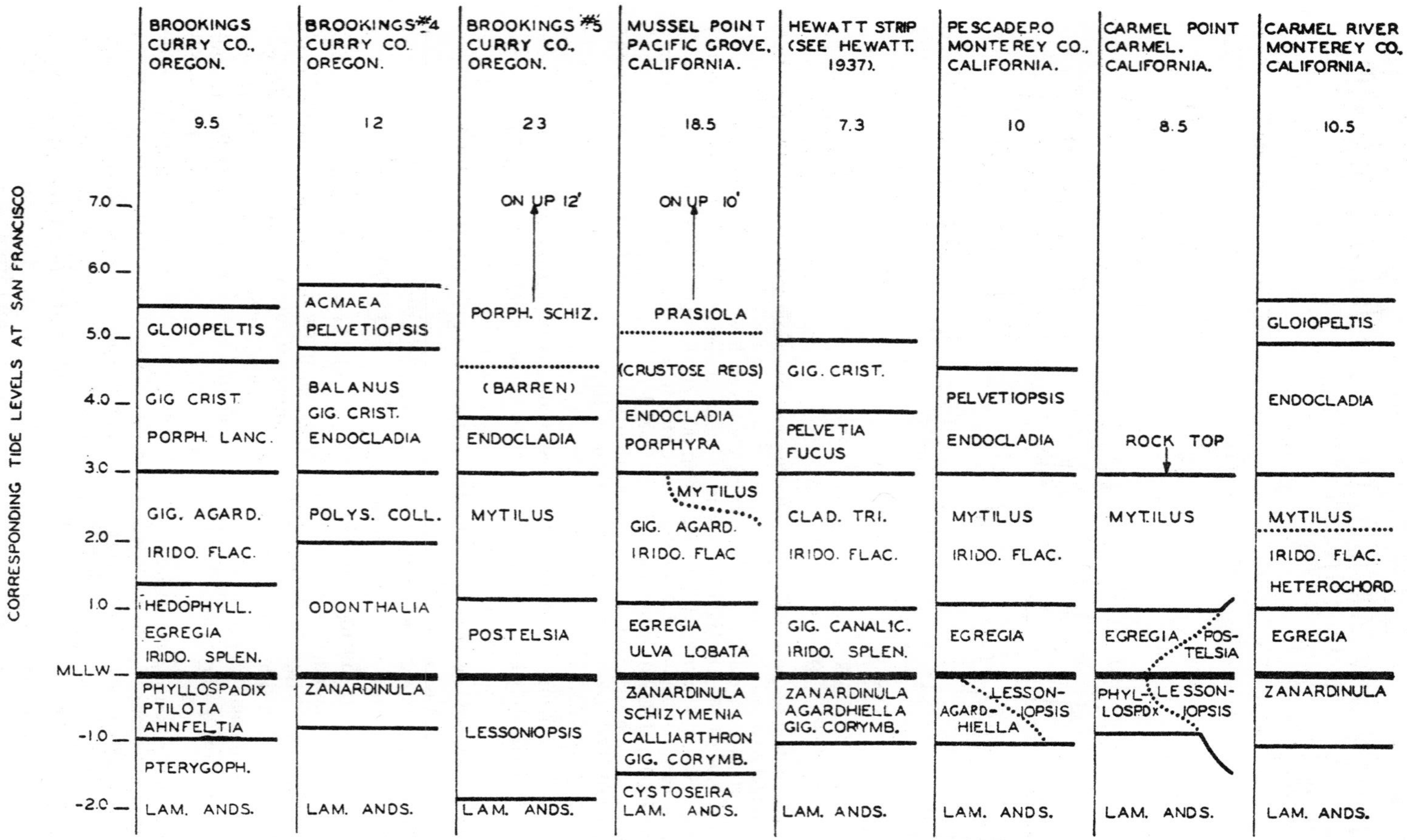

CORRESPONDING TIDE LEVELS AT SAN FRANCISCO: 7.0, 6.0, 5.0, 4.0, 3.0, 2.0, 1.0, MLLW, -1.0, -2.0

BROOKINGS CURRY CO., OREGON.	BROOKINGS #4 CURRY CO. OREGON.	BROOKINGS #5 CURRY CO., OREGON.	MUSSEL POINT PACIFIC GROVE, CALIFORNIA.	HEWATT STRIP (SEE HEWATT. 1937).	PESCADERO MONTEREY CO., CALIFORNIA.	CARMEL POINT CARMEL. CALIFORNIA.	CARMEL RIVER MONTEREY CO. CALIFORNIA.
9.5	12	23	18.5	7.3	10	8.5	10.5
		ON UP 12'	ON UP 10'				
GLOIOPELTIS	ACMAEA PELVETIOPSIS	PORPH. SCHIZ.	PRASIOLA	GIG. CRIST.	PELVETIOPSIS		GLOIOPELTIS
GIG CRIST PORPH. LANC.	BALANUS GIG. CRIST. ENDOCLADIA	(BARREN) ENDOCLADIA	(CRUSTOSE REDS) ENDOCLADIA PORPHYRA	PELVETIA FUCUS	ENDOCLADIA	ROCK TOP	ENDOCLADIA
GIG. AGARD. IRIDO. FLAC.	POLYS. COLL.	MYTILUS	MYTILUS GIG. AGARD. IRIDO. FLAC	CLAD. TRI. IRIDO. FLAC.	MYTILUS IRIDO. FLAC.	MYTILUS	MYTILUS IRIDO. FLAC. HETEROCHORD.
HEDOPHYLL. EGREGIA IRIDO. SPLEN.	ODONTHALIA	POSTELSIA	EGREGIA ULVA LOBATA	GIG. CANALIC. IRIDO. SPLEN.	EGREGIA	EGREGIA POSTELSIA	EGREGIA
PHYLLOSPADIX PTILOTA AHNFELTIA	ZANARDINULA	LESSONIOPSIS	ZANARDINULA SCHIZYMENIA CALLIARTHRON GIG. CORYMB.	ZANARDINULA AGARDHIELLA GIG. CORYMB.	LESSONIOPSIS AGARDHIELLA	PHYLLOSPADIX LESSONIOPSIS	ZANARDINULA
PTERYGOPH. LAM. ANDS.	LAM. ANDS.	LAM. ANDS.	CYSTOSEIRA LAM. ANDS.	LAM. ANDS.	LAM. ANDS.	LAM. ANDS.	LAM. ANDS.

FIG. 2. Several transects reduced to approximately same scale by adjusting mean lower low water levels (MLLW) and highest of higher low water levels (HHLW), arbitrarily selected as corresponding to 3.0 feet at San Francisco, to same height on figure. A few more conspicuous genera and species are indicated to illustrate consistency of zones from place to place in flora and fauna as well as in approximate proportion of total tide range. Actual height of each transect is given in feet just beneath its title.

rise and fall of tides at several other stations, this break, recognizable by species limited at this level, has been found to be near the MLLW level all along the coast, i.e. near the 0.0 tide level, which is the datum plane or chart datum of tide books and other Pacific Coast navigational aids. At least as far as temperature and desiccation are concerned, wetting by waves every few seconds is practically equal to being constantly submerged. As a result this line of demarcation considered as MLLW is actually, except in places protected from waves, a few inches higher than the 0.0 tide level as measured from bench marks or from tide gages. For the same reason all algal zones are shifted upwards and, as a rule, broadened in situations exposed to wave action and are lowered and, as a rule, narrowed in protected situations such as within bays.

Some two score vertical strips or transects were chosen in all seasons of the year at various places between Boiler Bay, Oregon (44° 50′ N. Lat.) and Carmel Bay, California (36° 31′ N. Lat.). On these strips running from the highest rocks bearing marine algae to those exposed only by the lowest tides, the vertical distribution of the more conspicuous, widely spread, and definitely determinable algae was noted in feet above or below this effective MLLW datum point. In general, the longer-lived or perennial species are much more consistent in their distribution and so were chosen whenever possible over the shorter-lived, more ephemeral species. Charts of the data were made on cross-section paper. Several are shown in figure 2. There are regularly at least six (or the uppermost divided into two and so seven) zones showing in each strip, unless the higher tides completely cover the area.

The approximate vertical height (in feet) is given at the top of each strip in figure 2. Consideration of these numbers shows that strips prepared by measuring the actual heights in feet above or below a given level are not directly comparable and will not give much clue to the factors underlying the regulation of vertical distribution. Similarly, it is clear that such a method of recording these facts in floristic studies intended to cover any broad geographic range would be inadequate. If the rough strip data are reduced proportionally to an arbitrary range, there is a good correlation between relative positions of the zones as identified by their flora, in regard to both the order in which they appear and the levels through which they grow. This additional adjustment has been made in figure 2 by aligning two rather distant clearly recognizable breaks. Owing to space limitations in the figure it has been impossible to indicate but very few of the many genera and species considered during the assembling of data for this project. In the field actually about 120 species were considered all the time, and so the homogeneity of the strips as indicated by the sampling process at different stations was even more striking than this figure could illustrate.

For the sake of brevity the breaks between zones are cited by the abbreviations LLLW, MLLW, LHLW, HHLW, LLHW, and LHHW (see figs. 1, 3). The lowermost level is the lowest level to which tides recede, the lowest lower low water (LLLW) level. The next level above this is the mean of the lower of the two low waters of each day or mean lower low water (MLLW). Next is the highest level uncovered only once each day, at the plane of the lowest of the higher low waters (LHLW). Above this the next level is near mean sea level and is the level of the highest higher low water (HHLW). The next, lowest lower high water (LLHW), is the highest level covered and uncovered at least twice each day. The uppermost level, lowest higher high water (LHHW), is the highest level covered at least once each day.

Results of correlation of field study and critical tide factors

Nature of the critical tide factors.—At many of the levels pointed out above there

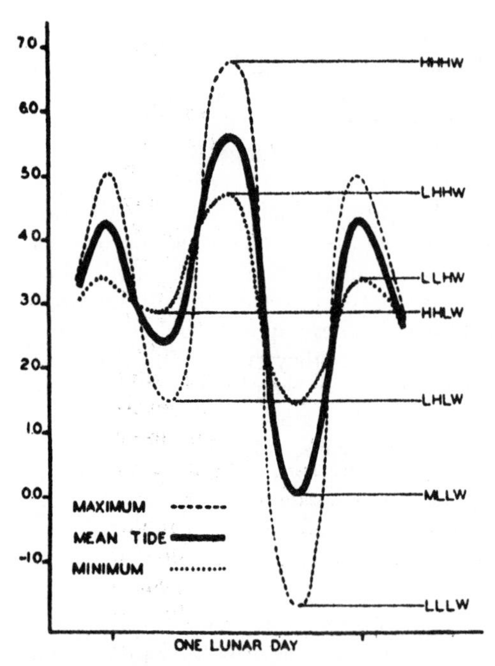

FIG. 3. Tide levels in feet above or below mean lower low water level (MLLW) at San Francisco are given along vertical axis and time across horizontal axis. Figure illustrates limits in variation, in order from left to right, for lower highwater of preceding lunar day and then higher low water, higher high water, lower low water, and lower high water of current lunar day. A great many levels, such as mean levels, could be recognized and pointed out on this figure, but only those factors or phenomena discussed further in the current paper are indicated in a vertical column along right-hand side of figure.

are sudden increases or decreases in the duration of exposure to the various factors in the environment. These may be as great as abrupt two- or three-fold, or greater, changes in the environment. As a rule two- or three-fold increases in time of exposure[2] over that necessary to injure or kill a few thalli are sufficient to kill all thalli. Figure 4 illustrates this point in regard to the critical levels below mean sea level. This figure indicates sudden two- to three-fold increases in the length of maximum single exposure to submergence during the year. As would be expected these are near 3.5 feet, 3.0 feet, and 1.0 foot and are in accord with the levels of LLHW, HHLW, and LHLW, respectively illustrated in figure 1. Exactly similar curves can be drawn for the maximum single periods of exposure to emergence, with breaks in the curve representative of the critical factors (sudden great increase in total single emergence time) correlated with the LLHW and LHHW levels. Levels at which such abrupt changes in the length of exposure occur are closely correlated with the breaks between the major algal zones, and with the upward or *downward* limits of distribution of those species not extending over the whole breadth of a zone. Following is a more detailed account of a number of these levels at which tide factors usually provide such sudden changes of maximum single exposure, and a brief sketch of their effects.

Zoning effects of the critical tide factors. —The highest portions of the "spray zone" extend upwards from the HHHW levels. There are three distinct communities that may be evident as high as twenty feet above the MLLW level; each composed of a single dominant species. *Rhodochorton purpureum*[3] occupies such vertical heights within caves often seemingly quite remote from any ordinary sort of wetting even by spray. · *Porphyra*

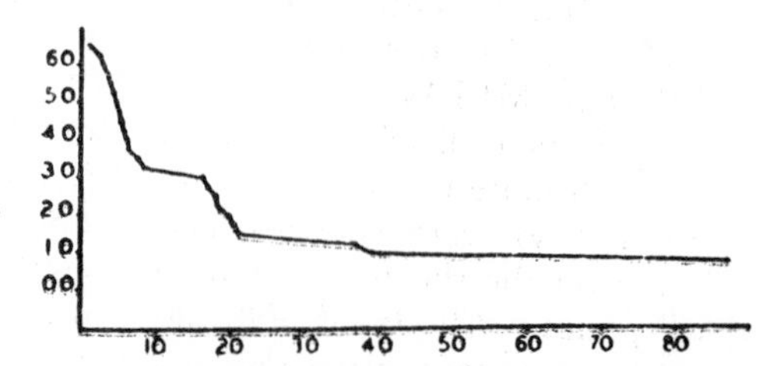

FIG. 4. Length of maximum single submergence in hours plotted along horizontal axis in correlation with tide levels as shown on vertical axis. Sudden increases in length of maximum single submergence at levels indicated for tide factors illustrated in figure 1 are shown clearly near levels of 3.5, 3.0, and 1.0 feet above mean lower low water (MLLW).

[2] Exposure here is defined as subjection to change in the influences of the environment.

[3] Authorities for the scientific names of algae are listed in figure 6.

schizophylla has been found on exposed rocky bluffs where it extends up as high as eleven feet above the very highest tide levels. In June and July most thalli are very much bleached and appear senescent. The third conspicuous community is that formed by *Prasiola meridionalis* accompanied by *Gayella constricta.* This last community is conspicuous as a green belt or patch in places extending at least ten feet above HHHW on most rocks covered above by bird droppings. These communities of the spray zone are present only in situations either exposed directly to spray or very well protected from desiccation; therefore, they are often completely absent. Further, since these species often extend down to the LHHW level, these communities may be considered merely as exceptionally strong developments of the next lower zone, rather than a separate zone unto themselves.

Between the HHHW levels and LHHW levels there is usually little of algal nature except occasional quasi-marine lichens, the above species, and, in shade or in crevices, coverings of the crustose *Hildenbrandia occidentalis.* This latter alga often runs down considerably lower than LHHW. This is a region of the shore often wet for only a few days in each month as a study of figures 1 and 5 will disclose. *Lophosiphonia villum* inhabits cave walls in these levels but may actually be a member of a lower zone. The marine animals are also zoned, though a bit less definitely, and their zonation (Hewatt, '37) is correlated with algal zonation. Such animals as the gastropods (Keen and Doty, '42; Hewatt, '37) *Acmaea scabra* (Gould), *A. digitalis* Eschscholtz, and *Littorina planaxis* Philippi, along with the isopod *Ligyda occidentalis* (Dana) in central and southern California, or *L. pallasii* (Brandt) in Oregon and on north, are common to this zone, though the isopods run up into the highest portion discussed above. At Brookings, Oregon, *Gloiopeltis furcata* forms a belt that seems to be above the LHHW levels and so is a regular representative of this zone. Along the central California coasts this species, when present, appears with the highest members of the next zone to be discussed.

Investigation of the LHHW periods will reveal phenomena similar to those of June 14–16 illustrated in figure 1B. Here it is seen that just below the 4.7-foot level at San Francisco the substratum was covered every 10 to 23 hours while levels above 4.7 feet were continuously uncovered. On the 16th a slightly higher tide inundated the rocks up to 4.9 feet and broke this long period of constant emergence of levels above 4.7 feet. Analysis of other such periods (fig. 5) reveals that this level was not below 4.6 feet during 1945 and was usually at this same 4.7-foot level or near it. This is the lowest level uncovered and exposed for periods longer than about 24 hours, and the length of maximum single exposure suddenly increases to two or three days just above this level during almost every neap-tide series of the year. In 1945 between March 1, from about 3:00,[4] and March 6, about 3:00, there was a 120-hour exposure above the 4.7-foot level. Had the neap-tide series fallen nearer the equinox the exposure above this level would have been even longer. It has been experimentally determined that exposures even in the shade for such lengths of time will seriously damage even the adult algal thalli growing just below this level, and the juvenile stages are usually even more easily damaged.

The common algae indicating the zone between this level, LHHW, and the next critical level below are *Pelvetiopsis limitata, Gigartina papillata,* and *Endocladia muricata.* The littorine, *Littorina scutulata* Gould, is likewise characteristic of this zone.

The critical phenomenon occurring near the plus 3.5-foot level at San Francisco seems to be the lowest level of the lower high tides (LLHW). The characteristics

[4] Time reckoned from 0:00 to 24:00 (midnight).

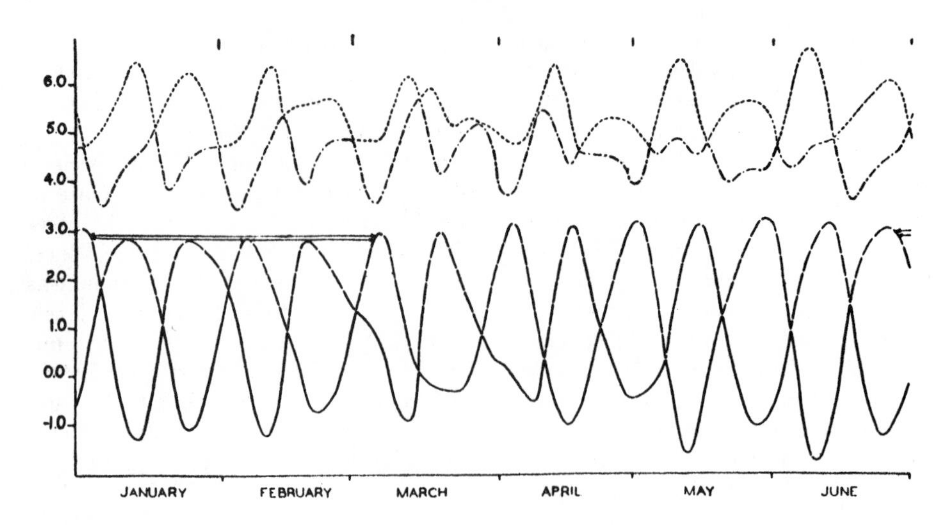

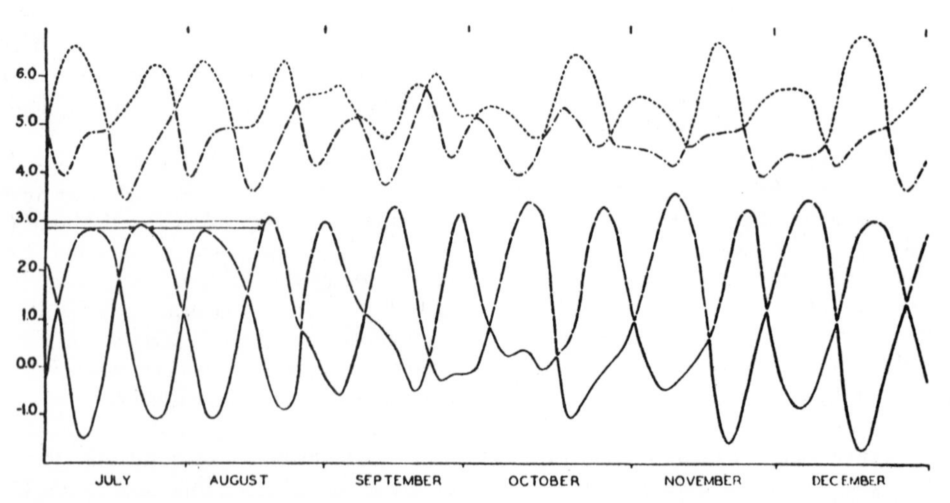

Fig. 5. Illustrating variations in tides at San Francisco during the year 1945. Solid line connects levels of lower low water of year; long-dashed line joins levels of higher low waters; line of alternating dots and dashes indicates levels of lower high waters; and short-dashed line levels of higher high waters.

of this phenomenon are outlined in figure 1C. It is seen that on June 19 below the level of plus 3.5 feet the shore is exposed for a period of 9¼ hours followed by a short period of submergence and then by another period of 8½ hours exposure, and that but a short distance above this level the shore was exposed for a continuous period of 18¾ hours. Here is seen another controlling, or critical, tide factor like that discussed just above in connection with the LHHW level.

Between this LLHW level and the next critical level below is a zone that is covered and uncovered by the ebbing and flowing tides twice each day. It is characterized principally by *Pelvetia fastigiata* in central and southern California but north through Oregon by *Endocladia muricata, Porphyra perforata, Gigartina cristata,*

Bangia vermicularis, and *Fucus furcatus.* The first of these last five species may more truly belong with the next zone above, and the last of these five is more truly characteristic of the lowermost half foot (at Monterey) of this zone. *Tegula funebralis* (A. Adams) is a characteristic and widespread gastropod in these levels.

Near plus 2.8 or 2.9 feet at San Francisco is the level (fig. 1B) of the highest higher low tides (HHLW) occurring during times of spring tide in the two months following both the winter and summer solstices. At this time there may be a period of 45–48 days during which this level is subject to semi-diurnal exposure for some hours to cold temperature, winds, rain, and sun in winter, or warm temperatures, winds, and bright sun in summer. At a slightly lower level this period of continual semi-diurnal exposures to such elements is reduced to less than one-half by periods when these levels are exposed but once each day, and then either early in the morning or late in the afternoon. Study of figures 1C and 4 shows that at this level there is a sudden increase in the length of maximum single submergence. While there are recognizable a number of other possibly controlling phenomena near this break, this latter seems to be the critical factor. Algae growing just above this level are uncovered by the tides over half the time; those below are submerged over half the time (for mean sea level at San Francisco is near the 3.0-foot mark). Frost killing and sun killing have been noted in different years below this level and less often above this level.

The algae indicative of the zone below HHLW are *Iridophycus flaccidum* and *Cladophora trichotoma.* On exposed prominences the mussel, *Mytilus californianus,* occupies this zone with *Iridophycus coriaceum* and *Polysiphonia Collinsii,* both of which may grow on the mussel shells or on rock. *Tegula brunnea* and *Tritonalia lurida* are gastropods abundant in these levels.

In some years a rather conspicuous break is found in the flora just above the plus 1-foot level in the San Francisco region. This may be defined as the highest level which is exposed only once each day but is referred to as the LHLW level. In figures 1C and 4 it is seen that there is a sudden increase in the length of maximum single submergence at this level. For while just above this level, as illustrated, the littoral population is covered during varying periods from but nine to twelve hours, a few tenths of a foot lower it is submerged for as long as one and one-half days without exposure. In addition to being correlated with LHLW, this level is also near mean low water level and the highest lower low water level.

Floristically, the LHLW level is recognized as the break between the distribution of *Iridophycus flaccidum* above and *Iridophycus splendens* below. Often it forms a distinct upper limit in the distribution of some or all of these following species: *Gigartina canaliculata, Ulva lobata* (and other Ulvaceae), *Rhodoglossum affine, Gelidium Coulteri,* and *Leathesia difformis.* In exposed situations *Postelsia palmaeformis,* while occurring at higher levels in respect to sea level, forms a community indicating the effective position of this zone. *Egregia Menziesii* and *Hedophyllum sessile* are the most certainly determinable and dependable indicators of this zone. As a gastropod indicator *Zizyphinus filosus* (Wood) (the *Calliostoma costatum* (Martyn) of authors), although ranging into the next zone below, has its upward limit here.

Near the MLLW (mean lower low water) level is the most conspicuous break in the flora. This is the sharp lower limit of the distribution of the species listed for the next zone above. There is no conspicuous algal species that regularly occurs above *and* below this level. The aspect is a sudden transition from the brownish phycocyanin-rich Rhodophyta and the Chlorophyta that come down to this level to the bright red Rhodophyta below. One flowering plant,

Phyllospadix Scouleri Hooker, has its upper limit of distribution here as have the algae *Schizymenia pacifica*, *Gigartina corymbifera*, *Zanardinula* spp.,[5] *Pikea* spp., *Calliarthron* spp., *Gymnogongrus* spp., *Ahnfeltia* spp., and *Agardhiella Coulteri*. On exposed slopes the top of the *Lessoniopsis littoralis* community indicates this break. This break is also indicated by the upward range limits of the sea urchin, *Strongylocentrotus purpuratus* (A. Agassiz), and the anemone, *Anthopleura xanthogrammica* (Brandt). Some species mentioned in this paragraph, while characteristic of the zone between MLLW and LLLW, extend on down into the region that is never exposed by receding tides.

Though this is the most conspicuous break in the flora, it is least well explained. No satisfactory sharp break in the environmental factors (fig. 1, etc.) provided by tides is known. Further study of the life history, particularly time of reproduction, of algae limited at this level will probably lead to disclosure of the controlling principles. At least it seems probable that the time of reproduction might give a clue as to when during the year to look for causal tide factors. Hewatt ('37:179) made the statement that above the MLLW level there is a sudden increase in the total length of exposure to atmospheric conditions during the year. It is to be remembered that exposure below this level occurs in either the evening or early morning when the humidity is relatively high, the temperature relatively low, and the absorption of light (relatively low in energy content at these times, at any rate) is least. During the late spring months when the air is relatively clear along the Pacific Coast there is much so-called sun-killing of algae growing at or just above this level.

No detailed study of tidepools is included here. If large enough, the tidepool flora tends to be the same as that found on unconfined shores below MLLW. If small or high in the intertidal zone, the species most susceptible to high temperatures and then those most susceptible to desiccation drop out first.

The lowest critical level (LLLW) to be discussed is the level below which the shore is never exposed by the low tides. At San Francisco this is, as figure 1C shows, 1.5 to 1.7 feet below MLLW. It is marked by the upward limit of distribution of *Cystoseira osmundaceae* and *Laminaria Andersonii*, two readily recognizable and wide-spread species. The upward limits of *Desmarestia herbacea* and *Pterygophora californica* seem to be at this level.

Discussion

Usually the zonation appears as outlined in the paragraphs above, but in extremely favorable situations, i.e. on plane surfaces equally exposed to the sea, as many as ten distinct zones have been observed. Such multiplication of the normal number of zones is seemingly due to one of two factors: (1) climax segregation of species, particularly perennial or biennial species; or (2) the result of the interlocking of the lens-like or narrowly cuneate margins of adjoining algal communities. To elaborate on the former: where the intertidal rocks expose a smooth plane surface in such a way that all levels for some horizontal distance are evenly exposed to the elements by tidal variations, almost each of the dominant individual species listed may form a sub-zone of its own.

Anomalous intermixtures of the species are not uncommon on rough shores, such as boulder fields. In such localities fewer than the normal number of intertidal zones are apparent. The spray zone communities are always missing except in situations either exposed to violent wave action or very well protected from desiccation. At times a zone will be absent or completely unrecognizable for some reasons other than those just given. Occasionally a barren zone three to six inches

[5] "Spp." in this paper signifies all species recognized during the study.

wide appears near mean sea level. Some species occasionally, even when distinctly sub-zoned, may be transposed within their own and more rarely transposed with members of neighboring zones. Such anomalies as these have for the most part yet to be explained. Since the algae and their zones are sometimes, but rarely, intermixed or arranged in these different orders, and since there may be bare gaps between zones, it seems unlikely that competition can be a factor in controlling vertical distribution.

Species whose occurrence is most consistent in regard to light incidence, exposure to wave action, etc., are those most variable in regard to the tide levels through which they appear. On the other hand, species most consistent in their vertical range appear, as a rule, with much less regard for incidence of light, wave action, etc. Fruiting of the species more variable in vertical range is very possibly controlled in large part by factors such as light, temperature, or oxygen tension, since these species show by their local distribution that they are very sensitive to such factors. This being true, then they would fruit at somewhat different times during the tidal series, from year to year, and from place to place, in correlation with current local weather and hydrographic features other than annual seasonal and tidal variation. The variation in levels at which they appear is then in correspondence with the different heights of the critical tide factors during the sensitive stages in their development or during the distribution of their reproductive bodies. Many species growing between HHLW and HHHW belong in this category, particularly the Rhodophyceae and Chlorophyceae noted as indicators of the upper limits of the MLLW to LHLW zone. The more consistent type is illustrated by the complex of forms recognized as *Zanardinula Lyallii*. It is found in light and shade and in warm water tidepools at high elevations but only exposed to the desiccation normal to levels below MLLW.

It must be pointed out that, because of the above and because the critical tide factors vary from year to year, the actual levels in feet above or below some permanent datum point will likewise vary. This has been found true by repeated observations from year to year on the same transects at the Hopkins Marine Station and elsewhere.

McDougall ('43:368), in discussing the mechanisms of vertical zonation among the invertebrates, emphasized the following three processes by which this may come about: "(a) larvae may settle over a wide vertical range, but those which do not chance upon a suitable location are subsequently killed off; (b) larvae, at the time when they are ready to settle, may be concentrated at a particular level in the water; (c) in the case of species with a motile adult stage individuals may enter and be confined to the optimum zone by characteristic reactions which prevent them from crossing its borders into unfavorable regions." Coleman ('33) in England and R. D. Northcraft (unpublished observations) at the Hopkins Marine Station confirmed the presence among the algae of the first method presented by McDougall. Gail ('18) observed species among the algae at Friday Harbor, Washington, which fall into the second of these patterns. The third mechanism discussed by McDougall would not apply to the macroscopic sessile adult thalli but without doubt could apply to the microscopic motile adult forms and to motile reproductory stages. Very likely, though, movements of the motile stages of algae would be rather futile in the open ocean along rocky shores but might be of significance within tidepools. No matter which of the above mechanisms, or some other, controls the presence or absence of these species, their vertical distribution is dependent upon such as the previously discussed tide factors for the sharp breaks between zones and upon the mean tide range for the vertical breadth of the zones.

A species having a certain genetic setup

can be expected to display the same valence toward the physical elements in its environment wherever it may occur. Further, the fact that it occurs in a locality speaks of the over-all effect of the physical conditions there, i.e. that the extremes are within the valence limits of the species. For these reasons, when we find an

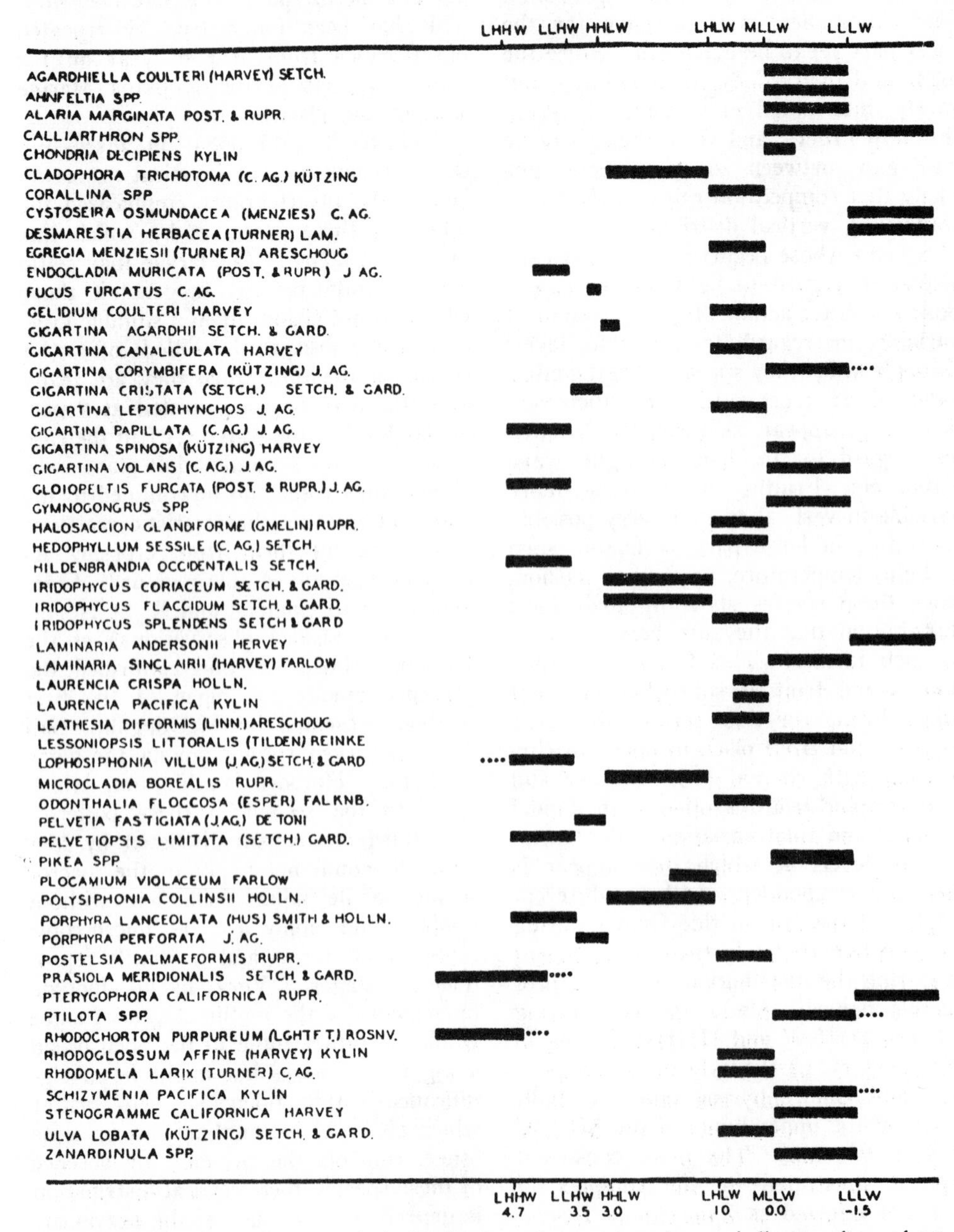

FIG. 6. Chart to indicate vertical distribution of more conspicuous indicator algae along central and northern Pacific Coast of United States. Corresponding tide heights (in feet above or below mean lower low water, MLLW) for San Francisco are given along lower axis as well as initials indicating critical tide factors correlated with breaks between zones along both horizontal axes.

algal species through the strips illustrated in figure 2, we may generally assume that it indicates the same set of primarily tidally-controlled physical conditions. That this is not always true is indicated perhaps by some of the occasional anomalies. Disregarding these occasional exceptions, there is no necessity for keeping the strips separate, and so they are fused and presented as figure 6. The figure lists the principal and conspicuous species of marine algae found from central California through northern Oregon and the tide factors and *effective* levels which characterize their vertical distribution. To aid in interpretation the heights of the tides for San Francisco are given along the side of the figure.

Summary

1. Using mean lower low water level as a datum point, the vertical distribution of all clearly recognizable macroscopic marine algae was recorded for the coasts of Oregon and northern California. Only incidental mention of horizontal distribution or association of the species is made.

2. It is observable along the Pacific Coast and elsewhere in the world that the intertidal flora and fauna are broken into a series of zones whose limits are correlated with the levels of certain critical (tide) levels. The vertical breadth of the intertidal zones is directly correlated with the tide range.

3. Study of critical tide levels reveals some peculiarities in the variations of the tides, referred to as tide factors. These tide factors for the most part provide sudden two- or three-fold increases in the exposure to changes in the environment. Changes of this magnitude are sufficient to account for the abrupt restrictions in vertical range observed. It is not yet understood whether the restriction is due to temperature, light, gas tensions, or other factors which are directly affected by the tide factors, or to desiccation and the consequent osmotic changes.

4. The vegetational cover is divided into zones between the following tide levels: below LLLW level; between the LLLW and MLLW levels; between the MLLW and LHLW levels; between the LHLW and HHLW levels; between the HHLW and LLHW levels; between the LLHW and LHHW levels; between the LHHW and HHHW levels and up to the upward limits of the spray.

5. Variation in vertical distribution seems to be correlated with daily, monthly, or annual variation in the levels at which the tidal phenomena occur and with variation in the time the algae reproduce, after account is taken of the local topography.

6. The zonation of animal and algal life agrees well except that the motile animals are often less sharply zoned.

Literature Cited

Broekhuysen, G. J. 1940. A preliminary investigation of the importance of desiccation, temperature and salinity as factors controlling the vertical distribution of certain intertidal marine gastropods in False Bay, South Africa. Trans. Roy. Soc. S. Africa **28**: 255–292.

Chapman, V. J. 1943. Zonation of marine algae on the sea-shore. Proc. Linn. Soc. Lond. 1941–42 (3): 239–253.

Coleman, John. 1933. The nature of intertidal zonation of plants and animals. Jour. Mar. Biol. Assoc. **18**: 435–476.

Delf, E. Marion. 1943. The significance of the exposure factor in relation to zonation. Proc. Linn. Soc. Lond. 1941–42 (3): 234–236.

Gail, F. W. 1918. Some experiments with Fucus to determine the factors controlling its vertical distribution. Publ. Puget Sound Biol. Sta. **2**: 139–151.

Hewatt, Willis G. 1937. Ecological studies on selected marine intertidal communities of Monterey Bay, California. Amer. Midl. Nat. **18**: 161–206.

Isaac, Wm. E. 1937. Studies of South African seaweed vegetation. I. West coast from Lambert's Bay to the Cape of Good Hope. Trans. Roy. Soc. S. Africa **25**: 115–152.

Johnson, Duncan S., and A. F. Skutch. 1928a. Littoral vegetation on a headland of Mt. Desert Island, Maine. I. Submersible or strictly littoral vegetation. Ecology **9**: 188–215.

——. 1928b. Littoral vegetation on a headland of Mt. Desert Island, Maine. II. Tide-pools and the environment and classification of submersible plant communities. Ecology 9: 307–338.

Keen, Myra, and Charlotte L. Doty. 1942. An annotated check list of the gastropods of Cape Arago, Oregon. Oregon State Monogr., Studies in Zool. no. 3. 16 pages.

Kemp, A. F. 1862. On the shore zones and limits of marine plants on the northeastern coast of the United States. Canadian Nat. 7: 20–34.

McDougall, K. D. 1943. Sessile marine invertebrates of Beaufort, North Carolina. Ecological Monogr. 13: 321–374.

Setchell, W. A., and N. L. Gardner. 1925. The marine algae of the Pacific Coast of North America. III. Melanophyceae. Univ. Calif. Publ. Bot. 8: 383–898.

Smith, G. M. 1944. Marine algae of the Monterey Peninsula, California. 622 pages. 98 plates. Stanford University Press. Stanford University, California.

Stephenson, T. A. 1944. The constitution of the intertidal fauna and flora of South Africa. Part II. Annals Natal Mus. 10: 261–358.

THE BIODEMOGRAPHY OF AN INTERTIDAL SNAIL POPULATION

Peter W. Frank
University of Oregon, Eugene, Oregon

Abstract. Numbers, deaths and growth rates were observed for three years in a delimited population of *Acmaea digitalis* whose members had been individually marked. The zone of vertical distribution of the species is determined by the behavior of these snails. In the fall and winter they ascend in the intertidal, and descend a smaller distance in the spring. This leads to a differential distribution of size classes. The oldest and largest animals are found highest in the intertidal zone. That the vertical movements are adaptive is suggested by death rates which are highest in the upper portions in summer, but rise in the lowest zone in winter. Probability of survival improves with age. Growth rate was remarkably consistent during the period of observation. It was fastest in the fall and winter months. During the month of July, and in part of June and August growth ceased, at least as measured by a change in shell length. Crowding decreased growth slightly but significantly. Growth rates in different areas suggest that time available for feeding may be a significant variable. Experimental manipulation of density indicates that emigration rates are density-dependent. The mechanism of density regulation is postulated as operating even at low overall densities because of the behavior pattern of the species.

The concept of population balance has a history of controversy (Nicholson 1933; Thompson 1939; Warren 1957). Although semantic difficulties are responsible for part of the problem, something important is at issue: whether and how commonly populations accumulate sufficiently so that they become limited by self-induced shortages in their resources. The wide common concern with the problem perhaps needs no documentation. Its pervasiveness may not be so evident, however, until one realizes that such disparate contributions as Varley's (1947) study of the knapweed gall fly, the comprehensive community analysis outlined by Elton and Miller (1954), MacArthur's (1960) and Preston's (1962) models of community structure and the short theoretical argument of Hairston, Smith and Slobodkin (1960) all bear directly on this point.

Although I was under no illusion that a single specifically oriented research effort could solve the general problem, work with the limpet *Acmaea digitalis* was begun with the rationale that empirical information gathered from natural populations was a requisite for progress towards a solution. This intertidal snail was thought to represent something of a limiting case, since it lives in what is generally conceded an extremely raw environment, yet occurs in seemingly dense populations. The apparent disadvantage that not all stages in the life cycle were accessible, since *Acmaea* has a planktonic larva, was regarded as a potential gain: a density-dependent birth rate was eliminated as a possible regulating factor, and the number of variables was thereby reduced.

At the outset certain limitations inherent in any feasible program were recognized. Except for cases where experimentation is possible, results represent correlations not clearly traceable to their causes. In cases where density can be manipulated, replication is still likely to be insufficient to assure the adequacy of controls. Sampling problems are bound to arise, since the rock surfaces on which the animals live are far from homogeneous. More fundamental than this is the question of the generality of the results: what does a limited though extensive set of observations and measurements, gathered over a relatively short time span and in a small portion of the species' range, signify regarding the performance of this and similar sorts of animals over their total area of distribution? These limitations apply not only to this, but to all work on natural populations conceived on a similar scale. They clearly imply that independent confirmation of significant conclusions is particularly important.

Animals and Their Habitat

Acmaea digitalis is the commonest limpet of the upper intertidal zone in Oregon, and ranges from the Aleutians to Baja California. Sexes are separate, and spawning occurs at least through winter and spring (Fritchman 1961). The length of planktonic life is unknown. These snails are common on virtually all hard substrates in a vertical zone that extends upwards from median high tide for as much as two to five m, depending on the degree of exposure to spray. They feed on microscopic algae, mainly blue-greens and diatoms, rasped off the substrate. Even casual observations of the sorts of areas where *Acmaea digitalis* is abundant suggest that the limpet lives in an extremely variable and hazardous milieu. At various times the animals are exposed to mechanical stresses from waves and suspended sand, to salinities ranging from that of sea to that of rain water, to large and abrupt temperature fluctuations and to prolonged desiccation.

The only other kinds of animals consistently seen locally with *A. digitalis* are the limpet *A. paradigitalis* (Fritchman 1960) with a narrower vertical range; another snail, *Littorina scutulata,* and the barnacle *Balanus glandula.* At the extreme lower limit of the vertical distribution of *A. digitalis,* a third limpet, *A. pelta,* begins to appear. It is only below this level that animal diversity increases greatly.

The area chosen for our observations lies inside Coos Head along the southern shore and near the entrance of Coos Bay. It is about 0.7 km from the Oregon Institute of Marine Biology, which served as a base of operations. The rocky coast here is partly protected from storms, particularly in summer when major wave trains come from the northwest. In winter, storms sweep up the bay, which opens to the southwest, and there is considerably more effect from rough water. Waves and swells more than a meter high are typical of winter but absent in summer. The difference in seasonal wave action is accentuated by the local situation. Soft sandstone emerges from a sandy beach whose level, during the period of these observations, fluctuated seasonally by about 1.5 m (Fig. 1). In summer the sand level is high, and the force of existing waves is largely broken even at high tide. In winter there is an absence of spray only at the lower tides; when tides are high much of the rock then receives the impact of relatively large waves. During the three years of our study, the beach sand gradually eroded around the base of the rocks in this area in October, and stayed low through March, when gradual accretion lasting into early May began. In 1963, presumably as a result of repair work on the south jetty of Coos Bay, the conditions of previous years were modified, and the winter situation prevailed for that summer.

Although other similar areas were used for subsidiary experiments, the main study site was a single rock that juts out from the main sea cliff (Fig. 1, 2). This soft sandstone rock is roughly conical and has habitable faces to the west, north and east. The cliff of which it is a part abuts to the south. At the two ends of the study area trickles of fresh water run down the rock surface virtually throughout the year. They become surrounded and overgrown by *Enteromorpha.* It was noted early in the preliminary phase of this investigation that limpets do not normally cross such areas, probably because of negative behavior similar to that described for *Patella vulgata* by Arnold (1957). We therefore selected this site

FIG. 1. The experimental rock. Left: typical summer conditions, June 1961. The white area at the upper right is a patch of bleached *Enteromorpha*. Right: typical winter conditions, January 1962. Note the stainless steel fences, which have loosened at this time. The holes in the rock reveal most clearly the differences in sand level.

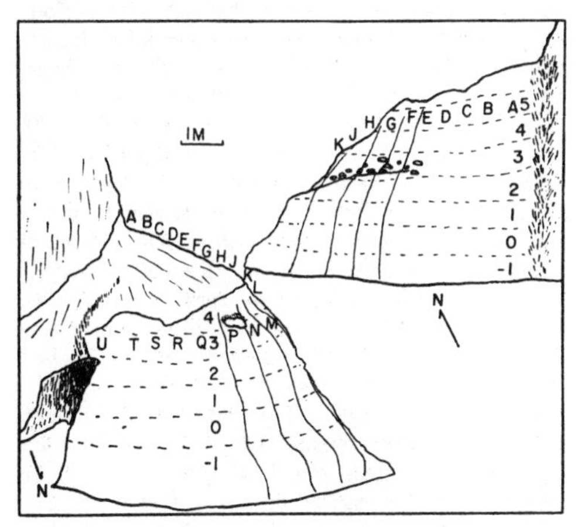

FIG. 2. Composite view of the experimental rock to indicate locations of designated squares, and of experimental fences.

and modified water flow by cutting channels into the rock, so that lateral emigration was prevented. The sand formed an effective lower barrier. Those parts of the rock lying within the vertical zone of distribution of *A. digitalis* are generally at an angle of 40 to 80 degrees from horizontal. Of considerable significance are the numerous small depressions and potholes eroded from the sandstone. Many of these are about 8 to 10 cm wide and 3 cm or more deep, and may hold small pools of water. In these areas and near other irregularities in the rock, aggregations of limpets very similar to those observed by Abe (1932) in *A. dorsuosa* often form.

The total area occupied by limpets varies somewhat seasonally, but encompasses something of the order of 40 to 45 m^2 on the experimental site. For convenience in locating animals, a grid of squares, approximately 65 cm along a side, was superimposed on the study area, so that there were 19 columns of such squares arranged in 5 to 7 rows, depending on precise location (Fig. 2). Since the experimental rock is roughly cone-shaped, the lower quadrats are actually a bit larger and the upper ones smaller than the specified size. Corners of these squares were permanently marked with stainless steel screws driven into the rock in 1961. Individual columns of squares are designated by letters A to U (omitting I and O), and levels from low to high are coded from -1 to $+5$. Mean high tide corresponds roughly to the top of row 1. Several gradients, caused by tidal, solar and local features, exist in the area. Aside from the obvious vertical zonation, a gradient of decreased wave exposure exists from column M towards both column A and U. The gradient of increased insolation is generally from U to A, although it is complicated by seasonal variation. The high cliff behind the rock cuts off a certain amount of mid-day sunlight in winter but not summer. Columns S to U receive fresh-water spray from above in the spring, and are generally moister than the rest of the area.

The seasonal moisture régime has a marked effect on algal cover. In summer when the rock remains dry above level 2 most of the time, only traces of algae, particularly bluegreens, can be found in rock scrapings, and the sandstone in the higher regions looks bare to the unaided eye. This is in contrast to a nearby cave area where, at the same level and with roughly comparable limpet densities, diatoms and other small algae form a dense cover. In winter when the experimental rock is continually moist, a heavy diatom blanket develops on it also, and larger algae gain a foothold. These include *Ectocarpus* sp., *Enteromorpha* sp., *Gelidium* sp., *Gigartina papillata, Polysiphonia* sp., *Porphyra perforata, Pterosiphonia robusta, Scytosiphon lomentaria* and *Ulva linza.*

Observational and Experimental Procedures

Individual recognition of limpets was basic to many of our observations. Smaller snails could not be handled effectively, but all limpets more than 5 to 6 mm in shell length were marked. First they were carefully removed from the rock and placed on moistened sheets of flexible plastic. In the laboratory the postero-dorsal shell was smoothed with a grinding tool, and an adhesive tape tag with an India ink code number was applied to the shell. Numerals to 1,999 and combinations of numbers and letters were used to designate individual animals. The adhesive tape was then covered with two coats of a quick-drying plastic adhesive (composed of a mixture of Butvar B-76 kindly furnished by Shawinigan Resins, and an acrylic resin 'Dekophane' available from Rona Pearl Corp.). Shell length was measured and recorded before the limpets were replaced, usually at the following low tide or about 12 hours after they were removed. Marked animals received an additional coat of the adhesive *in situ* a year after they were originally labelled. We made no attempt to replace limpets in the exact spot from where they had come, but, except during the first year, replaced them at the same vertical level. The marking procedure leads to losses of limpets from various causes, but these can be kept to about five per cent of those marked. For purposes of census marked limpets were disregarded until they had been recovered on the experimental rock at least once. The immediate losses from tagging were compensated by the addition of a similar number of limpets from near by. From the summer of 1960 on, all limpets 6 mm or more were kept individually tagged through the period of study.

Our marking procedure is related to two major shortcomings in our data: 1) We cannot account in any precise fashion for animals until they have been on the rock for approximately two to three months. Even were it possible to mark younger animals, it is difficult or impossible to identify limpets this small to species. One component of possibly density-dependent mortality is thus removed from precise analysis. 2) Tag losses occur, especially in winter at the lower levels. We can estimate tag losses both from the size of animals and from tag remnants that are normally visible for a period of weeks. However, corrections for tag losses made in the data make conclusions regarding some of the differences in death and emigration rates more tenuous. Over the total period of study, comprising over four years, the total number of limpets of all species marked was about 9,500. Marking was done almost exclusively in June, July and August, when student help was most available. Other observations were also conducted most intensively then. During the rest of the year information was gathered every two weeks, except when occasional severe storms upset this schedule.

Although it may seem simple to estimate density and mortality, the only two parameters that, superficially at least, are significant to the problem posed, one of the major emphases of this report is that, because of heterogeneity of the population and of the habitat, the crude total estimates are almost valueless. The main body of data includes initial numbers and sizes of animals, and periodically repeated measurements of recruitment, disappearance, growth and spatial position. Every two weeks, and oftener in summer, we tried to measure the shell length of a sample of 50 to 200 animals and to census the entire population. To avoid time-consuming duplication during counts, small areas were delimited during census by marking them off with a paste of starch and chalk squeezed from a plastic bottle. Except in summer data were read into a portable tape recorder, since the time available during low tides was rather limited. Efficiency of censuses can be checked, and errors can be corrected by comparisons with subsequent censuses. The number of limpets found on a later census that were missed on a prior one indicates, to a first approximation, the efficiency of the earlier census. The process of approximating the true number of snails present at the time of a particular census can be further corrected for the efficiency of the subsequent census and for the estimated intervening deaths. Usually this correction is of minor dimensions and of uncertain validity.

When conditions were favorable, censuses were about 95% effective. When weather was really inclement efficiency dropped to as low as 70%. Unless they were recovered on a subsequent census, animals that disappeared from our records were assumed dead at the time of their first disappearance. It is highly probable that these animals had died. Reasonably diligent searching of all habitats in the adjoining area—a total of about 200 m of shoreline—revealed only three marked animals outside the experimental area during three years. All of these were on immediately adjoining rocks, where they could have crawled directly

TABLE I. Frequency distribution of limpets according to location, July 3, 1961

Row	Column																		
	A	B	C	D	E	F	G	H	J	K	L	M	N	P	Q	R	S	T	U
5	0	0	0	0	0	3	1	0	0	1	0	0	4	1	0	0	0	0	0
4	0	0	13	1	4	11	5	4	2	0	0	2	5	1	1	0	0	2	0
3	0	0	12	1	2	12	11	17	29	25	1	3	2	20	5	20	0	7	3
2	16	0	31	1	10	22	41	58	34	69	31	42	49	69	8	23	2	2	8
1	9	8	13	18	19	15	4	6	24	12	26	92	43	4	7	3	2	2	1
0	0	2	0	4	7	2	12	1	3	3	2	11	9	2	0	0	0	0	0

from the main area. We repeatedly tried to see what would happen to an animal on sand. The foot invariably became coated with sand grains, and the snail was incapable of moving on its own.

The intended procedure of our study was to gather data for two years under essentially undisturbed conditions to serve as partial controls for density manipulations in the third year. Accordingly, columns F, G, H, M, N and P were fenced off from each other and from the rest of the rock with stainless steel mesh in the summer of 1962 (Fig. 1, 2). Although earlier trials on a nearby rock had indicated that the fences would stay in place and would effectively prevent immigration and emigration, wave action in late fall weakened the experimental fences which had been effective to that time. We therefore removed most of the ineffective remnants in winter (Jan. 7, 1963).

Aside from these observations on the experimental rock proper, other information was gathered from additional areas. Samples of limpets were removed at two-week intervals throughout one year to check on reproductive condition and to determine size-specific biomass. Growth and settlement on a north facing cave wall of about 30 m^2 was assessed by removal of its limpet population and continuing observations of the newly settled individuals. Separate experiments on movements were done on a steep rock face using animals that had been mass marked with various quick-drying paints. Essential details concerning other procedures are best reserved for those sections that deal specifically with them.

DISPERSION AND MOVEMENTS

Since this investigation centers about the problem of what limits animals to the zones occupied and the densities achieved, it becomes necessary to describe their spatial distribution. Here and elsewhere the data are too unwieldy to be easily summarized, and I must resort to representative sections of them. The gross pattern of dispersion can be readily illustrated by our quantitative data, as exemplified in Table I, in which numbers of limpets are listed by quadrat. The finer patterning of aggregations of a few to 30 or more limpets in a tight cluster, usually in small depressions or cracks in the rock is difficult to document quantitatively, as is the fact that one rarely finds one limpet on top of another, even where they are closely packed.

From Table I it is evident that the snails center about levels 1 and 2, and are generally more abundant in central than in peripheral columns. On closer inspection the table reveals some of the local differences for which limpet numbers can provide an index. For example, a cleft in column C characteristically contains more limpets than do the surrounding more exposed areas. Most of the heterogeneity illustrated in the table persists over the period of study. It may be worth noting at this point that the columns chosen for eventual density manipulation, G and P, were consistently similar to the adjoining columns that served as their controls. The difference in stratification between column P and columns M and N observable in the table is caused by a short term difference in distribution.

Although the main density pattern persists, there are seasonal shifts, as indicated by Table II.

TABLE II. Vertical distribution of *Acmaea digitalis* by season

Level	% of total *Acmaea:*	
	July 3, 1961	Dec. 8, 1962
5	0.9	11.6
4	4.6	17.1
3	15.3	22.6
2	46.4	28.2
1	27.7	17.1
0	5.2	3.4
Total No.	1,113	1,284

Characteristically a greater proportion of animals are found higher in the winter than in summer. Distribution of limpets is not random with respect to size either. As is true of *Patella* (Lewis 1954), the largest *Acmaea digitalis* live highest in the zone occupied by the species (Table III). This is in apparent contradiction to Shotwell's (1950)

TABLE III. Frequency distribution of *Acmaea digitalis,* by size and vertical zone, Aug. 1961

Tidal level	Size class in mm.																				
	5	6	7	8	9	10	11	12	13	14	15	16	17	18	19	20	21	22	23	24	25	26
5...........	—	—	—	—	—	—	—	—	—	1	1	—	—	3	—	—	—	—	—	—	—	—
4...........	—	—	—	—	1	—	—	—	—	1	—	3	3	7	3	5	4	1	—	2	1	—
3...........	—	—	—	1	—	—	—	2	3	8	11	9	21	20	10	8	4	2	1	1	—	—
2...........	—	—	7	5	10	15	9	14	23	16	37	35	35	33	20	26	6	7	2	—	—	1
1...........	4	5	26	29	14	16	14	16	8	12	10	19	5	12	7	1	2	—	1	—	1	—
0...........	—	1	3	1	—	3	1	1	—	1	2	—	6	3	1	—	—	—	—	—	—	—

findings among other members of the genus. A statistically significant regression line may be fitted to the data: Y (tidal level) = 0.143X (size in mm) — 0.22. The relation between size and position is readily explained when it is examined as a function of time. From the censuses, data on successive positions of individual animals are available. By disregarding all except the vertical zone where limpets are found, and restricting our examination to animals one or more years old, it is possible to assemble the information contained in Table IV. Individual animals that have moved upward from the beginning to the end of the period are scored +1, 2 etc., depending on the number of squares moved. Descent is scored as a negative integer. Quantitatively the averages mean very little, since the data reveal a tremendous amount of variation. This only means that not all animals behave alike in any one time interval, and that, at a time when some individuals move up, others may remain at the same level or may even descend. However, with the number of observations available, the reality of the ascent in fall and winter and the lesser downward movement in spring is clearly revealed by the averages, which are five or more times their standard errors. In 1961 to 1962 the upward movement was much less pronounced than in the following fall. Unlike the autumnal rise, which was immediately noticed from the obvious change in position of many limpets, the spring movement in the opposite direction was not detected until the data were analyzed. As a net result of these movements, larger and older animals tend to live higher than do younger ones. Similar observations have been made in littorines by Cranwell and Moore (1938), and in limpets by Abe (1932) and Lewis (1954). Lewis' paper bears a very striking set of resemblances to our observations; the genera *Patella* and *Acmaea* show a number of beautiful convergent adaptations. The general phenomenon of a change in size distribution with position in the intertidal seems relatively common among intertidal animals.

TABLE IV. Average net movement of all animals more than one year old between summer of 1961 and 1962

Time interval	Average movement, No. of squares	No. of *Acmaea*	SD^2	SE
Aug.-Oct........	+0.158	479	0.28	0.019
Oct.-Apr........	+0.242	426	0.41	0.031
May-July......	−0.184	255	0.34	0.037

In *Acmaea digitalis* the upward movement is presumably directly related to a behavioral response that anyone who has ever placed these animals in an aquarium will have noted. Within a few hours at most, these limpets will be found above the water surface. Similarly, as winter storms bring increased wave action on parts of the rock receiving only spray in summer, the limpets move upward. The downward movement in spring is not so readily related to a proximate cause. It may perhaps be likened to the presumed behavior of these animals in a hypothetical moisture gradient. At this time of year the upper parts of the rock become significantly drier.

Lateral movements tend to be smaller in extent than vertical ones. Except at the edges of the study site, no obvious barriers to lateral movement exist. Nevertheless the limpets tend to stay in the same or adjacent columns with a high degree of probability for long periods of time (Frank 1964). Especially in the upper parts of their zone of distribution these snails form aggregations of closely clustered individuals. Successive censuses may reveal some changes in relative position of the members of such a group, but suggest that few individuals move more than a few centimeters away for periods of several weeks or months. Strict homing does not occur. In summer those animals located high enough on the rock may be incapable of moving for long periods because the rock may remain dry. Under these conditions the foot of the snail becomes cemented to the underlying rock by dried mucus. Later, when the snail moves away or dies, a noticeable scar remains.

Two other observations are worth recording, since they may be of some interest to students of behavioral physiology. Several times we noted movements of several animals as in a group. The most striking instance was that of 16 animals, all of which were in square H_2 in November 1962. Following removal of the fence, they all appeared in square P_4 in January of the following year. Behavior such as this suggests that at times the limpets may be reacting rather simply to a single environmental stimulus. The second set of observations concerns the reactions of transplanted animals, and parallels more extensive observations of Segal (1956) with *Acmaea limatula.* In summer of 1961 we marked several groups of animals from high in their area of distribution with quick-drying paint to distinguish them from animals from lower in the intertidal, which were labelled with a contrasting color. Subgroups of both classes were repeatedly placed at high and low levels of a selected test area. The limpets that had occupied higher zones moved upwards rapidly as compared with those from lower zones, which tended to remain low. The converse result was not observed, probably because the rock higher up was not wet enough to permit crawling. We did not determine whether the differences are results of acclimation, as in *A. limatula,* or not.

One thus gains the general impression that these limpets move up and down as a result of a number of natural stimuli, but are otherwise conservative with respect to position. Although capable of moving more than a meter in a single tide period, such large scale movements are exceptional. Generally the area grazed by a single individual seems to be restricted to a few square centimeters in any one day. Although there is no return to the same spot, most animals apparently spend their lives, once they have settled, moving in an ambit that is no more than a meter wide. Unfortunately we were consistently unable to observe the limpets at high tide, the time when they feed and move most actively. There is therefore the possibility that effective density is reduced by the animals' dispersing temporarily. This certainly happens immediately around aggregations, but all our indications from records of positions of animals at successive low tides, sometimes for periods up to a month, suggest that grazing in far off areas followed by return to the same small area does not occur.

Numbers, Mortality and Settlement

Figure 3 summarizes census data. It describes total numbers of limpets on the experimental rock during three years and indicates death rates of *A. digitalis.* As may be seen, the greatest overall density existed at the outset. In June of 1960 there were a total of 3,841 limpets large enough to be marked. By June 1961 these had decreased to 1,092. The 1961 year class of recruits consisted

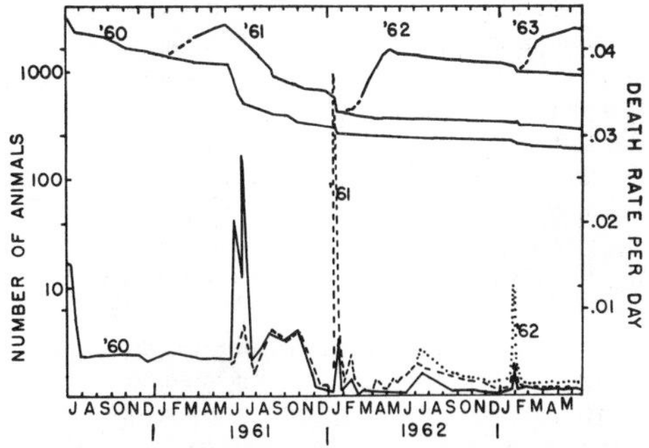

Fig. 3. Numbers and death rates of limpets on the experimental rock. The family of curves at the top indicate numbers of successive year classes as indicated. Note that numbers are graphed on a logarithmic scale. The curves near the bottom indicate instantaneous mortality rates. Those marked '60 apply for animals of all ages marked in 1960. The dashed and dotted lines refer to death rates of the 1961 and 1962 year classes respectively.

of 1,215 newly marked animals, of which, however, many died before the end of summer. The 1962 year class was augmented by 400 limpets that were imported from another area. The graph of total numbers is based on all species of *Acmaea* present on the experimental rock. It includes a few *A. pelta* and the relatively common *A. paradigitalis.* In contrast, the death rates of Figure 3 apply specifically to *Acmaea digitalis.* The ascending portion of the numbers curve of successive year classes is based on rather crude estimates. Despite attempts at total census of small, unmarked limpets, our data on numbers in a new year class are not meaningful until June of the year. Total numbers were undoubtedly somewhat higher in the spring than represented on the graph.

Inspection of that part of the figure that deals with mortality is probably more instructive. The indicated death rates were calculated from corrected census data according to the formula

$$d = \frac{\ln N_o - \ln N_t}{t}$$

where d is the instantaneous death rate, N_o the initial and N_t the final number of animals for a time interval t days long. Death rates are plotted separately for the total population of 1960, and for subsequent successive year classes.

For at least some of the deaths general causes can be assigned with some confidence. This applies particularly to the high death rates observed during the summers of 1960 and 1961 and during the winters 1961 and 1962. The two former years were characterized by relatively dry, sunny summers along the coast. As early as the latter part of June of both years, limpets could be seen that adhered to the substrate only by a narrow median piece of the foot, the rest having withdrawn and shrivelled. Some of these animals, when placed in fresh sea water, revived, but the majority did not. By the end of the summer of 1961 *A. digitalis,* especially on south-facing rocks and at the higher levels, had essentially disappeared all along this part of the Oregon coast, and shells of these limpets were washed up on the beaches at each tide.

In both summers a mouse was repeatedly seen feeding on limpets from the experimental site. It destroyed a sizable number, a dozen or so at a time, but removed only limpets so high up that they were already in an advanced stage of desiccation. It is improbable, considering the pull required to remove a healthy limpet, that the latter would be a source of food for mice. Predation by birds, observed only indirectly was only noted at this time. Twice empty shells were left on top of the site under conditions that made it obvious they had been eaten out by birds, probably gulls. The only other predator we have seen feeding on *A. digitalis* does not occur on the experimental rock, although it may be of some significance elsewhere. A polyclad flatworm, probably *Leptoplana,* was repeatedly seen feeding on limpets in nearby places during the late winter months. Often a trail of small limpet shells was left by the worms, some of which had ingested as many as five of these snails. The flatworms are limited to moist, shaded rock, and seem to be abundant among mussel beds. They rarely venture above the equivalent of zone 0 of the experimental area.

It is noteworthy that the deaths during 1961 struck young of the year only half as severely as they did the older year classes. This was probably true the previous summer, but cannot be documented for that period. The explanation lies in the previously described spatial distribution and in its relation to the moisture gradient: the older animals live higher and are thus more subject to desiccation. This is true despite the fact that they tolerate prolonged drying better than do smaller limpets.

The catastrophic mortality of the two winter periods is intimately associated with exceptionally severe frosts during these two years. The frosts seemed to have no direct effect on the limpets *per se.* A census the morning after temperatures fell to −7°C in the vicinity found the snails in good condition, although such temperatures are quite unusual in this area. However, the ice that formed in the top layers of the sandstone caused gradual but extensive exfoliation during the ensuing weeks. Animals attached to areas that exfoliated were carried off. For example, squares M_1 and M_2, which up to this time had had dense populations, lost their total populations in the winters of 1961 and 1962 respectively. Our notes show that in all areas limpet mortality correlates well during this period with the fraction of rock exfoliated. After removal of the top surface, the rock looks noticeably different and has no dense limpet populations for some months.

Aside from these incidents, a general, low level of mortality of about 0.001 to 0.004 per day prevailed. With rates this low, causation is hard to establish, nor is it necessarily particularly significant in regard to the major problem. The age specific differences in mortality are of some interest. Evidently young of the year are at a disadvantage in winter, whereas in summer relations are more variable and probably depend on the severity of drying. That the two freezes affected the young so much more than old animals may be of no general significance. It probably relates only to the nature of the particular rock surface. A survivorship curve based on our data, but disregarding the periods of obvious catastrophes thus would look concave, implying that the probability of survival increases with age at least for those ages for which sufficient data can be gathered. In view of the heterogeneity of the habitat and among the animals, I feel that specific survival probabilities given by such a curve have little meaning at this time. More useful is an analysis of mortality according to location on the experimental rock and to time of year. The data are insufficient to permit useful comparisons by individual quadrats. Too many quadrats lack animals at one time or another. The coarser grouping of Figure 4 illustrates the sort of information that can be gathered. There is a distinct contrast with respect to the time of year and location. The data for July 1961 indicate that very heavy mortality is then typical of the higher portions of the rock. In the fall there is excellent survival in equivalent regions. Little major upward movements had occurred in 1961 by this time, so that these data reflect deaths by vertical zones rather accurately. The bottom half of the figure illustrates in addition that probability of survival then decreases at the lower levels. A significant additional piece of information from a finer analysis is that animals at level 0 consistently, i.e. throughout the year, exhibit a higher death rate than those at level 1.

Except for the period from February 1 to 15 1962, the time of exfoliation, mortality experience similar to that of the lower part of Figure 4 prevails through the following winter and spring. One distinctive element is added: from February to June column group S to U has a higher death rate than does any other group of columns. This

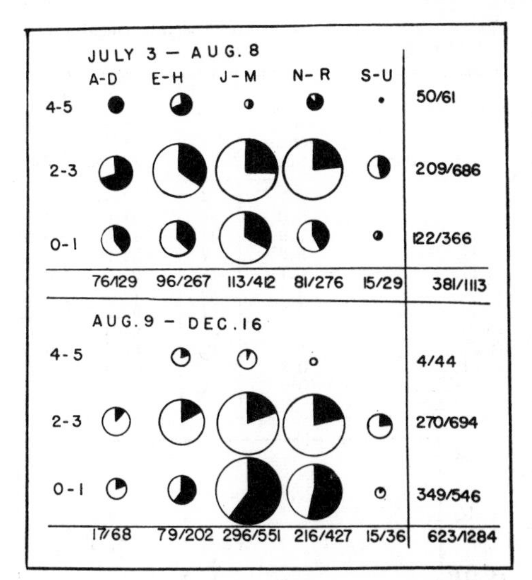

FIG. 4. Numbers and death rates during summer and fall 1961. The sizes of the circles are proportionate to the numbers of limpets in the quadrats indicated on the top and at left. The proportion of the area blackened indicates deaths occurring in the interval. The marginal totals indicate as the numerator the number of deaths out of a total number, given as denominator.

increase is correlated with, and undoubtedly caused by, the fresh water spray from the cliff above, which is then most intense.

Data of this sort can be subjected to a Chi square analysis to test additive effects of rows and columns and interaction. Significant differences in survival for rows and columns can usually be found. Moreover, after these effects have been eliminated (by statistical treatment), a statistically significant amount of heterogeneity often remains. One must be somewhat cautious with the data, since they have been corrected for census errors. In several instances, however, these effects are so large that they cannot be considered arithmetic artifacts. The residual heterogeneity was checked for possible correlation with density without significant results. So far these differences are interpretable as resulting from local attributes, but escape more searching analysis.

The view of this limpet population that one forms from these data on mortality and from the earlier section dealing with spatial distribution and movements bears some striking resemblances to the model presented for the grasshopper *Austroicetes cruciatus* by Andrewartha and Birch (1954). Quite evidently the snails are restricted to a relatively narrow region of the intertidal zone by behavior and by high death rates outside their even narrower and shifting optimum zone. Vertical zonation in this species of limpet has thus been "explained." Andrewartha and Birch argue further that, in the case of *Austroicetes* and generally, abundance is merely another aspect of the same phenomena that determine distribution. This is not the case here, as may be seen when one considers what would happen to our scheme were 10,000 limpets recruited annually rather than the number that are. In the absence of additional sources of deaths, no limit would be set to numbers. The population phenomena described for *A. digitalis* to this point give no evidence regarding the maximum densities likely to be achieved by these snails.

A priori, data on settlement are unlikely to be much help. If density of adults has an effect on settling rate of planktonic larvae, it is unlikely to be a density-regulating effect (Nicholson 1954), but rather may be density-disturbing (e.g. Knight-Jones 1951; Wilson 1952). A more likely place to look for negative feedback might be in animals too small to be marked, but already present on the rock. This group of animals may regulate numbers. At least this is a possibility that cannot be eliminated by any data we have. In an attempt to test the question we surveyed a variety of areas containing *Acmaea digitalis* in the summers of 1961 and 1962. Twelve vertical transects 50 cm to 1 m wide were stripped of limpets. Animals less than 6 mm long were separated from the larger ones, and numbers of these classes were then compared per unit area. The correlation coefficient between numbers of adults and young was positive ($r = +0.43$), but not statistically significant at the 95% level of confidence. The positive correlation may imply solely that an area that is suitable for dense populations of limpets one year is also likely to be attractive for settling in another. It certainly lends no credence to the hypothesis that density-dependent mortality of the young snails is an effective population regulator.

Settlement of small limpets could be detected on the rock through much of the year. There was a sharp peak in late April and early May. Spent gonads among animals from nearby rocks were first seen in March of the year. Gonads regressed during the summer months. Otherwise there was no time when a majority of animals in any local population was spawned out. Since it was impossible for us to identify newly settled limpets by species, we cannot be certain that settlement of *A. digitalis* ceased entirely in the fall. Virtually no limpets were recruited to the 6-mm size class during the early months of the year, however.

Growth

Estimates of growth were obtained primarily for the eventual determination of production rate. For present purposes they are suitable both as an index to the quality of environment and as an indicator of age. Since shell shape varies with growth rate (Orton 1933; Moore 1934), it is usually desirable to measure height of the shell. We were restricted to length measurements with vernier calipers, since otherwise the animals would have had to be removed from the rock. We have some evidence from dead animals that shell shape is not a significant variable in the experimental area, although it differs here from shape in some other nearby areas. To convert shell length to more generally useful measures, we made estimates of the relation between length, volume and dry weight exclusive of the shell. To determine volume, the shells of 47 animals were filled with paraffin, and the casts were weighed. The weights so obtained, when divided by the specific gravity of the paraffin, yield the volumes of Figure 5.

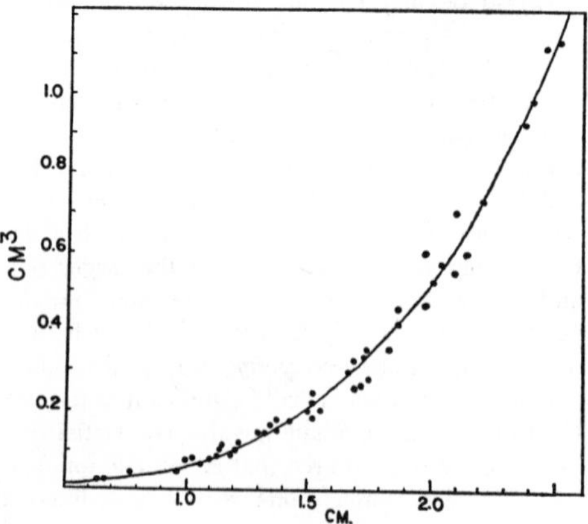

Fig. 5. The relation between shell length of limpets and their volume.

Dry weight estimates did not vary significantly between seasons or sexes, so that a simple conversion factor of 0.35 converts the volumes to dry weight.

Table V provides a summary of annual growth rates for June 1961 to 1962. The data are not separable by sex, but neither age specific size distributions nor a small number of growth data for sexed individuals suggest a significant difference in growth of the sexes. Growth rate is, however, quite variable, especially among the larger snails. From these data it is possible to construct the relation of age to size of Figure 6. For animals three years old or less the data are based on growth measurements for the entire period. For older animals an initial estimate of age was made on the basis of initial length.

Table V. A summary of size specific annual growth, 1961 to 1962

Initial size mm	Average yearly increase mm	No.	Coefficient of variation	SE
6	8.7	4	4.8	0.21
6- 8	8.0	46	21.6	0.26
8-10	7.6	44	17.6	0.20
10-12	5.6	7	33.2	0.70
12-14	4.1	18	21.5	0.21
14-16	3.5	29	28.2	0.19
16-18	2.8	34	52.2	0.25
18-20	1.8	24	64.3	0.23
20-22	1.8	12	57.7	0.27
22-24	0.8	4	48.0	0.18
24	0	1	—	—

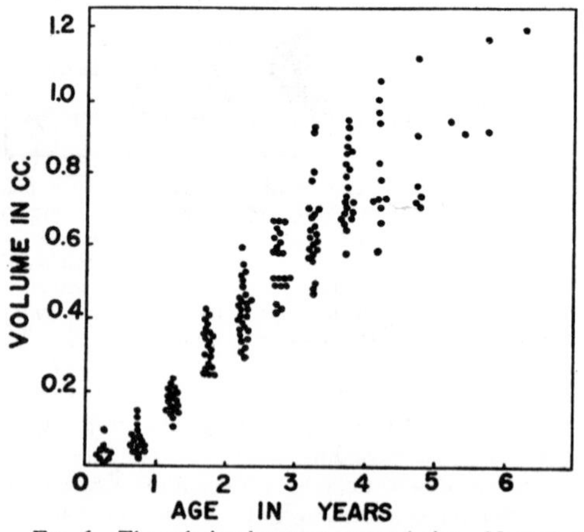

Fig. 6. The relation between age and size. Note that size is not a sufficiently good indicator of age to be used as a predictor. For animals over three years old, age is estimated by extrapolation. For animals less than four years, age and size were determined directly from marked animals.

With the exception of the summer months growth rates seem similar throughout the year. This information is conveyed, at least in part, by Figure 7, which is based on growth of 278 animals for the period June to July, 138 for February to May, and 233 for August to January. The graph is somewhat misleading. Snails less than 7 mm long in August or less than 11 mm in Februrary are probably atypical, i.e. are part of a sample selected unintentionally for low growth rate. Groupings by shorter time intervals yields no additional information, since numbers of animals become small, and the relative error of measurement becomes too high. Additional information regarding growth rates is best reserved for the section that deals with density manipulation.

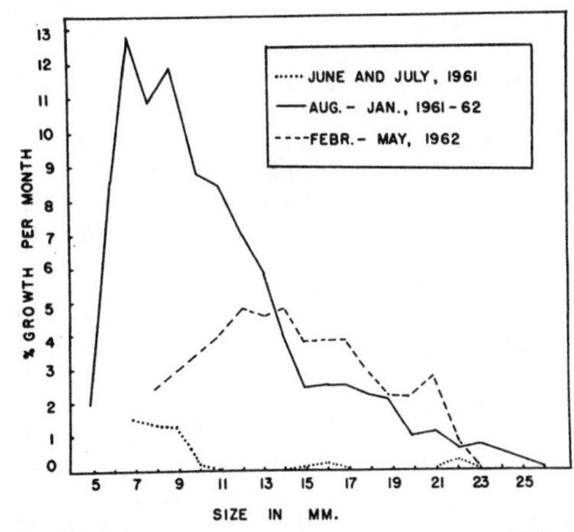

FIG. 7. Growth rate variations between seasons. Values for animals less than 8 mm in August and less than 12 mm in February are probably spurious (see text).

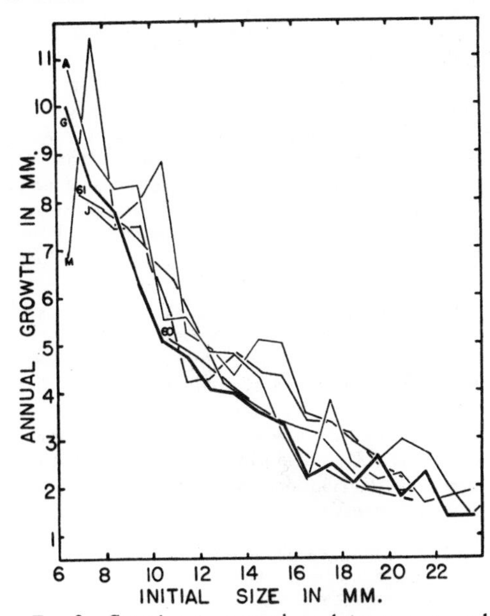

FIG. 8. Growth rate comparisons between years and between groups of columns. (A is growth rate of columns A to E; G applies to F to H; J to J to L; M to M to P; 60 is average growth for all animals measured for 1960-1961; 61 is the same information for the following year).

EFFECTS OF EXPERIMENTS

As a result of animals transplanted from one area of the experimental site to another and the addition of animals from elsewhere, columns G and P had significantly augmented numbers by August 10, 1962: column G had 636 limpets, of which 488 were *A. digitalis;* column P contained 475 snails, 393 of them *A. digitalis*. Total volume of limpets (exclusive of shell) in column G is estimated at 274 cc, and for column P at 180 cc. The area actually occupied by the limpets is difficult to estimate in any meaningful terms, but on the basis of where they were at that time, we have assigned an area of 2.8 m^2 to column G and of 2.1 m^2 to column P. This is equivalent to saying that average density per m^2 was 0.98 and 0.86 cc per m^2 respectively. Such an estimate seriously underrates the density in the more crowded middle squares. Although the fences that were to have kept the limpets confined to these columns disappeared or were removed in January, densities in these and in adjoining columns continued to reflect the crowding imposed the previous summer. Thus it is of some interest to determine whether an effect of crowding on growth rate is detectable. Comparison of growth by sub-areas for this year with growth over the area as a whole during the previous two years (Fig. 8) shows most evidently the consistent response of the population in this environment. The data may be subjected to an analysis of covariance, which is simplified by a logarithmic transformation of growth. The analysis supplies proper weighting to the means of the figure. Despite the similarity of the individual curves, the analysis leads to the conclusion that mean growth, adjusted for initial size, differs between them ($F = 9.28$; 99% confidence level is $F_{4,567} = 3.35$). Specifically the heavier line in the figure, which describes growth in columns F, G and H indicates a lower average. The other means do not differ significantly from each other. Since lability of growth rate in limpets is generally recognized and is undoubtedly real, the minor effect on growth was surprising. Fischer-Piette (1948) has supplied good evidence on large differences in growth rates of *Patella* in different environments, and we have observations on growth of *A. digitalis* in a shaded cave that also indicate that much greater variation is potentially possible. Apparently, however, conditions on the experimental site were such that increased crowding did not reduce growth rate significantly in one area (column P), and did so in the other (column G) only slightly. This is true despite the fact that at this time other areas on the rock had densities lower than any experienced since the start of the experimental period.

Because death rates in different columns and rows show so much heterogeneity, it is probably not legitimate to compare mortality except with adjoining squares that did not exhibit significant differences in previous years. Such comparisons must also be restricted to the period when the fences were effective, since after that time it is too difficult to assign death rates to the appropriate column. For column G, 437 of the original 488 limpets survived from August to December, as against 172 of an original 206 for columns F and H. For column P the equivalent values are 339 of 393, as against 282 of 321 for columns M and N combined. The biggest difference is one in the direction of increased survival in column G. However, none of the differences is significant when subjected to the appropriate Chi square test.

During this period there is a striking effect on spatial distribution. It is perhaps most clearly indicated by the fact that limpets for the only time during the period of study moved onto the more or less flat top of the experimental rock in column P (Fig. 2), that is, to an area corresponding to square P_5. The same is true of column G, where the distribution on December 7 is that shown in Table VI. It is quite evident that, either

TABLE VI. Positions of limpets on Dec. 7, 1962 in crowded and in adjacent, uncrowded squares

Vertical zone	Crowded squares G		Uncrowded squares F and H	
	No.	%	No.	%
5	51	11.7	3	1.7
4	88	20.1	14	8.1
3	134	30.7	84	48.8
2	97	22.2	55	32.0
1	49	11.2	11	6.4
0	18	4.1	5	2.9
TotalNo.	437		172	

as a result of the density increase or because of the addition of animals from elsewhere, the crowded quadrats had disproportionately more limpets in the highest and lowest quadrats. The movements thus exhibit a sort of density dependence that, over the year, presumably could effect control. Admittedly, because of the failure of the fences we are unable to prove this point conclusively. Curiously, moreover, tests for differences in rate of lateral dispersion after the fences were removed indicated no significant differences between rates of movement from crowded and from adjacent, less densely populated squares. This may be because density has already been adjusted downward by the vertical movements, and also because the animals may have established home ranges (Frank 1964).

Since these and earlier observations suggested space as important to the snail populations, a number of smaller scale tests of environmental features were made in the summer of 1963. Among these were two attempts to improve the environment: 1. two columns of the experimental site were continually kept moist by a trickle of seawater from a carboy placed on top of the rock and refilled daily; 2. small artificial shelters from insolation were added. These were made of slit inner tubes, anchored to the rock so that they formed a roof but were open at two ends. The effects of the first test were disappointing. Although some increase in algal growth was observed, wave splash this summer was heavier than in previous years. No extensive period of desiccation was thus experienced anywhere, and effects on death rate could not be established. Limpets did not actively aggregate in the moister areas; nor did the artificial shelters, located at levels 3 to 5, ever become significant centers of aggregation, although they usually harbored a few individuals. Apparently such areas do not serve as attractants over any distance, but rather allow limpets to remain when they reach the immediate vicinity. Presumably the snails cease to move in these darker or moister areas rather than actively seeking them out.

Discussion

Our results emphasize the role that behavior plays in determining the characteristics of an animal population. In this regard it complements and parallels studies by Lewis (1954) and by Evans (1951) rather specifically. The vertical level at which *A. digitalis* is found seems determined, at least in large part, by behavioral responses of the animals to their environment. That these responses are correlated with environmental features, i.e. are adaptive, is to be expected. Had we not found increased death rates outside the zone to which the limpets move, their movements would have been puzzling. The ultimate reasons for the upper limit are undoubtedly the physiological possibilities of the animals to tolerate desiccation. The lower limit is not so easily explained. Presumably the behavior of the limpet which determines this lower limit has evolved as a response to the increased predation (Connell 1961) or abrasion at lower levels. Potential competition with other species cannot be ruled out either, nor is there any reason why several factors may not operate together.

Behavior evidently is also involved in regulating abundance, probably in an interaction with space and food. As long as their preferred habitat is not saturated, limpets tend to remain where they are. We have a number of records of limpets that moved no more than a centimeter or two for periods of up to two weeks, when it would have been possible for them to move much larger distances. Rather than leaving, limpets tend to remain in the same small crack or cavity for prolonged periods. Presumably they begin to graze whenever local moisture conditions permit, and cease when they receive a set of signals that suggest that conditions locally are deteriorating. As an area of rock progressively dries, animals tend to be left in the moister spots, where movement finally stops. Since total contact of the foot with the substrate seems to be a necessary condition for normal inactivity, an otherwise favorable area which has a dense aggregation of limpets thus establishes its own upper limit. In a sense, these limpets behave similarly to the titmice studied by Kluyver and Tinbergen (1953), which established constant populations in the optimum habitats and had more variable numbers elsewhere.

Although food in this situation is unlikely to become absolutely limiting, relative scarcity (Andrewartha and Browning 1961) may be a common condition. The concentration of food depends on the production rate and on grazing. The highest areas have limited algal production, and opportunities for grazing there are also reduced. Elsewhere the amount of time available for grazing may determine its intensity. Time of activity as well as limpet numbers will determine how often a particular spot is covered. Limpets will not normally remain in the driest areas, as shown by the downward migration in spring. In other areas, time available for feeding rather than the existing food concentration may be most significant. This is suggested because the great increase in density in columns G and P had no greater effect on growth rate than it did, and because in other areas, e.g. a cave, growth is significantly faster. It is interesting to note that, where similar observations have been made among other limpets (Fischer-Piette 1948), high growth rates are associated with a shortened life span. This seems true in our animals also, and may be of some general importance. The relation of the phenomenon to adaptation and to density regulation is not at all clear, however.

A relative shortage of food resulting from its low concentration may sometimes operate. Castenholz (1961) estimates that a limpet volume of 0.8 cc per dm^2 can keep an area of high productivity free of algal mat. Since his limpet volumes are based on total displacement, they are about 20% higher, for the same biomass, than are ours. In any case algal mats should have been absent from the crowded columns, G and P, no matter which estimate is used. It is true that Castenholz's area was not equivalent, and that his observations stem from the summer and early fall primarily. Nevertheless the fact that during the fall of 1962 there were parts of the crowded quadrats that did have a visible cover of algae seems discordant. There is, however, no real discrepancy between our data and those of Castenholz. Limpets undoubtedly kept the main parts of our quadrats clear of algal mats. However, the habitat is heterogeneous, and as a result densities per m^2 or dm^2 are virtually meaningless, especially since the centers of activity of groups of limpets may shift. Our observations do indicate, however, that areas exist where food concentration is high enough that food is not limiting, even when limpet densities in the general vicinity are high. It is necessary to add that neither our data nor those of Castenholz explain the commonly observed algal mats seen in the upper intertidal in the winter months.

When our limpets were experimentally crowded, the effects on emigration rate were obvious, particularly since lateral movements were restricted. Although we were unable to demonstrate at that time that there were accompanying changes in death rates, these are clearly implied by other parts of our data, and their reality seems assured. Movements therefore can effectively regulate density. If a lower threshold exists below which density plays no part, it is likely to be very low, certainly lower than the existing density in areas normally considered reasonably good habitat for the animals. In a complex and shifting system of gradients, the limpets continually act to acclimate in an appropriate manner. This means that, even when overall densities are low, local crowding may occur. At times of catastrophic mortality the effects of such crowding may be obscured quite effectively. Ultimately, however, density regulation according to this scheme should be regarded as operating at virtually all densities, with increasingly severe, but never spectacular effects as progressively poorer habitats are invaded. This view, complex though it is, is undoubtedly an oversimplification. Different animals will possess different individual optima because of differences in age, history or genome. These individual differences will buffer the system of regulation.

Acknowledgments

Especially during the initial phases of this research, J. H. Connell provided much valuable advice; he and R. W. Morris were kind enough to criticize the manuscript. Discussions with R. W. Castenholz, H. K. Fritchman, A. J. Kohn and J. A. Shotwell were helpful. Several graduate students helped gather the data, and I am happy to be able to thank them here: David C. Coleman, Richard L. Darby and Vernie Fraundorf. The intensive observations and marking regimes pursued during summers would have been impossible without additional help, supplied largely by undergraduates, many of whom were supported by an undergraduate research participation program of the National Science Foundation: Harold Brown, Kenneth Graham, Jon Jacklet, Trudy Kofford, Roberta Langford, Robin Manela, Marjorie McFarland, Stephen Mohler,

John Palmer and Charlotte Patterson. I trust that they profited from the experience as much as they benefited the study. Expenses of the investigation were defrayed by National Science Foundation Grant G 11-3331, for which I am grateful.

Literature Cited

Abe, Noboru. 1932. The colony of the limpet (*Acmaea dorsuosa* Gould). Tohoku Imper. Univ. Sci. Reports, 4th Ser. (Biol.) **7:** 169-187.

Andrewartha, H. G., and L. C. Birch. 1954. The distribution and abundance of animals. University of Chicago Press. p. 782.

———, **and T. O. Browning.** 1961. An analysis of the idea of 'resources' in animal ecology. J. Theoret. Biol. **1:** 83-97.

Arnold, D. C. 1957. The response of the limpet *Patella vulgata* L. to water of different salinities. J. Mar. Biol. Assoc. U.K. **36:** 121-128.

Castenholz, R. W. 1961. The effect of grazing on marine littoral diatom populations. Ecology **42:** 783-794.

Connell, J. H. 1961. The influence of interspecific competition and other factors on the distribution of the barnacle *Chthamalus stellatus.* Ecology **42:** 710-723.

Cranwell, L. M., and L. B. Moore. 1938. Intertidal communities of the Poor Knights Island, New Zealand. Trans. Roy. Soc. New Zealand **67:** 374-407.

Elton, C., and R. S. Miller. 1954. An ecological survey of animal communities: with a practical system of classifying habitats by structural characters. J. Ecol. **42:** 460-496.

Evans, F. G. C. 1951. An analysis of the behaviour of *Lepidochitona cinereus* in response to certain physical factors of the environment. J. Anim. Ecol. **20:** 1-10.

Fischer-Piette, E. 1948. Sur les éléments de prospérité des patelles et sur leur spécificité. J. de Conchyliol. **88:** 45-96.

Frank, P. W. 1964. On home range of limpets. Am. Naturalist **98:** 99-104.

Fritchman, H. K. II. 1960. *Acmaea paradigitalis* sp. nov. (Acmaeidae, Gastropoda) Veliger **2:** 53-57.

———. 1961. A study of the reproductive cycle of California Acmaeidae (Gastropoda). Part III. Veliger **4:** 41-47.

Hairston, N., F. E. Smith and L. B. Slobodkin. 1960. Community structure, population control and competition. Am. Naturalist **94:** 421-425.

Kluyver, H. N., and L. Tinbergen. 1953. Territory and the regulation of density in titmice. Arch. neerland. de Zool. **10:** 265-289.

Knight-Jones, E. W. 1951. Gregariousness and some other aspects of the settling behaviour of *Spirorbis.* J. Mar. Biol. Assoc. U.K. **30:** 201-222.

Lewis, J. R. 1954. Observations on a high-level population of limpets. J. Anim. Ecol. **23:** 85-100.

MacArthur, R. 1960. On the relative abundance of species. Am. Naturalist **94:** 25-36.

Moore, H. B. 1934. The relation of shell growth to environment in *Patella vulgata.* Proc. Malacol. Soc. London **21:** 217-222.

Nicholson, A. J. 1933. The balance of animal populations. J. Anim. Ecol. **2,** Suppl.: 132-178.

———. 1954. An outline of the dynamics of animal populations. Austral. J. Zool. **2:** 9-65.

Orton, J. H. 1933. Studies on the relation between organism and environment. Proc. Liverpool Biol. Soc. **46:** 1-16.

Preston, F. W. 1962. The canonical distribution of commonness and rarity. Ecology **43:** 185-215; 410-432.

Segal, Earl. 1956. Microgeographic variation as thermal acclimation in an intertidal mollusc. Biol. Bull, **111:** 129-152.

Shotwell, J. A. 1950. Distribution of volume and relative linear measurements changes in *Acmaea,* the limpet. Ecology **31:** 51-62.

Skellam, J. G. 1951. Random dispersal in theoretical populations. Biometrika **38:** 196-218.

Thompson, W. R. 1939. Biological control and the theories of the interactions of populations. Parasitology **31:** 299-388.

Varley, G. C. 1947. The natural control of population balance in the knapweed gall-fly (*Urophora jaceana*). J. Anim. Ecol. **16:** 139-187.

Warren, K. B., Editor. 1957. Population studies: Animal ecology and demography. Cold Spring Harbor Symp. Quant. Biol. **22:** P. 437.

Wilson, D. P. 1952. The influence of the nature of the substratum on the metamorphosis of the larvae of marine animals, especially the larvae of *Ophelia bicornis* Savigny. Ann. Inst. Oceanogr. Monaco **27:** 49-156.

THE INFLUENCE OF INTERSPECIFIC COMPETITION AND OTHER FACTORS ON THE DISTRIBUTION OF THE BARNACLE *CHTHAMALUS STELLATUS*

Joseph H. Connell
Department of Biology, University of California, Santa Barbara, Goleta, California

Introduction

Most of the evidence for the occurrence of interspecific competition in animals has been gained from laboratory populations. Because of the small amount of direct evidence for its occurrence in nature, competition has sometimes been assigned a minor role in determining the composition of animal communities.

Indirect evidence exists, however, which suggests that competition may sometimes be responsible for the distribution of animals in nature. The range of distribution of a species may be decreased in the presence of another species with similar requirements (Beauchamp and Ullyott 1932, Endean, Kenny and Stephenson 1956). Uniform distribution is space is usually attributed to intraspecies competition (Holme 1950, Clark and Evans 1954). When animals with similar requirements, such as 2 or more closely related species, are found coexisting in the same area, careful analysis usually indicates that they are not actually competing with each other (Lack 1954, MacArthur 1958).

In the course of an investigation of the animals of an intertidal rocky shore I noticed that the adults of 2 species of barnacles occupied 2 separate horizontal zones with a small area of overlap, whereas the young of the species from the upper zone were found in much of the lower zone. The upper species, *Chthamalus stellatus* (Poli) thus settled but did not survive in the lower zone. It seemed probable that this species was eliminated by the lower one, *Balanus balanoides* (L), in a struggle for a common requisite which was in short supply. In the rocky intertidal region, space for attachment and growth is often extremely limited. This paper is an account of some observations and experiments designed to test the hypothesis that the absence in the lower zone of adults of *Chthamalus* was due to interspecific competition with *Balanus* for space. Other factors which may have influenced the distribution were also studied. The study was made at Millport, Isle of Cumbrae, Scotland.

I would like to thank Prof. C. M. Yonge and the staff of the Marine Station, Millport, for their help, discussions and encouragement during the course of this work. Thanks are due to the following for their critical reading of the manuscript: C. S. Elton, P. W. Frank, G. Hardin, N. G. Hairston, E. Orias, T. Park and his students, and my wife.

Distribution of the species of barnacles

The upper species, *Chthamalus stellatus*, has its center of distribution in the Mediterranean; it reaches its northern limit in the Shetland Islands, north of Scotland. At Millport, adults of this species occur between the levels of mean high water of neap and spring tides (M.H.W.N. and M.H.W.S.: see Figure 5 and Table I). In southwest England and Ireland, adult *Chtham-*

alus occur at moderate population densities throughout the intertidal zone, more abundantly when *Balanus balanoides* is sparse or absent (Southward and Crisp 1954, 1956). At Millport the larvae settle from the plankton onto the shore mainly in September and October; some additional settlement may occur until December. The settlement is most abundant between M.H.W.S. and mean tide level (M.T.L.), in patches of rock surface left bare as a result of the mortality of *Balanus*, limpets, and other sedentary organisms. Few of the *Chthamalus* that settle below M.H.W.N. survive, so that adults are found only occasionally at these levels.

Balanus balanoides is a boreal-arctic species, reaching its southern limit in northern Spain. At Millport it occupies almost the entire intertidal region, from mean low water of spring tides (M.L.W.S.) up to the region between M.H.W.N. and M.H.W.S. Above M.H.W.N. it occurs intermingled with *Chthamalus* for a short distance. *Balanus* settles on the shore in April and May, often in very dense concentrations (see Table IV).

The main purpose of this study was to determine the cause of death of those *Chthamalus* that settled below M.H.W.N. A study which was being carried on at this time had revealed that physical conditions, competition for space, and predation by the snail *Thais lapillus* L. were among the most important causes of mortality of *Balanus balanoides*. Therefore, the observations and experiments in the present study were designed to detect the effects of these factors on the survival of *Chthamalus*.

Methods

Intertidal barnacles are very nearly ideal for the study of survival under natural conditions. Their sessile habit allows direct observation of the survival of individuals in a group whose positions have been mapped. Their small size and dense concentrations on rocks exposed at intervals make experimentation feasible. In addition, they may be handled and transplanted without injury on pieces of rock, since their opercular plates remain closed when exposed to air.

The experimental area was located on the Isle of Cumbrae in the Firth of Clyde, Scotland. Farland Point, where the study was made, comprises the southeast tip of the island; it is exposed to moderate wave action. The shore rock consists mainly of old red sandstone, arranged in a series of ridges, from 2 to 6 ft high, oriented at right angles to the shoreline. A more detailed description is given by Connell (1961). The other barnacle species present were *Balanus crenatus* Brug and *Verruca stroemia* (O. F. Muller), both found in small numbers only at and below M.L.W.S.

To measure the survival of *Chthamalus*, the positions of all individuals in a patch were mapped. Any barnacles which were empty or missing at the next examination of this patch must have died in the interval, since emigration is impossible. The mapping was done by placing thin glass plates (lantern slide cover glasses, 10.7 × 8.2 cm, area 87.7 cm^2) over a patch of barnacles and marking the position of each *Chthamalus* on it with glass-marking ink. The positions of the corners of the plate were marked by drilling small holes in the rock. Observations made in subsequent censuses were noted on a paper copy of the glass map.

The study areas were chosen by searching for patches of *Chthamalus* below M.H.W.N. in a stretch of shore about 50 ft long. When 8 patches had been found, no more were looked for. The only basis for rejection of an area in this search was that it contained fewer than 50 *Chthamalus* in an area of about 1/10 m^2. Each numbered area consisted of one or more glass maps located in the 1/10 m^2. They were mapped in March and April, 1954, before the main settlement of *Balanus* began in late April.

Very few *Chthamalus* were found to have settled below mid-tide level. Therefore pieces of rock bearing *Chthamalus* were removed from levels above M.H.W.N. and transplanted to and below M.T.L. A hole was drilled through each piece; it was then fastened to the rock by a stainless steel screw driven into a plastic screw anchor fitted into a hole drilled into the rock. A hole ¼" in diameter and 1" deep was found to be satisfactory. The screw could be removed and replaced repeatedly and only one stone was lost in the entire period.

For censusing, the stones were removed during a low tide period, brought to the laboratory for examination, and returned before the tide rose again. The locations and arrangements of each area are given in Table I; the transplanted stones are represented by areas 11 to 15.

The effect of competition for space on the survival of *Chthamalus* was studied in the following manner: After the settlement of *Balanus* had stopped in early June, having reached densities of 49/cm^2 on the experimental areas (Table I) a census of the surviving *Chthamalus* was made on each area (see Figure 1). Each map was then divided so that about half of the number of

TABLE I. Description of experimental areas*

Area no.	Height in ft from M.T.L.	% of time sub-merged	Population Density: no./cm² in June, 1954			Remarks
			Chthamalus, autumn 1953 settlement			
			Undisturbed portion	Portion without *Balanus*	All barnacles, undisturbed portion	
MHWS	+4.9	4	—	—	—	—
1	+4.2	9	2.2	—	19.2	Vertical, partly protected
2	+3.5	16	5.2	4.2	—	Vertical, wave beaten
MHWN	+3.1	21	—	—	—	—
3a	+2.2	30	0.6	0.6	30.9	Horizontal, wave beaten
3b	"	"	0.5	0.7	29.2	" " "
4	+1.4	38	1.9	0.6	—	30° to vertical, partly protected
5	+1.4	"	2.4	1.2	—	" " " " "
6	+1.0	42	1.1	1.9	38.2	Horizontal, top of a boulder, partly protected
7a	+0.7	44	1.3	2.0	49.3	Vertical, protected
7b	"	"	2.3	2.0	51.7	" "
11a	0.0	50	1.0	0.6	32.0	Vertical, protected
11b	"	"	0.2	0.3	—	" "
12a	0.0	100	1.2	1.2	18.8	Horizontal, immersed in tide pool
12b	"	100	0.8	0.9	—	" " " " "
13a	−1.0	58	4.9	4.1	29.5	Vertical, wave beaten
13b	"	"	3.1	2.4	—	" " "
14a	−2.5	71	0.7	1.1	—	45° angle, wave beaten
14b	"	"	1.0	1.0	—	" " " "
MLWN	−3.0	77	—	—	—	—
MLWS	−5.1	96	—	—	—	—
15	+1.0	42	32.0	—	—	Chthamalus of autumn, 1954 settlement; densities of Oct., 1954.
7b	+0.7	44	5.5	3.7	—	

* The letter "a" following an area number indicates that this area was enclosed by a cage; "b" refers to a closely adjacent area which was not enclosed. All areas faced either east or south except 7a and 7b, which faced north.

Chthamalus were in each portion. One portion was chosen (by flipping a coin), and those *Balanus* which were touching or immediately surrounding each *Chthamalus* were carefully removed with a needle; the other portion was left untouched. In this way it was possible to measure the effect on the survival of *Chthamalus* both of intraspecific competition alone and of competition with *Balanus*. It was not possible to have the numbers or population densities of *Chthamalus* exactly equal on the 2 portions of each area. This was due to the fact that, since *Chthamalus* often occurred in groups, the *Balanus* had to be removed from around all the members of a group to ensure that no crowding by *Balanus* occurred. The densities of *Chthamalus* were very low, however, so that the slight differences in density between the 2 portions of each area can probably be disregarded; intraspecific crowding was very seldom observed. Censuses of the *Chthamalus* were made at intervals of 4-6 weeks during the next year; notes were made at each census of factors such as crowding, undercutting or smothering which had taken place since the last examination. When necessary, *Balanus* which had grown until they threatened to touch the *Chthamalus* were removed in later examinations.

To study the effects of different degrees of immersion, the areas were located throughout the tidal range, either *in situ* or on transplanted stones, as shown in Table I. Area 1 had been under observation for 1½ years previously. The effects of different degrees of wave shock could not be studied adequately in such a small area

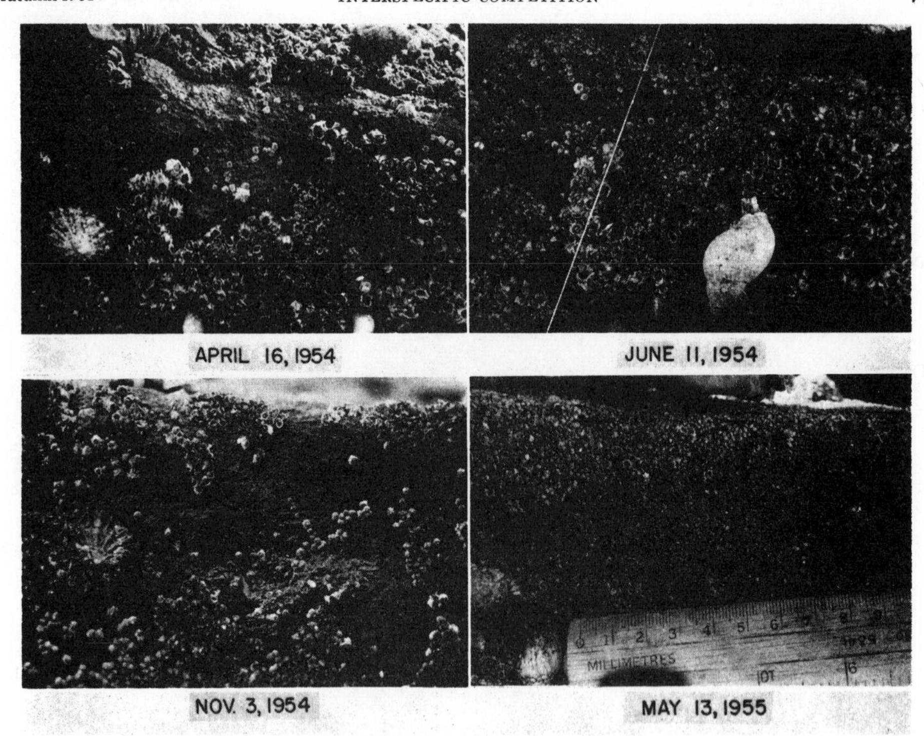

FIG. 1. Area 7b. In the first photograph the large barnacles are *Balanus,* the small ones scattered in the bare patch, *Chthamalus*. The white line on the second photograph divides the undisturbed portion (right) from the portion from which *Balanus* were removed (left). A limpet, *Patella vulgata,* occurs on the left, and predatory snails, *Thais lapillus,* are visible.

of shore but such differences as existed are listed in Table I.

The effects of the predatory snail, *Thais lapillus,* (synonymous with *Nucella* or *Purpura,* Clench 1947), were studied as follows: Cages of stainless steel wire netting, 8 meshes per inch, were attached over some of the areas. This mesh has an open area of 60% and previous work (Connell 1961) had shown that it did not inhibit growth or survival of the barnacles. The cages were about 4 × 6 inches, the roof was about an inch above the barnacles and the sides were fitted to the irregularities of the rock. They were held in place in the same manner as the transplanted stones. The transplanted stones were attached in pairs, one of each pair being enclosed in a cage (Table I).

These cages were effective in excluding all but the smallest *Thais*. Occasionally small *Thais*, ½ to 1 cm in length, entered the cages through gaps at the line of juncture of netting and rock surface. In the concurrent study of *Balanus* (Connell 1961), small *Thais* were estimated to have occurred inside the cages about 3% of the time.

All the areas and stones were established before the settlement of *Balanus* began in late April, 1954. Thus the *Chthamalus* which had settled naturally on the shore were then of the 1953 year class and all about 7 months old. Some *Chthamalus* which settled in the autumn of 1954 were followed until the study was ended in June, 1955. In addition some adults which, judging from their large size and the great erosion of their shells, must have settled in 1952 or earlier, were present on the transplanted stones. Thus records were made of at least 3 year-classes of *Chthamalus*.

RESULTS

The effects of physical factors

In Figures 2 and 3, the dashed line indicates the survival of *Chthamalus* growing without contact with *Balanus*. The suffix "a" indicates that the area was protected from *Thais* by a cage.

In the absence of *Balanus* and *Thais,* and protected by the cages from damage by water-borne objects, the survival of *Chthamalus* was good at all levels. For those which had settled normally on the shore (Fig. 2), the poorest survival was on the lowest area, 7a. On the transplanted stones (Fig. 3, area 12), constant immersion in a tide pool resulted in the poorest survival. The reasons for the trend toward slightly greater mortality as the degree of immersion increased are unknown. The amount of attached algae on the stones in the tide pool was much greater than on the other areas. This may have reduced the flow of water and food or have interfered directly with feeding movements. Another possible indirect effect of increased immersion is the increase in predation by the snail, *Thais lapillus,* at lower levels.

Chthamalus is tolerant of a much greater degree of immersion than it normally encounters. This is shown by the survival for a year on area 12 in a tide pool, together with the findings of Fischer (1928) and Barnes (1956a), who found that *Chthamalus* withstood submersion for 12 and 22 months, respectively. Its absence below M.T.L. can probably be ascribed either to a lack of initial settlement or to poor survival of newly settled larvae. Lewis and Powell (1960) have suggested that the survival of *Chthamalus* may be favored by increased light or warmth during emersion in its early life on the shore. These conditions would tend to occur higher on the shore in Scotland than in southern England.

The effects of wave action on the survival of *Chthamalus* are difficult to assess. Like the degree of immersion, the effects of wave action may act indirectly. The areas 7 and 12, where relatively poor survival was found, were also the areas of least wave action. Although *Chthamalus* is usually abundant on wave beaten areas and absent from sheltered bays in Scotland, Lewis and Powell (1960) have shown that in certain sheltered bays it may be very abundant. Hatton (1938) found that in northern France, settlement and growth rates were greater in wave-beaten areas at M.T.L., but, at M.H.W.N., greater in sheltered areas.

At the upper shore margins of distribution *Chthamalus* evidently can exist higher than *Balanus* mainly as a result of its greater tolerance to heat and/or desiccation. The evidence for this was gained during the spring of 1955. Records from a tide and wave guage operating at this time about one-half mile north of the study area showed that a period of neap tides had coincided with an unusual period of warm calm weather in April so that for several days no water, not even waves, reached the level of Area 1. In the period

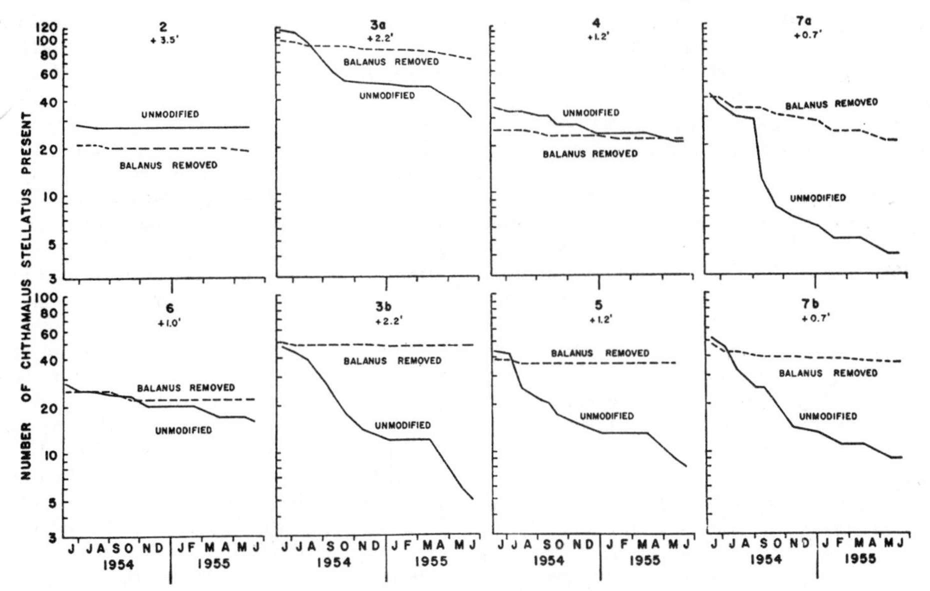

FIG. 2. Survivorship curves of *Chthamalus stellatus* which had settled naturally on the shore in the autumn of 1953. Areas designated "a" were protected from predation by cages. In each area the survival of *Chthamalus* growing without contact with *Balanus* is compared to that in the undisturbed area. For each area the vertical distance in feet from M.T.L. is shown.

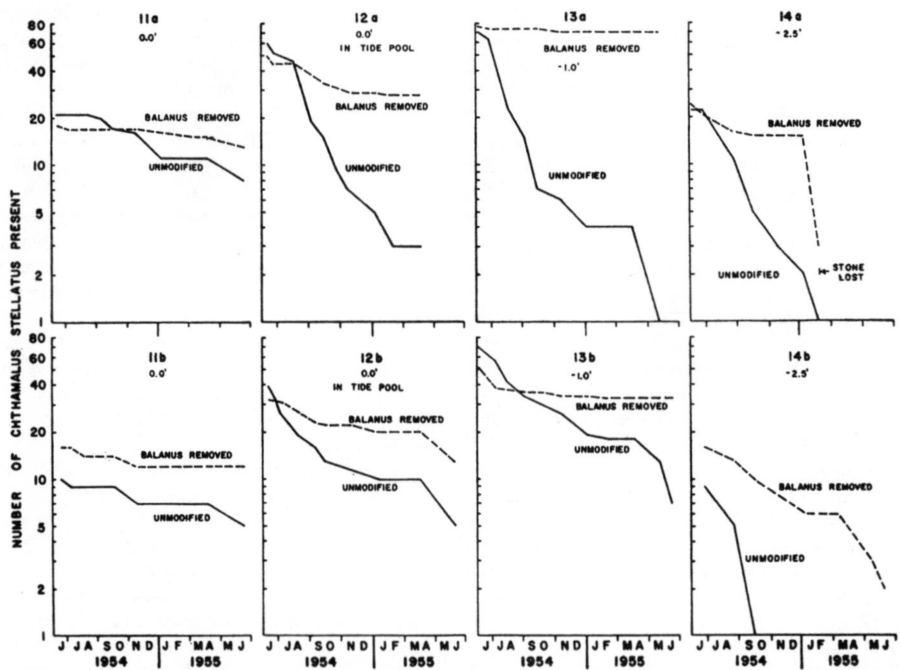

FIG. 3. Survivorship curves of *Chthamalus stellatus* on stones transplanted from high levels. These had settled in the autumn of 1953; the arrangement is the same as that of Figure 2.

between the censuses of February and May, *Balanus* aged one year suffered a mortality of 92%, those 2 years and older, 51%. Over the same period the mortality of *Chthamalus* aged 7 months was 62%, those 1½ years and older, 2%. Records of the survival of *Balanus* at several levels below this showed that only those *Balanus* in the top quarter of the intertidal region suffered high mortality during this time (Connell 1961).

Competition for space

At each census notes were made for individual barnacles of any crowding which had occurred since the last census. Thus when one barnacle started to grow up over another this fact was noted and at the next census 4-6 weeks later the progress of this process was noted. In this way a detailed description was built up of these gradually occurring events.

Intraspecific competition leading to mortality in *Chthamalus* was a rare event. For areas 2 to 7, on the portions from which *Balanus* had been removed, 167 deaths were recorded in a year. Of these, only 6 could be ascribed to crowding between individuals of *Chthamalus*. On the undisturbed portions no such crowding was observed. This accords with Hatton's (1938) observation that he never saw crowding between individuals of *Chthamalus* as contrasted to its frequent occurrence between individuals of *Balanus*.

Interspecific competition between *Balanus* and *Chthamalus* was, on the other hand, a most important cause of death of *Chthamalus*. This is shown both by the direct observations of the process of crowding at each census and by the differences between the survival curves of *Chthamalus* with and without *Balanus*. From the periodic observations it was noted that after the first month on the undisturbed portions of areas 3 to 7 about 10% of the *Chthamalus* were being covered as *Balanus* grew over them; about 3% were being undercut and lifted by growing *Balanus*; a few had died without crowding. By the end of the 2nd month about 20% of the *Chthamalus* were either wholly or partly covered by *Balanus*; about 4% had been undercut; others were surrounded by tall *Balanus*. These processes continued at a lower rate in the autumn and almost ceased during the later winter. In the spring *Balanus* resumed growth and more crowding was observed.

In Table II, these observations are summarized for the undistributed portions of all the areas. Above M.T.L., the *Balanus* tended to overgrow the *Chthamalus,* whereas at the lower levels, undercutting was more common. This same trend was evident within each group of areas, undercutting being more prevalent on area 7 than on area 3, for example. The faster growth of *Balanus* at lower levels (Hatton 1938, Barnes and Powell 1953) may have resulted in more undercutting. When *Chthamalus* was completely covered by *Balanus* it was recorded as dead; even though death may not have occurred immediately, the buried barnacle was obviously not a functioning member of the population.

TABLE II. The causes of mortality of *Chthamalus stellatus* of the 1953 year group on the undisturbed portions of each area

Area no.	Height in ft from M.T.L.	No. at start	No. of deaths in the next year	Percentage of Deaths Resulting From:			
				Smothering by *Balanus*	Undercutting by *Balanus*	Other crowding by *Balanus*	Unknown causes
2..........	+3.5	28	1	0	0	0	100
3a..........	+2.2	111	81	61	6	10	23
3b..........	"	47	42	57	5	2	36
4..........	+1.4	34	14	21	14	0	65
5..........	+1.4	43	35	11	11	3	75
6..........	+1.0	27	11	9	0	0	91
7a..........	+0.7	42	38	21	16	53	10
7b..........	"	51	42	24	10	10	56
11a..........	0.0	21	13	54	8	0	38
11b..........	"	10	5	40	0	0	60
12a..........	0.0	60	57	19	33	7	41
12b..........	"	39	34	9	18	3	70
13a..........	−1.0	71	70	19	24	3	54
13b..........	"	69	62	18	8	3	71
14a..........	−2.5	22	21	24	42	10	24
14b..........	"	9	9	0	0	0	100
Total, 2-7..	—	383	264	37	9	16	38
Total, 11-14..	—	301	271	19	21	4	56

In Table II under the term "other crowding" have been placed all instances where *Chthamalus* were crushed laterally between 2 or more *Balanus,* or where *Chthamalus* disappeared in an interval during which a dense population of *Balanus* grew rapidly. For example, in area 7a the *Balanus,* which were at the high population density of 48 per cm^2, had no room to expand except upward and the barnacles very quickly grew into the form of tall cylinders or cones with the diameter of the opercular opening greater than that of the base. It was obvious that extreme crowding occurred under these circumstances, but the exact cause of the mortality of the *Chthamalus* caught in this crush was difficult to ascertain.

In comparing the survival curves of Figs. 2 and 3 within each area it is evident that *Chthamalus* kept free of *Balanus* survived better than those in the adjacent undisturbed areas on all but areas 2 and 14a. Area 2 was in the zone where adults of *Balanus* and *Chthamalus* were normally mixed; at this high level *Balanus* evidently has no influence on the survival of *Chthamalus.* On Stone 14a, the survival of *Chthamalus* without *Balanus* was much better until January when a starfish, *Asterias rubens* L., entered the cage and ate the barnacles.

Much variation occurred on the other 14 areas. When the *Chthamalus* growing without contact with *Balanus* are compared with those on the adjacent undisturbed portion of the area, the survival was very much better on 10 areas and moderately better on 4. In all areas, some *Chthamalus* in the undisturbed portions escaped severe crowding. Sometimes no *Balanus* happened to settle close to a *Chthamalus,* or sometimes those which did died soon after settlement. In some instances, *Chthamalus* which were being undercut by *Balanus* attached themselves to the *Balanus* and so survived. Some *Chthamalus* were partly covered by *Balanus* but still survived. It seems probable that in the 4 areas, nos. 4, 6, 11a, and 11b, where *Chthamalus* survived well in the presence of *Balanus,* a higher proportion of the *Chthamalus* escaped death in one of these ways.

The fate of very young *Chthamalus* which settled in the autumn of 1954 was followed in detail in 2 instances, on stone 15 and area 7b. The *Chthamalus* on stone 15 had settled in an irregular space surrounded by large *Balanus.* Most of the mortality occurred around the edges of the space as the *Balanus* undercut and lifted the small *Chthamalus* nearby. The following is a tabulation of all the deaths of young *Chthamalus* between Sept. 30, 1954 and Feb. 14, 1955, on Stone 15, with the associated situations:

Lifted by *Balanus*	: 29
Crushed by *Balanus*	: 4
Smothered by *Balanus* and *Chthamalus*	: 2
Crushed between *Balanus and Chthamalus*	: 1
Lifted by *Chthamalus*	: 1
Crushed between two other *Chthamalus*	: 1
Unknown	: 3

This list shows that crowding of newly settled *Chthamalus* by older *Balanus* in the autumn mainly takes the form of undercutting, rather than of smothering as was the case in the spring. The reason for this difference is probably that the *Chthamalus* are more firmly attached in the spring so that the fast growing young *Balanus* grow up over them when they make contact. In the autumn the reverse is the case, the *Balanus* being firmly attached, the *Chthamalus* weakly so.

Although the settlement of *Chthamalus* on Stone 15 in the autumn of 1954 was very dense, 32/cm^2, so that most of them were touching another, only 2 of the 41 deaths were caused by intraspecific crowding among the *Chthamalus*. This is in accord with the findings from the 1953 settlement of *Chthamalus*.

The mortality rates for the young *Chthamalus* on area 7b showed seasonal variations. Between October 10, 1954 and May 15, 1955 the relative mortality rate per day × 100 was 0.14 on the undisturbed area and 0.13 where *Balanus* had been removed. Over the next month, the rate increased to 1.49 on the undisturbed area and 0.22 where *Balanus* was absent. Thus the increase in mortality of young *Chthamalus* in late spring was also associated with the presence of *Balanus*.

Some of the stones transplanted from high to low levels in the spring of 1954 bore adult *Chthamalus*. On 3 stones, records were kept of the survival of these adults, which had settled in the autumn of 1952 or in previous years and were at least 20 months old at the start of the experiment. Their mortality is shown in Table III; it was always much greater when *Balanus* was not removed. On 2 of the 3 stones this mortality rate was almost as high as that of the younger group. These results suggest that any *Chthamalus* that managed to survive the competition for space with *Balanus* during the first year would probably be eliminated in the 2nd year.

Censuses of *Balanus* were not made on the experimental areas. However, on many other areas in the same stretch of shore the survival of *Balanus* was being studied during the same period (Connell 1961). In Table IV some mortality rates measured in that study are listed; the *Balanus* were members of the 1954 settlement at population densities and shore levels similar to those of the present study. The mortality rates of *Balanus* were about the same as those of *Chthamalus* in similar situations except at the highest level, area 1, where *Balanus* suffered much greater mortality than *Chthamalus*. Much of this mortality was caused by intraspecific crowding at all levels below area 1.

TABLE III. Comparison of the mortality rates of young and older *Chthamalus stellatus* on transplanted stones

Stone No.	Shore level	Treatment	Number of *Chthamalus* present in June, 1954		% mortality over one year (or for 6 months for 14a) of *Chthamalus*	
			1953 year group	1952 or older year groups	1953 year group	1952 or older year groups
13b	1.0 ft below MTL	*Balanus* removed	51	3	35	0
		Undisturbed	69	16	90	31
12a	MTL, in a tide pool, caged	*Balanus* removed	50	41	44	37
		Undisturbed	60	31	95	71
14a	2.5 ft below MTL, caged	*Balanus* removed	25	45	40	36
		Undisturbed	22	8	86	75

TABLE IV. Comparison of annual mortality rates of *Chthamalus stellatus* and *Balanus balanoides**

Area no.	Height in ft from M.T.L.	Population density: no./cm^2 June, 1954	% mortality in the next year
	***Chthamalus stellatus*, autumn 1953 settlement**		
1	+4.2	21	17
3a	+2.2	31	72
3b	"	29	89
6	+1.0	38	41
7a	+0.7	49	90
7b	"	52	82
11a	0.0	32	62
13a	−1.0	29	99
12a	(tide pool)	19	95
	***Balanus balanoides*, spring 1954 settlement**		
1 (top)	+4.2	21	99
1:Middle Cage 1	+2.1	85	92
1:Middle Cage 2	"	25	77
1:Low Cage 1	+1.5	26	88
Stone 1	−0.9	26	86
Stone 2	"	68	94

* Population density includes both species. The mortality rates of *Chthamalus* refer to those on the undisturbed portions of each area. The data and area designations for *Balanus* were taken from Connell (1961); the present area 1 is the same as that designated 1 (top) in that paper.

In the observations made at each census it appeared that *Balanus* was growing faster than *Chthamalus*. Measurements of growth rates of the 2 species were made from photographs of the areas taken in June and November, 1954. Barnacles growing free of contact with each other were measured; the results are given in Table V. The growth rate of *Balanus* was greater than that of *Chthamalus* in the experimental areas; this agrees with the findings of Hatton (1938) on the shore in France and of Barnes (1956a) for continual submergence on a raft at Millport.

TABLE V. Growth rates of *Chthamalus stellatus* and *Balanus balanoides*. Measurements were made of uncrowded individuals on photographs of areas 3a, 3b and 7b. Those of *Chthamalus* were made on the same individuals on both dates; of *Balanus*, representative samples were chosen

	CHTHAMALUS		BALANUS	
	No. measured	Average size, mm.	No. measured	Average size, mm.
June 11, 1954	25	2.49	39	1.87
November 3, 1954	25	4.24	27	4.83
Average size in the interval	3.36		3.35	
Absolute growth rate per day x 100	1.21		2.04	

After a year of crowding the average population densities of *Balanus* and *Chthamalus* remained in the same relative proportion as they had been at the start, since the mortality rates were about the same. However, because of its faster growth, *Balanus* occupied a relatively greater area and, presumably, possessed a greater biomass relative to that of *Chthamalus* after a year.

The faster growth of *Balanus* probably accounts for the manner in which *Chthamalus* were crowded by *Balanus*. It also accounts for the sinuosity of the survival curves of *Chthamalus* growing in contact with *Balanus*. The mortality rate of these *Chthamalus*, as indicated by the slope of the curves in Figs. 2 and 3, was greatest in summer, decreased in winter and increased again in spring. The survival curves of *Chthamalus* growing without contact with *Balanus* do not show these seasonal variations which, therefore, cannot be the result of the direct action of physical factors such as temperature, wave action or rain.

Seasonal variations in growth rate of *Balanus* correspond to these changes in mortality rate of *Chthamalus*. In Figure 4 the growth of *Balanus* throughout the year as studied on an intertidal panel at Millport by Barnes and Powell (1953), is compared to the survival of *Chthamalus* at about the same intertidal level in the present study. The increased mortality of *Chthamalus* was found to occur in the same seasons as the increases in the growth rate of *Balanus*. The correlation was tested using the Spearman rank correlation coefficient. The absolute increase in diameter of *Balanus* in each month, read from the curve of growth, was compared to the percentage mortality of *Chthamalus* in the same month. For the 13 months in which data for *Chthamalus* was available, the correlation was highly significant, $P = .01$.

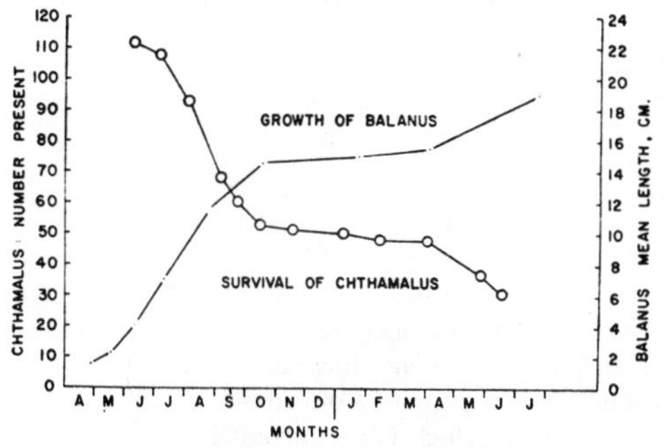

FIG. 4. A comparison of the seasonal changes in the growth of *Balanus balanoides* and in the survival of *Chthamalus stellatus* being crowded by *Balanus*. The growth of *Balanus* was that of panel 3, Barnes and Powell (1953), just above M.T.L. on Keppel Pier, Millport, during 1951-52. The *Chthamalus* were on area 3a of the present study, one-half mile south of Keppell Pier, during 1954-55.

From all these observations it appears that the poor survival of *Chthamalus* below M.H.W.N. is a result mainly of crowding by dense populations of faster growing *Balanus*.

At the end of the experiment in June, 1955, the surviving *Chthamalus* were collected from 5 of the areas. As shown in Table VI, the average size was greater in the *Chthamalus* which had grown free of contact with *Balanus;* in every case the difference was significant ($P < .01$, Mann-Whitney U. test, Siegel 1956). The survivors on the undisturbed areas were often misshapen, in some cases as a result of being lifted on to the side of an undercutting *Balanus*. Thus the smaller size of these barnacles may have been due to disturbances in the normal pattern of growth while they were being crowded.

These *Chthamalus* were examined for the presence of developing larvae in their mantle cavities. As shown in Table VI, in every area the proportion of the uncrowded *Chthamalus* with larvae was equal to or more often slightly greater than on the crowded areas. The reason for this may be related to the smaller size of the crowded *Chthamalus*. It is not due to separation, since *Chthamalus* can self-fertilize (Barnes and Crisp

TABLE VI. The effect of crowding on the size and presence of larvae in *Chthamalus stellatus*, collected in June, 1955

Area	Treatment	Level, feet above MTL	Number of *Chthamalus*	Diameter in mm		% of individuals which had larvae in mantle cavity
				Average	Range	
3a......	Undisturbed	2.2	18	3.5	2.7-4.6	61
"......	*Balanus* removed	"	50	4.1	3.0-5.5	65
4.......	Undisturbed	1.4	16	2.3	1.8 3.2	81
".......	*Balanus* removed	"	37	3.7	2.5-5 1	100
5.......	Undisturbed	1.4	7	3.3	2.8-3.7	70
".......	*Balanus* removed	"	13	4.0	3.5-4.5	100
6.......	Undisturbed	1.0	13	2.8	2.1-3.9	100
".......	*Balanus* removed	"	14	4.1	3.0-5.2	100
7a & b..	Undisturbed	0.7	10	3.5	2.7-4.5	70
" ..	*Balanus* removed	"	23	4.3	3.0-6.3	81

1956). Moore (1935) and Barnes (1953) have shown that the number of larvae in an individual of *Balanus balanoides* increases with increase in volume of the parent. Comparison of the cube of the diameter, which is proportional to the volume, of *Chthamalus* with and without *Balanus* shows that the volume may be decreased to ¼ normal size when crowding occurs. Assuming that the relation between larval numbers and volume in *Chthamalus* is similar to that of *Balanus*, a decrease in both frequency of occurrence and abundance of larvae in *Chthamalus* results from competition with *Balanus*. Thus the process described in this paper satisfies both aspects of interspecific competition as defined by Elton and Miller (1954): "in which one species affects the population of another by a process of interference, i.e., by reducing the reproductive efficiency or increasing the mortality of its competitor."

The effect of predation by Thais

Cages which excluded *Thais* had been attached on 6 areas (indicated by the letter "a" following the number of the area). Area 14 was not included in the following analysis since many starfish were observed feeding on the barnacles at this level; one entered the cage in January, 1955, and ate most of the barnacles.

Thais were common in this locality, feeding on barnacles and mussels, and reaching average population densities of 200/m² below M.T.L. (Connell 1961). The mortality rates for *Chthamalus* in cages and on adjacent areas outside cages (indicated by the letter "b" after the number) are shown on Table VII.

If the mortality rates of *Chthamalus* growing without contact with *Balanus* are compared in and out of the cages, it can be seen that at the upper levels mortality is greater inside the cages, at lower levels greater outside. Densities of *Thais* tend to be greater at and below M.T.L. so that this trend in the mortality rates of *Chthamalus* may be ascribed to an increase in predation by *Thais* at lower levels.

TABLE VII. The effect of predation by *Thais lapillus* on the annual mortality rate of *Chthamalus stellatus* in the experimental areas*

Area	Height in ft from M.T.L.	% mortality of *Chthamalus* over a year (The initial numbers are given in parentheses): a: Protected from predation by a cage			b: Unprotected, open to predation		
		With *Balanus*	Without *Balanus*	Difference	With *Balanus*	Without *Balanus*	Difference
Area 3..	+2.2	73 (112)	25 (96)	48	89 (47)	6 (50)	83
Area 7..	+0.7	90 (42)	47 (40)	43	82 (51)	23 (47)	59
Area 11..	0	62 (21)	28 (18)	34	50 (10)	25 (16)	25
Area 12 .	0†	100 (60)	53 (50)	47	87 (39)	59 (32)	28
Area 13..	−1.0	98 (72)	9 (77)	89	90 (69)	35 (51)	55

*The records for 12a extend over only 10 months; for purposes of comparison the mortality rate for 12a has been multiplied by 1.2.
†Tide pool.

Mortality of *Chthamalus* in the absence of *Balanus* was appreciably greater outside than inside the cage only on area 13. In the other 4 areas it seems evident that few *Chthamalus* were being eaten by *Thais*. In a concurrent study of the behavior of *Thais* in feeding on *Balanus balanoides*, it was found that *Thais* selected the larger individuals as prey (Connell 1961). Since *Balanus* after a few month's growth was usually larger than *Chthamalus*, it might be expected that *Thais* would feed on *Balanus* in preference to *Chthamalus*. In a later study (unpublished) made at Santa Barbara, California, *Thais emarginata* Deshayes were enclosed in cages on the shore with mixed populations of *Balanus glandula* Darwin and *Chthamalus fissus* Darwin. These species were each of the same size range as the corresponding species at Millport. It was found that *Thais emarginata* fed on *Balanus glandula* in preference to *Chthamalus fissus*.

As has been indicated, much of the mortality of *Chthamalus* growing naturally intermingled with *Balanus* was a result of direct crowding by *Balanus*. It therefore seemed reasonable to take the difference between the mortality rates of *Chthamalus* with and without *Balanus* as an index of the degree of competition between the species. This difference was calculated for each area and is included in Table VII. If these differences are compared between each pair of adjacent areas in and out of a cage, it appears that the difference, and therefore the degree of competition, was greater outside the cages at the upper shore levels and less outside the cages at the lower levels.

Thus as predation increased at lower levels, the degree of competition decreased. This result would have been expected if *Thais* had fed upon *Balanus* in preference to *Chthamalus*. The general effect of predation by *Thais* seems to have been to lessen the interspecific competition below M.T.L.

Discussion

"Although animal communities appear qualitatively to be constructed as if competition were regulating their structure, even in the best studied cases there are nearly always difficulties and unexplored possibilities" (Hutchinson 1957).

In the present study direct observations at intervals showed that competition was occurring under natural conditions. In addition, the evidence is strong that the observed competition with *Balanus* was the principal factor determining the local distribution of *Chthamalus*. *Chthamalus* thrived at lower levels when it was not growing in contact with *Balanus*.

However, there remain unexplored possibilities. The elimination of *Chthamalus* requires a dense population of *Balanus*, yet the settlement of *Balanus* varied from year to year. At Millport, the settlement density of *Balanus balanoides* was measured for 9 years between 1944 and 1958 (Barnes 1956b, Connell 1961). Settlement was light in 2 years, 1946 and 1958. In the 3 seasons of *Balanus* settlement studied in detail, 1953-55, there was a vast oversupply of larvae ready for settlement. It thus seems probable that most of the *Chthamalus* which survived in a year of poor settlement of *Balanus* would be killed in competition with a normal settlement the following year. A succession of years with poor settlements of *Balanus* is a possible, but improbable occurrence at Millport, judging from the past record. A very light settlement is probably the result of a chance combination of unfavorable weather circumstances during the planktonic period (Barnes 1956b). Also, after a light settlement, survival on the shore is improved, owing principally to the reduction in intraspecific crowding (Connell 1961); this would tend to favor a normal settlement the following year, since barnacles are stimulated to settle by the presence of members of their own species already attached on the surface (Knight-Jones 1953).

The fate of those *Chthamalus* which had survived a year on the undisturbed areas is not known since the experiment ended at that time. It is probable, however, that most of them would have been eliminated within 6 months; the mortality rate had increased in the spring (Figs. 2 and 3), and these survivors were often misshapen and smaller than those which had not been crowded (Table VI). Adults on the transplanted stones had suffered high mortality in the previous year (Table III).

Another difficulty was that *Chthamalus* was rarely found to have settled below mid tide level at Millport. The reasons for this are unknown; it survived well if transplanted below this level, in the absence of *Balanus*. In other areas of the British Isles (in southwest England and Ireland, for example) it occurs below mid tide level.

The possibility that *Chthamalus* might affect *Balanus* deleteriously remains to be considered. It is unlikely that *Chthamalus* could cause much mortality of *Balanus* by direct crowding; its growth is much slower, and crowding between individuals of *Chthamalus* seldom resulted in death. A dense population of *Chthamalus* might deprive larvae of *Balanus* of space for settlement. Also, *Chthamalus* might feed on the planktonic larvae of *Balanus;* however, this would occur in March and April when both the sea water temperature and rate of cirral activity (presumably correlated with feeding activity), would be near their minima (Southward 1955).

The indication from the caging experiments that predation decreased interspecific competition suggests that the action of such additional factors tends to reduce the intensity of such interactions in natural conditions. An additional suggestion in this regard may be made concerning parasitism. Crisp (1960) found that the growth rate of *Balanus balanoides* was decreased if individuals were infected with the isopod parasite *Hemioniscus balani* (Spence Bate). In Britain this parasite has not been reported from *Chthamalus stellatus*. Thus if this parasite were present, both the growth rate of *Balanus,* and its ability to eliminate *Chthamalus* would be decreased, with a corresponding lessening of the degree of competition between the species.

The causes of zonation

The evidence presented in this paper indicates that the lower limit of the intertidal zone of *Chthamalus stellatus* at Millport was determined by interspecific competition for space with *Balanus balanoides*. *Balanus,* by virtue of its greater population density and faster growth, eliminated most of the *Chthamalus* by directing crowding.

At the upper limits of the zones of these species no interaction was observed. *Chthamalus* evidently can exist higher on the shore than *Balanus* mainly as a result of its greater tolerance to heat and/or desiccation.

The upper limits of most intertidal animals are probably determined by physical factors such as these. Since growth rates usually decrease with increasing height on the shore, it would be less likely that a sessile species occupying a higher zone could, by competition for space, prevent a lower one from extending upwards. Likewise, there has been, as far as the author is aware, no study made which shows that predation by land species determines the upper limit of an intertidal animal. In one of the most thorough of such studies, Drinnan (1957) indicated that intense predation by birds accounted for an annual mortality of 22% of cockles (*Cardium edule* L.) in sand flats where their total mortality was 74% per year.

In regard to the lower limits of an animal's zone, it is evident that physical factors may act directly to determine this boundary. For example, some active amphipods from the upper levels of sandy beaches die if kept submerged. However, evidence is accumulating that the lower limits of distribution of intertidal animals are determined mainly by biotic factors.

Connell (1961) found that the shorter length of life of *Balanus balanoides* at low shore levels could be accounted for by selective predation by *Thais lapillus* and increased intraspecific competition for space. The results of the experiments in the present study confirm the suggestions of other authors that lower limits may be due to interspecific competition for space. Knox (1954) suggested that competition determined the distribution of 2 species of barnacles in New Zealand. Endean, Kenny and Stephenson (1956) gave indirect evidence that competition with a colonial polychaete worm, (*Galeolaria*) may have determined the lower limit of a barnacle (*Tetraclita*) in Queensland, Australia. In turn the lower limit of *Galeolaria* appeared to be determined by competition with a tunicate, *Pyura*, or with dense algal mats.

With regard to the 2 species of barnacles in the present paper, some interesting observations have been made concerning changes in their abundance in Britain. Moore (1936) found that in southwest England in 1934, *Chthamalus stellatus* was most dense at M.H.W.N., decreasing in numbers toward M.T.L. while *Balanus balanoides* increased in numbers below M.H.W.N. At the same localities in 1951, Southward and Crisp (1954) found that *Balanus* had almost disappeared and that *Chthamalus* had increased both above and below M.H.W.N. *Chthamalus* had not reached the former densities of *Balanus* except at one locality, Brixham. After 1951, *Balanus* began to return in numbers, although by 1954 it had not reached the densities of 1934; *Chthamalus* had declined, but again not to its former densities (Southward and Crisp 1956).

Since *Chthamalus* increased in abundance at the lower levels vacated by *Balanus*, it may previously have been excluded by competition with *Balanus*. The growth rate of *Balanus* is greater than *Chthamalus* both north and south (Hatton 1938) of this location, so that *Balanus* would be likely to win in competition with *Chthamalus*. However, changes in other environmental factors such as temperature may have influenced the abundance of these species in a reciprocal manner. In its return to southwest England after 1951, the maximum density of settlement of *Balanus* was 12 per cm^2; competition of the degree observed at Millport would not be expected to occur at this density. At a higher population density, *Balanus* in southern England would probably eliminate *Chthamalus* at low shore levels in the same manner as it did at Millport.

In Loch Sween, on the Argyll Peninsula, Scotland, Lewis and Powell (1960) have described an unusual pattern of zonation of *Chthamalus stellatus*. On the outer coast of the Argyll Peninsula *Chthamalus* has a distribution similar to that at Millport. In the more sheltered waters of Loch Sween, however, *Chthamalus* occurs from above M.H.W.S. to about M.T.L., judging the distribution by its relationship to other organisms. *Balanus balanoides* is scarce above M.T.L. in Loch Sween, so that there appears to be no possibility of competition with *Chthamalus*, such as that occurring at Millport, between the levels of M.T.L. and M.H.W.N.

In Figure 5 an attempt has been made to summarize the distribution of adults and newly settled larvae in relation to the main factors which appear to determine this distribution. For *Balanus* the estimates were based on the findings of a previous study (Connell 1961); intraspecific competition was severe at the lower levels during the first year, after which predation increased in importance. With *Chthamalus*, it appears that avoidance of settlement or early mortality of those larvae which settled at levels below M.T.L., and elimination by competition with *Balanus* of those which settled between M.T.L. and M.H.W.N., were the principal causes for the absence of adults below M.H.W.N. at Millport. This distribution appears to be typical for much of western Scotland.

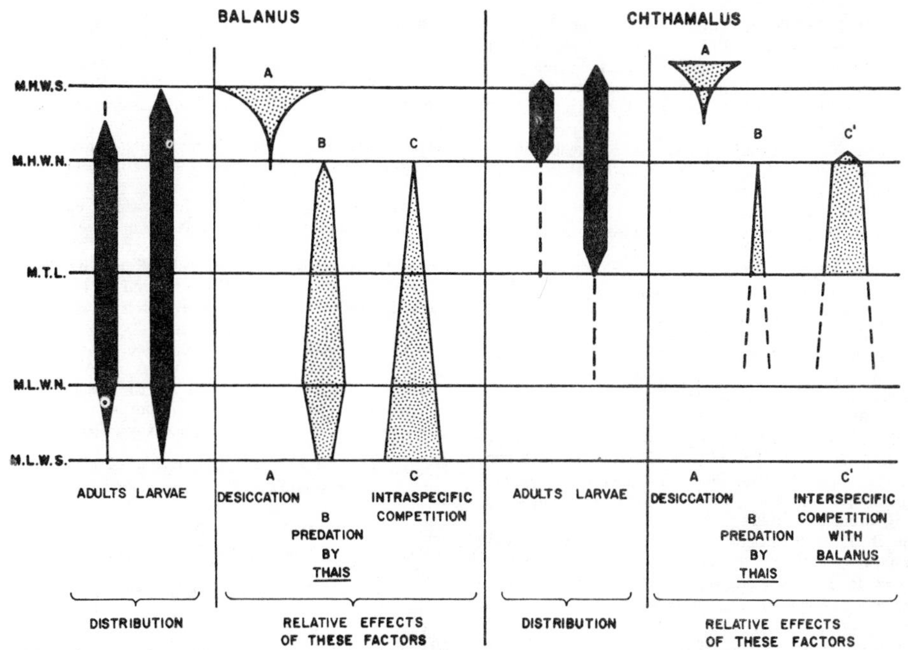

FIG. 5. The intertidal distribution of adults and newly settled larvae of *Balanus balanoides* and *Chthamalus stellatus* at Millport, with a diagrammatic representation of the relative effects of the principal limiting factors.

SUMMARY

Adults of *Chthamalus stellatus* occur in the marine intertidal in a zone above that of another barnacle, *Balanus balanoides*. Young *Chthamalus* settle in the *Balanus* zone but evidently seldom survive, since few adults are found there.

The survival of *Chthamalus* which had settled at various levels in the *Balanus* zone was followed for a year by successive censuses of mapped individuals. Some *Chthamalus* were kept free of contact with *Balanus*. These survived very well at all intertidal levels, indicating that increased time of submergence was not the factor responsible for elimination of *Chthamalus* at low shore levels. Comparison of the survival of unprotected populations with others, protected by enclosure in cages from predation by the snail, *Thais lapillus*, showed that *Thais* was not greatly affecting the survival of *Chthamalus*.

Comparison of the survival of undisturbed populations of *Chthamalus* with those kept free of contact with *Balanus* indicated that *Balanus* could cause great mortality of *Chthamalus*. *Balanus* settled in greater population densities and grew faster than *Chthamalus*. Direct observations at each census showed that *Balanus* smothered, undercut, or crushed the *Chthamalus;* the greatest mortality of *Chthamalus* occurred during the seasons of most rapid growth of *Balanus*. Even older *Chthamalus* transplanted to low levels were killed by *Balanus* in this way. Predation by *Thais* tended to decrease the severity of this interspecific competition.

Survivors of *Chthamalus* after a year of crowding by *Balanus* were smaller than uncrowded ones. Since smaller barnacles produce fewer offspring, competition tended to reduce reproductive efficiency in addition to increasing mortality.

Mortality as a result of intraspecies competition for space between individuals of *Chthamalus* was only rarely observed.

The evidence of this and other studies indicates that the lower limit of distribution of intertidal organisms is mainly determined by the action of biotic factors such as competition for space or predation. The upper limit is probably more often set by physical factors.

REFERENCES

Barnes, H. 1953. Size variations in the cyprids of some common barnacles. J. Mar. Biol. Ass. U. K. **32:** 297-304.

———. 1956a. The growth rate of *Chthamalus stellatus* (Poli). J. Mar. Biol. Ass. U. K. **35**: 355-361.

———. 1956b. *Balanus balanoides* (L.) in the Firth of Clyde: The development and annual variation of the larval population, and the causative factors. J. Anim. Ecol. **25**: 72-84.

——— **and H. T. Powell.** 1953. The growth of *Balanus balanoides* (L.) and *B. crenatus* Brug. under varying conditions of submersion. J. Mar. Biol. Ass. U. K. **32**: 107-128.

——— **and D. J. Crisp.** 1956. Evidence of self-fertilization in certain species of barnacles. J. Mar. Biol. Ass. U. K. **35**: 631-639.

Beauchamp, R. S. A. and P. Ullyott. 1932. Competitive relationships between certain species of fresh-water Triclads. J. Ecol. **20**: 200-208.

Clark, P. J. and F. C. Evans. 1954. Distance to nearest neighbor as a measure of spatial relationships in populations. Ecology **35**: 445-453.

Clench, W. J. 1947. The genera *Purpura* and *Thais* in the western Atlantic. Johnsonia **2**, No. 23: 61-92.

Connell, J. H. 1961. The effects of competition, predation by *Thais lapillus,* and other factors on natural populations of the barnacle, *Balanus balanoides.* Ecol. Mon. **31**: 61-104.

Crisp, D. J. 1960. Factors influencing growth-rate in *Balanus balanoides.* J. Anim. Ecol. **29**: 95-116.

Drinnan, R. E. 1957. The winter feeding of the oyster-catcher (*Haematopus ostralegus*) on the edible cockle (*Cardium edule*). J. Anim. Ecol. **26**: 441-469.

Elton, Charles and R. S. Miller. 1954. The ecological survey of animal communities: with a practical scheme of classifying habitats by structural characters. J. Ecol. **42**: 460-496.

Endean, R., R. Kenny and W. Stephenson. 1956. The ecology and distribution of intertidal organisms on the rocky shores of the Queensland mainland. Aust. J. mar. freshw. Res. **7**: 88-146.

Fischer, E. 1928. Sur la distribution geographique de quelques organismes de rocher, le long des cotes de la Manche. Trav. Lab. Mus. Hist. Nat. St.-Servan **2**: 1-16.

Hatton, H. 1938. Essais de bionomie explicative sur quelques especes intercotidales d'algues et d'animaux. Ann. Inst. Oceanogr. Monaco **17**: 241-348.

Holme, N. A. 1950. Population-dispersion in *Tellina tenuis* Da Costa. J. Mar. Biol. Ass. U. K. **29**: 267-280.

Hutchinson, G. E. 1957. Concluding remarks. Cold Spring Harbor Symposium on Quant. Biol. **22**: 415-427.

Knight-Jones, E. W. 1953. Laboratory experiments on gregariousness during setting in *Balanus balanoides* and other barnacles. J. Exp. Biol. **30**: 584-598.

Knox, G. A. 1954. The intertidal flora and fauna of the Chatham Islands. Nature Lond. **174**: 871-873.

Lack, D. 1954. The natural regulation of animal numbers. Oxford, Clarendon Press.

Lewis, J. R. and H. T. Powell. 1960. Aspects of the intertidal ecology of rocky shores in Argyll, Scotland. I. General description of the area. II. The distribution of *Chthamalus stellatus* and *Balanus balanoides* in Kintyre. Trans. Roy. Soc. Edin. **64**: 45-100.

MacArthur, R. H. 1958. Population ecology of some warblers of northeastern coniferous forests. Ecology **39**: 599-619.

Moore, H. B. 1935. The biology of *Balnus balanoides.* III. The soft parts. J. Mar. Biol. Ass. U. K. **20**: 263-277.

———. 1936. The biology of *Balanus balanoides.* V. Distribution in the Plymouth area. J. Mar. Biol. Ass. U. K. **20**: 701-716.

Siegel, S. 1956. Nonparametric statistics. New York, McGraw Hill.

Southward, A. J. 1955. On the behavior of barnacles. I. The relation of cirral and other activities to temperature. J. Mar. Biol. Ass. U. K. **34**: 403-422.

——— **and D. J. Crisp.** 1954. Recent changes in the distribution of the intertidal barnacles *Chthamalus stellatus* Poli and *Balanus balanoides* L. in the British Isles. J. Anim. Ecol. **23**: 163-177.

———. 1956. Fluctuations in the distribution and abundance of intertidal barnacles. J. Mar. Biol. Ass. U. K. **35**: 211-229.

EFFECTS OF COMPETITION, PREDATION BY *THAIS LAPILLUS*, AND OTHER FACTORS ON NATURAL POPULATIONS OF THE BARNACLE *BALANUS BALANOIDES*

JOSEPH H. CONNELL
Department of Biological Sciences, University of California at Santa Barbara, Goleta, California

TABLE OF CONTENTS

INTRODUCTION

This study represents an attempt to measure some of the major factors which affected the recruitment and mortality of a natural population of *Balanus balanoides* (L.) during a period of 2 2/3 years (1952-1955). Most emphasis was placed on the investigation of the effects of certain biological interactions such as intraspecific competition and predation by *Thais lapillus* L. (*Nucella, Purpura;* see Clench 1947 for synonymy). Other associated animals were studied, although in less detail; food and parasites were not studied. The size of the study area was purposely kept small in order to reduce the variability caused by differences in wave action, salinity, temperature, etc. Weather and tide records, however, afforded some information as to the effect of the physical factors of the environment. The results are presented as they occurred in the life of the barnacles, beginning with their attachment to the rock.

Barnacles possess certain advantages for this sort of study. The survival of individuals can be determined very accurately by simply mapping the positions of all the members of a group and then following the same individuals by regular censuses. As Deevey (1947) has pointed out, this method is as accurate as that used with laboratory populations, and is much superior to those methods which use the age at death or the differences between the numbers of successive generations in a sample. Furthermore, field experimentation is facilitated by the small size of barnacles, their dense concentrations and intertidal location.

I would like to express my thanks to C. M. Yonge for his advice and encouragement throughout this work. Other members of the Zoology Department of the University of Glasgow provided technical assistance and helpful discussions. My sincere thanks also go to the staff of the Marine Station, Millport, for their help and forbearance, during my stay there. T. B. Bagenal made observations of the study area after I had left and throughout the study provided many stimulating discussions. I would like also to thank Charles Elton and other members of the Bureau of Animal Population, Oxford, for much encouragement and enlightening discussion. E. W. Fager read the complete manuscript and his suggestions, especially on statistical matters, are gratefully acknowledged. Finally I wish to thank my wife for her constant encouragement and help.

METHODS

The area of study was located on the tip of Farland Point, Isle of Cumbrae, in the Firth of Clyde, Scotland. The shoreline at this point faces south and consists of a series of ledges of Old Red sandstone, dipping downward to the west. These create a series of parallel ravines oriented at right angles to the shoreline. The east wall of each ravine has a slope of about 30° to the horizontal, while the west wall is almost vertical, varying in height from 2 to 6 ft. In the ravines loose pieces of rock varying from a few inches to 3 ft in diameter occurred. Most of the smaller pieces were composed of bostonite, a harder basaltic rock derived from larger dikes in the vicinity.

Detailed studies were made in three areas, located within a 50-ft stretch of this shore (Fig. 1). Area 1 was located on the vertical west wall of a ravine. On this wall, which was about 6 ft high, study plots were located at four levels. At each level small squares, of about 25 cm² each and spaced a few inches apart, were marked by drilling shallow pits at the corners. The positions of all the barnacles in each square were mapped. At each level one or more of the squares was covered with a cage of stainless steel wire netting to protect the barnacles from predators.

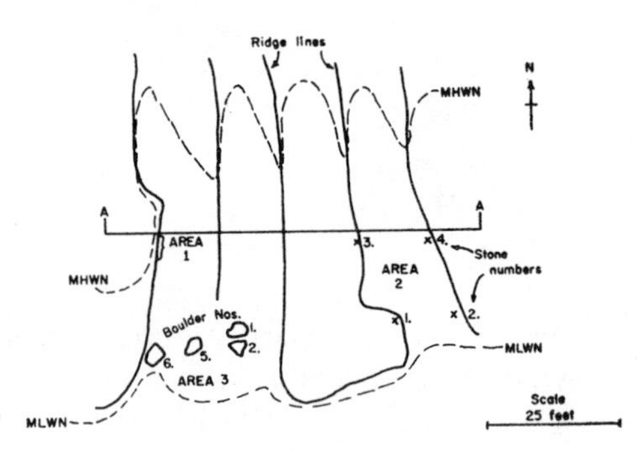

FIG. 1. Sketch map of the study area with approximate contours. Stones 7 and 8 are located on Area 1.

During the period of this study, moderately strong wave action was observed during gales each year, though probably extreme wave action, such as that on the shores exposed to the open Atlantic, does not

TABLE 1. Descriptions of the areas upon which the periodic censuses of *Balanus balanoides* were carried out.

Area No.	Subdivisions of the area	(Standard tidal levels)	Height in feet, relative to M.T.L.	% of time exposed to air	Descriptions of the small areas at each level where the censuses were made, with the names used for each in the text and figures. The date of first census of each area is given; all were continued until June, 1955.
1.		(MHWS)	+4.9	96	Jan. 1953: 3 adjacent squares, mixed *Balanus* and *Chthamalus stellatus*: each 25 cm.²
	Top level		+4.2	91	
	High level	(MHWN)	+3.1	79	July 1953: Only censuses of *Thais* were made at this level
	Upper level		+2.6	75	Nov. 1952: Cage 1, Cover, Control 1: each 29 cm.² Dec. 1953: Control 2, Cage 3: each 50 cm.² July 1954: Cage 2: 66 cm.²
	Middle level and stones 7 and 8		+2.1	69	Nov. 1952: Cage 1, Cover, Control 1: each 29 cm.² April 1953: Stones 7 and 8: 5 to 6 cm.² Nov. 1953: Control 2, Cages 2, 3, 4: Control, 47 cm.², Cages, 92, 98 and 98 cm.², respectively
	Lower level		+1.5	63	Nov. 1952: Control 1 and Control 2: 29 and 25 cm.² respectively Oct. 1953: Cage 1: 61 cm.² Feb. 1954: Control 3: 52 cm.²
2.	Stones 3 and 4 Stones 1 and 2		+1.1 −0.9	59 42	April 1953: 5 to 6 cm.² April 1953: 5 to 13 cm.²
3.	Boulder 1 Boulder 2 Boulder 5 Boulder 6		−0.9 −0.9 −1.8 −1.9	42 42 35 34	June 1953: 25 cm.² in 1953, 47 cm.² in 1954-5 June 1953: 25 cm.² in 1953, 53 cm.² in 1954-5 July 1953: 50 cm.² July 1953: 50 cm.²
		(MLWN) (MLWS)	−3.0 −5.1	23 4	

NOTE: The areas of census on the stones were different in each settlement season; the sizes are included in the ranges given.

occur in the Clyde sea area. However, the alga *Alaria esculenta* (L.) Breve, an indicator of fairly strong wave exposure (Lewis 1954b), occurs only at this point on the Isle of Cumbrae. *Ascophyllum nodosum* (L.) Le Jol., an alga which does not thrive in wave-beaten areas, occurred only in the upper sheltered parts of the ravines. *Fucus spiralis* (L.) and *Fucus vesiculosus* L. were the dominant large algae on Areas 1 and 2, *Gigartina stellata* (Stockh.) Batt. on Area 3.

In addition to these squares which were mapped *in situ*, two flat stones of bostonite about 4 inches in diameter were fastened at the "middle" level, 2.1. feet above mid tide level; these were numbered 7 and 8 (Table 1). They could be removed and replaced during the low tide interval and were used to follow the pattern of the spring settlement. All the squares and stones in the three lower levels of Area 1 were located in a space of about 1 m^2.

Area 2 consisted of two pairs of removable stones located in a ravine about 50 ft east of Area 1. One pair was fastened about 1 ft above mid tide level, the other about a foot below, further down the ravine. Although both members of a pair were at the same level, one was fastened to the sloping east side of the ravine so that it faced upward, while the other member was fastened underneath an overhanging ledge on the west side so that it faced downward. In the lower pair, stone 1 faced down, stone 2 up; in the higher pair, stone 3 faced down, stone 4 up. All these stones were similar in size and composition to those attached to Area 1.

Area 3 consisted of four boulders, each about 3 ft in diameter, situated in the lower part of the same large ravine in which Area 1 was located. The relative heights are given in Table 1. Squares were mapped on the sloping tops of these boulders. A few counts and experiments were made on other rocks in the same area; these will be described as necessary.

The three areas spanned almost the entire intertidal distribution of *Balanus balanoides*. Only scattered individuals occurred above the highest level of Area 1, and below the lowest boulder of Area 3.

The stones and cages were fastened to the rock by means of a stainless steel screw inserted in a plastic tube, sold as a screw anchor, which was fitted into a 1" deep hole, 1/4 inch in diameter, drilled into the rock. One such screw with suitable stainless steel and plastic washers was enough to hold a small stone or cage in place. These remained in place for over two years with no losses. The cages were made from stainless steel wire netting, No. 22 gauge wire, eight meshes to the inch; this netting has an open area of 60%. They consisted of a floorless enclosure about 6 x 6 inches, with walls about 1 inch high, so that the roof was well above the barnacles. In two instances, a piece of netting was stretched out from the side of a cage as a roof over an area of barnacles, but with no sides. This was done in an effort to create a physical environment similar to that in the cage, yet allowing access to predatory whelks; these were termed "covers."

The surface area used to calculate the population densities given in this study was that of the rock surface projected on to a plane surface. The population density of the youngest age group was calculated on the basis of the "available area" for that group; i.e., the area of free rock surface unoccupied by older barnacles.

Initially, the mapping of each square was done on graph paper, a grid of threads of 1 cm^2 opening being used as a guide on the rock surface. Later a more efficient method using a piece of thin glass (lantern slide cover glass) was developed. The glass was held over the barnacles and their positions were marked directly on it with glass-marking ink. A paper copy was made from this so that notes could be made after each census. For the first six months after settlement, the individual barnacles were not mapped, but successive counts were made on a small portion of each area. Yellow water color paint was used as an aid to avoid duplication in counting; it washed off during the next few high tides.

In the first census, November 1952, every individual was mapped, and the settlement of that year, then six months old, was distinguished from the rest. It was possible to distinguish these from the older year groups by their smaller average size and the appearance of the shell. Above mid tide level, barnacles aged 6 months had quite thin shells and the upper edges of the wall plates or parieties were still uneroded; the surface was fairly smooth and white. The previous year group, then aged 1½ years, had undergone much erosion so that the upper edges of the shell were thick and rounded, no new shell having been added there. (Growth in the compartments of a barnacle, according to Darwin (1854), occurs only along the basal and lateral edges of each wall plate; erosion of the top edge is not replaced.) The surface was darker and often pitted in these older individuals, probably due to boring algae (Parke & Moore 1935).

The differences in appearance were more reliable than size in separating the youngest age group from all the rest, since there was generally some overlap in the size range at this time. This overlap was noted in measurements made from photographs of the plotted squares which had been under observation for two years; two year classes had settled in this period and at the time of the photograph were 6 and 18 months old, respectively. The same sort of overlap in the extremes of size in the 6 and 18 month age groups was evident in some data kindly provided by Mr. H. T. Powell, from the study of growth rate published in Barnes & Powell (1953). An additional check was provided by the fact that the largest 6 months old barnacles were the most different from the 18 month group in general appearance. Besides having smooth thin shells, these fast growing individuals had increased their basal area faster than their opercular opening, resulting in a low barnacle with a proportionally smaller opercular opening.

No reliable way of distinguishing the various year groups of age 1½ years and older was found. Kuznetzov & Matveeva (1949) state that barnacles from the Arctic coast of Russia have definite annual growth rings. Such rings were not obvious at Millport, perhaps because the seasonal variations in climate were not so great as in Russia.

Copies of the original data of this study have been placed in the Bureau of Animal Population, Department of Zoological Field Studies, University of Oxford, Oxford, England.

Abbreviations for the intertidal shore levels used in this paper are as follows:

MHWS: Mean high water of spring tides.
MHWN: Mean high water of neap tides.
MTL: Mean or Mid tide level.
MLWN: Mean low water of neap tides.
MLWS: Mean low water of spring tides.

The absolute heights and the percentage of the time in which these levels are exposed to the air are given in Table 1.

SETTLEMENT

METHODS

To determine both the rate of attachment and the mortality during the settlement season, the fate of individual cyprids was followed. On each detachable stone, one small area was examined daily and the positions of the newly settled cyprids were mapped. At every other low tide the stones (from Areas 1 and 2: see Table 1) were brought into the laboratory and examined with a dissecting microscope. A glass slide ruled with a grid was placed over a marked area of the stone and all changes, such as new attachments, losses, and metamorphoses were marked on an enlarged map of the area.

The stones were always returned to the shore before the tide rose again and when in the laboratory were kept outside on a window ledge, except during the brief period when they were being examined. Thus the conditions were little different from those during a normal low tide. That this treatment did not markedly increase the mortality can be seen from the records of Stone 3, which was subjected to the same amount of handling as the other stones. This stone evidently was in a protected place on the shore since the cyprid mortality between attachment and metamorphosis during the 1953 and 1954 settlement periods was only 5%. Some of this mortality probably occurred on the shore, so that the amount resulting from the examinations in the laboratory must have been small.

After the position of a cyprid had been plotted it was assigned a grid number and records of its metamorphosis and/or death were kept. The fate of about 8,000 individuals was followed in this manner over three seasons.

There are several possible sources of error in this method: (1) if any cyprids had been clinging temporarily to the surface between tides instead of being permanently attached, their absence at the next examination would have been erroneously recorded as mortality; (2) some cyprids may have attached and been killed between successive examinations without a record having been made; and (3) cyprids, having cemented themselves down, might be detached and then reattach themselves elsewhere. Evidence that temporary clinging occurred only very rarely was obtained during the last two seasons by picking the cyprids off an area of 4 cm² on each stone with a needle. Fewer than 1% were not cemented down, and these were usually recognizable by their position, lying on the side. The loss of cyprids between examinations seems likely to have been small, since many cyprids did not become detached from the rock for several days after their death. A later study of this species at Woods Hole, Massachusetts, showed that if attached cyprids were removed and placed in dishes of seawater, they could metamorphose and become reattached. The possibility of this occurring in the turbulent water of the intertidal area is small, however, since the antennules by which the initial attachment is effected would be encased in a mass of dried cement.

During the first two settlement seasons individuals were followed on all six stones. In 1955 such records were only made on stones 7 and 8; on the other four, daily counts of small areas were made, but the cyprids were not followed individually.

The stones were kept in the same positions on the shore for the three seasons. Some of the individuals of the 1953 year group were still present at the beginning of the 1954 season so that at this time a small area on each stone adjacent to that followed in 1953 was cleared for observation. A third area was cleared in 1955. Thus the 1953 and 1954 year groups were followed without disturbance until June 1955. The disadvantage of this procedure was that the adjacent areas followed each year may have differed slightly in surface contour. Only on stone 1 was the same area used in consecutive years, 1953 and 1954; this was possible since the 1953 year group disappeared completely during its first winter on this stone.

THE PATTERN OF SETTLEMENT

Fig. 2 shows the numbers of attached cyprid larvae and metamorphosed barnacles as they accumulated on each stone in each settlement season studied. In all cases the pattern is similar; an initial period of low and slightly increasing rate of attachment, followed by a period of 5 days or so in which the rate of attachment was high and finally a return to a low rate approaching zero. This sort of pattern might be expected to result if there were individual variations in speed of development of the nauplii in the plankton following a sudden massive liberation of larvae. A few fast-developing larvae would be the first to arrive, followed by the great bulk of those which had developed at the average rate and the

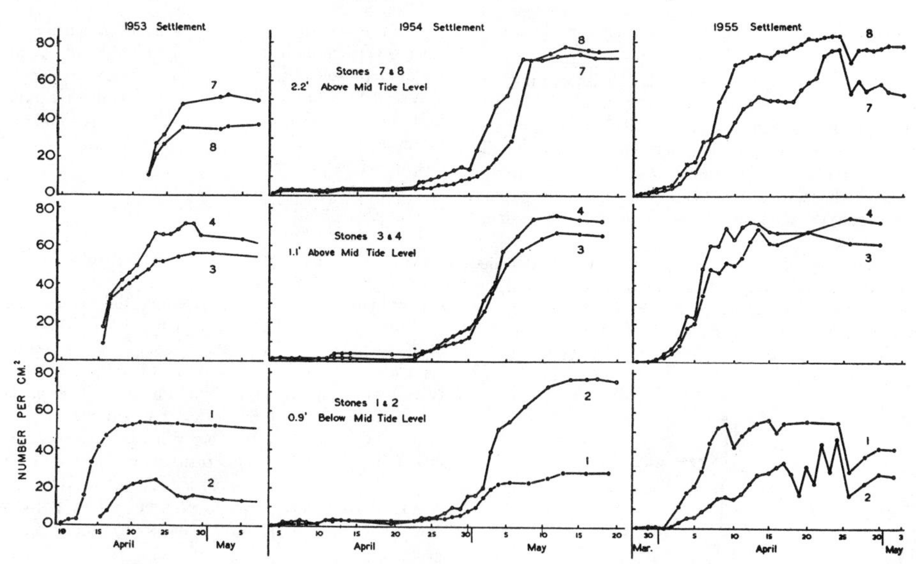

Fig. 2. Settlement pattern of *Balanus balanoides* on the same stones in three seasons at three levels. The density represents all living attached cyprid larvae and metamorphosed barnacles.

attachment rate would decline as the slowest-developing larvae arrived.

Another explanation of the gradual increase in rate of settlement has been suggested by Knight-Jones & Crisp (1953). They pointed out that if previously settled cyprids stimulate others to settle, (as shown by Knight-Jones 1953), the rate of settlement would be proportional to the numbers already attached. No actual data were given although the authors stated that they had repeatedly witnessed such a phenomenon. A decision between the two explanations cannot be made on the basis of information obtained in the present study.

The gradual decrease in settlement rate as the maximum density is approached might also be explained by a gradual diminution in numbers of larvae in the plankton. However, a decrease in settlement rate occurred on stone 1 in 1953 long before the number of planktonic larvae available for settlement had decreased. This is shown in the curves for 1953 in Fig. 2, where the maximum rate of settlement on stones 7 and 8 occurred after April 20, at a time when very few individuals were attaching to stone 1 which had reached its maximum density a week before.

This was investigated during the settlement seasons of 1954 and 1955 by picking the cyprids off small areas of each stone at each examination. In 1954, the settlement on all the stones had reached the maximum by about May 10. After this time the rate on the natural areas dropped to less than one cyprid attaching per cm^2 per tide. On the cleared areas the rates stayed above 4 cyprids/cm^2/tide until May 25. In 1955, the same thing was observed on Stones 7 and 8; the rate on the undisturbed areas approached 0 after May 1, while that on the cleared areas stayed above 3 until May 11. In most instances, the rates of settlement on the cleared areas were much higher than those observed on the undisturbed ones (in 1954, the highest average rates for the 6 stones were: undisturbed, 4.5 cyprids/cm^2/tide; cleared, 9.9 cyprids/cm^2/tide). This difference may have been due to the presence of body fluids and bits of cement left behind in the process of clearing. These have been shown to stimulate settling (Knight-Jones 1953). The rates from the two types of treatment are not, therefore, comparable. But the fact that the rates of settlement did not decrease appreciably on the cleared areas until long after the rates on the natural areas were approximately 0 indicates that cyprids were available in great numbers in the plankton and that the cessation of settlement shown in the upper parts of the curves in Fig. 2 was not due to an inadequate supply of larvae.

Chipperfield (1948) counted the accumulating settlement of *Balanus balanoides* on pier piles at Liverpool, and also on *Pecten* shells exposed at intervals to study the variations in intensity of settlement. These counts showed that after the settlement had stopped at all levels on the piles it continued on unoccupied shells placed out later. A similar pattern of accumulation of Bacteria and Protista on plates exposed to the sea is given on page 42 of "Marine Fouling and its Prevention," Woods Hole Oceanographic Institution (1952). Other examples of laboratory and natural populations undergoing sigmoid growth are given in Allee *et al.* (1949).

OBSERVATIONS OF THE BEHAVIOR OF ATTACHING CYPRIDS

The first cyprids attached in the hollows and concave portions of the surface. An analysis of two small portions of Stone 7 in the 1955 settlement season is given in Fig. 3. The settlement began earlier on the concave portion and in the first main period of cyprid abundance it became almost fully occupied. On the convex portion the settlement became dense only during the second increase in the numbers of planktonic cyprids. The cyprids evidently attached to the convexities only after the hollows were filled. This preference of cyprids for grooves and depressions in the surface has been often noted in the literature on barnacles and Crisp & Barnes (1954) have described it in detail.

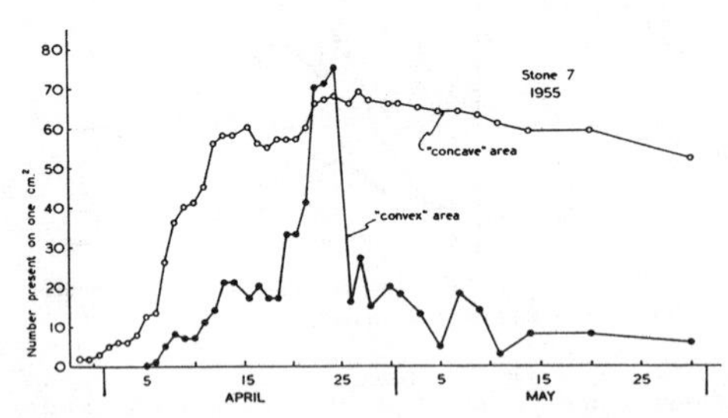

FIG. 3. The settlement of *Balanus balanoides* on two small areas about 5 cm. apart on stone 7 in 1955. The concave area was a hollow adjacent to some adults; the convex one was on a smooth raised area.

Some information was gathered concerning the amount of space needed by a cyprid to effect settlement and the reasons for avoiding densely occupied surfaces. It was noted frequently in the plotted areas that at high population densities the loss of an individual was usually followed by a new attachment on the same spot, usually in the next few days. The searching cyprid may have been attracted by some substance left behind by the damaged individual, or may have detected the open space on the otherwise occupied surface. Underwater, a densely occupied area is covered by a mass of beating cirri which probably would make searching rather difficult.

To test whether cirral activity keeps cyprids from attaching to densely occupied surfaces, the following observations were made. A surface at mid tide level was scraped clean during the 1954 settlement season. At each subsequent examination the cyprids or newly metamorphosed barnacles on one portion of the area were killed with a needle but not removed. The attachment cement held them in place. This created an area of occupied surface having no cirral activity. Another part of the area was kept clear by picking off the cyprids, while on the final part living barnacles were allowed to accumulate. After about two weeks the settlement had stopped on the two occupied areas bearing either living or dead individuals but continued on the cleared area.

It appears that a bare space on a suitable surface, even though it is only the size of a newly metamorphosed barnacle, is the only requirement for a cyprid to attach. Prompt replacement of a missing individual in dense settlements of living barnacles indicates that the surface must be constantly searched, despite cirral activity.

VARIATIONS IN SETTLEMENT

Variation in rate of settlement on closely adjacent areas. To gain information on possible variations in settlement rates in apparently similar situations, the records of the areas cleared on the stones at each examination were analyzed. On each pair of stones at the same level each day's counts were compared separately, as a "matched pair" of values. By doing this, daily variations in the numbers of cyprids in the plankton would not affect the comparison. The data were analyzed by the Wilcoxon matched-pairs test (Siegel 1956). The results are shown in Table 2. Of the six comparisons, significant local variation occurred in three. Variation occurred in both 1954 and 1955 but on different pairs of stones in each year.

TABLE 2. Comparison of the rates of settlement on adjacent areas cleared of cyprids at each examaination. Counts were made every 1-3 days in 1954, daily in 1955.

		MAY 1-24, 1954			APRIL 2-MAY 2, 1955		
Stone No.	Height from MTL	No. of observations	Average Set No/cm²/Tide	Probability	No. of observations	Average Set	Probability
7	+2.1	17	4.3	.001 ***	27	15.8	.316
8		17	7.1		27	15.5	
3	+1.1	15	7.1	.01 **	28	16.6	.102
4		15	8.9		28	14.6	
1	−0.9	14	5.3	.05	28	16.1	.001 ***
2		14	5.5		28	7.5	

NOTE: The data were analyzed using the Wilcoxon matched-pairs test, Siegel (1956).

Since the periods between observations were short, mortality on the shore would probably not account for the observed variation. Nor would it be expected that that patchiness in the distribution of larvae in the plankton would cause such variation between nearby surfaces. The most likely cause of local variability is a difference in surface contour between the two areas. A surface with slightly more irregularity, with more grooves and hollows, would be more attractive to searching cyprids than a smoother surface. It is also possible that such irregularities of the surface may have resulted in the two squares of the pair being of slightly different surfaces area. Each square measured 2 x 2 cm, but the actual area enclosed was impossible to measure accurately. It was taken as 4 cm² in all cases, so that a difference in surface contour of the two stones would cause an error. In the present instance this error is judged to be of much lesser significance than the former reason, in bringing about the observed differences.

Differences in settlement between shore levels. As seen in Table 2, differences occurred in the rates of settlement between shore levels on the stones. An analysis of variance was made of this data, between and within levels, and a significantly greater difference ($p = .05$) was found between the levels. However, the difference in positions of areas 1 and 2 might have created differences other than those due to shore level alone.

On areas of the shore other than the stones, much greater variations in settlement at different levels were observed as follows. On May 3, 1954, when the density of new settlement on the stones had reached about 35/cm^2 and the adjacent areas were moderately settled, it was noticed that only a very few cyprids were present on the boulders of area 3. (These boulders had been bare of barnacles since winter, as will be described later). The first metamorphosed barnacles appeared on these boulders on May 6, and by May 16 the densities were 16 to 20/cm^2.

The delayed settlement on these bare areas at a lower level may be explained in several ways. The effect of surface texture must be ruled out, since the boulders were composed of the same type of rock as existed higher up. The most likely explanation for this delay in settlement on large bare surfaces is one based on the work of Knight-Jones (1953). He has shown that larvae of this species and of other sessile marine animals are stimulated to settle by the presence of individuals of their own species. In the situation outlined above, the cyprids attached first to the areas where members of the previous year groups had persisted. Once the space there had become occupied they began to colonize the other surfaces. Barnes & Powell (1953) have suggested that settlement is stimulated by the draining away of thin films of water, such as occurs when the tide rises or falls past a surface; this would occur more often between neap tide limits than above or below. This would not be an alternate explanation for the delay in settlement on Area 3, however, since this area is above MLWN, and so was crossed by the same number of tides as the areas near MTL.

Other authors have observed differences in the density of settlement of *Balanus balanoides* at various shore levels. Fischer-Piette (1932) found the greatest set at low water. Moore (1935b, 1936a) also found the greatest numbers at low water in most localities, but not invariably, a few having the highest density at MTL. Hatton (1938) found that at three locations in two years, the greatest densities were at MTL three times and at LWN three times.

Rice (1935), at Friday Harbor, Washington, counted the settlement density of three species of barnacles. For *Balanus cariosus* (Pallas) the number of adults was greatest at MTL, as was the settlement. *Balanus glandula* Darwin, with no adults present, settled rather evenly, with slightly more at LWN. *Chthamalus dalli* Pilsbry, with only a few adults present, had equal densities of spat at HWN and MTL and a lower density at LWN.

Thus there does not appear to be any particular level at which the settlement density is maximum. It appears from the study of Rice (1935) that when adults are more abundant at one shore level, settlement is higher there.

Variations in settlement within one season. In 1954, a light set occurred in early April, obviously separate from the main settlement which began in late April (Fig. 2). In 1955, as shown in Fig. 3, two distinct maxima in the rate of settlement on the areas cleared daily occurred in the second and fourth weeks of April, respectively.

Chipperfield (1948) found three maxima in intensity of settlement at Liverpool, spaced 16 and 12 days apart, two at times of spring tides, one at neaps. Pyefinch (1948b), sampling from a pier during the 1947 season at Millport, found that the proportion of total larvae represented by each larval stage varied in a regular manner. Each stage showed three maxima, spaced 15 and 18 days apart. Although his method of collection was not sufficiently quantitative to give accurate estimates of population density, it is probably adequate for estimates of the relative proportions of naupliar stages, especially of the older ones.

These variations probably reflect not the variable behavior of planktonic larvae but successive liberations of nauplii from the parent stock. Crisp & Davies (1955) have shown that the same individual of *Elminius modestus* Darwin may produce several successive broods in one season. The evidence for *Balanus balanoides* is that each individual produces only one brood per year. Pyefinch (1948b) at Millport and Bousfield (1954) in Eastern Canada examined adults before and after the larvae first appeared in the plankton. They found that only part of the population had released their larvae in the first liberation.

Variation in the sizes of cyprids settling early and late in the settlement season. In 1954, the cyprids which attached in the second week of the settlement season appeared to be much larger than those which had attached in the first week. Measurements of newly metamorphosed individuals, shown in Fig. 4, confirm this. The measurements of the individuals which attached in the first week were actually made during the second week; even so they were much smaller than most of the ones settling that week. The population of small individuals which began to settle in the first week continued into the second, at which time a group of larger barnacles began to arrive; no settlement occurred in the third week, but the size in the fourth week was similar to that in the second. No further measurements were made, but the barnacles which attached in later weeks appeared to be similar in average size to those of the fourth week.

Measurements of attached cyprids and newly metamorphosed barnacles were also made during the other

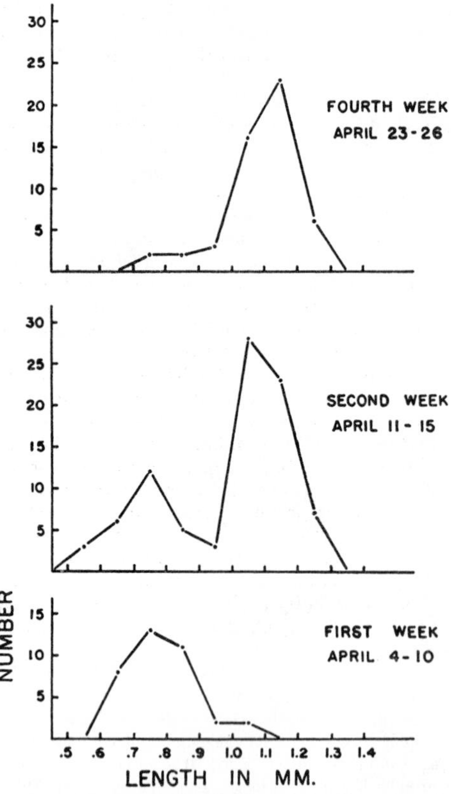

FIG. 4. Frequency distributions of the sizes of newly metamorphosed *Balanus balanoides* which settled during three weeks of the 1954 season.

settlement seasons. These measurements are summarized in Table 3, together with those made by other authors. In 1955 there was no indication of the presence of a distinctly smaller population during the first week such as occurred in 1954. The average size was slightly below that of later settlers, but not significantly so.

The measurements of the other authors appear to indicate a trend to smaller average size toward the southern end of the geographical range. These differences may be misleading, however, since an almost equal amount of variation was found within one season in the present study.

The cyprids which had metamorphosed into the smaller-sized barnacles in the first two weeks were darker and more sharply ridged dorsally than the larger, later, group. The only other species of barnacle which was likely to be settling at this time was *Balanus crenatus* Brug; its cyprids, being much paler (Pyefinch 1948a), did not resemble those discussed above. Cyrids identified as those of *B. crenatus* were found on the stones occasionally, but they never survived.

TABLE 3. Average lengths of cyprids and newly metamorphosed barnacles of *Balanus balanoides* at different periods in the settlement seasons studied. Measurements of other authors are given for comparison.

Present Study, Millport

Week of Season	Attached Cyprid Larvae			Newly Metamorphosed Barnacles		
	No. Meas.	Av. Length (mm.)	Range of Meas.	No. Meas.	Av. Length (mm.)	Range of Meas.
1953						
4 and 5......	59	1.0	0.9 -1.2	60	1.0	0.7 -1.2
1954						
1............	6	0.85	0.81-0.89	37	0.76	0.64-1.05
2............	101	1.04	0.72-1.25	98	0.98	0.56-1.21
4............	75	1.12	0.77-1.25	44	1.08	0.72-1.21
1955						
1............	29	1.03	0.94-1.14	20	0.96	0.86-1.07
4............	9	1.11	1.02-1.22	10	1.00	0.76-1.05

Other Authors, Planktonic Cyprids

Place	Av. Length (mm.)	Range of Meas. (mm.)	Authority
Herdla, Norway........	1.20		Runnstrom, 1925
Altane Fjord, Sweden...		1.02-1.24	Barnes, 1953a
Millport, Scotland......		0.84-1.20	Pyefinch, 1948a
Millport, Scotland......		0.82-1.22	Barnes, 1953a
Liverpool, England.....	1.09	0.90-1.25	Chipperfield, 1948
Plymouth, England.....	0.94		Bassindale, 1936

During the second week of 1954, measurements were made on each individual cyprid and then on the barnacle into which it metamorphosed. The cyprid measurements were grouped into 1/10 mm classes and the average sizes of the cyprids and the barnacles which had metamorphosed from them were computed. These values are given in Table 4. Cyprids of less than 1.0 mm in length developed into relatively smaller barnacles than did the larger cyprids.

Barnes (1953a) found that measurements of cyprids of *Balanus balanoides* taken from the plankton showed

TABLE 4. The relation of size of cyprid to the size of the barnacle into which it developed. (April 11-15, 1954 only)

Size range of cyprids in each group (mm.)	.70-.79	.80-.89	.90-.99	1.00-1.09	1.10-1.19	1.20-1.29
Number of individuals	1	10	12	37	43	11
Average length of cyprids..........	0.72	0.86	0.98	1.07	1.13	1.22
Average length of the barnacles developing from these cyprids..........	0.56	0.72	0.84	0.99	1.11	1.14
Length ratio, Barnacle/Cyprid........	0.78	0.84	0.85	0.93	0.97	0.93

two modal sizes. Stage I nauplii removed from adults aged 1 yr collected at a low intertidal level were slightly smaller than those from adults of the same age at a high level. Barnes suggested that the two size groups of cyprids may have come from adults of different shore levels. Although this suggestion remains a possibility, the effects of other factors such as size of parent and conditions during the planktonic stages need to be studied. There is some evidence that planktonic conditions were abnormal in 1954 at Millport. Barnes (1956) found that the phytoplankton bloom which normally develops in early March failed to do so in 1954. The larvae released early into the plankton may not have had sufficient food for proper development, so that the few cyprids which developed from them might have been small.

Annual variations. Records of the occurrence of barnacle larvae in the plankton and on the shore at Millport are available for the years from 1944 to 1955, except for 1948 (Pyefinch 1948b, Barnes 1956, present study).

In every year except 1946, cyprids first appeared in the plankton in early April, although the time of the main settlement varied from early April to early May. In the present study these extremes were encountered in 1955 and 1954, respectively (Fig. 2). Cyprids never settled earlier than March 21, and were never present in large numbers until near the end of March. The subject of annual variation will be treated further following the discussion of mortality during the settlement season.

MORTALITY DURING SETTLEMENT

In most instances it was easy to decide whether an individual barnacle had died in the interval between observations. Missing individuals, those with broken shells and gaping or deeply sunken opercular plates, and cyprids with withered shells, were all recorded as having died since the last observation. In some cases, however, cyprids which never metamorphosed gradually became darker, but remained attached for two weeks or more, so that the exact date of death could not be determined. During the 1955 settlement season, the time between attachment and metamorphosis of 951 cyprids was recorded in daily observations; the average time was 1.5 days. Therefore, the cyprids to which it was impossible to assign a specific date of death were assumed to have died 1.5 days after attachment.

Individuals missing since the previous observation were also assumed to be dead. Detachment in natural conditions is probably accompanied by bodily damage, and reattachment is unlikely to occur, as discussed earlier.

DIFFERENCES IN MORTALITY BETWEEN THE EARLY AND LATE SETTLERS

To study the differences in mortality between groups settling at different periods, the settlement period was divided into weeks and the barnacles which had attached in each week were considered as a group. Fig. 5 illustrates the survival of the weekly groups in each year, the barnacles on all the stones being combined. (In 1954 there was no set in week 3).

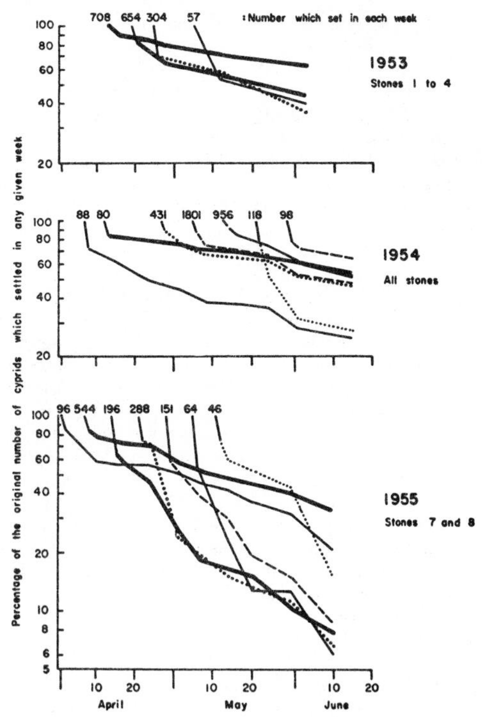

FIG. 5. Survival of *Balanus balanoides* during the settlement season; each curve represents all those which attached during a single week of the season. The first portion of each curve, a thinner line, denotes the mortality of cyprids before metamorphosis. In 1953, the first curve represents the second week of settlement, no records having been kept for the first week. In 1954 there was no set in week 3.

On each of the weekly survival curves (except the first and last of 1953 when adequate records were not kept), the first segment represents the mortality of cyprid larvae, plotted as if it had occurred in one and half days, as explained previously. Thus the differences in larval mortality are shown by the different slopes or lengths of the first part of each curve. No consistent relationship seems to exist between cyprid mortality and time of attachment. In 1954 the early and late settlers fared worse than the ones during the middle weeks, while in 1955, the reverse was true.

Once metamorphosed, the barnacles which attached in the early weeks appeared to survive better than those attaching in the later weeks. This is particularly clearly shown by the good survival of the barnacles of the second week in all three years; the slope of the curve indicates the relative mortality rate.

The barnacles attaching in the first week in 1954 showed poorer survival. In this year, as discussed previously, the barnacles settling in the first week were of smaller average size than the later settlers.

In 1955 the first cyprids to settle was similar in appearance to the later ones, and their average size was only slightly smaller (Table 3). The survival of the group settling in the first week was poorer during the first two weeks after which it was about the same as that of the second week (Fig. 5). Thus, aside from the obviously different group of the first week, 1954, the survival of the earlier settlers seemed to be consistently better than that of the later ones.

This was especially evident during some of the later periods of heavy mortality, as for example during a gale in the last week of May, 1954. Most of the populations suffered increased mortality at this time, as can be seen in Fig. 5. The percentage mortality for each week on each stone was calculated for this period. These data are shown in Table 5, where percentage mortalities of all the barnacles which attached to the stones in the three early weeks are compared to those of the three later weeks. The percentages for the early weeks were compared to those of the later ones using the Mann-Whitney U Test (Siegel 1956); it indicated that the later settlers suffered significantly greater mortality during the gale ($U = 63$, $p. = .025$).

TABLE 5. Mortality during a gale, May 24-26, 1954. A comparison has been made between the mortalities suffered by barnacles which settled early and those which settled later in the settlement season. The figures represent percentage mortality over an interval of three high tides.

Stone No.	Early Settlers			Later Settlers		
	Weeks No.			Weeks No.		
	1	2	4	5	6	7
1	0	8	22	31	61	78
2	—	—	6	22	28	0
3	0	9	2	1	8	0
4	—	—	19	32	37	—
7	54	14	26	28	39	87
8	—	0	0	3	5	33

NOTE: Only those weekly groups which contained six or more individuals were used.

The barnacles which settled in the eighth week of the 1954 season survived well for the first few weeks, as can be seen in Fig. 5. However, the later survival of the barnacles of this week was poorer than that of any of the other weeks; except on Stone 1, they were all dead within six months. The later survival of those attaching in weeks 1 and 7 was only slightly better. The main bulk of the population present after the end of the settlement season was composed of those settling in the second to sixth weeks.

As discussed earlier, attaching cyprids tended to settle in concavities such as grooves, pits, or other hollows of the surface. Once these places were filled by the early settlers the later ones were forced to settle on the rest of the surface, plane or convex. The differences in mortality on the two types of surface were investigated in the following manner. Two small portions of the area studied on stone 7 in 1955 were chosen, one a concave area where barnacles had persisted, the other a convex one which was almost completely bare at the end of settlement. From the individual records the curves of settlement were constructed and are shown in Fig. 3. The settlement started a week later on the convex area and during the main settlement period in early April only reached one-third the density of the concave area. An increase in planktonic cyprids about April 20, as indicated by increased attachments on an area of stone 7 cleared of cyprids daily, resulted in a rapid colonization of the convex area when the concave areas were almost completely occupied. The gale of April 25th caused great destruction on the convex area, but not on the concave one. The behavior of cyprids which leads them to attach in hollows evidently has great survival value.

This same analysis was used to determine whether the cyprids which attached earlier survived better by virtue of their greater age, when presumbably they would have a somewhat thicker shell. The same two areas on stone 7 were divided into the weekly sets and the death rate during the April 25th gale calculated. The percentage mortalities for barnacles which attached during the second, third and fourth weeks are 0, 7 and 8 for those on the concave area and 62, 60 and 92 for those on the convex area. Thus the differences in death rates were much greater between areas than within them. There is some suggestion that within each area the earlier settlers could better withstand the effects of a gale. However the protection offered by the concavity was much more important, indicating that in general the earlier settlers will survive better because they tend to occupy the more protected positions.

MORTALITY AND WEATHER

Since the weather often changed radically from day to day, the 1955 settlement season was chosen for analysis, observations having been made daily in that year. As pointed out earlier, the precise date of death of a cyprid was usually harder to decide than that of a metamorphosed barnacle, so that it would have been difficult to determine any relationship between the mortality of cyprids and daily changes in the weather. Thus for the cyprid mortality, only two periods were analyzed, one being a period of unusually warm, calm weather which persisted for five days, and the other a severe gale. During the warm weather there was no wind, so that there was little lapping of small waves; this prolonged the period of exposure to air.. As shown in Table 6, for both these periods the percentage mortality per day of cyprids was much higher than average for the whole season.

TABLE 6. Mortality of cyprids and metamorphosed barnacles during the 1955 settlement season in relation to daily weather conditions.

Daily Weather Type	No. of Days in Each Type	Av. % Mortality per Day of Barnacles		Av. % Mortality per Day of Cyprids
		Stone 7	Stone 8	Stone 7
Warm, calm	5	2.2	0.7	57.0
Warm, light wind	3	1.2	0.3	—
Sunny, cold, windy	8	4.0	1.7	—
Partly cloudy, cool	13	3.6	1.4	—
Rain, over .05 in./day	10	2.5	2.1	—
Gales	3	16.4	7.1	54.0
Whole season, March 28-May 11	42	3.7	1.6	29.0
Number of organisms	—	883	452	116

The effect of gales on metamorphosed barnacles was shown in another analysis. The weather occurring in the 1955 season was classified into six types, as listed in Table 6. Then the average percentage mortality per day of metamorphosed barnacles for all the days in each type was calculated. It was unusually great only during the gales. After a gale many barnacles were missing or broken, probably due to debris being thrown about by the waves.

The effect of increased exposure to air at higher shore levels was difficult to ascertain from these data, since the stones were not strictly comparable in surface roughness or angle of attachment. However some comparisons seem justified; the percentage of cyprids which metamorphosed after attachment varied from 95 to 70%. In any one season there was no systematic variation, and the stones at the highest level did not have the highest mortality. The percentage survival from initial attachment of the cyprid to the end of the settlement season varied from 88 to 24%. The highest mortalities occurred on different stones in 1953 and 1954, stones 2 and 4 respectively. Again the stones at the highest level did not suffer the greatest mortality. The erratic nature of these results indicates that chance damage was the main cause of death; stone 3, protected under a small overhang, always had the lowest mortality. Once metamorphosed, the most dangerous time for newly settled barnacles seemed to be when they were immersed.

Mortality Associated with Other Animals, Including Intraspecies Relationships.

During the 1955 season, broken barnacles were observed associated with regular grooves, which appeared to be radula marks, on the surface. The most likely cause for these marks was the grazing of the limpet, *Patella vulgata* L., which was very common in the area. An experiment performed during the 1954 settlement season to demonstrate the effect of limpets is illustrated in Fig. 6. At the beginning of barnacle settlement, two cages were attached to a slope of bare rock where radula marks indicated heavy grazing by limpets. One cage was attached over two limpets, the adjoining cage having none. After the settlement season was over, the photographs in Fig. 6 were taken. It is possible that the restriction by the cages may have resulted in more intense grazing inside the cage than on the open shore. But it seems obvious that the limpets can reduce the density of barnacle settlement appreciably. Inside the cages on Area 1 (described in the next section) the density of settlement in the narrow spaces between the tall older barnacles was very high. This suggests that the limpets present in the cages were not able to graze in these places. Hatton (1938) cleared the limpets away from the area surrounding his settlement sites to prevent the destruction of the young barnacles. He states that the limpets did not damage older barnacles; this was confirmed in the present study by the good survival of older barnacles in cages with limpets on Area 1. Lewis (1954a) noted that the activities of a population of *Patella vulgata* removed the newly settled spat of *Balanus balanoides* above MHWN. Those spat which settled in pits or among older barnacles survived. An indirect effect of *Patella* on *Balanus* has been suggested by Southward (1956). Limpets had been removed from a strip of intertidal shore on the Isle of Man and the recolonization studied (Jones 1948, Burrows & Lodge 1950, Southward 1953). Following the limpet removal, algae colonized the strip, covering it almost completely; after this there was good survival and growth of newly settled limpets under the algal canopy. Their grazing presumably prevented further recruitment of algae so that by the sixth year following the start of the experiment the algae had declined to the original degree of coverage. The density of *Balanus balanoides* (at MTL but not at HWN) declined when the algae increased. Southward (1956) ascribes this to the barrier effect of algae preventing the cyprid larvae from reaching the rock surface. However, several alternate explanations are possible; the increased limpet population may have destroyed more of the settlement than usual. Also, the period of decline in the *Balanus* population coincided with years of poor settlement of this species over much of England, including the Isle of Man (Southward & Crisp 1954). In addition, if more *Thais* occurred under the protective algal canopy this would partly explain why the decrease occurred at MTL where *Thais* was common but not at HWN, where it was never recorded by Southward (1953).

In a visit to Millport in the summer of 1958, similar changes were found; these are summarized in Table 7. *Patella* had decreased, and the cover of larger algae had increased. The algae on Area 1, mainly *Fucus vesiculosus* L., were 10 to 12 cm in length in late June, 1958. Hatton (1938) found that this species grew 20 cm per year at St. Malo, and in Denmark, Lund (1936) found that it grew 9-15 cm per year. Thus the algae on area 1 had probably attached during the summer or autumn of 1957. The 1958 *Balanus* settlement was less dense under the *Fucus* than in the open, possibly due to the barrier

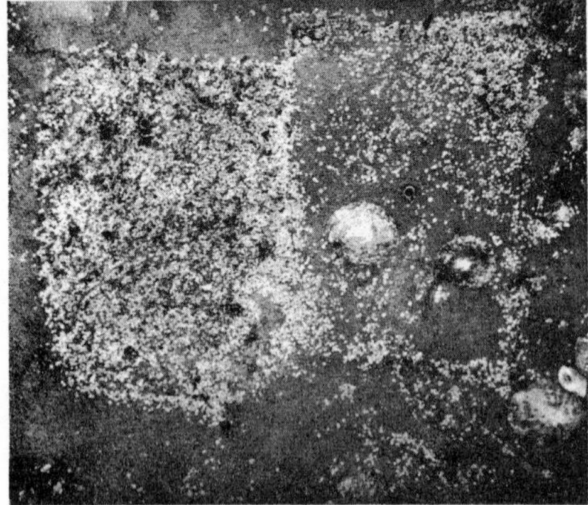

FIG. 6. The effect of grazing by the limpet, *Patella vulgata,* on newly settled *Balanus balanoides.* Both photographs were taken on June 22, 1954, after the cages had been attached for 10 weeks during the settlement season. The left cage excluded limpets, the right enclosed two. Each cage was about 9 x 6 inches in size.

effect of the algae as suggested by Southward (1956), or to the decreased water circulation under algae as suggested by Hatton (1938). The alternative explanations given above would not apply to the 1958 settlement under the *Fucus* since the limpet density was low and, as will be shown later, *Thais lapillus* did not feed much on young barnacles if older ones were available. *Thais* was never observed to feed on newly settled barnacles; possibly very small *Thais* might do so but none were observed at this time.

To investigate the other animals associated with the barnacles, all the organisms on small areas of barnacle-covered rock were collected on several occasions in the summer and autumn of 1953. This material was brought into the laboratory and examined. Collections were made from above and below MTL. The principal groups represented were Turbellarians, Nematodes, Oligochaetes, small Gastropods, Lamellibranchs, Ostracods, and Hydracarina. A few Cope-

TABLE 7. Summer observations on the abundance of *Patella vulgata* and *Fucus.*

Patella vulgata, numbers

		Area m.^{2}	July-Aug. 1953	August 1958
1. Area 1		1.00	14	5
2. Reef below MTL	Large limpets	5.80	276	22
	Medium limpets	5.80	76	8
	Small limpets	5.80	13	14
3. Area 3	Boulder 1	0.33	20	3
	Boulder 2	0.30	20	7
	Boulder 6	0.37	27	11

Fucus spiralis and *F. vesiculosus;* % Cover

		1953-1955	August 1958
Area 1	Upper level	5	50
	Middle level	5	60
	Lower level	10	60

pods, Amphipods, Isopods and insects, *Lipura* (= *Anurida*) *maritima* Guerin and Chironomid larvae, were found, plus a few polychaetes of the species *Eulalia viridis* (O. F. Muller). Only the last species would be likely to be a predator, the others all being very small. Moore & Kitching (1939) observed *Eulalia* opening a barnacle. Several observations of foraging by *Eulalia* were made in the present study; the worms seemed to be searching in all the crevices and empty barnacles. One was seen capturing a large chironomid larva among barnacles, but no sign of any attack on live barnacles was observed. Blower (1957) has reported observations of a centipede, *Scolioplanes maritimus* (Leach) feeding on *Balanus balanoides* at night on the Isle of Man. In the few night visits made during the present study, no centipedes were observed.

To study the larger animals, a fish trap was constructed over an area of barnacles so that any fish or crabs foraging over the surface at high tide would be caught. No bait was used. Six species of fish and three species of crabs were caught during operation of the trap for 19 tides in July and August, 1953. The list of species caught follows:

Fish

Labrus bergylta Ascanius
Ctenolabrus rupestris (L)
Centrolabrus exeoletus (L)
Gadus virens L.
Cottus bubalis Euphrasen
Onos mustela (L)

Crabs

Portunus puber (L)
Carcinus maenas (Pennant)
Cancer pagurus L.

Only *Ctenolabrus rupestris* was caught regularly. A few stomachs of each of the above species were examined, and traces of barnacles were found only in those of *Labrus* and *Portunus*. No blennies, *Blennius pholis* L., which have been found to feed heavily on barnacles, (Qasim 1957) were caught.

The droppings of the Purple Sandpiper, *Calidris maritima* (Brunnich) and the Black-headed gull, *Larus ridibundus* L. were examined but no barnacles were found in them. Rock pipits, *Anthus spinoletta* (Mont.), were common but the observations of Gibb (1956) indicate that *Littorina, Idotea,* and dipterous larvae are its principal food. Rats and rabbits were also observed on the shore but no stomach contents were examined.

From this limited survey it would seem that none of these possible predators in their present abundance could cause mortalities among newly attached barnacles comparable with those associated with gales.

From the preceding analysis, the following tentative conclusions may be drawn. The first settlers, by selecting the hollows for settlement, had a better chance of survival than those which arrived after the concavities had been occupied. During severe wave action the convexities on the surface of the rock might be rubbed clean, although older individuals with thicker shells appeared to withstand this. Once metamorphosed, the barnacles suffered little mortality during warm, dry weather as compared to that of the cyprids. Thus most of the early mortality of metamorphosed individuals may have occurred when the barnacles were under water. Limpets were the only animals which appeared to contribute greatly to this early mortality. The average mortality did not increase with height except at levels above MHWN. In 1954 the first group of cyprids was small and dark; their survival was poor. Possibly these were liberated shortly before a failure of the phytoplankton population and so did not develop normally (Barnes 1956).

Some previous work has been concerned with this period of barnacle mortality. Deevey (1947), using the data of Weiss (1948b), suggested that survival was poorer at times of maximum settlement of *Balanus improvisus* Darwin on continually submerged glass plates in Florida. These data were obtained by comparing the final numbers of barnacles surviving on plates exposed for a month with the sum of the numbers of cyprids attaching each day for a month to a plate wiped daily. At certain times, fewer attached over a month to the plate cleaned daily than survived after a month's exposure on the other plate. This suggests that the wiped plate was less attractive to cyprids than the one which had both a bacterial slime and previously settled barnacles, both of which have been shown to increase the numbers attaching (Miller *et al.* 1948, Knight-Jones 1953). In addition, the maximum number on the panel exposed for a month depended not on the area but on the season, since during the summer the barnacles grew faster and covered the panel with about a third of the number that could find room to attach in winter. Thus from this study no accurate relation between mortality and density can be derived.

There seems to be no other evidence that cyprid mortality is associated with population density. As seen in Fig. 5 there was no correlation between the weekly cyprid mortality and population size. However, the population density may indirectly affect survival since the cyprids which attach when the surface hollows are filled have a smaller chance of surviving.

FACTORS DETERMINING THE POPULATION DENSITY AT THE END OF THE SETTLEMENT SEASON

The density at the end of the settlement season is the difference between the numbers which attached and those which died during the season. In the previous work on barnacle settlement, no measurements of the mortality during the settlement season were made, so that it is difficult to decide which factors determined the densities observed.

In some instances, it was obvious that a low settlement density was the result of a limited supply of larvae from the plankton. Hatton (1938) noted that the density of settlement for *Balanus balanoides* at St. Malo was lower in calm areas than in those with moderate currents or waves. He attributed this to the fact that the numbers of larvae brought to the shore were fewer when the circulation was slower. In places where large algae (*Fucus*) grew densely the settlement density was lower; he removed the *Fucus* in patches and found that the settlement was heavier on the bare rock. His explanation was that the *Fucus* reduced the circulation close to the surface. No data were available on mortality; however, it seems logical that less circulation would result in fewer larvae being brought to the shore area.

Bousfield (1955) found that *Balanus balanoides* colonized only the lower portions of a large estuary in eastern Canada because the planktonic larvae remained near the surface where they would be carried seaward. Another species, *Balanus improvisus* Darwin, whose larvae were found near the bottom, colonized the upper estuary, probably being carried there in the landward current near the bottom. This is a particularly clear demonstration of the control of local shore distribution by events which occurred during the planktonic phase.

Another method of approach to the question of the importance of planktonic supply in determining settlement density is that of comparing annual variations in both planktonic and shore populations. Studies of this nature have been carried out at Millport since 1944 in connection with work on fouling.

Pyefinch (1948b) found great annual variations in the numbers of planktonic cyprids at Millport in four successive years. In April, 1945, the number of cyprids of all species in the plankton was only 1/10 that of 1944. Yet the density of newly settled cyprids in 1945 was 74/cm^2, similar to the maximum densities shown in Fig. 2 of the present study. Moreover, counts of young barnacles in April showed that the density in 1945 was about 3/4 of that in 1944. These figures suggest that great variations in the planktonic supply were not reflected in similar variations in densities on the shore. As discussed earlier, Pyefinch's methods of sampling plankton were only roughly quantitative, but probably were sufficiently accurate to reveal differences as great as in the two years cited.

Barnes (1956) studied the planktonic history of *Balanus balanoides* at Millport, and has summarized his own and earlier records for nine years from 1944 to 1954. In four of these years, termed "failure" years, very few late-stage nauplii and cyprids were found one month after the liberation of larvae into the plankton in early March. In the other five "normal" years, moderate to great numbers of later larvae were collected in late March and early April.

In three of the four failure years the phytoplankton bloom of early March disappeared after a short time while in the normal years it persisted through March. This seemed to be the most likely cause for the failure of the early planktonic larval population. Other influences such as egg viability, predation, deleterious water factors, etc., appeared to have no correlation with these larval variations.

As has already been pointed out, all the adults do not liberate their larvae at the same time. Barnes (1956) stated that during a failure year, a later liberation might develop successfully if there was a late diatom increase such as occurred in 1954. Barnes (personal communication) has also pointed out that the proportion of the adult population which liberates larvae in the first outburst of a failure year, will determine whether there are enough left for a second liberation of sufficient size to populate the shore.

Unfortunately, few data are available on the settlement density in these years. For the normal years of 1944, 1945, and 1949, the settlement density was equal to or greater than that found to be maximal in the present study (Pyefinch, 1948b, Barnes & Powell, 1950). For the failure years, no quantitative data were given although Pyefinch stated that the late liberation in 1946 led to a "light settlement."

For the three years of the present study, plankton data were available to Barnes for only one, 1954; it was classified as a failure year. In 1955, very heavy settlement occurred at the beginning of April; although no plankton data are available for 1955, this early settlement gives a strong presumption for classifying it as a normal year. If it were a failure year, the first liberation would have had to occur far in advance of that of any of the previous nine years. Also, the settlement continued into the third week of May, a normal length of season. 1953 also appeared to have been a normal year since cyprids began attaching in the first week of April, with the heavy settlement in the second week.

As shown in Fig. 2, the maximum density reached was similar on some stones in all three years. The average rates of settlement (no/cm^2/tide) on the areas cleared daily ranged from 4.3 to 8.9 in 1954 and from 7.5 to 16.6 in 1955. The maximum rates of settlement recorded in daily observations for the six stones ranged from 5.0 to 10.6 in 1954, and from 10.2 to 20.0 in 1955. The average and maximal rates observed in 1954 were thus about half those in 1955. Thus the second liberation in 1954, although producing a less intense settlement, filled the space available (Fig. 2) and as shown by the settlement on areas cleared daily, continued after the space was filled. The success of the second liberation indicates that the habit of this species of liberating its larvae in successive bursts has survival value in an environment of fluctuating food supply.

Densities determined at the end of each settlement season during the present study are given in Table 8. The light settlement in 1958 was undoubtedly limited by the supply of planktonic larvae. Differences between the three earlier years were smaller.

TABLE 8. Annual variations in the densities of *Balanus balanoides* (current year-group only) at the end of settlement at Millport and at St. Malo, France, from Hatton, (1938).

Millport

Area	Feet from MTL	No./CM.^{2} 1953	1954	1955	1958
Area 1, Top........	+4.2	34	17	18	0.8
Stones 5 - 6........	+2.5	16	14	18	—
Area 1, Upper......	+2.6	67	42	56	2.0
Area 1, Middle.....	+2.1	47	31	48	1.0
Stones 7 - 8........	+2.1	34	52	35	—
Area 1, Low........	+1.5	24	15	42	0.2
Stones 3 - 4........	+1.1	40	38	57	—
Stones 1 - 2........	−0.9	34	52	30	—
Area 3.............	−1.4	17	10	25	0.5
Average........		35	30	37	0.9

St. Malo

Level	No./CM.^{2} 1930	1931	1932
Ouest II (HWN)........	8.0	5.8	—
Est II (HWN)..........	4.0	5.2	—
Cité II (HWN).........	10.0	11.4	—
Ouest between II and III	23.0	20.0	19.5
Est between II and III	12.0	12.5	13.0
Cité between II and III	22.0	19.5	18.5
Ouest III (MTL).......	23.2	20.2	—
Est III (MTL).........	12.1	12.9	—
Cité III (MTL).........	22.1	19.9	—
Ouest IV (LWN).......	15.3	17.5	—
Est IV (LWN).........	15.0	13.1	—
Cité IV (LWN)........	23.8	13.9	—

In four instances the densities were much lower in 1954, in two, higher, and in the other three only slightly lower than in 1953 and 1955. Comparing the final densities on the stone with the maxima shown in Fig. 2, it is obvious that much mortality had occurred. Since most of the mortality during the settlement season was caused by scouring of the convex portions of the surface, it is evident that differences in the surfaces chosen from year to year could account for much of the annual variation shown in Table 8.

After I had left Millport, Mr. T. B. Bagenal continued to make counts and photographs on the study areas. Some of his observations on Area 3, together with those which I made in a return visit in July, 1958, are summarized in Table 9. The 1956 settlement was heavy, the 1958 one light. Although no counts were made in 1957, the settlement then must have been at least moderately heavy. This is shown by the densities on Area 1 in July, 1958. The mean densities (no./100cm^2) of the 1958, 1957 and 1956 year groups were 74, 270 and 2, respectively; thus the 1957 settlement after one year was much greater than that of the 1958 group soon after settlement. A moderate settlement must also have occurred in 1952, judging by the density after a year, as shown in Table 9. Thus in the seven seasons from 1952 to 1958, the settlement was light in only one. In the four earlier years when some settlement data are available, only one, 1946, had a light settlement.

TABLE 9. Observations of the population densities (no./cm.^{2}) of *Balanus balanoides* on Boulder 1, from 1952 to 1958.

Year group	Density after the end of the settlement season (in late June or early July)	Density the following year, at age 12-13 months
1952.........	—	1.0
1953.........	19.5	0
1954.........	14.0	2.5
1955.........	24.0	0.5
1956.........	32.0	—
1957.........	—	0.1
1958.........	1.3	—

Hatton (1938) observed the density at the end of the settlement seasons of three years. These are given in Table 8. Little difference in density was found between the years. This is especially interesting, since the settlement season of 1930 started three weeks later than that of 1931, the two seasons being 6 and 9 weeks long, respectively. This suggests a pattern similar to that described by Barnes (1956) at Millport. In neither place were the fluctuations in the length of the settlement season correlated with the settlement density at the end of the season.

From all these studies it appears that while the density observed at the end of the settlement season may occasionally be severely limited by the supply of planktonic larvae at Millport it was usually determined by occurrences on the shore, there being a vast oversupply of larvae. This situation may also apply to other similar areas, but may not apply on open coasts where larvae may be carried away. At Millport, the mortality was greatest on the exposed convex surfaces so that the population density could be regarded as varying directly with the proportion of "concave" surface, including in this category the spaces between older barnacles.

It is sometimes tacitly assumed that variations in the supply of planktonic larvae will be reflected in similar variations in the population of adults on the bottom. Thorson (1950) has suggested that in the life of sedentary marine invertebrates which possess a long planktonic larval phase, the most "critical" period is the planktonic stage. Many authors have shown that adults during the breeding season and developing larvae often have narrower tolerances to environmental factors than during other periods in their life cycle. As evidence for his hypothesis, Thorson showed that greater annual fluctuations in biomass of dredged material occurred in three species of bivalve molluscs with a long planktonic larval life than in two species which had short free-swimming phases. However, since no study was made of such factors as the spatial distribution, mortality, and growth, which also might affect the estimates of the biomass of these bottom invertebrates, this hypothesis can only be regarded as tentative.

MORTALITY FROM CROWDING

FACTORS DETERMINING CROWDING

When the settlement density was high, as it was in the present study, the barnacles after a short period of growth soon began to touch one another. In the following discussion, crowding will refer to the process in which barnacles grow while in contact with each other. The degree of crowding is thus determined by the rate of growth, population density and average size.

The growth of *Balanus balanoides* has been well documented in Western Europe. The results of these studies of most relevance to crowding are as follows. Growth is more rapid when there is more water movement, such as on wave-beaten points or in tidal currents, (Hatton 1938, Moore 1935b, Chipperfield 1948). These authors all believe that this is the result of more food being brought to the population by the increased circulation.

Moore (1935b) noted that barnacles packed closely in a groove grew more slowly than adjacent isolated individuals. He suggested that the food in the water flowing over the surface is shared among more individuals at a higher population density. This suggestion is supported by the observations of Chipperfield (1948), also at the Isle of Man. On the other hand, Kuznetzov & Matveeva (1949) suggest that growth is faster at higher densities. Collections were made at three locations in east Murman (Arctic Ocean) at densities of 5, 3 and 2/cm^2, respectively. Age was determined by growth rings on the wall and opercular plates; the maximum ages were 7 years at the two higher densities and 12 years at the lowest. At the same age, barnacles at the higher densities were larger than those growing at low

TABLE 10. Growth of *Balanus balanoides* in its later years at Millport and St. Malo.

	No. of Barnacles Meas.	SECOND YEAR (10-22 mos.) Ave. Length (mm)	SECOND YEAR Mean Specific gr. rate per day x 100	SECOND YEAR Mean Absolute gr. rate per day x 100	THIRD YEAR (22-34 mos.) Ave. Length (mm)	THIRD YEAR Specific gr. rate	THIRD YEAR Absolute gr. rate	FOURTH YEAR (34-46 mos.) Ave. Length (mm)	FOURTH YEAR Specific gr. rate	FOURTH YEAR Absolute gr. rate
Millport—Mid Cage 1+2.1 ft. above MTL										
1952 set	9	3.1	.147	.456	4.6	.086	.395	—	—	—
Pre-1952 set	4	—	—	—	5.5	.071	.391	6.7	.041	.275
Millport—(Barnes & Powell, 1953)										
Panel 3: +1′ above MTL	31	17.0	.088	1.496	—	—	—	—	—	—
Panel 7: −6′ below MTL	26	19.5	.078	1.520	—	—	—	—	—	—
St. Malo—(Hatton, 1938)										
Decolle Ouest, Level II, HWN	50	4.4	.154	.678	6.0	.025	.150	—	—	—
Decolle Est, Level II, HWN	50	4.6	.136	.626	6.2	.039	.241	—	—	—
Decolle Est, Level III, MTL	50	4.2	.135	.567	5.3	.015	.079	—	—	—
Cite, Level II, HWN	50	4.5	.102	.459	5.7	.033	.188	—	—	—
Cite, Level III, MTL	50	4.1	.080	.328	—	—	—	—	—	—
Cite, Level IV, LWN	50	3.9	.089	.347	—	—	—	—	—	—

NOTE: All measurements were made in mid-February, except for those in Barnes & Powell (1953) where measurements were made in January and October, at ages of 9 and 18 months. The Pre-1952 set in the present study may have included some individuals older than three and four years.

density. Whether the larger sizes found at higher densities were the result of the differences in density or in location is difficult to decide.

Moore (1934, 1935a), Hatton (1938) and Barnes & Powell (1953) recorded seasonal variations in growth rate. Growth was fastest in spring and early summer, decreasing later to a very slow rate in winter. Most of the growth occurred during the first and second seasons; after the age of 18 months the growth was very slow.

The only data published on the growth of *Balanus balanoides* which were followed after the second year are those of Hatton (1938). In the present study, some barnacles were followed through their fourth year on Area 1. Photographs had been taken of the Middle Cage 1 in February in three successive years, 1953-1955. A few individuals were in a clear position for measurement in all three photographs. Only those individuals which could be followed through the entire two year period were included. For the 1952 settlement, nine individuals were measured, at the ages of 10, 22 and 34 months; for the pre-1952 settlements, four individuals were measured at the ages of 22, 34 and 46 months (although some of these may have been even older). The mean specific growth rate (increase in length per unit length per day x 100) and the absolute growth rate (increase in length per day x 100) were calculated for each group for each year. Similar rates were calculated from the data of Hatton over the same periods, and from those of Barnes & Powell (1953) for the period between the ages of 9 and 18 months. These data are given in Table 10.

Although only a few barnacles were measured in this study, the values indicate that growth continues, at a declining rate, during the fourth year; it was greater at Millport at this age than at St. Malo during the third year. One of the barnacles of the pre-1952 group reached a length of 10 mm after at least four years. In comparing the growth rates at different levels in the intertidal zone, these authors were in agreement that the rate is faster lower on the shore during the first year. At the end of the second season Hatton (1938) found that the differences in average size between levels were reduced; the smaller barnacles at higher levels appeared to have grown faster and caught up with those at low levels. Barnes & Powell (1953) found that the mean specific growth rate was slightly greater at high levels during the second season; the absolute growth rate was about the same, however, so that the barnacles on their lowest panel were still larger than those on the highest panel at the end of the second year. Moore (1935b) found that in the second year the larger barnacles occurred at high levels in some localities, low in others. Evidence will be given later that the results of Moore and Hatton may be explained by predation by *Thais lapillus,* which selects the larger barnacles at lower levels. This selection inevitably reduces the average size of the barnacles at these levels.

From all these data it appears that for barnacles of the same size and age, growth is closely associated with food supply, being fastest in the spring when phytoplankton is most abundant and, during the first year, faster at lower levels on the shore where the time of immersion for feeding is longer. After the first year, the growth rate decreases with size and age, irrespective of location.

Crowding varies directly with population density, when barnacles of the same size and rate of growth are compared. Fig. 7 shows such a comparison of stones 1 and 2, which were located close together at the same shore level. The intense crowding at high population density on Stone 2 is evident.

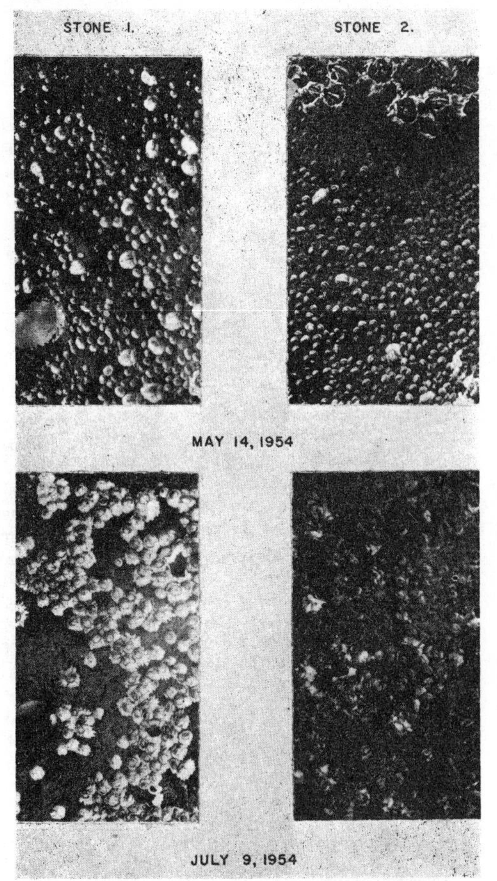

FIG. 7. Photographs of Stones 1 and 2 after most of the settlement of *Balanus balanoides* had occurred and later after some growth had taken place. Each photograph represents an area with dimensions of about 4.5 x 2.5 centimeters.

In populations of the same density, more contacts between barnacles would occur in those composed of large barnacles than in those of smaller ones. However, the increased number of contacts would not necessarily imply greater crowding, since the growth rate of the larger barnacles might be slower, either because of a lower growth rate, or because of a less favorable location.

PHYSICAL EFFECTS OF CROWDING

Of the many possible consequences of crowding, three will be discussed: (1) changes in growth form, (2) mortality, and (3) impedance of cirral activity. The most obvious results of crowding were the changes in body form of the barnacles. At low levels, or at higher ones where water movement was very rapid, barnacles grew very rapidly; at moderate population densities the barnacles had no room to expand laterally, so grew upward, developing into cylindrical or trumpet shapes. This development was not usually uniform over the surface, local groups developing into "hummocks." The barnacles were attached only at the relatively small base and so were often detached in heavy wave action.

This growth form was ascribed to crowding by Darwin (1854) in England and by Pilsbry (1916) in the United States. Pilsbry states that in *Balanus balanoides* this form "is in no sense a race, as it is commonly found in the same group with patelliform individuals." Darwin (1854), Pilsbry (1916) and Moore (1934) described occasional elongate individuals which were found isolated; no explanation of this was given. This phenomenon has been observed in this species at many other places: in Germany by Trusheim (1932) and Schafer (1948), in France by Hatton (1938), at Millport by Barnes & Powell (1950) and on the arctic coast of Russia by Sokolova (1951). I observed it at Woods Hole, Massachusetts in 1955. It has also been described for many other species by Darwin (1854), Pilsbry (1916), Rice (1935), and Barnes & Powell (1950).

The present observations of the process agree with those of Barnes & Powell (1950). In their study, the settlement densities on the shore were over 80/sq cm, and the barnacles reached a maximum height of 1.3 cm. Hummocking was observed in the present study at densities of 44/sq cm and 16/sq cm; maximum heights at these densities were 2.1 and 1.3 cm, respectively. On Area 3, hummocking occurred only at densities greater than 16/sq. cm. Its widespread occurrence, together with the fact that it was observed in each year on the present study area, indicate that it is a common, rather than unusual, phenomenon.

In areas where growth was less rapid, as at the higher shore levels, a variety of body forms resulted from the less extreme crowding. Some individuals retained the normal "patelliform" shape, while others developed cylindrical or low trumpet-shapes and so fitted into the spaces between the "normal" individuals. This elongated condition is shown on stone 2 in Fig. 7, where it is compared with an area of lower density on stone 1 where most of the barnacles were shaped normally.

Besides the effects on growth form, some observations were made on the killing of barnacles by what appeared to be the direct effects of crowding. A series of photographs taken at short intervals of a dense population of barnacles just after the season of settlement is shown in Fig. 8. During the period shown, 10 of the 24 deaths which occurred were probably a direct result of crowding. Some of these are marked on the photographs; some were flattened laterally, (Nos. 1, 2, and 4) while others were undercut and tilted by the growth of their neighbors (Nos. 3 and 6). Once tilted they often dried out, probably because the basal membrane was exposed. A few small individuals situated between larger ones became almost buried by the growth of the neighboring barnacles.

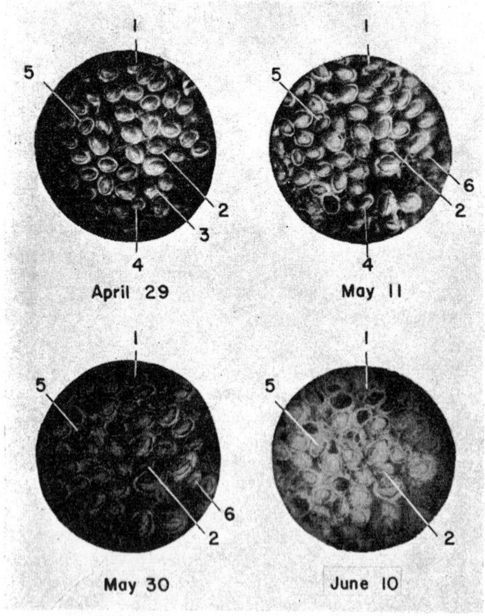

FIG. 8. Photomicrographs of the same area on stone 7 taken at intervals following a dense settlement of *Balanus balanoides*. Some individual barnacles which were being crowded are indicated. Each photograph represents an area of about one square centimeter.

At the higher levels where several year-classes occurred together another effect of crowding was observed. Young barnacles attached themselves to the upper parts of the shells of older ones, and then grew over the opening of the older barnacle, partly occluding it. The older individual usually died, presumably because it was unable to feed. All of these observations were made while following individual barnacles in periodic censuses.

In the present study, the term crowding has been restricted to situations in which crushing, displacement, smothering or distortion of growth form have resulted from the growth of contiguous barnacles. Other authors have used the term in a less restricted sense to refer simply to barnacles existing at a high population density or forming a complete coverage of the rock. It is probable that these other usuages are justified, since certain effects other than those described in this paper may occur in dense populations. One such effect was observed on a stone bearing a dense stand of young barnacles. The stone had been placed in a dish of sea water to observe the feeding movements of the cirri. Three barnacles were situated in a row, the slightly smaller middle one oriented in the opposite direction to the ones on either side so that the direction of beat of its cirri was in the opposite direction to that of its two neighbors. Its beating often coincided with that of one or both of the others, and the continual opposition of its outer cirri with those of its neighbors on either side appeared to inhibit the activity of the middle individual, since its beating was much more irregular and less vigorous than that of either of the other two.

Hatton (1938) in his discussion of the interaction between *Balanus balanoides* individuals, made the following observations. When two young barnacles settled next to each other, one invariably died; the victor was thought to be either the faster grower or the "stronger" of the two. He especially emphasized the effect of the growth of older barnacles in displacing young ones. Only if a young barnacle came into contact with an older one after a sufficient period of growth could it withstand the undercutting of the latter. In the present study, displacement or crushing of a young barnacle by an older one was often observed, especially during the settlement season. However, probably because of the faster growth rate and greater number of the younger barnacles, crowding between them was much more frequently observed than that involving older ones. Hatton also observed a single case of a young barnacle growing over the opercular opening of a three-year-old. The older one died and the younger one survived another year. He also noted that *Balanus* could undercut and remove fronds of the alga *Fucus vesiculosus*. A settlement of *Balanus* removed all the *Fucus* from an area in one growing season. Between individuals of *Chthamalus stellatus* (Poli), he never observed displacement. He suggests that the growth of *Chthamalus* was slower and more continuous than *Balanus* so that the young of *Chthamalus* had time to establish themselves before they could be displaced by the older ones. No observations of the interaction between *Balanus* and *Chthamalus* were given by Hatton. In the present study it was observed that any *Chthamalus* in contact with *Balanus* below MHWN were crushed, smothered, displaced or sometimes lifted on the shell of the *Balanus*. Weiss (1948a) noted in Florida that barnacles on continuously submerged panels were smothered by growths of colonial tunicates and encrusting bryozoa.

MORTALITY ASSOCIATED WITH CROWDING

No method for integrating growth and density into a "crowding index" was devised in the present study. Therefore, to illustrate the effects of crowding on mortality in the following discussion, the observed mortality rate has been considered in relation to variations in growth rate and in population density separately.

At the same population density and average size, faster growth might be expected to lead to more severe crowding, with consequent greater mortality. Seasonal variations in growth would be expected to be reflected in variations in crowding and mortality.

During the first year of life the survival curves often showed a decrease in the relative mortality rate in the winter (Figs. 9, 10, 16, 17), at the time of decrease in the growth rate.

The curves in Figs. 12, 16, 17 and 18 show that barnacles older than two years, if protected from predation by a high shore situation or by experimental means, had a very low mortality rate. The growth of these older barnacles was relatively slow, although they were mingled with younger ones which were still growing rapidly. Evidently the crowding among the younger barnacles had little effect on the older ones, except for the occasional smothering as described earlier.

In Figs. 9 and 10 are shown the survival curves of populations of *Balanus balanoides* at St. Malo, France, from Hatton (1938), as compared to those at Millport. The greater mortality at Millport at MTL, shown in Fig. 10, was probably due to the greater densities of settlement. Near LWN, as shown in Fig. 9, the densities at Millport (Area 3) and at St. Malo were similar, yet the mortality was greater at Millport. Extreme crowding leading to hummock formation had been observed at Millport on boulder 1 in 1953 and on boulder 5a in 1954, followed by complete destruction of the populations. If the

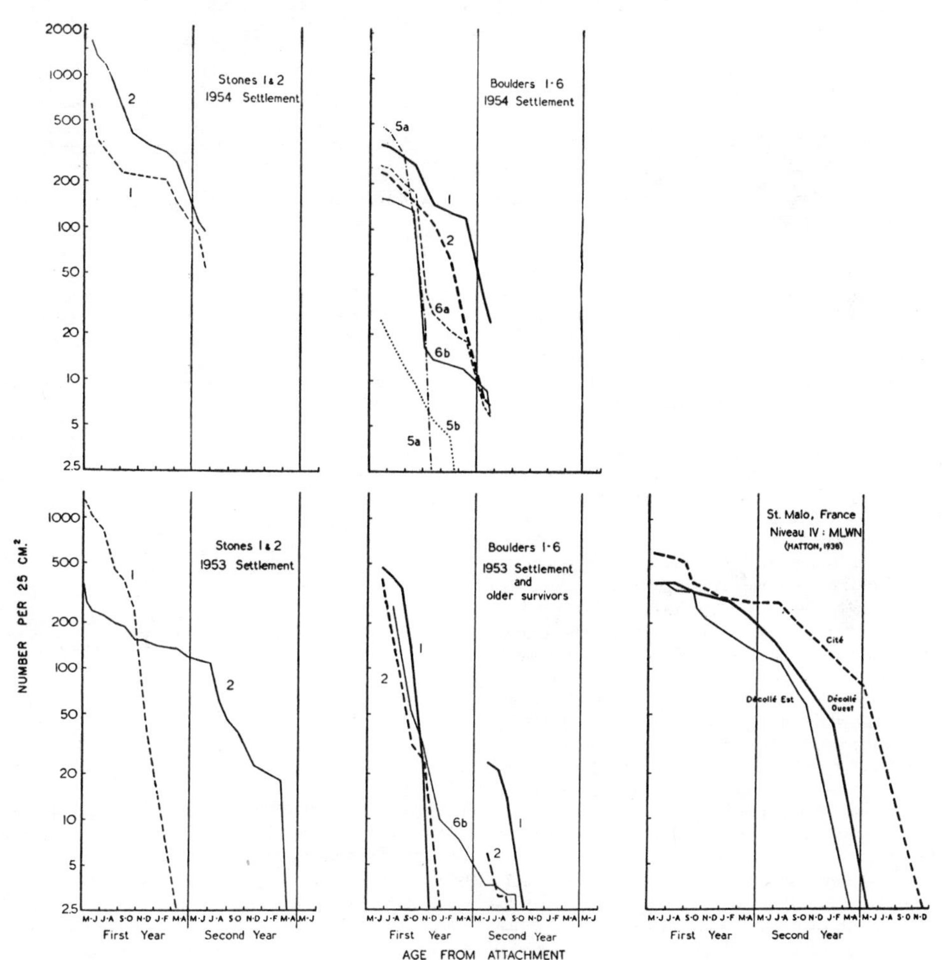

FIG. 9. Survival curves for *Balanus balanoides* below MTL at Millport and in Brittany, the latter replotted from Hatton (1938).

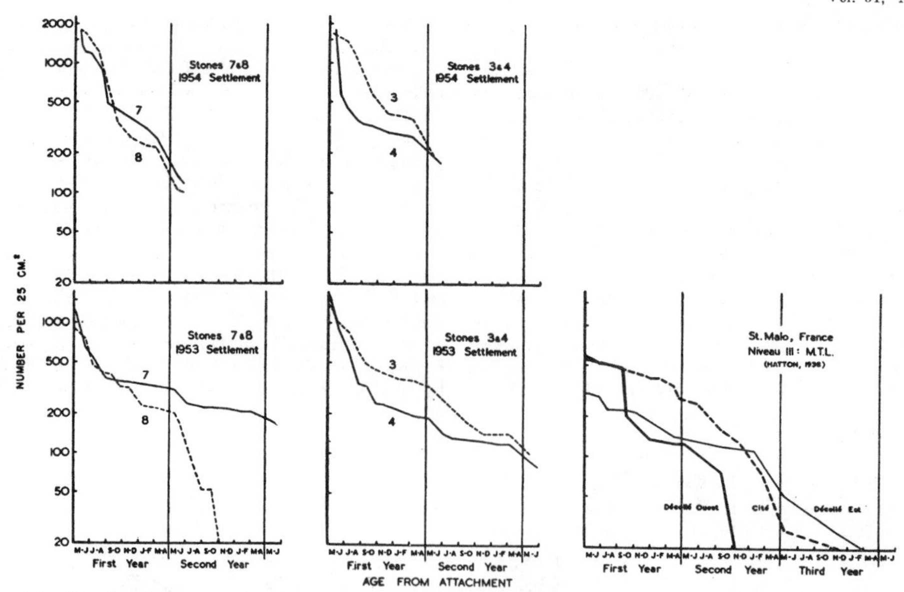

FIG. 10. Survival curves for *Balanus balanoides* just above MTL at Millport and MTL in Brittany, the latter replotted from Hatton (1938).

rate of growth were slower at St. Malo, it would account for the much lower mortalities found by Hatton at similar population densities. Growth rates at Millport during the first growing season were calculated from measurements made of isolated individuals in photographs of the same area near MTL, taken in June and November, 1954. To obtain the growth rate from Hatton's data, the sizes late in August, 1930, were taken from his curves of level III, MTL. This was the date when the maximum size was reached at St. Malo; after this the curves decreased, probably because of selective predation. All the growth rates were calculated as beginning at the same average size as that found in June, 1954, at Millport, 1.9 mm. The time interval from Hatton's curves was taken between the date when the average size was 1.9 mm and the date of maximum size in late August. These readings and the calculated mean specific growth rates are given in Table 11. This method, using only the period of apparent maximum growth at St. Malo while taking a longer span for the Millport measurements, probably tends to underestimate the Millport rates as compared to St. Malo. However, the growth rate was still greater at Millport. At the same population density and shore level, crowding would be expected to be more intense at Millport.

The foregoing analysis has dealt with the mortality during the growing season. However, it must be kept in mind that while some barnacles are killed during the growth period, others may die during the winter as a result of being crowded earlier. This delayed mortality is illustrated by the mass destruction in storms in late autumn and winter of populations which had developed the "hummocks" described earlier. Even at levels above MTL, where hummocks rarely formed, crowding produced some individuals of unstable form, squeezed between normally-shaped ones. An analysis of the mortality during the winter on Area 1 is given in Table 12. For those areas at the same level which had higher population densities at the end of settlement, and had thus experienced

TABLE 11. Comparison of the growth rate of *Balanus balanoides* at Millport and St. Malo during the first growing season. All barnacles were growing without contact on intertidal rocks.

Millport, just above and below MTL.

JUNE 11, 1954		NOV. 3, 1954		Mean specific growth rate per day x 100	Absolute growth rate per day x 100
Number measured	Ave. length in mm.	Number measured	Ave. length in mm.		
39	1.9	27	4.8	0.59	1.93

St. Malo, data from Hatton (1938). Level III, just below MTL. About 50 individuals in each measurement.

Location	Date when average length was 1.9 mm.	GREATEST AVERAGE SIZE REACHED IN THE FIRST SIX MOS.		Mean specific growth rate per day x 100	Absolute growth rate per day x 100
		Date	Length		
Decolle Ouest	April 15	Aug. 25	3.4	0.44	1.136
Decolle Est	May 10	Aug. 25	3.0	0.42	1.027
Cite	April 18	Aug. 20	3.5	0.48	1.240

TABLE 12. Mortality during the winter months, November through February, in relation to the population density a month after the end of settlement. Only barnacles on Area 1 in cages, protected from predation, were considered. (1954 settlement in their first winter.)

Level of Area 1	Height above MTL in feet	Cage No.	Population Density July 5, 1954	% Mortality Nov., 1954 through Feb., 1955
Upper.....	2.6	1	70	69
Upper.....	2.6	2	27	29
Middle....	2.1	1	57	55
Middle....	2.1	2	23	27
Middle....	2.1	3	24*	27
Middle....	2.1	4	15*	20
Low......	1.5	1	21	52

* These densities indicate values for July extrapolated back from August 1, when the first counts on these two areas were made.

greater crowding, the mortality during the winter was also higher.

The effect of variations in population density on mortality during the first growing season was investigated by calculating the percentage mortalities for all the squares on Area 1 and for the experimental stones for the six months after the end of the settlement seasons of 1953 and 1954. The relation between these mortalities and the population densities at the end of settlement is shown in Fig. 11. The effect of differences in growth rate is minimized if points representing areas at the same shore level are compared. Except for a few of the areas at the lowest levels, mortality varied directly with density. Above 50 per cm^2 there was little increase in mortality.

To study the effect of population density on mortality later in life, those populations of Area 1 protected from predation by cages were used. For each year class the percentage mortality was calculated over the same six-month periods as were used in the calculations for Fig. 11. No consistent increase in mortality with increasing density was found in the older groups. Although growth continued in the second season, the densities were low, so that the effects of crowding were reduced within this year group. In the third and fourth summer seasons both the densities and rates of growth were low, as was the mortality.

Another study has been made of the effect of crowding on mortality by Deevey (1947), using the data of Hatton (1938) from St. Malo. He calculated a "crowding coefficient," which was the number of binary contacts per sq cm, a function of population density and average size. When the coefficients were compared to the average expectation of life for the various populations no consistent correlation was found. When the average expectation of life was calculated starting at settlement, populations with from 7 to 76 binary contacts per sq cm showed about the same expectation of life; only the population with a coefficient of about 1.0 had a better expectation.

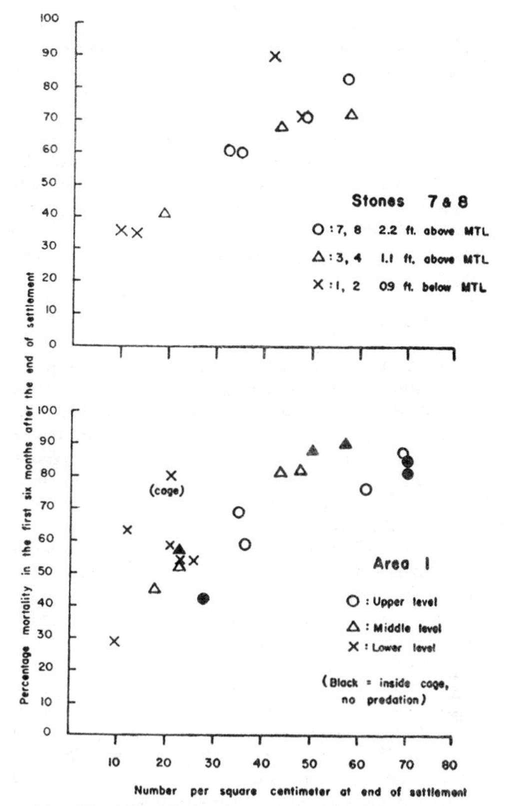

FIG. 11. The effect of population density of *Balanus balanoides* at the end of settlement on mortality during the first growing season. Both the 1953 and 1954 settlements were included.

From the age of 6 months onward, barnacles at high shore levels had high expectation and low coefficients; at other levels there was no consistent trend. These findings agree with those of the present study, that the mortality of older barnacles was not correlated with crowding.

An analysis of Hatton's data was made for comparison with that in Fig. 11. The percentage mortality for the first six months after settlement (May 15-November 15) was not correlated with the density at the end of settlement. This agrees with Deevey's calculation when the first six months were included. The reasons for the differences between these findings and those of the present study shown in Fig. 11 are probably that at St. Malo the densities were low (4 to 24 per cm^2) and the growth rate slower, both tending to decrease crowding.

One possible defect in Deevey's "crowding coefficient" is that it ignores the fact that barnacles of similar size may grow at different rates at different shore levels. Thus with the same "crowding coefficient," physical crowding might be expected to be more intense at low shore levels; hummocks are seldom encountered above MTL, even though densities may be great.

In an attempt to estimate the proportion of the total mortality which might have been due to crowding, a series of photographs of the area enclosed by the Middle Cage 1 were studied. This area was protected from predation for over 2½ years; photographs taken at intervals during this period are shown in Figure 19. When the area was first mapped in November, 1952, the younger age group, aged 6 to 7 months, was distinguished from the barnacles 18 months of age and older, as described in the introduction.

Of the 62 individuals in the younger group, 48 died within a year; 9 of these were attached to older individuals which died and fell off, carrying the younger ones with them. Eight died in situations in which no crowding was apparent. The remaining 31 deaths, 65% of the mortality, may possibly have resulted from crowding. Some of these occurred in barnacles situated between others which grew appreciably during the year. Where young barnacles were attached high on the side of older ones they often grew rapidly. When this led to their overhanging other young individuals, the lower ones often died, possibly from impedance of their cirral activity or reduction in the amount of water circulating over them; they might even have been crushed by the upper one.

Of the 63 barnacles in the older groups in this cage, 21 died in two years. Nine, or 43%, resulted from crowding. Four were smothered by the growth of young ones attached near the top; five were of small size and may have been crushed by slight growth of their neighbors. Two were carried away when the shell of a large dead one fell off. Ten died without apparently having been crowded. Some of these were very large and may have died of old age, probably a rare occurrence in natural populations. However, at the upper shore limit of distribution of this species, where predation by *Thais lapillus* is absent, very large barnacles occur commonly. In this situation, deaths from old age may be a regular occurrence; the protection given by the cages on Area 1 seems to have produced a similar situation at a lower level.

When a small population is examined in detail in this manner, it is impossible to assign a definite cause of death in many cases. However, the good correlation of higher mortality with faster growth and higher population density indicates that crowding was one of the important influences in the populations studied at Millport. As growth slowed with increasing size and age, the importance of crowding decreased as that of predation increased.

MORTALITY AT HIGH SHORE LEVELS

THE UPPER LIMIT OF *Balanus balanoides* DISTRIBUTION.

Near HWN the upper limit of distribution of *Balanus balanoides* merged into the bottom of the narrow *Chthamalus stellatus* zone. Little crowding occurred at this level since the population density was low and growth slow. Three squares were mapped above MHWN on the rock face above Area 1. A cage and netting "cover" were attached at first, but since no *Thais* were observed at this level and since the survival was the same after a year on all the squares, the cage and cover were removed. The results of these counts are given in Fig. 12, all three squares being combined. This will be referred to as the Top level, Area 1.

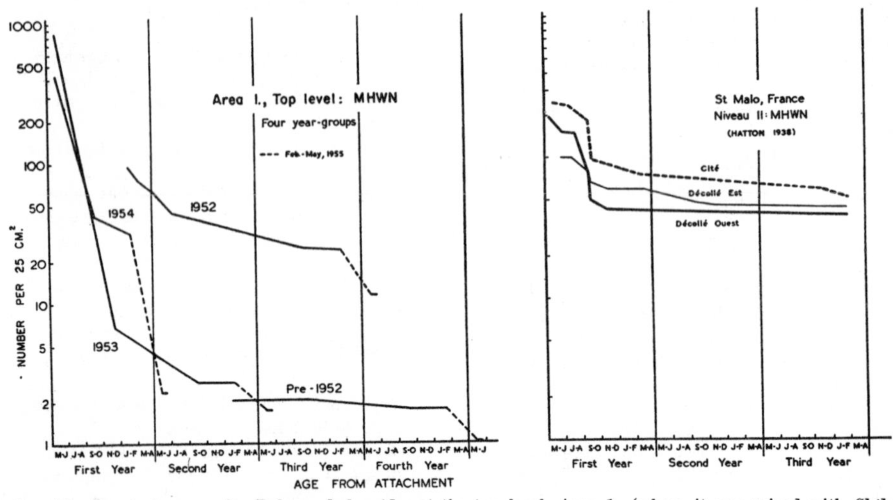

FIG. 12. Survival curves for *Balanus balanoides* at the top level, Area 1, (where it was mixed with *Chthamalus stellatus*); curves from a similar level in Brittany were replotted from Hatton (1938).

In the first mapping in December, 1952, the 1952 year class was distinguished from the previous settlements by its smaller size and the smoother appearance of the shell surface. As explained earlier this method is probably valid for distinguishing the youngest age class from older classes. Because of the slow growth at this level the average size difference between the first and second year individuals was less than that of the populations low on the shore. For the small number of individuals where uncertainty existed, the numbers were divided equally between the 1952 and the older group.

The data from the equivalent level at St. Malo from Hatton's (1938) study are also shown in Fig. 12. At both places the greatest mortality occurred in the first six months after attachment. The survival was good after this except for the spring of 1955 at Millport; this heavy mortality will be discussed later.

The curves in Fig. 12 were arranged so that the survival of the different year classes could be compared at the same age. Only the 1952 year class survived well after the first year of life, and except for a short time after each settlement season, outnumbered the other year classes for the whole study period. The almost complete destruction of the 1953 class soon after settlement eliminated that group; for the 1954 class the early mortality was less but the mortality in 1955 at eleven months of age greatly reduced the density. The 1952 group probably had an unusually favorable period in early life. It and the pre-1952 and 1953 groups were better able to withstand the heavy mortality in the spring of 1955, apparently by virtue of their greater ages.

At this level on the shore, where some individuals survived for at least four years, one year class outnumbered the rest. The occurrence of a "dominant year class" is common in fish populations. Most workers in fisheries biology who have studied this problem have concluded that it had resulted from variations in the survival of the planktonic larvae (cf. Hjort 1926, Sette 1943, Carruthers *et al.* 1951). In the present barnacle populations the recruitment from the plankton was remarkably uniform from year to year. The "dominance" of the 1952 year class shown in Fig. 12 was the result of variations in mortality within the first year after settlement. At lower levels where the length of life was rarely more than two years, there was obviously no opportunity for dominant year classes to exist. Where areas had been protected for a long time from predation, members of the older year classes occupied most of the space (cf. Fig. 19), when normally they would probably have been eliminated by predation.

AN UNUSUAL OCCURRENCE OF HIGH MORTALITY IN THE UPPER SHORE LEVELS.

Sometime between the censuses of mid-March and early June, 1955, great mortality occurred at and above the upper level of Area 1, as shown in Figs. 12 and 16. This mortality might have been expected among the barnacles which were still growing and crowding each other but its occurrence among older barnacles at the top level and inside the cages at the upper level was very unusual. The mortality rate was much lower during the period before the next census in late June.

Predation was not involved since the caged populations also suffered this mortality; no instances were observed of older barnacles being smothered by younger ones. The mortality during this period of the *Chthamalus stellatus* on the top level of Area 1 is compared to that of *Balanus balanoides* in Table 13. All ages of *Balanus* suffered heavy mortality, while only the youngest *Chthamalus* died in great numbers. A similar but lesser mortality occurred during this period at the middle level but almost none at the lower level of Area 1 (Figs. 17 and 18). This progressive lessening of mortality at lower levels suggested that the cause might be related to some deleterious effect of exposure to air.

TABLE 13. Mortality of *Balanus balanoides* and *Chthamalus stellatus* from February 13 to May 28, 1955 on Area 1, Top level (MHWN).

	Age Group			
	Older than 1952	1952	1953	1954
Number present, Feb.: Balanus	5	70	8	93
Chthamalus	173	42	143	159
% Mortality by May: Balanus	40	53	37	92
Chthamalus	0	0	5	62

A tide gauge had been operating for a year previously at Keppel Pier, 1/2 mile north of the study area. With the aid of the tidal records, an analysis of the degree of exposure to air of the Upper level on Area 1 was made.

Measurements on Area 1 indicated that the height of the upper level was 8.6 ft above Ordinance Datum, and that the vertical amplitude of the waves was 2.5 times that recorded by the tide gauge. The ability of the gauge to record waves was very important, since it was necessary to determine how long the level concerned remained dry. The gauge recorded the waves caused by the frequent steamer passages which would wet levels above high tide even on very calm days. Each such passage resulted in a series of waves lasting less than one minute.

The records from March 1954 to May 1955 were examined and all high tides which did not reach 8.6 ft were noted. The only periods of exposure longer than three high tides during this time occurred in March and April, 1955, when such periods of 4 and 6 tides respectively, occurred. The weather in April was warm and calm, whereas during the period of exposure in March it was cooler and rainy.

The period in April was investigated in detail; between April 16 at 0540 and April 18 at 2110 the level was untouched by any wind-driven continuous waves. Between these times waves caused by steamers passing close to the shore wet the level only six times, occurring in three of the four high tides.

This exposure of about 60 hours with only six short wettings appears to have been fatal to many of the barnacles. Previous exposures of shorter duration did not cause such high mortality. It would probably be safe to say that no such long exposure had occurred in the past 2 1/2 years, since no similar mortality of older barnacles had occurred on the top or the upper level of Area 1 (Figs. 12 and 16). In this locality the combination of calm warm weather and the very small amplitude of neap tides is evidently fairly rare.

At the top level of Area 1, the effects of these rare periods are probably the main causes of death for those individuals which survive the period immediately after settlement. The survival curves in Fig. 12 indicate that very little mortality normally occurs after the first year of life so that accidents must be very rare; death from old age may occur in a few cases.

Since the unusual mortality of 1955 occurred within and just below the zone of *Chthamalus,* it may have provided space for a downward extension of *Chthamalus* in its autumn settlement. The settlement of *Balanus* in late April filled the existing spaces but the delayed detachment of dead individuals during the summer would probably provide more bare areas than in years when no such mortality occurred.

Hatton (1938) dealt at some length with the settlement, growth and mortality of *Balanus balanoides* at high shore levels. The density at the end of settlement and the growth rate at his level II, HWN, were lower on surfaces which faced south than on those which faced other directions. If shade was provided, the density was the same at all orientations. Since all the surfaces dried out quickly in the shade, Hatton considered that the observed differences were a result of the effects of direct sunlight.

Young barnacles were observed to settle during spring tide periods up to 1 1/2 meters above the adults; the young usually all died during the next period of neap tides. Hatton placed rocks bearing newly attached barnacles above the tidal range; all were dead after a week of continuous exposure to air.

Hatton transplanted rocks bearing barnacles from his level II at HWN to level I at HWS, where no adults occurred. In an observation a week later it was found that the very young barnacles were all dead while those aged two months and any older adults lived four to six months longer. In contrast, young *Chthamalus stellatus* moved from levels II to I suffered only 26-40% mortality after nine days; adult *Chthamalus* occurred normally at level I. Settlement densities and growth rates of *Chthamalus* at these levels were similar on surfaces facing south or north.

Hatton performed two other interesting experiments. He fixed a basin of sea water at HWS with a tiny hole so that the water dripped slowly out. Young *Balanus* survived and grew for three months only where the rock was wet, even though the salinity of the water varied. He also transplanted rocks bearing barnacles two years of age from levels IV to II, LWN to HWN. Barnacles of this age left at LWN lived only a few more months, but the ones moved to HWN lived there for two more years, resembling those which had lived at this level all their lives. Southward (1958) found that *Chthamalus stellatus* could withstand higher sea water temperatures than *Balanus balanoides* in the laboratory. Both species endured water temperatures greatly in excess of those which occur in their natural environment. The relative tolerances of the two species confirm the present findings and those of Hatton that *Chthamalus* lives higher on the shore because of its ability to withstand greater extremes of heat or desiccation. These factors, probably in combination, determine the upper limit of *Balanus* distribution, acting more severely on the barnacles in their first year.

PREDATION BY *THAIS LAPILLUS* ON *BALANUS BALANOIDES*

This interaction was studied by determining the seasonal variations in population density, movement and feeding rates of *Thais,* while at the same time recording the mortality of *Balanus* populations either protected from *Thais* by cages or open to predation.

POPULATION DENSITY OF *Thais*

At each visit to the study area, counts were made of the numbers of *Thais* on certain areas, such as the boulders of Area 3. Also, a map of Area 1 was made showing the position of rock irregularities and cages; the positions of the whelks were plotted on copies of this map at each visit. These records were used to calculate the population densities at different levels on Area 1, corresponding to those levels where the barnacles were being studied. In 1953 and 1955 some of the *Thais* were numbered with India ink on a small part of the shell which had been filed clean. When dry, the number was covered with a mixture of "Distrene," a transparent plastic, dissolved in xylol, as described by Quayle (1952).

It was assumed that those whelks which occurred on areas covered with barnacles were actively feeding. By slowly tilting one of these whelks away from the surface, it could usually be observed to be feeding on a barnacle, whose opercular valves would often be gaping open. Of about 400 whelks thus observed on Area 3 in May, 1955, two-thirds were found to be actively feeding. The remaining third were clinging to the rock or to a barnacle, but not feeding; it is probable that these whelks were exposed by the tide while moving from one barnacle to the other. In crevices, on the under sides of boulders, or among algae, whelks were observed which were not feeding.

The average density of *Thais* at various levels on Area 1 is given in Table 14; each level was divided into seaward and landward halves. No *Thais* were

TABLE 14. Distribution of *Thais lapillus* on Area 1, at different levels and seasons. For the "coefficient of dispersion" values, 1.0 indicates random occurrence; those in brackets did not deviate significantly from random. All the others show aggregation.

Side and *Level*	AVERAGE NO. PRESENT PER OBSERVATION ON 0.1 M²				FISHER'S "COEFFICIENT OF DISPERSION" ON EACH AREA OVER THE STATED PERIOD		
	All seasons	Autumn 1953 & 1954	Winter 1954 & 1955	Spring 1954 & 1955	Autumn 1953 & 1954	Winter 1954 & 1955	Spring 1954 & 1955
Seaward Side							
High level	1.9	4.2	0.3	0.7	2.7	(1.5)	(1.3)
Upper level	4.7	7.8	0.8	2.7	10.5	(1.5)	2.9
Middle level	4.8	7.5	1.0	4.4	6.4	(1.4)	2.4
Lower level	8.3	12.5	3.7	6.2	4.7	2.4	2.9
Landward Side							
High level	0.8	0.7	0.6	0.4	1.6	(0.9)	(1.2)
Upper level	3.8	4.8	0.6	2.2	(1.4)	2.0	4.6
Middle level	3.4	3.0	0.9	3.7	3.5	2.0	3.0
Lower level	3.2	6.5	1.7	2.7	3.2	3.4	(1.3)
No. of observations in each season	141	34	46	61	34	46	61

ever seen at the "top" level of Area 1. The whelks were more abundant on the seaward side at each level. The density at the highest level was markedly less than that lower down, but there was little difference between the other levels, except that the seaward side of the lowest level supported a much higher density than the others. It seems that the inhibiting effect of increased air exposure began to be important only at the high level of Area 1, just below MHWN.

In Table 15 the population densities of *Thais* are compared between the low level of Area 1 and three of the boulders of Area 3. The monthly means were used, since the number of observations in each month varied; the period between January and June, 1954 was excluded from the calculations since Area 3 was bare of barnacles at that time, while Area 1 had barnacles. Boulder 6 had significantly fewer *Thais* than the other two boulders, but no other differences of this magnitude were found. The *Thais* population on boulder 6 showed the greatest fluctuations (Fig. 13) and the lowest density. As shown in Fig. 9, this boulder also had the fewest barnacles.

TABLE 15. Population densities of *Thais lapillus*, Areas 1 and 3. The values are averages of the monthly means (see Fig. 13 and 14), excluding the months of January to June, 1954.

Area	Ht., feet above or below MTL	No. of monthly means used	Mean no./m²
Area 1, Low level only	+1.5	14	54.5
Area 3, Boulder 1	−0.9	14	76.5
Area 3, Boulder 2	−0.9	14	88.5
Area 3, Boulder 6	−1.9	13	56.1

Using the Mann-Whitney U test, the monthly means were compared between pairs of each of the four areas. The only significant differences (p=.05) found were those between boulders 1 and 6 and between boulders 2 and 6.

Several other authors have made observations on the distribution and abundance of *Thais*. Moore (1938b) on the island of Skye, counted the maximum number, 200/m², at a level between MTL and HWN; the density decreased above and below. *Thais* did not extend upward as high as did *Balanus;* whelks of less than 0.5 cm height occurred only below MTL. Barnes & Powell (1950) counted the whelks during the summer of 1949 at Millport at three levels, MTL, LWN and below LWN. More *Thais,* to a maximum number of 150/m², occurred at MTL than at the lower two levels. The lowest area never had more than 12/m². As the barnacles were stripped off by wave action in the autumn, the numbers of *Thais* also decreased to densities of 3—25/m² in November. Southward (1953), at two localities between the neap tide levels on the Isle of Man, gave average densities of 1.0 and 0.5 per m², with maxima of 31 and 12 per m². These *Thais* were all feeding on barnacles. From all these observations, it appears that on shores covered with barnacles, *Thais* is most abundant at mid shore levels. The upper limit is probably a result of decreased tolerance to increased exposure to air, while the decrease toward the lower levels may be a reflection of the decrease in the population density of barnacles, as on Boulder 6, Area 3.

Whelks were sometimes observed in dense groups rather than scattered over the area at any one level. This was sometimes due to the presence of a concentration of their prey, as when on two occasions cages temporarily were removed (Fig. 16), so exposing barnacles which had been protected for some time. In both cases, these barnacles were much larger than those on the surrounding area; since *Thais* selects larger barnacles if given a choice, this was obviously an area on which they might be expected to feed. In both instances, the whelks congregated on the

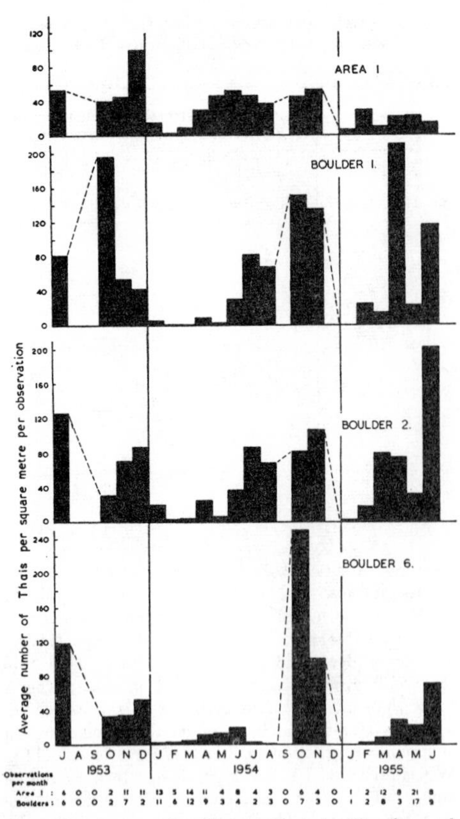

FIG. 13. Average densities of *Thais lapillus* for each month on Areas 1 and 3. When no observations were made during a month a dashed line crosses the space.

small area where the large barnacles occurred so as to almost completely cover it.

During any one season the numbers of whelks which were observed to be feeding on the barnacles varied greatly from day to day. This irregularity in occurrence was analyzed by the same method used to study spatial distribution. Fisher's "coefficient of dispersion" (Blackman 1942, Holme 1949), calculated by dividing the variance by the mean, was used. The formula used was:

$$\frac{\sum (x-\bar{x})^2}{\bar{x}(n-1)}$$

The limits of significance were calculated using the formula:

$$1 \pm 2\sqrt{\frac{2n}{(n-1)^2}}$$

The results of this analysis are given in Table 14. A coefficient of 1.0 indicates a random distribution; all the coefficients which were significantly different from a random distribution were greater than 1.0, indicating aggregation.

This variation is probably a result of short term changes in weather; during gales or periods of cold weather, whelks tended to congregate in crevices, on the lee side of boulders, or in algal growths. This variation in occurrence would tend to make estimates of density less precise than if the whelks occurred regularly or at random.

In Figs. 13 and 14 are shown the monthly mean population densities of *Thais* on Areas 1 and 3. The lowest densities were found from January to March, during which period the temperature of the water was at its minimum (Barnes, 1955). An exception to this occurred in February, 1955 at the lower levels of Area 1.

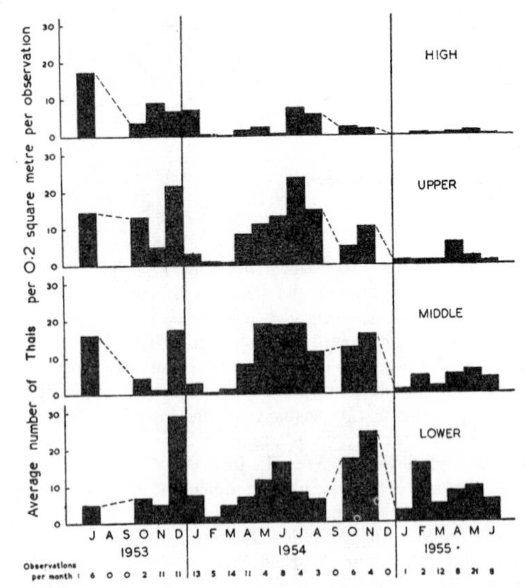

FIG. 14. Average densities of *Thais lapillus* for each month on each level of Area 1. When no observations were made during a month a dashed line crosses the space.

The few individuals which were observed feeding in winter were mostly small, with thin lips to the orifice of their shells. The higher proportion of young individuals among those feeding in the winter is shown by several counts made at different seasons, when large and small whelks were distinguished (Table 16). Moore (1936b) has shown that this thin-lipped type of shell indicates immaturity. Moore (1938a) showed that the tissue weight of mature *Thais* dropped sharply in February and March, and then began to increase. Immature whelks showed no such drop in weight in winter. Moore ascribed the weight decrease either to loss of reproductive products or cessation of feeding by mature whelks during the main spawning period, January to April.

TABLE 16. Proportion of immature *Thais lapillus* in the population in the different seasons.

Month	No. of observations	Total No. *Thais* observed	No. of *Thais* which were small, lips of orifice thin	% of total which were small
Area 2, MTL				
December, 1953	5	151	85	56
January, 1954	10	147	105	71
February, 1954	3	45	39	87
March, 1954	5	19	14	74
April, 1954 & 1955	4	24	7	29
Shore transect, HWN to LWN, most *Thais* found at MTL				
August, 1953	1	1344	576	43
Area 1, above MTL				
February, 1955	1	43	40	93
March, 1955	1	12	7	58

From the present evidence, it seems that the latter reason may account for much of the decrease.

The seasonal variation in numbers of *Thais* observed feeding at any one location may have been a result both of changes in the proportion of time spent feeding and of changes in the total population size. If it is assumed that in the same month in two different years *Thais* spent the same proportion of time feeding, a comparison of the numbers observed in the two months should reveal any annual changes in population size. In the records of *Thais* over two years those in the months of August, September, December and January were inadequate. For the other eight months an average density was available for each month in two successive years (Figs. 13 and 14).

On Area 1, as shown in Fig. 14, in many of the months of 1954 greater numbers of *Thais* were observed than in the corresponding months in 1953 and 1955. On Area 3 (Fig. 13), the opposite was the case except for the autumn season. Using the sign test (Siegel 1956), the monthly means of *Thais* density in 1954 were compared with those of 1953 and 1955. For Area 1 there were significantly greater numbers in 1954 ($p=.032$), whereas for Area 3 the numbers were significantly fewer ($p=.013$).

For the first months of 1954, Area 3 and the low shore in general was almost completely bare of barnacles, those of the previous year's settlement having been stripped off following "hummock" formation as shown in Fig. 9. With little food at the lower levels it might have been expected that the *Thais* population would shift upward to shore levels at which barnacles persisted, increasing the density of *Thais* at and above M.T.L. That such a shift probably occurred is shown by the low numbers on Area 3 and the high numbers on Area 1.

Thais were first seen feeding on the barnacles of the 1954 settlement on Area 3 in July, 1954. These barnacles, having settled less densely than those of the previous year, then survived for over a year. Presumably as a result of the reestablishment of this food supply at low shore levels, the density of *Thais* was greater on Area 3 from the autumn of 1954 through June, 1955 than in these months the previous year. On Area 1, changes in the opposite direction were occurring; in the autumn of 1954 the density decreased at the upper levels and increased at the lower levels. The density in the spring of 1955 was much lower at all levels of Area 1 than it had been in the spring of 1954. From these considerations it appears that the annual changes in the *Thais* population shown in Figs. 13 and 14 can be accounted for in large part by vertical redistribution of the population rather than by increases or decreases in the total population size.

There is little published information on mortality of *Thais lapillus*. Orton & Lewis (1931) reported a great mortality of *Thais lapillus* on oyster beds in southern England in an unusually cold winter. In a fecal pellet of the purple sandpiper, *Calidris maritima* (Brunnich), collected at Millport in January, were found 28 spires of small *Thais*. They corresponded in size to whelks 1-9 mm in length. Shells of *Thais* were found in a nest of the herring gull, *Larus argentatus* Pontopp., and oyster-catchers and thrushes have been seen eating *Thais* (Moore 1938b). A few *Thais* were found in the stomachs of two fish, one *Labrus bergylta* and one *Gadus virens*, caught in the intertidal fish trap at Millport in August, 1953. Moore (1938b) estimated that *Thais* may live at least four years on the Isle of Man. In the present study a marked *Thais* lived to an age of at least three years.

None of these observations help in judging whether the mortality rate of *Thais* is greater at any particular season at Millport. Predation by fish would be expected to decrease in winter, while that of birds might be relatively constant throughout the year. Gales occur at all seasons.

As shown in Table 16, the proportion of large older *Thais* was least in February, increasing in the spring as the numbers feeding also increased. This difference in both age structure and numbers was the most pronounced seasonal change in the *Thais* population during the year. One explanation might be that a great number of the large mature *Thais* died in January and that the younger ones grew through the winter so that they appeared in the spring as the mature members of the population. This would mean that few of the *Thais* lived more than a year after reaching maturity and that growth was rapid in winter and spring. Moore (1938a) found that on the Isle of Man *Thais* grew for about three years before attaining maturity, after which time growth stopped. Growth occurred at all seasons at an average annual rate of about 10 to 15 mm, but was somewhat faster in the warmer seasons. In a small collection of *Thais* made in July at Millport, the thin-lipped immature forms had a height range of 1.1 to 2.7 cm, while the thick-lipped mature forms showed a range of 2.5 to

3.4 cm. If the growth rates were similar to those on the Isle of Man, the *Thais* at Millport would become mature in about two years.

The appearance of many large mature *Thais*, often with heavily eroded shells, in the spring makes it seem unlikely that these all developed from the small whelks which constituted the winter feeding population. A more likely cause for the decrease of numbers observed in winter and the increase in spring is that *Thais* spent a smaller proportion of its time feeding in winter and increased the proportion in spring. This would account for the reduction in tissue weight of mature individuals in winter, as reported by Moore (1938a). For each season the density of the *Thais* population is correlated with the sea water temperature (Table 21); this might be expected if the differences in density were a result of changes in activity.

Thais were first counted when they were about one centimeter in height. It takes about 16 months after egg laying for the whelks to reach this size, four months being spent in development in the capsule. Since the main egg-laying season lasts for six months, with some laying all year, and since some individual variation probably occurs, some individuals are probably attaining the size of one centimeter at all seasons. There does not appear to be any particular season when young whelks enter the population in great numbers. Regarding the mortality of *Thais*, the evidence is sparse; no exceptionally cold periods occurred during the winters of 1954 and 1955, and predation did not appear to be heavier at any particular season. Thus the size of the *Thais* population seemed to be relatively steady, annual variations slight and the seasonal variations probably due mainly to the whelks spending less time feeding in colder periods.

MOVEMENTS DURING FEEDING

Since the positions of the marked *Thais* had been plotted on maps at each observation, movements could be measured. Only those observations made at consecutive low tide periods were used so that each record represents a movement during a single high tide period.

These records do not accurately represent the activity of *Thais* for several reasons. The whelks may not have moved in a straight line, as the distances were measured. In addition, all movements to and from shelter were excluded; these were probably longer journeys than those made while feeding. Thirdly, since the observations were made on a small area, usually about one square meter, the more sedentary individuals would tend to be recorded more often, while the more active ones, once they had left the area, would be missed.

The pattern was usually one of a series of short movements followed by one or more longer movements. For example, three whelks which had been observed for 8 or 9 consecutive low tides showed the following sequence of movements (distances in centimeters): No. 1: 2, 3, 20, 8, 34, 7, 5; No. 2: 2, 14, 11, 31, 6, 28, 17; No. 3: 2, 4, 8, 22, 1, 1, 8, 13. It

TABLE 17. Distances moved by *Thais lapillus* while feeding on *Balanus balanoides*, expressed as a frequency distribution. *Thais* were marked individually and observed at a series of low tides during the periods shown.

		Number of Movements in Each Distance Class						
		Low shore levels				Area 1		
Dates:		July 18-19 1953		July 28-Aug. 2 1953		July 27-30 1953	Dec. 7-16 1953	Apr. 25-May 28 1955
Area:		B	C	B	C			
Height from MTL:		−0.5	−0.9	−0.5	−0.9	+1.5 to +2.5		
No. of *Thais* observed:		16	28	12	29	4	9	7
No. of successive low tide observations:		3	2	12	9	7	7	16
Centimeters moved in one high tide.	0-4	20	6	13	65	13	10	3
	5-9	4	8	24	25	1	6	4
	10-14	6	2	18	7	2	4·	8
	15-19	3	4	5	4	2	3	4
	20-24	2	8	5	6	0	0	1
	25-29	1	3	1	2	0	0	2
	30-34	0	3	1	3	1	1	2
	35+	7	10	1	2	0	0	5
Total no. of observations		43	44	69	125	19	24	29
Mean distance in centimeters moved per tide		14	19	7	8	5	6	18

is probable that the whelks did most of their feeding during the more sedentary periods.

Most of the observations were made during the summer of 1953. A few others were made on Area 1 in other seasons. In Table 17, frequency distributions of the movements are given. At low shore levels on slightly sloping rocks (series B and C, Table 17) the distances moved during the first, shorter, summer period were much greater than those during the second longer one. This is probably because the more active marked animals, present at first, gradually left the area, leaving only the more sedentary ones to be recorded later. In the second period it was noticed that during some high tides most of the whelks had moved very little, while in others, a greater proportion of longer movements had been made. No correlation of these differences with weather was apparent.

On Area 1, a vertical surface at a higher level, fewer *Thais* were marked. The movements were shorter that at low levels. In December the movements were about the same as in July, but in the late spring of 1955 they were much longer. This difference is probably not due to selection for active individuals, as occurred in the low shore observations, since the period of observation was a long one. Evidently *Thais* is more active in the spring.

THE RATE OF FEEDING OF *Thais* ON BARNACLES

In any predator-prey study, the measurement of the feeding rate of the predator is obviously important. In order to do this, whelks were enclosed in cages with barnacles which were counted periodically. An adjacent empty cage served as a control. The physical conditions were similar in the experimental and control cages. In the later experiments several cages with *Thais* enclosed were used to provide replicates. Details of the experiments are as follows.

Experiment 1—July 22 to August 8, 1953, at MTL. Two cages were used, one with a *Thais* inside, the other empty as a control. The barnacles were very dense, much crowding occurred, and so many deaths occurred in both cages that the experiment was ended.

Experiment 2—September 10 to December 11, 1953, one foot below MTL. Here the barnacles were scattered and complete counts of living barnacles were made at each examination. Two cages were used, one with a single large *Thais* in it, until October 21; after this two more cages were attached with a *Thais* in each. In December many very small *Thais* began entering the cages so the experiment was ended.

Experiment 3—November 11, 1953 to August 8, 1954. Four cages were attached at the Middle level of Area 1, about one foot north of Cage 1. Two had one whelk each, two were kept as controls. These cages had a mesh of 1/4 inch, but the absence of small whelks above MTL rendered this large mesh satisfactory. In all the other experiments the 1/8 inch mesh was used.

Experiment 4—October 22, 1954 to May 31, 1955. Five cages were attached to a slightly sloping rock surface five yards east of Area 1 at the height of the "Lower" level of Area 1. In three cages two *Thais* were placed in each, in one cage four *Thais*, and one cage was kept as a control. All the whelks had thin lips to the orifice, indicating immaturity (Moore 1936b). One cage with two *Thais* contained

TABLE 18. Feeding rates of *Thais lapillus* (no. of *Balanus balanoides* eaten per day by one *Thais*). The size of *Thais* is the height in cm. See text for method of computing the relative volume eaten.

	Autumn	Winter	Spring	Summer
Experiment 1.				July 22-Aug. 8, '53
Thais size				*2.0*
Feeding rate, number				2.0
Feeding rate, rel. vol.				2.0
Experiment 2.	Oct. 21-Dec. 11, '53			
Thais size	*2.1 2.8 3.1 Ave.*			
Feeding rate, number	1.9 2.4 0.7 1.7			
Feeding rate, rel. vol.	3.3 4.1 1.2 2.9			
Experiment 3.	Nov. 11, '53-Jan. 26, '54	Jan. 27-Mar. 14, '54	Mar. 14-Aug. 8, '54	
Thais size	*2.0 2.5 Ave.*	*2.0 2.5 Ave.*	*2.0*	
Feeding rate, number	1.4 0.5 0.9	0.1 0.01 0.05	1.4	
Feeding rate, rel. vol.	1.9 0.9 1.4	0.2 0.03 0.12	2.5	
Experiment 4.	Oct. 23, '54-Jan. 6, '55	Jan. 7-Mar. 25, '55	Mar. 26-May 31, '55	
Cage no.	1. 2. 3. 4.	1. 2. 3. 4.	1. 2. 3. 4.	
Thais size (average)	*2.6 2.5 2.6 2.7 Ave.*	*2.6 2.5 2.6 2.7 Ave.*	*2.6 2.5 2.6 2.7 Ave.*	
Feeding rate, number	0.8 0.9 0.4 1.1 0.8	0.4 0.4 0.2 0.4 0.37	0.2 0.9 0.1 0.4 0.44	
Feeding rate, rel. vol.	1.9 3.8 2.6 1.4 2.4	0.6 1.6 0.7 0.5 0.80	0.4 2.1 0.7 0.6 0.90	
Experiment 5.				July 1-Aug. 16, '58
Thais size				*2.6 2.5 2.4 1.7 1.6 Ave.*
Feeding rate, number				1.8 2.3 2.2 1.5 1.6 1.9
Feeding rate, rel. vol.				4.0 5.1 5.0 3.3 3.5 4.7

NOTE: *In Experiment* 4, Cage 1 had four *Thais* in, while the other three cages had two *Thais* each. Cage 4 had only first year barnacles inside, while the other cages had three-year groups of barnacles. In all the other experiments there was only one *Thais* per cage.

only first-year barnacles as prey while the other four cages contained three year-groups.

Experiment 5—July 1 to August 16, 1958. Six cages were attached at the middle level, Area 1, at the same level as Experiment 3. A single *Thais* was placed in each of five cages, the sixth being left empty as a control. Two of the *Thais* were small, the others being average-sized (see Table 18). The barnacles were predominately one year of age.

Since two or three year-groups of barnacles of correspondingly different sizes were present in experiments 3 and 4, to compare the amounts eaten it was necessary to make an estimate of the volume of the barnacles consumed. This was calculated using the cube of the length, which is proportional to the volume. From photographs of experiments 3 and 4, the average basal length of barnacles of each age group was measured. For barnacles six months of age (in the autumn) this was 3.0 mm. The cube of the average length of each year group was expressed as a multiple of that of the six-month size. In addition, the ratios of cubes of length of *Balanus balanoides* at six months to those at 12, 18, 24 and 30 months of age, were calculated from the growth studies of Moore (1934), Hatton (1938), Barnes & Powell (1953) and Bousfield (1955); the ratios agreed well with those taken from the photographs in the present study, even though the absolute sizes differed. The calculated numbers of each age-group consumed on each area were multiplied by their respective relative volume and were then summed for all the groups.

The results of all the feeding experiments are given in Table 18. The average number of barnacles eaten in the autumn was about the same in experiments 3 and 4, but the volume consumed was greater in 4, owing to the larger size of the barnacles present. In experiment 4, a comparison of cage 4, which contained only barnacles of the most recent year group, with the other cages, reveals an interesting situation. The whelks in this particular experiment appeared to be able to open only about 1.1 barnacles per day in the autumn, fewer in the other seasons, regardless of the size of the barnacles. The whelks in cage 4 were consuming about two-thirds the volume of barnacles of the whelks in the other cages. Assuming that a greater volume means a greater amount of food, the whelks in cage 4 would be expected to have opened more barnacles if they could. There was never a shortage of barnacles in any cage. The implications of this finding are that it takes about the same amount of time for a whelk to open any barnacle, and that therefore it would obviously be advantageous for the whelk to select larger barnacles. Later in this paper evidence will be presented to show that *Thais* does select larger barnacles. The evidence that whelks can only open a limited number of barnacles cannot as yet be regarded as conclusive. In experiments 2 and 3 during the autumn, and in the other seasons the rates were higher. Also, the whelks in cage 4 may have been exceptionally slow feeders; additional experiments are required.

Imprisonment of whelks in cages might have resulted in abnormal feeding rates, though no evidence for this could be obtained. The whelks which were marked individually (Table 17) showed an average movement of about 10 cm per tide. The dimensions of the cages in the enclosure experiments were about 12 x 12 cm, so that an average whelk may not have been much restricted in its movements by being enclosed. Although these whelks could not retreat to a refuge to rest, the corner of a cage might be regarded as a protected place. However, individual variation might have been enough to produce the differences found between the cages in experiment 4.

The feeding of *Thais* was not continuous. In order to ascertain the proportion of time which *Thais* spent feeding, the records of the whelks which had been marked were analyzed. The total population of marked whelks present at any one observation was

TABLE 19. The fraction of the population of *Thais lapillus* which was observed to be feeding at any one time. (See text for explanation.)

	No. of *Thais* marked	No. of Low Tide Observations	Average Fraction of the Population Feeding at One Time	Range of Fractions Calculated from Daily Observations
Area 3, Series B July 18-30, 1953	19	14	0.61	.38 - .81
Area 3, Series C July 18-30, 1953	82	14	0.55	.24 - .72
Area 1, July 18-30, 1953	9	9	0.64	.50 - .78
Area 1, Dec. 7-16, 1953	11	8	0.73	.43 - 1.00
Area 1, Apr. 28-May 16, 1955	10	11	0.49	.12 - .90

taken to be the number observed feeding on the open rock plus those which, though not visible at the time, reappeared at a later observation during this period. The results of the calculations are shown in Table 19. In the summer the population was feeding about 60% of the time, in December somewhat more, in May, less.

Another analysis was made of observations taken over twelve consecutive low tides in the summer to discover the average lengths of the feeding and "resting" periods. A summary of these observations of 79 marked individuals is given in Table 20. It was found that in the summer the whelks only fed for about half the time. Most of the periods of feeding and of "resting" were short; 70% of them were of only one or two tides duration. A few whelks fed for relatively longer periods.

In contrast to the observations on 82 marked *Thais* made in the summer, the data for the other seasons shown in Table 19 are relatively sparse. Therefore, another estimate of the proportion of time spent feeding was made. In discussing the differences in

TABLE 20. Frequency distribution of, (a.) the number of consecutive low tide periods when marked individuals of *Thais lapillus* were observed feeding, and (b.) the length of the interval between feedings. These observations were made during twelve consecutive low tide periods, of 79 marked whelks, series C, on Area 3, 0.9 feet below MTL. July 27-Aug. 2, 1953.

No. of consecutive tides	a. Observed Feeding Actively		b. Intervals Between Successive Periods of Feeding	
	No. of Observations	% of total no.	No. of Observations	% of total no.
1	103	60.0	74	57.4
2	28	16.3	19	14.7
3	21	12.2	11	8.5
4	5	2.9	9	7.0
5	4	2.3	6	4.7
6	7	4.1	6	4.7
7	0	0	1	0.8
8	2	1.2	1	0.8
9	0	0	2	1.5
10	0	0	0	0
11	1	0.6	0	0
12	1	0.6	0	0
Total:	172	100	129	100
Average number of consecutive tides:	2.0		2.2	

numbers of *Thais* observed it was shown that annual variations were small except in spring, 1954, when barnacles were absent at low shore levels and the *Thais* were concentrated at higher levels. In addition, neither the natality nor the mortality rates seemed to vary seasonally so that the seasonal differences in numbers probably reflected changes in the proportion of time spent feeding. Using the relative numbers observed in each season, and taking the proportion of time spent feeding in the summer as a base, the proportions in the other seasons were calculated (Table 21). The averages of the numbers observed in each season for two years were used to compensate for the annual variation, which was slight except for the spring. The estimates for the autumn and spring are lower than those shown in Table 19; the possibility of error in these latter estimates is great, however, since the number of marked *Thais* observed was very small. Therefore, it was decided to use the estimates from Table 21 for the later calculations of the proportion of barnacle mortality due to *Thais* predation.

Hanks (1957) has shown that a related whelk, *Urosalpinx cinerea* (Say), consumed clams at a lower rate at lower temperatures in the laboratory. It is not possible to determine from the data given whether the lower rates were due to less frequent periods of feeding or to slower activity while feeding. The low *Thais* feeding rate in spring was probably due to fewer feeding excursions rather than to a slower rate of opening barnacles. This is the spawning season,

TABLE 21. Proportion of time spent feeding by an average *Thais lapillus* in various seasons, based on the densities on Area 1, and the proportion of time spent feeding in the summer given in Table 19.

	July-Sept.	Oct.-Dec.	Jan.-March	Apr.-June
Population density (no/m²) 1953-54	66	51	13	49
1954-55	60	45	15	21
Average	63	48	14	35
Ratio to July-Sept. value	1.0	0.76	0.22	0.51
% of time spent feeding	60	46	13	30
1949-53 Mean Sea Temperature, °C	13.35	10.53	7.19	9.41

and Thorson (1958) has brought together evidence to show that many marine predators do not feed during the period when the gonads are swollen with egg or sperm. In *Thais lapillus* the present study indicates that the feeding rate was reduced at this time, although there is a possibility that feeding may have been stopped entirely for shorter periods.

EFFECT OF PREDATION BY *Thais* ON THE *Balanus* POPULATION.

As described earlier, cages were attached at three levels on Area 1 to exclude *Thais*. Some of these are shown in Figure 15. Occasionally, small *Thais*, 1/2 to 1 cm in length, entered the cages through gaps at the line of juncture of netting and rock surface. These gaps were probably caused by distortion of the cage in strong wave action or from crumbling of the rock. Since inspections were frequent and the likelihood small of a whelk escaping back through the gap, it is believed that all these entries were noted. To estimate the extent of these lapses in predator control, each occurrence was multiplied by half the time

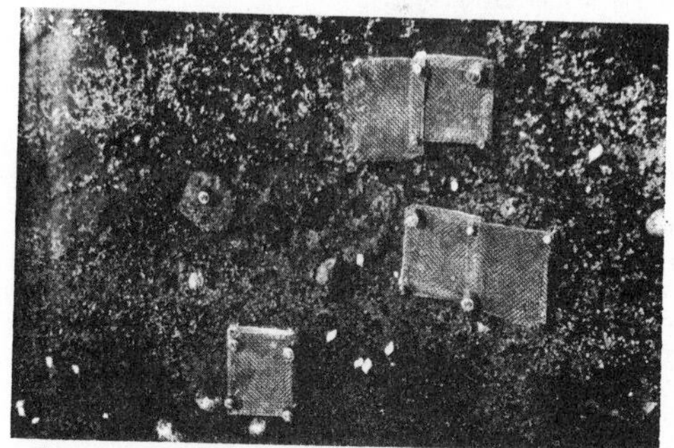

FIG. 15. Photograph of Area 1 made on April 6, 1954. The original cages and "covers" at the three levels are shown; stone 7 is attached to the left at the middle level. A true horizontal line would run slightly upward to the left, passing through both the center of the midde cage and stone 7. The dimensions of the area in the photograph are 1.1 x 0.7 meters.

since the last inspection. Summing these average times for all occurrences and dividing by the total time of attachment of the cages yields the value of 2.6% as the proportion of time when predation by small whelks occurred inside the cages. Since the cages gave complete protection against *Thais* of medium and large size, the method can probably be

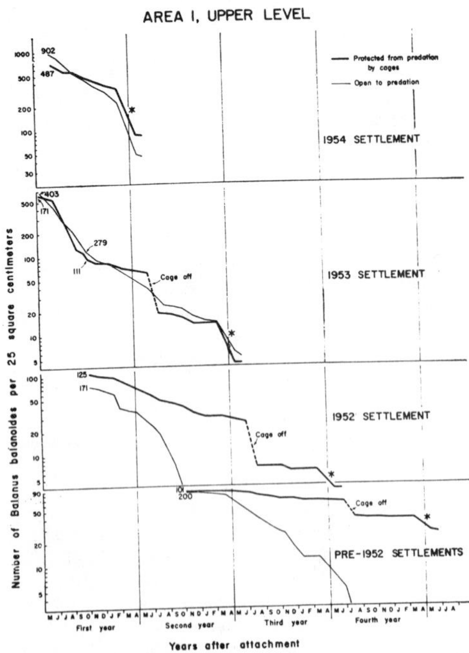

FIG. 16. Survival curves for *Balanus balanoides* with and without predation on Area 1, Upper level. The period of twelve days in July, 1954, when the cage was missing, is shown as a dashed line. The asterisk refers to the period in the spring of 1955 when the level was exposed to air for several days in warm calm weather. The initial number of barnacles counted is given with each curve.

judged effective. With infrequent inspections, modifications would be necessary. Very small *Thais,* able to pass directly through the 1/8-in mesh, were not recorded on Area 1, although their abundance low on the shore rendered ineffective there all attempts at control by cages. Only one instance of human interference occurred, in July, 1954, when the Upper Cage 1 was removed in the author's absence. A new cage was attached after twelve days of predation; the effect can be seen in Figure 16.

Besides excluding the whelks the cages gave protection from damage by water-borne objects. This protection completely altered the appearance of the barnacle population. In the two cages which were in place for 2½ years, barnacles of succeeding settlements attached to surviving adults and grew higher, sometimes bridging the spaces between the older individuals. Examples of this are illustrated in the series of photographs in Figure 19. These populations resembled the clusters of sublittoral barnacles such as *Balanus balanus* (L.) where three-year-groups were found attached to one older barnacle (Barnes 1953b). This condition did not exist on unprotected areas of *Balanus balanoides* even above the range of *Thais* predation probably because the uppermost barnacles would be very vulnerable to damage from waveborne objects.

Since this situation was so different after a year's time from the normal one outside the cages, new caged squares were established at intervals, beginning a year after the first cages had been attached as indicated in Table 1. These resembled the natural areas closely for at least a year or more following their enclosure. With frequent visits and the addition of new cages at about yearly intervals, this method of predator control was very satisfactory. Some algal fouling on the cages occurred, especially by *Porphyra umbilicalis* J. G. Agardh; it was easily removed and constituted no problem. This fouling was much heavier on other cages attached below MTL. Limpets were placed inside the cages in the summer of 1953 to control the growth of algae which had begun to grow abundantly in the cages

Figs. 16, 17 and 18 show the survival of barnacles on all the squares at each of the three levels of Area

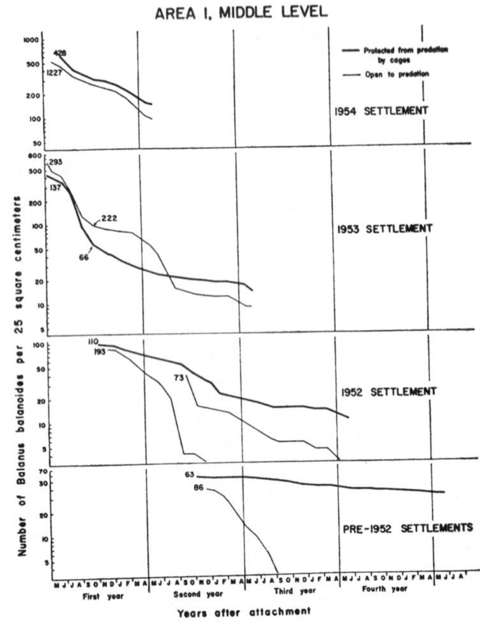

FIG. 17. Survival curves for *Balanus balanoides* with and without predation on Area 1, Middle level. The initial number of barnacles counted is given with each curve.

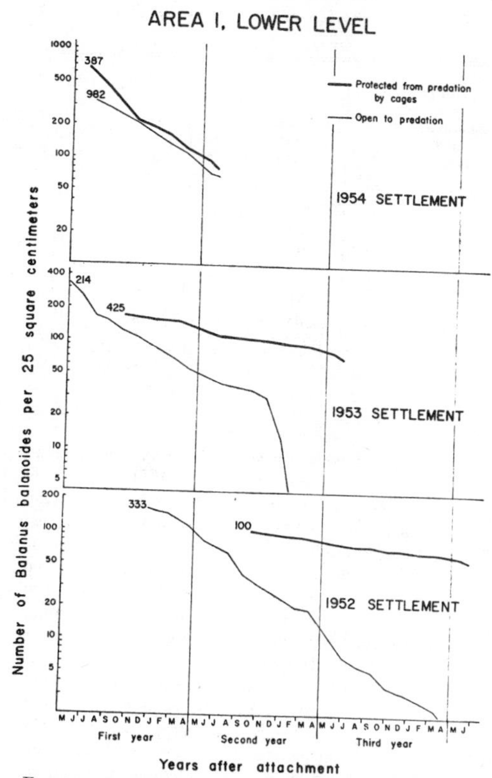

FIG. 18. Survival curves for *Balanus balanoides* with and without predation on Area 1, Lower level. The initial number of barnacles counted is given with each curve.

1. The density is indicated by the height of the curve, the relative death rate at any time by the slope. At the end of the 1953 settlement season the density of the newly settled barnacles was high at the upper and middle levels. At this time the first counts of the new set were made on smaller areas, 1.5 to 4.0 cm^2, within the original area. After six months the numbers had declined, and the counts of the young barnacles were made over the whole of the original area. For each such area, the end of the survival curve of the barnacles on the small area during the first six months was joined to the beginning of the curve of those on the larger, original area. The height of the curve was determined by the density of the later count made over the original area. Thus for the 1953 year group, at the upper and middle levels, the survival curves are composite ones. All others given in this paper were for the same group followed from beginning to end. The actual initial numbers are given with each curve, including those made at age six months as explained above.

Since each curve is the record of the survival of a group of animals rather than of samples from a population, the curve must pass through each point exactly. Since no attempt was made to smooth the curves, they appear quite angular in Figs. 16-18. Some of the changes in slope were probably even more abrupt in actual fact, since some factors such as wave damage or hot weather probably operate over very short periods.

The relation of mortality to density in the early growing stages has already been discussed. A comparison of the slopes of the survival curves of caged and uncaged populations indicates that predation was a minor factor in causing the mortality during the first year above MTL; this was also illustrated in Fig. 11. The slopes of the curves first began to diverge at the end of the first year; the unprotected populations then experienced an increase in relative death rate, while that of the protected ones either remained constant or decreased. The highest death rate due to predation, (indicated by the greatest differences in slopes of the two types of curves), occurred in the summer.

The variation in age structure of the *Balanus balanoides* population at different shore levels on the area studied was similar to that described by Moore (1934) for the Isle of Man. At the top level of Area 1, near HWN, the population consisted mainly of large older *Balanus balanoides* mingled with *Chthamalus stellatus*. Below HWN, the proportion of the older age groups of *Balanus* dwindled, until near LWN only the most recent year class was present except for a short period in the spring when the survivors of the previous year were mixed with the newly settled group. Fig. 9 illustrates the survival on Area 3, below MTL.

To test whether the mortality of barnacles at low shore levels was correlated with the abundance of *Thais*, data from Area 3 were analyzed. On four portions of this area the 1954 settlement was light, so that intraspecific crowding was probably not an important cause of barnacle mortality. For seven periods between June, 1954 and May, 1955 data on both barnacle mortality and *Thais* density were available. Correlations were tested using the Spearman rank correlation coefficient (Siegel 1956) as shown in Table 22; significantly positive correlations were found in two of the four areas. In all four areas, when the barnacle mortality was high, appreciable numbers of *Thais* were present. Although the analysis is based on few observations, it suggests that in the absence of crowding at low shore levels, predation is an important cause of mortality.

On Area 1, it was seen that the death rates for the first year of life were not related to predation. Lower on the shore, where growth was much more rapid and the new set constituted the main proportion of the barnacle population, predation began much sooner. In 1954, Boulders 1, 2, and 5 of Area 3 were bare before settlement so that a completely new set occurred at this low level. *Thais* were first observed on these boulders after July 10 when the barnacles

TABLE 22. Correlation between the percentage mortality of *Balanus balanoides* and the population density of *Thais lapillus* on Area 3.

Interval	No. of Thais Censuses	Boulder 1		Boulder 2		Boulder 6a		Boulder 6b	
		Balanus % mort./month	*Thais* no./m²	*Balanus* % mort./month	*Thais* no./m²	*Balanus* % mort./month	*Thais* no./m²	*Balanus* % mort./month	*Thais* no./m²
June 9-July 9, 1954	1	2	9	7	3	5	5	1	5
July 10-Aug. 24, 1954	4	10	89	12	93	13	1	6	1
Aug. 25-Oct. 1, 1954	0	9	—	13	—	15	—	7	—
Oct. 2-Nov. 5, 1954	7	27	162	16	66	62	221	70	221
Nov. 6-Dec. 13, 1954	2	19	74	13	61	21	9	1	9
Dec. 14, 1954-Feb. 3, 1955	2	7	12	24	11	14	3	4	4
Feb. 4-March 16, 1955	7	5	28	51	78	11	6		
March 17-May 17, 1955	16	36	77	32	45	31	12	12	12
Spearman Coeff. r_S		.786		.241		.714		.471	
P		$<.05$ $>.01$ *		$>.05$		.05 *		$>.05$	

were about 8 weeks of age. In 1955, boulders at various levels were scraped clean a month before settlement. These received their first settlers at the beginning of April; whelks were first observed feeding on the lower of these areas, adjacent to Area 3, on May 11, when the barnacles were only 5 to 6 weeks old. On similarly treated areas above MTL no feeding by *Thais* occurred.

The difference between the two years, 1954 and 1955, in the age at which predation began may possibly be explained by the differences in distribution of *Thais*. In Fig. 13 it is shown that many fewer *Thais* were present on Boulders 1 and 2 in May and June of 1954 than in these months in 1955. Also, there were adult barnacles present on these areas in 1955 but not in 1954 (Fig. 9). It appears that the presence of adult barnacles attracted large numbers of *Thais*, which then began to feed on the newly settled barnacles, even on adjacent areas from which all adults had been removed. In the following discussion it will be demonstrated that *Thais* does tend to feed more on adult barnacles so that the presence of adults would be expected to attract more *Thais*. Fischer-Piette (1935) noted that *Thais* did not attack *Balanus balanoides* until six months after attachment; no indication of shore level was given, however.

If, whenever it is "hungry," a *Thais* feeds on the barnacles it is then touching, it would be expected to encounter and feed on the various age groups in direct proportion to the area covered by each. From this hypothesis, some feeding would have been expected to occur on the newly settled group on Area 1. Since this did not happen to any extent, it must be concluded that the whelks ignored this young group and selected older individuals. These older barnacles, besides being larger, had rough shells, the upper edges of the wall and opercular plates being worn and rounded. This may have enabled *Thais* to distinguish older individuals.

Besides this evidence of selection some direct observations were made in a simpler situation where only two age groups were present. In May, 1955, some *Thais* were noticed feeding between MTL and LWN on the new set when it was 5-6 weeks old. Some were feeding also on survivors of the previous year, although these comprised only a small proportion of the total barnacle cover. The *Thais* thus had two groups of contrasting sizes to choose between. The proportions of the area occupied by old and young barnacles were determined by laying a meter square frame divided into 100 cm² squares on the rock, and for each small square separately, estimating by eye the total barnacle cover, the coverage of the older group (their yellow shells contrasted well with the white shells of the young barnacles) and the percentage of these older barnacles which were alive. By estimating each small square independently it was believed that a very close approximation to the true proportions was made. On each rock area, one horizontal and two vertical transects were made of ten such squares. Since at this stage very few young were dead, the coverage of living young ones was calculated as the total barnacle cover minus the coverage of older barnacles. This estimate of older barnacles included both dead and live individuals, so that it was multiplied by the percentage alive to give the area of living older individuals. The proportion of live old to live young was then calculated. Since some young which had settled inside the dead adult shells were missed and some adults killed by *Thais* still had their opercular plates in place, it was believed that the error in the measurement tended to overestimate the proportion of living adults.

On the assumption that no selection between the two groups by *Thais* was taking place, it would be expected that the *Thais* actually observed feeding would be distributed at random over the barnacles. In other words it would be expected that there would be a direct relationship between the numbers of *Thais* feeding on each age group and the areas covered by each group. To test this hypothesis, on these areas each *Thais* was carefully tilted away from the surface to expose the individual barnacle upon which it was

TABLE 23. The numbers of *Thais* observed to be feeding on two age groups of *Balanus*, compared to those numbers which would be expected if *Thais* occurred on each age group of barnacles in proportion to the area covered by it.

	Observed and Expected Numbers of *Thais* for Each Daily Observation.												Total Numbers	
	May 18		May 19		May 20		May 21		May 22		May 24			
	Obs.	Exp.	Obs.	Exp.	Obs.	Exp.	Obs.	Exp.	Obs.	Exp.	Obs.	Exp.	Obs.	Exp.
No. of *Thais* feeding on one year old barnacles:	49	12	28	8	25	6	30	7	36	9	28	8	196	50
No. of *Thais* feeding on newly settled barnacles:	10	47	11	31	6	25	8	31	9	36	11	31	55	201
Total *Thais* Observed:	59		39		31		38		45		39		251	
Chi-square (1 d.f.):	143.20		62.90		74.60		92.63		101.25		62.90		532.36	
Probability:	<.001		<.001		<.001		<.001		<.001		<.001		<.001	

feeding. When a *Thais* was feeding on a barnacle the opercular valves were usually opened out, with a white "cleaned" spot at the juncture of the valves. This spot appears to be made by the whelk in the process of opening the barnacle. Since it occurs in both old and new barnacles, it made the identification of the individual being eaten equally certain in both groups. In some instances the *Thais* was attached to the rock surface or to an individual barnacle which showed no sign of being attacked. These instances were noted as indeterminate, and were not used in the calculations. The results of these observations are given in Table 23. The total number of *Thais* feeding each day would be expected to be divided between the two barnacle groups in proportion to the areas occupied by them if there was no selection. Comparison of the expected and observed distributions (chi-square) showed that selection was certainly taking place; the whelks were consistently choosing the adult barnacles. The heterogeneity chi-square was 4.12, $p=0.50$, affirming this conclusion.

Some areas, adjacent to those just discussed, having been scraped clean before the 1955 settlement began, were covered only by newly settled individuals. The whelks feeding on these individuals, which were then 5 to 6 weeks old, seemed from casual observation to be eating the larger of these barnacles. To test this observation, on these areas each *Thais* was lifted as before and the individual being eaten was identified and measured. On the same day a piece of the rock was chipped off from the same area and was brought into the laboratory where all the individuals on a small area were measured to indicate the size variation in the total prey population. Measurements were made on six days. Combining all measurements, those representing the whole prey population yielded an average length of 2.0 mm, range 0.8 to 3.7 mm. Of those being eaten by *Thais*, the average length was 2.8 mm, range 1.7 to 3.8 mm. The difference between the two sets of measurements was found to be highly significant ($p = .00003$), using the Mann-Whitney U test (Siegel 1956). Since those being eaten were not apparently rougher or more eroded than the smaller individuals, size alone seems to be the stimulus to the whelk. Large barnacles in a dense settlement tend to stand apart as individuals in comparison to the many average-sized individuals around them, and this may create a spatial discontinuity acting as a tactile stimulus to *Thais*. This situation is illustrated in Fig. 7.

Little information was obtained concerning the feeding rate of different sized whelks. As shown in Table 18, the whelks which were less than 1.7 cm in height in experiment 5 ate at a slower rate than did the larger ones. In the other experiments, however, where the sizes were not so different, the correlation was less clear. It is obvious that there was much individual variation among whelks of about the same size.

There was no obvious correlation between the size of the whelk and that of its prey. In the observations on size selection of barnacles 5-6 weeks old, the size of some of the whelks was measured at the same time as that of the barnacles being fed upon. The average size of the 55 whelks measured was 27.1 mm, all but nine being between 25 and 30 mm. The average length of the killed barnacles was 2.8 mm. The very large or small whelks showed no consistent relation to the size of their prey. A scatter diagram of size of whelk vs. size of the barnacle being fed upon showed no trend, but only a cluster of points near the average sizes of the two animals. The extreme individuals showed irregular behavior.

One other aspect of whelk behavior which might have a bearing on the selection of larger barnacles is the habit-forming behavior described by Fischer-Piette (1935). Fischer-Piette described how the whelks have difficulty in changing their food from barnacles to mussels. The mussels were usually drilled through but barnacles were only rarely drilled. In the present study a few barnacles were found drilled through one opercular valve or, once, through the side wall; these were in a cage below MTL in which only very small whelks were present. More usually, the opercular valves were opened in the same way as the barnacle opens its own valves, that is to say, outward. As pointed out earlier, barnacles which have been opened by *Thais* usually show a small white area at the juncture of the two opercular valves, which is formed by the removal of the outer layer of shell so that a smooth "cleaned" spot remains.

The method of opening barnacles thus appears to be very different from the drilling of large mussels and may well require some "learning," as described by Fischer-Piette. In the present study, the only very large barnacles which were found below MTL together with dense *Thais* populations were in rock spaces kept clear by *Patella vulgata* on a reef otherwise covered with mussels. Here it might be that the whelks had formed the habit of eating mussels and so ignored the few barnacles.

It was decided to test the possibility that the choice of larger, eroded barnacles in preference to the smoother newly settled individuals was due to a habit, developed after feeding in late autumn and early spring on these older barnacles. The method of feeding was not different, as in the mussel-barnacle situation described above, but direct evidence was judged desirable. Therefore during observations in May, 1955, individuals feeding on older barnacles were marked with brown paint, and those feeding on newly settled barnacles with red. Of the 25 whelks marked as feeding on older barnacles, 1, 5, and 2 individuals were observed during the next three observations feeding on new barnacles; conversely, of the 21 whelks originally feeding on new barnacles, 5, 7, and 3 were found feeding on older barnacles in the next three observations, respectively. Thus about a quarter of the whelks changed the size of their prey over a period of three days. The preference shown for older barnacles does not appear to be due to a preformed habit.

The advantage to *Thais* of selecting large barnacles as prey is evident if it can only open a fixed number of barnacles in a given time, as the evidence from the feeding experiments indicated. Since *Thais* selects the larger barnacles, the danger from this predation increases with the age of the barnacle. Only in a narrow zone above MHWN are the barnacles free from predation by *Thais*. Thus the variation in the age structure of the *Balanus* population at different shore levels appears to result principally from the variation in distribution of *Thais*. This was shown in the caging experiments on Area 1, where the age structure of the population in the high zone above MHWN was reproduced inside cages lower down after two year's protection from *Thais* predation.

This selection by *Thais* of large sized barnacles may explain a conflict in the literature concerning the growth rate of *Balanus balanoides* at different shore levels. Moore (1935b, 1936a) collected barnacles at various localities and levels in summer, and separated them into two age groups, the current year's set, aged 2 to 3 months, and the previous years' sets of 14 months or older. Hatton (1938) followed the growth at three levels on the shore for 3 years, while Barnes & Powell (1953) measured individuals on intertidal panels attached to a pier for 1 1/2 years. The level at which the largest average size occurred at ages of 2 1/2 and 14 months respectively is given for each of these studies in Table 24; the largest barnacles in the first growing season were almost always low on the shore, but in the second season there was great variability. In the studies of Moore and Hatton, the lo-

TABLE 24. A comparison of the average sizes of *Balanus balanoides* at different intertidal levels in Great Britain and Northern France. Measurements of young and adults were made under various conditions of wave action and salinity.

Location and Authority	Collecting Stations	Environmental Conditions	SHORE LEVEL AT WHICH THE LARGEST AVERAGE SIZE WAS MEASURED		Relative abundance of *Thais lapillus* at each location
			Age 2-3 months	Age 14 months	
Plymouth, England. Moore, 1936a	Hen Point.	Brackish water	Low	Low	Absent
	Tinside, Drake Is., Misery Pt.	Normal salinity	Low or Middle	About the same at all levels	Common
Isle of Man, Irish Sea Moore, 1935b	Bradda Head	Heavy wave action	Low	Low	Few
	Dub Reef, Port St. Mary	Moderate to few waves	Low	High	Common
St. Malo, No. France Hatton, 1938	Decolle Ouest	Heavy waves	Low	Low	Few
	Decolle Est	Sheltered	Low	High	Common
	Cite	Sheltered, strong current	Low	High	Common
Millport, Scotland Barnes & Powell, 1953	Panels on Keppel Pier	Sheltered, strong current	Low	Low	Absent

NOTE: Size was expressed in tissue weight by Moore and in linear dimensions of the shell by the other authors.

cations with strongest wave action, or where the water was brackish, continued in the second season to have the largest barnacles at low levels, while in protected areas or at normal salinities, the situation was reversed. At Millport, the barnacles were always largest at low levels.

Moore thought that the growth of older barnacles might be retarded by a harmful factor in sea water, which would operate longer at low levels, but would be offset by the greater food supply in wave beaten places or in brackish estuaries where suspended matter was abundant. There is as yet no evidence for this hypothesis. Sokolova (1951) found that *Balanus balanoides* in Russia (Arctic Ocean) grew faster at low levels during their second summer.

Barnes & Powell (1953) showed that the specific growth rate was inversely proportional to the size of the barnacle. They suggested that the better growth which Moore and Hatton found at high levels in the second season resulted from the smaller barnacles growing faster because they were smaller at the start of the second season. This is undoubtedly true, but it does not explain why the high level ones should be larger than those lower down. Although the high level barnacles studied by Barnes & Powell showed a slightly higher specific growth rate in the second season's growth, the absolute growth rates were about the same on their highest and lowest panels, nos. 3 and 7, so that the difference between the average sizes on these panels was the same at age 18 months as it was at 9 months, the barnacles at the low level still being larger.

FIG. 19. The effects of protection from predation and damage for 2 1/2 years. Photographs of the Middle Control 1, left, and Middle Cage 1, right, were taken at intervals throughout the study. The dimensions of the area represented by the photographs are about 16 x 7 centimeters.

An alternate explanation might be as follows. Predation by *Thais* tends to eliminate the larger individuals and so would reduce the mean length. Since this predation occurs much more heavily at low than at high levels, the average size at the low levels might well be found to be smaller than at higher levels in the second season. Where predation is light or absent, the effect of longer feeding time would continue to favor growth at low levels so that at the end of the second season the average size would still be larger there. The rather larger decreases in average size found by Hatton (1938) at low levels were thought by him to be due to irregularities in shape or erosion; measurements of a single barnacle made by Hatton & Fischer-Piette (1932) showed this. However, Deevey (1947) suggested in reference to the data of Hatton (1938) that there may have been a greater mortality of older individuals from unknown causes. From the present evidence, this would appear to be the case, with predation the cause.

Moore (1935b, 1936b) observed that *Thais* was absent or less common in brackish water or in very wave beaten situations. Assuming that this was so at the three locations studied by Hatton, the relative abundance of *Thais* has been deduced and the estimates included in Table 24. The places where *Thais* was relatively more common were also those where the average size was smaller at low levels. Thus the differing results seem to have been due to variations in selective predation. When studying the growth rate of an animal it is therefore important to establish that any mortality which occurs is random with respect to size.

THE PROPORTION OF THE TOTAL MORTALITY OF *Balanus* CAUSED BY PREDATION AND OTHER FACTORS

With the data collected on the numbers and feeding rates of *Thais* over two years on Area 1, it was possible to calculate the number of barnacles which could have been eaten by the observed population of *Thais* per unit area and time. This was done for periods of three months' duration, regarding January through March as the winter season when the sea water temperature was lowest (Barnes 1955), and the other seasons accordingly.

Since the total mortality rate of the barnacles during each of these seasons was also known, the mortality which could be accounted for by predation by *Thais* was expressed as a proportion of the total rate. In addition, some mortality occurred inside the cages on Area 1, presumably from agents other than large predators or damage from floating objects. Intraspecific competition for space, disease, parasites and the effects of weather were possible causes of this mortality. This "caged" mortality rate was also expressed as a proportion of the total mortality rate occurring outside the cages. Thus the total mortality rate of *Balanus* during each season at each level was divided into three fractions: that which would have occurred even with protection by cages, that due to predation by *Thais,* and a remainder which may have been due to other predators or damage.

A correction had to be applied to the feeding rates of *Thais* given in Table 18 before they could be used in this calculation. The rates gained from *Thais* enclosed in cages included both periods of feeding and "resting." However, the *Thais* counted on the vertical surface of Area 1 were undoubtedly engaged in active feeding. For example, in the summer, the average feeding rate of *Thais* in cages was 1.9 barnacles per day. However, since *Thais* fed only about 60% of the time in summer, the rate during this active period was 3.17 per day; thus for each season the feeding rates were corrected using the proportion of time spent feeding as given in Table 21. These corrected rates are given in Table 25. As shown in this table, for both the autumn and winter, feeding rates from two separate experiments were available.

TABLE 25. Corrected feeding rates (No. of *Balanus* eaten by one *Thais* per day) at the level of Area 1. The rates from Table 18, experiments 3 to 5, are here recalculated to eliminate the periods spent "resting" between feeding excursions.

	Winter	Spring	Summer	Autumn
Average Feeding rates, from Table 18:				
Exp. 3..	0.05	1.45		0.9
Exp. 4..	0.37	0.44		0.8
Exp. 5..			1.9	
Mean of these rates:	0.21	0.44	1.9	0.85
Fraction of time spent feeding (from Table 21).	0.13	0.30	0.60	0.46
Feeding rate (per day) during the period of active feeding.........	1.62	1.47	3.17	1.85
Same rate per 90 days...........	145.	132.	285.	166.

For the summer, the rate from experiment 1, lower on the shore, was very close to that of experiment 5. Only one estimate is available for the spring, from experiment 4. The average of this rate with that of the summer from experiment 5 is 1.17; this average shows fair agreement with that of 1.45 over the spring and summer from experiment 3.

In making the calculation, the number of barnacles which could have been eaten by *Thais* was subtracted first from the total number which had died. Then the number which would have died even if the area had been protected by cages (calculated from the relative mortality of those in cages) was subtracted, leaving (sometimes) a remainder. Some of the barnacles which supposedly died from causes included under "caged" mortality might have been killed by the various deleterious factors, such as predation and damage, which operated only outside the cages. A correction was made for this, thereby increasing the proportion of the total mortality due to factors occurring outside the cages.

There are several possible sources of error in these calculations. Individual variations occurred in the feeding rates of *Thais,* as shown in Table 18. It is felt, however, that these were not extreme, and were probably not biased in one direction. All sizes of *Thais* which occurred in any abundance on Area 1 were used in the feeding experiments, and two independent estimates were available for each season except spring. The most divergent estimates were those in the winter. The proportion of time which *Thais* spent in feeding in the summer was estimated in two independent ways (Tables 19 and 20). The extension to the other seasons (Table 21) may have introduced some error, since the calculation was based on the assumption that seasonal differences in abundance were the result of differences in the frequency of feeding rather than of mortality. The estimates of population density of those *Thais* which were seen feeding in different seasons were based on many censuses, but these were not made with equal frequency in all seasons. In addition, the fact that *Thais* sometime occurred in groups rather than being distributed at random would reduce the accuracy of the censuses.

From all these considerations, it appears that the information about *Thais* is most accurate for the summer, and somewhat less so in the autumn. In the spring only one separate feeding rate is available, but as discussed previously, it is probably not greatly in error. The winter rates were quite different in the two experiments, but errors in this season were rendered less important since the numbers of *Thais* which were active at this season were small (Fig. 14) and so contributed little to the mortality of the barnacles.

By comparison, the estimates of the mortality of the barnacles were much more accurate. For the uncaged areas there were at least two replicates at each level. For the first year there was only one cage at each level; in later years more cages were added. The data from areas caged for more than about one year were not used, since the age structure of these populations had become very different after being protected for this length of time. The only exception to this procedure was at the lower level, where only one cage was present; the barnacles became densely crowded after the autumn of 1954. This is probably the only instance where the caged population is not representative of the natural population. In the rest of the estimates, the total mortality and the portion representing "caged" mortality are probably accurate ones, while the fraction representing mortality from *Thais* is a less accurate estimate. Only those barnacles aged 6 months and older were included in these calculations. As shown in Figs. 16-18, barnacles less than 6 months old were not fed upon by *Thais* on Area 1.

As shown in Fig. 20, the effect of *Thais* was greatest in the summer. At every level in both summers (except at the lower level in 1953), *Thais* could have accounted for all the mortality of *Balanus balanoides*. In some other seasons, especially when the amount of "caged" mortality was high, *Thais* also accounted for most of the remaining mortality. There was a "remainder," unaccounted for by *Thais,* in 5 of the 6 winter periods at all levels, and in certain of the other cooler seasons. This may perhaps be attributed to damage during severe gales.

The proportions given in Fig. 20 included all barnacles older than six months. If those aged between 6 and 18 months were considered separately from the older ones, differences in the proportions were found. As was shown in the survival curves of Figs. 16-18, the mortality inside the cages was very low after the age of 18 months. In these older barnacles the only seasons when the "caged" relative mortality rate exceeded 12% were the autumn of 1953 and the spring of 1955; mortality during the latter period has already been discussed. In contrast, the "caged" relative mortality rate of the younger group was generally high, the average mortality rate per season in the cages being 23.5% for the young group as compared to 9.8% for the older group.

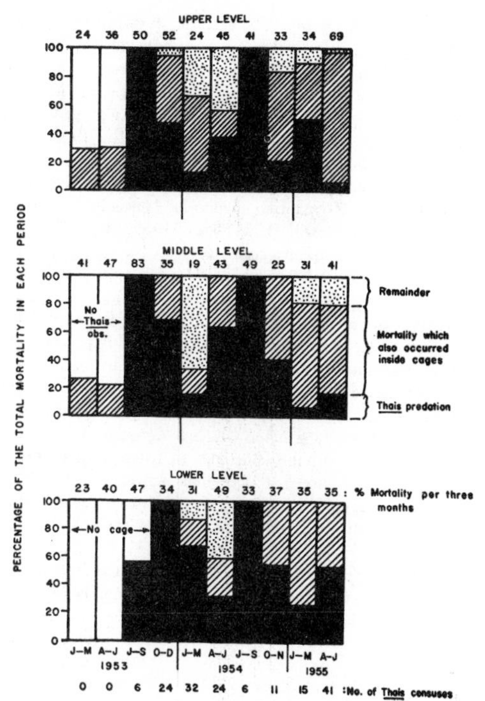

FIG. 20. Proportions of the total mortality of *Balanus balanoides*, over 6 months old, in each season on Area 1, which could be ascribed either to predation by *Thais lapillus* or to factors which operated inside the cages. See text for further explanation.

With regard to predation, the evidence that *Thais* selects larger barnacles as prey was used as follows. It was assumed that *Thais* attacked members of the older group first. For each season and level the number of barnacles calculated as having been eaten by *Thais* was compared to the number of the older group dead in excess of "caged" mortality. In all instances the former number was greater; in other words, *Thais* could have accounted for all the mortality of older barnacles in excess of the "caged" mortality. It is improbable that *Thais* accounted for every such death, but when it is considered that older barnacles with heavier shells are much less likely to be damaged by objects thrown about by waves, the likelihood is high that *Thais* is the main cause of death in older barnacles below MHWN level.

As discussed earlier, in the summer of 1958 the population density of *Patella vulgata* was much lower than it had been in 1955, and the cover of *Fucus* was greatly increased. The age structure of the *Balanus balanoides* population was much different also. The mean densities (no./100 cm^2) on Area 1 in July, 1958, for the 1958, 1957, and 1956 year groups were, respectively, 74, 270 and 2; obviously the 1958 settlement had been very light. The high proportion of the 1957 year group, (then just over one year old), to the older groups was also unusual. In the summers of 1953, 1954 and 1955, the densities of one year/older year groups were respectively, 223/77, 157/17 and 300/26, with an average of 228/40. Thus the value of 270/2 in the summer of 1958 shows a much lower proportion of barnacles aged two years or more. This low value for the older barnacles was not due to poor settlements, since the 1955 and 1956 settlements on other areas were good (Table 9). Hummocking was never seen on Area 1, so that the low density of these age groups was probably not due to intense intraspecific crowding. From the previous evidence that *Thais* selected larger barnacles it was natural to ascribe the low proportion of older barnacles to increased predation by *Thais*. The average densities of *Thais* on the same three levels of Area 1 in which the densities of *Balanus* were calculated, for the summers of 1953-1955 were 12, 14 and 4 per 0.2 m^2, respectively; for the summer of 1958 it was 16. The slightly higher density of *Thais* in 1958 may be indicative of greater predation, but no direct evidence is available on densities of *Thais* earlier in the year, when the mortality of the older barnacles must have occurred. Some indirect evidence exists, however, which suggests that the *Thais* population may have been higher on Area 1 since the previous winter.

In the censuses of *Thais* made in the summer of 1958 it was noticed that many *Thais* occurred on the barnacles beneath the cover of *Fucus*. To test whether the *Thais* tended to occur more densely under the *Fucus* canopy than in the open, the total number at each level in ten observations was divided into the proportions in which the surface was covered by *Fucus* or bare. These are the numbers which would be expected if the distribution of *Thais* bore no relationship to that of *Fucus*. These numbers were compared (chi-square) to those observed under the *Fucus* canopy and in the open. At every level more *Thais* were observed to be under the algae than would be expected in a random distribution. The total chi-square probability was less than 0.0005; the *Thais* evidently tended to occur more densely under *Fucus*. Most of the smaller *Thais* found on Area 1 in the summer of 1958 were under the *Fucus* canopy. Moore (1938b) found very young *Thais* only near the bottom of the shore; evidently young *Thais* are less tolerant of dry conditions. An increased coverage of *Fucus*, such as had occurred on Area 1 since the summer of 1957, would thus render the area more favorable for younger *Thais*, so increasing the total numbers of *Thais*. If this occurred, heavier predation on the older groups of *Balanus* would be expected to follow, with the resultant effects on the age structure of the *Balanus* population seen in the summer of 1958.

DISCUSSION AND CONCLUSIONS

While the various environmental factors measured in this study have been shown to cause much mortality, other "intrinsic" causes may have been important. Pearl & Miner (1935) cite the rate of living and genetic constitution as the two main endogenous causes of death. Neither from the present study nor from previous work on this species has any information been found concerning genetic factors. With regard to the rate of living, relative growth was found to be faster at low levels in the first year, at high levels later (Barnes & Powell 1953). Rate of activity, as represented by cirral beat, was studied by Southward (1955), for barnacles from different shore levels. *Balanus balanoides* gave ambiguous results; the cirral beat was faster in those from low levels soon after collection, but the differences disappeared after a short time in the laboratory. In any case, cirral beat may not be a good measure of activity in natural conditions. Observations were made at Millport of *Balanus balanoides* just after the rising tide had covered them on a calm day. They extended their cirri stiffly facing the current, retracting after irregular intervals of up to twelve seconds, presumably after having caught a particle of food. Similar behavior under natural conditions has been described in *Chthamalus stellatus* (Crisp 1950). Segal, Rao & Thompson (1953) found that limpets and mussels taken from lower levels had faster heart beat and water propulsion, respectively, even after acclimation in the laboratory. Acclimation was achieved within a month when limpets were transplanted in the field. It is obvious that, as yet, the extrinsic and intrinsic causes of mortality in barnacles cannot be separated.

As a means of summarizing the changes in population density which occurred at various levels during this study, the cumulative curves of Figure 21 have been plotted. Three locations were selected to illustrate the differences between shore levels, these being the top and middle levels of Area 1 (both caged and uncaged areas at the middle level), and two boulders from Area 3. These represented the upper and lower limits of distribution (except for scattered individuals) of *Balanus balanoides*, together with a mid shore level where successful predator control had been exercised for the entire period of study.

Fig. 21 illustrates several points dealt with previously. Barnacles live shorter lives at the lower levels, never more than two year-groups being present together. At the top levels, early mortality may result in only a few barnacles surviving after a year, but these may then live for a long time. The "dominance" of the 1952 year-group is shown.

At the middle level, the survival of older barnacles protected from predation was almost equal to that at the higher levels, but the early survival was better, probably owing to the less extreme physical conditions. This resulted in more equal representation of

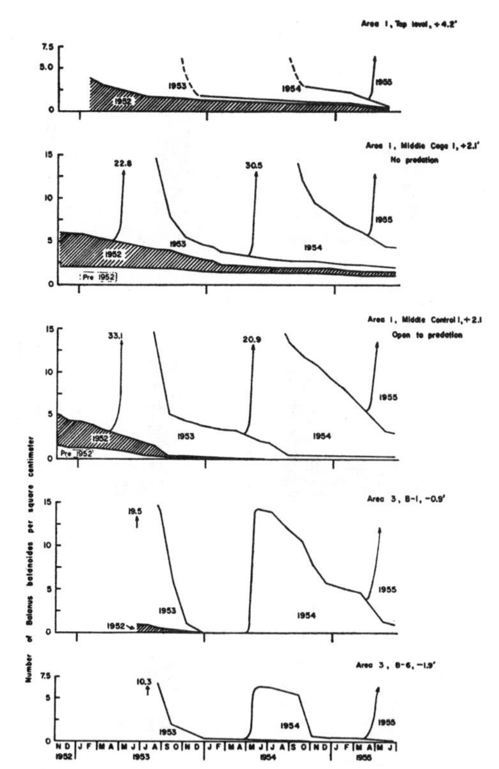

FIG. 21. Cumulative curves of the numbers of *Balanus balanoides* present at various shore levels; the height of each level is given, in feet above or below MTL. The upper portion of the curves just after settlement have been omitted in some cases; the densities reached have been indicated. For comparison, the 1952 settlement has been shaded in all instances.

each age group as compared with the higher levels. On the natural area unprotected from predation there were usually two year groups present, three at times, in contrast to the low levels.

In comparing the caged and uncaged areas at the middle level of Area 1, it is apparent that there was not much difference in the numbers of barnacles present, although the age structure of the population was very different. However, the older barnacles, being larger, occupied more surface space and represented a greater biomass per unit area owing to their greater height. To illustrate this quantitatively, measurements were made from a photograph (shown in Fig. 19), of these two areas made February 18, 1955, after the barnacles in Middle Cage 1 had been protected for 2 1/4 years. Surface areas covered by each age group were measured with a planimeter. As a relative measure of volume, the surface area of each age group was multiplied by the average height of the same group. This average height was determined by first obtaining the average length of the age group from the photograph. Then measurements of length and height of a group of *Balanus balanoides* collected from Area 1 in August, 1958, were made. From these data, an average height was determined for each age group, using the average length obtained from the photograph.

This calculation of volume can be used to compare different age groups only if the barnacles do not change shape as they grow. In the measurements made by Barnes & Powell (1953) of barnacles growing without crowding on panels, the barnacles became squatter as they grew, average length/height ratios of 2.4 and 3.3 being found at average lengths of 3.1 and 6.0 mm, respectively. In the present measurements of barnacles collected from Area 1, the barnacles changed shape only slightly: average length/height ratios of 1.7 and 1.6 were found with average lengths of 2.6 and 6.6, respectively. This difference in the reverse direction was probably the result of moderate crowding on Area 1. Since the barnacles change shape only slightly under these conditions, a valid comparison may be made between the calculated volumes of the different age groups.

The comparative measurements of number, surface area and relative volume for the different age groups with and without predation are given in Fig. 22. Without predation a greater volume (standing crop biomass) was supported by the same area while providing less bare rock surface for settlement of the new age group later in the spring (although some settlement may occur on the shells of the older barnacles). Since the number of larvae produced is proportional to the volume (Moore 1935a, Barnes 1953a), the effect of predation was to reduce greatly the larval output as well as the standing crop. Since there was usually a great oversupply of larvae seeking places to settle, the predation probably was not enough to limit recruitment. But by providing more space for new settlement and removing the older barnacles that grew more slowly, predation by *Thais* may have increased the rate of production of biomass. Therefore the predators, by selecting the larger barnacles, may have evolved a method of harvesting tending toward the "optimum yield." This provides an example from a natural population paralleling that of the artificial predation on fish by fishermen using nets which select the larger sizes. Fishermen, of course, strive for an optimum yield, with varying success.

In discussing the mortality at high shore levels, it was shown that annual variations in the mortality and strength of successive year classes were greater there than at lower levels. As can be seen in Fig. 21 more extreme variations occur at low than at middle levels. At the higher level the fluctuations were undoubtedly due to chance occurrence in the weather, such as occurred in the spring of 1955, although even less extreme conditions could probably cause destruction in young barnacles. At the low levels, overpopulation led to the formation of unstable "hummocks" since growth was faster at these levels. In addition, the numbers of *Thais* showed greater fluctuations at low than at mid shore levels, as shown in comparing areas 3 and 1, Fig. 13. Thus fluctuations in the physical factors decrease toward lower levels while variations in the biotic factors, such as crowding and predation, decrease at higher levels. Since mortality at high levels is due mostly to physical factors (there being little evidence of predation from the land, by mammals, birds, etc.), while mortality at low levels is mainly due to biotic factors, the damping of the fluctuations in both these as mid shore levels are approached tends to result in less variability in the mortality here. The intertidal zone of *Balanus balanoides* is thus bounded by physical factors at the top and biotic ones at the bottom, with fluctuations in mortality caused by these factors being greatest at the limits and decreasing toward the mid tide region.

SUMMARY

The accessibility and sessile nature of intertidal barnacles allowed very accurate records to be made on recruitment and survival under natural conditions.

Recruitment occurs as settlement after a planktonic larval period. The population density of *Balanus balanoides* after settlement was determined by the space available and the mortality during or just after settlement, rather than by a limited supply of larvae from the plankton, in most of the years studied. Attached larvae were killed during warm weather, but once metamorphosed, most early mortality was a result of damage from wave-borne material. Those young barnacles in hollows suffered least, so that the settlement density on a surface was partly a function of the proportion of protected sites. Larvae were liberated by adults into the plankton at intervals. Some of the first settlers were smaller and did not survive well, while those arriving later in the season were forced to settle in less favorable sites and suffered heavier mortality. Limpets damaged newly settled barnacles, but no other animals appeared to cause mortality at this time.

When the barnacles began to grow they soon touched each other at moderate population densities. Crowding then ensued, and barnacles were killed when they were undercut and displaced, smothered, or crushed by growing neighbors. In the first growing season, mortality was greater when population density was higher. At low shore levels the relative growth rate was faster and crowding and mortality were consequently greater. Crowding also resulted in some barnacles assuming unstable growth forms. These barnacles were detached in later storms and so there was sometimes a lag in the mortality due to crowding. Growth was slower in later years, and older barnacles were affected by crowding only when young barnacles attached themselves to the older ones and smothered them. When barnacles were protected from predation in cages, the mortality de-

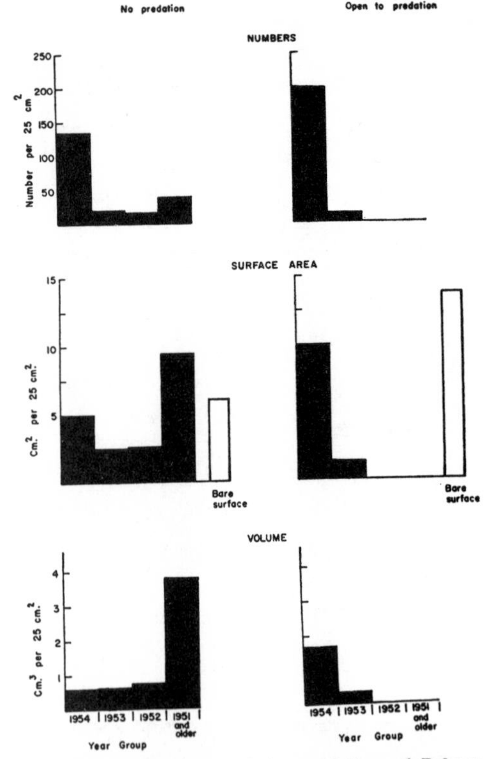

FIG. 22. A comparison of the populations of *Balanus balanoides* on two adjacent areas, Middle Cage 1, left, and Control 1, right, after the former had been protected from predation for 2 1/4 years. The data were taken from a photograph made on February 18, 1955, shown in Figure 19.

creased during the first winter and increased slightly during the next spring and summer, decreasing in the following years. This is also the pattern of growth, and indicates that intraspecific competition for space had a great effect on survival during the first year of life.

The upper shore limit of distribution was probably set by adverse weather conditions. Most of the mortality there occurred in the first year of life. Occasional periods of warm, calm weather may kill most of the younger barnacles at high levels; the erratic nature of this factor produces dominant year classes there.

The most important predator was a gastropod, *Thais lapillus*. It did not occur as high on the shore as *Balanus*, and was scarce below the lower limit of *Balanus*. It was less abundant during the winter and spring, which is its spawning season. In summer it fed about 60% of the time, and moved about 10 cm per tide, with much variation. Its feeding rate on barnacles was greatest in summer, probably due to the greater frequency of feeding excursions at that time.

From experiments where *Thais* was excluded from small populations of *Balanus* it was discovered that *Thais* selected the larger barnacles as prey; direct observations confirmed this finding. This behavior accounts for the changes in age structure of the barnacle population at different shore levels. Above the upper limit of *Thais* distribution several year groups of barnacles occurred together. Lower on the shore the older barnacles were selected by *Thais* and the number of concurrent age groups was less there. Near the lower limit of *Balanus* distribution where fewer older barnacles occur, predation began earlier in the life of the barnacles.

Previous studies of the growth rate of *Balanus* made by measuring samples of the population resulted in erroneous conclusions, partly because the mortality at lower levels (with *Thais* predation) was not random with respect to size.

Calculations of the proportion of the total mortality caused by predation by *Thais* showed that in the summer it could account for all of the mortality of barnacles older than six months. In other seasons it was less important. Annual variations in the intensity of predation occurred.

The evidence from the measurement of feeding rate of the predators suggested that they could open only a certain number of barnacles per day, regardless of size. If so, it provides evidence for the selective advantage to *Thais* of a behavior trait for selecting larger barnacles as prey. In addition, the evolution of such a trait would tend to increase the productivity of the barnacle population, by providing proportionately greater space for new settlers which grew relatively faster. Thus this behavior tends to produce the "optimum yield" for the predator population.

The barnacle zonation on the intertidal shore seems to be bounded at the top by deleterious physical factors (weather), at the bottom by biological ones (competition for space and predation). The fluctuations in these factors were greatest at the upper and lower limits; the recruitment and mortality were less variable at the middle shore.

LITERATURE CITED

Allee, W. C., A. E. Emerson, O. Park, T. Park & K. P. Schmidt. 1949. Principles of Animal Ecology. Philadelphia: W. B. Saunders Co.

Barnes, H. 1953a. Size variations in the cyprids of some common barnacles. J. Mar. Biol. Ass. U. K. **32:** 297-304.

———. 1953b. Orientation and aggregation in *Balanus balanus* (L.) Da Costa. J. Anim. Ecol. **22:** 141-148.

———. 1955. Climatological and salinity data for Millport, Scotland. Glas. Nat. **17:** 193-204.

———. 1956. *Balanus balanoides* L. in the Firth of Clyde: the development and annual variation in the larval population and the causative factors. J. Anim. Ecol. **25:** 72-84.

Barnes, H. & H. T. Powell. 1950. The development, general morphology and subsequent elimination of barnacle populations, *Balanus crenatus* and *B. balanoides*, after a heavy initial settlement. J. Anim. Ecol. **19:** 175-179.

———. 1953. The growth of *Balanus balanoides* (L.) and *B. crenatus* Brug. under varying conditions of submersion. J. Mar. Biol. Ass. U. K. **32:** 107-128.

Bassindale, R. 1936. The developmental stages of three English barnacles, *Balanus balanoides* (Linn.), *Chthamalus stellatus* (Poli) and *Verruca stroemia* (O. F. Muller). Proc. Zool. Soc. Lond. 1936: 57-74.

Blackman, G. E. 1942. Statistical and ecological studies in the distribution of species in plant communities. I. Dispersion as a factor in the study of changes in plant populations. Ann. Bot. **6:** 351-70.

Blower, J. G. 1957. Feeding habits of a marine centipede. Nature, Lond. **180:** 560.

Bousfield, E. L. 1954. The distributions and spawning seasons of barnacles on the Atlantic Coast of Canada. Bull. Nat. Mus. Can. **132:** 112-154.

———. 1955. Ecological control of the occurrence of barnacles in the Miramichi estuary. Bull. Nat. Mus. Can. **137:** 1-69.

Burrows, E. M. & S. M. Lodge. 1950. Note on the interrelationships of *Patella, Balanus* and *Fucus* on a semi-exposed coast. Rep. Mar. Biol. Sta. Pt. Erin. **62:** 30-34.

Carruthers, J. N., A. L. Lawford, V. F. C. Veley & B. B. Parrish. 1951. Variations in brood-strength in the North Sea haddock in the light of relevant wind conditions. Nature, Lond. **168:** 317-19.

Chipperfield, P. N. J. 1948. Unpublished Ph. D. thesis, Liverpool University, England.

Clench, W. J. 1947. The genera *Purpura* and *Thais* in the western Atlantic. Johnsonia 2, No. 23: 61-92.

Crisp, D. J. 1950. Breeding and distribution of *Chthamalus stellatus*. Nature, Lond. **166:** 311.

Crisp, D. J. & H. Barnes. 1954. The orientation and distribution of barnacles at settlement with particular reference to surface contour. J. Anim. Ecol. **23:** 142-162.

Crisp, D. J. & P. A. Davies. 1955. Observations *in vivo* on the breeding of *Elminius modestus* grown on glass slides. J. Mar. Biol. Ass. U. K. **34:** 357-80.

Darwin, C. 1854. A monograph of the sub-class Cirripedia. Vol. II. London: Ray Society.

Deevey, E. S., Jr. 1947. Life tables for natural populations of animals. Quart. Rev. Biol. **22:** 283-314.

Fischer-Piette, E. 1932. Repartition des principales especes fixees sur les rochers battus des cotes et des iles de la Manche de Lannion a Fecamp. Ann. Inst. Oceanog. **12:** Fasc. 4.

———. 1935. Historie d'une mouliere. Bull. Biol. **69:** 152-177.

Gibb, J. 1956. Food, feeding habits and territory of the rock pipit *Anthus spinoletta*. Ibis. **98:** 506-530.

Hanks, J. E. 1957. The rate of feeding of the common oyster drill, *Urosalpinx cinerea* (Say) at controlled temperatures. Biol. Bull. Woods Hole **112:** 330-335.

Hatton, H. 1938. Essais de bionomie explicative sur quelques especes intercotidales d'algues et d'animaux. Ann. Inst. Oceanogr. Monaco **17:** 241-348.

Hatton, H. & E. Fischer-Piette. 1932. Observations et experiences sur le peuplement des cotes rocheuses par les cirripedes. Bull. Inst. Oceanogr. Monaco **592.**

Hjort, J. 1926. Fluctuations in the year classes of important food fishes. J. Cons. Int. Explor. Mer **1:** 1-38.

Holme, N. A. 1950. Population-dispersion in *Tellina tenuis* Da Costa. J. Mar. Biol. Ass. U. K. **29**: 267-280.

Jones, N. S. 1948. Observations and experiments on the biology of *Patella vulgata* at Port St. Mary, Isle of Man. Proc. Liverpool Biol. Soc. **56**: 60-77.

Knight-Jones, E. W. 1953. Laboratory experiments on gregariousness during setting in *Balanus balanoides* and other barnacles. J. Exp. Biol. **30**: 584-598.

Knight-Jones, E. W. & D. J. Crisp. 1953. Gregariousness in barnacles in relation to the fouling of ships and to anti-fouling research. Nature, Lond. **171**: 1109.

Kuznetzov, W. W. & T. A. Matveeva. 1949. The influence of the density of the populations on certain biological processes in *Balanus balanoides* (L.) from Eastern Murman. C. R. Acad. Sci. U. R. S. S. **64**: 3. (In Russian)

Lewis, J. R. 1954a. Observations on a high-level population of limpets. J. Anim. Ecol. **23**: 85-100.

———. 1954b. The ecology of exposed rocky shores of Caithness. Trans. Roy. Soc. Edin. **62**: 695-723.

Lund, S. 1936. On the production of matte and the growth in some benthic plants. Rep. Danish Biol. Sta. **61**.

Miller, M. A., J. C. Repean & W. F. Whedon. 1948. The role of slime film in the attachment of fouling organisms. Biol. Bull. Woods Hole **94**: 143-157.

Moore, H. B. 1934. The biology of *Balanus balanoides*. I. Growth rate and its relation to size, season and tidal level. J. Mar. Biol. Ass. U. K. **19**: 851-68.

———. 1935a. The biology of *Balanus balanoides*. III. The soft parts. J. Mar. Biol. Ass. U. K. **20**: 263-277.

———. 1935b. The biology of *Balanus balanoides*. IV. Relation to environmental factors. J. Mar. Biol. Ass. U. K. **20**: 279-308.

———. 1936a. The biology of *Balanus balanodies*. V. Distribution in the Plymouth Area. J. Mar. Biol. Ass. U. K. **20**: 701-16.

———. 1936b. The biology of *Purpura lapillus*. I. Shell variation in relation to environment. J. Mar. Biol. Ass. U. K. **21**: 61-89.

———. 1938a. The biology of *Purpura lapillus*. Part II. Growth. J. Mar. Biol. Ass. U. K. **23**: 57-66.

———. 1938b. The biology of *Purpura lapillus*. Part III. Life history and relation to environmental factors. J. Mar. Biol. Ass. U. K. **23**: 67-74.

Moore, H. B. & J. A. Kitching. 1939. The biology of *Chthamalus stellatus* (Poli). J. Mar. Biol. Ass. U. K. **23**: 521-541.

Orton, J. H. & H. M. Lewis. 1931. On the effect of the severe winter of 1928-29 on the oyster drills of the Blackwater estuary. J. Mar. Biol. Ass. U. K. **17**: 301-314.

Parke, M. W. & H. B. Moore. 1935. The biology of *Balanus balanoides*. II. Algal infection of the shell. J. Mar. Biol. Ass. U. K. **20**: 49-56.

Pearl, R. & J. R. Miner. 1935. Experimental studies on the duration of life. XI. The comparative mortality of certain lower organisms. Quart. Rev. Biol. **10**: 60-79.

Pilsbry, H. A. 1916. The sessile barnacles (Cirripedia) contained in the collections of the U. S. National Museum; including a monograph of the American species. Bull. U. S. Nat. Mus. **93**: 1-366.

Pyefinch, K. A. 1948a. Methods of identification of the larvae of *Balanus balanoides* (L.), *B. crenatus* Brug. and *Verruca stroemia* O. F. Muller. J. Mar. Biol. Ass. U. K. **27**: 451-63.

———. 1948b. Notes on the biology of Cirripedes. J. Mar. Biol. Ass. U. K. **27**: 464-503.

Qasim, S. Z. 1957. The biology of *Blennius pholis* L. (Teleostei). Proc. Zool. Soc. Lond. **128**: 161-208.

Quayle, D. 1952. The rate of growth of *Venerupis pullastra* (Montagu) at Millport, Scotland. Proc. Roy. Soc. Edinb. B. **64**: 384-406.

Rice, L. A. 1935. Factors controlling arrangement of barnacle species in tidal communities. (in: Some marine biotic communities of the Pacific Coast of North America.) Ecol. Monogr. **5**: 293-304.

Runnstrom, S. 1925. Zur biologie und entwicklung von *Balanus balanoides* (Linne). Univ. Bergen Arg. naturv. R. **5**: 1-46.

Schafer, W. 1948. Wuchsformen von Seepocken (*Balanus balanoides*.) Natur u. Volk **78**: 74-78.

Segal, E., K. P. Rao & T. W. James. 1953. Rate of activity as a function of intertidal height within populations of some littoral molluscs. Nature, Lond. **172**: 1108-09.

Sette, O. E. 1943. Biology of the Atlantic mackerel (*Scomber scombrus*) of North America. Part 1: Early life history, including growth, drift, and mortality of the egg and larval population. Fish. Bull., U. S. **50** (38): 147-237.

Siegel, S. 1956. Nonparametric statistics for the behavioral sciences. New York: McGraw-Hill.

Sokolova, M. N. 1951. Influence of the conditions of the environment on the density of the colony and the shape of the shell in *Balanus balanoides*. C. R. Acad. Sci. U. R. S. S., N. S. **78**: 1227-1230. (In Russian.)

Southward, A. J. 1953. The ecology of some rocky shores in the south of the Isle of Man. Proc. Lpool. Biol. Soc. **59**: 1-50.

———. 1955. On the behaviour of barnacles. II. The influence of habitat and tide-level on cirral activity. J. Mar. Biol. Ass. U. K. **34**: 423-33.

———. 1956. The population balance between limpets and seaweeds on wave-beaten rocky shores. Rep. Mar. Biol. Sta. Pt. Erin, No. **68**: 20-29.

———. 1958. Note on the temperature tolerances of some intertidal animals in relation to environmental temperatures and geographical distribution. J. Mar. Biol. Ass. U. K. **37**: 49-66.

Southward, A. J. & D. J. Crisp. 1954. Recent changes in the distribution of the intertidal barnacles *Chthamalus stellatus* Poli and *Balanus balanoides* L. in the British Isles. J. Anim. Ecol. **23**: 163-177.

Thorson, G. 1950. Reproductive and larval ecology of marine bottom invertebrates. Biol. Rev. **25**: 1-45.

———. 1958. Parallel level-bottom communities, their temperature adaptation, and their "balance" between predators and food animals. In: Perspectives in marine biology, Univ. Calif. Press, 67-86.

Trusheim, F. 1932. Palaontologisch Bemerkens wertes aus der Okologie rezenter Nordsee-Balaniden. Senckenbergiana **14**: 70-87.

Weiss, C. M. 1948a. The seasonal occurrence of sedentary marine organisms in Biscayne Bay, Florida. Ecology **29**: 153-172.

———. 1948b. Seasonal and annual variations in the attachment and survival of barnacle cyprids. Biol. Bull. Woods Hole **94**: 236-243.

Woods Hole Oceanographic Institution. 1952. Marine fouling and its prevention. U. S. Naval Institute, Annapolis, Maryland. (W.H.O.I. contribution No. 580).

Reprinted from the Proceedings of the NATIONAL ACADEMY OF SCIENCES
Vol. 45, No. 4, pp. 617–622. April 1959.

*COMPARISON OF POPULATION ENERGY FLOW OF A HERBIVOROUS AND A DEPOSIT-FEEDING INVERTEBRATE IN A SALT MARSH ECOSYSTEM**

BY EUGENE P. ODUM AND ALFRED E. SMALLEY†

UNIVERSITY OF GEORGIA, ATHENS

Communicated by G. Evelyn Hutchinson, February 16, 1959

Very frequently the autotrophic and heterotrophic components of an ecosystem are partially separated in space in that they are stratified one above the other (vegetation-soil on land, phytoplankton-sediments in water). Also, the basic functions are usually partially separated in time in that there may be a considerable delay in the heterotrophic utilization of a large portion of the net production of autotrophic organisms. Consequently, between the first and second trophic levels the energy flow of the community is often divided into two broad streams resulting in two types of primary consumption: (1) direct and immediate utilization of living plant tissues by herbivores and plant parasites, and (2) delayed utilization of dead tissues and stored food by other consumers. In most ecosystems many species of primary consumers rely exclusively on one or the other types of food source while other species utilize both, or shift from one to the other seasonally.

In addition to the problem of time-lag in utilization, one of the difficulties in developing a satisfactory comparative population ecology of heterotrophs is the fact that different species, and also life history stages of the same species, associated together in a given area of the earth's surface differ greatly in size and rate of metabolism. In general, numbers tend to overemphasize and biomass to underemphasize the importance of small organisms in the community, while the reverse tends to be true with large organisms. However, if changes in numbers and biomass can be integrated and the population energy flow determined, then quite diverse populations may be compared in so far as their true impact on the community is concerned. While the general pattern of community metabolism has been worked out for a number of ecosystems,[1, 2] energy flow at the population level has as yet received very little attention. At the Sapelo Marine Institute a concerted effort by several investigators is being made to work out the details of energy flow in tidal marshes where the limited taxa and uniform habitat simplify sampling. In this paper we wish to present a comparison of the annual pattern of population energy flow of two invertebrate species, a grasshopper and a snail, which represent, to a partial extent at least, the two basic types of primary consumers described above. The grasshopper, *Orchelimum fidicinium*, is a strict herbivore, both nymphs and adults feeding exclusively on living marsh grass, *Spartina alterniflora* (especially growing grass tips); it is thus closely associated with the autotrophic layer of the community and reaches its greatest abundance in the taller grass (1–2 meters) of the low marsh. There is but one generation per year. The snail, *Littorina irrorata*, lives on the surface of the sediments and on the stems of marsh grass (especially dead stems); it is a detritus-aufwuchs feeder and is abundant in the high marsh where *Spartina* is dwarfed (one-half meter tall or less).

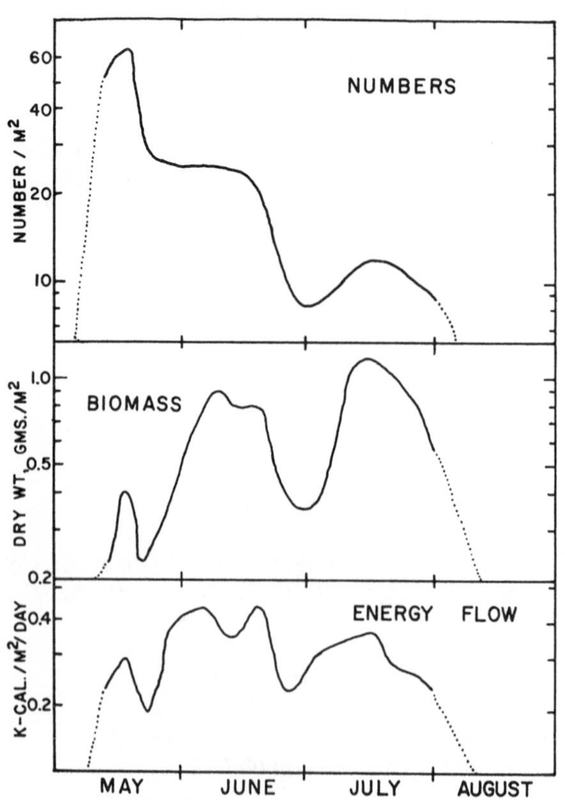

FIG. 1.—Numbers, biomass (dry weight), and energy flow per square meter in a population of salt marsh grasshoppers (*Orchelimum fidicinium*) living in the low *Spartina* marsh, Sapelo Island, Georgia.

Details of methods and the tabular data are presented elsewhere.[3] Briefly, the procedures were as follows: The numbers and biomass per square meter were determined at frequent intervals (every 3–4 days for the rapidly changing grasshopper population and monthly for the more stable snail population). From these data, the population growth, or production, was determined by adding the increase in weight of survivors to the growth of individuals which died during the census interval. Production was then converted to Calories/M²/day. Oxygen consumption (respiration) of different age or size groups in relation to temperature was then determined in the laboratory. From these data population respiration of the average standing crop during each interval was calculated and adjusted to the actual temperature of the natural environment. Oxygen consumption was converted to Calories by means of an oxycaloric coefficient.[4] The total population assimilation rate, or energy flow, was then obtained by adding production to respiration.

The annual cycles of numbers, biomass (as dry weight, not including shell of the snail), and energy flow of the grasshopper and snail populations are shown in Figures 1 and 2. Numbers of grasshoppers were at a peak in late May, when numerous small nymphs hatched out from overwintering eggs, and then declined rapidly during the season. Two periods of heavy mortality were followed by accelerated growth of survivors, and also by a small amount of recruitment (new eggs hatching in early summer, immigration in late summer). A definite homeostasis or self-adjustment in the population was evident since energy flow fluctuated only 2-fold while numbers and biomass varied 5- to 6-fold. Since metabolism-per-gram of small nymphs was several times greater than that of adults, the high number but small biomass population of spring was about equal to the low number but large biomass population of late summer. Note also that the peak in energy flow (and hence the time of greatest consumption of marsh grass) did not correspond with either maximum numbers or maximum biomass (although more nearly with the latter than the former), but occurred when the population was composed of a medium number of medium-sized nymphs all growing rapidly.

The snail population comprised two distinct components, small young snails with

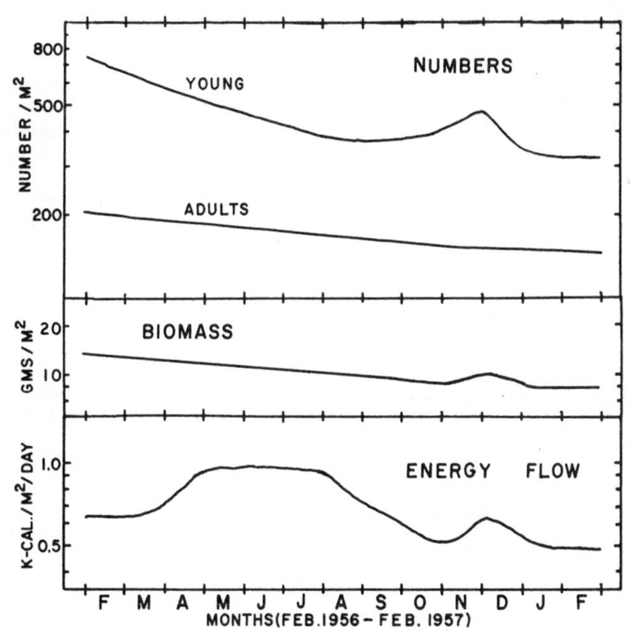

FIG. 2.—Numbers, biomass (dry weight, not including shell), and energy flow per square meter in a population of periwinkle snails (*Littorina irrorata*) living in the high *Spartina* marsh, Sapelo Island, Georgia.

a shell length of 1–6 mm, and large adults with a shell length of 16–21 mm; individuals of intermediate size were scarce. The young snails were growing but suffering heavy mortality while the adult group was characterized by little or no growth and a low mortality rate. During 1956 the young snail component was dominated by the very large 1955 year class recruited from the plankton in the autumn of that year. The 1956 year class was very small producing only a temporary increase in density of young snails (Fig. 2). The numbers of adult snails remained relatively constant throughout the year. It was evident from the population structure that an occasional massive "set" of young snails (perhaps occurring only once in several years) replenishes the population of long-lived adults, and that very few, if any, snails from a small recruitment (such as that of 1956) would survive to adult size. Despite the large numbers and high metabolism-per-gram the young snails made but a minor contribution to population energy flow because of their very small total biomass and low survival. The snail population was thus metabolically, but not numerically, dominated by the stable population of large snails. Consequently, the annual pattern of energy flow was largely regulated by temperature, assimilation being greater during the warmer months (Fig. 2). Only about 10 per cent of the annual total energy flow of the snail population was production (i.e., 90 per cent was respiration) while more than a third of the grasshopper energy utilization was production.

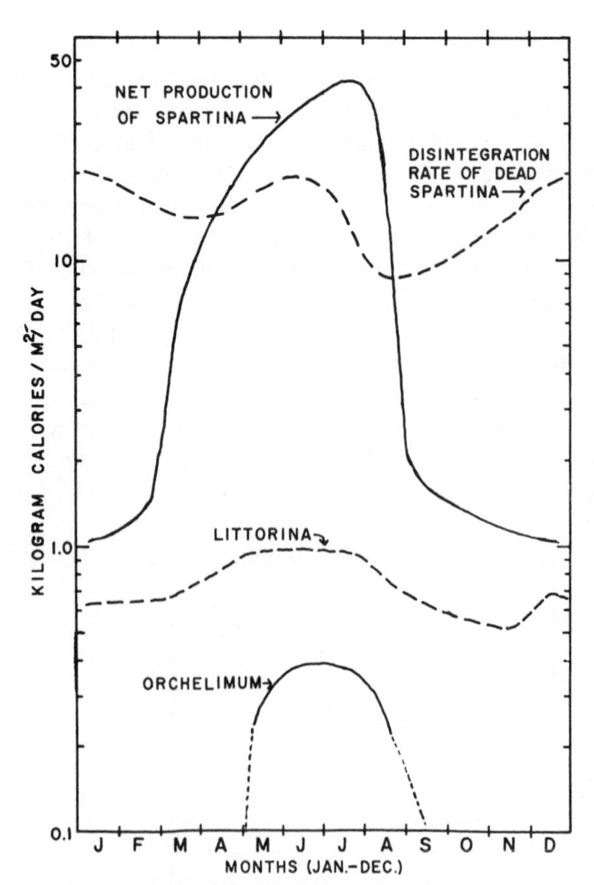

FIG. 3.—Comparison of the annual pattern of energy flow in *Littorina* and *Orchelimum* populations in relation to certain potential food sources. Net production of *Spartina* in the low marsh is the sole source of food energy for *Orchelimum* while the disintegration of dead *Spartina* through the entire marsh and the subsequent transport of detritus (including associated microflora) to the high marsh provides one potential food source for *Littorina*.

In Figure 3 the broad annual patterns of energy flow of the grasshopper and the snail populations are compared. Shown also are measurements of net production of the low marsh *Spartina* on which the grasshopper feeds, and our preliminary estimate of the rate of formation of *Spartina* detritus which comprises at least a part of the food of the snail population. Actually we do not have a good estimate of detritus "production," but the rate of disappearance of dead grass from quadrats may provide a rough estimate—probably an overestimate since some decomposed grass is undoubtedly lost to estuaries and deeper sediments. The point to emphasize is that detritus is produced more or less continually throughout the year; more dead grass is present in the colder months, but this is counterbalanced by the more rapid decomposition during the warmer months.[5] *Littorina* undoubtedly also consumes algae and other aufwuchs; the proportion of different food items actually utilized in nature is unknown. Pomeroy[6] has shown that net production of "mud algae" (algae living on and in the soft sediments alternately exposed and covered by tides), while not as great as that of marsh grass, makes a major contribution to the total primary productivity of the salt marsh ecosystem. The rate proved to be almost constant the year around with production occurring in summer chiefly when the sediments are covered by the high tide, and in winter when the sediments are exposed and quickly warmed by the sun. Consequently, *Littorina* has a more or less constant supply of both detritus and algae available throughout the year. It is evident from Figures 1–3 that the population structure of both *Littorina* and *Orchelimum* is such that the pattern of population energy flow of each species is well adjusted to the utilization of the particular sources of food potentially available to each population.

Certain ratios or "ecological efficiencies" based on the total annual energy flow are shown in Table 1. Ingestion was estimated by adding the caloric value of feces to assimilation. The most rapidly growing segment of the *Littorina* population (the 1955 year class) proved to have a higher assimilation efficiency but a lower growth efficiency as compared with *Orchelimum*. If all the snails in the population are considered the growth efficiency is even lower. Thus, the slow-growing, long-lived snails apparently assimilate a considerable part of ingested food but use a relatively small amount of the assimilated matter for growth. In contrast, the grasshoppers—which emerge, grow and die within a short period—assimilate a smaller percentage of ingested food but convert nearly 40 per cent of the assimilated material into protoplasm.

TABLE 1

Ecological Efficiencies of *Littorina* (1955 Year Class Only) and *Orchelimum* Populations on an Annual Basis

	Assimilation Efficiency: Assimilation/Ingestion, %	Growth Efficiency: Production/Assimilation, %
Littorina	45	14
Orchelimum	36	37

A high population growth efficiency would presumedly be of survival value for a population which depends on a seasonal "bloom" of primary production, whereas a high assimilation efficiency would presumedly benefit a population whose food is produced at a lower but more continuous rate. Whether an increase in assimilation efficiency tends to result in a decrease in growth efficiency (and vice versa) is, as yet, an unanswered question.

The total annual energy flow of the grasshopper population was estimated to be about 28 kg. calories per square meter per year. During the summer the population assimilated less than 1 per cent of the net production of grass, but ingested three times as much, that is, about 84 Calories or 21 gm. dry weight of grass (4 Calories/

gm.) per square meter. The nonassimilated material (feces) does not change trophic levels but is available to other primary consumers such as bacteria or detritus feeders. The annual energy flow of the snail population was estimated to be about 290 Calories/M^2/year, but during the summer the average standing crop of 700 snails/M^2 assimilated only about twice as much as the average population of 10–20 grasshoppers/M^2 (Fig. 3). While it would appear that both populations utilize only a small percentage of the total net primary production of the ecosystem it should not be assumed that food is nonlimiting. The maximum portion of the net production which is actually available to the populations has not yet been determined, nor has the utilization of competing primary consumers been considered. The fact that growth rate per individual in snails declined with increasing density in the field suggests that the rate at which nourishment can be ingested and assimilated on a square meter of marsh is not unlimited.

It is evident that populations which differ greatly in life history characteristics, age structure, and metabolic rate cannot be compared on the basis of numbers and biomass (i.e., "standing crop") alone. However, through the common denominator of energy flow valid comparisons can be made, and the true role of the populations in the community can be evaluated. Energy flow analysis is the logical first step in finding out how populations really function in nature.

The authors wish to thank Drs. R. A. Ragotzkie, J. M. Teal and L. R. Pomeroy of the University of Georgia Marine Institute for collaborative data and suggestions.

* Contribution No. 14 from the University of Georgia Marine Institute, Sapelo Island, Georgia. This research was supported by funds from the Georgia Agricultural and Forestry Research Foundation and from the National Science Foundation.

† Present address. Department of Zoölogy, University of Kentucky, Lexington.

[1] Odum, Howard T., *Ecol. Monogr.*, **27**, 55 (1957).

[2] Teal, John M., *Ecol. Monogr.*, **27**, 283 (1957).

[3] Smalley, Alfred E., Ph.D. Thesis, University of Georgia, 1959.

[4] Ivlev, V. S., *Biochem. Ztschr.*, **275**, 49 (1934).

[5] Burkholder, P. R., and G. H. Bornside, *Bull. Torry Bot. Club*, **84**, 366 (1957).

[6] Pomeroy, L. R., *Limnol. and Oceanogr.*, **4** (in press).

ENERGY FLOW IN THE SALT MARSH ECOSYSTEM OF GEORGIA[1]

JOHN M. TEAL

Woods Hole Oceanographic Institution, Woods Hole, Massachusetts

INTRODUCTION

Along the coast of the United States from northern Florida to North Carolina runs a band of salt marsh bordered on the east by a series of sea islands and on the west by the mainland. The Marine Institute of the University of Georgia was established on one of these islands, Sapelo, and has tended to focus attention on the marsh. Several studies have provided data from which it is now possible to construct a picture of the energy flow through the organisms of this marsh.

Reasonably detailed studies of the energy flow, or trophic level production have been limited to a few natural ecosystems. These include Cedar Bog Lake, reported in the pioneer work of Lindeman (1942), and 2 fresh-water springs (Odum 1957, Teal 1958). There have been a number of studies of the energetics of laboratory populations (Richman 1958, Slobodkin 1959), and some theoretical comments upon energetics of populations and ecosystems (e.g. Patten 1959, Slobodkin 1960), but work on even the broad details of energy flow in natural ecosystems has lagged.

The present paper draws heavily upon the work of others. The authors are cited in the appropriate places but I wish here to express my appreciation for their cooperation.

[1] Contribution No. 38 from the University of Georgia Marine Institute, Sapelo Island, Georgia. This research was supported by funds from the Sapelo Island Research Foundation and by N.S.F. grant G-6156.

The physical and chemical features of the marsh have been described (Teal 1958, Teal & Kanwisher 1961) but I will briefly define 5 regions into which the marsh was divided in many of these studies (Figure 1).

Creek bank: muddy and/or sandy banks of tidal creeks between low water and the beginning of *Spartina* growth.

Streamside marsh: an area 1-3 m wide of closely spaced, tall *Spartina* located just above the bare creekbank.

Levee marsh: *Spartina* of intermediate height spacing atop the natural levees bordering the creeks.

Short-*Spartina* marsh: flat areas behind the levees with short, widely spaced *Spartina*.

Salicornia marsh: sandy areas near land where plants other than *Spartina* occur, among which *Salicornia* is conspicuous.

The relative areas of these various marsh types were measured on aerial photographs (Table V) to enable calculation of averages for the marsh as a whole.

The marsh fauna

Animals living in the marsh must be able to survive or avoid the great changes in salinity, temperature and exposure. Salinity of water flooding the marshes varies from 20 to 30 o/oo with values as low as 12 o/oo recorded in heads of creeks just after heavy rains. Salinity of water in the mud may be 5 o/oo in isolated areas where fresh water drains from the islands and 70 o/oo in isolated low areas during rainless summer periods. An average aquatic or soil animal must be able to withstand variations from 20 to 30 o/oo but probably escapes greater extremes by burrowing in the mud and/or migrating short distances.

The limited number of animals which have adapted to these extremes are relatively free from competing species and enemies. For example, mussels living in the marsh are bothered by neither snails nor echinoderms, which take great toll of the estuarine bivalves living only a few meters away. Ants and grasshoppers are each represented by only one common species, *Crematogaster clara* and *Orchelimum fidicinium* respectively, which is quite abundant in marsh areas optimal for it. Once adapted to the marsh, the lack of competition from similar animals has perhaps allowed them to occupy a broader niche and

TABLE I. Known macro-fauna of a Georgia salt marsh listed by groups according to distribution and origin

(1a) Terrestrial species living in marsh
Orchelimum fidicinium Rehn & Hebard
Ischnodesmus sp.
Prokelisia marginata (Van Duzee)
Liburnia detecta Van Duzee
Tabanus spp.
Culicoides canithorax Hoffman
Dimicoenia spinosa (Loew)
Plagiopsis aneo-nigra (Loew)
Parydra vanduzeei (Creeson)
Chaetopsis aenea (Wiedermann)
Chaetopsis apicalis Johnson
Haplodictya setosa (Coquillett)
Mordellid sp.
Crematogaster clara Mayr
Camponotus pylartes fraxinicola M. R. Smith
Hyctia pikei Peckham
Seriolus sp.
Lycosa modesta (Keyserling)
Philodromus sp.
Grammonata sp.
Hyctia brina (Hentz)
Eustala sp.
Singa keyserlingi McCook
Tetragnatha vermiformis Emerton
Dictyna sp.
Rallus longirostris Boddaert
Termatodytes palustris (Wilson)
Ammospiza caudacuta (Gmelin)
A. maritima (Wilson)
Oryzomys palustris (Harlan)
Procyon lotor (Linne)
Mustela vison Schreber

(1b) Terrestrial or fresh-water species only on landward edge of marsh
Pachydiplax longipennis Burmeister
Pantala flavescens Fabrisius
Erythrodiplax verenice Drury
Anax junius Drury
Erythemis simplicicollis Say
Orphylella sp.
Platunus cincticollis (Say)
Kinesternum s. subrubrum (Lacepede)
Lutra canadensis (Schreber)

(2a) Estuarine species limited in marsh to low water level
Bouganvillia carolinensis (McCrady)
Campanularid sp.
Oerstedia dorsalis burger
Nolella stipata Gosse
Eteone alba Webster
Autolytus prolifer (O. F. M.)
Polydora ligni Webster
Heteromastis filiformis (Claparede)
Crassostrea virginica (Gmelin)
Mercenaria mercenaria (Linne)
Tagelus plebeius Solander
T. divisus Spengler
Mulinia lateralis Say
Epitomium rupicolum (Kurtz)
Balanus improvisus Darwin
Microprotopus maculatoides Shoemaker
Paracaprella sp.
Crangon heterochelis (Say)
Clibanarium vittatus (Bosc)
Molgula manhattensis (DeKay)

(2b) Estuarine species in streamside marsh
Nassarius obsoletus Say
Chthamalus fragilis Darwin
Neomysis americana (S. I. Smith)
Leptochelia rapax Harger
Cassidisca lumifrons (Richardson)
Gammarus chesapeakensis Bousfield
Melita nitida Smith
Callinectes sapidus Rathbun
Panopeus herbstii Milne-Edwards
Eurypanopeus depressus (Smith)
Malaclemys terrapin centrata (Latreille)

(2c) Estuarine species occurring well into marsh
Neanthes succinea (Frey & Leuckart)
Laeonereis culveri (Webster)
Streblospio benedicti Webster
Capitella capitata (Fab.)
Orchestia grillus (Bosc.)
O. Platensis Kroyer

(3a) Aquatic marsh species with planktonic larvae
Manayunkia aestuarina (Bourne)
Modiolus demissus Dillwyn
Polymesoda caroliniana Bosc
Littorina irrorata (Say)
Littoridina tenuipes (Couper)
Eurytium limosum (Say)
Sesarma reticulatum (Say)
S. cinereum (Bosc)
Uca pugilator (Bosc)
U. minax (LeConte)
U. pugnax (S. I. Smith)

(3b) Aquatic marsh species living entirely within marsh
Oligochaetes-3 spp.
Melampus bidentatus Say
Cyathura carinata (Kroyer)
Orchestia uhleri Shoemaker

be more abundant than would otherwise be possible.

Table I lists the marsh fauna divided into several groups: (1) typically terrestrial insects and arachnids subdivided into those occurring throughout the marsh and those confined to the landward edge, (2) the aquatic species with their center of abundance in the estuaries and 2a confined to regions near low water, 2b occurring in the streamside marsh, or 2c occurring throughout the marsh, (3) marsh species derived from aquatic ancestors with their centers of distribution within the marsh which are subdivided into those with planktonic larvae and those that spend their entire life cycle in the marsh.

The list shows that of the aquatic species, 33 or 60% are in groups 2a and 2b, estuarine forms that have managed to colonize the lowest portion of the salt marsh. They can survive only where periods of exposure at low tides are short. The individuals in the marsh are living at one edge of their species' distribution and their numbers are maintained by migrations from the surrounding waters. Those living above the mud are especially subject to damage by extremes of weather, and species that have penetrated farthest into the marsh are burrowers. The remaining aquatic species are either tolerant enough to inhabit the entire marsh although they are most common in the estuaries, group 2c, or are most common in the marshes themselves. But even among the latter, only 6 do not spend part of their life cycle in the estuaries. These are isopods, amphipods, oligochaetes and the pulmonate snail, *Melampus*, the last 2 of which are derived from fresh water or terrestrial rather than marine ancestors.

That part of the fauna derived from the land is at present the least well known, but the species so far encountered comprise nearly half of the marsh animals. They are, however, far less important in the energetics of the community than their aquatic counterparts, as will be seen below.

Most of the terrestrial species have made only slight adaptation to the marshes. They breath air and resist salinity changes and desiccation by means of their impervious exoskeleton. Most of the larger forms climb the marsh grass to escape rising tides, but can climb under water to seek refuge from birds. Some insects, such as the ant, *Crematogaster*, are easily drowned, but live within *Spartina* stems and can effectively plug the entrance to their nests.

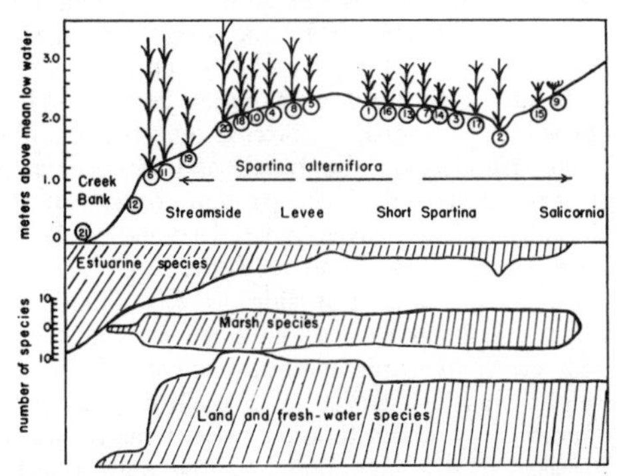

FIG. 1. Representative section of a Georgia salt marsh with horizontal scale distorted non-uniformly. Sample sites indicated by circled numbers. Site 2 represents the beginning of a drainage channel, not an isolated low spot. Symbols for grass are drawn to correct height for average maximum growth at those sites. The number of species of animals of 3 groups listed in Table I are plotted against sample sites. Names of marsh types used herein are also indicated.

The distribution of marsh species can be seen in Figure 1. The data are from the samples reported in this paper taken at the sites indicated as well as other samples and general collecting. Most of the insect and spider samples are from Smalley (unpublished). Estuarine species occur mostly near low water and in drainage channels and are reduced on the levees which may dry out between spring tides. The aquatic marsh species are distributed relatively evenly throughout the marsh. Terrestrial species are common throughout the grassy areas, with more species present on the higher ground and in the taller marsh-grass.

The food web

The herbivorous faunas of many ecosystems can be divided into 2 groups, those which feed directly on living plants and those which feed on plants only after the plants have died and fallen to the ground (Odum and Smalley 1959). The marsh fauna may be grouped in a similar manner (Figure 2).

A group of insects lives and feeds directly upon the living *Spartina*: *Orchelium*, eating the tissues, and *Prokelisia*, sucking the plant juices. These and their less important associates support the spiders, wrens, and nesting sparrows. A different group lives at the level of the mud surface and feeds on the detritus formed by bacterial decomposition of *Spartina* and on algae. These mud dwelling groups function mostly as primary consumers, although the detritus also contains animal remains and numbers of the bacteria that help break the *Spartina* into small pieces. The carnivores preying on the algal and detritus group are principally mud crabs, raccoons, and rails.

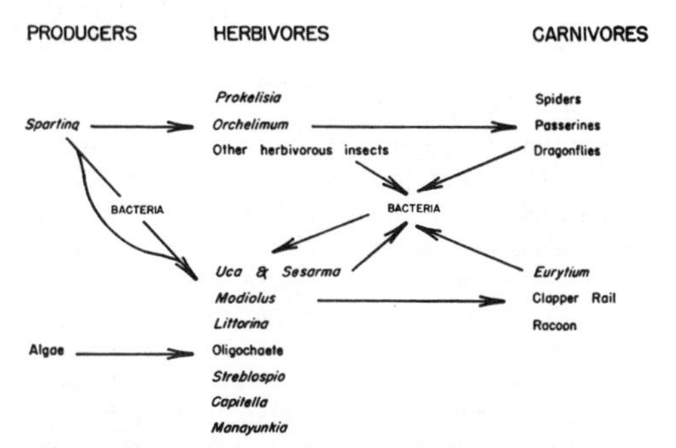

FIG. 2. Food web of a Georgia salt marsh with groups listed in their approximate order of importance.

The species of the detritus-algae feeding group that are important in the economy of the marsh are the fiddler crabs, oligochaetes, *Littorina,* and the nematodes among the deposit feeders, and *Modiolus* and *Manayunkia* among the suspension feeders. Thus, the community consists of 2 parts, one deriving its energy directly from the living *Spartina* and the other deriving its energy from detritus and algae.

Energy Flow by Trophic Groups

Methods

Some populations were sampled completely enough that production could be measured directly, e.g. *Spartina* and grasshoppers. In other cases production could not be determined from the sampling but was estimated either from turnover time, in which case production equals one maximum population per turnover period, or by assuming that the ratio between respiration and production for the group in question is 0.25 to 0.30 as has been found for other groups (Teal 1958, Slobodkin 1960). Respiration in air was the measure of energy degradation. Some of the mud dwelling forms live in completely anaerobic conditions (Teal & Kanwisher 1961), e.g. nematodes, and the energy degradation under anaerobic conditions is assumed to be equal to that in air. This is supported only by the observation of Wieser and Kanwisher (1961) that nematodes from anaerobic muds are as active under anaerobic conditions as under aerobic.

Primary production

The only higher plant of importance on the salt marsh is *Spartina alterniflora.* It grows over the entire marsh, is eaten by insects, then dies, decomposes and as detritus furnishes the food for much of the remaining fauna of the marsh. Smalley (1959) measured production of *Spartina* by harvesting and weighing plants at monthly intervals. Teal and Kanwisher (1961) measured respiration. Since the net production was determined by short-term harvesting, it is necessary to add the 305 kcal/m² yr consumed by the insects (see below) to arrive at the true net production. Table II shows that net production of *Spartina* comes to only 19% of gross production. There are indications (Odum 1961) that the production values are underestimated, but not the standing crops upon which respiration values are based. Furthermore, the *Spartina* used to measure respiration was collected in spring. If there is appreciable acclimation to temperature by this species, then summer values will be lower and winter values higher than indicated, but as summer contributes twice as much (4 vs. 2 months), an adjustment would lower the total figure for respiration. But assuming perfect summer acclimation would bring net production to only 24% of gross production.

TABLE II. Data for *Spartina* in Georgia salt marhes. Production figures from Smalley (1959), respiration rates from Teal and Kanwisher (1961)

Season		Short Spartina 42% of total area	Levee-Streamside 58% of total area
Winter 2 mo at 10°	Standing crop Respiration	300 fresh g/m² 235 kcal/m²	750 fresh g/m² 580 kcal/m²
Spring 3 mo at 17.5°	Standing crop Respiration	600 g/m² 1250 kcal/m²	1350 g/m² 2800 kcal/m²
Summer 4 mo at 26°	Standing crop Respiration	705 g/m² 6450 kcal/m²	3225 g/m² 29600 kcal/m²
Autumn 3 mo at 20°	Standing crop Respiration	900 g/m² 3240 kcal/m²	1800 g/m² 6480 kcal/m²
Production		2570 kcal/m² yr	8970 kcal/m² yr

Marsh average: net product on = 6580 kcal/m² yr
respiration = 28000 kcal/m² yr

gross production = 34580

In addition to *Spartina,* the algae living on the surface of the marsh mud contribute 1800 kcal/m² yr gross production and not less than 1620 kcal/m² yr net production (Pomeroy 1959).

Decomposition of *Spartina*

Before the *Spartina* is available to most of the marsh consumers it must be broken down by bacteria. Part of the *Spartina* crop decomposes in place on the marsh, especially that portion in the Short Spartina areas not subject to strong tidal currents. *Spartina* from the streamside marsh, however, is swept off by the water and carried back and forth until it is either decomposed in the water, stranded, or carried out to sea. Material stranded on the beach and representative of that carried out of the system consists mostly of stalks. The leaves have decomposed before leaving the marsh-estuarine system.

Burkholder and Bornside (1957) found that when marsh grass was confined in cages in a tidal creek over the winter, one-half of the dry weight was broken up and washed away after 6 months, by which time only the stems remained.

Pieces of dead standing *Spartina* were collected in mid-winter when decomposition was starting. Algae were not present, hence oxygen consumption was a measure of bacterial activity. From the initial rate of decomposition measured at 15° C, it was estimated that leaves submerged at every tide would be completely consumed in 2 months, stems in 3½ months. Leaves and stems that dry

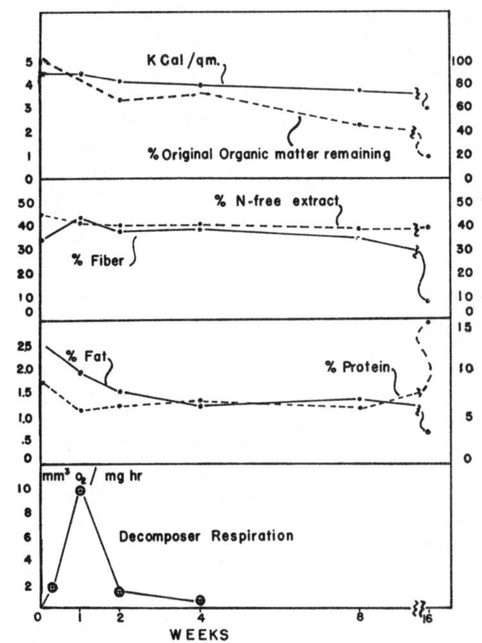

FIG. 3. Decomposition of marsh grass in sea water showing changes in oxygen consumption and composition of grass-bacteria mixture. Line between 8-week point and break shows correct slope.

out in periods between spring tides would last about twice as long.

To find out what changes in composition occur during the breakup of *Spartina,* 10 g of finely-dropped air-dried marsh grass were placed in 500 ml flasks with 200 ml sea water, inoculated with 1 ml of marsh mud and placed on a shaker in the dark at 20° C. The oxygen content of the flasks was not measured but in no case did the material go completely anaerobic. After periods of 0, 2, 7, 14, 28, 56 and 112 days samples were removed. Oxygen consumption of 2 samples was measured and the material from 4 flasks was lumped for analyses of moisture, fat, protein, crude fiber, nitrogen-free extract, ash and caloric content made by Law and Company, Atlanta, Georgia. The results are shown in Figure 3.

Respiration in the flasks rapidly reached a maximum as bacteria grew on the material liberated from crushed cells by the chopping. In 2 weeks this phase passed and bacterial action remained low for the remainder of the experiment. The initial phase is reflected in the decrease in fat and protein percentages but, while fat continued to decline slowly, protein concentration increased gradually until at 16 weeks it was twice as high as at the beginning. At the same time carbohydrates (N-free extracts) remained constant but fiber, principally cellulose, declined to less than ¼ of its initial value. The caloric content declined by 33%. During the period 82% of the organic matter was consumed.

If this is representative of what happens to the *Spartina* as it is changed from standing marsh grass to detritus then, although the total amount of material is decreased, the animal food value of what remains is increased. Bacteria attack the grass substances and convert a portion into bacterial protoplasm and in this process cellulose in the bacterial-detritus mixture decreases most swiftly and protein least swiftly.

The magnitude of the bacterial metabolism was calculated with figures for respiration of plankton in the estuarine waters and bacteria in the marsh sediments and on the standing *Spartina.* Ragotzkie (1958) found that plankton respiration averaged 1600 kcal/m² yr in estuarine waters. Since the turbidity of the water is high, there is very little phytoplankton and I assumed that all of this respiration represents bacterial action upon *Spartina* detritus. Since a planimetric survey of charts and aerial photos of the region showed that there is twice as much marsh as estuarine area in the system, this represents 800 kcal/m² yr of marsh. From Teal and Kanwisher (1961) we find that the bacteria in the marsh sediments degrade 2090 kcal/m² yr. The average respiration of bacteria on standing, dead *Spartina* is about 60 mm³/gm hr, which, multiplied by the biomass of that *Spartina* (Smalley 1959) comes to 1000 kcal/m² yr. Thus the activities of bacteria account for 3890 kcal/m² yr averaged over the marsh area. This amounts to 59% of the available *Spartina.*

Besides bacteria, colorless blue-green algae are also active in the degradation of *Spartina.* Bits of partly decomposed *Spartina* were often found within the mud which was usually black just around them. Within these bits were often numerous filaments of what were apparently *Thioploca* and *Beggiatoa* or *Oscillatoria.* The algae were alive and active, living in a lightless, highly reduced environment as has been discussed by Pringsheim (1949).

Herbivorous insects

The salt marsh grasshopper and the plant hopper are the only important animals in this category. The grasshoppers respire 18.6 kcal and produce 10.8 kcal of tissue per m² per year which adds to an assimilation of 29.4 kcal/m² yr (Smalley 1960). The corresponding plant hopper figures are: respiration 205 kcal/m² yr, produc-

TABLE III. Summary of energy-flow for detritus-algae feeders in Georgia salt marsh

	Respiration	Production	Assimilation	Production / Assimilation
Crabs	171	35	206	17%
Annelids	26	9	35	25%*
Nematodes	64	21	85	25%*
Mussels	39	17	56	30%
Snails	72	8	80	10%
Totals	372	90	462	19.5%

* Assumed as means of calc. P.

tion 70 kcal/m^2 yr and assimilation 275 kcal/m^2 yr (Smalley 1959). Production of plant hoppers was not measured but calculated on the assumption that the ratio of production to assimilation equals 25%. The grasshoppers assimilate only about 30% of what they ingest which means they produce nearly 70 kcal/m^2 yr of fecal matter which can probably serve as food for some of the other marsh inhabitants.

Detritus-algae feeders

Table III gives a summary of the results for the various groups of animals that feed at the surface of the mud-eating Spartina-detritus, algae and to a lesser but unknown extent each other. The groups are considered in turn below.

Crabs

Uca pugilator, U. pugnax, and *Sesarma reticulatum* are the most conspicuous consumers in the marsh. One or more of these species is present in all parts of the marsh. They feed on the surface of the mud for the most part, picking up clawfuls of mud, sorting it with their mouthparts, spitting the rejected material into a claw and depositing it back on the mud and swallowing the remainder. If the spit is compared with the undisturbed mud surface, it is apparent that most of the sorting consists of rejecting larger particles. Spit from *U. pugilator* which live on sand consists of sand from which the smaller particles, the algae and detritus, are gone. In spit from *U. pugnax* feeding on mud rich in diatoms and nematodes most of the larger diatoms and nematodes were no longer distinguishable but the average particle size of the spit was larger than that of the mud and contained many bits of diatom shells. Apparently the large diatoms and nematodes were crushed and then the finer particles were swallowed and the larger ones rejected.

By comparing the amount of feces produced in a few hours by freshly collected animals with the normal rates of respiration and growth, the portion of the ingested food actually assimilated was estimated. Four measurements during the winter on groups of from 5 to 16 animals gave values from 23 to 31%. One measurement on 57 crabs in August gave a value of 75%. Algae and detritus are scarcer in summer than winter on the areas where these crabs were collected and apparently they assimilate a larger part of the digestable material when it is scarce. A parallel situation is found in copepods (Marshall and Orr 1955).

In areas of dense crab populations the entire surface of the marsh is worked over between successive high tides. The feces produced by the crabs feeding on muddy substrates contain about one-third more calories per gram than the mud; feces produced by crabs on sand about 10 times as many calories as the sand. Both by the working over of the marsh surface and the concentration of organic matter in their feces the crabs will have considerable influence upon other organisms, especially the nematodes, annelids and bacteria.

The crab populations were sampled by placing metal rings, 30 cm high, on the marsh while the tide was in and the animals were in or near their burrows, returning at low tide and removing

TABLE IV. Crab populations in g/m^2 in Georgia salt marshes followed by standard errors. Sizes: s=0-150 mg; m=150-500 mg; 1=>500 mg

Marsh type	Size	Winter	Spring	Summer	Autumn	Species
Creek bank	s	0	?	4.89± 1.04	0	*Uca pugilator*
	m	0	0	3.67±0.75	0	
	l	0	0	33.1 ±5.6	0	
Stream-side	s	0	0	4.8 ±2.8	0	*Sesarma reticulatum*
	m	0.7 ±0.4	3.65± 1.15	7.35± 1.15	*	
	l	10.2 ±1.5	*	*	*	
Stream-side	s	0	0	2.6 ±1.6	0	*Eurytium limosum*
	m	0	0	2.85±0.45	*	
	l	8.7 ±1.44	*	*	*	
Levee	s	0.5 ±0.3	?	0.8 ±0.8	0.41±0.14	*Sesarma reticulatum*
	m	0.5 ±0.5	0.2 ±0.2	2.34±0.45	1.4?	
	l	12.4 ±4.35	12.4 ±4.33	22.1 ±0.8	17.0?	
Levee	s	0.81±0.36	2.27±0.68	5.74±0.93	2.47±0.67	*Uca pugnax*
	m	3.0 ±1.5	9.8 ±2.1	7.14±0.65	5.0?	
	l	16.7 ±3.4	32.5 ±9.0	54.3 ±9.5	35.5?	
Levee	l	15.6 ±2.9	*	*	*	*Eurytium limosum*
Short Spartina	s	1.0 ±0.45	0	3.16±0.71	1.0 ±0.45	*Uca pugnax*
	m	2.30±0.60	1.95±0.65	5.00±0.85	3.65?	
	l	16.25±3.90	9.5 ±2.0	12.2 ±1.70	14.25?	
Salicornia-Distichlis	s	?	?	1.91±0.43	8.13±1.17	*Uca pugilator*
	m	3.6 ±0.5	*	*	*	
	l	114.9 ±20.4	*	*	*	

? indicates no samples were taken, numbers if entered are interpolations. When data for several seasons were pooled, the mean appears, followed by asterisks for other seasons involved in the average.

TABLE V. Respiration of Georgia salt marsh crabs by marsh type. Values are kcal/m²/season

Marsh type	Species	Winter	Spring	Summer	Autumn	Total	% by adults
Creek bank.	U. pl.	0	0	69.5	0	69.5	66%
Streamside..	S. r.	2.9	10.5	66.6	17.4	97.3	37%
Levee......	S. r.	3.8	8.6	49.4	17.0	78.7	81%
	U. px.	5.9	50.0	139.6	51.2	246.6	67%
Short spartina..	U. px.	5.5	11.6	48.0	23.3	88.4	56%
Salicornia...	U. pl.	37.3	80.6	170	100	388	90%

everything within the ring, separating, counting and weighing the crabs. Rings of $\frac{1}{5}$ m² were used for adults which were picked out by hand but rings of only 0.018 m² were used for the young, which had to be separated by sieving. Crabs of more than 150 mg were ignored in the small samples and vice versa.

The sampling results are shown in Table IV. In general the biomass of crabs follows the same distribution as the numbers of species of marsh animals (Fig. 1). The *U. pugilator* on the creek bank are apparently completely killed during the autumn and replaced the following spring. It seems unlikely that they migrate as they are confined to sandy substrates (Teal 1958) and the Streamside, Levee and Short Spartina marshes are uniformly muddy (Teal and Kanwisher 1961). The values for the Salicornia marsh must be divided by 4 since only about $\frac{1}{4}$ of the area is occupied by plants and crab burrows, the rest being open sand flats not included in the samples where the crabs feed but do not live.

Table V lists values for respiration of the crabs by marsh types. Respiratory rates are from Teal (1959). The last column gives an idea of the relative importance of the large and small individuals to the population's energy degradation.

By assuming that the crab populations replace themselves annually the energy flow figures in Table VI were calculated.

TABLE VI. Energy flow of detritus-eating crabs in a Georgia salt marsh. Data in kcal/m²/yr

Marsh type	Relative area	Respiration	Production
Creek-bank......	10%	70	28
Streamside.......	10%	97	19
Levee...........	35%	325	65
Short Spartina....	40%	88	14
Salicornia........	5%	97	20

Average respiration	171
Average production	35
Average assimilation	206
Production efficiency	17%

TABLE VII. Summary of annelid sampling in Georgia salt marsh

Marsh type	NOV-DEC SAMPLE		JUL-AUG SAMPLE	
	N	g/m²	N	g/m²
Creek bank..........	8	2.0 ±2.0	7	0.95 ±0.25
Streamside..........	15	2.9 ±0.6	10	1.0 ±0.53
Levee..............	35	1.8 ±0.28	38	2.2 ±0.32
Short Spartina.......	39	2.2 ±0.95	8	0.5 ±0.12
Salicornia...........	5	0.2 ±0.2	2	0.0
Marsh average.......		2.01 ±0.46		1.16 ±0.14

Annelids

The annelids in the marsh are mostly deposit feeders, feeding either from fixed burrows or working their way through the sediments like the oligochaetes, although *Manayunkia* is a filter-feeder. Although *Neanthes,* because of its jaws, might be thought to be predaceous, guts examined at different seasons revealed only diatoms, detritus, and mud and sand.

The annelids were sampled once in November-December and once in July-August. Five samples were taken with a plastic coring tube at spots chosen by taking a pair of random numbers, one indicating the distance north and the other, the distance east of a stake marking the southwest corner of a square meter plot selected at random from the marsh as a whole. The sites are indicated in Fig. 1 and may be seen on an aerial photograph of the marsh in Teal and Kanwisher (1961). Each core was divided into 3 parts at 2 cm intervals and the portion below 6 cm discarded. The annelids were removed by gentle washing in sea water in a sieve with 16 meshes/cm. All samples were examined within 24 hours of collection. Biomass was calculated by multiplying the average weight of each species by the number found in the sample. Average weights were determined for the more common species by weighing on a quartz helix as well as by measuring length and width and calculating the weight based on a specific density equal to sea water. The methods agreed within 15%. Insect larvae collected with the annelids are included in the figures (Table VII).

Table VIII shows the numbers of the most common annelids and insect larvae from selected representative sites. *Capitella,* the oligochaetes, *Streblospio* and *Manayunkia* made up most of the biomass, usually in that order. Two of these, *Capitella* and *Streblospio,* are characteristic not of the marshes but of the estuaries, indicating that the annelids have had to make relatively little adaptation to marsh life. In general they are most

TABLE VIII. Numbers of selected annelids and insect larvae/0.01 m^2 in representative marsh types in a Georgia salt marsh

Site	21	6	19	10	8	4	1	7	3
Winter Series									
Capitella capitata	0	10	30	70	50	30	0	10	0
Steblospio benedicti	50	150	80	40	50	0	20	30	0
Neanthes succinea	0	10	0	0	0	0	0	0	10
Manayunkia aestuarina	0	0	40	90	0	290	0	0	0
oligochaete	10	10	170	80	30	80	40	40	0
dipteran larvae	0	0	10	0	10	10	0	0	0
Summer Series									
Capitella capitata	30	26	14	56	136	38	12	30	0
Streblospio benedicti	14	38	40	6	4	0	4	0	0
Neanthes succinea	0	0	0	0	2	0	0	0	0
Manayunkia aestuarina	8	10	12	52	0	60	0	0	0
oligochaete	78	10	60	88	88	66	30	32	20
dipteran larvae	2	0	0	0	0	2	0	0	0

numerous in the most productive parts of the marsh as are other animals, except that they are somewhat scarcer on the highest and thus the driest parts of the levees.

Energy flow for the annelids was calculated on the basis of an average respiration rate of 400 mm^3/gm hr (Zeuthen 1953) and a production equal to 25% of assimilation. The latter results in a turnover time of 1.6 months which is reasonable since the animals are between 20μ gm and 200μ gm in weight.

Nematodes

Using the relative areas of various marsh types from Table V, and the weights of nematodes from Teal and Wieser (1961) I calculate that there are about 2.76 g fresh weight/m^2. The samples were all taken in spring so it must be assumed that the nematode biomass does not change appreciably throughout the year. Wieser and Kanwisher (1961) found that there was only slightly more than a 2-fold variation in a marsh at Woods Hole, Mass. where the climate is considerably colder and more variable. Using the average respiratory rate of 540 mm^3/fresh gm/hr (Teal and Wieser 1961) the nematodes would respire an average of 64 kcal/m^2 yr. Assuming their production to equal 25% of assimilation, production would be 21 kcal/m^2 yr which would amount to a turnover of the population every 1.6 months. This may be compared with turnovers of 1 year and 1 month quoted by Wieser and Kanwisher (1961) and Nielsen (1949) respectively.

Snails and Mussels

The mussels (Kuenzler 1961) respire on the average 39 kcal/m^2 yr and produce 17 kcal/m^2 yr. The small snails of the current year class had a production equal to 14% of their assimilation but Smalley's population data did not provide a production value for the adults which grew very slowly. Since there is some growth of adult snails as well as the formation of gametes, which amounts to $1/6$ of the total production of the mussels (Kuenzler 1961), the snail production must be something less than 14%, and 10% is used here.

Secondary consumers

The population data for the mud crab, *Eurytium limosum,* were presented in Table III. An average respiration value of 21.9 kcal/m^2 yr was derived from data of Teal (1959) and a production of 5.3 kcal/m^2 yr was calculated in the same manner as for the other crabs. For the Clapper Rails, a production of 0.1 kcal/m^2 yr was calculated from the data of Oney (1954) for population and mortality, and an average respiration of 1.6 kcal/m^2 yr was calculated on the basis of the weight-metabolism curve of warm blooded animals in Hemmingsen (1950). So figured the total assimilation of the carnivores comes to 30.6 kcal/m^2 yr divided into 25.1 kcal/m^2 yr respiration and 5.5 kcal/m^2 yr production. The raccoons in the marsh have not been studied but on the basis of general observation during 4 years when I was in the marsh nearly every week, they are considered to have an assimilation equal to that of the rails.

As yet there is no study of the carnivorous birds and spiders that feed on the marsh insects. For purposes of this calculation they are assumed to take the same portion of their prey as do the predators feeding on the detritus-algae eaters.

COMMUNITY ENERGY FLOW

Figure 4 shows the energy flow for the marsh system calculated in the preceding sections. The value for light input is from Kimball (1929). The amount of light that is actually intercepted by the plants is unknown but it is obvious from colored aerial photographs that a considerable portion of the mud is exposed especially in the parts of the marsh not close to a creek. Nevertheless, gross production is 6.1% of the incident light energy. This may be compared with values of from 0.1 to 3.0% reported for various fresh-water and marine areas (Odum and Odum 1959, 1955). The salt marsh occupies a highly advantageous position where nutrients are plentiful and circulation is supplied by the tides.

However, as noted above, a large part of the gross production is metabolised by the plants themselves and net production over light is a little less than 1.4%. This is still high compared to other

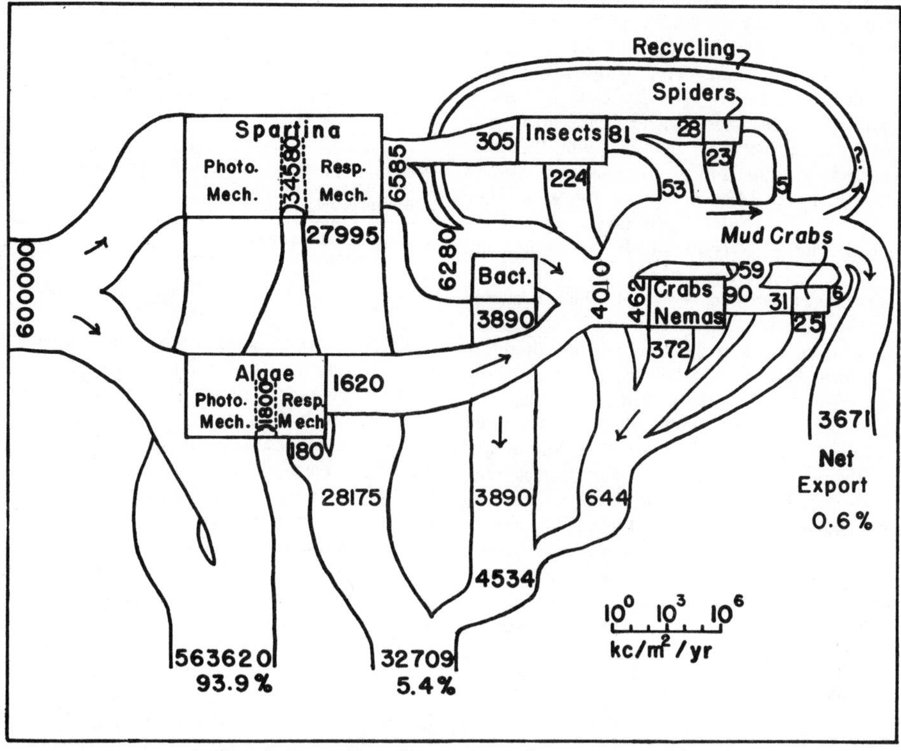

FIG. 4. Energy-flow diagram for a Georgia salt marsh.

systems although values as high as 6.0% have been reported for pine plantations (Ovington 1959).

The herbivorous insects assimilate 4.6% of their potential food, the net production of *Spartina*. They eat the plants directly and there is no significant time lag between production and consumption, since Georgia is far enough south that the *Spartina* grows to some extent throughout the year. The grasshoppers feed on it during the summer whereas the leaf hoppers are most abundant during the cooler seasons.

The relation between production and consumption for the algae-detritus feeders is more complex. Mud algae production is rather constant throughout the year (Pomeroy 1959) and algal turnover is much more rapid than *Spartina* turnover. When the algae-detritus feeders utilize algae there is little or no time lag between production and primary consumption, but when they feed on *Spartina* detritus and the associated bacteria, there is a definite time lag. Detritus is produced throughout the year as the older leaves die and are broken up, but most of the *Spartina* dies in autumn and winter after seed formation. Low winter temperatures retard formation of detritus and spread the supply out into the spring. At the beginning of summer, a new supply forms as the leaves of the spring growth die and decay. Whatever the actual time relationships may be, it is certain that there is considerable delay before detritus feeders can use the *Spartina* and the longer the delay, the less food remains.

These animals compensate for the variations in detrital supply by eating algae. The most conspicuous and abundant consumers of the marsh, the fiddler crabs, have perhaps the most omnivorous food habits. They can survive on detritus, algae, bacteria or animal remains (Teal 1958), i.e. they have a very unrestricted diet.

Odum and Odum (1959) and MacArthur (1955) have suggested that the fluctuations characteristic of certain communities may be correlated with the presence of few species as in arctic or desert areas. In the salt marsh we have a system with few species but one which seems to exhibit considerable stability. In the 5 years during which these studies were carried on there were no noticeable changes in population size in any of the important animals. (The microfauna of the soil is probably relatively little affected by the weather extremes which make life difficult for the larger animals and are not considered in this argument.)

Stability is a valuable asset for an ecosystem as it minimizes disturbance which might lead to partial or total extinction. Stability will therefore have selective value and ecosystems will tend to develop more stable configurations with time (Dunbar 1960). Salt marsh faunas and floras have had long periods in which to develop and although they have been greatly affected by the considerable changes in sea level which occurred during the Pleistocene they may not be considered as youthful as arctic areas. There has therefore been sufficient time for stability to develop and for species to adapt to the marsh conditions.

MacArthur (1955) shows that a community may achieve stability by having either many species with restricted diets or fewer species with broad diets. The former alternative will permit greater efficiency and, other things equal, will be the one selected. The salt marsh has, however, the 2nd alternative. Among the detritus-algae feeders there are only a few important species and they all have a very unrestricted diet. There are also only a few species among the carnivores that prey on the detrital feeders and they also have an unrestricted diet. Among the insects there are only 2 important species and though they feed only on *Spartina,* this is the only higher plant growing on most of the marsh and so the only food available. These 2 insects are not especially restricted in other ways; they feed on various parts of the plants and in various degrees of exposure. The situation among the spiders and carnivorous insects has not been adequately investigated but there seem to be more species in these groups and perhaps more specialization in their niches.

There are 2 principal reasons why the salt marsh should have the less efficient alternative of the 2 paths to community stability. There is only one higher plant on the marsh and consequently a lack of variety of possible niches such as could be found in a forest at the same latitude. Possibly even more important is the restriction of biomass by the removal of much of the marsh production by the tidal currents. As Hutchinson (1959) has pointed out, if the total biomass is restricted, "then the rarer species in a community may be so rare that they do not exist." With these 2 limitations of the possible numbers of species that can survive in the marsh community, the only road to stability is the development of broad, unrestricted food habits such as is found in the marsh.

TABLE IX. Summary of salt marsh energetics

Input as light	600,000 kcal/m²/yr
Loss in photosynthesis	563,620 or 93.9%
Gross production	36,380 or 6.1% of light
Producer Respiration	28,175 or 77% of gross production
Net Production	8,205 kcal/m²/yr
Bacterial respiration	3,890 or 47% of net production
1° consumer respiration	596 or 7% of net production
2° consumer respiration	48 or 0.6% of net production
Total energy dissipation by consumers	4,534 or 55% of net production
Export	3,671 or 45% of net production

Table IX gives a summary of the energy flow for the system. The producers are the most important consumers in the marsh followed by the bacteria which degrade about 1/7 as much energy as the producers. The animals, both primary and secondary consumers, are a poor 3rd degrading only 1/7 as much energy as the bacteria. As far as the consumers are concerned, the situation in the salt marsh is not very different from that in other systems. But the high consumption by the producers is unusual. In a stable system such as a so-called "climax forest" consumption equals production and there is no accumulation of organic matter, but the trees are relatively unimportant consumers. Ovington (1957) gives data indicating that in mature pines respiration is something less than 10% of production. The fact that salt marsh *Spartina* respires over 70% of its production may be associated with existence in an osmotically difficult situation.

In spite of the high rate of producer respiration, net production in the salt marsh is 1.4% of incident light which is higher than in most systems studied (Teal 1957). Table IX shows that 45% of this production is lost to the estuarine waters. The fauna of the estuaries has not been quantitatively studied but the numbers of shrimp and crabs taken by the local fishery give evidence of their abundance. Since the waters of these estuaries are so turbid and well mixed that the phytoplankton spend most of their time in the dark and their net production is zero (Ragotzkie 1958), the estuarine animals must be living on the exported marsh production. There is about 1/2 as much estuarine as marsh area behind the sea islands and since 45% of the marsh production is exported to the estuaries, there can be 1.6 times as much consumer activity in the latter region as in the former.

The tides are of supreme importance in controlling the environment of the salt marshes. They limit the number of species that can occupy the system and so make it simple enough to be studied in the detail reported here. They are responsible for the high production of *Spartina,* as witnessed by the luxuriant growth along the tidal creeks as compared with that on the Short Spartina areas. At the same time the tides remove 45% of the production before the marsh consumers have a chance to use it and in so doing permit the estuaries to support an abundance of animals.

References

Burkholder, P. R., and G. H. Bornside. 1957. Decomposition of marsh grass by aerobic marine bacteria. Bull. Torrey Bot. Club **84:** 366-383.

Dunbar, M. J. 1960. The evolution of stability in marine environments. Natural selection at the level of the ecosystem. Amer. Nat. **94:** 129-136.

Hemmingsen, A. M. 1950. The relation of standard (basal) energy metabolism to total fresh weight of living organisms. Rep. Steno. Mem. Hosp. **4:** 1-58.

Hutchinson, G. E. 1959. Homage to Santa Rosalia or Why are there so many kinds of animals? Amer. Nat. **93:** 145-160.

Kimball, H. H. 1929. Amount of solar radiation that reaches the surface of the earth on land and on the sea, and methods by which it is measured. Mon. Weather Rev. **56:** 393-398.

Kuenzler, E. J. 1961. Structure and energy flow of a mussel population in a Georgia salt marsh. Limnol. and Oceanogr. **6:** 191-204.

Lindeman, R. L. 1942. The trophic-dynamic aspect of ecology. Ecology **23:** 399-418.

MacArthur, R. 1955. Fluctuations of animal populations and a measure of community stability. Ecology **36:** 533-536.

Marshall, S. M., and A. P. Orr. 1955. Experimental feeding of the copepod *Calanus finmarchicus* (Gunner) on phytoplankton cultures labelled with radioactive carbon. Pap. Mar. Biol. Oceanogr., Deep Sea Res. Suppl. **3:** 110-114.

Nielsen, C. O. 1949. Studies on the soil microfauna. II. The soil inhabiting nematodes. Natura Jutlandica **2:** 1-131.

Odum, E. P. 1961. Personal communication.

———, **and H. T. Odum.** 1959. Fundamentals of Ecology. Philadelphia: Saunders.

———, **and A. E. Smalley.** 1959. Comparison of population energy flow of a herbiverous and a deposit-feeding invertebrate in a salt marsh ecosystem. Proc. Nat. Acad. Sci.

Odum, H. T. 1957. Trophic structure and productivity of Silver Springs, Florida. Ecol. Monogr. **27:** 55-112.

———, **and E. P. Odum.** 1955. Trophic structure and productivity of a windward coral reef community on Eniwetok Atoll. Ecol. Monogr. **25:** 291-320.

Oney, J. 1954. Final report, Clapper rail survey and investigation study. Georgia Game Fish. Comm.

Ovington, J. D. 1957. Dry-matter production by *Pinus sylvestris* L. Ann. Bot. N. S. **21:** 287-314.

Patten, B. C. 1959. An introduction to the cybernetics of the ecosystem: the trophic-dynamic aspect. Ecology **40:** 221-231.

Pomeroy, L. R. 1959. Algal productivity in the salt marshes of Georgia. Limnol. and Oceanogr. **4:** 367-386.

Pringsheim, E. H. 1949. The relationship between bacteria and myxophyceae. Bact. Rev. **13:** 47-98.

Ragotzkie, R. A. 1959. Plankton productivity in estuarine waters of Georgia. Inst. Marine Sci. **6:** 146-158.

Richman, S. 1958. The transformation of energy by *Daphnia pulex*. Ecol. Monogr. **28:** 273-291.

Slobodkin, L. B. 1959. Energetics in *Daphnia pulex* populations. Ecology **40:** 232-243.

———. 1960. Ecological energy relationships at the population level. Amer. Nat. **94:** 213-236.

Smalley, A. E. 1959. The growth cycle of Spartina and its relation to the insect populations in the marsh. Proc. Salt Marsh Conf. Sapelo Island, Georgia.

———. 1960. Energy flow of a salt marsh grasshopper population. Ecology **41:** 672-677.

Teal, J. M. 1957. Community metabolism in a temperate cold spring. Ecol. Monogr. **27:** 283-302.

———. 1958. Distribution of fiddler crabs in Georgia salt marshes. Ecology **39:** 185-193.

———. 1959. Respiration of crabs in Georgia salt marshes and its relation to their ecology. Physiol. Zool. **32:** 1-14.

———, **and J. Kanwisher.** 1961. Gas exchange in a Georgia salt marsh. Limnol. and Oceanogr. **6:** 388-399.

———, **and W. Wieser.** 1961. Studies of the ecology and physiology of the nematodes in a Georgia salt marsh. Ms.

Wieser, W., and J. Kanwisher. 1961. Ecological and physiological studies on marine nematodes from a small salt marsh near Woods Hole, Massachusetts. Limnol. and Oceanogr. **6:** 262-270.

Zeuthen, E. 1953. Oxygen uptake as related to body size in organisms. Quart. Rev. Biol. **28:** 1-12.

J. exp. mar. Biol. Ecol., 1968, Vol. 2, pp. 1–23; North-Holland Publishing Company, Amsterdam

FAUNA AND SEDIMENTS OF AN INTERTIDAL MUD FLAT: A MULTIVARIATE ANALYSIS

R. M. CASSIE and A. D. MICHAEL [1]

Department of Zoology, University of Auckland, New Zealand

Abstract: On Karore Bank, an intertidal mud flat near Auckland International Airport, New Zealand, 40 faunal quadrats 0.5×0.5 m were dug, 21 of these being submitted to a grain-size analysis of the sediment. Twelve species of invertebrates (40 samples) were arrayed in 4 communities using principal component analysis. A reduced list of 8 species (21 samples) was grouped in the same manner and the communities correlated with sediment grade by multiple regression analysis. The two most clearly defined communities were characterized by *Chione stutchburyi* and *Macomona liliana*, and by *Halicarcinus cookii* and *Owenia fusiformis*. These were negatively correlated and associated respectively with coarse and fine sediments. The definition of the communities and their correlation with sediments was consistent with those which would have been obtained by conventional subjective methods, but the former extended the range of the conclusions. An analysis of the same data by canonical correlation appeared to produce less meaningful results, though this technique may well repay further investigation.

The data presented were collected in connection with a more comprehensive project sponsored by the Wildlife Division, New Zealand Department of Internal Affairs, to investigate the habits of sea birds which constitute a hazard to flying at Auckland International Airport, Mangere, Auckland (Fig. 1). Invertebrate fauna of the adjacent intertidal mud flat are a potential source of food to these birds. A quantitative survey of the benthic fauna on the Karore Bank, Manukau Harbour was carried out by one of us (A.D.M.) as a thesis project for the degree of M. Sc. While the numbers of samples is relatively small, there is at least sufficient information for an exploratory investigation of the potentialities of multivariate analysis as a technique for objectively identifying the benthic communities and relating them to the properties of the sediments. Since the sampling area was exposed at low tide, the usual quantitative difficulties encountered in grab or dredge sampling were not present. The computer programming and analysis were carried out by R.M.C. on the IBM 1130 computer at the University of Auckland, using programmes written in Fortran IV.

SAMPLING METHODS

Stations on the Karore Bank were located in a systematic pattern as indicated in Fig. 3. At each of the 40 stations a 0.5×0.5 m quadrat was dug to a depth of 6–10 cm, and the fauna separated using a sieve of 2.5 mm linear aperture. At alternate stations, 21 in all, a 250 cm^2 sample of the substratum to a depth of 5 cm was taken.

[1] Present Address: Institute of Oceanography, Dalhousie University, Halifax, Nova Scotia

These were subjected to a grain size analysis by washing successively through square meshed sieves of 20 cm (8 in.) diameter with linear apertures 1.67, 0.853, 0.422, 0.251, 0.124, 0.064 and 0.031 mm. In order to preserve as far as possible the original *in situ* characteristics, the sediments were not pre-treated in any way and were processed within a few hours of collection. An analysis of organic carbon content

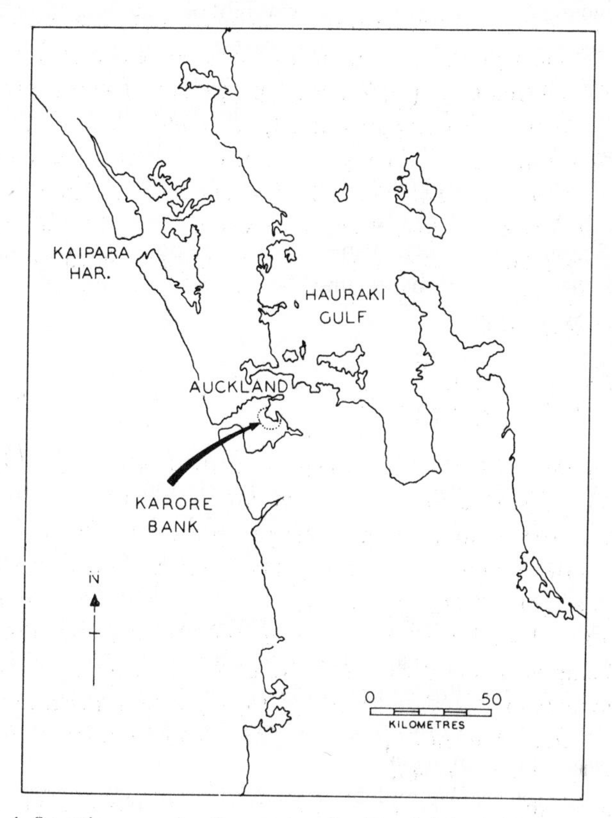

Fig. 1. Location map (northern extremity, North Island, New Zealand).

was carried out using a modified Walkley and Black method (Morgans, 1956). Of the species recorded, twelve were present in sufficient abundance for numerical analysis of the full 40 stations. This number was reduced to eight for correlation with sediment properties in 21 stations.

Multivariate Analysis

Two types of multivariate analysis have been employed, principal component analysis (see, Cassie, 1963; Seal, 1964) and canonical correlation (see, Quenouille, 1952; Kendall, 1957). (Note, however, that 'canonical correlation', though mathematically related, is a different procedure from 'canonical analysis' as described, *e.g.*, by Seal, 1964). For both methods used here, the initial processing of data is the same. The table of raw data is subjected to the following transformations:

$Y_{ij} = \ln(C_{ij}+1)$, where C_{ij} is the count of the jth species in the ith quadrat.

$X_{ik} = \arcsin\sqrt{P_{ik}}$, where P_{ik} is the proportion by weight of the kth sediment grade ($k = 1, \ldots, 8$) in the ith quadrat.

$X_{i9} = \sqrt{P_{i9}}$, where P_{i9} is the proportion by weight of organic carbon in the sediment.

The combined sediment grades and organic carbon content (collectively designated X) may be briefly referred to as the 'sediment properties'. The three transformations are necessary to satisfy, at least approximately the multivariate model in which it is assumed that all variates are normally distributed and, as a consequence, linearly related one to another (see Appendix). The convention of X (independent variable) and Y (dependent variable) is based upon regression analysis, and implies that the distribution and abundance of organisms is at least in part controlled by the sediment properties, and that the relationship may be expressed in one or more meaningful linear equations of the type:

$$\sum_{j=1}^{p} y_{ij}a_{jl} = \sum_{k=1}^{q} x_{ik}b_{kl}+\varepsilon_{il}, \quad i = 1, \ldots, n, \quad l = 1, \ldots, m, \tag{1}$$

where a_{jl} and b_{kl} are constants appropriate to the jth species, kth sediment property and lth equation, ε_{il} is an error term for the ith station and lth equation, p and q are the number of species and sediment properties respectively, n is the number of stations, and m is the number of equations. The lower case y and x are standardized values corresponding to X and Y, *e.g.*, $y_{ij} = (Y_{ij}-\bar{y}_j)/s_j$, where $\bar{y}_j$ and s_j are the mean and standard deviation of Y_{ij}. The variates, x and y, and the error term, ε, are treated as being normally distributed. Note, however, that, except in the simple principal components model of equation (5), the regression model in which only the y's are subject to error does not apply, and (1) should be modified to include error terms on both sides of the equation.

From the entire table of X's and Y's a correlation matrix is computed and partitioned as follows:

$$\left[\begin{array}{c|c} \mathbf{A} & \mathbf{B} \\ \hline \mathbf{B}' & \mathbf{C} \end{array}\right]$$

where **A** represents $x \times x$ correlations, **B**, $x \times y$, **C**, $y \times y$ and **B**$'$ is the transpose of **B**.

Using the principal components approach, the a's in (1) are the elements of one of the latent vectors (or eigenvectors) of **C**. The y-function, left-hand side of equation (1), defines a new variable or principal component:

$$z_{il} = \sum_{j=1}^{p} y_{ij}a_{jl} \tag{2}$$

which has a variance λ_l, the latent root (or eigenvalue). The b's in the x-function right-hand side of equation (1), are now derived by multiple regression analysis, with z as dependent variable and the x's as independent variables. The number of equations possible is p, but in practice only the m ($< p$) of these with the largest latent roots are retained. The z's have the two special properties, first that they are uncorrelated one with the other, and secondly that they may be arranged in descending order of variance, each successive z having the maximum possible variance after removing the contribution of previous z's to the matrix **C**.

The vector elements, a, have a further use in that they are scores (in fact, correlation coefficients) indicating the relative contribution of species to any given principal component. Species with large scores of the same sign may be regarded as members of a common 'community', though it remains to be seen whether this classification coincides with that of a community which might be defined by conventional non-mathematical means. Large scores of both negative and positive sign occurring in the same vector define two negatively correlated communities, the one tending to replace the other in space. Since different vectors are associated with different variances (as measured by their roots), the comparison of scores between vectors is facilitated by the scale transformation:

$$f_{ij} = \lambda^{\frac{1}{2}} a_{ij}. \tag{3}$$

The symbol f (factor loading) is chosen since this is the usual convention for scaling vectors in factor analysis. Vector elements are presented in the form f_{ij} in Table VI and as a_{ij} in Table VII.

In canonical correlation the objectives are different even though the results are superficially similar to those of principal component analysis. Instead of deriving a y-function based on matrix **C** (which represents the community structure as presented by the biota alone), the matrices **A**, **B** and **C** are processed simultaneously in a manner analogous to multiple regression analysis. The x and y functions of the first equation are computed to have maximum (canonical) correlation, and each successive correlation is likewise maximized, subject only to the restriction that any two functions other than those of a pair must have zero correlation. If there are p x's and q y's, the number of equations and of canonical correlations is p or q, whichever is less. The a's are no longer indices of community structure in the previous sense, and in practice these coefficients are usually difficult to interpret, if not meaningless. On the other hand, the equation will reflect more efficiently (in the mathematical sense) the relationship between biota and environment. Of the two procedures, it is possibly a matter of judgment rather than *a priori* reasoning which is the more useful ecological tool.

Since the a and b coefficients derived by canonical correlation have no absolute scale, it is convenient to standardize them so that:

$$\sum_{j=1}^{p} a_{ij} = 1. \tag{4}$$

The Data: Biota and Sediments

The twelve species of invertebrates which were enumerated are listed in Table I, together with comments on their feeding habits and mobility. The order in the table together with the reference numbers is retained throughout the presentation. Species marked* (6–9) were not sufficiently abundant to be useful in the correlation of biota with sediments and appear therefore, only in the initial (12 × 12) principal component analysis. *Hemiplax hirtipes* is a burrowing crab, and *Halicarcinus cookii* a spider crab. *Owenia fusiformis* and *Pectinaria australis* are tubiculous polychaetes. The remaining eight species are lamellibranch molluscs.

Table I

The biota.

Species	Feeding habit	Mobility
1. *Chione stutchburyi* (Gray)	suspension	low
2. *Macomona liliana* Iredale	deposit	high, deep-burrowing
3. *Amphidesma australe* (Gmelin)	suspension	low, (juveniles only were present)
4. *Nucula hartvigiana* Pfeiffer	deposit	high, non-burrowing
5. *Hemiplax hirtipes* (Jacquinot)	deposit and scavenger	high
6. *Cyclomactra ovata* (Gray)*	suspension	low, deep-burrowing
7. *Pectinaria australis* (Ehlers)*	deposit	nil
8. *Solemya parkinsoni* E. A. Smith*	deposit	high
9. *Soletellina siliqua* Reeve*	deposit	high
10. *Leptomya retiaria* (Hutton)	suspension	high, shallow-burrowing
11. *Halicarcinus cookii* (Filhol)	deposit and scavenger	high
12. *Owenia fusiformis* della Chiaje	deposit	nil

It was evident from field observations and from inspection of the raw data that at least two distinct communities were present, the first dominated by *Chione* (a suspension feeder) and *Macomona* (a deposit feeder), and the second by *Owenia* (a deposit feeder) and *Halicarcinus* (a scavenger). The *Chione-Macomona* community favours the coarser sediments and *Owenia-Halicarcinus* the finer. Tentatively, *Amphidesma* and *Nucula* would be classified with *Chione-Macomona* and *Leptomya* with *Owenia-Halicarcinus*. It was also evident that the two suspension feeders, *Chione* and *Leptomya*, tend to be mutually exclusive, since they occur together at only three stations and in each of these the density of both was low. The remaining five species are apparently distinct from the *Chione-Macomona* community, but there is no clear evidence whether they should be associated with *Owenia-Halicarcinus* or placed in one or more separate communities.

Eight particle sizes, based on the linear dimensions of the retaining sieve, are recorded in descending order: 1.67, 0.853, 0.422, 0.251, 0.124, 0.064, 0.031, and < 0.031 mm. The last variate of the sediment data table is the estimated organic carbon content. All figures are initially expressed as percentages by weight.

Correlation Matrices

The species × species correlation matrix for $p = 12$ (species), $n = 40$ (stations) is given in Table II together with the means ($\bar{y}$) and standard deviations (s_y) for the species. The order of tabulation has been revised to correspond with the principal components arrangement, though it may be noted that a similar, if not identical, arrangement might have been made from the correlation matrix itself. When the matrix is partitioned (broken lines), coefficients in the upper left and lower right sub-matrices tend to be high-positive, and those in the remaining sub-matrices low-positive or

Table II

Species × species correlation matrix, 40 stations.

	1.00	0.66	0.24	0.44	−0.12	−0.16	−0.13	−0.33	−0.38	−0.47	−0.47	−0.48
		1.00	0.44	0.68	0.04	0.00	−0.25	−0.14	−0.11	−0.25	−0.29	−0.59
			1.00	0.32	−0.27	−0.20	−0.16	−0.14	−0.21	−0.29	−0.34	−0.28
				1.00	0.29	−0.06	−0.20	0.05	−0.05	−0.11	−0.08	−0.40
					1.00	0.14	0.18	0.12	0.13	0.18	0.07	−0.28
						1.00	0.20	0.01	0.03	0.06	0.06	0.12
							1.00	0.02	0.04	0.03	0.05	0.28
								1.00	0.17	0.26	0.27	0.39
									1.00	0.28	0.39	0.16
										1.00	0.37	−0.06
											1.00	0.50
												1.00
$\bar{y}$	(2.08	2.85	0.50	2.46	0.76	0.18	0.16	0.27	0.18	0.86	2.12	0.98)
s_y	(2.28	1.80	0.97	2.07	0.92	0.48	0.52	0.61	0.44	1.42	1.34	1.72)

negative. The reduced species × species matrix, **C**, with $p = 8$, $n = 21$ (Table III), shows comparable features to the larger matrix, and may be partitioned in the same way. The species × sediments matrix, **B′**, with $p = 8$, $q = 9$, $n = 21$ (Table IV), indicates that species 1 and 2 are both correlated positively with coarse sediments (left-hand side of the first two rows) and negatively with fine sediments. The op-

Table III

Species × species correlation matrix, 21 stations.

1.00	0.62	0.08	0.39	−0.14	−0.48	−0.47	−0.40
	1.00	0.40	0.68	0.06	−0.27	−0.43	−0.49
		1.00	0.22	−0.23	−0.28	−0.43	−0.25
			1.00	0.20	−0.02	0.07	−0.38
				1.00	0.44	0.07	−0.29
					1.00	0.25	−0.20
						1.00	0.60
							1.00

TABLE IV

Species × sediment grades (rows × columns) correlation matrix, 21 stations (coarse grades to the left).

									R
0.28	0.08	0.12	0.21	−0.00	−0.16	−0.34	−0.02	0.10	0.79
0.25	0.08	0.16	0.27	0.40	−0.30	−0.46	−0.44	−0.31	0.75
−0.08	−0.49	−0.27	−0.20	0.31	−0.12	−0.22	−0.33	−0.27	0.79
0.39	−0.02	−0.01	−0.01	0.11	0.18	−0.16	−0.25	−0.11	0.59
−0.07	0.05	0.38	0.38	−0.35	0.15	0.47	0.21	0.22	0.78
0.05	0.25	0.12	−0.04	−0.25	−0.03	0.21	0.32	0.03	0.53
−0.04	0.32	0.05	−0.18	0.00	0.39	0.15	−0.08	0.06	0.77
−0.24	0.02	−0.13	−0.21	0.17	0.12	−0.03	−0.16	0.05	0.45

TABLE V

Sediment grade × sediment grade correlation matrix, 21 stations.

	1.00	0.55	0.27	0.21	−0.29	0.09	0.04	0.18	−0.22
		1.00	0.66	0.36	−0.25	−0.05	0.01	0.26	−0.25
			1.00	0.84	−0.14	−0.01	0.03	−0.00	−0.42
				1.00	−0.06	−0.15	−0.07	−0.05	−0.40
					1.00	−0.39	−0.77	−0.91	−0.34
						1.00	0.43	0.09	0.01
							1.00	0.63	0.30
								1.00	0.43
									1.00
$\bar{x}$	(0.12	0.05	0.06	0.09	0.99	0.39	0.16	0.24	0.83)
s_x	(0.09	0.03	0.06	0.07	0.23	0.10	0.10	0.21	0.26)

posite trend might have been expected with the last two species, but this is not evident. The column of figures at the right-hand side of the matrix is the multiple correlation coefficient, R, for species × (combined sediment properties), which ranges from 0.45 to 0.79. These are moderately high values and suggest that appropriate functions of the sediment properties would have some value in predicting the abundance of some species. Table V shows the sediments × sediments correlation matrix, **A**. This can be partitioned in a similar manner to species × species.

PRINCIPAL COMPONENT ANALYSIS

Of the twelve possible latent roots found, using the approximate test of Lawley (1956) it is found that eight can be effectively differentiated at the 5 % level of significance. In the present instance, however, it seems possible to place a useful interpretive value only on the first three roots. It is not uncommon to find all roots significantly different (*e.g.*, Cassie, 1963), so that some subjective procedure of rejection seems necessary, otherwise the principal components will be just as complex as the original data. The first three latent roots, λ, and vectors, $\boldsymbol{f}$, of the 12×12 correlation matrix (Table II) are shown in Table VIa. The largest absolute vector element for each row of the matrix (*i.e.*, for each species) is in bold type. The rows (species) have been arranged in descending order of the elements in the first vector. Although this is an objective procedure, it may not necessarily be the best general method of rearrangement (as, for example, were the procedure to be written into the computer programme).

TABLE VI

Latent roots, λ, and vectors f, derived from correlation matrices, Table II (VIa) and Table IV (VIb).

	VIa			VIb			
	(3.62	1.80	1.36)	(3.08	1.80	1.17	0.91)
	0.83	0.06	−0.10	**0.77**	−0.16	−0.23	−0.49
	0.80	−0.38	0.07	**0.87**	0.13	−0.25	0.09
	0.59	0.13	0.35	0.54	−0.26	0.27	**0.72**
	0.55	**−0.65**	0.17	0.59	0.40	**−0.60**	0.26
	−0.08	**−0.67**	−0.50	−0.04	**0.83**	0.02	−0.11
	−0.20	−0.11	**−0.54**				
f	−0.32	0.12	**−0.58**				
	−0.42	−0.30	0.33				
	−0.44	−0.40	0.20				
	−0.49	−0.46	0.04	−0.39	**0.76**	0.24	0.13
	−0.62	−0.36	0.32	**−0.69**	0.13	−0.65	0.19
	−0.71	0.37	0.22	**−0.68**	−0.50	−0.38	0.09

It is seen that all but one of the twelve species is accounted for by a bold figure. For species 8, the 5th vector has an element, $f_{85} = 0.48$, which is only a little larger than $f_{81} = 0.42$, so that only a minor inaccuracy would be introduced by assigning this species to the first vector. On this basis we are now able to group the species into four communities:

1. Species 1–3, and possibly also 4, if we permit some overlap between communities. This we may call the *Chione-Macomona* community after the two species with the highest coefficients, which had already been classified as a community by inspection of the data.
2. Species 8–12 which also predominate in the first vector but have coefficients of the opposite sign. This may be called the *Owenia-Halicarcinus* community, the same remarks applying as for *Chione-Macomona*.

These first two communities are obviously complementary, one tending to exclude the other, just as has already been noted to be the case with *Chione*, and *Leptomya*.

3. Species 4 and 5. A community intermediate between 1 and 2, and tending to share one of its members, *Nucula*, with community 1. It will be named the *Hemiplax* community, since this seems to be the more exclusive member of the pair.

4. Species 6 and 7. Another intermediate community; its affinities, are if any, with community 2. It will be named *Cyclomactra-Pectinaria* since the two species are of about equal status.

The degrees of confidence in assigning these communities varies. It will be seen that the vectors (Table VIb from the 8 × 8 correlation matrix) assign some species differently, though presumably less reliably. It is difficult to obtain any objective measure of the relative reliability of the four communities as assigned. A measure of the average information contributed by a single species within a community is obtained by calculating the mean square of the vector elements for the dominant species (*i.e.*, the bold elements in Table VIb) in each community. For the four communities, these figures are: 1, 0.56; 2, 0.33; 3, 0.43; 4, 0.32. On this basis, the order of reliability is: 1, 3, 2, 4. On the other hand, from other subjective observations on Karore Bank and elsewhere, it seems that community 2 is a satisfactory grouping and almost equals community 1 in reliability, while communities 3 and 4 will remain tentative until evidence is available from further data.

Table VII shows the first 4 latent roots, λ, and vectors, $\boldsymbol{a}$ from the 8 × 8 species correlation matrix, and the partial regression vectors, $\boldsymbol{b}$, which are required to complete the four equations (1). (However, Lawley's test indicates that the fourth root is not quite 'significant' at the 5 % level.) The first of these equations, for example, is:

$$0.44y_{i1}+0.50y_{i2}+0.31y_{i3}+0.33y_{i4}-0.02y_{i5}-0.22y_{i10}-0.39y_{i11}-0.39y_{i12}$$
$$= 0.25x_{i1}-0.60x_{i2}+0.22x_{i3}-0.29x_{i4}-2.50x_{i5}-0.78x_{i6}-0.82x_{i7}-1.79x_{i8} \quad (5)$$
$$-0.12x_{i9}+\varepsilon_{i1}.$$

The multiple correlation coefficients, R, for the first equation is 0.73, or $R^2 = 0.52$, indicating that 52 % of the variation of community 1 may be predicted from the sediments.

One of the problems which has beset benthic ecologists has been to obtain a single statistic which best describes the distribution of particle sizes in a sediment as affecting the biota living in the sediment. This would be a simple matter if particle sizes had a unimodal distribution approximating, at least after transformation, to a normal distribution, but this is seldom the case. Typical procedures are to calculate a median or a mean particle size. For the median, correlation with the y-function (5) is -0.22 and for the mean (*cf.*, Holme, 1949) -0.49. Squaring the correlation for the mean, R^2 is0. 24, a little less than half the value for equation (5). Thus the x-function is, for the data in hand, about twice as efficient as the mean. It is seen, however, that the x-function bears some resemblance to a mean in that the coarse sediments give a high, and the fine, a low value.

One of the most useful functions of a principal component analysis is to present the information available in the smallest possible array of numbers. Full information concerning the present biota requires 8 equations, but the information content

TABLE VII

Latent vectors, ***a***, partial regression coefficients, ***b***, and multiple correlation coefficients, R, for 21 stations.

	0.44	−0.12	−0.21	−0.52		0.25	0.51	0.14	0.40
	0.50	0.10	−0.24	0.09		−0.60	−0.61	−0.90	−0.79
	0.31	−0.19	0.25	**0.76**		0.22	1.08	0.77	0.77
a	0.33	0.30	**−0.55**	0.27	***b***	−0.29	−0.16	−0.28	−0.81
	−0.02	**0.62**	0.02	−0.12		−2.50	2.03	1.44	1.53
	−0.22	**0.57**	0.22	0.14		−0.78	0.48	−0.07	0.23
	−0.39	0.10	−0.60	0.20		−0.82	0.51	0.24	0.42
	−0.39	−0.37	−0.35	0.10		−1.79	1.68	1.82	1.04
						−0.12	0.26	−0.33	−0.48
					R	(0.73	0.52	0.69	0.75)

decreases in successive equations, so that those which appear later in the series may be abandoned with minimum loss of information. If each species is allotted one unit of variance (as is implied in computing the correlation matrix), the progressive gain in information as successive equations are added may be summarized as in Fig. 2. The upper curve (cumulative λ) represents the y-functions, and the lower (cumulative λR^2) the x-functions. Eight y-functions carry 8 units of variance and thus describe the biota as faithfully as the original correlation matrix: however, the last four functions carry only one unit of variance, so that the first four contain 87 % (7/8) of the original information. In Table VIa we found that three y-functions, defining four communities, were sufficient to represent all twelve species, and in terms of Fig. 2

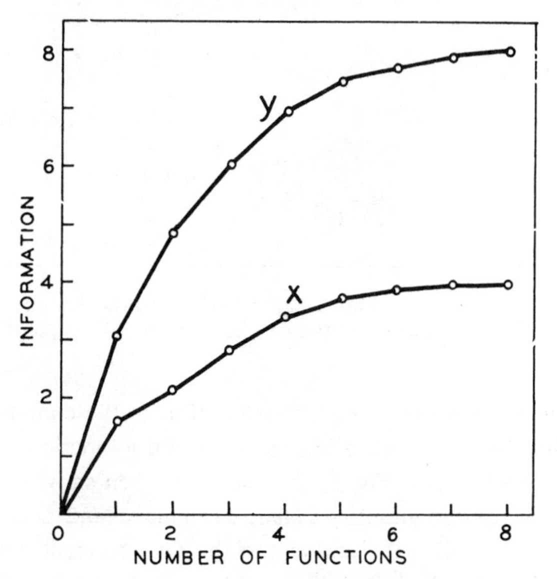

Fig. 2. Cumulative plot of information against number of functions, equation (1); x = x-functions, y = y-functions; one unit of variance per species.

this would account for 75 % (6/8) information. The residual 25 % is not non-significant in the statistical sense, but is relatively meaningless ecologically, so that the numerical relationship of the original eight species can be described reasonably faithfully in terms of three functions.

The x-functions give approximately 50 % as much information as the corresponding y-functions, the trends of the two curves being similar throughout. Thus three x-functions give 40 % of the total information, or 84 % of the information available from sediments alone. On the present data, it seems that the only biologically meaningful information is that given by the first function which describes the level of 'coarseness' of the sediments.

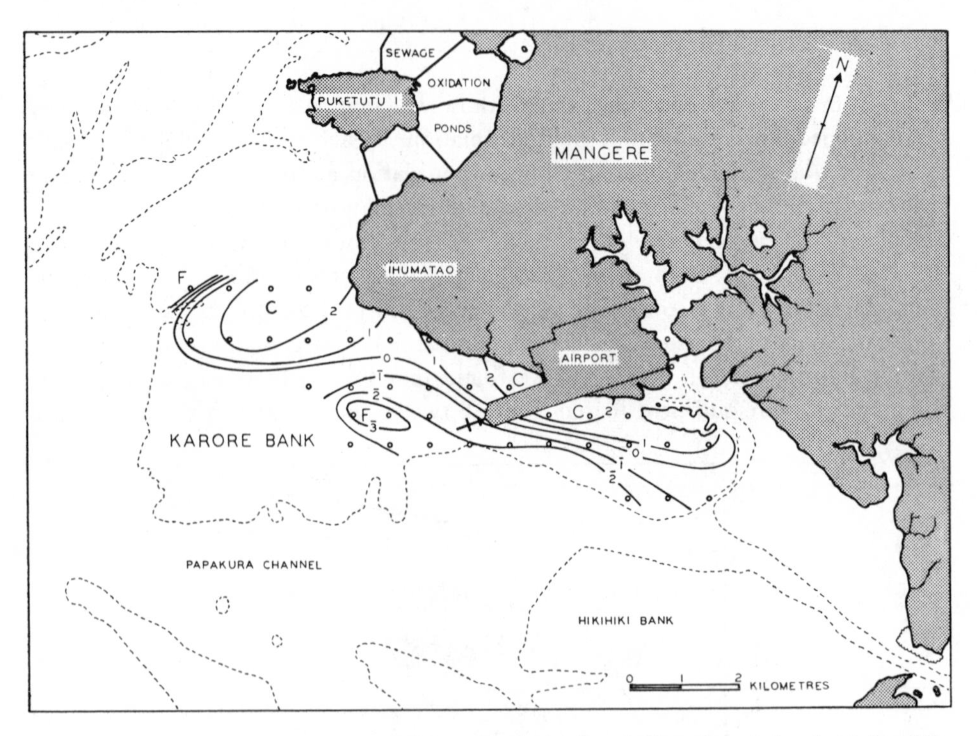

Fig. 3. The sampling area, Mangere and Karore Bank; Stations indicated as circles; isopleths of the y-function, left-hand side of equation (5), derived by principal component analysis.

A relatively comprehensive and simple presentation of the community pattern in space will be obtained by plotting station values of the y-function derived from the first column of Table VIa, as in Fig. 3. The isopleths then show the relative quantitative distribution of the communities *Chione-Macomona* and *Owenia-Halicarcinus* or, as they may now be called, the *coarse* and the *fine* sediment communities. The maxima are accordingly labelled C and the minima F. In general, the coarse community is closer to the shore-line and the fine closer to the tidal channel.

A possible extension of the method is to compute principal components from the sediments × sediments correlation matrix. This permits particle sizes to be ordinated into 'sediment components' comparable to 'communities', and also provides a new set of variables with which the communities may be correlated. Latent roots and vectors (***f***) of the sediment matrix are shown in Table VIII. This reveals that the

TABLE VIII

Latent roots, λ, and vectors, ***f***, derived from the correlation matrix for sediments (Table V).

λ	(2.98	2.75	1.07)
f	0.35	0.51	−0.02
	0.36	0.72	−0.21
	0.20	0.89	0.05
	0.06	0.81	−0.02
	−0.98	0.06	0.01
	0.41	−0.15	0.84
	0.81	−0.23	0.22
	0.88	−0.15	−0.34
	0.34	−0.66	−0.39

largest component of variability is in the relative abundance of medium (0.12–0.25 mm) and finest (< 0.06 mm) particles, while the second largest component is concerned with the largest (< 0.25 mm) particles. In contrast with the biota, all latent roots of the sediments prove 'significant' by Lawley's test. Table IX shows the correlation between the first three communities (rows) and the first three sediment components (columns). The testing of significance for coefficients of such complex

TABLE IX

Correlation between the first three communities (rows) and the first three sediment components (columns).

−0.26	0.34	0.11
0.15	0.16	−0.13
−0.08	0.01	−0.24

derivation is problematical, but it is clear that none of these would in any case reach even the 5% level of significance and that the information content is low when compared with the multiple correlation coefficients, R, in Table VII. Thus, in the present example, the sediment principal components have been of little assistance.

CANONICAL CORRELATION

The canonical vectors, ***a*** and ***b***, and correlations, R, are presented in Table X, in similar fashion to Table VII, except that the latent roots are no longer appropriate.

The most obvious features are the relatively high values of the canonical correlations, particularly the first, which is 0.97. (Of the 8 particle-size percentages only 7 are independent, so that 7 rather than 8 degrees of freedom are lost.) These are, however, less remarkable when one considers that 16 variates have been fitted to 21 samples, thus reducing the number of degrees of freedom to 5. The successive coefficients of determination, R^2, may be compared for the two methods of analysis:

Principal components: 0.63, 0.56, 0.62, 0.35, 0.62, 0.28, 0.60, 0.20
Canonical correlation: 0.95, 0.87, 0.77, 0.59, 0.46, 0.30, 0.14, 0.03

The principal component coefficients are distributed about a mean of approximately 0.5 and are uncorrelated with their position in the series, while the canonical coefficients have been derived in such a manner that they descend continuously from very high values nearly to vanishing point. Dominant *a*'s are shown in bold face as for the latent vectors, though in contrast to the principal component vectors these coefficients lead to no useful community classification. The *a*'s are in effect coefficients

TABLE X

Canonical vectors, ***a*** and ***b***, and canonical correlation coefficients, *R*, for 21 stations.

	a					b			
	0.57	0.43	**−0.75**	−0.04		−0.08	−0.50	0.34	−0.18
	−0.19	−0.27	0.00	**0.72**		−0.60	−0.82	−0.65	−0.22
	0.35	−0.14	−0.12	0.03		0.61	−0.53	0.52	−0.04
	−0.04	−0.33	0.10	**−0.58**		−0.44	0.48	0.23	0.15
	0.47	0.35	0.36	0.06		−2.22	−1.19	2.70	−0.31
	0.20	−0.22	**−0.46**	−0.04		−0.66	0.02	0.55	−0.26
	−0.30	**0.58**	−0.26	0.37		−0.60	−0.02	0.99	0.04
	0.40	−0.33	−0.13	−0.09		−1.40	−1.10	1.92	−0.54
						0.09	0.44	0.01	0.01
					R	(0.97	0.93	0.88	0.77)

for partial regression of the *x*-functions on the *y*'s. It is well known that partial regression coefficients often undergo a considerable change in value, and even a reversal of sign, as variables are added or deleted from the regression equation. One would expect much the same to happen if canonical analysis were carried out with more or fewer species.

On the other hand there is an obvious correspondence between at least the earlier *x*-function coefficients derived from the two methods. The correlation between the first two sets of coefficients is 0.98, for the second 0.83, the correlation diminishing progressively for subsequent vectors. This is perhaps to be expected in that the *x*-function coefficients in both methods have been derived by a regression-type method. Canonical correlation is, in effect, no more than a more general case of multiple regression and the two processes become identical if the number of variables on either side of the equation is reduced to one. This also helps to explain the apparently capricious nature of the canonical correlation *y*-coefficients, as well as the apparent failure of the first set of x-coefficients in both methods to relate more than approximately to what would be expected of a mean particle size statistic (using the term 'mean' in its broadest sense and not necessarily referring to an arithmetic mean). It is generally recognized (*cf.*, Quenouille, 1952) that partial regression coefficients may not give the best estimate of the true underlying relationship between variables, unless the independent variables are free from error. Regression coefficients will always provide the best *prediction* of dependent from independent variables, but at the same time will not necessarily indicate faithfully the relative contributions of the independent variables.

It would perhaps be useful to plot and contour one or more of the canonical y-functions for comparison with Fig. 3, but this seems best reserved for a larger series of data, remembering that 40 stations are available for principal component analysis, but only 21 for canonical correlation. In view of the high correlation between the b's of the first equation of each method, one might predict some similarity between the two sets of contours.

Conclusions

Of the two analyses, principal component analysis has proved the more versatile in that it permits both diagnosis of the community structure and at least a plausible contouring of two of these communities in space. Subsequent multiple regression analysis with 'communities' as dependent and sediment properties as independent variables produces regression coefficients more or less equivalent to the corresponding coefficients obtained by canonical correlation, and there seems to be little to chose between the two methods in approximating the most appropriate statistic for predicting community distribution from sediment distribution. The first x-function from either method would undoubtedly be meaningful, but judgement must be reserved concerning the second and third, since the communities to which they refer may well be related to environmental properties not revealed by the particle-size analysis. Principal component analysis of the particle sizes will also produce x-functions which are appropriate for a physical characterization of the sediments but, at least for the present example, these seem to have little biological significance. Although canonical correlation and principal components of particle size have been less useful, this is possibly attributable in part to the smaller number of samples available for these analyses, and the two methods will probably repay further investigation.

While the present analysis must be regarded as exploratory, and in some respects inconclusive, the principal component analysis has been consistent with what is known of the habits of the organisms concerned. As was predicted, *Owenia-Halicarcinus* and *Chione-Macomona* are ranked as the dominant members of two negatively associated communities. Other partly subjective observations on similar fauna elsewhere in the Auckland region may be quoted. Adult *Amphidesma* is commonly associated with *Chione-Macomona.* While the crabs *Hemiplax* and *Halicarcinus* are commonly found on the same shores, the former is usually at lower levels, or at least in flat regions with overlying surface water at low tide. *Cyclomactra* and *Pectinaria* both appear to be associated with sediments of higher clay content, a feature which would not necessarily appear in the particle-size analysis. Adult *Chione* are commonly associated with a coarse substrate, even though immature individuals are sometimes found in heavily silted regions. No detailed observations were made of surface contours or tidal channels, but there is general agreement between the isopleths in Fig. 3 and subjective observations of tidal movement. It would be expected that, within any one community, species would occupy different niches and that this would be reflected in feeding habits, mobility, burrowing depth, *etc.* This last hypothesis is a difficult one to assess objectively, but at least there are no obvious inconsistencies apparent when the characteristics of species (Table I) are related to the principal component analysis (Table VIa). *Soletellina* and *Solemya* (community 2) have some superficial similarity in that both are highly mobile filter feeders, but there is an obvious functional difference between the sharp, knife-like foot of *Soletellina* and the blunt plug-like foot of *Solemya.*

Computation time for all the procedures outlined above was approximately 20 minutes. The mathematical procedures are essentially combinations of standard matrix algebra subroutines – multiplication, inversion, solution of simultaneous equations and solution of eigenvalues and eigenvectors.

Acknowledgements

The authors wish to thank Professor J. E. Morton and Dr M. C. Miller of the Zoology Department, University of Auckland, for advice on the ecology of the invertebrate fauna.

Appendix

The Use of Transformations in Multivariate Analysis

BY R. M. CASSIE

Throughout the above paper, it has been assumed that the basic mathematical model is the multivariate normal distribution. While the manuscript was in draft, it became apparent that this model is not accepted by all users of multivariate analysis and that, in particular, the transformations of raw data for the purpose of ensuring approximate normality is regarded with some suspicion. Though there is little specific reference to 'non-normal' analysis in the literature, it seems desirable to discuss the mathematical theory underlying the multivariate normal model, the reasons for choosing these particular transformations, and the possible consequences of failing to transform.

The assumption of normality is implicit in most, if not all texts on multivariate analysis, but not all are equally explicit, so that it is possible to learn the computational procedure and much of the theory of, say, principal component analysis without actually encountering the word 'normal' at all. Perhaps the most useful introductory text is that of Seal (1964) who discusses the multivariate normal distribution (Chapter 6) and the dependency of various techniques on departure from this model (p. 170).

Equations such as (1) and (2) show clearly that multivariate analysis deals with linear models and is thus an extension of the linear regression equation:

$$y = \alpha + \beta x + \varepsilon, \tag{6}$$

where α and β are constants, x and y are variables and ε, the error term, is the only variate (where a variate is defined as a random variable). Nonlinear models are of course possible, but the theory would be formidable, so that it is expedient to convert the raw data to a form which can be handled by linear methods. In multivariate analysis, random variation is not confined to the error term. For example, the x's and y's in (1) may be regarded as variates which, though each contains a random element, are in some measure interdependent. The first task of the multivariate analyst is to estimate the variability of individual variates (the variances) and the components of this variability common to every possible *pair* of variates (the covariances). Subsequent analysis may be simplified by assigning to each variate the arbitrary variance of 1, so that the covariances become correlation coefficients. Since these estimates apply, not merely to the sample, but also to the population from which the sample is drawn, it is essential to define the distribution of the variates (*i.e.*, the shape of the histogram or other frequency diagram). It is almost universal practice to assume that this is multivariate normal, although other distributions, such as the multinomial (Anderson, 1958, p. 1) are possible. However, in the words of Kendall (1957, p. 86): "Practically everything that is known about exact distributions in the multivariate case depends on the assumption that the parent distribution is multivariate normal". For example, the product-moment correlation coefficient becomes a maximum likelihood estimate if based upon two normal variates. Thus normality is required even before multivariate analysis commences. Since the multivariate normal distribution is a linear model, linearity need no longer be specified as a separate requirement.

Having derived the correlation (or variance-covariance) matrix, the investigator then applies various *linear* transformations to the matrix (as opposed to the nonlinear transformations of individual variates), with the object of estimating components of variance common, not only between pairs, but between groups of variates. Geometrically this amounts to *rigid rotation* of the multi-dimensional scatter diagram of the transformed data (though this becomes an abstraction if the dimensions are more than three). Rigid rotation implies that the properties of multivariate normality remain undisturbed so that when new variates (such as principal components) are defined they retain the same statistical properties as the old. The process of rotation is discussed and illustrated by Cassie (1963) and Seal (1964).

Some confusion has probably arisen in that the geometric operation of rotation, corresponding to the algebraic operation of, say, principal components, is a non-statistical operation quite independent of the underlying distribution (*e.g.*, Anderson, 1958, p. 273). If, however, the distribution of the input variates has not been specified, it is not possible to draw any statistical inferences from the output. This will apply, not only to tests of significance, as has been implied, *e.g.*, by Reyment (1963, p. 453), but also to estimates of community structure (from the latent vectors) and to any subsequent statistical operations, such as regression or correlation analysis.

In developing the transformations used in this paper, it is convenient first to base our arguments upon the less restrictive requirement, linearity, and to show then that the same transformations are also appropriate for normality. In the ideal situation, where all variates are truly homogeneous and random, there can be only one transformation appropriate to each variate, and the normal criterion is absolute. In natural populations, this ideal is seldom realized, and some relaxation of statistical rigour may be necessary. Optimal transformations (*e.g.*, Kleczkowski, 1949) are not necessarily helpful, since it is usually only possible to optimize one criterion. Thus, for example, one might eliminate skew, but fail to remove correlation between standard deviation and mean, or curvilinearity in relation to other variates. Fortunately, there are many situations in which all three criteria can be approximately satisfied by simple transformations which are reconcilable with reasonable hypotheses concerning animal behaviour as related to the environment. Further developments of statistical theory may permit some refinements of these transformations, but it seems likely that the gain in efficiency would be slight, except in dealing with very small samples.

The relationship between c, the numerical abundance of an organism, and x, the value of an environmental factor, may frequently be expressed in the exponential form:

$$c = Ae^{\beta x}E, \tag{7}$$

where A and β are constants and E is a multiplicative error term. Transforming so that $y = \ln c$, $\alpha = \ln A$, and $\varepsilon = \ln E$, this reduces to the linear form of (6). (As we shall see below, it may be necessary to make x also a transformation of the original environmental measurement.) Every organism will be influenced by many environmental factors, and summing over all of these, (6) becomes:

$$y = \alpha + \sum_i \beta_i x_i + \varepsilon, \tag{8}$$

where ε is now a Poisson variate. Since the expression in x_i is a linear compound of a number of random variables, y will tend to be normally distributed. The Poisson term, ε, will not disturb this normality provided the mean of all y's is large. It is commonly found that sample counts of organisms are indeed approximately log-normal, *i.e.*, the transformation, $y = \ln c$, is normal. When some sample counts are zero, it is conventional to transform to $\ln(c+1)$ in order to give finite logarithms to the zeros. While this is an arbitrary expedient, it has some theoretical justification in that it tends to correct for departures from normality arising from the Poisson error component which causes pronounced skew when the mean count is small. The theory of this model has been discussed in more detail by Cassie (1962).

If x influences the abundances, c_1 and c_2, of two organisms:

$$c_2 = A' c_1^{\beta_2/\beta_1} E', \tag{9}$$

where A' is a constant, E' an error factor, and β_1, β_2 are the exponents appropriate to c_1 and c_2. This is also a curvilinear relationship, but includes the special case ($\beta_2 = \beta_1$) which is linear. In practice, it may be possible for the linear case to be realized, even for an entire correlation matrix, so that a principal component analysis of untransformed species counts might give approximately the same latent roots and vectors as for transformed counts. This is possibly the reason why some workers have been led to assume that transformations are unnecessary. It can by no means be assumed, however, that $\beta_2 = \beta_1$ under all circumstances, and if principal component analysis is to be followed by analysis of regression or correlation on the x's, the curvilinear equation (7) still remains to be negotiated. Even when the principal component vector solution is regarded as an end in itself, it is *implied* that the new variates so defined are linearly related to some external factors though these may not have been identified.

Fig. 4 is a computer simulation of the situation in equation (9), with c_2 plotted against c_1 in a scatter diagram, the raw data in Fig. 4a, and the transformations $\ln c_2$, $\ln c_1$ in Fig. 4b. Both sets of points are enclosed by a 99 % probability envelope. The equations for the line A-B are:

$$c_2 = 1.5\, c_1^{0.8} \qquad \text{(Fig. 4a)}, \tag{10}$$

$$\ln c_2 = 0.4 + 0.8 \ln c_1 \qquad \text{(Fig. 4b)}. \tag{11}$$

In Fig. 4a, the main axis A-B is curvilinear, and the points are mainly clustered toward the lower left extremity of the envelope, while toward the upper right the scatter (variance) of points increases considerably. In Fig. 4b, A-B is linear and the points are distributed symmetrically within the envelope, with their greatest density toward the centre. Furthermore, if the points in Fig. 4a are projected upon, say, the X-axis, their distribution will be highly skewed, whereas in Fig. 4b their distribution will approximate to normal. (In fact, they would be *exactly* normal if an infinite population of points had been generated by the computer.) In Fig. 4b, normality is preserved in the projection, no matter how the scatter diagram is rotated in relation to the axes, but in Fig. 4a the degree of skew will change according to the degree of rotation, so that the distribution following linear transformation is unpredictable. Further, if the coefficient, A', in (9) is changed, say, from 1.5 to 2.0, corresponding to a change in absolute abundance of species c_2 without change in relative distribution, the line A-B will be replaced by the broken line, which has a different slope and curvature. A corresponding change will take place in the shape of the 99 % probability

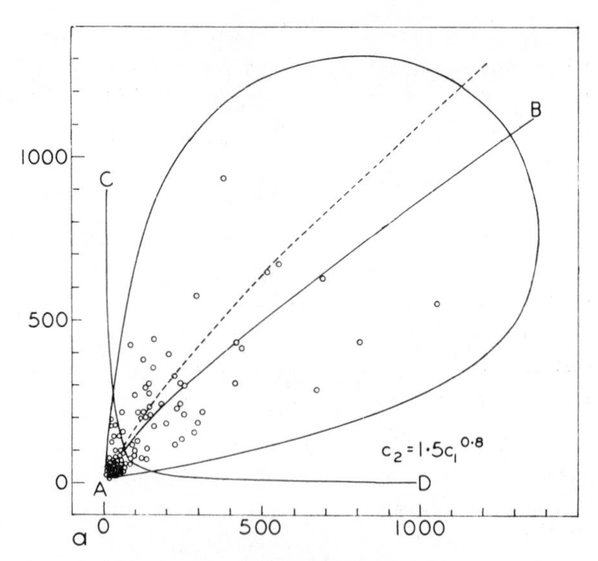

Fig. 4a. Scatter diagram of the raw counts, c_1 and c_2 of two species related by the equation: $c_2 = 1.5c_1^{0.8}$ (line A–B). A–B and C–D correspond to the lines similarly labelled in b. The closed curve is the 99 % probability envelope, and the broken line represents the equation: $c_2 = 2.0c_1^{0.8}$: the population coefficient of linear correlation is $\rho = 0.70$.

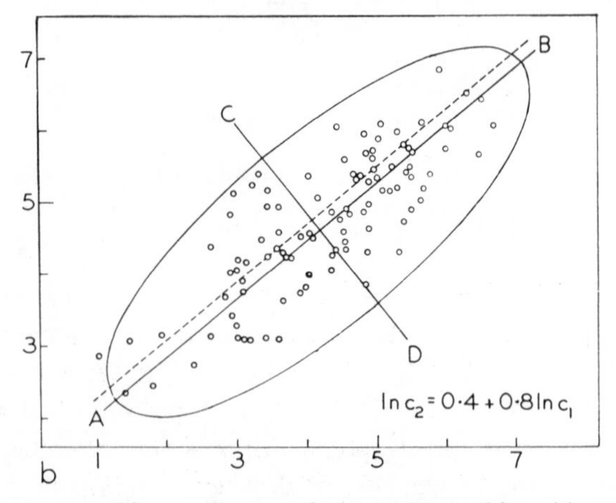

Fig. 4b. The same diagram with c_1 and c_2 transformed to natural logarithms: the 99 % probability envelope becomes an ellipse with major and minor axes A–B and C–D, and the correlation coefficient is increased to $\rho = 0.80$.

envelope which is not shown. In Fig. 4b, A-B, together with the elliptical envelope, is merely displaced upward without any change in form. Since we are concerned with variances (shape of the diagram) rather than means (position of the diagram), the model is effectively unchanged. Thus, the log-transformed analysis provides a generalization which is appropriate for the same two species at different places or seasons, or even under different sampling techniques, provided none of these changes affect the ratio β_2/β_1. Without transformation no such generalization is possible. Finally, the (linear) correlation coefficient for Fig. 4b is 0.8 and for Fig. 4a is 0.7, so that a greater amount of information has been extracted from the data, *i.e.* c_2 can be predicted more precisely from c_1 or *vice versa.* As we shall see below, in the multivariate case, higher correlation permits more information to be expressed in any given number of principal components.

The assumption that $y(= \ln c)$ is linearly related to x is a reasonable one, provided x does not in its measured range approach an upper or lower limiting value. Such a limiting value could occur in one or other of two ways. As an example of the first, the temperature of water has lower and upper limits of 0 °C and 100 °C outside which levels water (in the common sense) no longer exists. A 'warmth-loving' organism, however, might cease to react favourably to rising temperature at a very much lower level than 100 °C, say 30°C, thus giving a second kind of limit, which does not concern us at present (appropriate transformations for this case are discussed by Cassie, in press).

An instance of the first kind of limiting values occurs with environmental properties commonly expressed as proportions or percentages, *e.g.*, the proportion of a certain sized particle in a sediment. Let us imagine that a certain organism tends to avoid coarse sediments. This is essentially the same as seeking fine sediments, so that we might describe this particular property either as p, the proportion of particles above a specified size, or q, the proportion below this size. It is convenient to make $p+q = 1$, though the argument is easily modified to apply to percentages. If the average values for a given environment are $p = q = 0.5$, the organism would probably be relatively slightly affected by a change to, say, $p = 0.4$ or $p = 0.6$. On the other hand, if average $p = 0.05$ ($q = 0.95$), a change to $p = 0.15$ ($q = 0.85$) will have a substantially greater effect, and $p = 0.0$ ($q = 1.0$) even greater again. Thus the response of the organism may be measured by a coefficient, β', which plays the same role as β in (7) except that it is no longer constant, but inversely proportional both to p and to q:

$$\beta' = \mathrm{d}y/\mathrm{d}p = k/\sqrt{(pq)}, \tag{12}$$

where k is a constant.

Referring to Fig. 5a, p and q are represented as areas of squares on the sides of a right-angled triangle. If the hypotenuse is of unit length, $p+q = 1$, as specified above. The angle opposite p is designated x, so that: $p = \sin^2 x$
and

$$\frac{\mathrm{d}p}{\mathrm{d}x} = \frac{\mathrm{d}\sin^2 x}{\mathrm{d}x} = 2\sin x\cos x = 2\sqrt{(pq)}. \tag{13}$$

Then, from (12) and (13):

$$\frac{\mathrm{d}y}{\mathrm{d}x} = \frac{\mathrm{d}y}{\mathrm{d}p}\frac{\mathrm{d}p}{\mathrm{d}x} = 2k = \text{a constant}. \tag{14}$$

Thus we may replace p by the new variate, x, which is linearly related to y:

$$y = \alpha + \beta x, \tag{15}$$

where $\beta = 2k$ = a constant, α = a constant and $x = \arcsin \sqrt{p}$.

The curve representing this transformation is seen in Fig. 5c, where p is the ordinate and x the abscissa. While an angle may seem a strange substitute for an area, the effect is easily visualized in Fig. 5b, where the areas p and q are replaced by sectors in a quadrant which has the same area as the square on the hypotenuse of Fig. 5a. In practice it is unimportant whether the sum of transformed p and q is unity or some

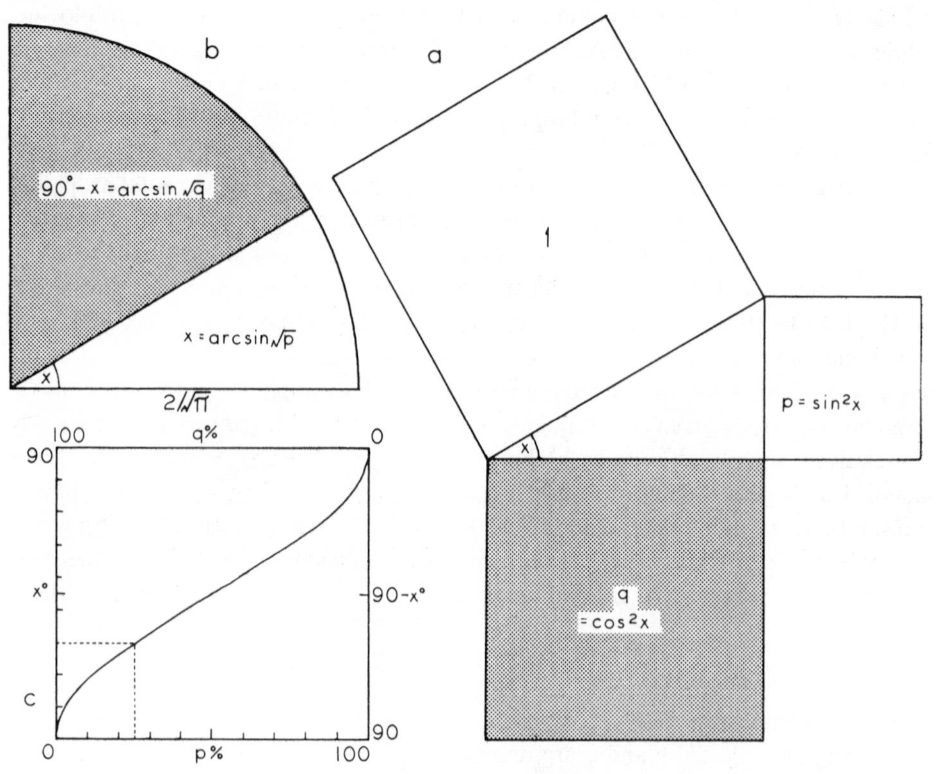

Fig. 5. Derivation of the angular transformation, $x = \arcsin \sqrt{p}$: in all figures $p = 25\,\%$ (0.25), $x = 30°$ ($\frac{1}{6}\pi$).

other constant. In statistical tables the 'angular' transformation is usually given in degrees (0–90°) while computer subroutines give an output in radians (0–$\frac{1}{2}\pi$). (The usual library subroutine is arctan rather than arcsin, so that the transformation becomes: $x = \arctan \sqrt{(p/q)}$.) When all percentages are small (as for organic carbon content), as $p \to 0$, $q \to 1$, and, substituting 1 for q in (12):

$$\beta' \sim k/\sqrt{p} \tag{16}$$

so that the simpler transformation, $x = \sqrt{p}$ may be substituted. The angular transformation also has the effect of eliminating the skew, positive for small and negative for large values, inherent in the distribution of percentages, and is a well-known expedient for removing the dependence of variance on mean in the binomial distribution (*e.g.*, Rao, 1952). Thus, once again the transformation selected for linearity produces at least some of the criteria of normality. Goodall (1954) has appropriately used the same transformation for a principal component analysis of percentage plant cover.

To illustrate the consequences of failure to transform, the analysis in this paper was repeated on the raw data. The species × species correlation matrix (*cf.* Table II) is shown in Table XI. Many, but not all of the correlation coefficients are reduced

TABLE XI

Species × species correlation matrix, untransformed data, 40 stations (*cf.* Table II).

1.00	0.65	−0.06	0.30	−0.17	−0.10	−0.14	−0.19	−0.18	−0.18	−0.30	−0.27
	1.00	−0.16	0.42	−0.07	−0.16	−0.23	−0.20	0.01	−0.16	−0.36	−0.42
		1.00	0.21	−0.20	−0.12	−0.10	−0.11	−0.11	−0.11	−0.24	−0.18
			1.00	0.30	−0.11	−0.14	−0.10	0.18	−0.04	−0.06	−0.27
				1.00	0.09	0.26	0.05	0.21	0.13	−0.08	−0.22
					1.00	0.17	0.00	−0.03	−0.03	0.00	0.05
						1.00	−0.03	−0.03	−0.01	−0.05	−0.11
							1.00	0.07	0.06	0.15	0.33
								1.00	0.03	0.20	0.05
									1.00	0.23	−0.08
										1.00	0.57
											1.00

in value, indicating varying degrees of non-linearity, and the sum of squares of off-diagonal elements is 2.93 as compared with 4.99 for the transformed matrix, representing a loss of information of about 40 %. Loss of information alone will result in a less definitive diagnosis of 'community' structure, but an even more serious deficiency is that the loss is unevenly distributed through the matrix, and the diagnosis is correspondingly distorted.

In the principal component analysis, the first four latent roots are:

Transformed:	3.62	1.80	1.36	1.11
Untransformed:	2.88	1.61	1.45	1.09

Clearly the transformed analysis is more efficient in that the larger roots account for more of the variance. Thus, for example, three 'transformed' roots have a total variance of 6.78, which is almost as much as 7.03 for four 'untransformed' roots.

Other comparisons are necessarily partly subjective. Without transformation, interpretation was on several counts inconsistent with what would be predicted by conventional ecological procedures. Thus, for example, the *Chione-Macomona* community appears to be associated with fine rather than coarse sediments, while *Owenia* and *Halicarcinus* are allocated to different communities. It might, of course, be argued that the subjective observations are wrong, but most ecologists would be reluctant to accept a mathematical analysis which cannot be reconciled with the more *obvious* distribution patterns they see in the field. Even if it were not possible to apply well-established statistical criteria, it would seem legitimate to develop a computer programme which not only reproduces the judgement of an experienced ecologist, but also extends the scope of this judgement to what would otherwise be marginal situations.

Although the method has not, as yet, been applied extensively, present indications are that it has some degree of generality. Two further mud-flat surveys recently completed in the Auckland region show that wherever there is an overlap in species composition, the computed community structure is consistent with the transformed, but not with the untransformed Karore Bank analysis. Thus, at Hobson Bay, *Chione*, *Macomona*, and *Nucula* appear in a single community correlated with the coarser sediments, and *Halicarcinus* in a separate community correlated with finer sediments. At Whangateau, where the sediments are uniformly coarse, only one community is found, dominated by *Chione* and *Macomona*, but with *Amphidesma* and *Nucula* also present.

References

Anderson, T. W., 1958. *An introduction to multivariate statistical analysis.* Wiley, New York and London, 374 pp.

Bartlett, M. S., 1950. Tests of significance in factor analysis. *Br. J. Statist. Psychol.*, Vol. 3, pp. 77–85.

Cassie, R. M., 1962. Frequency distribution models in the ecology of plankton and other organisms. *J. Anim. Ecol.*, Vol. 31, pp. 65–92.

Cassie, R. M., 1963. Multivariate analysis in the interpretation of numerical plankton data. *N.Z.J. Sci.*, Vol. 6, pp. 36–59.

Cassie, R. M., (in press). Sampling and statistics. In, *Methods for assessment of secondary production in freshwater*, IBP Handbook, edited by W. T. Edmondson.

Goodall, D. W., 1954. Objective methods in the classification of vegetation. III. An essay in the use of factor analysis. *Aust. J. Bot.*, Vol. 2, pp. 304–324.

Holme, N. A., 1949. The fauna of sand and mud banks near the mouth of the Exe Estuary. *J. mar. biol. Ass. U.K.*, Vol. 28, pp. 189–237.

Kendall, M. G., 1957. *A course in multivariate analysis.* Griffin, London, 185 pp.

Kleczkowski, A., 1949. Transformation of local lesion counts for statistical analysis. *Ann. appl. Biol.*, Vol. 36, pp. 139–155.

Lawley, D. N., 1956. Tests of significance for latent roots of covariance and correlation matrices. *Biometrika*, Vol. 43, pp. 128–136.

Morgans, J. F. C., 1956. Notes on the analysis of shallow-water soft substrate. *J. Anim. Ecol.*, Vol. 25, pp. 367–387.

Quenouille, M. H., 1952. *Associated measurements.* Butterworth, London, 242 pp.

Rao, C. R., 1952. *Advanced statistical methods in biometric research.* Wiley, New York and London, 390 pp.

Reyment, R. A., 1963. Multivariate analytical treatment of quantitative species associations: an example from palaeoecology. *J. Anim. Ecol.*, Vol. 32, pp. 535–547.

Seal, H. L., 1964. *Multivariate statistical analysis for biologists.* Methuen, London, 207 pp.

Tropical Ecology

TROPHIC STRUCTURE AND PRODUCTIVITY OF A WINDWARD CORAL REEF COMMUNITY ON ENIWETOK ATOLL[1]

HOWARD T. ODUM[2] and EUGENE P. ODUM[3]

TABLE OF CONTENTS

INTRODUCTION

coral reef communities of the world are tre-usly varied associations of plants and animals ıg luxuriantly in tropical waters of impover- plankton content. Under intense equatorial ion the plants apparently grow rapidly and ten rapidly. Save for fluctuations the reef unchanged year after year, and reefs appar- persist, at least intermittently, for millions of With such long periods of time, adjustments ganismal components have produced a biota successful competitive adjustment in a rela- constant environment. The reef community is s for its immense concentrations of life and its exity.

ıaps in the structure of organization of this ely isolated system man can learn about optima ilizing sunlight and raw materials, for man- great civilization is not in steady state and its n with nature seems to fluctuate erratically and ously. What, then, is the relationship between c productivity, energetic efficiency, and the ng crop structure of a coral reef community? re steady state equilibria such as the reef eco- self adjusted?

Since nuclear explosion tests are being conducted in the vicinity of these inherently stable reef communities, a unique opportunity is provided for critical assays of the effects of radiations due to fission products *on whole populations and entire ecological systems in the field.* In the present paper some results are presented of a variety of measurements made on an Eniwetok Atoll reef which as yet has been little affected by nuclear explosions. These measurements represent both a multiple approach to the problem of obtaining practical assays of total function which will aid future comparisons between the normal and the irradiated reef ecosystems, and also a continuing effort by many to answer the questions posed in the preceding paragraph.

THE PROBLEM OF RELATING STANDING CROP AND PRODUCTION

In recent years rapid advances in technique and approach have permitted the measurement of the metabolism and productivity (rate of production) of aquatic communities and their components. Ingenious methods such as used by Sargent & Austin (1949, 1954) building on the work of Mayor (1924), Yonge (1940), and others, have permitted estimates of the productivity of coral reefs. Intense postwar interest in the tropical Pacific by geologist and biologist alike has led to new and detailed popu-

ntribution from the Atomic Energy Commission Marine at Eniwetok.
partment of Zoology, Duke University, Durham, N. C.
partment of Zoology, University of Georgia, Athens, Ga.

Reprinted from ECOLOGICAL MONOGRAPHS, Vol. 25, July, 1955

lation and zonational study of reef fauna and flora such as by Ladd, *et al.* (1950), Tracey, *et al.* (1948), Emery, *et al.* (1954), Wells (1951), Doty & Morrison (1954), and Cloud (1952, 1954). It seems now time to determine the relationship between the standing crop, defined as the dry biomass of existing organisms per area, and productivity, defined as the rate of manufacture of dry biomass per area.

It has long been felt that the productivity of the various trophic levels of a community is very roughly proportional to the standing crop being maintained although the reason has not been entirely clear. Many have confused these entirely different properties of ecosystems. The distributions of standing crop may be represented graphically by trophic level so as to form block diagrams in the shape of pyramids. For a discussion of pyramids and production see Odum, E. P. (1953). In another communication by Odum & Pinkerton (1955) theoretical reasoning based on the second law of thermodynamics is presented to show that systems of many types when in open steady state tend to adjust to maximum output of energy consistent with available input energy and a corresponding low but optimum efficiency. If steady state systems tend to be similarly self adjusted regarding efficiency of energy utilization between trophic levels, then there is theoretical reason for expecting pyramids of biomass to be similar for components with similar metabolic rates. It is pertinent that somewhat similar pyramids have been found in Silver Springs, Florida, a rich constant-temperature aquatic community in slightly pulsing steady state (Odum, H. 1953), and in successional terrestrial communities the Savannah River Atomic Energy Commission ar of South Carolina (Odum, E. P. 1954).

ACKNOWLEDGMENTS

This study was supported by the U. S. Ato Energy Commission through a contract (No. AT(2)-10) between the AEC and the University of Ge gia. We are grateful for the courtesies and a provided by AEC personnel at Eniwetok, especial T. Hardison, E. Wynkoop, J. C. Kyriacopulos, Taft, and D. Ellif. It is a pleasure to acknowled the many stimulating discussions, help, and enco agement provided us during the course of this stu by experienced reef scientists: R. W. Hiatt, M. Doty, and T. S. Austin in Hawaii; L. Donaldson, Lohman, A. Weylander of the Univ. of Washingto D. W. Strasburg, Duke Univ.; L. R. Blinks, Stanf Univ.; and H. Ladd, U. S. G. S. in Washingt D. C.

We are also indebted to L. D. Tuthill, Departm of Zoology and R. W. Hiatt, Director of the Mar Laboratory of the University of Hawaii for provid facilities for the work done in Hawaii.

Identifications of organisms were provided by F. Bayer (corals), M. Doty (algae), C. E. Cuttr (anemones), R. Hiatt (invertebrates), and A. W lander (fish). Code numbers have been assigned unidentified corals along with tentative names specimens deposited with the U. S. National Museu

FIG. 1. Aerial view (July 1954) of the Japtan reef looking northeastward into the trade winds. The study transect is delimited by the two arrows. The insert shows the position of the Japtan reef in Eniwetok Atoll with wind direction indicated.

GENERAL PROCEDURES

)uring a six weeks' period at the Atomic Energy ımission Eniwetok Marine station a transect of drats marked by iron stakes was established on a tively undisturbed and fairly typical inter-island ' shown in Figure 1. Many varied sampling proıres were combined to estimate the standing crop the major component groups of the reef biota. n chemical methods were used upstream and nstream to estimate the primary production and l respiratory metabolism of the reef. From these ding crop and productivity estimates, the turn- was obtained. Productivity data were combined calculated light intensities to obtain an estimate nergetic efficiency.

ı this approach it was imperative that a wide ety of methods be used all at the same time on the e area. Thus, fewer replications were made than ld be required to obtain maximum accuracy from method. Therefore it is the orders of magnitude ch mainly emerge, but care is taken to base conions only on large, probably significant differs. Details of methods used are outlined in approte sections which follow.

he taxonomic composition of the reef community emendously varied from spot to spot whereas we eve the biomass per area is more constant. Weight nates by trophic level based on our few quadrats thus probably representative, but the quadrats ıld not be considered as population estimates by ies. Many more quadrats would be required to nate the densities of individual populations.

he 20 ft by 20 ft quadrat maps in Figure 5 were e in the field with pencil on acetate boards so that vings could be made underwater using face masks. rs. R. W. Hiatt and M. S. Doty, who visited the , agreed with the authors that the quadrats ped were fair samples regarding percentage of l and general physiography.

THE WINDWARD REEF COMMUNITY, ZONATION

ıe study reef (Fig. 1) is a part of the ring of ıerged reefs which connects the small islands that e up Eniwetok Atoll. The transect is located ¼ north of Japtan Island (Lady Slipper or Muti ıd on some maps), on the eastern and windward of the atoll where the reef is 1500 ft wide (455 Most previous study transects of atoll reef comities have been made on island reefs where the r must break onto the reef and then return in same path as an undertow. In contrast, this sect, like that of Sargent & Austin (1949), is ss an inter-island reef where the water moves in one direction from east to west with the wind; is, from the open sea into the lagoon. On Enik Atoll, the inter-island reef is actually the imınt predominant type.

; pointed out by Cloud (1954), many island reefs ılly represent eroding reefs which were elevated e the water surface by a six-foot fall in ocean which began at the end of the postglacial ther-

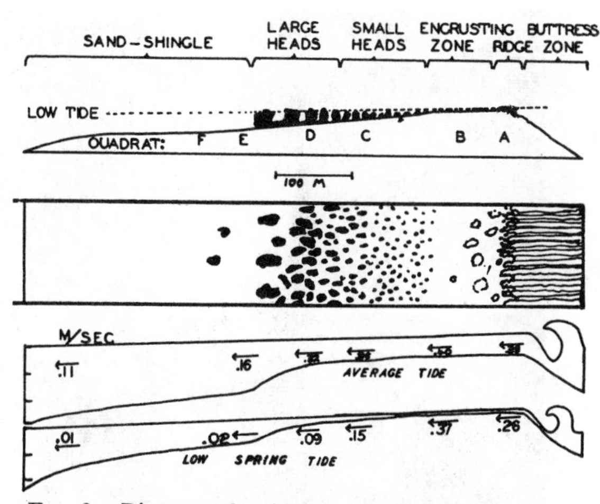

FIG. 2. Diagram showing the physiographic zones of the reef in surface and cross section view, and the average current velocities in m/sec. The approximate location of the 6 quadrats is indicated in the upper diagram.

mal maximum about 3000 years ago. These reefs are now in the process of being worn down. Only the front edge of present-day island reefs of Eniwetok support actively growing reef-building corals and algae, whereas many of the inter-island reefs support a vigorous community throughout. Consequently, the inter-island reef more nearly represents the "climax" or "steady state" community under present water level conditions. From the standpoint of productivity and metabolism it is quite clear that the inter-island reef is much more typical of an active coral community than is the half-dead, decadent island reef. These latter reefs, of course, may again become active when sufficiently worn down or if there is a future rise in sea level.

The transect of stakes was erected in a line parallel to the steady current traversing the reef. At no time during our observation were there tide pools with entirely stationary water. The zonation is very distinctive and apparently regulated by current velocities that decrease downstream as the depth increases. The descriptions of island reef zonation by Cloud (1952), Banner & Randall (1952), and Wells (1951) show very few similarities with the inter-island zonation on this reef, which is not surprising in view of points raised in the preceding paragraph. Tracey, Ladd & Hoffmeister (1948) class this type of interisland reef as type IA.

As diagramed in Figure 2 and shown in aerial photographs and horizontal views in Figures 3-5, there are six physiographic zones as described briefly below.

WINDWARD BUTTRESS ZONE

The only information on the leading front of the reef comes from the aerial photograph in Figure 4 which shows the surge channels and buttresses running at least 200 ft (66 m) to seaward. It was not possible to sample this area because the algal ridge breakers could not be crossed.

Brief glimpses from a helicopter suggest that there is about a half coverage of coral in this zone.

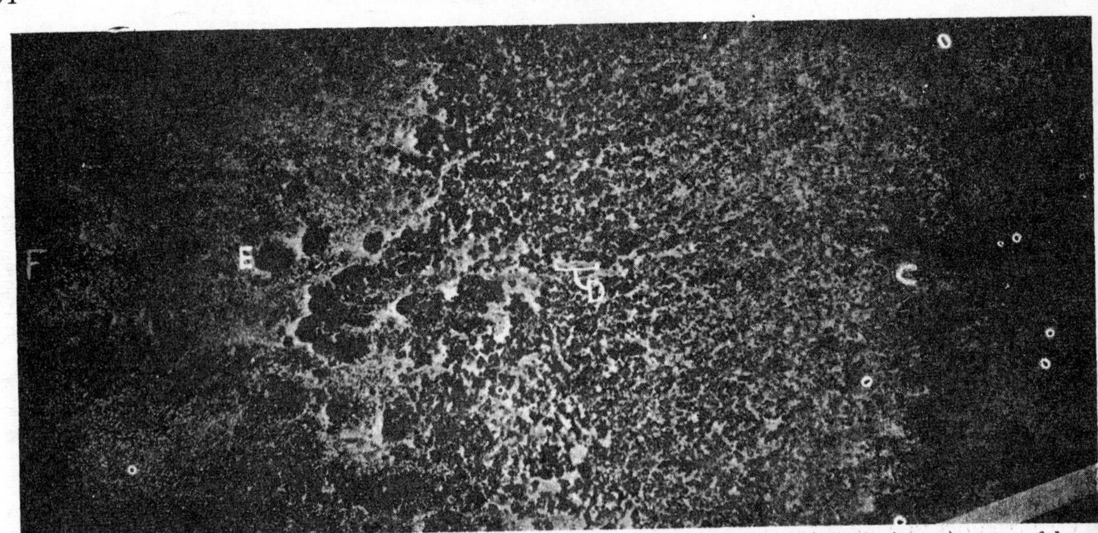

FIG. 3. Aerial photograph showing the zone of small heads (right one-third of picture), zone of large heads (center), and zone of sand-shingle of the down stream reef (left). Water flows from right to left. Zones of Quadrat C-F are marked.

Mayor (1925) found primarily species of *Acropora* in such an area.

CORAL-ALGAL RIDGE

In contrast to the massive algal ridges which have been described for windward reefs of Bikini and other atolls (Taylor 1950), the front ridge on the inter-island reef is a low, narrow ill-defined strip of limestone about 50 ft wide. The irregular surface is covered largely with corals and soft algae. The calcareous red algae (chiefly Porolithon), which are so prominent on Bikini reefs, are found mainly in small patches below the other algae or down in crevices. Our reef is an example of the principle (Tracey, *et al.* 1948; Ladd, *et al.* 1950; Cloud 1954) that inter-island reefs do not have so extensive an algal ridge or so elaborate a tunnel structure as do the island reefs. This is reasonable because the undertow on the inter-island reef is less on the front since much of the water goes over rather than back through. However, even the island reefs on Eniwetok have less calcareous superstructure than Bikini although much more than on our inter-island transect area. The lack of extensive superstructure may make the crossing of the algal ridge by a swimmer more difficult. The foam and combers on algal-coral ridges diminish the light penetration significantly and possibly cause the red algae to dominate.

The yellow encrusting *Acropora palmerae* (B-1),* small clumps of Pocillopora (A-1), and an encrusting form of *Millepora platyphylla* (B-2) are the chief corals which may cover up to half of the area of the ridge zone. Otherwise the ridge is covered with a thick mat of fleshy algae such as *Dictyosphaeria intermedia, Zonaria variegata,* Ceramium, Dictyota, and *Caulerpa elongata.* A characteristic feature is large purple sea urchins (*Heterocentrotus trigonarius*) which wedge themselves into holes under the pou[illegible] ing surf. Rotenone sampling revealed a prevale[illegible] of small blennies and groupers. Sampling work [illegible] the ridge was possible only for a half hour du[illegible] each low spring tide. The ridge is marked by w[illegible] surf in Figures 1 and 4. Some idea of the distr[illegible] tion of coral and the much folded algal surfaces [illegible] be obtained in the diagram of Quadrat A in Figur[illegible]

ENCRUSTING ZONE

The first 200 ft (66 m) downstream from the r[illegible] is a high, gently sloping plateau that at low sp[illegible] tide is covered with only 6 in. of water. It is [illegible] tively the smoothest area with corals being eithe[illegible] a flat encrusting growth form or restricted to [illegible] rounded "heads" but little raised above the ge[illegible] reef surface. The range between tops of heads [illegible] ridges and the bottoms of depressions is only a[illegible] one foot. As on the coral-algal ridge zone shee[illegible] yellow Acropora and Millepora are conspicuous. [illegible] addition, there are scattered low, rounded head[illegible] *Porites lobata* (B-6) and several species of f[illegible] (B-3, B-4, B-5). As shown in Figure 5, quadr[illegible] living coral colonies on these low heads are [illegible] crescent or doughnut shaped probably becaus[illegible] higher center portions are killed periodically b[illegible] posure during exceptionally low spring tides.

Filamentous red, brown, green, and blue-[illegible] algae form heavy encrusting mats over all of the [illegible] which is not covered by coral, there being no are[illegible] white sand as in the back reef zones. Small sea a[illegible] ones are abundant, occurring in clusters throu[illegible] the algal mat. These belong to the genus *Acti[illegible] ton* (Carlgren 1938), and apparently represe[illegible] undescribed species according to Charles E. [illegible] ress who is currently working on this material. [illegible] anemones are remarkable in that they coat them[illegible] with calcareous sand grains which are permeate[illegible]

* Code numbers refer to specimens deposited in the U. S. National Museum where final or more complete identification will be made when taxonomy of Marshall Island corals becomes better known.

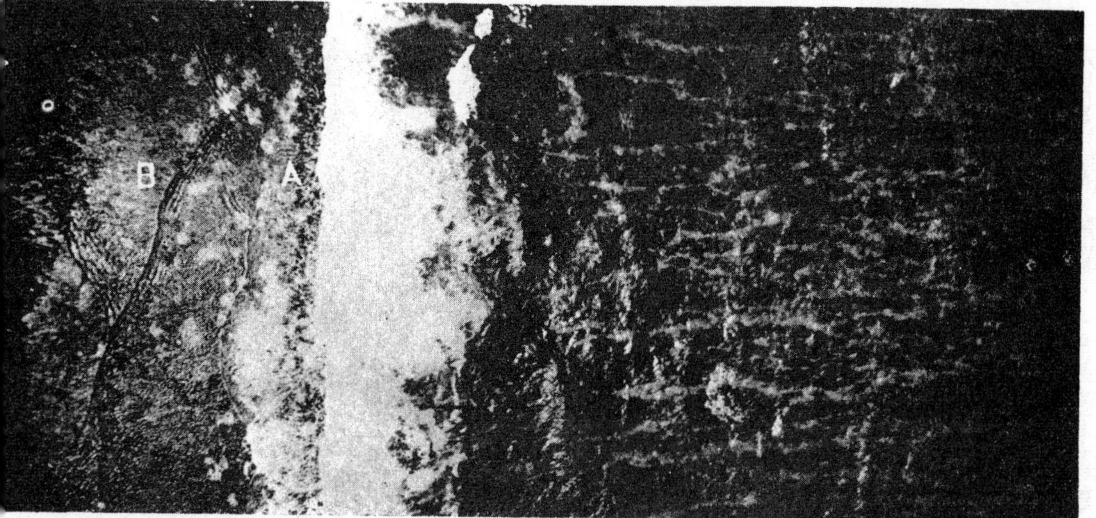

FIG. 4. Aerial photograph showing the windward buttresses on the east (right), the surf zone, the algal ridge (just to left of white line of surf), and the encrusting zone (left of ridge and covered with light-olored splotches of encrusting coral). Water leaving the zones pictured here flows downstream across zones pictured in Figure 3. Quadrats A and B are marked.

nentous algae of the same type as found in the etons of corals.

orals cover much less than half of the surface (Fig. 5, quadrat B). From the air the zone a wine-red color (algae) splotched with yellow als). The zone receives pulses of foam-water as breakers throw rolls of water up on the plateau. e there is a distinct slope the current is always ng even at low spring tide when the water pours dily across like a broad mountain stream rippling a rocky bed. Visual observation indicates that are not numerous in this shallow, rough-water , although schools of parrot fish were observed ross the area and small fish were found in the crevices which are available. The encrusting zone sible just back of the line of breakers in Figure 4.

ZONE OF SMALLER HEADS

airly abruptly beyond the sloping plateau (en-ting zone) the water begins to deepen and the ent diminishes accordingly. Coral heads become r and more numerous but are still only a foot or n diameter and height. The non-coral surfaces me lighter in color with less algal matting and sand and more fragments of calcareous skele-(rubble and shingle). As one moves downstream, heads become larger and are coalesced into com-nd heads, often composed of several species. lrat C, Figure 5, is located near the lagoonward of this zone. The numerous small heads and the ation of larger heads are well shown in this e. Encrusting forms of Acropora, so prominent he previous two zones, are absent. Massive, ded heads of large calyx favids (C-2, C-3 *Favia da,* C-4, C-5 *Cyphastrea serailia,* C-7) reach mum abundance in this zone. *Porites lobata* nues to be an important species, while colonies e short, branching forms of Acropora first make appearance in numbers (Figure 5). Large branched Acroporas (*A. gemmifera,* C-10), which become more important in the next zone, are present in small numbers.

Small fishes are numerous in this zone and large fish come into the area when the current is not too strong. Two individuals of the poisonous stone fish, *Scorpaena gibbosa,* were found resting on the top of dead portions of heads when quadrat C was being mapped. So well did these fish blend with the background that one was at first sketched in as part of the reef structure before disturbance caused it to change its position!

The zone of small heads is well shown in the aerial photograph in Figure 3.

ZONE OF LARGER HEADS

As the water depth increases and the current becomes much less, the heads become massive compound structures, 2 to 4 ft or more in height and 2 to 20 ft in diameter, with channels of white sand and cobble floor in between the heads (quadrat D, Figure 5). Branching corals predominate, such as *Acropora gemmifera* (D-7), *A. cymbicyathus* (D-9), Pocillopora, Stylophora, and others, but massive types such as Porites and Montipora are present. There is a distinct Millepora zone composed of *M. platyphylla* (E-1) and the "stinging coral" *M. murrayi* (E-2) at the back edge of the zone of larger heads. The blue coral, *Heliopora caerulea* (E-3), is fairly common, while large heads of *Turbinaria mesenterina* (E-4) represent the last important coral formations as one passes into the next zone in the lagoon. Quadrat D lies in the front part of the zone of large heads where Milleporas are less prominent.

Mounds of coral shingle (dead coral fragments, usually permeated with living filamentous algae) form the central mass upon which smaller live heads form wreaths. At low spring tide the branching type corals

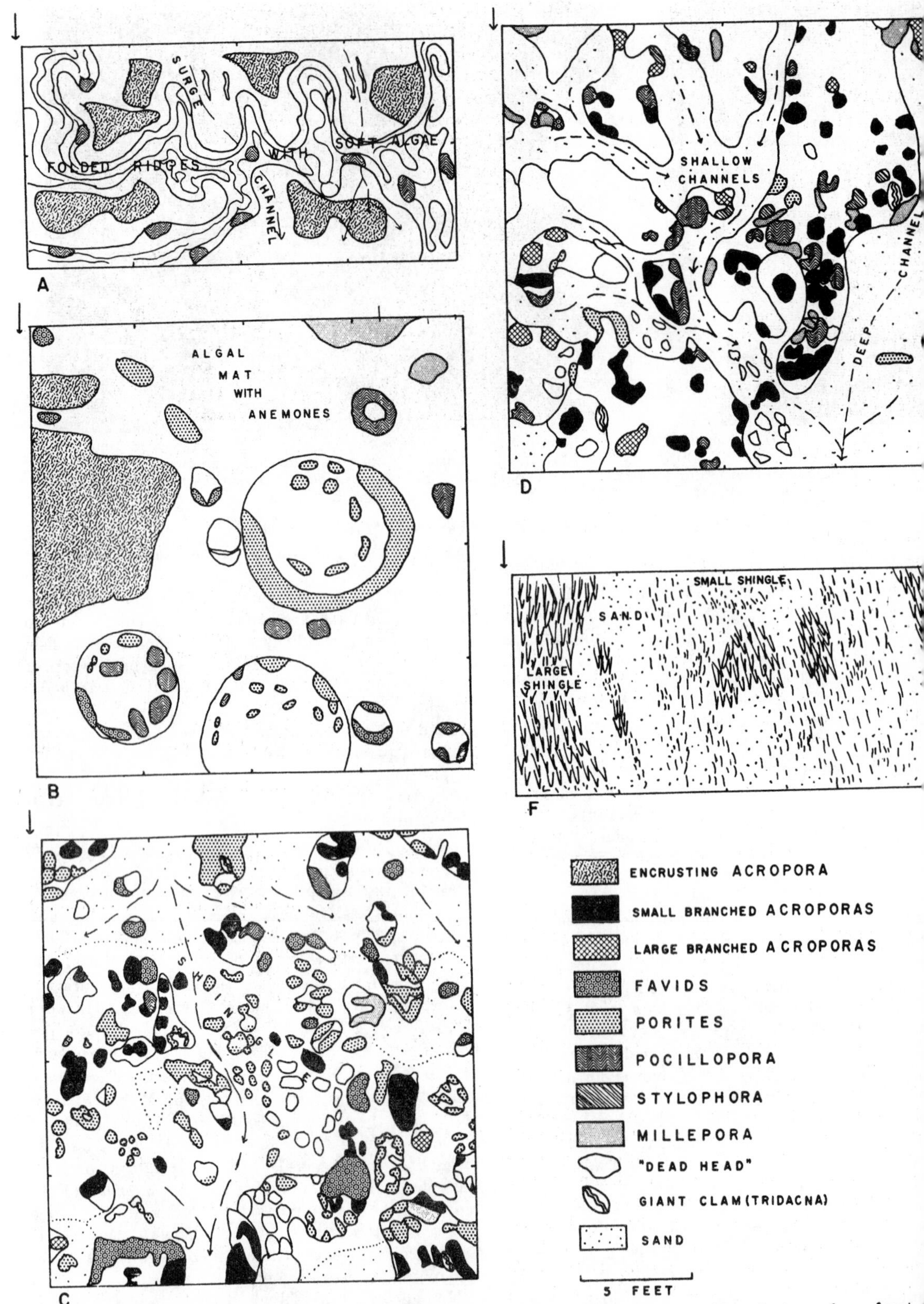

FIGURE 5. Quadrat maps showing distribution of corals and other major habitat features from front to back of the reef (see Fig. 2, upper diagram, for the approximate location of quadrats on the reef). Quadrat A, coral-algal ridge; quadrat B, encrusting zone; quadrat C, zone of small heads, quadrat D, zone of large heads; quadrat F, zone of sand and shingle.

the tops of these compound heads break the sur-
ɜe, whereas the reef floor is 3 ft submerged. Swim-
ng becomes easy in this zone. The majority of
rrot fishes and surgeon fishes browse and school
re. In superficial visual appearance, this zone is a
utiful jungle of live coral and fishes although as
own later, in terms of live protoplasm the plant
rld nevertheless dominates as usual. This zone as
n in Figure 3 is about 400 ft wide.

ZONE OF SAND AND SHINGLE

With gradually increasing depth and diminishing
rent, both the small and the larger coral heads,
e and dead complexes, come to an abrupt end. A
g flat shelf slopes lagoonward for 500 ft or more,
ered with coarse and fine sand made up of frag-
nts carried downstream from the front reef. Fora-
nifera of large types are abundant (Table 7). The
es decrease except for the schools of sardine-like
es that feed on the downstream drifting fragments
pseudoplankton and a few larger carnivorous
es, including sharks, which cruise here. The area
predominantly white except for filamentous algae
the larger coral shingle fragments. Beyond the
ft of this long bare shelf there is a sharp steepen-
of slope for over 100 ft (33 m) down to the
gular lagoon floor. As seen from a helicopter
steepening of the slope is sharply marked by a
der of the green turbidity of the lagoon water as
floor drops out of visible range.

he over-all change from the dark color of encrust-
filamentous algae of the front reef to the white
r of the back reef, where the filamentous algae are
nly within the dead and porous calcareous frag-
ts, suggests a transition from a water-filtering
rce of nutrients up front to a sub-surface decom-
ition source of plant requirements on the back
e. It will be shown that the quantitative totals of
r-all plant protoplasm per area are similar in
er of magnitude.

CURRENT VELOCITIES

n the course of the study, 26 dye current measure-
ts were made in different zones and time. Fluo-
ein dye (from air-sea rescue kits) was released
the water by one observer and the time required
ravel a measured distance to a second observer
determined. The maximum current measured
1.44 m/sec across the reef during a high water
p tide. Currents probably twice this velocity were
untered in incoming spring tides when the ob-
ers were too busy hastening to shore to get a
surement. The lowest velocity measured on the
t reef encrusting zone was 0.18 m/sec; the lowest
he back reef lagoon shelf zone was 0.009 m/sec.
mean of eleven measurements on the encrusting
was 0.49 m/sec. The 5 ft-deep water of high
ng tides probably permits strong flow over the
although no measurements were made at this
. The larger fish which are unable to hide behind
ll bumps and coral heads apparently could not
vse on the middle reef when the current was run-
ning over 0.3 m/sec, for no larger fishes were observed in these zones at this time.

As a rough estimate of comparative and average currents, measurements made of water transport at a neap tide at one station are converted by calculation of depth effect into velocities for the different zones and reported in Figure 2. Currents change so rapidly with time in the tidal cycle that this is the only way to make a synoptic comparison. As indicated above it was difficult to obtain measurements during high water spring tide without being washed off the reef.

TROPHIC STRUCTURE

PLANT AND ANIMAL COMPONENTS IN LIVE CORAL; THE PREDOMINANCE OF PLANT PROTOPLASM

Early taxonomists classified corals as plants, as in John Ray's system (Nordenskiold 1932), because of their vegetative appearance. Later, corals were found to be coelenterates and thus classified as animals. Yet their ecological roles that resemble plants remained a point of interest such as their predominance in the community and their practice of laying down carbonaceous substance in quantity sufficient to maintain the community substrate. Then, when the symbiotic zooxanthellae were found in the tissues of the animal polyp, it became evident that, metabolically, corals might be part-plant and contribute to the primary production of the community. Yonge, Yonge & Nicholls (1932); Kawaguti (1937); and others have shown that corals do indeed produce an excess of oxygen over carbon-dioxide during the daytime, although most measurements show that production does not quite equal respiration over a 24-hr period.

On ecological grounds, zooxanthellae, to match coral respiration, must either carry out photosynthesis many times faster than corals respire or exceed the coral animal protoplasm several times if a pyramid of mass should exist as required by the second law of thermodynamics for most systems. Yet corals have been shown to come close to achieving a balance between photosynthesis and respiration while possessing seemingly only a small amount of plant tissue in the form of scattered single algal cells restricted to the endoderm of the polyp.

However, there is yet a second plant component characteristic of corals, the significance of which seems to have been overlooked. When one breaks open a fresh, live coral head, conspicuous green bands are seen in the skeleton of the living polyp zone and also in the concentric layers in the older skeleton well below the zone of animal tissue as shown in Figures 6 and 7. These bands are not due to the yellow-brown, rounded zooxanthellae cells, but to a network of bright green filamentous algae growing in the pores of the inert skeleton (Fig. 7E). Although sometimes located as much as 2 or 3 cm below the surface, these algae growing within the translucent aragonite skeleton are, nevertheless, within light range of the intense penetrating tropical sun. All of the species of hard aragonite corals, examined at Eniwetok including hydrozoan, octocoral, and hexacoral groups contained these filamentous green algae in abundance. Only Dendrophyllia, growing in the shade in Hawaii, was different in not possessing either the zooxanthellae or the filamentous greens of the skeleton.

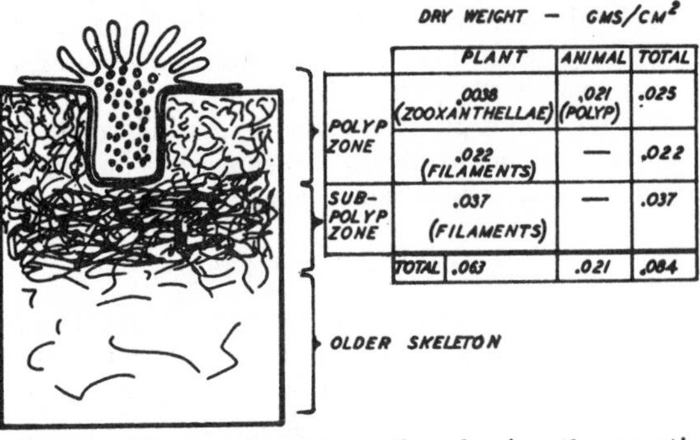

FIG. 6. Diagram in cross section showing the quantitative distribution of plant (algal) and animal (coral polyp) tissue in a generalized live coral head. Data on plant and animal biomass are from Table 4.

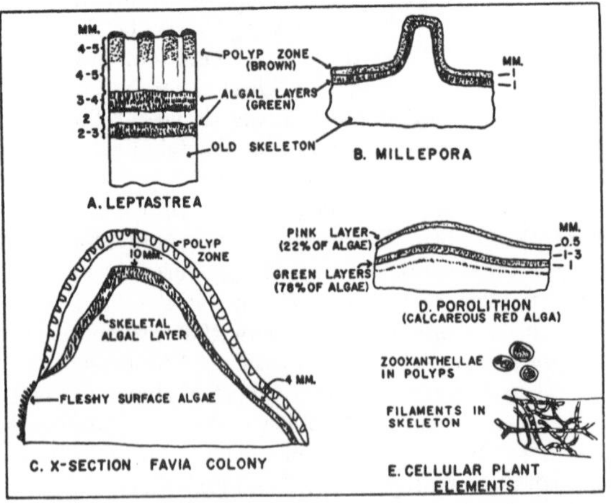

FIG. 7. Sketches from fresh material showing: A to C, the relation of the bands of skeletal algae to the polyp zone in three genera of corals. D, sub-surface green bands in the skeleton of calcareous red algae. E, the two major types of symbiotic plant material in coral colonies.

While zooxanthellae have long been considered symbiotic or mutualistic with coelenterate polyps, the possibility that the filamentous skeletal algae might also be symbiotic or at least important to the nutrition of the coral colony has been little considered. Indeed, most writers refer to these conspicuous algae as "boring algae" and consider them as parasitic agents which weaken the skeleton and hinder the growth of the coral colony. Duerden (1905) found two kinds of filaments, one with cross walls, one without. As one of the first to describe the filamentous skeletal algae, he stated that they "invade the corallum, weakening it if having no other effect." This view seems but little challenged although Edmondson (1929) stated that his evidence was not conclusive that these algae check the metabolism since he found that some "heavily infested" colonies showed good coral growth in the laboratory. It may well be, that the boring filamentous algae of live corals are be ficial, and at least under conditions existing on study reef, contribute to the survival and ra growth of major reef builders. Thus, there is a sh contrast between the boring algae in live corals the different species boring in dead coral. The spe of algae which have been described as boring in careous substrates are listed by Utseumy (1942).

The evidence that there is a predominance of p ducing plant protoplasm rather than coelente polyp protoplasm in a live coral head is based high chlorophyll content found in the non-polyp p of live coral heads relative to dry weight estimate animal polyps. The dry weight of producing pl tissue was determined from the chlorophyll va with the graph in Figure 8 on the assumption algae in the coral skeleton and polyps have a chl phyll-dry weight ratio similar to free-living al Details of the method used are given below in the tion on primary producers. Rough quantitative mates of the dry weight of the animal-polyp c ponent of a coral head were obtained by estima the volume of the coral head occupied by polyps assuming that the dry weight to volume ratio anemones represents that of coral polyps. The p volume was estimated by a vaseline method. Fur details on estimation of polyp volumes and weights are given in the section on coral pol Rough estimates of zooxanthellae dry weights obtained from chlorophyll extraction of isol polyps of a large polyp species (Lobophyllia) from histological sections pictured by Yonge, Y & Nicholls (1931) and further described in the lowing section on producers. The data on plant animal components are given in Tables 1, 2 and

The estimate of .075 gm/cm² residue after t ment with 20% nitric acid (Mayor 1924) comp well with the finding of .062 gm/cm² mean los ignition (Table 2) as an estimate of total biomas

The diagram in Figure 6 and the mean estim of protoplasmic components in Table 4 summariz data and present a general picture of the coral h which is almost a whole ecological community in i with producer, herbivore, and carnivore roles a one. The essential components thus appear to (a) animal (non-photosynthetic) tissue, (b) zoo thellae, (c) filamentous green algae in the sub face skeleton and between polyps, (d) bacterial ponents which are not estimated in this study which may be of importance.

The important quantitative conclusion from T 4 is that the total plant protoplasm exceeds the mal biomass (about 3 to 1) and the filamentous g algae have a greater biomass than the zooxanth (about 16 to 1). If the filamentous skeletal alga considered an integral part of the coral colony with the zooxanthellae, a reasonable biomass pyr is obtained which is in line with the high phot thetic activity shown by most coral colonies.

Comparison of live and dead corals provide direct evidence of symbiotic relationship bet

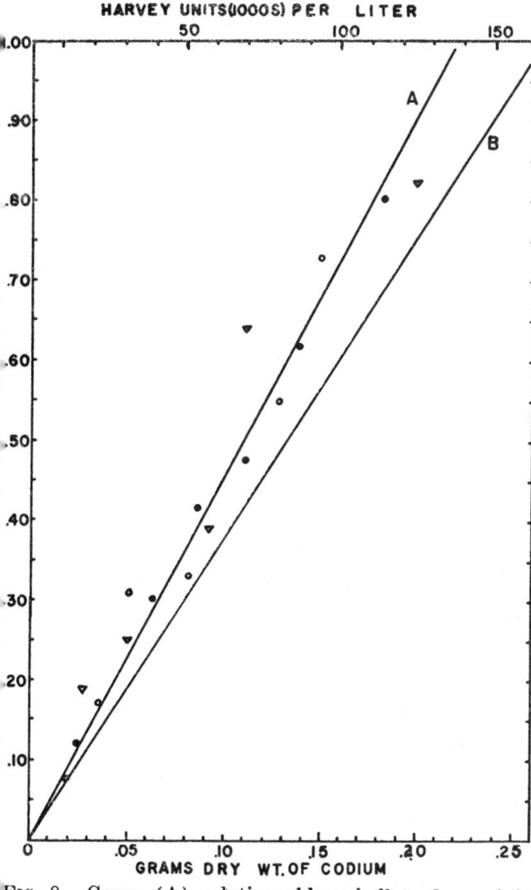

FIG. 8. Curve (A) relating chlorophyll to dry weight *Codium edule.* This graph was used as a means of taining rough estimates of the dry plant protoplasm coral heads and other calcareous substrates. The rve is based on three replications as indicated by the ee types of symbols. B is correction for average loss color on drying. See text for further explanation.

al animals and filamentous algae in the skeleton. gae occur in characteristic bands under living lyps (Figs. 6 & 7) and the bands show patterns ich are characteristic for a given species. These -surface bands disappear if the coral animals ove them died to be replaced by other algae which w on and near the surface of the dead skeleton. us, the bands are not present on the sides of the d where there are no live polyps (Fig. 7C). Al- ugh no attempt was made to classify the filaments, microscopic appearance of algae in bands under polyps is clearly different from that of algae wing on and in dead coral material. Dead corals luded boring greens, reds, and blue-greens, whereas live corals observed contained only greens.

These observations suggest the hypothesis that letal algae, as well as zooxanthellae, have a sym- tic relationship with coral animals. Nutrients of al metabolism may readily diffuse through the ous skeleton to the algae. The coral skeleton pro- es matrix and some enclosed nutrient, and the yp-protoplasmic sheet on the outside protects the icate filaments from competition, browsing, and

Table 1. Quantitative distribution of algae in corals.

	Percent of Total Algae	
	Polyp Layer	Sub-polyp Layer
A. Vertical Distribution of Algae in Massive Corals		
Leptastrea sp. (B-3)	40	60
Favia sp. (C-3)	34	66
Porites sp. (B-6)	40	60
Porites sp. (B-6)	46	54
Turbinaria mesenterina (E-4)	43	57
Mean	41	59

	Percent of Total Algae	
	In Branches	In Basal Zones
B. Distribution of Algae in Branching Corals		
Acropora sp. (D-9)	59	41
Acropora sp. (D-9)	57	43
Pocillopora sp. (C-8)	68	32
Lobophyllia sp.	66	34
Mean	62	38

C. Zooxanthellae in Coral Animal Tissue

Percent zooxanthellae in a coral planula*	27%
Percent zooxanthellae in a coral polyp (estimated from histological cross section)*	14%
Mean of extraction of two *Lobophyllia* polyps	16%

	Individual No. 1		Individual No. 2		Mean	
D. Biomass Distribution in *Lobophyllia*, Gms/cm²						
Biomass in Polyps:						
Zooxanthellae in polyps	.013	13%	.018	19%	.016	21%
Animal tissue in polyps	.041	97%	.077	81%	.059	79%
Total biomass of polyp	.054	100%	.095	100%	.075	100%
Comparison of Zooxanthellae and Algal filaments outside Polyps:						
Zooxanthellae in polyps	.013	31%	.018	18%	.016	19%
Algal filaments	.029	69%	.084	82%	.067	81%
Total plant biomass	.042	100%	.102	100%	.083	100%

Distribution of Plant Biomass:		
Zooxanthellae in polyps	.013	31%
Filaments around polyps	.015	36%
Filaments below polyps	.014	33%
Total plant biomass	.042	100%
Comparison of Animal and Plant Biomass:		
Animal biomass (from Table 4)	.072	46%
Total plant biomass	.083	54%
Total plant and animal biomass	.155	100%

*Calculated from data of Yonge et al. (1931).

intense sunlight (note from figure 7C that algal bands are deeper in the skeleton at the apex where light intensity is greatest).

From the quantities of algae present relative to coral protoplasm, it would be supposed that coral animals benefit from diffusion of organic substances from the algae but direct evidence of this was not obtained. Coral animals would still need to obtain some food and critical nutrients such as nitrogen by ingesting plankton, since there is not enough plant material to completely support the coral and since the coral requires a higher nitrogen content. Also, it is possible that the sub-surface algae raise the pH

TABLE 2. Quantitative estimates of animal tissue in coral polyp zones gm/cm².

Species	Depth of Polyp Zone cm	Total Biomass (Loss on Ignition*)	Plant Biomass (Extract Methods; Tables 1 & 5)	Animal Biomass in Gm/cm²			
				By subtraction (Total Minus Plants)	By Estimate of Polyp Volumes† (See Table 3)	By Direct weighing of isolated polyps‡	Mean Animal Biomass
Pocillopora (C-8).........	.1	.035	.036	−.001	.0015	....	.007
Millepora (B-2)..........	.15	.044	....	....	.007	....	.007
Porites (D-2)...........	.2	.043	.016	.027	.009	....	.018
Heliopora (E-3)..........	.3	.043	....	....	.0144	....	.014
Leptastrea (B-5).........	.4	.062	.019	.043	....	....	.043
Astreopora (M-30).......	.5	.045	....	....	.015	....	.015
Turbinaria (E-4).........	.8	.062	.026	.036	.019	....	.027
Favia (C-3)............	1.0	.050	.019	.031	.029	.015	.025
Lobophyllia.............	2.2	.173	.081	.092	.064	.059	.072
Mean of corals..........	.64	.062	....	....	....	....	.025
Anemone-sand grain—algae complex........	.5	.086	....	....	....	.034	....

*Loss on ignition in furnance at 600°C.
†Volume of polyp zone assumed to be filled with protoplasm like that of sea anemones growing on the reef with a density of about 1 and a 15.1% dry weight o[f] weight. (Similar dry weights of wet weights are summarized for other Coelenterates by Vinogradov, 1953.) Estimates with 15% error of underestimation due t[o] tissue in pores possibly cancelled by pore space occupied by plant filaments and zooxanthellae.
‡It is not possible to separate all of the polyp out, even in large polyp species. Possible compensation comes from using weights rather than loss on ignition of sepa[rate] polyps.

TABLE 3. Density, porosity, loss on ignition of skeletons.

Species and Zone		Skeletal Density gm/cc	Gross Dry Density gm/cc	Porosity in % (vaseline method)			Loss on Ignition (of dry
				A	B	Mean	
FRESH DRIED LIVING CORALS							
Pocillopora.........	branch	2.09	1.86	10.0	12.4	11.2	4.9
Millepora..........	polyp zone	....	1.51	....	....	33.3	8.9
	whole	2.27	1.42	....	....	37.6	....
Porites.............	polyp zone	2.38	1.66	30.4	30.4	30.4	12.7
	Algal zone	1.98	1.50	17.4	26.5	22.0	4.7
	layer 3	2.08	1.70	15.4	21.0	18.2	4.5
	layer 4	2.33	1.72	24.4	26.0	25.2	4.4
	layer 5 (eroded)	2.24	1.68	20.1	32.2 (hole)	26.1	3.9
Leptastrea (B-3)....	polyp zone	2.38	1.82	....	....	23.3	4.1
	algal zone	2.28	2.12	....	....	7.0	2.6
	sub-algal	2.47	2.18	....	....	11.6	1.7
Heliopora..........	polyp zone	2.89	1.63	....	....	31.9	8.9
	whole	2.42	2.24	....	....	7.1	....
Astreopora.........	polyp zone	2.51	1.99	22.6	18.1	20.3	4.8
Turbinaria.........	polyp zone	2.44	1.90	....	....	22.0	4.5
	whole	....	1.50	....	....	38.7	....
Favia (C-3)........	polyp and algal zone	2.33	1.83	22.2	20.2	21.4	5.4
Lobophyllia.........	polyp zone	2.18	1.64	25.1	24.7	24.9	4.8
FRESHLY DRIED ALGAL SKELETON							
Porolithon.........................		2.38	2.20	....	....	7.8	4.5
SKELETONS FROM REEF MASS, 20 FT DEEP IN DYNAMITE HOLE							
Porolithon.........................		2.64	2.53	....	....	3.2	3.9
Favia?.............................		2.38	2.22	....	....	6.7	2.9
MEAN OF POLYP ZONES (7)...........		2.44	1.73	....	....	26.6	7.0
MEAN OF SUB-POLYP ALGAL ZONES (2)..		2.13	1.81	....	....	14.5	7.3
MEAN OF SUB-ALGAL LAYER ZONES (3)		2.29	1.87	....	....	18.3	3.5

TABLE 4. Summary of components of biomass in als, mean data. Figures given: in gm/cm² dry ight and in percent of total biomass.

		Plants Tissue	Animals Tissue	Total Biomass
p	In Polyps	.0038* (zooxanthellae) 4.5%	.021† 25%	.025 30%
;	Between Polyps	.022‡ (filaments) 26%	....	.022 26%
	Total in Polyp Zone	.026‡ 31%	.021 25%	.047‡ (.062§) 56%
polyp Zone		.037** (filaments) 44%	....	.037 44%
l Biomass Outside 'olyps		.059 70%	....	.059 70%
l of all Layers		.063†† 75%	.021 25%	.084 100%

Mean of zooxanthellae extracted from Lobophyllia (13% and 19% of polyps ble 1) and calculated from Yonge (14% of polyps in Table 1). 15.3% of ms/cm² polyps (from Table 2).
025 gms/cm² polyps (from Table 2) minus zooxanthellae estimate .0038 cm².
Total plant estimate for the polyp zone minus zooxanthellae estimate.
41% (from Table 1) of Plants are in Polyp zone; mean plant estimate per of coral .063 gms/cm² from Table 5.
Sum of plants and animal estimates.
Total organic matter estimate based on loss on ignition at 600°C for com- on. (from table 2)
59% of plants in sub-polyp algal zone (from Table 1); mean plant estimate rea of coral, .063 gms/cm² from Table 5.
From Table 5.

d thus slow skeletal decomposition) or actually tribute directly to the skeleton formation of the al complex. Certainly evidence is lacking that the sence of algae in any way weakens the live skele- so long as the coral-algal complex is intact. The a strontium content of aragonite coral skeletons ng in the same or higher ratio to calcium as the sea er (9.23 atoms per 1000 atoms) is unusual for areous animals. The similar high ratio in the n calcareous alga Halimeda suggest some simi- ty in the deposition process of corals and Hali- la that might be explained by a role of the green nentous algae in deposition in the coral. Since green filaments are tightly enclosed within the l colony, any organic matter produced by them as wth or surplus diffusable products cannot escape out going through the enclosing polyp zone. re is no room for such growth except as the whole ny grows and there is no visible accumulation of nic products. The situation leads to the supposi- that either the plants are growing close to the pensation point or are supplying coral animal ps with organic materials. Since there are often ral bands of healthy green filaments, it is possible , if the deepest band were at the compensation t, the top bands would be making an excess which being used by the corals and deeper algae.

rom the standpoint of the entire reef ecosystem it not matter how much food made by algae within live coral head is used by coral animals directly. y by considering the large amount of producer e in a coral head is it possible to explain the

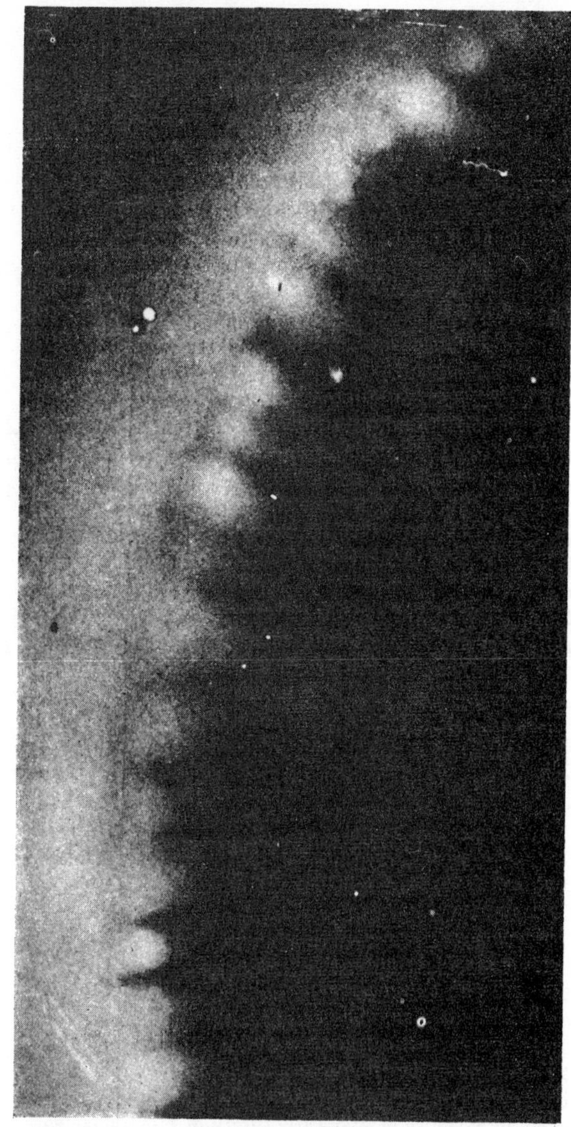

FIG. 9. A positive print of an autoradiogram of a section of a coral head (Goniastrea) after two days' exposure; light areas indicate exposed and dark areas unexposed film. Note that the surface polyp zone, but not the subsurface algal bands (see Fig. 7), exhibited considerable radioactivity.

great preponderance of organisms classed as animals. Thus the coral reef is like most known self-sufficient ecosystems in having a much greater weight of plant biomass than animal biomass.

An internal use and reuse of nutrients in a living coral is suggested by the autoradiogram (positive) in Figure 9 which is typical of ten made from corals from the generally (low level) radioactive Japtan reef. The radioactivity is entirely restricted to the animal polyp zone. Whereas algae on dead reef surfaces were intensely radioactive, these forms beneath the corals are apparently not in nutrient contact with the outside water except through the coelenterate tissues. This may be interpreted as an evidence that the coral algal filaments receive nutrients from the corals. By this view the radioactive elements in the polyps are not of the type which are released by the corals during their daily metabolic cycles. Cerium and praesodymium have been strongly implicated as main components to previous reef radioactivity (Blinks 1952; Donaldson *et al.* 1950). Apparently the calcophile elements like calcium and strontium are not retained but rapidly exchanged away by the high non-radioactive concentrations in the flux of sea water over the reef.

Although the filaments and zooxanthellae are small in size, their photosynthetic rates on a dry plant weight basis (Mayor 1924) are not nearly as high as free algae growing in nature as quoted by Verduin (1952); thus, for the same production per area, more biomass is required than in a plankton population, for example.

If the filamentous algae are truly mutualistic with corals then there may be no need to create separate taxonomic names for them. The coral-algal complex could be considered as a single species entity as is the fungal-algal complex of lichens, and the established names for the corals used to cover both elements.

PRIMARY PRODUCERS

In the previous section the complex association of plants and animals in a live coral head was discussed with a new viewpoint. In this section all of the main groups of primary producers (plants) are described and quantitative estimates are presented in Table 5.

The main primary producers grouped ecologically are as follows:

(1) filamentous algae in live corals.

(2) zooxanthellae in coral polyps.

(3) algae matted as an encrustation on and in the dead rigid porous reef substrate surface in the swift current zone.

(4) Encrusting fleshy green types such as Dictyosphaeria, Zonaria, and Caulerpa attached to the irregular surface, with encrusting calcareous reds such as Porolithon and Lithophyllum mostly beneath.

(5) Small algae in and on loose coral shingle (broken coral pieces) lying in channels and in areas between coral heads.

(6) Small filamentous algae in and on "dead heads." (A dead head is defined as a coral formation no longer containing living coral polyps but still standing erect on the surface of the reef.)

(7) Large conspicuous bunches of branching algae attached around dead heads including genera such as Codium, Asparagopsis, and Halimeda.

(8) Algae in the coarse, white, calcareous sand which covers inter-coral areas of the back reef.

(9) Zooxanthellae and filamentous algae in animals other than coral polyps such as sea anemones and giant clams (Tridacna, etc.).

(10) Planktonic algae derived from the open sea and the much larger quantity of pseudoplankton breaking off the other sessile masses. Sargent & Austin (1949) showed the relative insignificance of the plankton production in the reef's metabolism. The relative sparsity of true plankton is suggested the absence of attached plankton feeding mollus hydroids, tunicates, and ectoprocts on the glass slid as reported in Table 6. The attachment aufwuc was entirely autotrophic.

A reef community resembles a complex, tropic terrestrial community in that it is not dominated one or two producer species, but there is considera diversity and variation from place to place. On t reef, for example, the myriads of tiny blue gre

TABLE 5. Primary producers of the reef.

1. ALGAE IN LIVING CORALS (filaments in skeleton and zooxanthellae in polyps)	Quadrat	Gms. D Wt. p cm²
(a) Massive Scleractinia		
Leptastrea (B-3)	B-C	0.04
	B-C	0.03
	B-C	0.04
	B-C	0.03
	B-C	0.03
	B	0.07
Favia (C-3)	B-C	0.05
	B-C	0.06
	B-C	0.05
Favia (B-4, green species)	B-C	0.11
	C	0.13
Porites lobata (C-1)	C-D	0.04
	C-D	0.04
	B	0.04
Porites lobata (in deep water)	E	0.01
Turbinaria mesenterina (E-4)	E	0.06
Acropora palmerae (B-1, encrusting form)	B	0.03
Mean of Massive Scleractinia	..	0.05
(b) Branching Scleractinia		
Acropora cymbicyathus (D-9)	D	0.03
	D	0.04
Pocillopora (C-8)	C	0.07
	C	0.03
	A	0.04
Lobophyllia	*	0.12
	*	0.05
Mean of Branching Scleractinia	..	0.06
(c) Octocorallia and Milleporina		
Heliopora caerulea (E-3)	D	0.05
Millepora platyphylla (B-2), encrusting section	D	0.02
vertical branch	E	0.09
Millepora murrayi (E-2), encrusting section	E	0.07
encrusting section	E	0.08
encrusting section	E	0.06
encrusting section	E	0.04
vertical branch	E	0.11
vertical branch	E	0.07
Mean of Octocorallia and Milleporina	..	0.07
Mean of all corals (33 specimens, 12 species)	..	0.06

TABLE 5. Primary Producers of the Reef (Continued).

2. ALGAE IN REEF SUBSTRATES	Quadrat	Gms. Dry Wt. per cm^2
Reef floor—with fleshy algal mat	A	0.062
Reef floor—with fleshy algal mat	A	0.041
Reef floor—*Porolithon* encrusted	A	0.042
Reef projection, upper plus under surface	A	0.092
Reef projection, upper plus under surface	A	0.080
Reef floor—with fleshy algal mat	B	0.072
Reef floor—with anemones on surface	B	0.070
Reef floor—very porous	C	0.122
Algal mat separated from reef surface	B	0.008
Algal mat separated from reef surface	B	0.007
Algal mat separated from reef surface	C	0.004
Algal mat separated from reef surface	C	0.003
Porolithon separated from reef surface	A	0.030
Rubble (Shingle), new, non-porous	C	0.029
Rubble (Shingle), old, porous	C	0.047
Rubble (Shingle), old, porous	D	0.066
Rubble (Shingle), old, porous	D	0.060
Rubble (Shingle), old, porous and eroded	D	0.141
Rubble—buried 25 cm deep in dead head	D	0.028
Rubble (Shingle), old, porous	F	0.110
Deadhead—surface and internal algae	D	0.049
Deadhead—internal algae only, surface "grazed" by fish	D	0.042
Deadhead—*	D	0.061
Deadhead projection, upper plus under surface	D	0.160
Sand—deep channel	D	0.0019
Sand—shallow channel	D	0.017
Sand—deep channel	F	0.009
Mean—reef floor	A-C	0.073
Mean—old shingle	C-F	0.085
Mean—deadheads	D	0.078
Mean—sand	D-F	0.009

3. ALGAE IN AMINALS OTHER THAN CORALS	Quadrat	Gms. Dry Wt. per cm^2
Anemones—sand complex; zooxanthellae in polyps and filamentous algae in attached sand grains.		
Sample 1—zooxanthellae in polyps plus filamentous in attached sand	B	0.037
Sample 2—zooxanthellae in polyps plus filamentous in attached sand	B	0.025
Sample 3—zooxanthellae in polyps only	B	0.030
filamentous in attached sand	B	0.017
Total	B	0.047
Sample 4—zooxanthellae in polyps only	B	0.045
filamentous in attached sand	B	0.019
Total	B	0.064
Giant clam (Tridacna)		
Algae in mantle exposed to sun when shell fully open	D	0.052
Algae in calareous shell of same specimen	D	0.031

*From lee reef at Rigili Island

TABLE 6. Reef fouling on glass slides submerged 21 days. (Each figure unless otherwise indicated is the mean of 5 counts; slides submerged July 3 to July 24, 1954; counts in numbers of individuals per cm^2.)

	Front Reef Encrusting Zone	Back Reef Zone of Large Heads	Slides imbedded in coarse bottom sand
Station designation (quadrat)	B-1	D	C
Depth of water over slides at low-water spring tide in cm	10	30	20
Larger algae (greater than 5 microns) individuals/cm^2:			
Greens (*Enteromorpha*, siphonaceous)	97	33	0
Browns (*Ectocarpus* and others)	290	53	133
Reds (filaments and calcareous pieces)	2	26	0
Blue-Greens	47	39	3
Diatoms	0	693	20
Smaller algae (less than 5 microns)			
Greens	1.2×10^5	9.0×10^5	15.4×10^5
Blue-Greens	2.4×10^5	$.2 \times 10^5$	6.6×10^5
Encrusting diatoms	52.8×10^5	5.4×10^5	$.4 \times 10^5$
Bacteria	4.4×10^8	2.4×10^8	1.2×10^8
Ciliates, nematodes, other similar sized soft bodied animals	600	132	15
Foraminifera (small sized types)	.6	.4	0
Dry weight of algal protoplasmic biomass by chlorophyll method in mg/cm^2	2.5	1.6	
Plankton feeders such as barnacles, molluscs, ectoprocts, hydroids, tunicates	0	0	0

brown, red, and green algal filaments embedded in live and dead calcareous materials have been little studied, yet they make up a large part of the food making biomass of the reef community. Some idea of this component can be gained from counts of attached algae on glass slides (Table 6). In spite of the great diversity in species there seems to be a fairly uniform distribution of producer biomass as estimated with the chlorophyll method. Particularly since the living primary producers are so completely interwoven with animal material and in dead skeletal material, the chlorophyll extracting method seemed to be the most practical way of obtaining quantitative estimations readily comparable with other communities. To permit comparison with other areas the chlorophyll methods are outlined in some detail. It is recognized that chlorophyll is not a perfect measure of the active plant protoplasm and that it varies considerably with species, physiologic age, and growth conditions.

All values for irregular shapes are reported in gms dry weight per area of the horizontal plane covered. (This surface is much less than the surface area of the irregular object.) For corals a small block was cut out with a hacksaw and the area of the block's projection on the horizontal surface was measured. The block was cut deep enough to include all visible green material (25 mm usually sufficed for massive corals, deeper sectioning was required for branching forms). It was later found that some algae occur even below the visible green zones but the amount is so small as to be scarcely measurable with the methods used. The block was pulverized with a hammer, ground up in a mortar and extracted with successive washings of acetone until all the chlorophyll had been removed. The extract was filtered into a test tube, adjusted to 20 ml and read photometrically. Reef substrates, rubble and dead heads were also cut into blocks with a measured surface area and treated in the same way, care being taken not to lose the algae growing on the surface. Sand from a measured area was removed into a bottle and then ground and extracted in the same manner. It was found that materials must be extracted fresh, preferably not more than 12 hours after collection. Oven-dried materials invariably gave lower values than comparable fresh materials. For corals and most reef materials a piece with a surface area between 1 and 4 sq cm was required for extraction with 20 ml of acetone.

In the massive corals, algae are concentrated in layers just beneath the surface (Fig. 7) while in branched corals it is distributed along the branches with less concentration at the base and with relatively little at the exposed tips of branches. In fact in some branched forms the algae are so diffuse that the intensely green solutions derived from pale pieces of coral are surprising. A "block" of a branching coral type such as Acropora or Pocillopora, as used for extraction, consisted of a vertical branch together with the basal section from which the branch arose. The cross sectional surface area of such a block (including all branches in the imaginary prism directly above the measured area on the horizontal) was considered comparable with the surface area of flat encrusting corals, since the amount of light per square centimeter of horizontal surface should be similar.

To convert photometric readings into biomass of producing tissue, a calibration curve was determined from acetone extracts of known dry weights of an arbitrarily chosen alga abundantly available in the field. *Codium edule* was selected as a standard and a relatively constant relation was found between chlorophyll content and dry weight (Fig. 8). A small portable colorimeter was used at first until a Coleman Spectrophotometer Model 6 became available. Codium used for calibration was freshly collected and paired, duplicate pieces were respectively dried and extracted for chlorophyll. Since a 20% loss of chlorophyll was found in pieces oven dried at 100° C for 6 hr, all extractions based on oven dried materials must be corrected for loss on drying. Figure 8 shows the calibration curve for optical density at 670 millimicrons as a function of Codium dry weight based on 3 replications on 3 different batches of material. A straight line results with monochromatic light. With the colorimeter first used, a reproducible curved line was obtained for absorption as a function of dry weight of Codium. The line allowing 20% correction for loss of chlorophyll on drying is also drawn in Figure 8. A series of nickel-chromium solutions was made as reference standards and readings equivalent to dry weights of Codium were included in Figure 8 (upper scale). 10,000 Harvey units/1 were made up with 4.3 gm/1 nickel sulfate and .25 gm/1 potassium dichromate. Since the readings reported in Figure 8 were made with a spectrophotometer as 670 m microns instead of visually as originally defi (Harvey 1934), these are not Harvey units as usu used as a measure of chlorophyll but considera different. Measurements made later on the nic chromium solution with a Beckman model DU sp trophotometer indicate that the optical densities Figure 8 were made with an optical path of ab 2 cm. From Richards (1952) these densities in cate an order of magnitude of .7 mg chlorophyll/ dry Codium.

The values for algal dry weights obtained with above methods are given in Tables 1 and 5 arran according to the general producing types previo listed. These dry weights may be overestimates s the small filamentous strands, so important in m reef materials, are smaller and thus likely to hav lower dry protoplasmic weight-chlorophyll ratio t that characteristic of the standard used, Codi No correction was made for ash in Codium.

The following tentative conclusions are indica from these data about primary producers:

1. There is a striking similarity between values tained with different species of corals. When pressed in terms of cross-sectional area projected the horizontal, the branching types as compared flat or rounder massive types were little if any hig in algal content even though much more calcare matter was extracted and even though the actual face area of branching forms exposed to the w was much greater. The branching life-form may an advantage to the animal part of the coral in ca ing plankton from deep water and possibly usefu plant components in obtaining nutrients. The f tional plant producers, however, seem to be regul by the available light and are more widely dispe in the branching forms. Thus, in general, the chl phyll, per area perpendicular to the sun, is su ingly uniform. The chlorophyll per area of a ste state community may be expected to have a unifc ity and greater significance than in transient bl populations and laboratory cultures. It may b better measure of producers and productivity u these more constant conditions.

2. While the algal content of different specie corals was of the same general order of magnit distinct species differences are indicated. among the massive corals an unidentified specie Favid (code #B—4) had about twice the algal co of other species of favids (Table 5). This cor very green in appearance with abundant green a and zooxanthellae in the surface in the polyp zo well as in the subsurface zone. Among the branc types, Millepora appears to be high and in both cies (*M. platyphylla, M. murrayi*) the tall ve branches were found to contain more algae per cross section than flat portions of the same colony wide 100 fold range of production and respir values had been established for corals by Kaw (1937).

3. In so far as our small number of samples sh

there was little evident difference in producer content of live corals in different zones of the reef which range from ½ to 4 ft deep at low spring tide. Similarly, the corals collected in Kaneohe bay in Oahu, Hawaii although from more turbid water gave similar orders of magnitude of chlorophyll content for comparable species.

4. The white sand area of the back reef was the only major area of the reef which had a definitely lower biomass of producer protoplasm.

5. From the data on glass slide attachment in Table 6, about twice the growth of encrusting algae was obtained in the front reef as on the back reef correlating with the predominance of encrusting forms up front with boring forms in back.

CORALS AS CONSUMERS

Estimates based on mapped quadrats (Figure 5) indicate, along with cursory survey, that most of the reef surface is between 16% and 50% covered with live coral. Although the living part of a coral is more plant than animal there is, nevertheless, an important total weight of animal coral. It has been repeatedly shown (Mayor 1924, Edmondson 1929, Yonge 1930) that coral polyps are in part carnivores in trophic classification, since they catch zooplankton, especially at night. And if the inferences of the previous sections and of previous authors (reviewed in Yonge 1930, Kawaguti 1937) are correct, a coral animal polyp is very much an herbivore because of nutrition received from symbiotic algae. Thus, the animal part of a coral is partly divided in trophic classification between two trophic levels, herbivores and carnivores. Three procedures were used to obtain an estimate of the dry weight of the consumer fraction in live coral.

First, the volume of the polyps was estimated by filling the pores of a slice of coral from the polyp zone with melted vaseline. From the amount of vaseline filling the pores, the polyp zone porosity is determined as an upper limit to the polyp volume. One source of error, the inclusion of pores not occupied by polyps, may be cancelled by the error of not including pores partly blocked by animal and plant dry-residues. In drying, shrinkage of polyps may be expected to be greater than shrinkage of algal filaments within the skeleton. Thus, a rough upper limit figure for primarily animal volume (with enclosed zooxanthellae) may be obtained. The steps in this procedure are as follows: (1) cut a slice of polyp zone with hacksaw, measure surface area, dry, and weigh; (2) place in melted vaseline in oven 6 hours until permeated, remove, cool, wipe all excess vaseline off the outside, weigh; (3) the figure for vaseline in pores obtained by subtraction of weights, should be divided by the density of the vaseline to obtain the volume occupied; (4) multiply the volume by the dry weight equivalent for anemones and divide by the area to obtain final figure for dry weight polyp per square centimeter.

For two species with large polyps, it was possible to tease out the protoplasm and obtain a dry weight directly.

In the procedure for the third method of estimating polyp biomass, the chlorophyll-based estimate of the plant part of the polyp zone is subtracted from the total loss on ignition (600°C). This method assumes that most of the loss on ignition of the polyp zone is due to ashing of live plant and animal tissues.

Some estimates for polyp weights from the three methods outlined above are given in Table 2, where they are in sufficient agreement to permit some confidence in the order of magnitude at least. The predominance of plants over the animal component seems clear.

The role of current in limiting coral distribution is supported by their distribution relative to currents at low tide (Fig. 2). Values for the quadrats show a decrease in coral coverage as the current decreases. Since ample light for photosynthesis penetrates the clear waters of the back reef it would seem likely that current is a major factor in the decrease of coral coverage.

CONSUMERS OTHER THAN CORALS

Although the trophic relationships of most of the higher organisms on the reef are very imperfectly known, an attempt has nevertheless been made to make rough groupings by trophic level as to herbivores, carnivores, and decomposers. Drs. Hiatt and Weylander generously gave their help on this, drawing on studies of food relationships in preparation.

Each of the groups required a suitable means of obtaining a weight estimate per area. It must be realized that the great clarity of water permits face mask work with as great visual intimacy as on a terrestrial quadrat. In Tables 7-12 are presented the results of the various estimates by methods briefly listed with trophic levels below. Where an organism eats the matting of algae with included small invertebrates, an omnivore classification might be correct except that by weight most of this material is plant. So, such consumers are classed as herbivores. For example, a negligible biomass estimate was obtained for microcrustacea in algal mats.

The following are the groupings and methods used for estimating trophic components. Since considerable doubt exists as to the trophic relationships, the groupings are kept separate in the presentation to

TABLE 7. Benthic Foraminifera; counts of individuals /cm². (Counts of representative algal mats and sand patches have been multiplied by the coverage of these areas in the quadrats.)

Quadrat	Coverage of sands or mats containing Foraminifera	Small Forams .01 cm size	Large Forams .1 cm size
B (Front, encrusting)......	70%	25	0
C (Small Heads).........	34%	2	5
D (Large Heads).........	34%	2	32
E (Sand-Shingle).........	67%	3	54

TABLE 8. Dry biomass estimates on quadrat A on the algal-coral ridge.

Biomass Component	Quantity measured, calculation	Mean Biomass averaged over the Quadrat gms/m²
PRODUCERS		
Algae in corals	Coral coverage estimated 50% (Fig.5) Algae in non branching coral .063 gms/cm² (Table 5)	315.
Fleshy and calcareous Algae in crust and subcrust	Algae coverage estimated 50% (Fig. 5) .064 gms/cm² (Table 5 average Quad. A reef floor)	320.
Total Estimate of Producers		635.
HERBIVORES PREDOMINATELY		
Slate pencil urchins (*Heterocentrotus trigonarius*)	Quadrat count: 6(5, 5, 8) individuals/9 m² 59.3 gms loss on ignition/individual	39.4
Gammarids and other small crustacea	Mean of 2 methods A. 16 individuals/9 cm²; .0004 gms loss on ignition/individual B. .086 gms/.04 m² collected sample	4.7
Animal tissue in corals (partly carnivore but classed as predominately herbivore because of symbiotic algae)	50% coverage (Fig. 5); .021 gms/cm² (Table 4)	105.
Parrot fish	Visual counts: .4 individual/28 m²; 9.3 gms loss on ignition/individual	0.1
Total Estimate of Herbivores		149.
CARNIVORES PREDOMINATELY		
Annelids, mostly of Nereid type	.65 gms/.04 m² sample	16.1
Small crabs and other similar sized crustacea	.28 gms loss on ignition/.04 m²	7.
Total Estimate of Carnivores		13.1
Total Biomass H/P .24; C/H .09.		807.

permit rearrangements as further knowledge becomes available.

HERBIVORES

Small herbivorous fishes, including primarily surgeons and damsels, were counted on the 20 ft quadrats visually and converted to dry weight using the mean dry weight per fish found in a rotenoned sample. A similar method of visual census of coral reef fish has been recently described by Brock (1954). Three species, namely, *Acanthurus elongatus, Pomocentrus jenkensi,* and *P. vauili,* made up a large percentage by weight of small herbivorous fishes at quadrat C and D as shown by rotenone samples.

Large herbivorous fishes, including especially surgeons, damsels, parrot fish, and butterfly fishes were rapidly counted with 360° underwater vision. The area of this sample was estimated from horizontal

Table 9. Dry biomass on quadrat B on the encrusting zone.

Biomass Component	Quantity Measured, basis for calculation	Mean Biomas averaged over the Quadrat gm/m²
PRIMARY PRODUCERS		
Slab of reef rock surface containing substrate boring algae and mats of encrusting algae	70% coverage; (Fig. 5) Chlorophyll extract estimate: .072 gm/cm²	804.
Slab of reef floor covered with anemone—algal permeated grain complex	Coverage 7.0%; chlorophyll extract estimate .070 gm/cm²	49.
Halimeda clumps	Coverage 1%; .036 gm/cm² loss on ignition of clump	4.
Algae in corals	Coverage of corals 22% Mean biomass of algae in non-branching corals. .063 gm/cm² (Table 5)	139.
Total Estimate of Primary Producers		696.
HERBIVORES PREDOMINATELY		
Animal tissue in corals (partly carnivores; classed as herbivore due to symbeiosis with algae)	Coverage of corals 22% (Fig. 5) .021 gm/cm² animal tissue in coral (Table 4)	46.
Snails (*Thais*)	Counts: 16.5 (16, 17) individuals/1.44 m²; .094 gm dry tissue/individual	1
Sedentery annelids in reef floor	Visual count: 65/1.44 m² of non coral area; 70% coverage; .12 gm dry/individual	4
Cucumbers	Count: 3/36 m²; 2.4 gm/indiv. loss on ignition	
Parrot fishes	.4 individuals/28 m²; 9.3 gm loss on ignition/individual	
Ophiuroids in coral heads	Coverage of suitable heads 8.6%; 1.59 gm loss on ignition in sample head .04 m²	[illegible]
Anemones (also carnivores; partly herbivores because of symbiotic algae)	.043 gm loss on ignition/individual 7.0% coverage of anemone complex; 17 individuals/120 cm² of anemone area	
Total Estimate of Herbivores		5
CARNIVORES PREDOMINATELY		
Nereid type annelids in coral heads	Coverage of heads 8.6%; 1.86 gm loss on ignition/.04 m² head	
Small crustacea, crabs in coral heads	.22 gm loss on ignition/.04 m²; coverage of heads 8.6%	
Total estimates of carnivores other than corals and anemones		

was determined from a sample of 12 speared fis the same general size. *Scarus sordidus, S. erithro Chaetodon auriga, C. ephippium, C. trifasciatus, tropyge flavissimus, Naso lituratus, Acanthurus vaceusm,* and *Ctenochaetus striatus* were impo

TABLE 10. Dry biomass estimates on quadrat C on the ne of smaller heads.

Biomass Component	Quantity Measured, basis for calculation	Mean Biomass averaged over the Quadrat gm/m²
MARY PRODUCERS		
lgae in live coral	Coverage of coral 19% (Fig. 5) .062 gm/cm² dry algae in coral (Table 5)...	118.
oral shingle permeated nd encrusted with algae	Coverage of shingle 47%; .038 gm/cm² dry algae in shingle..............	178.
lgae in and on hard eef floor	Coverage of algal encrusted floor-rock 23%; .122 gm/cm² dry algae.........	286.
lgae in and on small lead-heads (Quadrat C lead-heads like reef floor n respect to algae)	Coverage of dead-heads 11%; .122 gm/cm² dry algae...................	134.
iant clam algae	.2% coverage; .052 gm/cm² of exposed mantle photosynthetic surface...	1.
al Estimate of Producers..	....	717.
RBIVORES PREDOMINATELY		
mall cucumbers in dead-eads	3 individuals/0.26 m² of head; 2.4 gm loss on ignition/individual; 11% coverage of dead and live coral heads..	31.
mall cucumbers around orals and in shingle	20 individuals/18 m²; 4.8 gm loss on ignition/individual..................	5.3
arge cucumbers	14 individuals/18 m²; 17.7 gm loss on ignition/individual..................	14.
mall urchins in dead eads (*Echinothrix*)	1.01 gm (4 individuals) loss on ignition in 260 cm²; 11% coverage of dead heads	4.2
arge urchins *Echinothrix*)	56 individuals/18 m²; .61 gms loss on ignition per individual...............	1.9
ridacna (small) erbivorous because of ymbiotic algae)	3 individuals/36 m²; 12.7 gm dry/individual minus 1 gm /m²plant in clam (see above)....................	.1
nnelids in dead heads	20 gm dry (19, 21)/100 cm² dead head; 11% coverage of dead heads; loss on ignition 57%; 70% herbivores..	9.0
edentary annelids on ard reef floor	.126 gm loss on ignition/30 cm² of reef flat; coverage 23%..............	10.4
ponges	10 cc volume/36 m²; 7.9% of wet is loss on ignition......................	.02
mall gastropods *Thais* and *Cowry*)	6 individuals/18 m²; .089 gm loss on ignition/individual..................	.03
nimal tissue in corals	19% coverage of coral; .021 gm/cm² animal tissue in coral................	40.
maller fishes	Visual counts: 25 (21, 23, 24, 25, 34, 34, 17, 33) fish/36 m²; 2.42 gms dry/individual and 61% herbivores based on poisoned sample.................	1.0

TABLE 10 (continued).

Biomass Component	Quantity Measured, basis for calculation	Mean Biomass averaged over the Quadrat gm/m²
Larger fishes	Visual counts: 52 (35, 40, 65, 75, 43)/692 m² area of horizontal visibility in all directions. 120 gm dry weight/individual; 90% herbivorous; large fishes absent from area 1/3 of time during maximum currents............	5.0
Total Estimate of Herbivores.	....	122.
CARNIVORES PREDOMINATELY		
Mollusca	6 individuals/18 m²; .09 gm loss on ignition/individual..................	.03
Small starfish	11 individuals/18 m²; 1.0 gm loss on ignition/individual..................	.6
Large starfish	1 individual/36 m²; 106 gm loss on ignition/individual..................	3.0
Smaller fishes	39% of fish counted (see herbivorous fishes above).......................	.65
Larger fishes	10% of fish counted (see herbivorous fishes above).......................	.7
	1 stone fish/36 m² (100 gm dry) (A stone fish was twice taken from the quadrat area during two days' work)	2.8
Annelids	Estimated 30% of annelids in dead heads (see herbivorous annelids above)	4.0
Total Estimate of Carnivores.	....	11.2

It can be noted that attempting to poison out quads or other measured areas with rotenone proved to a poor census method in itself, but was a valuable adjunct to visual counts. Many of the larger herbivorous fishes, which abound in coral reefs, travel in active schools and quickly move out of a limited poisoned area. But after the active fishes were censused by repeated counts rotenoning the quadrats revealed the hidden element of small fishes which could then be added to the population estimate.

Herbivorous molluscs, sea urchins, sea cucumbers, brittle stars, and other large invertebrates were counted by hand in subquadrats as the observer carefully took the superstructure of the reef apart, lifting dead material, and breaking open heads with a hammer. The loss on ignition value (600°C) for an average sized organism was used as a rough estimate of live protoplasm to convert numbers of each phyletic type into biomass. Herbivorous molluscs were primarily Thais, Cypraea, and Tridacna.

In the *Herbivorous annelids* were somewhat arbitrarily included all those annelids in sedentary tubes and all those without pharyngeal jaws. These were estimated from sample heads of measured horizontal area coverage carried back to the laboratory and broken open. Allowing the head to stand in stagnant water was found effective in inducing the annelid component to crawl out into the surrounding water prior to death. With estimates of the coverage of

TABLE 11. Dry biomass estimates for quadrat D on the zone of complex larger heads.

Biomass Component	Measurement; Basis for Calculation	Mean Biomass averaged over the Quadrat gms/m²
PRIMARY PRODUCERS		
Algae in live coral	Coverage of live Coral 16% (Fig. 5) .062 gm dry algae/cm² coral (Table 4)	100.
Algae in shingle (dead coral fragments)	Coverage 10%; .089 gm/cm² algae in shingle (Table 5)	89.
Algae in tall complex dead heads, encrusting and permeating	Coverage of dead heads; 40%; .049 gm/cm² algae in and on dead heads	197.
Halimeda clumps in dead heads	Coverage of dead heads 40%; .013 gms loss on ignition *Halimeda*/cm² head	52.
Fleshy algae on lower dead parts of tall live heads	Coverage of live heads 16%; 7.5 gms loss on ignition/400 cm² live head	30.
Algae permeating and encrusting dead cobble under live heads	Coverage of live heads 16%; .028 gm/cm² algae in sample 25 cm deep in head	45.
Algae in white sand in channels	Coverage of sand 34%; .009 gm/cm² Algae in sand (Table 5)	31.
Fleshy algae (Codium) on dead-heads along their lateral slopes	Coverage of dead-heads 40%; Estimated fraction of dead-heads with fleshy algae 30%; 36.5 gm dry algae/400 cm² sample	107.
Algae in *Tridacna*	Coverage of *Tridacna* .3%; .040 gm/cm² dry algae in *Tridacna*	1.2
Total estimate of primary producers		652.
HERBIVORES PREDOMINATELY		
Sponges in dead-heads	Coverage of dead-heads 40%; 4.0 gm (4, 9, 3, 2) loss on ignition/.04 m²	40.
Sponges in live heads	Coverage of tall live heads 16%; 7.0 gm loss on ignition/.04 m²	28.
Midget cucumbers in dead-heads	.27 gm loss on ignition/.04 m²; coverage of dead heads 40%	2.7
Small cucumbers in and around heads	2.4 gm loss on ignition/individual; 4 individuals/2.25 m²	4.3
Small urchins in dead-heads	.13 gm (.19, .06) loss on ignition/.04 m²; dead-head coverage 40%	1.3
Ophiuroids and urchins in live heads	.53 gm loss on ignition/.04 m²; live-head coverage 16%	2.1
Urchins around heads	27 individuals/2.25 m² subquadrat; .61 gm loss on ignition/individual	7.3
Ophiuroids around heads	36 individuals/2.25 m² subquadrat; .40 gm loss on ignition/individual	6.4
Herbivorous annelids in dead-heads	.10 (.09, .11) gm loss on ignition/.04 m²; coverage of dead heads 40%	1.0

TABLE 11 (continued).

Biomass	Measurement; Basis for Calculation	Me Bior avera over Qua gms/
Herbivorous annelids in live heads	Coverage of live heads 16%; .14 gm loss on ignition/.04 m²	
Small herbivorous crustacea in dead-heads	Coverage of dead heads 40%; .17 gm/.04 m² loss on ignition	
Small mollusks in dead heads (herbivores)	Coverage of dead heads 40%; .55 gm loss on ignition/.04 m²	
Small mollusks in live heads	.09 gm loss on ignition/.04 m²; Coverage of live heads 16%	
Animal part of corals	Coverage of live coral 16%; .021 gm/cm² animal tissue in live coral (Table 4) Surface of branching corals about 3 times area covered horizontally	9
Small herbivorous fishes	Counts on quadrats: 71 individuals/36 m²; 2.42 gm dry/individual	
Large herbivorous fishes	Counts per 600 m² horizontally visible area; 30 (24, 25, 33, 27, 42) fishes; estimated 3/4 herbivorous; 120 gm dry/fish	
Tridacna	.163 gm/cm² dry	
Total Estimate of Herbivores		12
CARNIVORES PREDOMINATELY		
Annelids and nemerteans in dead-heads.	.07 (.05, .08) gm loss on ignition/.04 m²; coverage of dead-heads 40%	
Annelids and nemerteans in live heads	.07 gm loss on ignition/.04 m²; coverage of live heads 16%	
Small crabs and shrimp in live heads	Coverage of live heads 16% 1.05 gm loss on ignition/400 cm²	
Small crabs around heads	4 individuals/2.25 m²; .15 gm loss on ignition/individual	
Small crabs in dead-heads	.07 gm loss on ignition/.04 m²; coverage of dead-heads 40%	
Carnivorous mollusks around heads	5 individuals/2.25 m²; .06 gm loss on ignition tissue/individual	
Carnivorous mollusks in dead-heads	.5 gm loss on ignition/.04 m²; coverage of dead heads 40%	
Small carnivorous fishes	5.3 individuals/36 m²;2.42 gm dry/individual	
Larger carnivorous fishes	1/4 of fishes counted in visible horizontal area (See herbivorous fishes above)	
Total estimate of carnivores		

the type of head counted from quadrat maps, t head counts for live and dead head-types were verted into over-all weights per area using loss ignition values. A correction by counting was n

TABLE 12. Dry biomass estimates on quadrat E—F on e sand-shingle zones of the back reef. (Data comned from open areas of Stations E and F.)

Biomass Component	Measurement; Basis for Calculation	Mean Biomass averaged over the whole Quadrat gm/m²
IMARY PRODUCERS		
Algae in sand	Coverage of Sand 67%; .009 gm/cm²; dry algae in sand (Table 5)..........	60.
Algae in shingle	Coverage of shingle 33%; .110 gm/cm²; dry algae in shingle (Table 5)....	331.
tal estimate of producer biomass...............................		391.
RBIVORES PREDOMINATELY		
Small herbivore fishes	Count 23 (31, 15) individuals/36 m² quadrat; 2.42 gm dry/average sized fish..............................	1.5
Schools of sardine-herring shes	Count of schools 1.2 (1, 1, 2, 1, 1) per 600 m² horizontal visible area; About 100 fish/school; 1 gm/fish...........	.2
Large herbivore fishes	Count 16 (16, 14, 17, 15) individuals per 600 m² horizontal visible area; 240 gm dry weight per fish..........	6.4
tal estimate of herbivores..	....	8.1
RNIVORES PREDOMINATELY		
Larger fishes other than harks	3.2 (4, 0, 6, 3) individuals counted per 600 m² horizontal visible area; 240 gm dry weight/fish..............	1.3
Sharks observed while walking across back eef zones	Counts per 20 minutes observation: 1.6 (1, 1, 0, 5, 1) individuals per 600 m² visible area; 90 degrees visibility at one time; each individual in sight about 30 seconds; Weight per shark about 50 lbs wet or 4540 gm dry (20% of wet) (Vinogradov 1953)......	prorated on time and area basis 1.2
al estimate of Carnivores.	....	2.5
COMPOSERS		
oraminifera	Counts: 54/cm² area 1.33 x 10⁴ gm loss on ignition/individual (13.5%) 67% coverage	48.

r tube worms imbedded in the hard base reef of e front quadrats.

Small *Herbivorous crustacea,* including mainly rimps and gammarids, were estimated in the same nner as herbivorous annelids.

Micro-crustacea in the algal encrusting mats were imated from some counts in samples scraped from asured areas. Only 237 (361, 120, 180, 288) were ınd per 28 cm². Estimating .02 cm x .01 cm x .01 as the size of these small species and allowing % dry tissue in wet volume (Vinogradov 1953), a gligible biomass of .022 gms/m² is found.

CARNIVORES (OTHER THAN CORALS)

Small carnivorous fishes, including mainly wrasses, oupers, and small moray eels, were estimated by means of counts and rotenone as described for herbivorous fishes. *Gymnothorax buroensis, Thalassoma quinquevittata, Epinephalus hexagonatus, E. spilotoceps, E. merra, Amblycirrhites arcatus, Scorpaena parvipinnis,* and *S. gibbosa* were examples of fishes in this ecological group.

Large carnivorous fishes, including a variety of species with no one type predominating, were very roughly estimated by counts as with the herbivorous fishes. A rough estimate of shark biomass was obtained as follows: The time sharks were in view during the 15 minute underwater walk to and from the area across the back reef zone was recorded. The fraction of the time when a shark was observed was assumed to be the fraction of one shark's range in view. The area of visibility of an observer looking from side to side through a face mask was assumed as ¼ of 360 deg. The average shark had an estimated weight of about 50 lbs. wet. Moray eels were estimated from the rotenone samples on the surely underestimating assumption that all the morays had climbed out into the channels to die. A rough estimate of area effectively rotenoned was used.

Carnivorous annelids including mainly the nereids were estimated as described for herbivorous annelids.

Carnivorous crustacea including mainly small crabs were estimated as described for herbivorous crustacea.

Carnivorous molluscs primarily Conus were estimated as described for herbivorous molluscs.

One series of night counts was made to estimate the larger night invertebrates. The basis for estimation was the number seen in walking a known distance where estimated visibility for bright eye reflections with an underwater flashlight is about 2 ft as a band along the path. A count of 6 individuals (4 spiny lobsters, 2 large crabs)/720m² was obtained. One individual weighed 150 gm wet with about 35 gms organic matter as estimated by loss on ignition. Thus about .3 gms/m² was observed.

DECOMPOSERS

Decomposers are here identified as that trophic group that subsists on the leakage from other food chains of dead organic matter no longer clearly assignable to a living group of producers, herbivores or carnivores. Included in this group are the bacteria, blennies, and foraminifera. Many others act as decomposers in part of their diet in nature, as when sea cucumbers eat sand, but in most of these cases a majority of the nutrition is from living algae. All algae are considered producers even though some may erode calcareous skeletons. The following were the only efforts made to assess the decomposer part of the community.

Bacteria. Counts of bacteria on glass slides suspended three weeks indicated an 80% coverage. What proportion of these were autotrophs of photosynthetic or chemosynthetic type is not known. As a possible upper limit, the surfaces of all reef objects (about 3 times horizontal area in complex zones) may be assumed to be bacterial covered to the same extent.

Allowing .05 gm/cm³ dry weight and a one micron thickness of bacteria an estimate of their possible biomass of .1 gm/m² results. We have no idea of the bacterial populations within skeletons.

Foraminifera. The foraminifera of the front reef algal mats are very small forms (.1 mm) characteristic of plankton and probably maintained by influx of oceanic water in the strong flow in this area. The forams of the sandy back reef are large benthic types like Calcarina (1. mm). Rough counts (Table 7) and loss on ignition values for these components permit rough estimate of their contribution to the biomass. The nutritive source for the large foram biomass (Table 12) on the back reef is not known.

Blennies. Following food studies by Strasburg (1953), the blennies which are numerous in rotenone samples from the algal ridge zone can be grouped as partly decomposers because of their eating of precipitated detritus (leptopel).

These exploratory estimates were mainly incidental to our study. An understanding of the trophic relationships of the bacteria and foraminifera on reefs is urgently needed. From estimates available, these components do not represent a large biomass in comparison to the other trophic levels. However the high metabolic rates of the small decomposers life system greatly magnifies the effect of a small biomass.

BIOMASS PYRAMIDS

Finally, the quantitative trophic structure of the reef community can be set out as a pyramid of mass. The estimates of biomass by trophic level as estimated in the previous sections are combined in the graphs of Figure 10. In spite of the various errors in the necessarily crude estimates, a general pyramid structure clearly results in all cases as predicted by ecological theory. Furthermore, these pyramids are not quantitatively too different on a weight basis from quadrat to quadrat even though entirely different types of reef community components are represented. Thus the combined mean estimates in the composite pyramid (Fig. 10) gives a reasonable picture of relationships of standing crops. Even if any one of the minor estimates were as much as two fold in error, the general shape of the pyramids would be unchanged. If the chlorophyll to organic matter ratio (Fig. 8) has been underestimated by using Codium, the correct pyramids may be steeper than shown in Figure 10. The ratios of standing crop between trophic levels is H/P 18.9%; C/H 8.3%. Decomposer estimates do not include all components and are left out of the pyramids. Although the reef in gross appearance is what is usually described as a coral reef (rather than an algal reef), and although even the front breaker ridge is in gross appearance half coral, the pyramids show that on a live protoplasmic basis the usual predominance of producer algae exists. This is partly due to the prevalence of plant protoplasm in coral and partly due to large concentrations of matted algae in and on all the reef surfaces.

The pyramids of biomass structures show up even within some taxonomic groups. In the fishes, fo example, there is a striking predominance on a weigh basis of herbivorous parrot fishes, surgeons, damse and butterfly fishes in comparison to wrasses, group ers and other carnivores. The numerous, beautif schools of brilliant herbivorous fishes are indeed th "cows" of the reef.

The single coral is first a producer, to a lesser e tent (in many cases) an herbivore, and somewhat carnivore, thus giving something of a pyramid with one coral head. Indeed the isolated coral heads grow ing in Eniwetok atoll lagoon practically constitute whole community since the plankton is scarce and much of the metabolism is internally complete, th fitting community definition. This need not be tr of all coral heads everywhere and is most certain not the case for clusters of non-photosynthetic De drophyllia growing in deeper waters at Bikini or shadows of ledges in Hawaii.

COMMUNITY METABOLISM

Having demonstrated roughly the trophic structu of the coral reef, consideration may be given the rat at which the community is operating, its productivit its metabolism, its turnover, and the efficiency of i primary production. Although there is consens that individual corals are not quite inherently s sufficient in production, the work of Sargent Austin (1949, 1954) suggests that the whole reef d subsist on its own primary production. They show using black bottles, that the production values water in both the open sea and lagoon side of t reef were far too small to be of significance in co parison to the production of the whole reef, althou this did not prove that plankton passing over the re was not quantitatively an important source of n trition. Their production measurements only mea that production by the plankton while passing o the reef was small relative to the attached commun below. Whether the large volume of water filte by the reef was contributing appreciable ene sources from organic matter previously accumula in the water was not settled. Whether the reef li entirely on its own production or not, it is lik that it derives critical nutrients from the strong fl over the community.

In this study, to assess the contribution of the flowing water to the reef metabolism, measureme of several variables were made in incoming sea wa (represented by water in the windward channel so of Japtan island), in water crossing the front r after passing through the breaker zone, and in wa leaving the reef over the back reef zone (Table 1 The general water characteristics are summarized Table 14. Discussion follows on the significance the changes observed as indicated in Figure 11

Sargent & Austin (1949) interpreted their h values of organic matter on the front reef as due the trajectory of the water in passing through seve turbulent eddys in the breaker zone at which t some of the production of the buttress zone was c

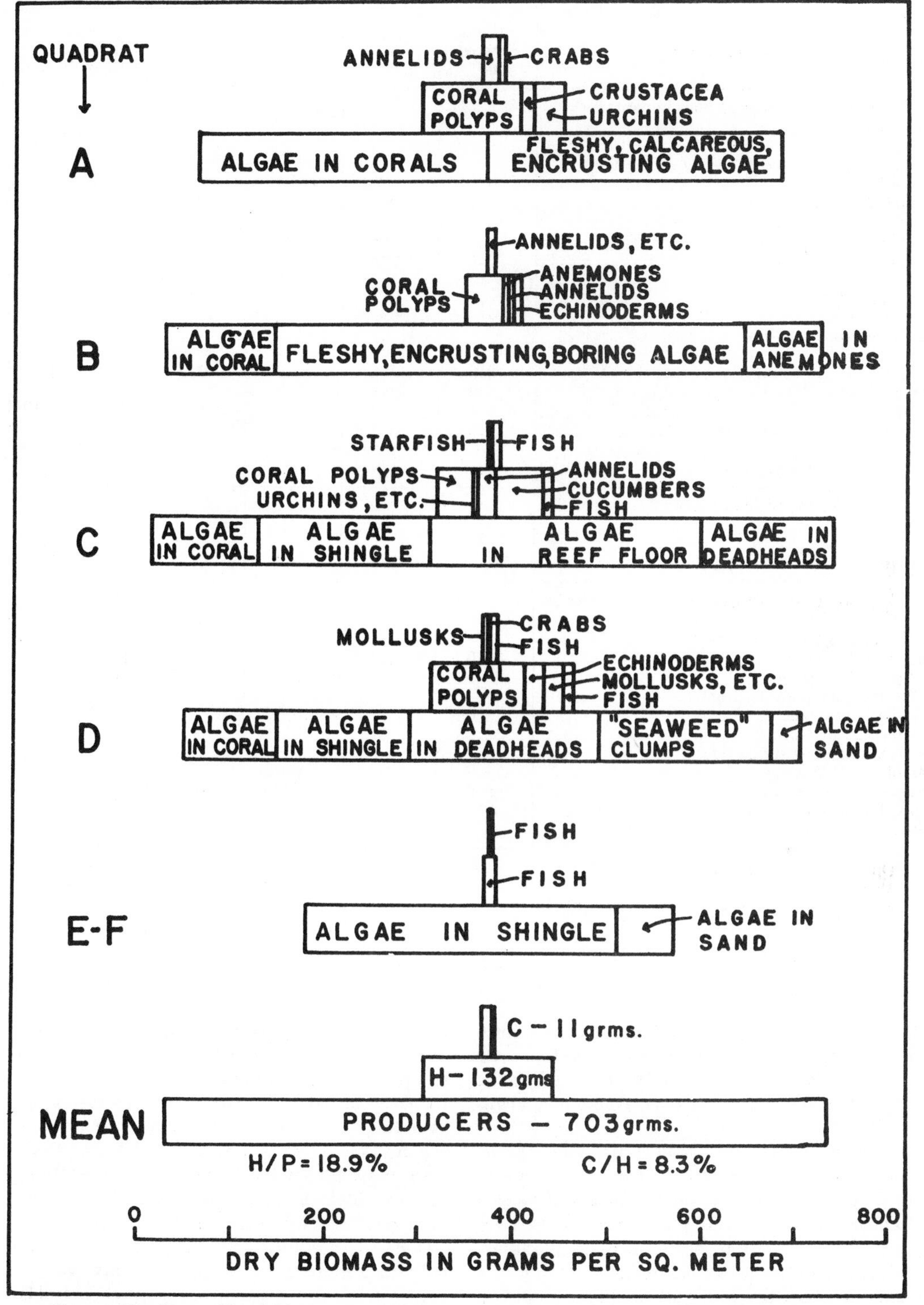

FIGURE 10. Pyramids of biomass resulting from estimates of the dry weight of living materials (excluding, of course, dead skeletal materials associated with protoplasm). For each quadrat, A-F, the weight of "producers" (bottom layer of pyramid), the "herbivores" (H) (middle layer), and the "carnivores" (C) (top layer) is shown, and also the average dry biomass for the reef.

TABLE 13. Plankton characteristics across the reef, July 1954.

Station, circumstance	Volume Filtered m^3	Dry Weight gm/m^3	Loss on ignition gm/m^3	Washed ash (dry) gm/m^3	Plant Fraction (Extract method) %	Radio-activit Thsds counts min/m
OCEANIC WATER (taken from M boat in Japtan—Parry channel)						
10:30 p.m. (night), July 19, falling spring tide:						
Sample 1	6.1	.0120	.0041	....	5.5%	....
Sample 2	6.1	.0121	.0042	.0059	....	13.5
2:00 p.m. July 17, rising spring tide:						
Sample 1	6.1	.0177	.0064	....	6.5%	....
Sample 2	6.1	.0099	.0036	.0054	....	24.1
Mean Oceanic Water	....	.0129	.00457	.0057	6.0%	18.8
WATER CROSSING ALGAL-CORAL RIDGE						
Midnight, July 19, turning low spring tide	3.6	.057	.030	.023	41.2%	188.
2:10 a.m. (night), July 19, rising spring tide	5.0	.063	.033	....	29.7%	....
Noon, July 17, turning low spring tide	4.4	.358	.175	.181	28.7%	1081.
10:45 a.m. July 27, falling spring tide	9.0	.039	.021	.0168	57.2%	73.
Mean of water crossing algal ridge	....	.129	.064	.074	39.2%	447.
WATER CROSSING END OF ENCRUSTING ZONE (Station B-2)						
4:00 p.m., July 13, falling neap tide	21.8	.079	.022	.0236	8.1%	....
WATER CROSSING ZONE OF LARGE HEADS (Station D)						
July 13, turning high neap tide:						
Sample 1, 3:00 p.m.	30.0	.024	.0079	.0034	32.3%	....
Sample 2, 3:20 p. m.	30.0	.027	.0099	.0054	30.6%	....
July 19, rising spring tide, Night:						
Sample 1, 1:00 a.m.	1.68	.054	.0327	....	3.3%	....
Sample 2, 1:37 a.m.	2.65	.034	.0199	.0081	....	52.
July 17, turning low spring tide:						
Sample 1, 10:53 a.m.	11.9	.0129	.0052	....	20.7%	....
Sample 2, 12:40 a.m.	5.9	.035	.0139	.0171	....	77.
July 27, falling spring tide:						
9:45 a.m.	3.0	.037	.0184	.0125	....	34.
Mean of water crossing large head zone	....	.032	.0154	.0093	22.0%	54.
WATER CROSSING ZONE OF SAND AND SHINGLE OF THE BACK REEF SHELF (Station E)						
July 27, falling spring tide, 8:50 a.m.	4.0	.022	.0119	.0042	....	35.

tinually being added. The plankton data suggests this picture to be correct for coarse plankton. Organic matter data are inconclusive as to whether the far larger dissolved organic-matter fraction changes in crossing the reef.

As the water passes the shallow portions of the reef, much plankton is removed. That which remains is exposed to settling in the quieter back waters and to the schools of small anchovy type fish of the back reef zone. The presence of some open-sea plankters in the samples of the back reef indicate that a few individuals do cross the reef without being removed. In general the reef is a highly efficient filter even though the water crosses the reef in 15 to 20 minutes.

Because the tidal cycle creates a variation in the current the reef plankton is highly variable. At very low tides the breakers up front pull off usual quantities of algae but since little water is being thrown and over the reef, they accumulate in the brea eddies until the incoming tide at which time plankton is unusually heavy in the water first com over the reef. Some such variation may account the organic values of Sargent & Austin (1949) wh were too high on the front reef relative to the b reef to match the known respiration of the reef. high plankton values in Sample 1, 2:00 p.m. July (Table 13) are due to the effect described above. data in Figure 11 show similar patterns across reef for loss on ignition, chlorophyll extracts of plankton, and plankton radioactivity.

FLUX OF LARGER PLANKTON

Plankton samples were made with a #10 net. the reef the net could be set on a stake to permit

TABLE 14: Chemical levels in Eniwetok waters.

Component	Analyses	Mean	Range
ganic Matter, alkaline permanganate method, in mg/l	13	.96	.74—1.41
trate nitrogen, strychnidine method, in mg atoms/m³	24	.44	.06—1.0
organic phosphorus, ammonium molybdate method in mg atoms/m³	29	.32	.26—.64
Total phosphorus, acid digested, in mg atoms/m³	6	1.7	0—3.4
ssolved oxygen, Winkler method, in mg/l.			
(1) Incoming Ocean Water (from channel)	8	6.54	6.38—6.68
(2) Algal-coral Ridge	12	6.50	6.09—6.97
(3) Back Reef zone of Large Heads			
Daytime	19	7.31	6.22—8.59
Night	6	5.37	4.89—6.29
(Beckman Model G)			
Daytime			
Incoming Ocean Water (from channel)	5	8.21	8.19—8.22
Algal-coral ridge	5	8.21	8.18—8.24
Back Reef zone of large heads	5	8.32	8.30—8.33
Night			
Incoming Ocean Water (from channel)	2	8.19	8.18—8.19
Algal-coral ridge	2	8.16	8.14—8.17
Back Reef Zone of large heads	2	8.10	8.10—8.10
mperature in degrees Fahrenheit			
Incoming ocean water (from channel)	2	82.6	82.6—82.7
Algal-coral ridge	2	82.9	82.7—83.0
Back reef zone of large heads			
Daytime	3	84.1	83.5—84.6
Nighttime	1	82.2	....

rong current to flow through. The channel samples ere taken from a boat. The volume of water passg through the net was determined by placing a drop dye (air-sea-rescue dye marker, fluorescein) in the outh of the net and counting the seconds required r the dye to wash through the net. The current on e reef was always sufficient to give a satisfactory mple of plankton after 10 to 40 minutes. The curnt outside the net was simultaneously determined ith the dye method in order that computations of tal plankton flux could be made. The average time r five dye spots to cross a distance of 20 ft was ed to determine water velocity.

Qualitatively, net plankton of the incoming water as characterized by pteropods, calanoid copepods, diolaria, and tiny filamentous algae. The reef ankton after passing the surf zone was conspicusly different being made up of large fragments of amentous algae derived from the buttress-breaker ne of the reef. The data (Table 13, Fig. 11) clearly dicate a very large increase of large-sized plankton the water crosses the breaker zone and a rapid loss most of this plankton in crossing the rest of the ef. Thus the reef consumes its own pseudoplankn. Since the amount of plankton leaving the reef the back is about the same or slightly greater than at in the incoming water, it seems that the reef is deed energetically self-sustaining and deriving no t gain of larger planktonic material from the inwing water. The data, however, do not entirely iminate the possibility of a gain from nannoplankn and dissolved organic matter.

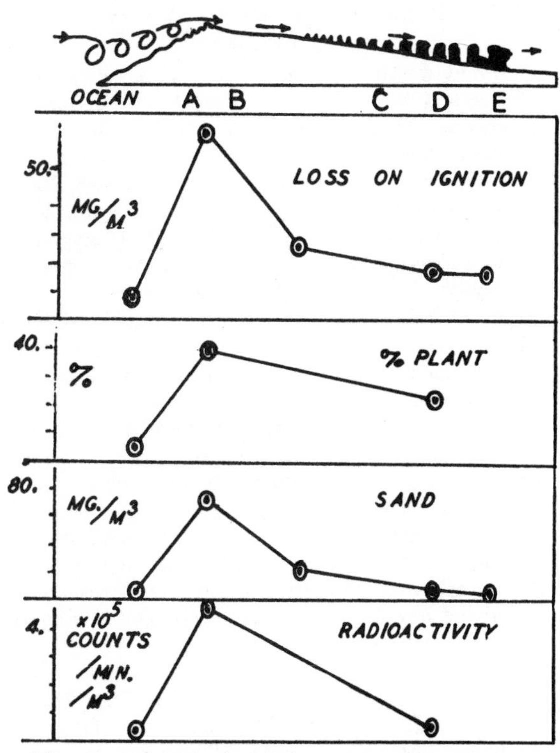

Fig. 11. Changes in water content as it crosses the reef in regard to: loss on ignition of net plankton and seston; green plant content (estimated by chlorophyll content) of net plankton; suspended sand; radioactivity.

TOTAL ORGANIC MATTER

A few measurements to determine the order of magnitude of the total organic matter were made with the alkaline permanganate method of Benson & Hicks (1931). Samples (100 cc) were digested 30 minutes at 95°C with standardized permanganate and sodium hydroxide. The ferrous sulfate was added in an amount equivalent to the permanganate. Oxidation-reduction conditions were adjusted with manganous salts and phosphoric acid to prevent oxidation of the chloride. The excess ferrous sulfate was titrated with more permanganate to determine the permanganate lost during digestion. A mean value of about 1 mg/l of oxygen consumed was found (Table 14). The values obtained with this very rough procedure were fairly consistent and probably give the general order of magnitude of dissolved organic matter. Significantly, the values were similar to those obtained by Johnstone with BOD determinations (Sargent & Austin 1949). We did not find the great difference between front and back reef found by Sargent & Austin although our permanganate method should probably not be relied upon as accurate enough to delimit a difference of less than 1 ppm. As in other kinds of water, the dissolved organic matter is much greater, though less conspicuous, than the particulate matter. Motoda (1938) found 2.0-3.1 mg/l, 35 day BOD, in open sea water at Palao and 5.5 mg/l in the bay.

PRODUCTION AND RESPIRATION BY THE FLOW METHOD

Sargent & Austin (1949) used an ingenious flow-rate method to measure the over-all production and respiration of the coral reef at Rongelap Atoll, Marshall islands. A similar method has been used by H. T. Odum in Silver Springs, Florida (1953, 1954). The oxygen content of the water upstream and downstream is measured simultaneously. The oxygen increase between stations during the day is the net photosynthetic production of the community. The oxygen decrease between stations during the night is the total respiration of the community. By taking a series of measurements over the daily cycle, one obtains the course of production during the day. Measurement of the current transport permits calculation of total reef metabolism. The respiration at night plus the net production during the day gives the total production. By comparing the area of the graph between the day curve and the zero line with areas of the graph under the zero line at night one can obtain an indication of what part of the excess production during the day is used up by respiration during the night.

A series of such measurements was made on several different days of typical cloud cover of from 1/10 to 3/10 small cumulus and 1/10 to 4/10 high and middle cloudiness, and also at night. These values are expressed on an area basis in Figure 12 following their conversion from depth and current measurements. The curve obtained by Sargent and Austin for their reef is also plotted in Figure 12.

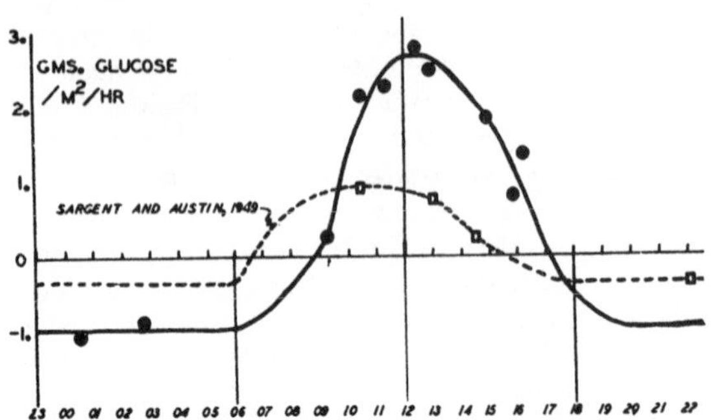

FIG. 12. Graph of primary production and respiration on the middle reef between the coral-algal ridge and quadrat D. A similar curve from Rongelap Atoll is recalculated from Sargent & Austin (1949).

Besides small errors due to inherent fluctuations and variability of oxygen samples and methods, current and depth measurements, and cloud cover changes, there is a major source of error that tends to cause values of production and respiration to be too small. This is the diffusion of oxygen from and to the atmosphere. This error is greatest when the displacement of the gaseous content of the water from equilibrium with the atmosphere is greatest. Thus the error in decreasing the production estimate is greater than that in the respiration since greater displacement from equilibrium occurs. Furthermore, during the day's production a carpet of bubbles of oxygen is observed to form over the algal-mat surfaces particularly on the front reef. These bubbles are continually breaking off and reach the surface without dissolving so that some of the oxygen is lost, further lowering the estimate of production below the correct figure. The diffusion error is further discussed in the section on over-all balance sheet for the reef.

Our curve of reef production for the middle reef (Fig. 12) permits the following interpretations: (1) The productivity of the reef is very high, greater than 24 gms/m²/day or 74,300 lbs/Acre/yr. (2) The productivity is greater than that of Silver Springs Florida, which is very nutrient-rich with a relatively constant temperature, and supports a similarly autotrophic community with a production of 50,000 lbs/Acre/yr (Odum, H. T. 1953-54). Silver Springs has a higher summer production rate but a smaller annual total due to the small insolation of the winter The reef production is greater than other marine localities reported. In comparison to the open tropical waters such as that flowing onto the reef (.2 gms/m² day; Sargent & Austin 1949), this reef is 120 time more productive. (3) The Japtan reef has twice th production value per area by the flow method as th Rongelap reef studied by Sargent & Austin. Th larger production of the reef per area is matched b a larger respiration per area in comparison to th reef studied by Sargent & Austin and therefore suggests a greater biomass per area. Sargent & Austi (1954) describe their reef as relatively barren. (4 There is a lag between the sun's light intensity an the oxygen production as measured in the water abov the reef each day. It is likely that part of this la is due to the location of many of the primary pro ducers down in the calcareous reef surface, sands, an below the coral polyps.

In view of the evidence for a lag in gaseous ex change, a symmetrical morning and afternoon pro duction curve seems incorrect even though Yonge Nicholls (1931) showed that isolated single corals i the sun may reach maximum photosynthesis in th morning. Thus Sargent & Austin's production curv drawn through three afternoon points, probabl shows too much productive area. It seems likely th their reef possessed little or no excess of productio over respiration and may not have been depositin stored organic biomass.

PRODUCTION EXPERIMENTS IN SUBMERGED BELL JARS AND PLASTIC BAGS

Following experience in Silver Springs, effor were made to enclose reef components with glass jar and to sample the water before and after periods production and respiration. After many difficulti a diaphragm of inner-tube rubber placed outside the glass jar was found to make a seal between t jar and the hard irregular reef floor. Sand a boulders heaped over the ring-shaped rubber bord prevented active circulation of the water outside t jar with the interior water. A rubber tube with

ening inside the jar and one outside the jar per-
itted sampling. Water samples were drawn out
th a type of sampler developed at Silver Springs
which the observer draws water through two bot-
s on a stick by sucking on a tube. The bottle near-
t the bell jar source is used for Winkler oxygen
terminations after the second bottle has filled.

The unexpected results given in Table 15 for ex-
riments on the reef seem to demonstrate a lag effect
also evidenced in the flow measurements discussed
the previous section. When black covers were put
er corals or reef surfaces during the day, oxygen,
esumably from sub-surface bubbles, continued to
evolved for over an hour. Supporting this inter-
etation is the observation that myriads of little
bbles rise from the interior of Millepora or Acro-
ra fingers when they are broken during the day.
he hard bumps of the reef surface that do not even
ok green externally, as well as the algal mats of the
ont reef, are coated with bubbles during the middle
the day. These bubbles on bare looking places are
robably coming from the subcrustal green algal
yers previously described in the primary producer
scussion.

To determine whether reef components had an
er-all net production or respiration, plastic bag ex-
eriments were conducted in the field. Following the
ggestion of Dr. Max Doty, 6 x 6 in. plastic bags
ere tied around small coral and algal heads with
bber bands and observed a day later. The bubbles
hich accumulated in the bags are assumed to have
en initially formed of oxygen because release of
her gases seems unlikely. Carbon-dioxide would
t be released in a gaseous form through basic sea
ater. Since no temperature changes could occur in
is field arrangement, air components would not be
leased from solution. Once a gaseous phase was
rmed other dissolved gases might enter. No exten-
ve decay is suspected that might produce large
antities of hydrogen, methane or other gas. As
sitive evidence that these bubbles contain consider-
le oxygen, water drawn from bags with bubbles
as super-saturated whereas water drawn from bags
ithout bubbles was undersaturated. However, after
gaseous phase had been formed, other dissolved
ses would diffuse in. Visual estimates were made
the volume of bubbles in the bags after one day.
ote (Table 16), that the dead heads with encrusting
gae have a large net bubble-production whereas the
rals had slight or no production even though 2 hrs

TABLE 15. Lag in gas exchange between reef and
ater. (Oxygen changes in dark and light bell jars on
ef substrates.)

Surface	Preceding Light Regime	During Measure-ments	Time Lapse Minutes	Oxygen Change gms/m²/hr
nd-shingle	7 hr light	dark	31	+.73
nd-shingle	31 min dark	light	30	−.24
ead-heads	7 hr light	dark	31	+.085
ead-heads	3 hr light	dark	73	+.125

TABLE 16. Volume of bubbles accumulating in plastic bags over reef heads. Bubbles estimated at noon.

	Time exposed hrs	Volume of bubbles cc
Dead heads permeated and encrusted with algae	26	15
	74	12
	26	25
Halimeda bunch	26	2
Bag of foraminiferal sand	26	.1
Live corals		
Millepora	26	.5
Porites	26	1.0
Heliopora	26	1.0
Acropora (read at 9:00 a.m.)	70	0
Acropora	50	.1
soft coral (*Lobophytum*)	26	.1
Control bag	26	.1

of sunlight of a second day were included in more than one day's measurement. This seems to indicate that photosynthesis in situ does not exceed respiration in many corals although it does match a large fraction of the respiration. It is possible that the role of photosynthesis is greater in the clear waters of the Marshalls than in more turbid waters in some other areas. Perhaps fortunately for corals, decreasing light, due to greater turbidity is often accompanied by increasing plankton content.

PHOSPHORUS AND NITROGEN

To establish the general nutrient level on the reef, a few analyses were made of inorganic phosphorus, organic phosphorus, and nitrate nitrogen. Methods used were rough adaptations of the ammonium molybdate-stannous chloride method for phosphorus (Robinson & Kemmerer 1930) and the strychnidine method for nitrate (Zwicker & Robinson 1944). Samples were necessarily determined 1 to 6 hrs after collection. Some loss due to uptake by bottle walls may have occurred. As expected for a central tropical ocean the incoming waters had extremely low values of both nutrient elements. Although the concentrations were in the lower limits of sensitivity of the methods, the values reported in Table 14 possibly are of valid order of magnitude.

Considering the productivity, as established, the required nutrients can be accounted for as follows. If the mean nitrogen content of the producing algae is 3.0% (½ protein), then .72 gms/m²/day nitrogen is required. Considering the width (Table 17) and volume transport of water (3. x 10^4 m³/day), only .6 gms/m²/day nitrogen could be supplied from inorganic nitrate even if it were all captured and used. It is not all used since it was detected in a concentration of the same order of magnitude in back-reef water. Using open-ocean values (Sverdrup, *et al.* 1946) .1 gm/m³ total organic nitrogen is found in water with about 2.0 gms/m³ total organic matter. Using this nitrogen/organic matter ratio of 5% and a total organic matter content of 1 gm/m³ (as determined, Table 14), about 4.7 gms/m²/day could be

TABLE 17. Balance sheet for the Japtan inter-island reef in July. From algal-coral ridge to the end of the zone of large heads, this zone is 322 m^2 long.

	gm/m²/day
INCOME*	
Planktonic organic matter (Table 13) from breaker zone	2.0‡
Primary production (measured as oxygen, calculated as glucose:	
Net (uncorrected) daytime production	14.0
Respiration during daytime	10.0
Total income	26.0
LOSSES*	
Planktonic organic matter lost to lagoon (Table 13)	0.4‡
Total respiration in 24 hr	24.0
Total outgo	24.4

*Dissolved Organic Matter; 0.96 gm/m³ (Table 14) (no significant difference between influx and outflux; analytical method not precise enough, however, to delimit.)

‡The mean water flux during plankton sampling was 425m³/hr across a band of reef 1 m wide.

acquired by the reef if all the nitrogenous organic matter were taken up. To meet the need of .72 gms/m²/day, therefore, much of the incoming organic matter would have to be taken out. Our few analyses showed no evidence of this magnitude of organic-matter uptake although Sargent & Austin had some evidence of large uptake. It seems equally likely that there is cyclic re-use of nitrogen along with some nitrogen fixation by the abundant blue greens of the front reef. That the surface encrusting algae are so definitely correlated with high current velocity, however, suggests the need for taking out some nutrients of organic or inorganic nature from the low nutrient water. As indicated in Table 5, however, the surface encrusting algae, although conspicuous, are relatively unimportant on a biomass basis in comparison to the algae permeating the calcareous substrates, living and dead. They may, however, have higher metabolic rates than the imbedded algae.

Similar calculations suggest that phosphorus is more abundant relative to needs than nitrogen.

The back reef production is accomplished largely by algae within the calcareous dead parts of corals and other back-reef components. The coverage of coral decreases from nearly 50% of the front reef to 16% of the back reef and finally 0%. Apparently, nutrients caught and stored by the algae and coral of the front reef are passed to the back-reef producers in the form of skeletal fragments. The front reef encrusting-producers, being in shallower water with much swifter currents (Fig. 2), are exposed to greater volume per individual so that the energy of nutrient gathering is partly supplied by the flow system. Thus an adequate nutrient source in water and plankton is available at the front. The nutrient and current regimes are thus entirely different for the front and back-reef producers. The front-reef producers need catch only enough nutrients to balance that leaking off the back reef to maintain the e system. The general habit of the boring produc in all the reef surfaces is most favorable to nutri conservation. One imagines that at the time o typhoon or other circumstance bringing richer wa to the surface of the ocean, the reef would capt critical nutrients in an efficient manner for fut use. Certainly the reef has concentrated radioacti of multiple types in water that is barely above ba ground as it drifts in 200 mi. from Bikini. Three sa ples of water, evaporated down without filtering, g mean radioactivity of 182 counts/min/l (determi by K. Lohman). This was roughly of the same m nitude as the dried-plankton radioactivities per v ume (Table 13) estimated with a Beta Gamma sur meter (AEC MOD-SGM-2B). Coral and algal s faces registered about 300 counts/min/cm² of s face. Thus, plankton radioactivity of a liter of wa has an order of magnitude equivalent to that o cm² of coral surface.

The reef surfaces, living coral, living calcare algae, and dead skeletons all act as a kind of soil that they conserve phosphorus nutrients and per plant biota to burrow in to reach these nutrients. indicated in Vinogradov (1953), fresh coral skelet have 1% phosphorus and .01 to .1% aluminum a iron. Calcareous algae contain much less. Gr layers of boring algae are found just under the surfaces of calcareous red algae as well as under corals (Fig. 7D).

The N/P ratio by atoms is roughly 2.1 (usin nitrogen value inferred from organic matter on assumption that most is protein), which indica particularly sparse nitrogen conditions. The la proportion of small blue green algae may sugges low nitrogen environment.

REEF DEPOSITION, REEF EROSION, pH CHANGES

Although the production and respiration measu ments suggest that respiration is not far from a b ance with production, there is little clear evide about the balance between skeletal reef deposit and erosion of any one reef yet studied. Certai the reef atoll as a whole can maintain its relations to sea level for long periods of time. As caref postulated by Mayor (1924), a reef seems to have mechanism for self regulating its balance of dep tion and erosion at about 6 in. below mean low wa spring tide. The excess of production or respirat does not in itself indicate whether deposition or e sion is in excess. For example, animals like oyst with entirely respiratory metabolism may neverthe relegate energy to deposition. Most marine pla with a primarily production-type metabolism do deposit a skeleton which their metabolism would t to precipitate in water almost saturated with resp to calcium carbonate, like that at Eniwetok. T might involve an energy expenditure. It is not c whether reefs, by a succession, destroy themselves becoming a terrestrial community or whether t form a climax at the 6 in. level below low wa spring tide.

Sargent & Austin attempted to estimate reef depo- tion on the assumption that most of the reef biomass as coral and that their respiration measurements for e whole reef would be extrapolated into a deposi- on rate, assuming a deposition to respiration ratio pical of corals as measured by previous workers. owever, it is clear from the pyramids (Fig. 10) that e majority of the biomass is not coral-animal tissue. Nor is the metabolism predominately coral. By lculation from Mayor's value for metabolism per ving biomass of .43 mg/gm/hr (mean of 4 species) d our figure of .062 gms/cm^2 total biomass in rals, and a coverage of 20% coral, one obtains a spiratory contribution of coral of 53 mg/m^2/hr. his is a relatively small part of the total respiration the reef determined with the flow method which is out 1.0 gm/m^2/hr.

Sargent & Austin's possible overestimation is rtly counteracted by the use of the wrong density. ven so, their over-all estimate of maximum material dded (1.4 cm) seems too high, on the basis of their lculations.

As a variation of Mayor's calculation the quadrat timates of coral coverage were used to estimate rate coral deposition with a figure for coral growth rived from Mayor (1924). Slightly lower growth tes are found in colder waters (Tamura & Hada 32, Ying 1934). By recalculating on an area basis d averaging 18 growth values for corals from ayor, an annual skeletal growth rate of 8.0 cm was tained. This involved using a dry gross density of ral of 1.9* gm/cc. Most of our reef seems to be ilt of coral (not calcareous algae) if one can judge y blocks of the reef from the dynamite holes else- here on the atoll (Parry island) or from the present mposition of most of the reef. Much of the deposi- on, therefore, comes from the 20% coverage of ral. Therefore, the 8 cm skeletal growth in the eas of coral is spread over the whole area in the rm of shingle and dead heads to form a net addition material of growth of 1.6 cm. Although this rate increment is almost identical with the one esti- ated by Sargent & Austin, their apparently less roductive reef may actually have a lesser deposi- onal rate.

On the basis of growth rate and coverage estimates, d a density of 1.8, Mayor estimated 0.8 cm annual position and simultaneous erosion on his study reef Samoa.

For the middle reef the estimate of 1.6 cm cal- reous deposition amounts to 3.05 gm/cm^2/yr. The ver-all income of 26.0 gm/m^2/day (Table 17) is equivalent to .95 gms/cm^2/yr glucose. Thus the bio- mass initially deposited is only ⅓ of the calcareous deposition. With the water near the inorganic de- position point, little energy is likely to be required for this calcareous deposition. Just how much is not yet known.

This coral increment being added is very likely being eroded just as fast by current abrasion and the complex of bacteria and boring algae that charac- terize the coral shingle fragments that dominate much of the back reef zones, so neither the above calcula- tion nor Sargent & Austin's evidence is at all indica- tive of over-all net reef growth or erosion.

Some idea of the magnitude of abrasion taking place on the front buttress zone, which must be bal- anced by growth to maintain the reef and must be exceeded to produce a reef-growth laterally into the wind, may be obtained from the sand in the reef plankton. The plankton ash in the samples collected at the front and back reef was washed with water leaving a residue consisting mainly of fine calcium carbonate sand that had been suspended in the water passing through the plankton net. The change in this sand fraction in crossing the reef is depicted in Figure 11. Strikingly, the sand content rises in cross- ing the breaker zone and as the current diminishes on the back reef the sand content falls, thus demonstrat- ing the action of front-reef growth in filling in and cementing the back reef. Since no measurements of sand in plankton were made during the strong cur- rents of high tide, the magnitude of deposition and erosion of reef sand is uncertain.

BALANCE SHEET FOR THE REEF COMMUNITY

Having made various measurements and estimates of photosynthetic rates and metabolic processes, we may now consider the data as a whole to see how nearly balanced are the gains and losses of organic matter on one reef section. In Table 17 the sources of energy storage gain are estimated, including pri- mary photosynthetic production and influx of or- ganic matter in the water. The losses of energy are also listed, including respiration, and outflow of or- ganic matter. The gains and losses are only 4% apart. In view of the rough nature of some of the estimates it is not certain whether this is a significant difference or whether the community is in a perfect steady state with losses matching gains. With the 15-20% lower total insolation in winter (insolation tables, Kennedy 1949) at this latitude than when these measurements were made in July, a lower pro- duction but relatively unchanged respiration may be expected to make an annual balance between produc- tion and respiration.

This vigorously productive reef is possibly one of those that Cloud (1954) thinks is now in slightly deeper water than the equilibrium depth because of a sea level rise starting about 100 yrs ago. Accord- ing to this idea the reef may be experiencing a net growth of calcareous matter. However, there is no definite evidence from this study to indicate that the

* In Table 3 are given density measurements of two types. ne is the density of the dried skeletal material obtained by eighing wet while suspended in water and by weighing dry. his density which is of interest mineralogically is the weight r volume of the component skeletal septa. The density of a y gross block including the empty pore spaces is considerably ss and is obtained by correcting for pore space. Although the eletal density (mostly aragonite) is about 2.3 gm/cc (Table , with a pore space of 16% the dry blocks of coral have a oss density of only 1.9 gm/cc. This latter density should have en used by Sargent & Austin rather than 2.5 gm/cc in esti- ating growth increment from weight increases. This error for- nately partly counteracted their error of overestimation of coral pulations discussed above. When the pore space is filled with ater, the gross wet density is 2.1 gm/cc, a sometimes useful antity.

reef is not in balance with respect to organic matter. Organic deposition and calcareous deposition are not necessarily in phase.

Similarly, as dicussed in the section on production measurements, a balance between gains and losses may have existed on Sargent & Austin's reef also. The biggest uncertainty in both these studies is still the question of changes in dissolved organic matter in the vast flow of water crossing the reef.

We may tentatively conclude, at least, that the Japtan reef is a true climax community, in the ecological sense, under present ocean level conditions, since there is little if any net increase in living biomass.

Efficiency

From tables of insolation reaching the ground (Kennedy 1949) a figure for insolation reaching the water surface can be obtained, taking into account approximate cloudiness. About ½ of this total insolation is in the visible range (Sverdrup, *et al.* 1946). With a Weston photographic light meter enclosed in a plastic bag a reading above the surface of 1500 fc was obtained compared to 800 fc 50 cm below the surface at quadrat D. Therefore about half of the surface light reaches the average reef depth of 2 ft. From these approximations the energy available to production on the reef can be estimated. For the latitude of Eniwetok in August, these approximations indicate about 1650 KCal/m^2/day incoming energy reaching the community. Relative to the 96 KCal/m^2/day (24 gm x 4.0 KCal/gm) primary production estimated from oxygen measurements in Table 16 this is about 5.8% efficiency of primary production.

That this efficiency is a low one in comparison to some laboratory experiments (Rabinowitch 1951) and yet higher than average terrestrial agriculture is an important result. Here is an ecosystem which has had millions of years to evolve an effective composition, which is built for a low efficiency. This may be support for the hypothesis (Odum & Pinkerton 1955) that there is an optimum but relatively low efficiency that produces the most effective trophic structure whose survival is based on a high primary productivity.

RELATIONSHIP BETWEEN TROPHIC STRUCTURE AND COMMUNITY METABOLISM

It seems clear that the vast coral reef community is highly productive and not far from a steady state balance of growth and decay. As a community of unquestionable durability and ancient origin, it may be postulated that some kind of optimum adjustment has been evolved. The evolution may be stated in terms of the stability principle (Holmes 1948) as follows: As an open system, the construction of self regulating interactions has led by selective process to the survival of the stable.

In the previous sections a standing crop of living biomass of about 700 gm/m^2 was found and a total primary productivity (glucose) of 24 gm/m^2/day or 8760 gm/m^2/yr. The ratio of annual primary pro duction to standing crop is therefore about 12.5 to 1 This ratio can be called the turnover. If there i an underlying relationship of primary production an standing crop under steady-state conditions, an an nual turnover value similar to the 12.5 for our ree may be found in other systems where there are simi lar temperatures, and similar supplementary energie supplied as water currents, and similar sized organ isms with similar metabolic rates.

SUMMARY

1. During a mid-summer 6-week period (1954), (1 the standing crop biomass of "producer" and "con sumer" organisms of a windward, inter-island cor reef on Eniwetok Atoll was estimated, (2) primar production and total respiration were determined b upstream and downstream chemical measurement and (3) from these data the turnover and energet efficiency of the reef ecosystem were calculated.

2. The reef, which has not as yet been directly di turbed by nuclear explosions, exhibited 6 distin zones as follows: windward buttress zone, coral-alg ridge, encrusting zone, zone of smaller heads, zone larger heads, zone of sand and shingle (Figs. 2-4 Zonation of this inter-island reef (with its one-dire tion current system) is very different from zonati on island reefs so abundantly described in the liter ture. Quadrats were mapped in 5 of the zones (Fi 5) and standing crop biomass determined for ea

3. Because producers (algae) are so intimately i terwoven with animal and dead skeletal material t chlorophyll extraction method appeared to be t most feasible means of estimating producer bioma Algal dry weight was determined by relating spect photometrically, chlorophyll content with known d weights of a reference species, *Codium edule*. Va ous methods were used to estimate major animal co ponents as described in appropriate sections of t paper.

4. On a horizontal surface area basis, the avera living coral colony proved to contain three times much plant as animal tissue, or .063 gm/cm^2 (weight of algae as compared with .021 gm of anim polyp (Table 4 & Fig. 6). Zooxanthellae (in coelenterate polyps) comprised only about 6% of total plant portion, filamentous green algae embed in the skeleton making up the bulk of plant mater The evidence indicates that these skeletal algae, of considered "parasitic" or "boring" by previous wo ers, may be actually mutualistic. The algal-coel terate complex, therefore, comprises a highly in grated ecological unit (comparable to the algal-fun complex of a lichen) which permits cyclic use reuse of food and nutrients necessary for vigor coral growth in tropical "desert" waters having very low plankton content. The coral is thus c ceived to be almost a whole ecological unit in it with producer, herbivore (utilizing food from s biotic algae), and carnivore (plankton feeding night) aspects.

5. The quantitative coverage of coral (50-16

surface encrusting algae decreased while the
ınt of subsurface algae increased from front to
across the reef correlated with mean current
ities suggesting a transition from a water filter-
source of nutrients up front to a subsurface deposition source of plant requirements on the back
s.

Although species differences were indicated, the
content of 33 samples of 12 species of corals
rather uniform when calculated as dry biomass
projected horizontal surface area basis (Table
Branching corals contained about the same
ınt of algae as massive and encrusting corals.
osed reef surfaces, "deadheads," and shingle con-
d comparable or somewhat larger amounts of
ucer tissue; only the calcareous sands of the back
gave low values.

While the kinds of primary producers were
variable from place to place and zone to zone
major types are listed in the ecological classifi-
n), the reef, whether covered with coral or not,
ed to have a rather uniform content of algae.
ther words, the algal standing crop was of a simi-
rder of magnitude (between 0.050 and 0.1 gm/
throughout, a situation certainly not evident on
rficial examination (because a large amount of
green plant material is subsurface).

The sessile part of the community is primarily
trophic with relatively few plankton feeders
than coral polyps; fouling on glass slides was
st entirely algal.

In all zones of the reef a trophic structure with
ramid of biomass was found (Fig. 10). Although
ely different taxonomic components were present
ifferent zones, similar biomass figures were ob-
d. The mean standing crop for the reef as a
e in gm/m^2 was: producers, 703; herbivores,
and carnivores, 11. The ratio between standing-
trophic levels was H/P, 18.9%, and C/H, 8.3%.

. A very high total production of about 74,000
cre/year was obtained with the flow rate method.
represents a turnover (the ratio of annual pri-
production to average standing crop) of about
times per year of existing biomass. The figures
the production curve (Fig. 12) provide quantita-
criteria for assaying the future effect of nuclear
osions, continued low-level radioactivity, or other
rs on the community as a whole.

. The production on the reef seems to about bal-
the respiration on the reef (Table 17). The
s do not constitute a dominant part of the whole
bolism. It is concluded that the reef community
nder present ocean levels, a true ecological climax
pen steady-state system.

. The efficiency of primary production computed
rms of the visible light reaching the underwater
nunity is about 6%. This is support for the
cated theory that steady state communities ad-
to a moderately low efficiency as a necessary
pensation for high total productivity.

. The reef does not derive a net gain from the larger components of plankton in the water crossing the reef under the stress of the trade winds. Whether a dissolved organic-matter gain is obtained is still uncertain.

14. Individual corals *in situ* in nature, like those in laboratory experiments of other workers, produce an excess of oxygen in the daytime but not over the course of 24 hours (Table 16). The coral with its 3:1 ratio of plants to animals is apparently just about "balanced" in gaseous exchange.

15. The location of the sub-surface boring algae leads to a time lag in gas diffusion with the water crossing the reef and an afternoon community-production maximum (Fig. 12).

16. The nutrient levels of nitrogen and phosphorus are very low (Table 14). Some evidence exists that nitrogen is more scarce relative to plant needs than is phosphorus and must be conserved, fixed, and recirculated.

17. Measurements of low-level radioactivity which was present (Table 13, Fig. 11) provided further evidence of nutrient conservation, and autoradiograms of corals (Fig. 9) provided additional evidence for symbiosis between corals and their skeletal algae.

18. Plankton and eroded sand broken from the front-reef breaker zone is recaptured on the middle-reef zones.

19. Estimates indicate 1.6 cm of calcareous deposition per year but there is no evidence that this is not eroded almost equally rapidly.

20. The Japtan inter-island reef at Eniwetok is primarily a coral reef rather than a calcareous algal reef in the geological reef-forming sense. But like other communities studied (whether aquatic or terrestrial), the Japtan reef has a large predominance of living plant biomass, even though organisms classified as animals are more conspicuous.

LITERATURE CITED

Banner, A. H. & J. E. Randall. 1952. Preliminary report on marine biology study of Onotoa atoll, Gilbert Islands. Atoll Res. Bull. **13**: 1-62.

Benson, H. K. & J. F. G. Hicks, Jr. 1931. Proposed modification of oxygen consumed method for determination of sea water pollution. Indust. Engin. Chem. Anal. Ed. **3**: 30-31.

Bonham, K. 1950. Invertebrates. In Donaldson: Radiobiological survey of Bikini, Eniwetok, and Likiep Atolls. July-August, 1949. U. S. Atomic Energy Commiss. Tech. Inform. Serv. Bull. **AECD-3446**: 119-128.

Blinks, L. R. 1952. Effect of radiation on marine algae. Jour. Cell. & Comp. Phys. **39**: suppl. 2, 11-18.

Brock, V. E. 1954. A preliminary report on a method of estimating reef fish populations. Jour. Wildlife Mangt. **18**: 297-308.

Cloud, P. E. 1952 Preliminary report on geology and marine environment of Onotoa Atoll, Gilbert Islands. Atoll Res. Bull. **12**: 1-73.

———. 1954. Superficial aspects of modern organic reefs. Sci. Monthly **79**(4): 195-208.

Donaldson, L. R., et al. 1950. Radiobiological Survey of Bikini, Eniwetok and Likiep Atolls—July-August

1949. U. S. Atomic Energy Commis. Tech. Inform. Serv. Bull. **AECD-3446**: 7-145.

Doty, M. S. & J. P. E. Morrison. 1954. Interrelationships of the organisms on Raroia aside from man. Atoll Res. Bull. **35**: 1-61.

Duerden, J. E. 1905. Recent results on morphology and development of coral polyps. Smithsn. Inst. Misc. Collect. **47**: 93-111.

Edmondson, C. H. 1929. Growth of Hawaiian corals. Bernice P. Bishop Mus. Bull. **45**. 43 pp.

Emery, K. O. 1948. Submarine geology of Bikini atoll. Geol. Soc. Amer. Bull. **59**: 855-860.

Emery, K. O., J. I. Tracey, Jr. & H. S. Ladd. 1954. Geology of Bikini and nearby Atolls. U. S. Geol. Survey Prof. Paper **260-A**: 1-265.

Fox, D. L., J. D. Isaacs & E. F. Corcoran. 1952. Marine Leptopel, its recovery measurement and distribution. Jour. Mar. Res. **11**(1): 29-46.

Harvey, H. W. 1934. Measurement of phytoplankton population. Mar. Biol. Assoc. Jour. **19**: 761-773.

Hiatt, R. W. 1953. Instructions for marine ecological work on coral atolls. Atoll Res. Bull. **17**: 100-108.

Holmes, S. J. 1948. The principle of stability as a cause of evolution. Quart. Rev. Biol. **23**: 324-333.

Johnson, M. W. 1949. Zooplankton as an index of water exchange between Bikini Lagoon and the open sea. Amer. Geophys. Union Trans. **30**: 238-244.

Kawaguti, S. 1937. On the physiology of reef corals. I. On the oxygen exchanges of reef corals. Palao Tropical Biol. Station Studies **2**: 187-198.

———. 1937. On the physiology of reef corals. II. The effect of light on colour and form of reef corals. Palao Tropical Biol. Station Studies **1**(2): 199-208.

Kennedy, R. E. 1949. Computation of daily insolation energy. Amer. Met. Soc. Bull. **30**(6): 208-213.

Ladd, H. S., J. I. Tracey, Jr., J. W. Wells & K. O. Emery. 1950. Organic growth and sedimentation on an atoll. Jour. Geol. **58**(4): 410-425.

Marshall, S. M. 1932. Notes on oxygen production in coral planulae. Sci. Repts. Great Barrier Reef Exped. **1**: 252-258.

Mayor, A. G. 1924. Structure and ecology of Samoan Reefs. Carnegie Inst. Wash. Dept. Mar. Biol. Papers **19**: 1-25, 51-72.

Motoda, S. 1938. Quantitative studies on the macro plankton of coral reef of Palao. Nat. Hist. Soc. Sapporo Trans. **15**(95): 242-246.

———. 1940. Organic matter in sea water of Palao, South Sea. Nat. Hist. Soc. Sapporo Trans. **16**(2): 100-104.

Nordenskiold, E. 1932. The History of Biology. London: A Knopf. 699 pp.

Odum, E. P. 1953. Fundamentals of Ecology. Philadelphia: Saunders. 384 pp.

———. 1954. Ecological survey of the Savannah River area. Progress Rep., March 1953-1954 to Atomic Energy Commission. Mimeo, 42 pp.

Odum, H. T. 1953., 1954. Productivity in Florida Springs. First, Second, and Third Progress Reports to the Office of Naval Research (ditto).

Odum, H. T. & R. C. Pinkerton. 1955. Times speed regulator, the optimum efficiency for maximum output in physical and biological systems. Amer. Sci. **43**: 331-343.

Odum, H. T. & E. P. Odum. 1953. Principles and concepts pertaining to energy in ecological systems. In Odum, Fundamentals of Ecology (Chap. 4). Philadelphia: Saunders. 384 pp.

Rabinowitch, E. I. 1951. Photosynthesis and re processes. Vol. II. New York: Interscience. 120

Richards, F. A. with T. G. Thompson. 1952. The mation and characterization of plankton popula by pigment analyses. II. A spectrophotometric me for the estimation of plankton pigments. Jour. Res. **11**: 156-172.

Robinson, R. J. & G. Kemmerer. 1930. Determin of organic phosphorus in lake waters. Wis. Sci. Arts Letters Trans. **25**: 117-181.

Sargent, M. C. & T. S. Austin. 1949. Organic ductivity of an atoll. Amer. Geophys. Union T **30**(2): 245-249.

———. 1954. Biologic economy of coral reefs. B and Nearby Atolls, Part 2. Oceanography (biolo U. S. Geol. Surv. Prof. Paper **260-E**: 293-300.

Strasburg, D. W. 1953. The comparative ecology of Salarin Blennies. Ph.D. Thesis, Univ. of Ha 266 pp.

Sverdrup, H. U., M. W. Johnson & R. H. Fleming. The Oceans. New York: Prentice-Hall. 1060 p

Tamura, T. & Y. Hada. 1932. Growth of reef buil corals inhabiting the south sea islands. Sci. R Tohoku Imperial Univ. (4th series) **7**: 433-455.

Taylor, W. R. 1950. Plants of Bikini and Other N ern Marshall Islands. Ann Arbor: Univ. Mich Press. 227 pp.

Tracey, J. I., H. S. Ladd & J. E. Hoffmeister. Reefs of Bikini, Marshall Islands. Geol. Soc. A Bull. **59**: 861-878.

Utseumy, F. 1942. Lime boring algae. South Sea (Kagaku Nanyo) **5**(1): 123-128.

Verduin, J. 1952. The volume based photosynt rates of aquatic plants. Amer. Jour. Bot. **39**: 157

Vinogradov, A. P. 1953. The elementary chemical position of marine organisms. Sears Foundation Marine Research, Memoir II, 647 pp., Yale Univ.

Wells, J. W. 1951. The coral reefs at Arno Marshall Islands. Atoll Res. Bull. **9**: 1-14.

Yonge, C. M. 1930. Studies on the physiology of co I. Feeding mechanisms and food. Sci. Reports of Great Barrier Reef Expedition **1**: 13-57.

———. 1940. The biology of reef-building corals. Reports, Great Barrier Reef Exped. **1**: 353-389.

Yonge, C. M. & A. J. Nicholls. 1931. Studies on physiology of corals. V. The effect of starvatio light and darkness on the relationship between c and zooxanthellae. Scientific Reports of the G Barrier Reef Expedition **1**: (7): 177-212.

Yonge, C. M., M. J. Yonge & A. G. Nicholls. Studies on the physiology of corals. IV. The struc distribution and physiology of the zooxanthe Scientific Reports of the Great Barrier Reef Ex tion **1**: (6): 135-176.

———. 1932. Studies on the physiology of corals. The relationship between respiration in corals and production of oxygen by their zooxanthellae. Sci. ports, Great Barrier Reef Exped. **1**: (8): 213-251.

Ying, T. H. Ma. 1934. On the growth rate of Japa corals and the sea water temperatures in the Japa islands during the latest geological times. Sci. ports Tohoku Imp. Univ., Sendai, Japan **16**(3): 189.

Zwicker, B. M. G. & R. J. Robinson. 1944. The ph metric determination of nitrate in sea water wi strychnidine reagent. Jour. Mar. Res. **5**: 214-232

THE ECOLOGY OF *CONUS* IN HAWAII*

ALAN J. KOHN[1]
Department of Zoology, Yale University, and Hawaii Marine Laboratory, University of Hawaii

TABLE OF CONTENTS

INTRODUCTION

Members of the gastropod genus *Conus* (Prosobranchia: Conidae) are among the most conspicuous invertebrates on the coral reefs and marine benches that fringe the Hawaiian Islands. At least 21 species of *Conus* are known to occur in these habitats.

Investigation of these natural populations was stimulated by the existence of such a large number of closely related species in a restricted environment. This phenomenon is not unique to *Conus*, for many other genera of marine invertebrates are also characterized by large numbers of sympatric species in tropical regions. The gastropod genera *Cypraea, Mitra,* and *Terebra* are represented by 30-50 species in Hawaii (Edmondson 1946). A non-molluscan example is the snapping shrimp genus *Alpheus*, represented by 30 species in Hawaii (Banner 1953). The evolution of such genera has contributed to a marked enrichment of the tropical littoral epifauna. Here the number of species approaches ten times that of temperate regions (Thorson 1956).

Although these assemblages are well-known to systematists, no previous comparative ecological studies are known to the present writer. The objective of the study reported here was to describe the ecological niches of the species, to determine the extent of isolation between ecologically similar species, and thus to elucidate the mechanisms that permit a large number of closely related species to survive and retain their identity in a narrow environment.

* Contribution No. 113, Hawaii Marine Laboratory.
[1] Present address: Department of Biological Sciences, Florida State University, Tallahassee, Florida.

This paper is the first in a projected series reporting the results of ecological observations on natural populations of *Conus* in different areas, with emphasis on the Indo-West Pacific region. The research reported here was carried out while the author was a fellow of the National Science Foundation. Financial aid was also received from the Higgins Fund and the Director's Fund, Sheffield Scientific School.

Gratitude is expressed to G. E. Hutchinson, Sterling Professor of Zoology, Yale University, for inspiration and guidance. Special thanks for the collection of material are due Dr. A. H. Banner, Mr. C. E. Cutress, Mr. R. M. Gray, Mr. E. C. Jones, Miss Alison Kay, Mr. R. A. McKinsey, Mr. C. M. Stidham, Mr. R. Sheats, and Miss Shirley Trefz. Miss Marian Adachi assisted in many phases of the work. Members of the staff of the Hawaii Marine Laboratory rendered much helpful assistance. Appreciation is expressed to Dr. J. L. Brooks, Dr. E. S. Deevey, Dr. W. D. Hartman, Dr. R. H. MacArthur, and Dr. G. A. Riley for helpful discussion and criticism of the manuscript, and to Dr. E. Mayr for examination of preliminary data. All of the algae mentioned were identified by Dr. A. J. Bernatowicz. The writer is also indebted to Mr. W. G. VanCampen for translating the paper by Takahashi (1939). Mechanical analyses of substratum samples were carried out at the Connecticut Agricultural Experiment Station, where Dr. T. Tamura, Dr. P. Waggoner, and the staff of the Department of Soils gave much assistance. The help of Mrs. Nancy Kimball in the preparation of the figures is gratefully acknowledged. The author is especially grateful to Dr. Olga Hartman for identification of the polychaetes discussed in the section on food and feeding. Without her willing and patient attention to a collection of partly digested, fragmentary remains of polychaetes, the food analyses presented here would not have been possible.

THE GENUS *CONUS* IN HAWAII

Of 45 species of *Conus* previously reported from the Hawaiian area, 32 are known from two or more specimens collected alive and are thus considered to be valid constituents of the Hawaiian marine fauna. Ecological observations on 25 of these, listed below, will be reported in this paper. Eighteen species were collected by the writer in the subtidal coral reef (noted by *) and intertidal marine bench (noted by +) habitats to be discussed in detail below, The specific names used are those given by Kohn (1959).

Conus abbreviatus Reeve*+
Conus catus Hwass *in* Bruguière+
Conus chaldaeus (Röding)*+
Conus distans Hwass *in* Bruguière*+
Conus ebraeus Linné*+
Conus flavidus Lamarck*+
Conus imperialis Linné*
Conus leopardus (Röding)
Conus lividus Hwass *in* Bruguière*+
Conus marmoreus Linné*
Conus miles Linné*+
Conus moreleti Crosse
Conus nussatella Linné
Conus obscurus Sowerby
Conus pennaceus Born*+ (Fig. 1)
Conus pertusus Hwass *in* Bruguière
Conus pulicarius Hwass *in* Bruguière
Conus quercinus Solander
Conus rattus Hwass *in* Bruguière*+
Conus retifer Menke+
Conus sponsalis Hwass *in* Bruguière*+
Conus striatus Linné*
Conus textile Linné
Conus vexillum Gmelin*+
Conus vitulinus Hwass *in* Bruguière*

Most of the species listed are widely distributed throughout the Indo-West Pacific region. One species, *Conus abbreviatus,* is believed to be endemic to the Hawaiian archipelago. It is closely related to *Conus coronatus* Gmelin, which occurs in other areas of the central and western Pacific.

Conus nanus Broderip is here regarded as conspecific with *C. sponsalis,* but the Hawaiian populations probably constitute a valid subspecies. All specimens of *C. marmoreus* known from Hawaii agree with the description of *C. bandanus* Hwass *in* Bruguière. Most systematists consider the latter a variety of *C. marmoreus,* but the Hawaiian populations probably constitute a valid subspecies.

THE HABITAT OF *CONUS*

Marine benthic communities may be separated into two principal types, "those that tolerate or require exposure to the atmosphere and occupy stable and usually hard substrata exposed to the full force of the air with each important fall of tide," and "those that do not tolerate exposure to the atmosphere and are practically always submerged in water" (Clements & Shelford 1939: 323).

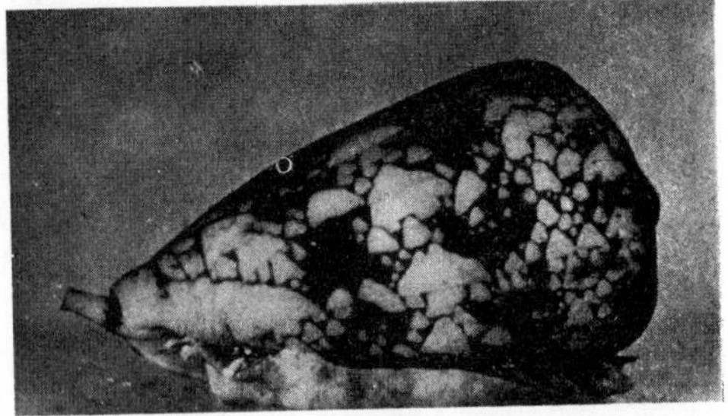

FIG. 1. Photograph of *Conus pennaceus* Born. Length of shell about 35 mm. Photograph by C. E. Cutress.

The present paper deals primarily with the ecology of natural populations of *Conus* which occupy habitats of these two types, marine benches and coral reef platforms, respectively. Notes on species occupying other habitats will also be included.

THE MARINE BENCH HABITAT

The geomorphology of emerged marine benches in the Hawaiian Islands has been discussed in detail by Wentworth (1938, 1939). These benches result from the single or combined action of processes designated by Wentworth as solution benching, water-level weathering, ramp abrasion, and wave quarrying.

SOLUTION BENCHES

In the Hawaiian Islands, solution benches occur where the shoreline is composed of reef rock and calcareous sandstone. On Oahu, this type of shore comprises 52 miles, or 31% of the coastline (Wentworth 1938). A detailed description of the characteristics and formation of solution benches is given by Wentworth (1939). A typical shore profile is shown in Fig. 2A.

Unbroken units of bench are ordinarily a few hundred feet long and 5-70 ft in width. The bench platforms are very flat. "The normal bench surface commonly shows variations of elevation of not over three to six inches in an area fifty feet wide by one hundred feet in length." (Wentworth 1939.) The outer edge of the solution bench rises more or less steeply from the water. The bench surface is ordinarily a few inches to 3 ft above mean sea level, and there is no raised rampart at the seaward margin. The bench itself is usually reef limestone, which consists of the firmly lithified skeletons of coral and calcareous algae. On the sloping outer edge, calcareous algae and, to a lesser extent, corals, contribute active building of the bench out from shore.

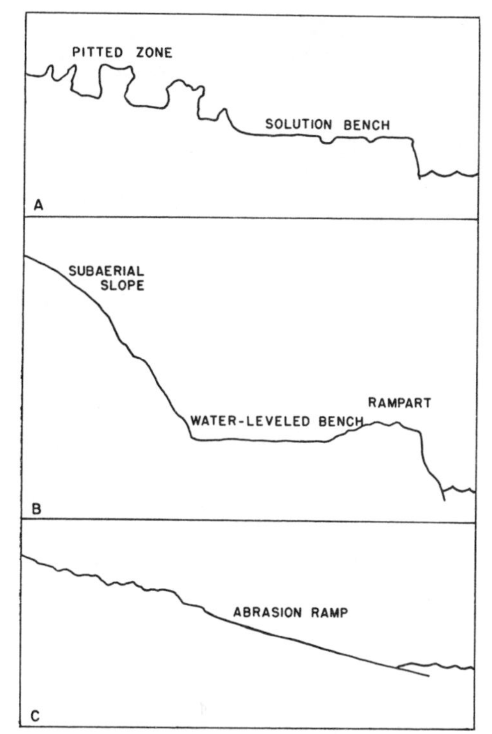

FIG. 2. Profiles of Hawaiian marine benches. A. solution bench. B. Water-leveled bench. C. Abrasion ramp bench. Modified from Wentworth (1938, 1939).

Other organisms, chiefly echinoids, are a destructive force in the same region.

The shoreward edge of the bench is sharply delimited by the front of a zone of pitted limestone (Fig. 3), ordinarily 10-50 ft in width and rising to a height of one to several ft above the bench surface. Wentworth (1939) presents evidence to show that this type of bench is formed chiefly by dissolution of the shoreward zone of pitted limestone by rain water, which collects in its pools; wave quarrying may finally produce the flat surface.

Details of solution benches selected for study are as follows:

Kahuku, Oahu (Sta. 5). The solution bench at Kahuku (Figs. 3, 4) is typical of the formation and was thoroughly investigated. It extends for several hundred feet along shore and is 40-80 ft wide. The bench surface is about one foot above mean sea level and is completely exposed for periods of up to four hours at tides of +0.2 feet[1] or less when seas are fairly calm. The coast here is exposed, and trade winds are ordinarily quite strong (about force 5). At high tide, the bench is strongly awash, and observations on it are not possible. The bench platform,

FIG. 3. Photograph of solution bench at Kahuku, Oahu (Station 5). The width of the bench (left to right) is about 60 ft. The seaward margin of the zone of pitted limestone is visible at the left.

which is of solid reef limestone with a few potholes, is covered by a well developed algal turf. Zonation of algae across the bench is present. The landward portion is characterized by *Laurencia* sp., *Sargassum polyphyllum* J. Agardh, and *Microdictyon setchellianum* Howe; the central portion, by *S. polyphyllum, S. echinocarpum* J. Agardh, and *Halimeda discoidea,* Decaisne; and the seaward portion, by *S. echinocarpum, H. discoidea* and *Dictyosphaeria cavernosa* (Forskål) Børgesen. *Lyngbya majuscula* Harvey ex Gomont and *Cladophoropsis membranacea* (C. Agardh) Børgesen are also of common occurrence.

Nanakuli, Oahu (Sta. 11). On the leeward coast of Oahu, a limestone shore with solution benches extends, with interruptions, from Nanakuli Beach to the northern end of Nanakuli town, near Maile Head (Fig. 4). The bench is generally 50 ft or less wide, although at the end of one section it is about 100 ft. This is a very short section, however, and the bench there occupies less than one-half acre. The bench platform is about one foot above sea level, but it is less often exposed than is the bench at Station 5, described above. This is due to a prevailing heavy swell in the region, so that even at low tide large waves breaking over the bench may make collecting impossible. The algal turf is as well developed here as at Station 5. *Valonia aegagropila* C. Agardh is the dominant species. *Jania capillacea* Harvey, *Sargassum* sp., and *Padina* sp. are common. Zonation of algae was not studied, but it was not obvious.

Gastropods other than *Conus* common on solution benches are *Mitra litterata* Lamarck, *Haminoea aperta Pease,* and *Cypraea caputserpentis* Linné. A number of sea anemones are abundant. The most conspicuous Crustacea are xanthid crabs, snapping shrimp, and hermit crabs. The microfauna is especially rich. Prominent are amphipods, isopods, harpacticoid copepods, polychaetes, and Foraminifera.

WATER-LEVELED BENCHES

A formation typical of palagonite tuff and weathered basalt shores in Hawaii is the water-leveled bench, the characteristics and origin of which are dis-

[1] All tidal data are referred to O datum = mean lower low water and are from Coast and Geodetic Survey Tide Tables.

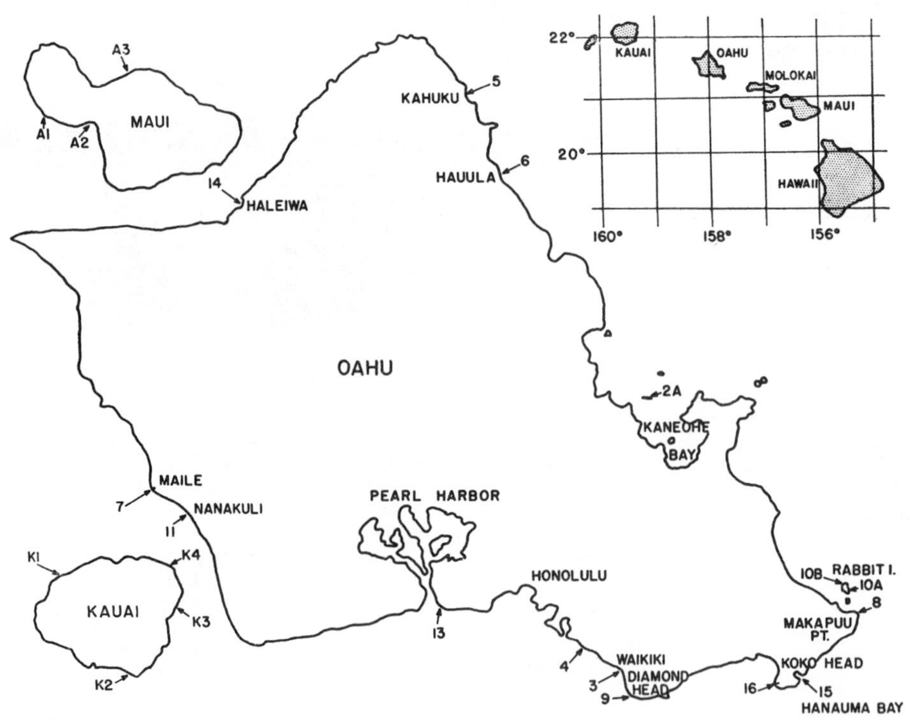

FIG. 4. Maps of the Hawaiian Islands showing location of stations. Stations lacking locality data on maps are: 2A, Sand (Ahuolaka) Island; 4, Ala Moana; 13, Ft. Kamehameha; A1, Olowalu; A2, Maalaea Bay; A3, Lower Paia; K1, Milolii; K2, Poipu; K3, Kapaa; K4, Moloaa.

cussed in detail by Wentworth (1938). Benches of this type may be 10-12 ft above sea level on exposed coasts, but are only 2-3 ft above sea level in sheltered places. Most of them are less than 100 ft wide. A steep subaerial slope landward of the bench is characteristic (Fig. 2B).

Unlike the solution bench, water-leveled benches are characterized by a well-developed rampart (Fig. 2B) of slightly higher rock at the seaward edge. The outer face of this zone is a steep cliff, characteristically occupied by echinoids. Landward of the rampart is the water-leveled bench proper. Its surface may be quite smooth or bear vertical irregularities of the order of a few inches to a foot. These are often due to differences in hardness of the dipping beds of tuff on which this type of bench is most often found.

Water-leveled benches which were given special study were as follows:

Lower Paia, Maui (Sta. A3). A fairly typical water-leveled bench near Lower Paia, Maui (Figs. 4, 5) was visited on 5 August 1956. The shore profile at this station is similar to that shown in Fig. 2B. The seaward face of the rampart zone is steep. The rampart zone (Fig. 5) averages about 15 ft wide. Its surface is irregular and is covered by a luxuriant algal turf, composed of many species. The water-leveled part of the bench (Figs. 2B, 5) is about 15 ft wide and quite smooth. It bears a low algal turf which binds some sand. The tide was extremely low (—0.2 ft) during the period of observation, and this region was quite dry. Landward of it are large tide pools 1-2 ft deep (Fig. 5).

FIG. 5. Photograph of water-leveled bench at Lower Paia, Maui (Station A3) from subaerial slope. The 100-sq ft quadrat is outlined. Tide pools and rampart are also visible.

Milolii, Kauai (Sta. K1). Two broad marine

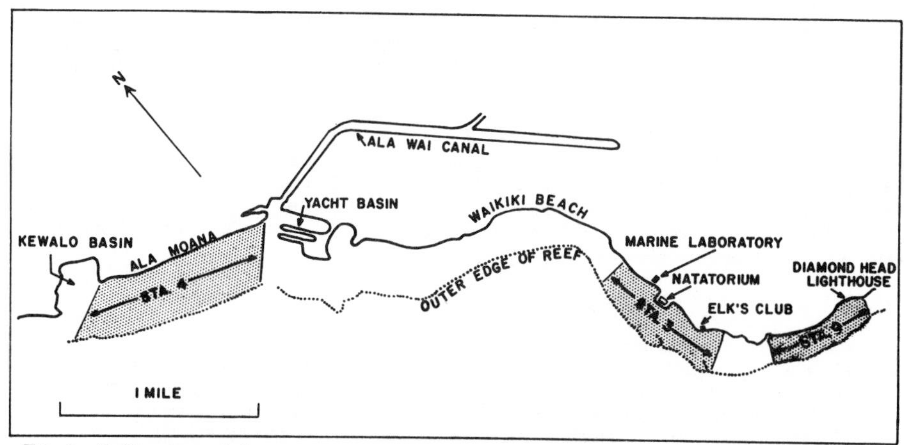

FIG. 6. Outline map of the south shore of Oahu, showing the nearly continuous fringing reef, Stations 3, 4, and 9, and landmarks.

benches fringe the shore at each end of a small beach, just east of Milolii Valley on the Napali (north) coast of the island of Kauai (Fig. 4). This region is accessible only from the sea. Four days were spent in the area in August 1955. The surface of the bench studied is about one foot above mean sea level. At minus tides it is completely exposed to the air. The benches reach a maximum width of about 200 ft. The algal turf of the platform is very low and does not present a "healthy" appearance. The dominant species is *Wurdemannia miniata* (Lamarck and DeCandolle) Feldmann and Hamel. Other species observed were *Dictyosphaeria* sp., probably young *D. cavernosa* (Forskål) Børgesen, *Gelidiella* sp., *Jania capillacea,* and *Valonia aegagropila.* At about 110 ft from shore a low pink encrusting alga, probably *Porolithon* sp., is present, and it continues to the seaward edge. The differences in flora between Station Kl and the solution benches may be due to the absence of heavy surf over most of the former.

ABRASION RAMP BENCHES

The distribution and formation of abrasion ramp benches on Hawaiian shores has not been treated in detail, but Wentworth (1938) includes a photograph of one. The formation is found on limestone and/or tuff shores. Abrasion is caused by washing of sand and gravel back and forth across the bench by waves. The result is a rather smooth sloping surface extending from below the low tide line often to several feet above it (Fig. 2C). A dense but low algal mat, usually of a varied flora, is typically present. Abrasion ramp benches are limited in size and constitute only a minor portion of shorelines. Three such benches were visited, Station 10A, on the southern shore of Rabbit (Manana) Island; Station 16, near Kawaihoa Point, on the eastern end of Maunalua Bay, Oahu; and Station A2, west of Kihei on Maalaea Bay, Maui (Fig. 3).

THE SUBTIDAL CORAL REEF PLATFORM HABITAT

More than half of the shoreline of Oahu and comparable portions of some of the other Hawaiian Islands are fringed by rather narrow coral reefs. These reefs are characterized by a predominantly sandy substratum. Living coral, patches of bare limestone and coral rubble also comprise varying proportions of the substratum, but the most actively growing corals and coralline algae are typically found at the outer edges and on the reef slopes. The reef platforms are typically subtidal and variable in depth, being usually 2-10 ft below mean sea level. Occasionally portions of the reef platform are exposed at low spring tides. The platform of some reefs is raised at the outer edge, but this rampart is often not well developed. A general discussion of Hawaiian coral reefs is given by McCaughey (1918). A typical reef fringes the south shore of Oahu, interrupted only by dredged channels and the drainage of streams. Investigation was concentrated on sections of this reef near the laboratory, noted as Stations 3 and 9 (Fig. 6). Additional collections were made at Station 4.

Brief descriptions of Stations 9 and 3, based on field notes, follow. They emphasize, respectively, the characteristics of the reef normal to the shore and parallel to the shore. The ecological notes on *Conus* which are included will serve as an introduction to the quantitative data presented below.

Diamond Head, Oahu (Sta. 9). The fringing reef on the Honolulu side of Diamond Head may be roughly divided into four zones. Zone (1) is an intertidal area of moderate surf which breaks over a substratum of detrital limestone. Landward of this zone, above the high tide limit, is a narrow sand beach. A dense growth of many species of algae occurs in zone (1), except in the bare limestone surge channels and tidal pools. This region takes the form of an abrasion ramp bench, 30-50 ft wide, which slopes to seaward. No *Conus* were observed here, but this is perhaps due to the convenience of the area to shell collectors.

Zone (2) comprises the broad submerged reef platform, 2-6 ft below 0 tide datum. The substratum is characterized by areas of coral rubble, coral heads and sand, more or less intermingled. This region comprises most of the width of the reef, which reaches a maximum of about 1,500 ft. The areas with substratum of coral rubble are extensive. Conspicuous benthic algae are absent. These areas are also barren of large gastropods, probably because of lack of shelter provided by the pieces of rubble, which are readily moved back and forth by wave action. Scattered coral heads, usually bearing only a small colony of living coral, harbor a variety of invertebrates, including *Conus pennaceus* on the sand beneath. *C. rattus* and *C. imperialis* are occasionally epifaunal on dead coral. Sandy areas interspersed with reef limestone outcrops form the typical habitat of *Conus flavidus* and *C. lividus,* which are usually epifaunal. *C. abbreviatus* is also found, usually burrowing in sand in these areas.

Zone (3), which is variable in width, is characterized by large areas of dead coral reef, which appear to be eroding. The surface areas of these regions are often near the 0 tide datum and hence are often dry at low tide. At high tide, surf over these areas is heavy. *Conus rattus* is occasionally found, and *C. ebraeus* sometimes occurs on the vertical edges of the eroding coral areas, or in small crevices of sand below. Channels 4-6 ft deep and normal to the shore separate these areas near the outer edge of the reef and broaden into extensive areas with sand substratum just inshore. *Conus pulicarius* is found typically beneath the surface of the sand. Reef fishes abound about the steep edges.

Zone (4) is the zone of heavy surf at the outer edge of the reef, where coral flourishes. Environmental conditions precluded extensive observations.

The most common alga at Station 9 was *Lyngbya majuscula. Sargassum, Hypnea, Codium,* and others were also common.

Waikiki, Oahu (Sta. 3). The wide fringing reef at Station 3 appeared to offer a wide variety of microhabitats. The collecting area extended about 1,000 ft north, and about 2,500 ft south, from the Waikiki Branch of the Hawaii Marine Laboratory (Fig. 6). The width of the reef is about 600-1,000 ft. Inshore areas of the southern portion, to about the Elks Club, are characterized by more or less abundant coral rocks set in sand or sand-rubble substratum. *Conus lividus* and *C. flavidus* are often common. *C. abbreviatus* sometimes occurs in the larger sandy areas between coral rocks. Much of the reef area to seaward is eroding dead coral reef, as at Station 9, and *Conus* is generally absent. At the seaward edge of the reef are sandy areas, with some limestone outcrops. The reef slope is gentle and quite sandy. Although the area appears suitable, *Conus* occurs only occasionally. This area can be visited only on calm days, and the usually heavy surf may make this portion of the habitat unsuitable for the snails. *C. flavidus* and *C. distans* have been found however. The latter typically occurs at reef edges and generally in rougher water than the other species of *Conus.*

From the Elks Club (Fig. 6) to just north of the laboratory, the substratum is quite different. An inner reef, shoreward of a dredged channel, is limestone bench and rubble and devoid of *Conus.* The outer reef is generally deeper, being 3-8 ft below MLLW. Sand is the dominant substratum, but coral heads are abundant and there are some rubble areas. *C. lividus, C. flavidus,* and *C. ebraeus* occur but are rather sparse. *C. distans* is sometimes found at the outer edge.

The northern portion of the area sampled supported a richer fauna. The substratum was of coral heads and rocks, reef limestone, and rubble areas. A detailed discussion of this region was given by Edmondson (1928). *Conus pennaceus* occurred under rocks in sand. Epifaunal species included *C. rattus, C. flavidus, C. lividus, C. ebraeus,* and *C. abbreviatus,* none of which was uncommon. This area especially provided sites for attachment of *Conus* egg capsules. Subsequent to the investigation, however, much of this area was dredged, and the inshore portion covered by a sand beach, to create an area for swimming.

Other reefs studied were essentially similar to the two described. Helfrich & Kohn (1955) and Kohn & Helfrich (1957) discussed the characteristics of Station K3. Extensive collecting was also carried out at reefs on Oahu at Maile (Sta. 7) and Ala Moana (Sta. 4). The location of the more important reef stations is shown in Fig. 4.

LIFE HISTORY OF *CONUS*

An account of spawning and larval development of *Conus* in Hawaii is in preparation and will be published elsewhere. Therefore, only information of ecological importance will be presented here.

Spawning Site and Season

Egg capsules of at least 12 species of *Conus* have been collected in Hawaii, chiefly by Ostergaard (1950) and by the writer and colleagues.

Coral reef platforms, but not marine benches, provide suitable attachment sites for egg capsules of *Conus.* Of 36 egg masses collected in the field, 29 were recorded from reef platforms. An almost complete absence of records from marine bench habitats suggests that spawning is unsuccessful there. This is probably due to the absence of protected pools in which egg capsules may be deposited without being subject to desiccation at low tide and/or torn away by heavy surf at high tide. Recruitment of bench populations is probably from pelagic veliger larvae which have been carried from other areas and are washed onto marine benches in condition to settle and assume the benthic mode of life.

All of the capsules were found between the months of February and August, although search for them was not confined to, or emphasized during, this period. The data suggest that most species of *Conus* spawn during about the same part of the year. The spawning season of most species for which more than one egg mass has been collected is rather extended over the period between the months cited. The most complete data are for *C. pennaceus,* of which 12 egg masses were collected, all in the months of May, June, July and August. The data are probably sufficient to establish the breeding season as continuing through these months.

REPRODUCTION AND LARVAL DEVELOPMENT

As is typical in the Prosobranchia, the sexes are separate in *Conus.* The male possesses an extensible penis. Copulation was not observed. In spawning, eggs are released from the genital aperture and pass ventrally over the foot in a temporary groove to the prominent aperture of the nidamental gland on the sole. There the capsular material is extruded, enclosing a number of eggs. The capsule is attached to a hard substratum, typically under a coral rock, or to the underside of the rock itself. Illustrations of the egg capsules of *Conus* are given by Ostergaard (1950). A number of capsules (3-78, in 12 species studied) are deposited to form a cluster. The number of eggs per capsule varied from 40 to 11,400 in 5 species studied.

In 4 species studied (*C. vitulinus, C. abbreviatus, C. imperialis, C. quercinus*), the trochophore stage is entered at 2-6 days, and the veliger stage at 6-10 days, after spawning. Larvae hatch as veligers about two weeks after spawning. These observations are in agreement with those of Ostergaard (1950), who also reported development of 4 other species, which hatched 12-16 days after spawning. Almost all of the eggs in a capsule develop completely, and no nurse eggs were observed.

With the exception of one species, the length of the pelagic stage could not be determined. The maximum survival time of free-swimming veligers in the writer's laboratory was 9 days. Metamorphosis was observed only in *Conus pennaceus,* which has an extremely short free-swimming stage of less than one day. On the second day after hatching, metamorphosis is virtually completed and the young snail begins to craw about on its foot. These juveniles survived for periods of up to 20 days, but no significant growth was observed after hatching. The nature of the food at this stage is unknown. Protozoa abounded in the cultures. Thorson (1946) concluded that all prosobranch larvae known from the Oresund feed on phytoplankton, and he calculated the theoretical maximum diameter of the food to be 5-45μ. The mouths of *Conus* veligers measured were of about the same diameter as the esophagus of the smaller larvae measured by Thorson. Thus the larvae of *Conus* probably depend for food on phytoplankton, nannoplankton and detritus. Examination of squash preparations of *C. pennaceus* a few days after settling revealed the presence of radula teeth. These differ in form from the adult teeth, being shorter in proportion to the thickness, and they are probably not functional. Thus neither the method of feeding nor the food is known at this stage of the life history.

POST-LARVAL DEVELOPMENT AND GROWTH

It was possible to study post-larval development only in *Conus pennaceus.* At hatching, the larvae of this species are several times as large as those of other species, measuring about 1.3 mm in shell length.

A rough estimate of the rate of post-larval growth was obtained in the following manner. Four clusters of egg capsules of *Conus pennaceus* were collected in a large tide pool adjacent to Station 9 on 13 August 1955, and additional clusters were observed but not removed. Adults collected at the same time ranged from 33 to 37 mm in shell length. On 30 November, 3.5 months later, 8 specimens of *C. pennaceus* were collected in the same tide pool. Of these, 6 were probably hatched from egg capsules the previous summer (Fig. 7). On 27 December, 41 specimens of *C. pennaceus* were collected in this tide pool.

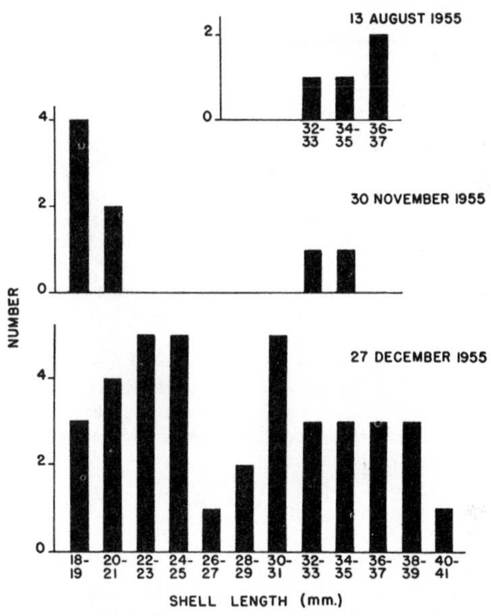

FIG. 7. Length-frequency distribution of *Conus pennaceus* at Station 9, 13 August-27 December 1955.

The length-frequency distribution of the December population is also shown in Fig. 7. Although bimodality suggesting two age classes is evident, it is quite possible that all of the specimens were spawned the previous summer. The minimum at 26-27 mm may not indicate separation of two age classes, because of the large number of specimens of greater length collected. If these were older specimens, most of them probably would have been collected from the tide pool on previous occasions. As noted above, the breeding season is long. The mean shell length of first age-class individuals in November was 19 mm, with S.D. = 6.3 mm. In December, mean shell length was 28 mm, with S.D. = 7.7 mm. The mean mean growth rate was thus 5-6 mm/month during the first 3.5 months. During the next month the mean increment was 5 mm.

It was not possible to obtain any other growth data of this type, since such isolated populations are exceptional. Shells of a number of specimens of *Conus ebraeus, C. abbreviatus,* and *C. sponsalis* from a marine bench (Sta. 5) were marked with a diamond point vibrator and returned to their natural habitat. Of these, three specimens, all *C. ebraeus,* were recovered after 133-221 days following release. The growth increments ranged from 0.3-0.6 mm shell length and 0.2-0.4 g wet weight per month. Although only these data are available, they presumably give the correct order of magnitude of growth of older individuals.

If the growth rates cited hold for other species, it may be concluded that, in species comparable in size to *Conus pennaceus* and *G. ebraeus* (Fig. 9), several millimeters per month in shell length are added during the first few months, the rate later falling off to a few tenths of a millimeter per month.

ABUNDANCE AND POPULATION DENSITY OF *CONUS*

Darwin (1859: 319) pointed out that most species of animals are characterized by being rather rare. Since species of *Conus* are not exceptions, and because time available for collecting was limited, several of the less abundant species are represented by rather small samples. In order to determine the ecological relationships of the species, the data presented in this report have therefore been subjected to appropriate statistical analyses.

As MacArthur (1957) has shown, the expected abundance of the rth rarest species in the community of a single habitat which has been adequately sampled, and in which ecological niches are nonoverlapping and continuous, is

$$\left(\frac{m}{n}\right) \sum_{i=1}^{r} \left(\frac{1}{n-i+1}\right) \qquad (1)$$

where m = the total number of individuals, n = the number of species, and i = the species rank. All of the data on abundance of *Conus* have been presented in comparison with the distribution expected according to this theory.

THE MARINE BENCH HABITAT

Solution Benches. The population of *Conus* at Station 5 was the densest stable population studied on Oahu. Eight species were collected, of which *C. ebraeus* was the most abundant. In 61 quantitatively sampled 100-sq ft quadrats, four species were collected. The relative abundance of species is shown in Fig. 8. If only data from the quantitatively sampled areas are included (Fig. 8A) a homogeneous population or single community is indicated by agreement

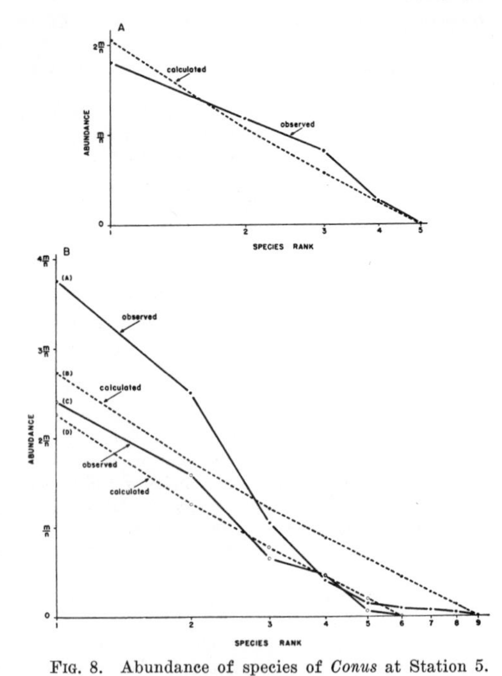

FIG. 8. Abundance of species of *Conus* at Station 5. A. Data from quantitatively sampled quadrats. m = 136, n = 4. Species rank: 1 = *ebraeus,* 2 = *abbreviatus,* 3 = *sponsalis,* 4 = *chaldaeus.*
B. Curves (A) and (B), data from all collections at Station 5. m = 500, n = 8. Curves (C) and (D), data from all collections, but with exclusion of three species characteristic of other habitats. Explanation in text. m = 484, n = 5. Species rank: (A) and (B) 1 = *ebraeus,* 2 = *abbreviatus,* 3 = *sponsalis,* 4 = *chaldaeus,* 5 = *lividus,* 6 = *flavidus,* 7 = *rattus,* 8 = *pennaceus;* (C) and (D) 1 = *ebraeus,* 2 = *abbreviatus,* 3 = *sponsalis,* 4 = *chaldaeus,* 5 = *rattus.*

of the observed curve with the expected distribution for the community of a uniform habitat.

Curve (A) in Fig. 8B is based on summation of results of all 18 collecting trips made to Station 5. The slope is steep, since common species are too common and rare species too rare. The curve is therefore not in good agreement with the theoretical distribution (Curve B), and a heterogeneous population, or the inclusion of occupants of more than one habitat, is indicated. The heterogeneous aspect of the total population may be explained as follows: Three of the species included occur more typically in habitats other than the solution bench platform. All three, *Conus pennaceus, C. flavidus,* and *C. lividus,* are typically subtidal species, which are rarely exposed by receding tide, and they occur much more commonly where such a habitat is provided, as will be shown below. If the abundance curve is plotted without these species, the resulting line (C) is considerably straighter and approaches the theoretical curve (D) more closely. It is therefore apparent that the sparse occurrence of three species which are more typical residents of a different habitat contributes to the heterogeneity shown by the curve (A) for total abundance at Station 5.

The mean density of *Conus* on this bench is 2.2 individuals/100 sq ft, based on the 61 quantitatively sampled quadrats. The mean population density of species censused in quadrats is shown in Table 1A.

TABLE 1. Population Density of Species of *Conus* at Two Marine Bench Stations.

Species	Mean number per 100 sq. ft. (=9.3m.²)
A. Station 5	
Conus ebraeus	1.02
Conus abbreviatus	0.66
Conus sponsalis	0.41
Conus chaldaeus	0.15
	2.24
B. Station K1	
Conus abbreviatus	0.85
Conus ebraeus	0.55
Conus sponsalis	0.48
Conus catus	0.24
Conus chaldaeus	0.18
Conus rattus	0.09
Conus flavidus	0.03
Conus retifer	0.03
	2.44

The mean biomass of all species was calculated to be about 0.6 g dry organic matter/100 sq ft, or 0.065 g/m².

The species differ in size. Length-frequency distributions are shown in Fig. 9. The population is essentially an adult one, and juvenile specimens are rarely found. Despite numerous collecting trips at all times of year, no egg capsules of *Conus* were ever found.

At Station 11, nine species of *Conus* were collected, of which *C. sponsalis* was the most abundant. In eight 100-sq ft quadrats, quantitatively sampled in September, four species were present, with mean density of 5.5 individuals/100 sq ft and abundances as shown in Fig. 10. The curve shows that the observed number of *C. sponsalis* (38) is much higher than that expected in a homogeneous population containing the observed numbers of the other species.

An even denser population was observed on 29 November 1955. At this time, counts of two areas, each of but one sq ft, were 3 and 7 individuals of *Conus sponsalis*. No other species were present. This abundance (= 500/100 sq ft) was present only on the extremely wide area of bench described above.

This high density of *Conus sponsalis* is believed to be related to the fact that the individuals were much smaller (mean length 13.5 mm) than elsewhere (mean length 22.1 mm at Sta. 5). Length-frequency distributions of specimens collected at Station 11 in September and November are shown in Fig. 11A. The September population is unimodal. However, the

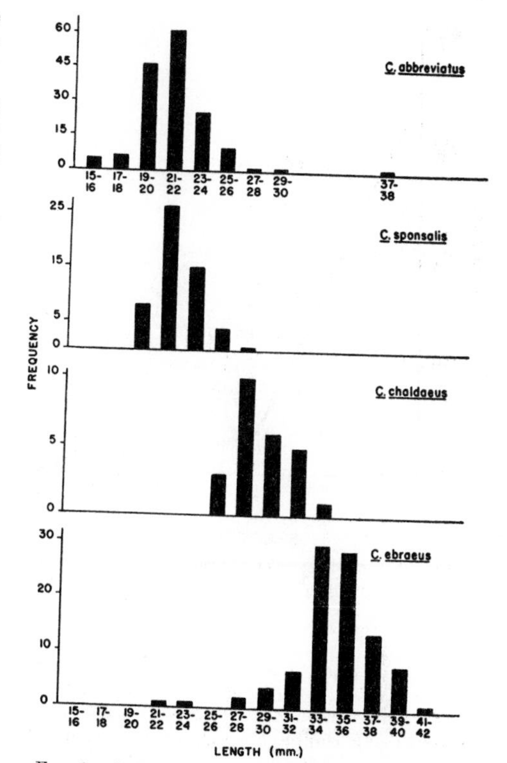

FIG. 9. Length-frequency histograms of the dominant species of *Conus* at Station 5.

November population is bimodal, with an absence of individuals of 12-13 mm. This suggests that a new age class of 6-11 mm individuals has been added to the population. Unfortunately, no information is available on the reproductive cycle of *C. sponsalis*. However, egg capsules of many species of *Conus* are found in summer and if eggs were laid in August, hatching might take place in September, with settling of pelagic veliger larvae in October giving rise to a dense population of juvenile individuals in November. Although no egg capsules were found at Station 11, the bench is interrupted by rather deep, somewhat protected pools, which may provide suitable spawning sites. Alternatively, larvae may arrive in numbers from other areas in settling condition.

Population density varied greatly in the area of Station 11. Particularly densely populated sections were quantitatively sampled by the transect method. In these regions, two transects of eight quadrats gave the mean density of 5.5 individuals (of four species)/100 sq ft. The maximum observed density of about 500 juveniles/100 sq ft has been mentioned.

On other occasions, searches of 0.63 and 1.25 man-hrs resulted in only one and three specimens, respectively. Since the time required to sample a 100-sq ft quadrat was usually 5 minutes (0.08 hr), this

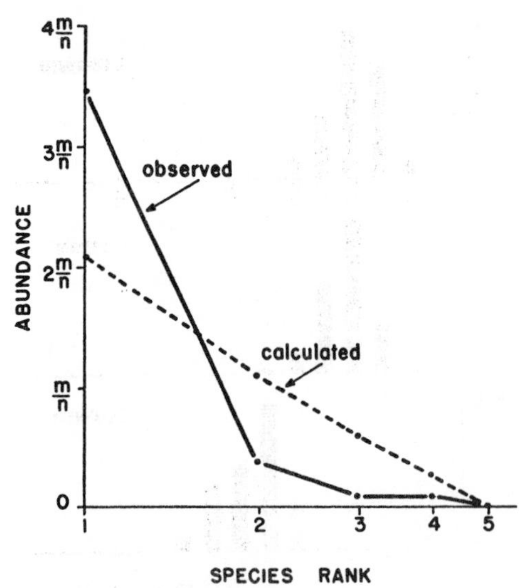

FIG. 10. Abundance of species of *Conus* at Station 11. Data from quantitatively sampled quadrats. m = 44, n = 4. Species rank: 1 = *sponsalis,* 2 = *chaldaeus,* 3 = *ebraeus,* 4 = *rattus.*

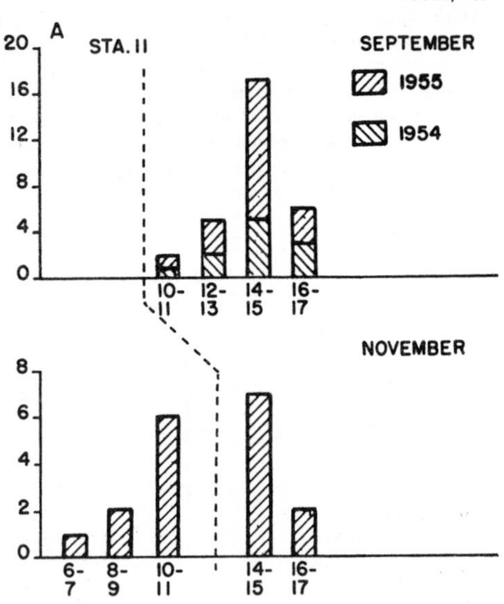

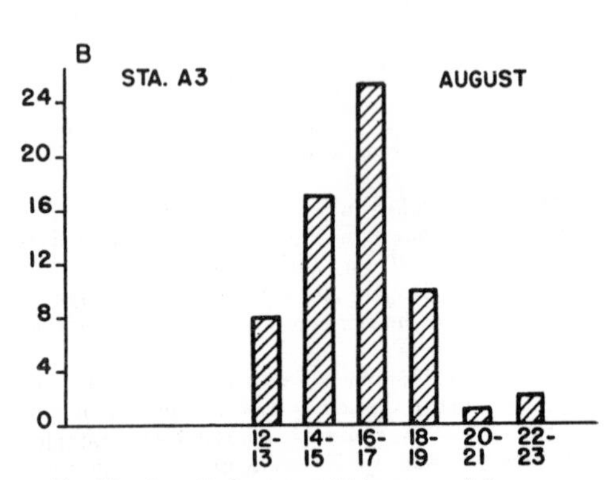

FIG 11. Length-frequency histograms of *Conus sponsalis.* A, Station 11. B, Station A3.

factor can be used to convert time-relative to space-relative density. A search of 0.63 man-hr would thus cover 7,900 sq ft and 1.25 man-hrs 15,600 sq ft. Corresponding densities are 0.01 and 0.02 individuals/100 sq ft, respectively.

Water-leveled Benches. Six species of *Conus* were collected at Station A3, of which *C. sponsalis* was by far the most abundant. The abundance of species is plotted in Fig. 12. Fig. 12A represents total abundance, and 12B the abundance in a single quantitatively sampled 100-sq ft quadrat on the water-leveled part of the bench (Fig. 4). Both curves show an obvious inflection point, which is caused by the relatively great abundance of *C. sponsalis.* Disagreement with the calculated theoretical distribution suggests a heterogeneous population.

Density on the water-leveled bench at Station A3 was second only to that in the two 1-sq ft quadrats sampled at Station 11. At Station A3, only the single quadrat was sampled. However, this represented a considerable fraction of the total available area (Fig. 4). One specimen of *Conus flavidus* and 25 *C. sponsalis* were present in this area.

The length-frequency distribution of *Conus sponsalis* at Station A3 is unimodal (Fig. 11B). The population is probably an adult one, although the individuals are not as large as those at Station 5. The relative superabundance of *C. sponsalis* cannot therefore be ascribed to an influx of first age class juveniles. Since the observations were made on 5 August, during the probable spawning season, this is even less likely.

Marine benches do not in general provide suitable sites for deposition of egg capsules of *Conus.* However, a specimen of *C. catus* was collected at Station A3 in spawning condition, as evidenced by the deposition of an egg capsule which was attached to the shell after collection.

Station K1. At Station K1, all *Conus* present in 33 quadrats, each of 100 sq ft, were counted. Eight species were present in the area sampled, of which the most abundant was *C. abbreviatus.* Relative abundance of speciės is shown in Fig. 13. The distribution calculated from Equation (1) is also included, and the data are seen to be in excellent agreement with it. This supports the observation that the habitat is a rather uniform one, with a homogeneous population, or single community, of *Conus.*

The solution bench at Station 5 was the only area

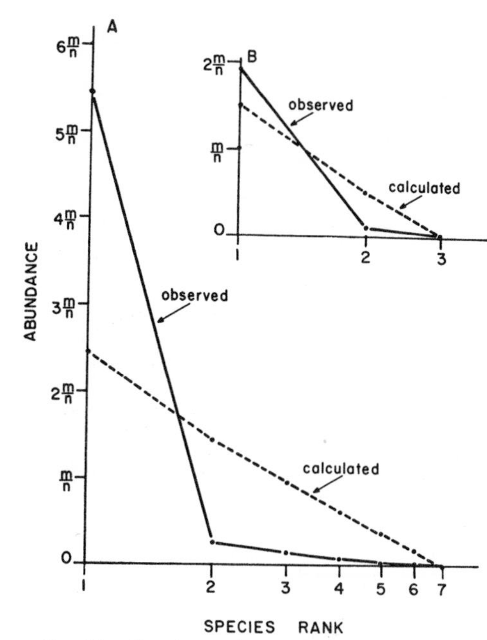

FIG. 12. Abundance of species of *Conus* at Station A3. A, Total data. m = 166, n = 6. B, Data from quantitatively sampled quadrat. m = 26, n = 2. Species rank: A. 1 = *sponsalis,* 2 = *rattus,* 3 = *catus,* 4 = *flavidus,* 5 = *abbreviatus,* 6 = *chaldaeus;* B. 1 = *sponsalis,* 2 = *flavidus.*

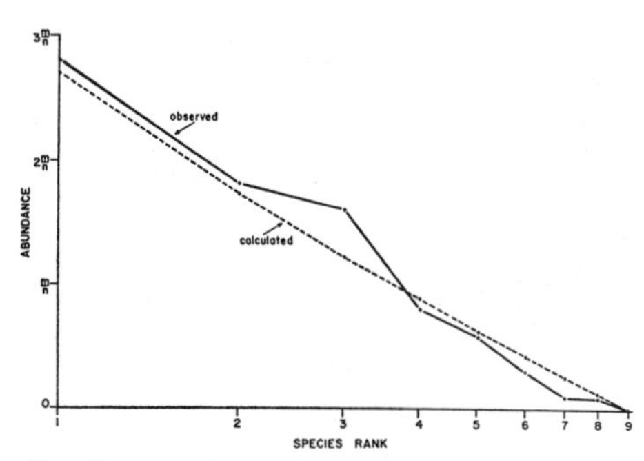

FIG. 13. Abundance of species of *Conus* at Station Kl. Data from quantitatively sampled quadrats. m = 81, n = 8. Species rank: 1 = *abbreviatus,* 2 = *ebraeus,* 3 = *sponsalis,* 4 = *catus,* 5 = *chaldaeus,* 6 = *rattus,* 7 = *flavidus,* 8 = *retifer.*

sampled more intensively than Station K1. The three most abundant species are the same at the two sites, although the order is different. The mean density of all *Conus* species at Sta. K1 is 2.44 individuals/100 sq ft, based on the 33 quantitatively sampled quadrats. The mean density of each species is shown in Table 1B. The mean density of *C. abbreviatus* is higher, while that of *C. ebraeus* is lower, than at Station 5. The densities of *C. sponsalis* and *C. chaldaeus* are about the same at the two stations.

Abrasion Ramp Benches. The rather smooth sloping surface and attendant wave action of abrasion ramp benches provide a rather unfavorable habitat for large gastropods. However, some shallow crevices provide shelter at high tide. A total of but 47 specimens of seven species of *Conus* were collected in five field trips to abrasion ramp benches. On all such benches studied, *C. rattus* was the most abundant species. *C. sponsalis* and *C. abbreviatus* were relatively common, and *C. chaldaeus, C. flavidus,* and *C. lividus* were also present.

SUMMARY

Data on abundance and population density of *Conus* on all marine benches studied are summarized in Table 4. Between six and nine species are found in such habitats. However, the figures in the fourth column of Table 2 probably indicate the number of species comprising a homogeneous population, at least at the more thoroughly studied stations. On very narrow water-leveled and solution benches, *C. sponsalis* is the dominant species, and only 1-3 other species may be present. Wider benches are occupied by proportionally greater numbers of species. As will be shown below, the species are then distributed non-randomly across the bench platform from shore to seaward edge.

TABLE 2. Summary of Abundance and Population Density of *Conus* on Marine Benches

Station	Width of Bench (ft.)	Total Number of Species	Number of Species in Quadrats	Number of Quadrats Sampled	Density (No./100 sq. ft.)	Most Abundant Species
Solution Bench Stations						
5	60	8	4	61	2.24	*ebraeus*
11	40	9	4	8	5.5 (0.01*-500)	*sponsalis*
Water-leveled Bench Stations						
K1	200	9	8	33	2.44	*abbreviatus*
A3	15	6	2	1	26	*sponsalis*
Abrasion Ramp Bench Stations						
Total	...	6	..	0	1.0*	*rattus*

*Calculated from time-relative density.

The species of greatest abundance is variable among benches of similar, as well as different, geological origin. Abrasion ramp benches were not studied in detail, hence data from them are combined in Table 2. However, *Conus rattus* was the most abundant species at all three such benches visited, and it may be termed typical of this formation. This species was of minor importance on both solution benches and water-leveled benches.

Sampling of a large number of 100-sq ft quadrats at Stations 5 and Kl indicated that a fairly stable population of 2-2.5 individuals/100 sq ft may be expected on solution benches and water-leveled benches. Other areas sampled quantitatively were those in which extremely large populations were observed by inspection. Conversion of time-relative to space relative density suggests a lower limit of population density on marine benches of about 0.01 individual/100 sq ft.

THE SUB-TIDAL REEF PLATFORM HABITAT

Inspection of a number of coral reef platforms led to the conclusion that the abundance and distribution of *Conus* in these habitats are characterized by patchiness. This not unexpected phenomenon must be considered in interpreting the data presented in this section and in the following one on distribution.

In Figs. 14-17, data on abundance at the four thoroughly sampled reef stations are presented. No one species is most abundant on all of the reefs. At each station, however, either *Conus flavidus* or *C. lividus,* or both, represent one or both of the two most abundant species. This is in marked contrast to marine bench stations where, as has been shown above, these two species occur infrequently.

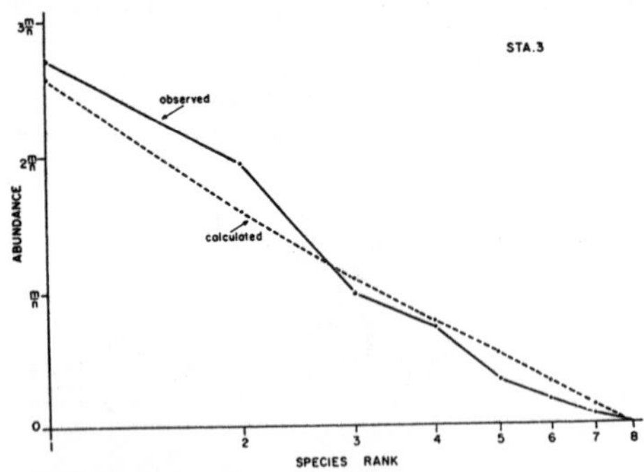

FIG. 14. Abundance of species of *Conus* at Station 3. m = 182, n = 9. Species rank: 1 = *flavidus,* 2 = *lividus,* 3 = *ebraeus,* 4 = *abbreviatus,* 5 = *rattus,* 6 = *imperialis,* 7 = *striatus.*

At only one of the adequately studied reef stations is the most abundant species neither *Conus lividus* nor *C. flavidus.* At Station 7, the most abundant species was *C. sponsalis.* This is attributed to substratum factors, which will be discussed in detail below.

At all stations, agreement of the results of analysis of abundance data with the curve calculated from Equation (1) is sufficient to justify the conclusion that the subtidal reef platform constitutes a single rather than composite habitat, which supports a homogeneous community or interspecific population of *Conus.* The striking agreement of the observed and calculated curves for Station 7, which is the closest of any census thus far analyzed in this manner (MacArthur, personal communication), is probably fortuitous.

Population Density. Population density of *Conus* was more difficult to study directly on subtidal reef

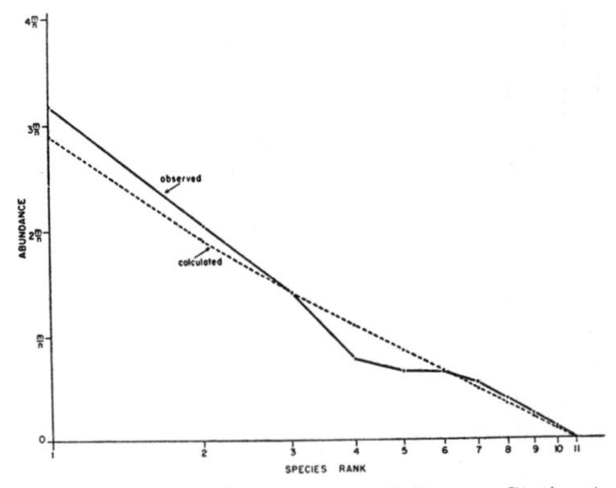

FIG. 15. Abundance of species of *Conus* at Station 4. m = 78, n = 10. Species rank: 1 = *lividus,* 2 = *flavidus,* 3 = *abbreviatus,* 4 = *imperialis,* 5 = *ebraeus,* 6 = *rattus,* 7 = *vitulinus,* 8 = *sponsalis,* 9 = *chaldaeus,* 10 = *vexillum.*

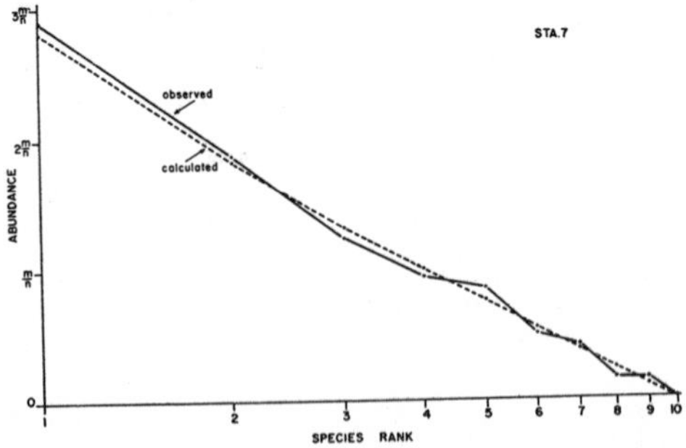

FIG. 16. Abundance of species of *Conus* at Station 7. m = 182, n = 9. Species rank: 1 = *sponsalis,* 2 = *lividus,* 3 = *ebraeus,* 4 = *flavidus,* 5 = *pennaceus,* 6 = *abbreviatus,* 7 = *rattus,* 8 = *distans,* 9 = *chaldaeus.*

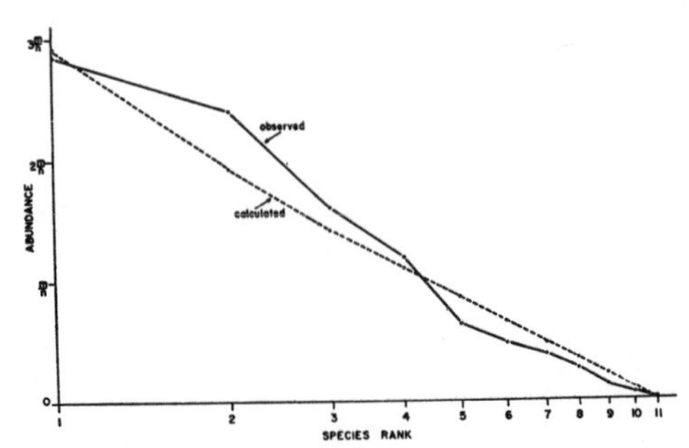

Fig. 17. Abundance of species of *Conus* at Station 9. m = 187, n = 10. Species rank: 1 = *flavidus,* 2 = *pennaceus,* 3 = *lividus,* 4 = *abbreviatus,* 5 = *pulicarius,* 6 = *ebraeus,* 7 = *imperialis,* 8 = *vexillum,* 9 = *rattus,* 10 = *marmoreus.*

platforms than on marine benches. This was due both to technical difficulties and to the obvious patchiness of the populations. During most field trips,

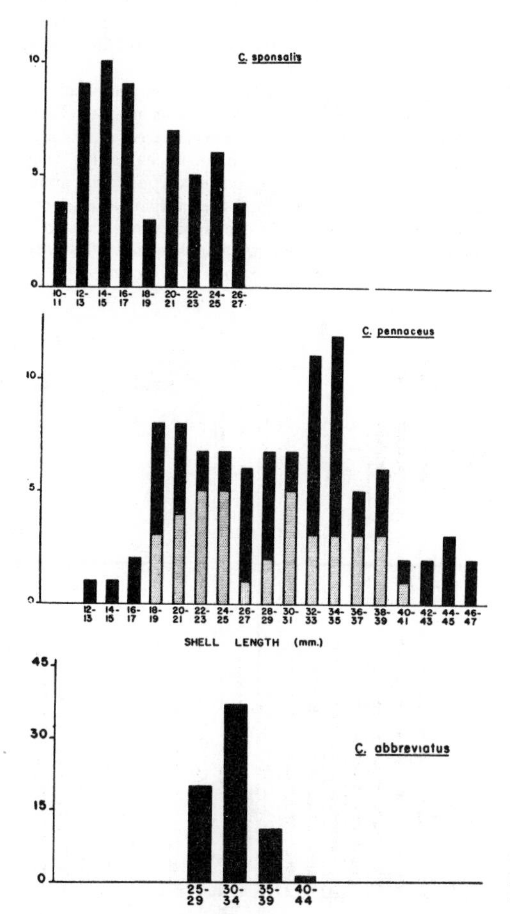

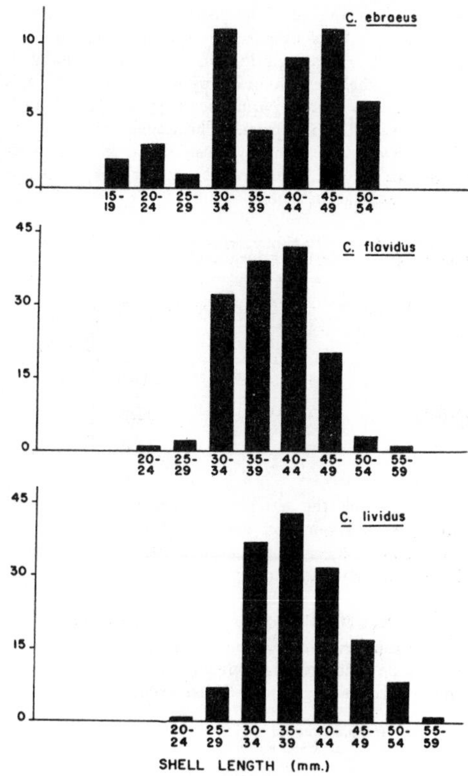

FIG. 18. Length-frequency distribution of *Conus* species at four reef stations. Stippled portion of *C. pennaceus* histograms represent collection at Station 9, 27 December 1955. See Fig. 7.

collecting efforts were of necessity concentrated in areas of greatest abundance. Conversion of time-relative to space-relative density would thus give values considerably in excess of the mean density on the entire reef platform.

At one reef station (Sta. K3, see Helfrich & Kohn 1955) transects of known area were sampled quantitatively. The results (Table 3) show that the mean population density of all species in the area sampled was 0.16 individual/100 sq ft. This is lower by an order of magnitude than the density on marine benches. This relationship is probably generally valid, although favorable parts of reef platforms may support local populations which approach those of marine benches in density.

The mean biomass of *Conus* in the quantitatively sampled areas at Station K3 was calculated to be 0.06 g dry weight organic matter/100 sq ft (0.0065 g/m^2). Although the population density on marine benches is about 15 times as large as on reef platforms, the biomass of *Conus* is only ten times as great. Although the comparisons are extremely rough, the discrepancy is probably real. Comparison of Figs. 9 and 18 clearly shows the prevalence of larger individuals in the populations occupying reef habitats. The mean shell length of two species common in both habitats is compared in Table 4. These values are 13% and 54% larger in reef platform than in marine

TABLE 3. Population Density of *Conus* at Station K3.

Area Sampled (sq. ft.)	Species	Density (No./100 sq. ft. $=9.3$ m.^{2})
3,000........	*C. ebraeus*	0.07
	C. abbreviatus	0.07
	C. chaldaeus	0.03
	C. flavidus	0.03
		0.20
3,000........	*C. ebraeus*	0.10
1,200........	*C.* (species not noted)	0.17

Mean Density of All Species=0.16

bench populations of *C. ebraeus* and *C. abbreviatus*, respectively. This may reflect the more equable environmental characteristics of the former habitat. The mean shell length of *C. sponsalis* on reefs was 17.8 mm. In bench populations, mean shell length ranged from 13 to 22 mm. The explanation may be that *C. sponsalis* occupies those parts of reef platforms where conditions most closely approximate those of marine benches.

TABLE 4. Mean Shell Length of Reef and Bench Populations of *Conus ebraeus* and *C. abbreviatus*.

	LENGTH IN MILLIMETERS			
	All Reefs		Sta. 5 (bench)	
	Mean	S.D.	Mean	S.D.
C. ebraeus. . .	39.0	9.7	34.6	3.2
C. abbreviatus	32.7	3.1	21.3	3.5

Except for the growth studies reported above, collection data were not analyzed for seasonal variations, since observations were made only during a single complete annual cycle.

COMMUNITY DIVERSITY AND HETEROGENEITY

The similarity of the species composition of different populations may be conveniently measured by an index of diversity given by Koch (1957):

$$I = \frac{t-n}{n(P-1)} \quad (2)$$

where n = the total number of species represented, P = the number of populations or communities sampled, and t = the arithmetic sum of n_1, n_2, n_3, ... n_p, which are the numbers of species in each population or community. If I is low, the species composition differs greatly from population to population. I approaches unity with increasing similarity of the populations compared.

This index was calculated separately for the four most thoroughly sampled reef stations (Sta. 3, 4, 7, and 9) and the four most thoroughly sampled bench stations (Sta. 5, 11, K1, and A3). In both cases, I was 0.58. Comparison of total collections from all bench stations with those from all reef stations gave I = 0.61. The three indices are so similar that there would appear to be no greater ecological difference between the reef and bench habitats than among different reefs, or among different benches. This is misleading, however, since the index used measures only similarity of qualitative species composition.

Consideration of quantitative data concerning the relative abundance of species in different habitats suggests a different interpretation. It has been shown that individual reefs and benches generally support homogeneous populations of *Conus* (Figs. 8, 10, 12-17). Inspection of these graphs shows that three species, *C. sponsalis, C. ebraeus,* and *C. abbreviatus,* are rather consistently the most abundant species on marine benches, while two entirely different species, *C. lividus* and *C. flavidus,* are dominant at almost every reef station.

A comparison of the relative abundance of all species at all reef and bench stations is shown in Fig. 19. Of the 18 species considered, 6 are proportionally more abundant on marine benches than on reef platforms. In 4, the discrepancy in abundance between the two types of habitats is at least one order of magnitude. Twelve species are relatively more abundant on reefs than on benches. The discrepancy is at least one order of magnitude in 8 of these.

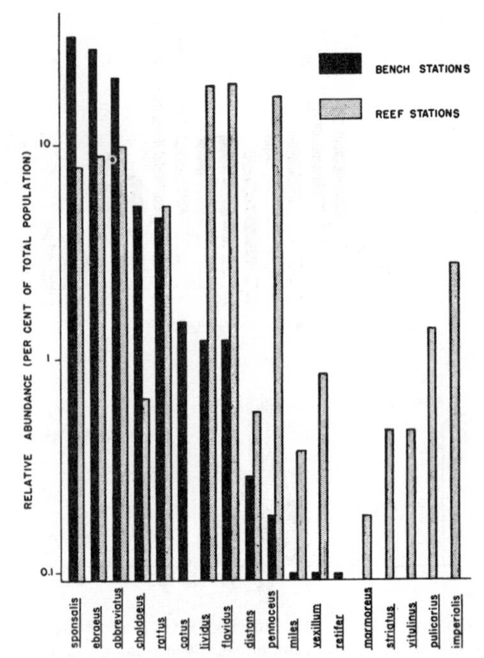

FIG. 19. Relative abundance of *Conus* species at all bench stations (solid histograms) and all reef stations (stippled histograms).

A more quantitative measure of community heterogeneity, H′, based on information theory, is given in Equation (3), which is modified from Margalef (1956):

$$H' = I_{m(AB)} - \left(\frac{I_{m(A)} + I_{m(B)}}{2}\right) \quad (3)$$

in which

$$I_m = \frac{1}{m} \log \frac{m!}{m_1! \cdot m_2! \cdot \ldots \cdot m_s!} \quad (4)$$

where m = the number of individuals of all species, m_1 = the number of individuals of the most abundant species, and m_s the number of the rarest species. The subscripts are equivalent to the values on the abscissas of Figs. 8, 10, and 12-17 and the values can be determined from data given in the graphs.

In Equation (3), A and B signify different communities. Equation (3) can be expanded for application to more than two communities. Equation (5) gives H′ for four communities, A, B, C, and D:

$$H' = I_{m(ABCD)} - \left(\frac{I_{m(A)} + I_{m(B)} + I_{m(C)} + I_{m(D)}}{4}\right) \quad (5)$$

Application of Equation (5) to the abundance data presented (Figs. 14-17) for populations of *Conus* at the four most thoroughly sampled reef stations gave H′ = 0.09. For the four most thoroughly sampled bench stations (Figs. 8, 10, 12, 13), H′ = 0.19.

Application of Equation (3) to summations of abundances at the four most thoroughly sampled reef stations (combined as community "A") and the four most thoroughly sampled bench stations (combined as community "B") gave H′ = 0.84.

The low heterogeneity of the *Conus* communities at individual stations has been demonstrated above. In addition, a low value resulted when the measure of heterogeneity introduced by Margalef (1956) was calculated for summed abundance data from four reef stations. Calculation from summed data from four benches gave a similar low value. However, comparison of summed data from four reefs with summed data from four benches resulted in a high value, indicating marked heterogeneity between the communities of *Conus* in these two kinds of habitats.

To summarize, two types of habitats, reef platforms and marine benches, were distinguished on the basis of observational data presented in the previous section. The data reported in the present section are interpreted as justifying this separation by indicating its significance to the gastropods under consideration, as well as to the investigator.

LOCAL DISTRIBUTION OF *CONUS*

MARINE BENCHES

Distribution of *Conus* across bench platforms was studied by the transect method. Two parallel lines, marked at 10-ft intervals and 10 ft apart, were secured across the bench from inshore edge to near the seaward margin. Ten-foot square areas of the platform were thus delimited. In each transect series, all of the *Conus* visible in each 100-sq ft area were counted. Counts were made at night, when the gastropods were actively crawling about on the bench. For convenience, it was assumed that all of the snails in the study areas were visible from above and none were buried in the algal turf or under stones or coral, as is often the case in the daytime.

Solution Benches. Distribution of *Conus* species, based on data from transects made at Station 5, is shown graphically in Fig. 20. Only the four most abundant species at Station 5 were observed in the transects. The populations of these species are not randomly distributed across the bench platform from pitted zone to outer edge. Interspecific differences are apparent from Fig. 20. Although *C. abbreviatus* is the more abundant, it and *C. sponsalis* are similarly distributed (Wilcoxon test: P ≫ .05). Most of the populations of these two species occupy a strip within 20 ft of shore, independently of the width of the bench. In contrast, the peak density of *C. ebraeus* occurs about halfway across the bench. Lower densities occur near shore and near the outer edge. The density peak of the least abundant species, *C. chaldaeus,* is nearer the outer edge, but the edge itself is not occupied. This species rarely occurs in the shoreward zone occupied by *C. abbreviatus* and *C. sponsalis.* Wilcoxon tests showed that the distribution of *C. ebraeus* differs significantly from that of *C. sponsalis* (P < .05) and from that of *C. chaldaeus* (P < .01).

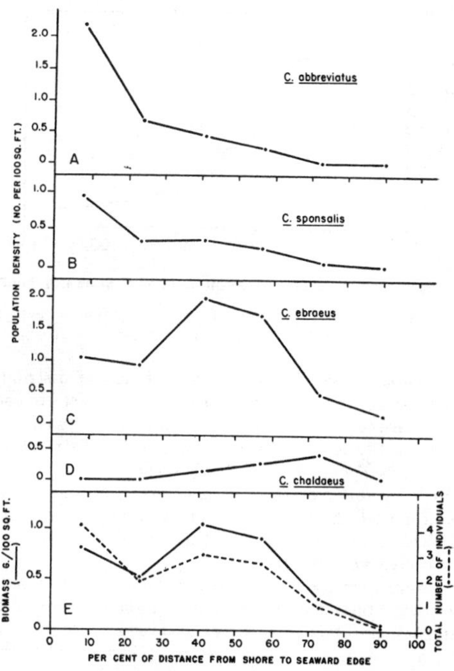

FIG. 20. Distribution of species of *Conus* across bench at Station 5. Each point to the left of 90% represents the average of 8 or 9 transects. The points at 90% represent the average of 5 transects. If width of bench varied from 60 ft at transect site, data were adjusted to a width of 60 ft.

The biological significance of this pattern of distribution will be discussed below. The total density of all species of *Conus* tends to decrease toward the outer edge. Fig. 20E shows this distribution in density as well as in terms of dry weight, excluding shells.

At Station 11, the distribution of *Conus sponsalis* across the bench is shown in Fig. 21 to be essentially similar to that at Station 5. Most of the population occupies the more protected shoreward portion of the bench. The striking difference between the populations at Stations 5 and 11 is the virtual absence of *C. ebraeus* and *C. abbreviatus* at the latter (Fig. 10), although the type of habitat afforded seems suited to these species.

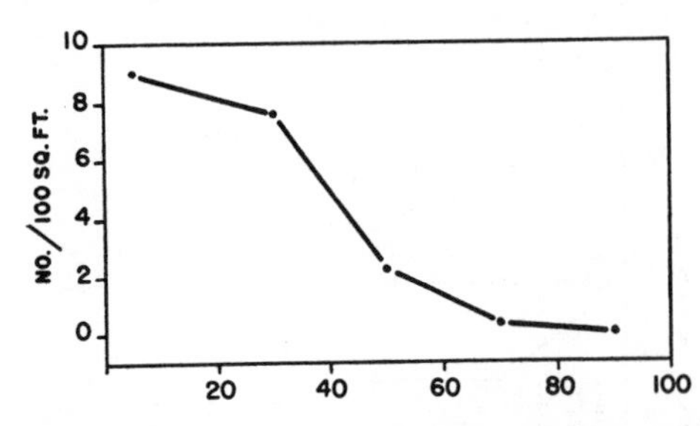

FIG. 21. Distribution of *Conus sponsalis* across bench at Station 11. Data from transects.

Water-leveled Benches. The patterns of distribution across the bench at Station K1, as determined by transecting (Fig. 22), are also quite similar to those at Station 5. As at Station 5, the distribution of *Conus sponsalis* and *C. abbreviatus* are not significantly different from each other ($P \gg .05$). The distribution of *C. ebraeus* differs significantly from that of *C. sponsalis* ($P < .01$) as well as from that of *C. chaldaeus* ($P < .01$).

A striking difference between the distribution patterns at Stations 5 and K1 is that although the bench at Station K1 is three times as broad as Station 5, the four dominant species occur in bands which are only about twice as wide as those at Station 5. Since population density is similar, the density peaks of *C. abbreviatus* and *C. sponsalis* are therefore considerably higher at Station K1 (5.8 and 3.5, respectively) than at Station 5 (2.3 and 0.9, respectively). Thus, in contrast with Station 5, a broad area of bench platform is present which is occupied only by *C. chaldaeus* of the four dominant species under consideration.

In this region, four other species were found. The most abundant was *Conus catus*. Densities of the others, *C. rattus, C. flavidus,* and *C. retifer,* were very low. Thus three of the species characteristic of the solution bench habitat do not extend to the seaward portion of the broad water-leveled bench at Sta. K1 but are replaced there by four other species, only one of which is common.

The density distribution of all species is shown in Fig. 22G. The density peak is relatively nearer shore than at Station 5. Total density decreases toward the seaward edge at both stations.

Station A3. Search of the rampart zone at Sta. A3 revealed no specimens of *Conus*. In addition to *Morula tuberculata,* which is the dominant gastropod, the only other common large gastropod was *Mitra*

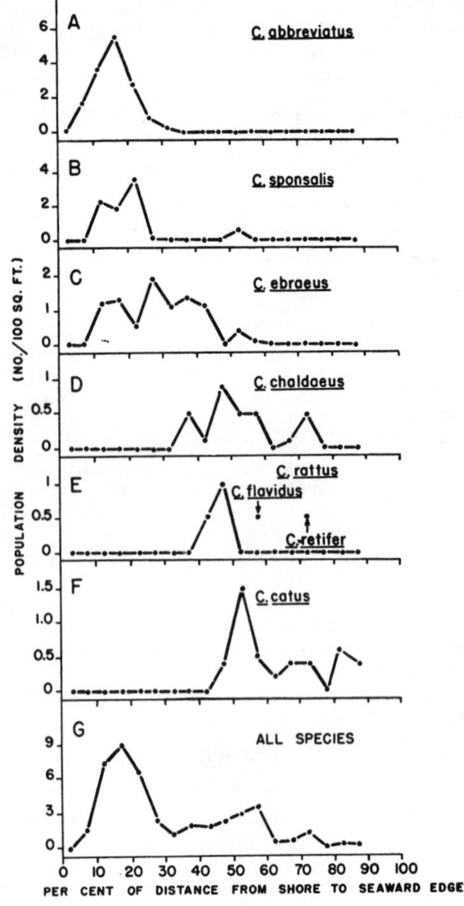

FIG. 22. Distribution of species of *Conus* across bench at Station K1. Data represent the average of two transects.

litterata. The factor which excludes *Conus* from this microhabitat is not known. Almost all of the specimens of *Conus sponsalis* were collected on the flat water-leveled platform. They were found with the shells partly buried in the algae-sand turf, and they thus apparently avoid desiccation during the day. The other species, in contrast, were found more typically on the inner margin of the rampart and level bench platform, or in other sites where the irregular tuff provides shelter during the day. These species may be too large to obtain adequate protecttion from the thin algal mat. The heterogeneity shown by the abundance curve (Fig. 12) can be accounted for by these microhabitat differences. It is of interest that the single specimen of *C. flavidus* collected in the quadrat was only 13 mm long, less than one-half the normal adult length of this species.

REEF PLATFORMS

The most apparent characteristic of the distribution of *Conus* on reef platforms, patchiness, has been mentioned above. It is believed to be related primarily to the nature of the substratum, discussed in more detail in the following section.

Technical difficulties and low population density on reef platforms precluded the use of the transect method for determining patterns of distribution normal to the shoreline. The less exact method of noting the approximate site of individual specimens, in terms of per cent of the distance from shore to outer edge where the specimen was collected, was employed. These data were recorded in the field by the method described by Kohn (1956a). In this manner the approximate distribution of 274 specimens of 11 species at four reef stations was determined. Most of the data concerned five species at three stations, as shown in Table 5. Data for Station 7 are tabulated separately. At the other stations, samples were somewhat biased since collecting effort was greater on the inshore half of the reef than the offshore half. At Station 7, the effort was not so biased, and the similarity with data from the other stations suggests that the latter are valid. The evidence thus indicates that more intensive collecting on the offshore parts of other reefs would reveal *Conus* in low density, and this was confirmed by inspection.

TABLE 5. Distribution of Five Species of *Conus* Across Reef Platforms at Stations 3, 7, and 9.

Per cent of Distance From Shore to Outer Edge	*abbreviatus*		*flavidus*		*lividus*		*ebraeus*		*sponsalis*		Total
Sta.	7	T	7	T	7	T	7	T	7	T	
0- 30	4	8	6	26	12	23	7	10	2	2	69
30- 60	0	4	6	16	10	18	2	10	20	20	68
60- 90	3	3	3	5	4	5	7	8	29	29	50
90-100	1	2	0	0	3	4	1	2	2	2	10
	8	17	15	47	29	50	17	30	53	53	197

T=Stations 3+7+9

The distribution patterns normal to the shore of *Conus abbreviatus, C. flavidus, C. lividus,* and *C. ebraeus* do not differ significantly from each other in the samples observed ($P \gg .05$). The distribution of *C. sponsalis* at Station 7 may differ significantly ($P > .05$) from that of *C. ebraeus* and thus from those of all other species. This variation, however, is in the opposite direction from that observed on marine benches, where *C. sponsalis* characteristically occupies the shoreward zone (Figs. 20-22). On the subtidal reef platform at Station 7, the bulk of the population of this species occupies the central region of the reef (Table 5). This is probably due to the infrequency of suitable substratum nearer shore, as will be discussed below.

The distribution patterns of *Conus lividus* and *C. flavidus,* two closely related species which are also similar in size, are strikingly similar, but some of the predominance of inshore individuals may be due to sampling bias.

The total density of all species is similar over most of the reef platform from shore to outer edge (Table 5). The total density decreases as the breaker line is approached, but sampling bias is also in this direction. The downward trend in the first three figures in the right hand column of Table 5 is probably not significant.

It may be concluded, then, that over most of the area of reef platforms the distribution of *Conus* is patchy, or clumped, and is not related to distance from shore or breaker line. In the zone of surf and very close to shore, however, *Conus* is less abundant, although present in low density.

SUBSTRATUM

MARINE BENCHES

By definition, a habitat must possess uniformity with respect to an important quality (Andrewartha & Birch 1954: 28). In the case of marine benches, uniformity is in the physiography of the bench platform, which constitutes the substratum of the species under consideration. This substratum is hard, rather smooth, and covered with an algal turf which may vary markedly in density. The algae, if sufficiently dense, bind sand, which is of importance because its presence increases the amount of residual water retained on the bench at low tide, and because it provides a medium into which *Conus* may burrow in order to escape desiccation or heavy wave action.

The substratum is of importance not only to adult *Conus,* but also to the pelagic veliger larvae. The importance of certain attributes of the substratum in the settling of pelagic larvae of certain other benthic marine invertebrates has been demonstrated by Wilson (1952, 1955), and this is likely to be the case in *Conus* also. Peculiarities of the substratum of marine benches may attract larvae of some species and not of others and thus determine qualitatively the species composition of the population of this habitat. The possibility of obtaining direct evidence relevant to this hypothesis was remote. However, it is probably one of a number of density-inactive mechanisms (Nicholson 1955) which determine the specific composition of populations of *Conus* on marine benches.

The hypothesis that the microhabitats of different species differed with respect to substratum preferred by the adults is more amenable to study. The predominant substratum of marine benches is the hard, algal-matted reef limestone noted above. In addition to the zonation of distribution discussed in the previous section, variations of substratum exist which may further differentiate microhabitats. These are patches of bare limestone or tuff which do not support an algal turf, and patches of sand, which usually fill shallow depressions in the bench platform. In the latter, benthic algae may or may not be present. The association of various species of *Conus* with these regions is summarized in Table 6A.

TABLE 6. Proportion of Populations of *Conus* Species Associated with Different Types of Substratum.

A. Marine Benches

Species	Sample Size	Bench with Algal Turf, Binding ± Sand		Sand Pockets or Patches on Bench		Bare Limestone or Turf	
		No.	%	No.	%	No.	%
C. sponsalis	211	189	90%	10	5%	11	5%
C. abbreviatus	96	63	65%	25	26%	6	6%
C. ebraeus	97	63	65%	15	15%	17	18%
C. rattus	45	28	62%	5	11%	11	24%
C. chaldaeus	29	13	45%	8	28%	8	28%
C. catus	11	5	45%	2	18%	4	36%
	489	361	74%	65	13%	57	12%

B. Reef Platforms

Species	Sample Size	Sand		Reef Limestone with or without Algal Turf		Dead Coral and Coral Rubble	
		No.	%	No.	%	No.	%
C. pulicarius	12	12	100%	0	0%	0	0%
C. abbreviatus	73	57	78%	13	18%	3	4%
C. lividus	133	86	64%	29	22%	18	14%
C. ebraeus	54	33	61%	11	20%	10	19%
C. flavidus	130	73	56%	34	26%	23	18%
C. imperialis	23	11	48%	6	26%	6	26%
C. sponsalis	61	23	38%	24	39%	14	23%
C. rattus	37	8	22%	14	38%	15	40%
	659	405	61%	160	25%	94	14%

C. *Conus pennaceus* on Reef Platforms

	Sample Size	SAND		Reef Limestone, with or without Algal Turf	Dead Coral and Coral Rubble
		Visible From Above	Under Rocks		
day	79	2	70	3	4 (all under rocks)
night	57	30	0	26	1 (visible from above)
Total	136	32 (102)	70	29	5
Per cent		75%		21%	4%

Most individuals of all species are found on the most abundant kind of substratum, algal turf on hard bench platform. A certain proportion of each species "spills over" onto the other types of substratum available. This frequency is significantly less in *Conus sponsalis* than in the other species. *C. abbreviatus* utilizes sandy areas more often than *C. sponsalis* ($P < .01$). *C. ebraeus* utilizes bare regions of bench more often than *C. abbreviatus*, but the difference is probably not significant ($P = .08$). Substratum preferences of other species do not differ significantly from one another.

The particle size distribution in samples of sand from several stations was determined by the use of graded sieves, in order to detect possible interspecific differences in preference for sand of diffrent mechanical proprties. The methods used varied but slightly from those of Holme (1954).

Since sand particles are subject to sorting by wave action, larger particles might be expected to dominate in samples taken near the outer edge of benches. Application of Wilcoxon tests to data from Stations 5 and K4 showed that *Conus ebraeus* is associated with somewhat coarser sands than *C. abbreviatus* ($P = .05$). However, mechanical properties of the sand probably do not vary sufficiently to provide different microhabitats with respect to this factor.

Sandy areas are more common on reef platforms than marine benches. The small areas of sand on benches may limit the density of some species of *Conus*. *C. pennaceus* is found rarely on benches, but it is common in habitats where sand is abundant, as the animals typically remain in sand under rocks during the day. Other species of *Conus* which occur in low densities on marine benches (e.g. *C. lividus, C. flavidus*) may be limited indirectly by the absence of sandy substrata which are required not by them but by the species on which they prey (see section on Nature of the Food of *Conus* on Marine Benches).

REEF PLATFORMS

The predominance of a sand substratum and the absence of extensive living coral characterize the platforms of Hawaiian fringing reefs and distinguish them from those of typical atoll and barrier reefs.

The nature of the substratum associated with 659 individuals of the 9 most abundant species of *Conus* on reef platforms is summarized in Table 6B. Most of the specimens (61%) of all the species collected on reef platforms occupied sandy substrata. From 12% to 14% of the populations were found on each of the other types of available substrata: bare limestone, limestone with algal turf (combined in Table 6B), and coral (including coral rubble and dead coral). The relative abundance of the available types of substratum was not measured, but sand is by far the most prevalent. Thus, as is the case on marine benches, most of the population of *Conus* occupies the most abundant type of substratum.

Specific Differences of Substratum. *Conus pulicarius* is probably entirely restricted to a sandy substratum. Although the sample size in Table 6B is small, it is supported by similar unpublished data from other regions where this species occurs. *C. pulicarius* typically occurs in areas of reefs characterized by sand bottom and the absence of limestone outcrops and growing coral. Individuals are usually partly or completely buried during the day. At night, they actively crawl about through the sand, leaving a broad track which is visible from above and facilitates collection. In quiet water, these tracks are often visible the next morning, and they may lead to the discovery of completely buried individuals.

Conus flavidus and *C. lividus*, the two most abundant species on reef platforms, also occur predominantly on sand substrata but are commonest in areas of sand patches and smaller pockets among solid substratum. These two species do not differ significantly from each other with respect to the types of substratum utilized ($P = .15$; see Table 20). Both species are predominantly epifaunal, and individuals are rarely found even partly buried in the substratum during the day.

Conus abbreviatus is also primarily a sand dweller when it occurs on reef platforms. It will be recalled that this species also occupies sandy regions of marine benches significantly more often than do the other species present. *C. abbreviatus* also burrows in sand on reef platforms significantly more often ($P < .01$) than the other species, with the exception of *C. pulicarius.*

The proportions of different types of substratum utilized by *Conus ebraeus* on reef platforms do not differ significantly from those of the other predominantly sand-dwelling species just discussed (Table 6B). This species is also typically epifaunal, but some individuals are found partially buried in the substratum.

Conus pennaceus is also primarily associated with a sand substratum. However, in contrast to the other species, the substratum of *C. pennaceus* is altered by the diurnal activity cycle. The substratum of 79 specimens collected during the day and 57 collected at night is shown in Table 6C. During the day, specimens are characteristically found on (37%) or partly buried in (63%) sand under basalt or coral rocks. At night, the snails actively crawl about on the surface of the sand or on reef rock, and they are visible from above. The substratum of *C. pennaceus* thus differs qualitatively from those of the other sand dwellers, which are found uncommonly under rocks.

The number of specimens of *Conus sponsalis, C. imperialis,* and *C. rattus* found on other types of substratum exceeds the number found on sand. Most of the specimens of *C. sponsalis* listed in Table 6B were collected at Station 7, where it is the most abundant species of *Conus. C. sponsalis* occurs sparsely or not at all on the other reefs studied (Figs. 14-17). The substratum of a large area of Station 7, particularly where *C. sponsalis* is common, consists of rather smooth limestone, which supports an algal turf, and dead coral, which presents a rough surface but is apparently being smoothed by wave quarrying. The former type of substratum especially is similar to that of marine benches, where *C. sponsalis* is also abundant. It is therefore not surprising that, when such a substratum is available on an otherwise predominantly sandy reef platform, it should be occupied by a population of *C. sponsalis.* Only a few scattered individuals of *C. sponsalis* are found on reefs which lack extensive areas of this type of substratum.

Conus rattus is also found more often on lithified than on sandy substrata, and in this respect probably does not differ significantly from *C. sponsalis* ($P = .08$). The numbers of these two species found on each of the substratum-types listed in Table 6B are significantly different, however ($P = .02$), in the samples collected.

Analysis of Particulate Sediments. The five most abundant species of *Conus* on subtidal reef platforms are characteristically associated with the sand moiety of the substratum. It was therefore of interest to determine the properties of the particulate sediments and the distribution of *Conus* with respect to these properties.

The sediments are almost entirely biogenic, comprising chiefly Foraminifera tests, fragments of coral, mollusk shells, echinoid tests and spines, and calcareous algae. Small amounts of olivine are often present, especially at Station 9, located at the foot of Diamond Head (Fig. 6), which takes that name from the local term "Hawaiian diamonds" for olivine.

Sediments associated with seven species of *Conus* on reef platforms were analyzed. *C. lividus* was found to be associated with somewhat coarser sediments than *C. flavidus* (Wilcoxon test: $P = .05$). However, all species are associated with sediments with extremely variable particle size distribution in the sand-gravel range. It therefore seems likely that, on reef platforms as on marine benches, there is no niche diversification of *Conus* species with respect to the particle size distribution of the sand moiety of the substratum.

EFFECTS OF ENVIRONMENTAL STRESSES

As Prosser (1955) has recently stated, "Determination of the importance of specific variations requires (a) the careful observation of ecological niches of subspecies and species occupying overlapping habitats, and (b) physiological tests of the effects of environmental stresses."

In order to test the effects of some environmental stresses on the species of *Conus* that inhabit marine benches, experiments on the effect of strong water currents and low oxygen tension were carried out. In addition, observations on activity with respect to desiccation will also be reported. The environment of the subtidal reef platform habitats is in general more equable than that of intertidal marine benches.

TEMPERATURE

Water temperatures ranging from 22.8 to 28.3°C were recorded at marine bench stations. The extreme range is probably somewhat greater, especially in shallow pools at low tide. Seasonal fluctuations appeared to be extremely small. Sea temperatures ranging from 22.0 to 29.1°C were measured on reef platforms. More extensive data for Station 3 were published by Edmondson (1928), who reported extremes of 21.5 and 29.0°C.

Analyses of diurnal fluctuations of temperature at Station K3 have been published elsewhere (Kohn & Helfrich 1957). On successive days in summer, ranges of 24.3-29.1°C and 24.1-27.1°C were recorded. The amplitude of diurnal temperature fluctuation thus closely approaches the annual range of variation on at least some reefs.

Temperature is not a limiting factor to species tolerant of the prevailing range, and population densities do not affect, and are not affected by, temperature. The qualitative species composition of populations of *Conus* in Hawaii may, however, be determined at least in part by the temperature regimen. The Hawaiian Islands represent the highest latitudes at which most of the species present are known to occur. The fauna of islands at lower latitudes, where sea temperatures are higher, is richer. Some species may be excluded from the Hawaiian area by temperatures below their tolerance limits. However, most regions nearer the equator which have been studied lie closer to the Indonesia-Melanesia center of distribution, and distance from this center is also a factor of major zoogeographical importance.

Exposure to Air at Low Tide

Marine benches, but not reef platforms, are subject to periodic exposure to air at times of low tide. Therefore, only a limited range of depth of water is available on marine benches at low tide, ranging from 0 (hereafter noted as "exposed") to about a foot in tidal pools.

A certain fraction of the population of each species of *Conus* was found to be left exposed at low tide, when the bench platform may not be awash for periods of up to four hours (Table 7). The frequency of exposure is significantly higher in *C. sponsalis* than in all other species ($P < .01$), mainly because the sample contains the large population at Station A3, which was entirely exposed. This dense population, and the virtual absence of *C. abbreviatus*, has been noted above. *C. abbreviatus* may be unable to withstand the long periods of exposure to air required for habitation of Station A3.

TABLE 7. Proportion of Populations of *Conus* Species Exposed to Air at Low Tide.

Species	Sample Size	Number Exposed	Per Cent Exposed
C. sponsalis	144	79	55%
C. abbreviatus	83	29	35%
C. chaldaeus	31	10	32%
C. ebraeus	73	21	29%
C. catus	28	8	29%
C. rattus	47	8	17%

Data from all marine bench stations.

The low proportion of the population of *Conus rattus* which is out of water at low tide is probably due to the fact that this species occurs more often on abrasion ramp benches than on other types. Since these slope into the sea, they are more often awash at low tide than the horizontal platforms of solution benches and water-leveled benches.

All species of *Conus* observed by the writer tend to remain quiescent in the daytime. On marine benches, the algae-sand turf provides shelter from heavy wave action at high tide, when the entire bench is awash, and the residual water it retains at low tide reduces the danger of desiccation for the smaller species (*C. sponsalis, C. abbreviatus*) during the day. If the turf is dense enough and sufficient sand is present, the shells may be completely buried. Larger species (*C. ebraeus, C. chaldaeus*) find shelter in shallow crevices or under pieces of coral rubble in the daytime.

In order to ascertain whether different species occupy different microhabitats with respect to depth of water at low tide, the proportion of each species population partly buried or otherwise sheltered was determined. The samples were rather small and excluded completely buried individuals. Nevertheless, the frequency of burrowing in *C. abbreviatus* is significantly greater than in any of the other species ($P < .01$). This is probably due at least in part to the preference of this species for sandier regions of the bench platform, where burrowing can be more easily accomplished. Proportions of populations of other species buried during the day did not differ significantly from each other in the samples examined.

Although seemingly unsuitable shelters that are vacant can always be found on marine benches, the density of available sheltered sites, and the ability of *Conus* to locate them when needed, may be a limiting factor of population size in some cases.

Heavy wave action, a prominent environmental factor on marine benches, is generally absent from subtidal reef platforms except near the outer edge. Large waves are broken by the outer edge of the reef, and wave action on the platform is rarely so heavy as to interfere with collecting or observation, even at high tide.

Activity Rhythms

At night, especially at low tide, the snails are typically up on the surface of marine bench platforms, moving actively about. This activity rhythm persists when *Conus* is maintained in laboratory aquaria. Degree of expansion of the foot, movement of the siphon, and movement of the entire animal are much greater at night than in the daytime, even if the aquaria are illuminated at night. Rhythmicity of activity in nature is thus probably at least partly endogenous.

Alternating periods of light and darkness are important to many marine invertebrates because they synchronize rhythms of locomotor and other activities (Brown *et al.* 1953). In the field, no interspecific differences in activity with respect to light or time of day, which might lead to reducing the possibility of interspecific competition, were noted. Enhanced activity may be correlated with the immediate presence of food, as Ohba (1952) observed in *Nassarius*, and as was shown in *Conus striatus* in the laboratory (Kohn 1956b).

The relationship of tidal fluctuation to locomotor activity in *Conus* is probably complex. Observations were hampered by the fact that most marine benches are accessible only at low tide. Maximum activity was observed during low spring tides at night. At such times, waves may not break over solution benches for periods of several hours. The retention of much residual water by the algae-sand mat and the absence of solar radiation probably make the problem of desiccation negligible.

The initial movements of *Conus,* which begin after sunset, are mostly vertical, being directed out of the daytime hiding place up onto the bench platform. This behavior pattern is illustrated from field observations which comprised counts made in seven 100-sq ft quadrats at Station 5 at intervals over a 4-hr period in the evening. No *Conus* were visible on the bench platform at 1800 hrs, but by 2030, 17 specimens of four species had become active and were crawling about on the surface of the 700-sq ft area studied. By 2200, three additional individuals had also become active. Time of low tide (—0.2 ft) was 2023. Ohba (1952) noted activity peaks of *Nassarius* in tide pools at times when agitation of the water ceased on the receding tide and was renewed on the rising tide. In the present study, however, the bench had been exposed since before 1800. The rhythm is probably timed so that increased activity begins when two factors, darkness and absence of strongly flowing water, become favorable.

Since heavy wave action made observations on benches impossible except at low tide, experiments designed to measure the ability of *Conus* to withstand strong currents were carried out. Individuals of the four species dominant on marine benches were subjected to artificially created currents of 20, 100, 175 and 210 cm/sec. The ability to withstand the currents was found to be roughly correlated with size, in that *C. ebraeus,* which is larger than the other species (Fig. 9), is the most tenacious. The tenacity of *C. sponsalis, C. abbreviatus,* and *C. chaldaeus* is similar, although individuals of the last named species are somewhat larger (Fig. 9).

These data may offer a partial explanation of the observed pattern of distribution on solution benches (Fig. 20). The two smallest species are less able to withstand strong currents generated by waves breaking over the bench. By occurring chiefly on the landward portion, they are probably able to spend more time seeking food between high tides. The larger and heavier *Conus ebraeus,* on the other hand, being better able to withstand strong currents, may find a wider area of bench within its optimal habitat.

The ability of *Conus chaldaeus* to withstand strong currents in the experiments was comparable only to that of *C. sponsalis* and *C. abbreviatus.* In size, *C. chaldaeus* ranks between these species and *C. ebraeus* (Fig. 9). The experimental data thus do not help to explain the distribution pattern of *C. chaldaeus,* which characteristically occurs at the more seaward portion of marine benches. However, the data may suggest a possible explanation for the small population size of this species at Station 5, where wave action is more violent than at Station K1 and other collecting sites.

Reef Platforms. Diurnal activity cycles are difficult to observe on reef platforms. Most specimens were collected during the day, when they are typically quiescent. Little apparent difference in activity is observable in the field in the case of typically epifaunal species such as *Conus lividus* and *C. flavidus.* However, both of these species probably feed only at night, as will be shown below, and they are therefore presumably most active then.

Diurnal differences in the activity of *Conus pennaceus* are more readily observable, as alluded to in the previous section. This species is almost always found on or partly buried in sand under rocks during the day (Table 6C). At night, however, the snails leave that microhabitat to crawl about on the surface of the sand or on coral or limestone. Of 60 specimens of *C. pennaceus* collected at night, three were found crawling about out of water. Exposure to air is insignificant as an environmental stress on reef platforms, however. Exposed individuals of other species were found even more rarely or not at all.

OXYGEN REQUIREMENTS

Diurnal cycles of solar radiation and semidiurnal tidal cycles are factors leading to large fluctuations in oxygen concentration in the sea water on Hawaiian marine benches. Amplitude of fluctuation is greatest at low tide. At mid and high tides, oceanic water breaking over the bench in surf is probably always saturated. During daytime low tides residual water on the bench platform becomes supersaturated with oxygen due to photosynthesis by the dense mat of attached algae. At night low tides, oxygen content of residual water is reduced by respiration of both plants and animals.

Determinations of oxygen concentration *in situ* were made at three marine bench stations. Determinations were made by the Winkler method. The highest concentration measured was 7.68 ml $0_2/l$, and the lowest was 0.97 ml $0_2/l$, or only 19% of saturation.

It was thought that the nocturnal minimum might represent an environmental stress on *Conus.* In an attempt to determine the oxygen requirements of *Conus* and the ability to survive low oxygen tension, experiments on three of the four dominant species on marine benches were carried out in a simple respirometer. The results of three experiments, each with 14-20 specimens of *C. ebraeus,* and six experiments, each with 16-20 specimens of *C. chaldaeus,* are summarized in Fig. 23. Seven experiments with *C. abbreviatus* had erratic results, and they have not been included in Fig. 23. Indeed, all results were rather erratic, as indicated by the large standard errors. Some of this variation may be due to diurnal and tidal cycles in rate of respiration, such as have been reported in other gastropod genera by Sandeen, Stephens & Brown (1954).

The data in Fig. 23 suggest that *Conus ebraeus* and *C. chaldaeus* are probably physiological adjusters rather than regulators with respect to oxygen. That is, the rate of consumption varies with the environmental concentration rather than remaining relatively constant (Prosser 1955).

Three respiratory rate determinations of *Conus*

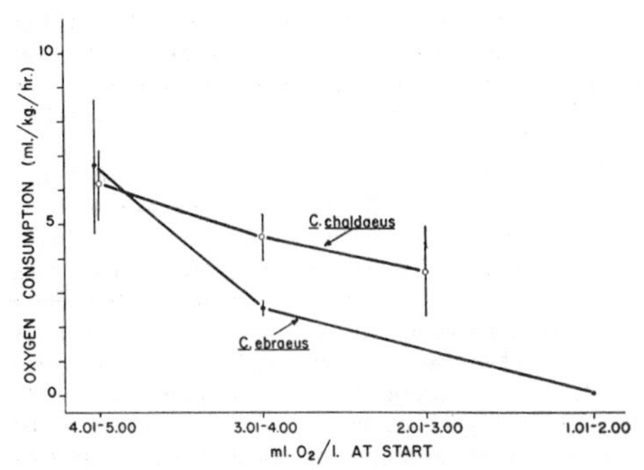

FIG. 23. Rates of oxygen consumption of *Conus* at different oxygen concentrations. Vertical lines indicate standard error of mean.

ebraeus in water containing 1.03-1.19 ml 0_2/l, near the lowest *in situ* concentration measured, showed no detectable uptake of oxygen over a period of 6 hrs. Nevertheless, all animals survived the experiment. On marine benches, oxygen concentration is probably never so low for so long a time. In addition, the typical activity cycle of *Conus* tends to obviate danger of hypoxia. When the snail crawls about out of water on the bench platform, a thin layer of water over the ctenidia ensures diffusion of oxygen from the air as the concentration in the water is reduced by passage of oxygen into the tissue.

Reef Platforms. Like temperature, oxygen concentration of the water over subtidal reef platforms undergoes diurnal fluctuation of lower amplitude than on marine benches. A graphic presentation of the diurnal cycle of oxygen concentration at Station K3 has been published (Kohn & Helfrich 1957). The minimum concentration recorded (at 0300) was 3.58 ml 0_2/l, equivalent to 77% saturation. Midday values ranged to 7.27 ml 0_2/l, or 168% saturation. A few determinations of oxygen concentration were also made at Stations 3, 7, and 9. The maximum value observed was 8.91 ml 0_2/l (197% saturation) recorded at Station 7.

Dissolved oxygen is probably never in short supply to *Conus* on reef platforms. However, individuals of several species burrow into the sand substratum. Below a centimeter or so, these sands are usually gray. This reducing environment probably does not adversely affect *Conus,* however, since the inspiratory organ, the siphon, projects above the water-sand interface and draws a stream of oxygenated water over the ctenidia.

FOOD AND FEEDING

THE FEEDING PROCESS

Members of the genus *Conus* are known to have a unique feeding mechanism (discussed in detail by Bergh 1896, and Hinegardner 1957, 1958) and are known to be predatory (Alpers 1932a, Kohn 1955). The feeding process in piscivorous species has been described elsewhere (Kohn 1956b). Alpers (1932b) studied the feeding process in *C. mediterraneus* Hwass *in* Bruguière, which feeds on polychaetes. He concluded erroneously that *Conus* ejects venom into the water in the vicinity of the prey, and he was not able to discern the function of the radula teeth.

Feeding of several of the vermivorous species which occur in Hawaii was observed in the laboratory. In *Conus abbreviatus* and *C. ebraeus,* the manner of injecting the radula tooth and accompanying venom do not differ from that described (Kohn 1956b) for *C. striatus.* However, the radula tooth is not held by the proboscis after injection. Rather, the proboscis retracts quickly, leaving the tooth in the prey. The mouth then expands and the paralyzed prey is engulfed. However, as will be noted below, other vermivorous species do retain the radula tooth within the proboscis and use it to draw the impaled prey into the mouth, as does *C. striatus.*

The method of feeding is somewhat different in *Conus pennaceus,* which, as will be shown, feeds on other mollusks. When the prey is stung, the radula tooth is completely freed from the proboscis. In contrast with the vermivorous species, not one, but up to six radula teeth may be injected into the same prey organism. If the prey is an opisthobranchiate mollusk with an internal shell, it is usually swallowed whole. The shell is presumably later regurgitated, since it is usually too large to pass into the intestine. When the prey is a prosobranch, or an opisthobranch with a large external shell, the shell is not swallowed. Rather the mouth of *C. pennaceus* is applied to the aperture of the shell of the prey after stinging. This position is maintained for 15 min-1 hr, following which the shell, now empty, falls away. Presumably the venom acts on the columellar muscle during this time, relaxing its attachment to the shell and allowing the soft body to be removed intact from the shell and swallowed. The feeding process was observed to be essentially identical in *C. textile.*

Digestion. After the prey has been completely swallowed, it lies in a large, distensible organ variously termed the crop (Clench & Kondo 1943) or esophagus (Speiseröhe of Bergh 1896; Ösophagus of Alpers 1931). Usually no digestion takes place in this organ, although enzymes may leak anteriorly from the intestine, causing some. Since there is no mechanism for trituration, prey in the esophagus is usually in a good state of preservation and identification is thus facilitated. The junction of the esophagus and intestine is marked by the entrance of two large ducts from the digestive glands. The prey is gradually moved from the crop into the intestine, where digestion and absorption occur. Fecal matter is not usually compacted into pellets but is excreted as undigested remains.

The piscivorous species represent an exception to the course of digestion just described, as noted by Kohn (1956b). Considerable digestion occurs in the anterior portion of the alimentary tract. The food swallowed is proportionately much larger than that eaten by the other species, and the lower parts of the tract are not very distensible.

TIME OF FEEDING

Feeding takes place at night and usually not during daylight hours. This was demonstrated in the following manner. In analyses of alimentary tract contents, the position of food in the tract was recorded. Since the time of collection was noted, the position of food in the tract could be plotted against time of day. Food frequency histograms for the three dominant species on marine benches are shown for pertinent times of day in Fig. 24. Data from all marine bench stations are included. Data for three species on reef platforms are presented in Fig. 25. At other times of day (during afternoon and early evening) the proportion of snails with empty alimentary tracts was so high that it was not profitable to collect and examine large numbers of them.

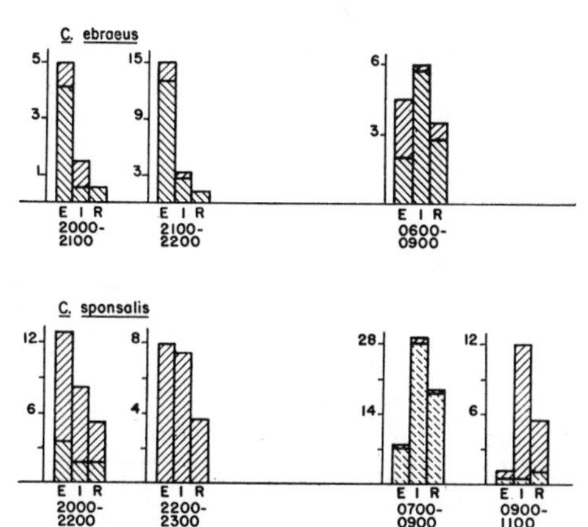

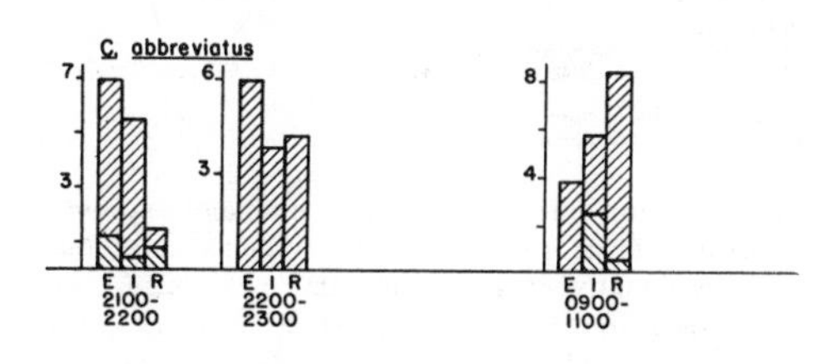

FIG. 24. Food in alimentary tracts of *Conus* collected at marine bench stations at different times of day. E = esophagus. I = intestine. R = rectum.

In Figs. 24 and 25, alimentary tracts are divided into esophagus (E), intestine (I), and rectum (R). The esophagus and intestine are quite distinct organs (Bergh 1896, Alpers 1931), while the rectum is somewhat arbitrarily considered as the region between the last curve of the intestine and the anus. The dat from marine benches show that, in all three species, the crops contain the largest number of food organisms at night. By mid-morning, food has largely passed from the esophagus into the intestine and rectum, and incidence of esophagus contents is greatly reduced.

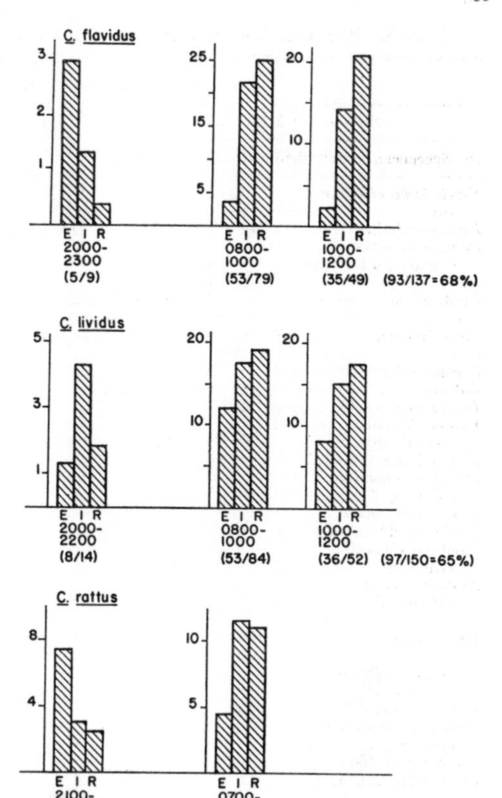

FIG. 25. Food in alimentary tracts of *Conus lividus*, *C. flavidus*, and *C. rattus* at different times of day. E = esophagus. I = intestine. R = rectum.

Of the dominant reef platform species, evidence of time of feeding was obtained for *Conus flavidus*, *C. lividus*, and *C. rattus*, as shown in Fig. 25. Nocturnal records for *C. flavidus* and *C. lividus* are probably too few to be meaningful. That feeding takes place at night in *C. flavidus* is plainly shown, however, by the low incidence of prey in the esophagus during the day. In *C. lividus*, the data are less clear-cut. This may be due to a slower rate of digestion, or some feeding may take place during daylight hours. *C. rattus* is plainly a nocturnal feeder. At night, prey organisms are found chiefly in the esophagus, while during the morning their remains occupy the lower regions of the alimentary tract.

The number of specimens examined which had food in the alimentary tracts is also indicated in Fig. 25. About two-thirds of the population of both *Conus*

TABLE 8. Prey Organisms Consumed by Vermivorous Species of *Conus* at Marine Bench Stations

	sponsalis				*abbreviatus*		*ebraeus*			*chaldaeus*			*rattus*	*lividus*			*flavidus*		*miles*	*distans*
Station...	5	11	A3	T	5	T	5	11	T	5	11	T	T	5	11	T	5	T	T	T
No. Specimens Examined	45	51	92	258	158	243	94	5	122	25	12	59	79	10	1	24	6	10	1	2
Nereis jacksoni Kinberg var.	—	6	5	30	—	—	—	—	—	—	—	—	—	—	—	—	—	—	—	—
Perinereis helleri Grube	15	14	—	32	19	20	104	3	115	—	—	—	13	—	—	—	—	—	—	—
Platynereis dumerilii (Audouin & Edwards)	2	2	6	10	2	3	—	—	—	17	16	45	—	—	1	1	—	—	—	—
Nereid sp. 350	1	2	1	4	—	—	—	—	—	—	—	—	—	—	—	—	—	—	—	—
Unidentified Nereidae	—	—	1	1	1	1	—	—	—	—	—	—	—	—	—	—	—	—	—	—
Total Nereidae	18	24	13	77	22	24	104	3	115	17	16	45	13	—	1	1	—	—	—	—
Lysidice collaris Grube	2	3	3	11	5	27	—	—	—	—	—	—	—	—	—	—	—	—	1	—
Palola siciliensis (Grube)	—	—	—	—	—	—	3	3	7	4	—	8	—	—	—	—	—	—	—	—
Eunice antennata Savigny	—	1	1	2	1	2	—	—	—	—	—	—	12	—	—	—	—	—	—	—
Eunice (Nicidion) cariboea (Grube)	—	5	30	36	8	10	—	—	—	—	—	—	1	—	—	—	—	—	—	—
Eunice filamentosa Grube	1	—	—	1	2	3	—	—	—	—	—	—	—	—	—	—	—	—	—	—
Marphysa sanguinea (Montagu)	—	—	1	1	1	3	—	—	—	—	—	—	—	—	—	—	—	—	—	—
Eunice afra Peters	—	—	—	—	—	—	—	—	—	—	—	—	—	—	—	—	—	—	—	1
Unidentified Eunicidae	1	—	—	1	1	1	—	—	1	—	—	—	—	—	—	—	—	—	—	—
Lumbrinereis sarsi (Kinberg)	2	3	8	14	10	11	—	—	—	—	—	—	—	—	—	—	—	—	—	—
Arabella iricolor (Montagu)	—	—	—	—	—	4	—	—	—	—	—	—	—	—	—	—	—	—	—	—
Total Eunicea	6	12	43	66	28	61	3	3	8	4	—	8	13	—	—	—	—	—	1	1
Nicolea gracilibranchus (Grube)	—	—	—	—	—	—	—	—	—	—	—	—	—	1	—	4	—	2	—	—
Terebellid sp. 837	—	—	—	—	—	—	—	—	—	—	—	—	—	—	—	—	—	3	—	—
Cirriformia semicincta (Ehlers)	—	—	—	—	—	—	—	—	—	—	—	—	—	—	—	1	—	—	—	—
Polydorid sp. 1500	—	—	—	—	—	—	—	—	—	—	—	—	—	—	—	—	—	1	—	—
Unidentified annelids	—	1	3	6	4	4	1	—	1	—	—	—	—	—	—	2	—	1	—	—
Total Annelids	24	36	59	149	45	89	108	6	124	21	16	53	26	1	1	8	—	7	1	1
Total Identified Food	24	36	59	149	54	89	108	6	124	21	16	53	26	3*	1	16†	—	7	1	1
Unidentified Food	1	1	1	4	1	3	—	—	—	—	—	—	—	—	—	3	—	1	—	—

Data from Stations 5, 11, and A3 are entered separately as noted. T=Data from all bench stations.
Numbers in body of table indicate numbers of polychaete species at left found in alimentary tracts of *Conus* species at top.
Chief references used in the identification of the polychaetes were Abbott (1946), Hartman (1940, 1944, 1948), Holly (1935), Fauvel (1927), Okuda (1937), and Treadwell (1906, 1922).
*Includes one specimen definitely identified, and one tentatively identified, as *Ptychodera flava laysanica* Spengel.
†Includes five specimens definitely identified, and three tentatively identified, as *Ptychodera flava laysanica* Spengel.

flavidus and *C. lividus,* and about one-half of the population of *C. rattus* succeed in capturing food each night.

Investigation of the rate of passage of food through the alimentary tract of *Conus abbreviatus* showed that the food starts passing into the intestine after about 1.5 hrs in the esophagus. After about 3 hrs, fecal matter is present in the rectum. From 12-24 hrs after feeding, the alimentary tract is completely emptied. Therefore, gastropods observed with empty tracts in the afternoon may or may not have fed the previous night.

NATURE OF THE FOOD OF *Conus* ON MARINE BENCHES

Solution Bench Stations. Of the eight species of *Conus* present at Station 5, five, including the four dominant species, were found to feed exclusively on polychaete annelids. Of the less abundant species, *C. flavidus* feeds predominantly on polychaetes but may occasionally take an unsegmented worm. *C. lividus* feeds on the enteropneust *Ptychodera flava* but consumes polychaetes as well. *Ptychodera* is a sand-dwelling worm and is uncommon on marine benches. Both *C. flavidus* and *C. lividus* prefer Terebellidae among the polychaetes. These worms are also typically sand dwellers, building tubes on the sides of rocks partly buried in sand. The low abundance of *Ptychodera* and terebellids on benches may limit the populations of *C. lividus* and *C. flavidus* in this habitat.

Conus pennaceus was earlier reported (Kohn 1955) to feed on other gastropods. Small species of the type on which *C. pennaceus* thrives (especially *Haminoea*) are abundant on benches, so the low density of *C. pennaceus* does not seem the result of limited food supply. Rather, *C. pennaceus* is probably limited by lack of sufficient sand in which to burrow during the day.

Contents of alimentary tracts of 343 specimens collected at Station 5 were determined by dissection of fixed specimens or collection of fecal matter from living individuals. From these, 210 prey organisms were identified, 199 of them to species. All eight species of *Conus* are represented in the sample, but food remains were not found in any alimentary tracts of the few *C. rattus* and *C. pennaceus* examined. The results are summarized in Table 8, which also includes totals from other bench stations for comparison.

These data show that at Station 5, *Perinereis helleri* is the primary food of the three most abundant species of *Conus*. *C. ebraeus* eats this species almost exclusively, while *C. abbreviatus* especially feeds on other polychaetes as well. Members of the superfamily Eunicea (only the families Eunicidae, Lumbrinereidae, and Arabellidae are represented) are eaten about as often as nereids. *C. sponsalis,* on the other hand, is more restricted to nereids. This may reduce the possibility of competition for food between these two species, which feed in the same zone of the bench. Most of the population of *C. ebraeus* is found feeding in the central portion of the bench (Fig. 20) and is thus seeking *P. helleri* in a different place from *C. sponsalis* and *C. abbreviatus.*

The food of the fourth commonest species, *Conus chaldaeus,* is strikingly different. *C. chaldaeus* also feeds exclusively on a single species of nereid, but its food is *Platynereis dumerilii,* which occurs predominantly toward the outer edge. A large substratum sample collected near the outer edge at Station 5, adjacent to two specimens of *C. chaldaeus,* contained the following Nereidae:

Platynereis dumerilii	16
Nereis jacksoni	14
Perinereis helleri	1
Unidentified epitokes	3

Thus *P. dumerilii* is an order of magnitude more abundant than *P. helleri* near the outer edge. The prey species of *C. chaldaeus* is thus correlated with the distribution pattern across the reef platform, as both prey and predator are less abundant near shore.

The three most abundant species of *Conus* exert an active demand on the local population of *Perinereis helleri.* If this demand were found to exceed the immediate supply of the prey species, it could be said that the three predator species compete with each other for this food, at least in areas of distribution overlap.

Food organisms were found in about 60% of all alimentary tracts of specimens examined from Station 5. Since specimens were collected at all hours, and since feeding has been shown to take place only at night, it is possible that some individuals had defecated remains of the previous night's meal before being examined. A mean of one polychaete per gastropod per night is a reasonable estimate of feeding rate.

The mean density of the *Conus* species that feed on *Perinereis helleri* is 2/100 sq ft. Since *P. helleri* constitutes about 74% of their diet, these species consume an average of 1.5 individuals of *P. helleri*/100 sq ft/night, or 0.17/m.²/night. Substratum samples taken about halfway across the bench contained the densities of nereids shown in Table 9. At the calculated feeding rate, about 28 years would be required to exhaust the observed population of *P. helleri,* considering no replacement.

TABLE 9. Population Density of Nereidae and Eunicea Halfway Across Station 5.

Species	No. counted in 625 cm.²	No./m.²	Eaten by *Conus* /m.²/day
Perinereis helleri....	121	1,940	0.17
Nereis jacksoni......	267	4,270	
Platynereis dumerilii.	7	112	
All Eunicea........	4/30 cm.²	1,300	

Although the smaller *Nereis jacksoni* was more abundant than *Perinereis helleri* at Station 5, it was not found to be eaten by *Conus.* Elsewhere it is eaten by *C. sponsalis* (Table 8).

The feeding habits of the species at Station 11 are shown in Table 8 to be essentially similar to those at Station 5. In addition, the few specimens of *Conus rattus* collected there were found to consume both nereid and eunicid polychaetes. The single specimen of *C. catus* collected had an empty alimentary tract.

No quantitative samples of polychaetes at Station 11 were analyzed. However, *Perinereis helleri* was observed to be common. Each of two samples, representing a few square centimeters of substratum surface, contained 9 polychaetes, of which 3 and 5, respectively, were *P. helleri.* Standing crop is probably of the same order as at Station 5.

Water-leveled Bench Stations. Alimentary tracts of 92 specimens of *Conus sponsalis* from Station A3 were analyzed. Of these, 90 were collected between the hours 0700 and 0900. Polychaete remains were found in 55 specimens. The frequency distribution of remains in alimentary tracts is shown in Fig. 24. The low frequency of esophagus contents indicates that feeding had ceased some time before collection, probably at or before dawn. Most of the remains are seen to be in the intestine and/or rectum. This is in essential agreement with data from other stations.

Fifty-two of the specimens examined contained remains of one polychaete in each, two contained remains of two polychaetes each, and one contained remains of three polychaetes. It may be concluded that most individuals succeed in capturing one polychaete per night.

The nature of the food of *Conus sponsalis* at Station A3, shown in Table 8, differs markedly from that of other bench stations. The primary prey organism is not a nereid, but the eunicid, *Eunice (Nicidion) cariboea*. *Perinereis helleri* was not found to be eaten at all (but its presence at Station A3 was not ascertained). This resulted in eunicids far exceeding nereids in the prey of *C. sponsalis* at Station A3, in contrast with other stations studied. This may be correlated with the fact that *C. abbreviatus,* which generally (Table 8) feeds on eunicids more often than nereids, is virtually absent from Station A3. It may be conjectured that where the two co-occur, they compete for eunicids, with the result that *C. abbreviatus* is the more successful, and *C. sponsalis* is forced to eat nereids, which are possibly less desirable as food. When *C. abbreviatus* is excluded for other reasons from a microhabitat where *C. sponsalis* does occur, the latter species would then be able to exploit eunicids as food.

As for other polychaete feeders at Station A3, remains of *Eunice antennata* were found in three of the seven specimens of *Conus rattus* which were examined, and the single *C. abbreviatus* had fed on a *Lysidice collaris.*

The food of *Conus catus* has previously been shown to be small fishes, chiefly blennies and gobies, and the feeding process has been briefly described (Kohn 1956b). Remains of fishes were found in two of the three specimens collected at Station A3. One of the fishes was identified as the goby, *Bathygobius fuscus.* The other was too poorly preserved to permit identification. The food of *C. catus* at bench stations is summarized in Table 10.

TABLE 10. Prey Organisms Consumed by *Conus catus* at Marine Bench Stations.

	Sta. K1	All Bench Stations
No. Specimens Examined	13	24
Bathygobius fuscus (Rüppell)	2	3
Istiblennius gibbifrons (Quoy and Gaimard)	3	3
Unidentified fishes	2	5
Total Fishes	7	11
Unidentified fecal matter		1

Only small samples of most species from Station K1 were examined for alimentary tract contents. The expected prey organism, *Perinereis helleri,* dominated in *Conus sponsalis* and *C. ebraeus.* Nine of 10 food organisms isolated from *C. abbreviatus* were *Lysidice collaris.* Thirty-eight specimens of *C. rattus* were analyzed. Eleven contained remains of *P. helleri,* one, *Eunice antennata,* and the remaining 26 were devoid of identifiable food. Results of analysis of alimentary tracts of 13 specimens of *C. catus* are shown in Table 10. The single specimen of *C. retifer* was found to have an empty alimentary tract. However, other members of its subgenus (*Cylinder*) feed exclusively on other gastropods, and it is likely that *C. retifer* does also.

Polychaetes did not appear as abundant as on solution benches, but no quantitative samples were collected. *Perinereis helleri* and *Lumbrinereis* sp. 239 were observed to be present. Three of the eight species of *Conus* collected feed primarily on *P. helleri.* It is not known whether the demand exceeds the immediate supply and, therefore, whether food is the requisite which governs population size. The possibility of competition for food between *C. catus* and the other species present is entirely precluded, as it feeds on fishes.

FOOD PREFERENCE

Of the species of *Conus* characteristic of the marine bench habitat, *C. sponsalis* and *C. abbreviatus* have been shown to be most similar to each other with respect to feeding habits and pattern of distribution. Both species feed exclusively on polychaetes. Frequency of different prey species found in alimentary tracts is shown in Table 8. If data from all marine benches studied are combined (Column T), polychaetes of the superfamily Eunicea comprise 68%, and Nereidae 26% of the diet of *C. abbreviatus.* The diet of *C. sponsalis* consists of 44% Eunicea and 52% Nereidae. Although feeding habits are rather similar, the difference between the two species is significant ($P = 10^{-3}$) in the samples analyzed.

Polychaetes of both groups are abundant (Table 9). The observed differences in feeding frequency may be accounted for by two alternative hypotheses: (1) *C. abbreviatus* is better adapted to feeding on eunicids, which burrow into limestone and coral, than on nereids, which occur epifaunally among the holdfasts of algae. Conversely, *C. sponsalis* is better adapted to feeding on nereids than eunicids. (2) *C. abbreviatus* exhibits active preference for eunicids over nereids, and *C. sponsalis* prefers nereids to eunicids.

Comparison of results of food studies at Station A3 with other stations provided at least some evidence in favor of the view that, where the two species co-occur, *C. abbreviatus* is the more efficient predator on eunicids.

The second hyopthesis was tested experimentally in a choice chamber, following the method of Van Dongen (1956). The chamber was a lead-sheathed wooden sea water table 125 × 58 cm in area and 8 cm deep (Fig. 26). Polychaetes were placed in Bull Durham bags, which were secured to the chamber floor. The water current thus passed over the polychaetes and then over the snails. All stimuli received by *Conus* from the polychaetes were thus of a chemical nature. Actual predation was prevented by the cloth bags, from which snails were unable to extract the polychaetes. In each experiment, one or two polychaetes were placed in each bag, and 12-59 specimens of a single species of *Conus* were placed at a distance of 88 cm from the goal (Line cd, Fig. 26).

Experiments were usually begun in late afternoon and allowed to continue overnight, to coincide with the snails' normal food-seeking regimen. A few of

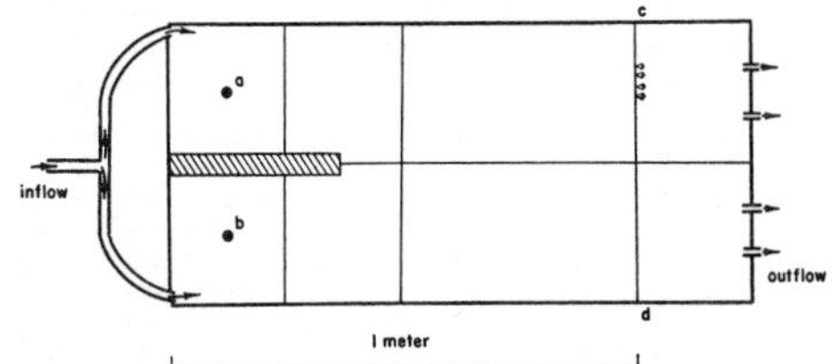

FIG. 26. Choice chamber for determination of food preferences of *Conus*. Lines were marked on floor of chamber as shown. During an experiment, fresh sea water was admitted to the chamber equally through the two inflow tubes. Outlets were located 1 cm above the chamber floor, so that a uniform depth of 1 cm of water was maintained in the chamber. A wooden partition (hatched) 40 cm in length partially separated the two portions of the chamber. This served to prevent mixing of stimuli and alteration of choice by snails. Food chambers are shown at a and b. Line cd is the starting line along which snails were placed at outset of an experiment.

the experiments with *Conus sponsalis* were allowed to run 40 hrs. At the termination of an experiment, snails were scored as having reached to within 25-50 cm of the goal, less than 25 cm, or adjacent to the food chamber. In the summary of results (Table 11), all three categories are summed.

TABLE 11. Summary of Results of Food Choice Experiments with *Conus sponsalis* and *C. abbreviatus.*

Number of Experiments	Number of Snails	Nereidae* Chosen	Eunicidae Chosen
C. sponsalis			
6	222	42	28**
2	61	22	2***
Total 8	283	64	30
C. abbreviatus			
2	87	17	35**

Figures in body of table indicate number of snails choosing polychaete listed above in choice chamber experiments.
*=*Perinereis helleri*
**=*Eunice antennata*
***=*Palola siciliensis*

Table 11 shows that, given equal choice of both, *Conus abbreviatus* chooses the eunicid (*Eunice antennata*) more often than it chooses the nereid (*Perinereis helleri*). *C. sponsalis* chooses the nereid (*P helleri*) more often than either of the eunicids tested (*E. antennata, Palola siciliensis*). The difference in preference for eunicids and nereids between *C. sponsalis* and *C. abbreviatus* is highly significant ($P < .01$).

Comparison of Food in Nature and Choice Experiments. A comparison between the experimental choices and food in nature is made in Table 12. Since the percentages given do not permit direct comparison between the experiments and nature, the ratios of eunicids to nereids for *Conus abbreviatus* and *C. sponsalis* are included. Variation of the relative frequency of predation on nereids and eunicids in nature from the frequency of choice in the experiments is significant for *C. sponsalis* ($P = .05$) but not for *C. abbreviatus* ($P \gg .1$).

TABLE 12. Comparison of Results of Choice Chamber Experiments with Food Habits of *Conus sponsalis* and *C. abbreviatus* in Nature.

	Food in Nature (all marine benches)	Choice in Experiments
C. sponsalis		
Nereidae	52%	23%
Eunicea	44%	11%
Ratio E:N	1:1.3	1:2.1
C. abbreviatus		
Nereidae	26%	20%
Eunicea	68%	40%
Ratio E:N	2.6:1	2.0:1

Comparison of data for nature and experiments shows that the food of *Conus sponsalis* and *C. abbreviatus* in nature is reflected in the choice experiments. The partition of the environment into different but overlapping microhabitats which occurs with respect to food may therefore be maintained by active preference of different prey species by the two predator species.

The hypothesis that *Conus sponsalis* is better adapted to feeding on nereids than eunicids was not amenable to experimental test. It has been suggested, however, that information bearing on this question might be gained from study of radula morphology (Peile 1939).

Although the anterior extremities of radula teeth show considerable variation throughout the genus *Conus,* the teeth may be grouped in two categories with respect to the posterior portion, or base. In the first, the base is simple and of more or less greater diameter than the shaft, thus forming a terminal knob. The base in the second group is characterized by the presence of a forward projecting cone (Fig. 27).

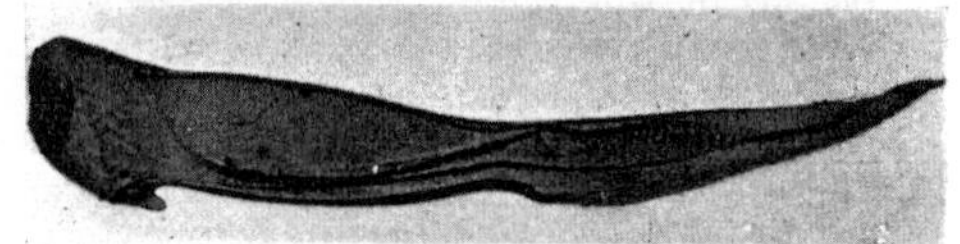

FIG. 27. Radula tooth of *Conus abbreviatus.* The forward projecting cone is visible at the left.

Peile (1939) made the plausible suggestion that this cone might serve to retain the tooth within the proboscis when prey is attacked. It was shown, however, that *C. striatus* which does not possess such a cone, retains the tooth in the proboscis in feeding (Kohn 1956b).

It will be shown below that the presence of the forward projecting cone is generally correlated with feeding on eunicid and other tube-dwelling polychaetes in nature. Eunicids, unlike nereids, live in burrows in coral and reef rock, and the basal cone may well aid the predator in extracting the worm from its burrow.

TABLE 13. Prey Organisms Consumed by Vermivorous Species of *Conus* at Subtidal Reef Stations. **Numbers in body of table indicate number of polychaete species at side found in alimentary tracts of *Conus* species at top.**

	sponsalis			*abbreviatus*				*ebraeus*				*chaldaeus*		*miles*		*rattus*			*distans*	
Station...	7	R	T	3	9	R	T	3	9	R	T	R	T	R	T	15	R	T	R	T
No. specimens examined....	64	72	330	19	24	99	342	15	8	55	199	4	106	11	20	18	55	149	19	21
Nereis jacksoni Kinberg var........	4	4	34	—	—	—	—	—	—	—	—	—	—	—	—	—	—	—	—	—
Perinereis helleri Grube...........	3	3	35	—	—	—	20	—	1	6	136	—	5	—	—	7	7	21	—	—
Platynereis dumerilii A. & E.......	3	3	13	—	—	—	3	—	—	—	—	2	98	—	—	—	1	1	—	—
Nereid sp. 350..................	1	1	5	—	—	—	—	—	—	—	—	—	—	—	—	—	—	—	—	—
Unidentified Nereidae............	1	1	2	—	—	—	1	—	—	—	—	—	—	—	—	—	—	—	—	—
Total Nereidae..................	12	12	89	—	—	—	24	—	1	6	136	2	103	—	—	7	8	22	—	—
Lysidice collaris Grube............	6	6	17	—	—	9	36	—	—	2	2	—	—	5	16	—	—	—	—	—
Palola siciliensis (Grube)........	—	—	—	1	—	1	1	14	8	29	44	—	14	—	—	—	—	—	—	—
Eunice antennata Savigny.........	1	1	3	5	10	20	22	—	—	—	—	—	—	1	1	3	17	29	1	1
Eunice (*N.*) *cariboea* Grube........	—	3	39	—	—	4	14	—	—	1	1	—	—	—	—	1	1	2	—	—
Eunice afra Peters...............	—	—	—	—	—	—	—	—	—	—	—	—	—	—	—	4	9	9	12	13
Marphysa sanguinea Montagu.....	—	—	1	1	1	3	6	—	—	—	—	—	—	—	—	—	—	—	—	—
Eunice filamentosa Grube..........	—	—	1	—	—	—	3	—	—	—	—	—	—	—	—	—	—	—	—	—
Unidentified Eunicidae...........	—	—	1	1	—	1	2	—	—	—	1	—	1	—	—	—	—	—	—	—
Lumbrinereis sarsi (Kinberg)......	2	3	17	—	—	—	11	—	—	—	—	—	—	—	—	—	—	—	—	—
Arabella iricolor (Montagu)........	—	—	—	—	—	5	9	—	—	—	—	—	—	—	—	—	—	—	—	—
Total Eunicea...................	9	13	79	8	11	43	104	14	8	32	48	—	15	6	17	8	27	40	13	14
Eurythoe complanata (Pallas)......	—	—	—	—	—	—	—	—	—	—	—	—	—	—	—	—	—	—	—	—
Unidentified Annelids.............	1	1	7	—	—	—	4	—	—	1	2	—	—	—	—	—	—	—	—	—
Total Annelids..................	22	26	175	8	11	43	132	14	9	39	186	2	118	6	17	15	35	62	13	14
Total Identified Food.............	22	26	175	8	11	43	132	14	9	39	186	2	118	6	17	15	35	62	13	14
Unidentified Food................	2	2	6	—	—	—	4	—	—	—	—	—	1	—	—	—	—	—	1	1

SUMMARY OF FOOD OF *Conus* IN THE MARINE BENCH HABITAT

Conus abbreviatus and *C. sponsalis,* both of which possess the small cone at the base of the radula tooth, feed on polychaetes which belong to different groups, the Nereidae and Eunicea, and differ ecologically as noted above. Although there is overlap (63%), the difference in numbers of the two groups of prey organisms eaten by *C. abbreviatus* and *C. sponsalis* is highly significant ($P = 10^{-3}$). Comparison by individual species of the food eaten by these two *Conus* species on all bench stations (Table 8) revealed overlap of only 25% and the probability of only 10^{-6} that the polychaetes preyed on represented random samples from the same population. Differences between the species composition of the food of *C. sponsalis* and *C. abbreviatus* are thus sufficiently great that competition for food is unlikely.

Of the other species which occur in the marine bench habitat, *Conus ebraeus* and *C. chaldaeus* do not possess the forward projecting cone on the radula tooth. Both species are extremely oligophagous. Nereids comprise 85% of the food of *C. chaldaeus* and 93% of the food of *C. ebraeus.* The two species do not differ significantly in the proportion of the diet comprised by nereids ($P = .24$). In striking contrast to this, however, is the specific nature of the prey eaten, as shown in Table 8. In the samples collected on all marine benches, the only nereid eaten by *C. ebraeus* was *Perinereis helleri,* while the only nereid eaten by *C. chaldaeus* was *Platynereis dumerilii.* Despite the fact that both predators are typically found on the same benches, and with overlapping distributions, they were never found to eat the "wrong" polychaete in the bench habitat.

Unfortunately, no choice experiments were carried out in order to test the possibility of differences in active preference of prey species by *Conus ebraeus* and *C. chaldaeus.* It is likely that the specific differences in prey are correlated with differences in the ecology of the prey species. In Table 9 *Perinereis helleri* was shown to be two orders of magnitude more abundant than *Platynereis dumerilii* about halfway across the bench at Station 5, where *C. ebraeus* is maximally abundant. Nearer the outer edge, where *C. chaldaeus* reaches its peak density, *P. dumerilii* is an order of magnitude more abundant than *P. helleri.*

R = All reefs; T = All stations; * = All from Station 4; + = Equals Total.

vexillum			*vitulinus*	*imperialis*	*pulicarius*		*flavidus*				*lividus*					
9	R	T	T*	R+	9	R+	3	9	R	T	3	7	9	R	T	Station
13	16	17	4	31	11	20	47	56	182	192	32	46	34	216	240	No. Specimens Examined
—	—	—	—	—	—	—	—	—	—	—	—	—	—	—	—	*Nereis jacksoni* Kinberg var.
—	—	—	—	—	—	—	—	—	—	—	—	—	—	—	—	*Perinereis helleri* Grube
—	—	—	—	—	—	—	—	—	—	—	1	5	—	11	12	*Platynereis dumerilii* A. & E.
—	—	—	—	—	—	—	—	—	—	—	—	—	—	—	—	Nereid sp. 350
—	—	—	—	—	—	—	—	—	—	—	—	—	—	—	—	Unidentified Nereidae
—	—	—	—	—	—	—	—	—	—	—	1	5	—	11	12	Total Nereidae
—	—	—	—	—	2	2	5	5	24	24	—	—	—	—	—	Capitellid sp. 1040
—	—	—	—	—	—	—	—	—	11	11	—	—	—	—	—	*Thelepus setosus* Quatrefages
6	9	9	—	—	—	—	13	5	22	22	—	—	—	1	1	*Polycirrus* sp. 660
—	—	—	—	—	—	—	—	—	2	5	—	—	1	1	1	Terebellid sp. 837
—	—	—	1	—	—	—	10	17	32	34	2	—	—	7	11	*Nicolea gracilibranchus* (Grube)
5	6	6	—	4	—	—	—	—	—	—	—	—	—	1	2	Unidentified Terebellidae
—	—	—	—	—												
—	—	—	—	—	—	—	23	22	67	72	2	—	1	10	15	Total Terebellidae
—	—	—	—	—												
—	—	—	—	—	—	—	—	—	5	5	—	—	—	1	1	Polydorid sp. 1500
					—	—	—	—	1	1	—	1	1	4	5	*Cirriformia semicincta* (Ehlers)
11	15	15	1	4	—	—	—	—	—	—	—	—	—	2	2	*Lygdamis nesiotes* (Chamberlin)
					—	—	—	—	—	—	—	—	—	3	3	*Sabellastarte indica* Savigny
—	—	—	—	11												
—	—	—	—	—	—	—	1	3	4	5	—	1	—	2	3	Unidentified Annelids
11	15	15	1	15	—	—	29	30	101	108	3	7	2	33	41	Total Annelids
					—	2	—	—	—	—	—	—	—	—	—	*Thalassema* sp.
					—	—	1	—	1	1	13	5	8	28	33	*Ptychodera flava laysanica* Spengel
					—	—	—	—	1	4	2	5	6	29	32	*P. flava laysanica* Spengel (tentative identification)
					—	—	—	—	—	—	—	—	1	1	1	*Octopus* sp.?
11	15	15	1	15	2	4	30	30	106	113	18	17	16	90	106	Total Identified Food
2	—	2	1	2	—	3	7	5	18	21	5	10	6	39	42	Unidentified Food

Conus rattus, the other vermivorous species characteristic of marine benches, does possess the basal cone on the radula tooth. It feeds predominantly on eunicids (Table 8) and in the proportion of Eunicea eaten does not differ significantly from *C. abbreviatus*. However, the specific nature of the food of these two species is significantly different ($P = 10^{-6}$). In fact, the nature of the food of each vermivorous species of the marine bench habitat differs significantly from that of all others at the 1% level. Thus these species, as well as *C. catus* which eats only fishes, avoid interspecific competition for food in the bench habitat.

NATURE OF THE FOOD OF CONUS ON SUBTIDAL REEFS

Of the 16 species of *Conus* collected by the writer on reef platforms, 10 were found to feed exclusively on polychaetes. These are *C. abbreviatus, C. ebraeus, C. sponsalis, C. rattus, C. imperialis, C. chaldaeus, C. vexillum, C. distans, C. vitulinus,* and *C. miles.* Polychaetes constitute more than 90% of the food of *C. flavidus,* which occasionally consumes an unsegmented worm. Polychaetes comprise about 50% of the food of *C. lividus,* which feeds also on an enteropneust. *C. pulicarius* feeds on polychaetes and echiuroids, as far as is known. *C. pennaceus* and *C. marmoreus* feed on other gastropods, and *C. striatus* feeds on fishes.

VERMIVOROUS SPECIES

Species Typical of Marine Benches Which Occur also on Reef Platforms. The species composition of the food of the vermivorous species of *Conus* which occur on reef platforms is given in Table 13 for a sample of 784 specimens of 13 species. From these, 360 prey organisms were identified, 254 of them to species. Most of the records are determinations of alimentary tract contents of fixed specimens. A few represent collection of fecal matter from living individuals. Total data for certain species include records from marine bench and deep water habitats for comparison.

The more diversified food on reefs in comparison with marine benches is made possible by the addition of the sand substratum microhabitat on reef platforms. All of the prey species listed in Table 13 but not in Table 8 are associated with the sand substratum. All but one of the polychaetes eaten by

Conus on marine benches were found to be present and eaten on reef platforms. The limestone outcrops and coral of the reefs provide abundant burrowing sites for eunicids. Nereids, however, are not common on the reefs, except where an algal turf is present on exposed limestone. This is especially reflected in the feeding habits of *C. sponsalis, C. abbreviatus, C. ebraeus,* and *C. rattus. Perinereis helleri* is the dominant prey of each of these species on marine benches (Table 8), but it is much less important to all four on reefs, where it is less abundant.

On reef platforms, *Conus sponsalis* is often found in regions of dead coral and reef limestone which support an algal turf and nereids, as well as eunicids. Both groups of polychaetes are eaten by *C. sponsalis,* in frequencies which do not differ significantly from those eaten by marine bench populations of this species.

No nereids were found in alimentary tracts of 99 specimens of *Conus abbreviatus* collected on coral reefs. This is as expected, since on reef platforms *C. abbreviatus* is not found on the algae-matted areas where some nereids occur. Eunicea, which commonly burrow into any available limestone on coral reefs (Hartman 1954), apparently constitute the entire diet of *C. abbreviatus.* The frequency of different polychaete species eaten differs significantly between bench and reef populations of *C. abbreviatus.* This is due largely to the predominance of *Eunice antennata* as prey in the latter habitat.

The food of *Conus ebraeus* in reef habitats also varies from that on marine benches in the direction of increased concentration on a species of eunicid. In contrast with the previous species, however, *C. ebraeus* feeds chiefly on *Palola siciliensis* and does not eat *Eunice antennata.*

The most obvious difference between the ecological niches of *Eunice antennata* and *Palola siciliensis* is that the latter species is typically found in more dense limestone. This observation is correlated with the fact that the mandibles are larger and stronger in *Palola* than in the other genera of Eunicea. (This has been discussed by Hartman (1954), who suggests that *Palola* may be the most destructive of the Eunicea to coral reefs. Following the same criterion, *Lysidice, Eunice, Marphysa, Lumbrinereis,* and *Arabella* would follow in order of decreasing destructiveness.) Filamentous algae were observed in intestines of *E. antennata.* These are perhaps the boring forms recently discussed by Odum & Odum (1955). The food of *P. siciliensis* was not determined. Takahashi (1939) concluded that several eunicids inhabiting coral reefs are omnivorous, consuming sand, diatoms, crustaceans, and algal fragments.

Conus chaldaeus occurs rather rarely on reef platforms. *C. rattus,* on the other hand, is about as common as it is on marine benches. As in other species common to both habitats, eunicids are consumed significantly more often than nereids ($P = .03$) by reef than bench populations of *C. rattus.*

In summary, there is virtually no qualitative difference in the specific nature of the prey eaten by those species of *Conus* common to both the reef and bench habitats. The frequencies of the different species eaten vary considerably, however, and are correlated with differences in substratum characteristics of the two habitats.

Typical Reef Platform Species. The most abundant species of *Conus* on subtidal reef platforms were shown (Figs. 14-17, 19) to be *C. flavidus* and *C. lividus.* Prey organisms recovered from alimentary tracts of about 200 specimens of each of these species are listed in Table 13. It is readily apparent that *C. flavidus* feeds almost exclusively on polychaetes, that the largest single item in the food of *C. lividus* is the enteropneust, *Ptychodera flava,* and that these species feed only occasionally or never on nereids and eunicids.

Tubicolous polychaetes comprise almost the entire diet of *Conus flavidus.* About two-thirds of these are members of the family Terebellidae. The other main prey species is an unidentified species of the family Capitellidae. The terebellids on which *C. flavidus* feeds live in tubes which are usually attached to the under sides of coral rocks or pieces of rubble resting in sand. The intestines of these polychaetes are usually filled with extremely fine sand particles. The terebellids are probably selective deposit feeders, selection apparently being effected by the tentacles.

Conus flavidus consumes enteropneusts only occasionally. In contrast, *Ptychodera flava* constitutes about 50% of the diet of *C. lividus.* Since it has no hard parts, *P. flava* is particularly difficult to identify in the partially digested state. The reddish eggs, which are apparently refractory to the digestive enzymes of *Conus,* aided identification of the females eaten. However, a number of *C. lividus* alimentary tract contents, which appeared to be remains of *Ptychodera,* could not be positively identified. These are entered separately in Table 13, but they are included in "total identified food."

The ecological niche of *Ptychodera flava* differs markedly from those of the terebellids on which *Conus lividus* feeds less often. Although all are deposit feeders in the same habitat, *P. flava* moves slowly about through the sand. It appears to be a nonselective feeder, ingesting in the manner of an earthworm. A wide range of sand particle sizes is found in its alimentary tract. As there is no apparent mechanism for trituration, the particle size distribution in the tract is presumably identical with that of the environment, but this was not determined.

The feeding of *Conus lividus* on *Ptychodera flava* was observed in the laboratory. When both were placed in a dish of sea water, the rostrum, but not the proboscis, of *Conus* was extended. When the mouth touched the worm, the latter was engulfed without being stung. During the entire engulfment process, which lasted only about 15 sec, the *Ptychodera* continued its normal peristaltic pulsations. This method of feeding may not be duplicated in nature. The observed presence of radula teeth in the alimentary tract with food remains indicates that *C. lividus* stings *Ptychodera* as well as polychaetes before feeding.

Other Reef Species. The remaining six reef-inhabiting vermivorous species together comprise only about 6% of the total population of *Conus* in this habitat. It was possible to collect and analyze only small samples, but the results obtained are of considerable interest and are included in Table 13 for completeness.

The most common of these species is *Conus imperialis.* Only two prey species were found in analyses of alimentary tract contents and recovery of fecal matter of 31 specimens. One of these was the eunicid, *Marphysa sanguinea,* which is also eaten to some extent by other species of *Conus.* Notes on the ecology of this polychaete were recorded by Abbott (1946), who stated that "this species has been found only in limited areas of the pond [Wailupe Fish Pond, Oahu] where the bottom is sandy and the salinity close to that of sea water. Here it occurs under rocks, and does not appear to burrow deeply into the sand. It has not been found in regions where a soft mud bottom prevails."

The more common prey of *Conus imperialis* is the amphinomid polychaete *Eurythoe complanata.* Although this species is the most conspicuous polychaete on Hawaiian coral reefs, it was never found to be eaten by any other species of *Conus. Eurythoe* possesses extremely abundant large setae, which easily penetrate human skin and cause a burning sensation. They may possess a venom (Halstead 1956). Nevertheless, intestines of *C. imperialis* were often observed literally packed with these setae. The polychaete is typically found under rocks or in crevices in coral. Its food is unknown.

On several occasions, specimens of *Eurythoe complanata* were fed to *Conus imperialis* in the laboratory. The stinging operation is typical. Like *C. striatus, C. imperialis* uses its radula tooth as a harpoon. The tooth is not freed from the proboscis, but the impaled prey is drawn into the mouth by rapid contraction of the proboscis.

Only four prey organisms were recovered from the alimentary tracts of 20 specimens of *Conus pulicarius* which were examined. These were sufficient to show that the food of this species is not restricted to polychaetes. Two of the food organisms were of the species of capitellid eaten by *C. flavidus,* but the other two were of the echiuroid worm, *Thalassema* sp. Both of the latter were regurgitated by the snails after capture. At least one of the echiuroids was alive when regurgitated. This suggests that *Thalassema* may not be stung before being swallowed by *C. pulicarius.*

All of the other four vermivorous species of *Conus* collected on the reefs may be restricted to Eunicidae for food, although the samples examined were small. The primary food of *C. distans* is *Eunice afra.* This polychaete was found most commonly completely buried in coral rocks or in coral or other calcareous encrustation on basalt boulders. *C. miles* feeds chiefly on *Lysidice collaris,* and *C. vexillum,* on *Eunice antennata* and *Marphysa sanguinea.* A single specimen of *E. afra* was recovered from the alimentary tract of *C. vitulinus.*

The large differences in the specific nature of the prey of the vermivorous species of *Conus* on reef platforms would seem to virtually preclude the possibility of interspecific competition for food. It was nevertheless of interest to obtain some information on the abundance of the prey species.

In order to determine the species and abundance of polychaetes associated with, or burrowing into, limestone, coral and coral ruble, sample blocks of substratum were removed from reefs and placed in sea water in sealed jars. When the oxygen tension decreased, the polychaetes were attracted out of their burrows and eventually fell to the bottom of the jar. After 2-3 days, the contents of the jar were fixed in 10% formalin. Cracking of pieces of coral revealed few or no polychaetes remaining within the blocks after this treatment. Dry weight, volume and surface (projection) area of the blocks were determined. Polychaetes equal to or greater than 0.3 mm in maximum diameter were identified and preserved.

The results of samples from Stations 3 and 7 treated in this manner are shown in Table 14. In suitable areas on reefs eunicids are seen to be extremely abundant. It is difficult, however, to assess the density of polychaetes over the entire reef platform, since the relative areas of different types of substratum could not be adequately measured. The data in Table 14 probably give the correct order of magnitude for areas with a predominantly lithified substratum.

MOLLUSCIVOROUS SPECIES

Two species of *Conus* found on subtidal reef platforms appear to feed exclusively on other gastropod mollusks. The species composition of their prey, as well as that of *C. textile,* found rarely on reefs, is shown in Table 15. In addition to specimens collected on reefs, total data include material collected in deeper water.

At least 13 species of prosobranchiate and opisthobranchiate gastropods were recorded from alimentary tract analyses of 146 specimens of *Conus pennaceus.* The most common food species is the bubble shell, *Haminoea crocata.* Second commonest is the small prosobranch, *Phasianella variabilis.* Since the shells of prosobranchs are not swallowed, the identification of partly digested remains was extremely difficult. About half of the prey organisms were identified to species, but this was not possible in some which were represented only by radulae and/or opercula.

Feeding of *Conus pennaceus* on *Haminoea crocata, Terebra gouldii* Deshayes, and *Cypraea maculifera* (Schilder) was observed in the laboratory. A specimen of *Cypraea moneta* Linné was not eaten, however, when left in a tank with two *C. pennaceus* for 24 hr. *C. pennaceus* did not prey on other species of *Conus,*

TABLE 14. Abundance of Polychaetes Associated with Hard Substrata on Reefs.

Sample No.	Station	Area (cm.^{2})	Species	Number Present
1269	7	50	*Eunice (Nicidion) cariboea*	6
			Cirriformia semicincta	2
			Eunice antennata	1
			Unidentified Polychaetes	6
				—
			Total	15
			Density of Large Polychaetes	3,000/m.^{2}
			Density of Eunicidae	1,800/m.^{2}
			(Number of Polychaetes <0.3 mm. diameter in sample	51)
1276	3	129	*Eunice antennata*	5
			Palola siciliensis	4
			Lysidice collaris	3
			Eunice (Nicidion) cariboea	2
			Cirriformia semicincta	1
			Eurythoe complanata	1
			Platynereis dumerilii	1
				—
			Total	17
			Density of Large Polychaetes	1,300/m.^{2}
			Density of Eunicidae	1,100/m.^{2}

although several were retained in the same aquaria for several months.

The most striking aspect of the food of *Conus marmoreus* is that it appears to consist entirely of other species of *Conus*. Remains of *Conus* species were found in alimentary tracts of five of the seven specimens of this rather rare species which were examined. On 31 March 1956, Mr. Charles Sueishi observed a specimen in the act of feeding on *Conus abbreviatus* near Station 11. Dissection revealed the remains of a second *C. abbreviatus* in the intestine. On 6 July 1956, the writer observed a specimen in the act of feeding on *C. lividus* at Station 9.

A specimen of *Conus textile* collected at Station 15 by Miss Valerie Lang contained the radula sheath, operculum, and other remains of a *Conus pennaceus* in its alimentary tract. Another specimen, which had just eaten a *Conus striatus*, was collected by Dr. C. M. Burgess near Makua, Oahu. One other gastropod, a *Morula ochrostoma* (Blainville), was identified from the alimentary tracts of *C. textile* examined. In the laboratory, specimens of *C. textile* were observed to sting and consume *C. abbreviatus*, *C. ebraeus*, *C. lividus*, *Cypraea caputserpentis*, *Cypraea moneta*, *Turbo intercostalis* Menke, Thais *aperta* Blainville, and *Drupa morum* Lamarck, but they did not sting *C. flavidus* or *Helcioniscus argentatus* Nuttall.

PISCIVOROUS SPECIES

Conus striatus occurs sparsely on Hawaiian reef platforms. Remains of fishes from the five specimens which were examined could not be identified. A goatfish (*Parupeneus* sp.) is known to be eaten by *C. striatus* in Micronesia (unpublished data). The other known piscivorous species, *C. catus* and *C. obscurus*, were not collected by the writer on reef platforms, but

TABLE 15. Food of Molluscivorous Species of *Conus* on Reefs.

	C. pennaceus	*C. marmoreus*		*C. textile*	
	All Reefs	All Reefs	Total	All Reefs	Total
No. Specimens Examined...	146	3	7	2	10
Haminoea crocata Pease	24	—	—	—	—
Haminoea sp. cf. *H. aperta* Pease	2	—	—	—	—
Haminoea sp. 1057	1	—	—	—	—
Phasianella variabilis Pease	12	—	—	—	—
Dolabrifera olivacea Pease	7	—	—	—	—
Gastropod sp. 1963	4	—	—	—	—
Gastropod sp. 1964	4	—	—	—	—
Trochus intextus Kiener	3	—	—	—	—
Turbo intercostalis Menke	2	—	—	—	—
Pleurobranchus sp. 1064	2	—	—	—	—
Gastropod sp. 988	2	—	—	—	—
Natica marochiensis Gmelin	1	—	—	—	—
Conus abbreviatus Reeve	—	1	3	—	—
Conus lividus Hwass *in* Bruguiere	—	1	2	—	—
Conus sp.	—	—	1	—	—
Conus pennaceus Born	—	—	—	1	1
Conus striatus Linne	—	—	—	—	1
Morula ochrostoma Blainville	—	—	—	—	1
Unidentified Gastropods	7	—	—	—	3
Total Gastropods	71	2	6	1	6
Unidentified Food	23	1	1	1	1

References consulted in the identification of the mollusks in this table were Edmondson (1946), Ostergaard (1955), Pease (1860), and Pilsbry (1917, 1920).

Ostergaard (1950) reported the former species from Station 3.

SUMMARY OF FOOD AND FEEDING IN THE SUBTIDAL REEF HABITAT

Modifications of radula teeth which can be correlated with increased frequency of feeding on tube-dwelling polychaetes are present in many of the vermivorous species of *Conus* characteristic of the subtidal reef platform habitat. A forward projecting cone on the base, which may aid in retaining the tooth in the proboscis while the prey is extracted from its tube, is present in the two dominant reef species, *C. flavidus* and *C. lividus*. The former preys almost exclusively on tube-dwelling Terebellidae. Polychaetes preyed on by *C. lividus* are chiefly terebellids, but the dominant food species is the enteropneust *Ptychodera flava*. In addition to the basal cone, both *C. flavidus* and *C. lividus* possess a backward projecting spur on the shaft of the radula tooth, about one-third of the length from the base. Peile (1939) suggested that this structure may serve to prevent the tooth from being forced back into the proboscis on the impact of the sting.

The chief differences in the radula teeth of *Conus lividus* and *C. flavidus* are that the latter is shorter and is finely serrate (in disagreement with Peile 1939), while the former has a long shaft and no serrations. There is no obvious adaptive significance to these differences; nevertheless, the difference in nature of the food of the two species is highly significant ($P < 10^{-6}$).

The characteristics of the radula teeth of *Conus*

sponsalis, C. abbreviatus, C. ebraeus, and *C. rattus* were discussed above. Differences in the feeding habits of these species on reef platforms are apparent from inspection of Table 13. The feeding habits of *C. abbreviatus* and *C. rattus* are most similar to each other, but even these differ highly significantly ($P = 10^{-6}$) in the samples examined.

Conus imperialis, C. miles, C. vexillum, and *C. vitulinus* all possess the forward projecting cone on the base of the radula teeth. All feed primarily on polychaetes which either burrow into coral (Eunicidae) or live under rocks or in crevices (*Eurythoe*). The presence of the cone is therefore correlated with the habit of feeding on burrowing polychaetes, and it may well serve the function of aiding in the extraction of the worm from its tube, as suggested by Peile (1939). Distinct differences in the nature of the food of all of these species is also apparent from Table 13, although the samples are rather small for rigorous statistical analysis.

Conus distans, which also feeds on burrowing eunicids, does not possess the basal cone on the radula tooth, but the terminal knob is extremely large and may have the same function. The food of *C. distans* is similar to that of *C. rattus,* but the differences between them are highly significant ($P < 10^{-3}$).

On reef platforms as on marine benches, the nature of the food of each vermivorous species of *Conus* differs significantly from that of all other species. Thus these species avoid competition for food among each other. In addition, they are completely ecologically isolated, with respect to food, from *C. pennaceus,* which feeds only on a large number of other gastropods. Two other molluscivorous species, *C. marmoreus* and *C. textile,* occur only very rarely on reef platforms. *C. striatus* is the only piscivorous species collected by the author on the reefs, although small numbers of *C. catus* also occur in this habitat.

AMOUNT OF FOOD EATEN

Thorson (1956) has recently called attention to the need for more information on the quantity of food consumed by predators in order to understand the dynamics of benthic communities. A few data on predatory benthic fishes show that the daily food consumption is equal to 3-5% of the weight of plaice (Dawes 1930, 1931). Smith (1950) calculated rates of 1.1-2.4% in benthic fishes of Block Island Sound. Among the gastropods, young, growing specimens of *Polinices duplicata* Say, feeding on *Gemma gemma* Totten, consumed about 7% of their own weight per day. The rate in older *Polinices* was about 5% (Turner 1951). Higher rates of 10-25% were calculated for other predatory gastropods by Thorson (1958).

In the data on *Conus* which follow, dry weight of the predator (excluding shell) is compared with dry weight of the prey. Dry weights of *Conus* were measured following heating at 100°C for 48-96 hr, or by estimation from "alcohol weight" using the conversion given by Holme (1953). Dry weights of polychaetes were determined by the method of Holme.

Enough specimens of *Perinereis helleri* were available to establish the relationship between maxilla length and body weight. Thus the weight of the prey organisms could be determined even if only the maxillae were found in the alimentary tract of the predator. Using data obtained with this method as well as from direct weighing of intact polychaetes found in crops, the amount of food (17 *P. helleri*) consumed by 17 adult *Conus ebraeus* was calculated to average 0.0049 ± 0.0005 g. Dry weight of the predators was 0.43 ± 0.02 g. Since the daily feeding rate was estimated to be one polychaete per snail, the daily food consumption of *C. ebraeus* is equal to about 1% of its own body weight. Average quantitative food consumption of the four dominant species of *Conus* on marine benches, in per cent of body weight per day, was:

C. sponsalis	4.6%
C. abbreviatus	3.4%
C. chaldaeus	1.2%
C. ebraeus	1.2%

The two smaller species consume proportionately more food per day than the larger species. The rates are somewhat lower than those cited above for other predatory gastropods. The reason is probably that all of the other figures are for temperate species, many of which do not feed in winter. Since the time available for feeding is shorter, the feeding rate during the season is likely to be higher than in a tropical gastropod such as *Conus,* which feeds at a constant rate throughout the year.

PREDATION ON *CONUS*

The extent of predation on *Conus* is difficult to evaluate. In the course of collecting trips, freshly dead fragments of *C. catus, C. rattus, C. abbreviatus,* and *C. flavidus* were observed on marine benches. In some cases, shells had been broken into many pieces so recently that most of the pieces were present within a few square centimeters. On reef platforms, shell fragments of freshly killed specimens of 11 species of *Conus* were collected. Fragments of the relatively thin-shelled *C. pennaceus* were found most often. Surprisingly, fragments of the extremely thick-shelled *C. flavidus* were second commonest. These cases are believed due to predation. The identity of the predators is not known, but parrot fishes (Scaridae) and the zebra eel, *Echidna zebra* (Shaw) are possibly responsible.

Other organisms, including the other species of *Conus* mentioned in the previous section, other gastropods (*Cymatium*), and starfish (*Asterope*), were observed to prey on *Conus* in the laboratory.

Cymatium nicobaricum Röding readily attacked *Conus* in laboratory aquaria. A large specimen (shell length 84 mm) devoured specimens of *C. ebraeus, C. abbreviatus,* and *C. catus.* Dead *C. ebraeus* were also eaten. In one case, a live *Conus ebraeus* was attacked by *Cymatium nicobaricum* about 20 min after the two were placed in the same aquarium. The predator introduced its proboscis into the aperture of the *Conus* shell, and apparently began to rasp off pieces of the foot with its radula. This position was maintained for 9 days, after which the empty *Conus* shell was released. *Cymatium,* which is represented by several species in Hawaii, is a likely predator of *Conus* in nature.

Although specimens of the starfish *Asterope carinifera* Lamarck were kept for months in aquaria with several species of *Conus,* predation was observed only rarely. It is not known whether *Asterope* preys on *Conus* in nature. The starfish is not very common, and its habitat is often not shared by *Conus.* Both *Asterope* and *C. ebraeus,* which the starfish ate in the aquarium, were however found at Station 13.

Xanthid crabs also attacked *Conus* in laboratory aquaria. The attacks usually resulted in the outer lip of the shell being broken off. Crabs were never observed to succeed in killing the gastropods, probably because the latter could retract farther into the shell, and the older portions of the shell were too thick for the crabs to break. However, laboratory observations were made only on two thick-shelled species, *C. lividus* and *C. flavidus.* Crabs may prey more successfully on some of the thinner-shelled species.

Two species of fishes, a wrasse, *Stethojulis axillaris* (Quoy and Gaimard) and a goby, *Chlamydes cotticeps* (Steindachner), were reported by Strasburg (1953) to feed on the eggs of *Conus* in Hawaii. The problem of predation on larvae remains unstudied. Large numbers of free-swimming veligers are undoubtedly consumed by carnivorous planktivores. Many newly settled larvae probably fall prey to brittle stars (*Ophiocoma* spp.) which abound on the reefs.

FOOD CHAINS AND TROPHIC STRUCTURE OF THE COMMUNITY

The polychaetes on which the dominant marine bench species of *Conus* feed are mainly herbivorous. At Station 5, *Perinereis helleri* was found to feed chiefly on the blue-green alga, *Lyngbya majuscula.* Alimentary tracts of specimens collected at night were often full of the filaments of this alga. Eunicids also feed on *Lyngbya* as well as on other algae.

Since predation on adult *Conus* may be assumed to be negligible, a short food chain of three steps is indicated. Since the numbers of species at the two higher trophic levels are large, increased efficiency associated with a restricted diet can be achieved without detracting from community stability (MacArthur 1955).

Data presented above for biomass of polychaetes and *Conus,* together with dry weights of algal samples collected about halfway across the bench at Station 5, were used to calculate the biomass pyramid for the algae-polychaete-*Conus* food chain shown in Fig. 28. The dominant alga in the sample was *Laurencia* sp. Since this does not appear to be eaten by the polychaetes under discussion, only *Lyngbya majuscula,* which represented 7% of the algae sample, is included in the pyramid.

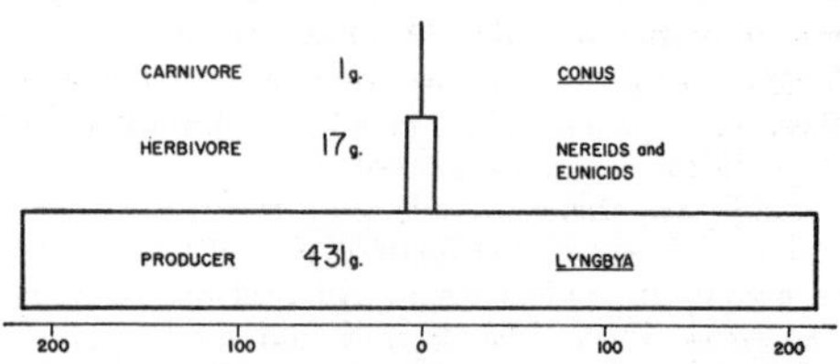

FIG. 28. Pyramid of biomass for the algae-polychaete-*Conus* food chain at Station 5.

For this food chain, the ratios of standing crop between trophic levels are herbivores/producers 4% and carnivores/herbivores 6%. It is to be noted that these ratios apply only to the single food *chain* considered and not to the entire community or food *web.*

ECOLOGICAL NOTES ON OTHER SUBTIDAL HABITATS

The most important habitats of *Conus* in the Hawaiian Islands are the intertidal marine benches and subtidal reef platforms which have been discussed in the previous sections. The relatively fragmentary ecological information which has been gained concerning species which occur chiefly in other habitats is summarized in the following paragraphs.

Conus quercinus. Certain limited regions off the shore of Oahu, usually in bays, are characterized by vast areas of sand substratum uninterrupted by coral heads or limestone outcrops. Wave action is usually comparatively light, and the salinity is often somewhat reduced by the proximity of streams. Portions of such areas may be a foot or two above the 0 tide datum, forming a sand spit. Station 2A, Sand (Ahuolaka) Island, in Kaneohe Bay (Fig. 4), is such a formation. The habitat of *C. quercinus* appears to be restricted to areas of this type, at least at certain seasons. The writer's observations are in agreement with those of Bryan (1915), who stated "they (*C. quercinus*) appear to prefer the muddy brackish water conditions at the harbor mouth to a life on the coral reef in the open sea." The species is often common where it occurs, but its distribution in shallow water is probably limited by the sparse occurrence of favorable habitats about the Hawaiian Islands and by seasonal differences in habitat.

Collections made by the writer confirmed a marked seasonality of *Conus quercinus,* which had been verbally reported to him by a number of collectors. Collecting trips to Station 2A made by the writer between July, 1954, and June, 1956, are noted by closed circles in Fig. 29. A marked seasonal fluctuation in abundance, with maxima annually in February and March, is apparent. Since no quantitative sampling of the area was done, the absolute amplitude of the maxima is not meaningful.

The biological significance of the fluctuation in numbers of *Conus quercinus* is in reproduction.

extension of the proboscis. In only one instance was a radula tooth ejected, however. The fish, a specimen of *Bathygobius fuscus,* escaped and was not swallowed but died a few minutes later, presumably from the effects of the venom.

Conus textile. Specimens were collected rarely on the reefs and at depths of 10-75 ft. Results of alimentary tract examinations are included in Table 15.

Conus miles. This species occurs occasionally on reefs and benches but is somewhat more common in deeper water. Individuals on reefs are usually larger, however. Specimens collected in 35-40 ft off leeward Oahu were found to have fed on *Lysidice collaris.* (Most of the specimens of *C. moreleti, C. obscurus, C. textile* and *C. miles* discussed in this section were collected by R. M. Gray, R. A. McKinsey and C. M. Stidham.)

DISCUSSION: ECOLOGICAL NICHES AND ECOLOGICAL ISOLATION

Odum (1953) satisfactorily defined ecological niche as "the position or status of an organism within its community and ecosystem resulting from the organism's structural adaptations, physiological responses, and specific behavior." The ecological niche is multidimensional.

In the preceding sections of this paper, the species of *Conus* inhabiting Hawaiian coral reefs and marine benches have been compared with respect to some of the dimensions of niches. The purpose of this discussion is to state concisely all comparative data. From these data, evidence of ecological isolation, or its reciprocal, the overlap of ecological niches, will be evaluated.

Formulation of the theory that ecological isolation is the result of interspecific competition is due to Volterra (1926; see also D'Ancona 1954). Since Gause (1934) clearly showed that this theory was applicable to experimental populations, the postulate that in a stable community each species occupies a different ecological niche and that two or more species with the same ecological requirements cannot coexist has become known as Gause's principle (Odum 1953), Gause's hypothesis (Gilbert, Reynoldson & Hobart 1952), or, more properly, the Volterra-Gause principle (Hutchinson 1953). In such experiments, only one species finally survives in a population by inhibiting the population(s) of the other species initially present more than its own.

Although the process of competition can thus be observed in experimental populations, it is very difficult to do so in natural populations. The reasons for this are that (1) if the process occurs in nature, its rate may be very slow, and (2) it is often difficult to make the necessary determination of the behavior of a species in the absence of its presumed competitor in completely natural populations. It should be pointed out that the demonstration of niches which overlap with respect to one or more dimensions does not prove the occurrence of competition. This seems to have been overlooked by some authors (e.g. Test 1945, Odum 1953). In its most satisfactory definition, competition requires common exploitation of a limited requisite. Furthermore, mechanisms by which even species whose niches overlap with respect to limited requisites can avoid severe competition have been pointed out by Hutchinson (1953). Severe competition is used here to mean competition leading to the elimination of the less successful species.

Despite the inherent difficulties, interspecific competition has been observed in nature, particularly in birds, on a few occasions (Mackenzie 1950, Pitelka 1951, S. D. Ripley, verbal communication). The rarity of such cases, together with the invocation of other factors, led Andrewartha & Birch (1954) to a general theory of population ecology without introduction of the concept of competition. A number of studies of natural populations containing ecologically similar species provide strong evidence of a less direct sort in favor of the operation of interspecific competition in nature. Additional evidence has been derived from studies of geographical replacement of species with "too slight ecological dissimilarity" (Svärdson 1949a, 1949b).

In all carefully studied stable populations containing two or more ecologically similar species, more or less subtle differences in the ecological niches of these species have been elucidated, thus demonstrating the validity of the Volterra-Gause principle in nature. Ecologically *similar* species are just that and not ecologically *identical.* It can be stated with Gilbert *et al.* (1952), that these observations support the hypothesis that "in a population of a species, mechanisms which reduce competition between it and populations of other species tend to persist."

In the experimental studies of the Volterra-Gause principle, avoidance of interspecific competition was shown to be the result of such competition. In many natural populations ecologically similar species coexist because of reduction or avoidance of competition. In this way, the evidence from natural populations favors the hypothesis that the process of competition, leading to the avoidance of competition, operates in nature, most probably as a selection pressure.

This evidence has been derived from studies of natural populations which contain ecologically similar species of many kinds, including flatworms (Beauchamp & Ullyott 1932), fruit flies (Da Cunha, Dobzhansky & Sokoloff 1951), copepods (Hutchinson 1951), mollusks (Test 1945), fishes (Daiber 1956), amphibians (Dumas 1956), birds (Lack 1945, 1947), and mammals (McCabe & Blanchard 1951, Johnson 1943).

Except for a few papers which consider only one dimension of the ecological niche, e.g. zonation (Fischer-Piette 1935, Eslick 1940), the work of Test (1945) is the only previous such study of marine gastropods known to the writer. It deals with herbivorous limpets, of the genus *Acmaea.* Test's work differs from most of the others listed, and is more similar to the present study, in that it deals with

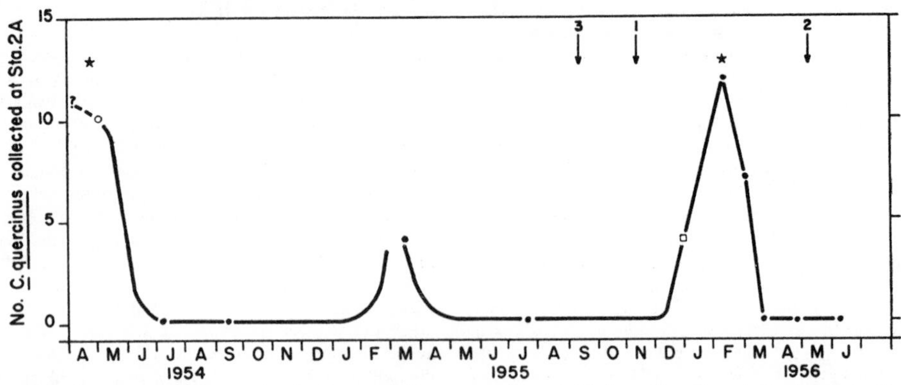

FIG. 29. Seasonal changes in abundance of *Conus quercinus* at Station 2A. Stars indicate observation of spawning of *C. quercinus* at Station 2A (1956) or in laboratory soon after being collected at Station 2A (1954). Closed circles indicate collecting trips by the author. Additional trips, during which no *C. quercinus* were seen, were made in August, 1954, and August, September, November, and December, 1955, but the exact dates were not recorded.

Explanation of vertical arrows: 1. Collection of one specimen in sampan channel, Kaneohe Bay, depth 10 ft, by A. H. Banner. 2. Collection of two specimens off Barbers Point, depth 160 ft, by R. Sheats. 3 Collection of two specimens in Bay ½ mile south of Hapuna Beach, Hawaii, depth 15-20 ft, by R. A. McKinsey.

Spawning was observed at Station 2A in February, 1956. A number of specimens collected at Station 2A in April, 1954, by staff members of the Hawaii Marine Laboratory, spawned in laboratory aquaria late the same month. The population apparently migrates to Station 2A from deeper water in early spring for spawning. The egg capsules are typically attached to the alga, *Acanthophora orientalis.* The adults then presumably return to deeper water. Specimens at Station 2A were collected in 2-10 ft of water and were most abundant in 2-5 ft. Other, "out of season" collection records from deeper water are also noted in Fig. 29.

Analysis of alimentary tract contents of 34 specimens of *Conus quercinus* revealed the commonest prey organism to be the enteropneust, *Ptychodera flava,* remains of which were found in 13 specimens. Two specimens contained remains of the sabellid polychaete, *Sabellastarte indica.* These were the only identifiable prey organisms. *Ptychodera flava* was extremely abundant at Station 2A; densities of several individuals per square foot were not uncommon.

Conus leopardus. The habitat of *Conus leopardus* is similar to that of *C. quercinus,* but the former species usually occurs in somewhat deeper water. It is rarely collected at depths of less than 10 ft. The deepest record known to the author is a specimen collected in 120 ft off Pearl Harbor, Oahu, by R. Sheats on 2 February 1956. Sand is the typical substratum, but specimens are also found on mixed sand-rubble bottoms.

No definitely identifiable food organisms were found in alimentary tracts of ten specimens which were examined. Remains in one specimen were tentatively identified as *Ptychodera flava.* The feeding of *C. leopardus* on *P. flava* was observed in the laboratory. Within a few seconds after the enteropneust was introduced into an aquarium containing a *C. leopardus,* the latter became active, extending its siphon and waving it about in the water. Several minutes later, the orange rostrum extended and began to engulf the *Ptychodera.* At no time was the proboscis visible. The prey was apparently never stung, since rhythmic contractions of the proboscis, collar and trunk regions persisted until engulfment was complete, some 18 min later.

Conus leopardus is the largest species of the genus in Hawaiian waters, the shell lengths of some specimens exceeding 200 mm. However, its radula and venom apparatus are extremely poorly developed. The radula teeth of a 160-mm specimen measured only 0.9 mm in length, comparable to those of an adult (40-mm) *C. lividus.* It is just possible that the radula, which is an extremely specialized apparatus in other species of *Conus,* is vestigial in *C. leopardus.* It must be recalled however, that some species with well developed radula teeth do not always sting the prey prior to feeding.

Conus moreleti. No identifiable food organisms were found in alimentary tract contents of 6 specimens of *Conus moreleti* which were collected in depths of 10-40 ft off the leeward (west) coast of Oahu. However, this species, which is rather similar to *C. lividus,* is probably vermivorous.

Conus obscurus. Nine specimens collected at depths of 15-35 ft off leeward Oahu were examined. Remains of an unidentified fish were present in one. Attempts to observe the feeding process in the laboratory were not completely successful. Introduction of a fish into the vessel with the snail usually evoked

TABLE 16. Summary of Ecological Characteristics of Species of *Conus* on Marine Benches.

	sponsalis	*abbreviatus*	*ebraeus*	*chaldaeus*	*rattus*	*catus*
Relative abundance at all bench stations	1	2	3	4	5	6
Population density on a solution bench (Sta. 5) (no./100 sq. ft.)	0.41	0.66	1.02	0.15	0.02*	0.00
Population density on a water-leveled bench (Sta. Kl) (no./100 sq. ft.)	0.48	0.85	0.55	0.18	0.09	0.24
Population density at abrasion ramp stations* (no./100 sq. ft.)	0.25	0.21	0.00	0.07	0.33	0.00
Distance from shore of density peak at Sta. 5 (% distance across bench)	8%	8%	50%	73%	—	—
Distance from shore of density peak at Sta. Kl (% distance across bench)	22%	18%	26%	48-70%	47%	53%
Per cent of population on substratum of algal turf on bench, binding ± sand	90%	65%	65%	45%	62%	(45%)
Per cent of population on sand patches on bench	5%	26%	15%	28%	11%	(18%)
Per cent of population exposed to air at low tide	55%	35%	29%	32%	17%	29%
Per cent of population partly buried in substratum during day	30%	63%	31%	42%	10%	(44%)
Ability to withstand strong water currents (exp't'l.) (arbitrary units)	+	+	++	+		
Active period in nature and in laboratory	All species actively crawl about at night, are quiescent during day					
Per cent of diet represented by Nereidae	52%	26%	93%	85%	50%	Eats Only Fish
Per cent of diet represented by Eunicea	44%	69%	6%	15%	50%	

*Calculated from time-relative density.

more than two ecologically similar sympatric species. As many as 17 species of *Acmaea* coexist in a broad region of the California coast. Test was able to show diversification of the ecological niches of these species, despite the demonstration of varying degrees of overlap in one or more dimensions of the niches. Only quite recently have similar studies been extended to large numbers of sympatric species in other groups, namely insects (Cooper & Dobzhansky 1956; Da Cunha, El-Tabey Shehata & de Olivera 1957) and birds (Betts 1955).

A total of 18 species of *Conus* were collected by the writer on Hawaiian coral reefs and marine benches. Of these, 16 were collected on reefs and 13 on benches, dispersed among these habitats as previously discussed. On marine benches, 6 of the 13 species present are characteristic of the habitat, and only these are included in the following discussion of ecological niches. Of the others, two represent unique collection records, and five are occasionally found on benches but are more typical of the reef habitat. The factors which probably limit their abundance on benches have been discussed above.

The ecology of the species of *Conus* inhabiting marine benches is complicated by the fact that the whole life cycle is not passed in the same habitat. Recruitment of all species is predominantly from pelagic veliger larvae which are washed onto benches from other areas of origin, particularly subtidal reefs, where conditions for spawning are more favorable.

The ecological requirements of pelagic larvae and newly metamorphosed young are unknown and are extremely difficult to study. Because of their small size, newly settled larvae are virtually impossible to observe in nature. Attempts to raise young from eggs in the laboratory generally failed. Veligers usually hatched and swam about freely but died before settling to the crawling mode of life. The food at these early stages is unknown. Veligers are probably plankton or seston feeders. It is not known whether or not newly settled larvae immediately assume the carnivorous habit, or at what stage the venom apparatus becomes functional.

Requisites which govern population density of pelagic veligers and newly-settled young stages may affect the observed adult population densities reported here. Predation may well be an important factor. These statements are, however, not based on positive evidence. The hypothesis was not amenable to investigation in the time available, so adult populations only were studied. The evidence to be summarized in Tables 16-20 suggests that adults of all species studied are sufficiently isolated ecologically that interspecific competition does not limit population densities.

In Table 16, all characteristics of ecological niches studied are summarized for the six most abundant species of *Conus* on marine benches. All of these species also occur in the subtidal reef habitat, although usually in considerably lower abundance. In Table 17, data for the 8 most abundant species on subtidal reefs are similarly summarized.

In order to determine the significance in niche differentiation of interspecific differences with respect to dimensions of niches, the results of statistical analyses of all observed ecological data relating to these species are sumarized in Tables 18-20. In the right column of these tables the category of relationship of *Conus* to a number of environmental factors is listed. The statistical test applied to the data and the probability of the samples of the two species having been drawn at random from the same population are given in the second and third columns, respectively. Probabilities in most cases are extremely low. However, it is not legitimate to attribute all

TABLE 17. Summary of Ecological Characteristics of Species of *Conus* on Subtidal Reefs.

	flavidus	*lividus*	*pennaceus*	*abbreviatus*	*ebraeus*	*sponsalis*	*rattus*	*imperialis*
Relative abundance at all reef stations............	1	2	3	4	5	6	7	8
Population density on a reef platform (Sta. K3) (no./100 sq. ft.).................	0.03	—	—	0.07	0.09	—	—	—
Per cent of population on sand substratum..............	56%	64%	75%	78%	61%	38%	22%	48%
Per cent of population on reef limestone substratum.....	26%	22%	21%	13%	20%	30%	38%	26 %
Per cent of population on coral, rubble and rough coral bench substrata..........	18%	14%	4%	4%	19%	23%	40%	26%
Dominant component of particulate sediments.....	coarse+very coarse sand		coarse sand	coarse+very coarse sand				fine to coarse sand
Per cent of population buried or under rocks during day.	15%	10%	100%	54%	27%	13%	14%	8%
Per cent of diet represented by Nereidae...............	—	—	—	—	15%	46%	23%	—
Per cent of diet represented by Eunicea................	—	12%	—	100%	82%	50%	77%	27%
Per cent of diet represented by Terebellidae............	64%	14%	—	—	—	—	—	—
Per cent of diet represented by all polychaetes..........	96%	39%	Eats only Gastropods	100%	100%	100%	100%	100%
Per cent of diet represented by enteropneusts............	4%	61%	—	—	—	—	—	—

of the differences solely to the dimensions of the niches. If the degree of overlap is large, chi-square (but not Wilcoxon) tests may indicate significant differences if the sample also is large. A hypothetical case is illustrated in Fig. 30. In both both A and B, the probability that the two samples were drawn at random from the same population is the same. The ecological significance of the two situations is, however, quite distinct. Because of differences in position of the curves on the abscissa, competition between N_1 and N_2 for the requisite represented is much less likely in B than in A. Competition does not necessarily take place in either case.

For this reason the degree of overlap of the species with respect to each dimension is also included in Tables 18-20. In entries where the interspecific differences are highly significant, and where per cent overlap is large, the biological significance is less than the low probabilities might imply. That is, the possibility of competition is not virtually precluded.

Inspection of Table 16 shows as a first approximation that *Conus sponsalis* and *C. abbreviatus* are ecologically more similar to each other than to the other species in the table. These species are also similar in absolute abundance and in size. They are closely related systematically, usually being placed in the same subgenus (*Virroconus*). The results of statistical analyses of all observed ecological data relating to these two species are summarized in Table 18. Population densities are similar at the two most thoroughly studied bench stations, but they differ significantly when all marine benches are considered. The great difference in abundance on different reefs is due to the fact that *C. sponsalis* is much more abundant than *C. abbreviatus* at Station 7 and less abundant at all other

TABLE 18. Statistical Analyses of Ecological Data: Comparison of the Ecological Niches of *Conus sponsalis* and *C. abbreviatus*.

Relation of *Conus* to Environmental Factor	Statistical Test	P	Per cent Overlap
1. Relative abundance at Stations 5 and Kl..............................	Chi-square	.8	95%
2. Relative abundance on three types of benches.........................	Chi-square	$<.01$	53%
3. Relative abundance on four reefs..	Chi-square	$<10^{-6}$	14%
4. Distribution pattern on a solution bench (Sta. 5)..................	Wilcoxon	$\gg.05$	80%
5. Distribution pattern on a water-leveled bench (Sta. Kl)	Wilcoxon	$\gg.05$	80%
6. Occupation of different types of substratum on marine benches.....	Chi-square	$<.001$	62%
7. Occupation of different types of substratum on reefs	Chi-square	10^{-5}	42%
8. Frequency of burrowing into substratum during day: marine benches	Chi-square	$<.01$	50%
9. Frequency of burrowing into substratum during day: reefs.........	Chi-square	$\ll.001$	42%
10. Frequency of exposure to dry air at low tide on marine benches........	Chi-square	$<.01$	67%
11. Nature of food: Frequency of Eunicea and Nereidae eaten on marine benches.........................	Chi-square	$<.001$	59%
12. Nature of food: Frequency of individual prey species on benches.........	Chi-square	10^{-6}	42%
13. Nature of food: Frequency of Eunicea and Nereidae eaten on reefs.......	Chi-square	$\ll.01$	34%
14. Nature of food: Frequency of individual prey species on reefs	Chi-square	$<10^{-6}$	25%
15. Food preference in choice experiments.	Chi-square	$<.001$	48%

Entries 8, 10, and 11, are dependent on 7, 12, and 13, respectively.

reef stations. Both species are almost identically distributed over the benches where both are abundant. Despite the fact that both are common in the same habitat, the two species differ at the 1% level of significance with respect to all other dimensions of the

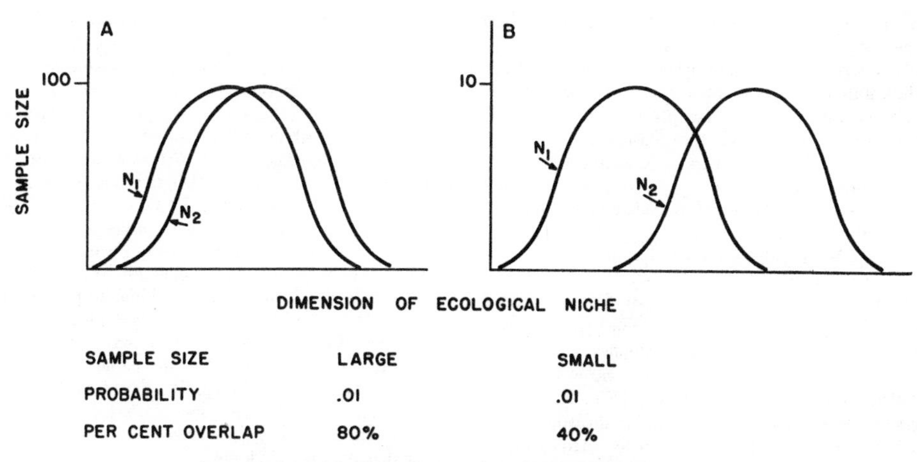

FIG. 30. Hypothetical case indicating the possibility of interspecific competition despite apparent statistically significant difference in niches. Explanation in text.

ecological niches which were investigated. Furthermore, the degree of overlap is small, hence ecological isolation is pronounced, with respect to all other factors.

Conus ebraeus and *C. chaldaeus* are the two species which are most closely related systematically. Both belong to the same subgenus (*Virroconus*) as *C. sponsalis and C. abbreviatus.* Many authors have considered *C. chaldaeus* to be a variety of *C. ebraeus.* This view has persisted among a few recent workers (e.g. Dodge 1953). Usually, however, both are accorded specific rank, and even opponents of this view have acknowledged the absence of intergrades. In Hawaii, the shells of these two species are quite distinct in appearance, much more so than in other parts of their range (Kohn unpublished.).

Furthermore, evidence discussed above and summarized in Table 19 indicates that the microhabitats of *Conus ebraeus* and *C. chaldaeus* are quite distinct, although the two species are typically found together in the marine bench macrohabitat. Since *C. ebraeus* is fairly common on reef platforms, but *C. chaldaeus* is virtually excluded, all of the comparisons in Table 19 are concerned only with marine benches. There the relative abundance, relation to substratum, frequency of exposure to air at low tide, and general nature of the food are very similar. However, striking differences in pattern of distribution across benches and especially in the specific nature of the food are apparent. The degree of overlap between the niches with respect to these two dimensions is extremely small. The two species are thus able to coexist and avoid interspecific competition.

Inspection of Tables 16 and 17 suggests that *Conus ebraeus* is ecologically about as similar to *C. abbreviatus* as it is to *C. sponsalis.* A similar comparison of the ecological data of *C. ebraeus* and *C. sponsalis* was also made. These two species differ

TABLE 19. Statistical Analyses of Ecological Data: Comparison of the Ecological Niches of *Conus ebraeus* and *C. chaldaeus* on Marine Benches.

Relation of *Conus* to Environmental Factor	Statistical Test	P	Per cent Overlap
1. Relative abundance at Stations 5 and Kl	Chi-square	.2	71%
2. Relative abundance on three types of benches	Chi-square	.03	80%
3. Distribution pattern on a solution bench (Sta. 5)	Wilcoxon	<.01	59%†
4. Distribution pattern on a water-leveled bench (Sta. K1)	Wilcoxon	<.01	14%*
5. Occupation of different types of substratum	Chi-square	.17	63%
6. Frequency of burrowing into substratum during day	Chi-square	.6	80%
7. Frequency of exposure to dry air at low tide	Chi-square	.6	94%
8. Occupation of particulate sediments with different mechanical properties	Wilcoxon	.05	75%
9. Nature of food: Frequency of Nereidae and Eunicea eaten	Chi-square	.22	83%
10. Nature of food: Frequency of individual prey species	Chi-square	$<10^{-6}$	0.5%

Entry 9 is dependent on entry 10.
†Overlap=3% if only 2nd and 3rd quartiles of the distribution shown in Figure 20 are considered.
*Overlap=0% if only 2nd and 3rd quartiles of the distribution shown in Figure 22 are considered.

significantly from each other with respect to almost all of the dimensions of ecological niches which were studied.

Inspection of Tables 16 and 17 suggests that *Conus rattus* is ecologically quite similar to *C. ebraeus. C. rattus* is placed in a different subgenus (*Lithoconus* or *Rhizoconus*) from the other bench species. An analysis similar to those of Tables 18 and 19 indicated that differences with respect to several dimensions of the niches of the two species are not significant and that there is considerable overlap. However, the two species are usually not found in the same place, for *C. rattus* dominates the abrasion ramp benches where *C. ebraeus* is absent. Where both occur, *C. ebraeus* is the more successful, judging from absolute numbers of both present on solution benches, water-leveled benches, and reef platforms. The difference between the food of these two species is also highly significant. On marine benches, the food of *C. rattus* is most similar to that of *C. sponsalis* ($P = .7$ in comparison of families of polychaetes eaten), but the individual species preyed on by *C. rattus* and *C. sponsalis* differ highly significantly ($P < 10^{-6}$), and the overlap is only 22%.

Of the species of *Conus* characteristic of marine benches, *C. catus* (subgenus *Chelyconus*) was the least abundant. This species was commonest at Station K1, where its greatest density was nearer the outer edge than any other species present (Fig. 22). Samples were too small to provide reliable information on most other characteristics of the niche. However, *C. catus* is completely ecologically isolated from all other species of *Conus* which inhabit marine benches by the nature of its food, which consists entirely of fishes.

Turning to the species characteristic of subtidal reef platforms, it is apparent from Table 17 that *Conus flavidus* and *C. lividus*, the two most abundant species, are very similar ecologically. Since both occur uncommonly on marine benches, only data from reef stations are presented in Table 20, which summarizes the comparative ecology of these two species.

TABLE 20. Statistical Analyses of Ecological Data: Comparison of the Ecological Niches of *Conus lividus* and *C. flavidus* on reefs.

Relation of *Conus* to Environmental Factor	Statistical Test	P	Per cent Overlap
1. Relative abundance on four reefs (Stations 3, 4, 7, 9)	Chi-square	.001	64%
2. Relative abundance on three reefs (Stations 3, 4, 9)	Chi-square	.08	75%
3. Distribution pattern on three reefs	Wilcoxon	$\gg .05$	88%
4. Occupation of different types of substratum	Chi-square	.15	77%
5. Association with particulate sediments with different mechanical properties	Wilcoxon	.05	42%
6. Nature of food: Frequency of major groups eaten*	Chi-square	$< 10^{-6}$	12%
7. Nature of food: Frequency of individual prey species eaten	Chi-square	$< 10^{-6}$	10%

*Major groups=Terebellidae, other polychaetes, and enteropneusts

On most reefs, the abundance of *C. flavidus* and *C. lividus* is similar. An exception is Station 7, where the latter species was much the more abundant. The significant difference in abundance in the first entry of Table 20 is due entirely to Station 7, as the second entry shows. The nature of, and relation to, the substratum is similar in both species. The most striking difference in the ecological niches of *C. flavidus* and *C. lividus* is in the nature of the food. Frequency differences in the samples examined are highly significant, and degree of overlap is exceedingly low. There is thus little possibility of competition for food. Extensive overlapping with respect to other environmental requisites may permit competition, however.

A number of vermivorous species occur on reefs in very low population densities. Three of these, *Conus distans*, *C. vexillum* and *C. imperialis*, attain a larger size than almost all of the other species found on the reefs. Specimens of *C. distans* collected ranged from 46 to 131 mm in shell length; the range in *C. vexillum* was 44-85 mm, and in *C. imperialis*, 52-88 mm. These species, as well as *C. miles* and *C. vitulinus*, feed on eunicid polychaetes. Some of these species may be restricted to eunicids for food, but *C. imperialis* also feeds on the amphinomid polychaete, *Eurythoe complanata*. Since the samples collected were small, little statistical information was obtained. However, the data presented in Table 13 suggest striking differences in the specific nature of the prey of these species and, therefore, in that dimension of their ecological niches. The factors influencing population density are not known.

Two salient features of the niche of *Conus pulicarius* serve to isolate this species ecologically from its sympatric congeners. *C. pulicarius* occurs most often in large areas of deep sand on reef platforms, while the more abundant *C. flavidus* and *C. lividus* occur most often on patches of thin sand on reef limestone, or in small sandpockets. Secondly, part of the diet of *C. pulicarius* consists of the echiuroid, *Thalassema*, which as far as is known is not exploited for food by any other species of *Conus*. Information on the ecology of *Thalassema* is not sufficient to determine whether these two aspects of the niche of *C. pulicarius* are related.

The ecological niche of *Conus pennaceus* differs qualitatively from those of the other reef species with respect to both space and food. This species typically remains under basalt or coral rocks on, or partly buried in, sand during the day. Other species are found only rarely in this microhabitat. Competition for food with the numerous vermivorous species is completely avoided, since *C. pennaceus* feeds solely on other gastropods. The two other molluscivorous species, *C. marmoreus* and *C. textile*, are exceedingly rare on the reefs. Data obtained by the author are too few to indicate the extent of overlap with respect to food. However, both *C. marmoreus* and *C. textile* feed at least partly on other species of *Conus* while *C. pennaceus* was never found to eat its congeners.

The wide variety of gastropods on which *C. pennaceus* feeds is further evidence that intraspecific competition is of much greater importance than interspecific competition to this species. For, as Svärdson (1949a) showed, dominant intraspecific pressure causes a species to approach more closely the tolerance limits of its niche. "In this case, the species may be said to go down the slopes of its adaptive peak" (Svärdson 1949a).

Both piscivorous species, *Conus striatus* and *C. catus*, are very rare on Hawaiian reefs. They are ecologically isolated from all of the other species by the nature of their food. Young *C. striatus* may feed on fishes of similar size to those eaten by adult *C. catus,* but the food of adult *C. striatus* is much larger.

The data presented in the preceding paragraphs indicate a high degree of ecological isolation among the species of *Conus* considered. Except where otherwise noted, each marine bench and subtidal reef habitat is a homogeneous one, containing a single community of *Conus,* as evidenced by agreement with the theoretical distribution given by MacArthur (1957). Although such agreement implies non-overlapping niches, this ideal is not completely realized in nature. As MacArthur (1957) showed, however, the distribution expected if niches overlapped randomly would fit the observed data much more poorly.

Ecological isolation results from fractionation of the habitat into microhabitats which differ especially with respect to zone occupied, relation to the substratum, and, especially, nature of the food. The first factor is of much greater importance on marine benches than on reef platforms. In this manner, interspecific competition severe enough to lead to the elimination of some of the species from the habitat is avoided. Such avoidance of competition is a likely result of the process of competition itself (Park 1954). But, as has been noted, direct evidence bearing on this is difficult to obtain from observation of natural populations (see also Mayr 1948).

Pertinent evidence obtained in the present study is that mentioned in connection with *Conus sponsalis,* which sometimes occurs where the ecologically similar *C. abbreviatus* is absent. Here the microhabitat is broadened to include, for example, more extensive use of eunicids for food. Also, *C. rattus* occurs in very low densities where *C. ebraeus* is abundant, but the former is the dominant species on the abrasion ramp type of marine bench, where the latter is absent. Restriction of the vermivorous species to apparently optimal regions of their possible habitats may also indicate the efficacy of interspecific competition, in accord with Svärdson (1949a).

In summary, density-inactive factors of the environment (Nicholson 1955) permit certain species of *Conus* to occur in the marine bench and subtidal reef habitats. Hydrographic characteristics and certain properties of the substratum are likely to be important factors of this type. The number of ecologically closely related species which may occupy a habitat is proportional to the amount of fractionation into microhabitats, which may overlap but are sufficiently distinct that severe interspecific competition is precluded. The process by which these microhabitats are established may be interspecific competition.

The population density is adjusted to certain governing requisites in the environment. Some of the species of *Conus* which occur on marine benches in Hawaii are believed to be limited by the extent of sand substrata suitable for burrowing (*C. pennaceus*) or required by prey organisms (*C. lividus, C. flavidus*). Six other species on marine benches may be termed dominants in this habitat. Of these, the population size of *C. catus* may be limited by the amount of available food. Population densities of the more abundant species on reefs may also be limited by the amount of available food. Governing requisites of the other species are probably not amount of adult food, space, or predators. Factors which are effective at a pre-adult stage in the life history, which were not amenable to study, are likely to be important.

SUMMARY

The gastropod genus *Conus* has contributed to the enrichment of the number of species of epifaunal marine invertebrates in tropical regions in that it is typically represented by many sympatric species. Ecological observations on 25 species which occur in the Hawaiian Islands are reported. Most of the data concern natural populations of 18 species studied on intertidal marine benches and subtidal coral reefs which fringe much of the coastline of the Islands. Investigations on marine benches were carried out at nine stations on the islands of Oahu, Kauai, and Maui. Eight subtidal reefs on Oahu and Kauai were studied.

Conus populations on marine benches are composed chiefly of adult individuals. A stable population density of about 2.5 individuals/100 sq ft (30/100m^2) may be expected on solution benches and water-leveled benches. Mean density of a few quantitative samples on a reef was 0.16 individual/100 sq ft (2/100m^2). Although populations are much denser on marine benches, spawning is usually unsuccessful there, presumably because of the absence of protected sites for the attachment of egg capsules. Recruitment is from pelagic veliger larvae which originate elsewhere but are carried to benches in condition to settle and assume the benthic mode of life.

Four species (*Conus sponsalis, C. abbreviatus, C. ebraeus,* and *C. chaldaeus*) are usually dominant on solution benches and water-leveled benches. *C. rattus* is the most abundant species on abrasion ramp benches. *C. catus* is also a typical inhabitant of marine benches. *C. distans, C. flavidus, C. lividus, C. miles, C. pennaceus, C. nussatella, C. retifer,* and juvenile *C. vexillum* were also recorded from marine benches.

Conus flavidus and *C. lividus* are the dominant species on subtidal reefs, although at one station *C. sponsalis* was most abundant. *C. pennaceus, C. abbreviatus, C. ebraeus,* and *C. rattus* are also common. Other species recorded from reef stations were *C. imperialis, C. distans, C. chaldaeus, C. marmoreus, C. miles, C. pulicarius, C. striatus, C. textile, C. vexillum,* and *C. vitulinus.*

Values of an index of diversity, which measured similarity of species composition of different populations, differed but little in comparisons a) among the several bench stations, b) among the several reef stations, and c) between reef and bench stations. Quantitatively, however, certain species are characteristically most abundant on benches, while others are most abundant on reefs. Calculation of a measure of heterogeneity gave low values when the *Conus* communities of reefs were compared among each other (H′ = .09) and when the communities of benches were compared among each other (H′ = .19). Comparison of summed reef populations with summed bench populations showed much greater heterogeneity (H′ = .84).

At the most thoroughly studied bench and reef stations, number of species and number of individuals are related in a manner which agrees with the theoretical distribution expected in an adequately sampled, homogeneous community of a single habitat, where niches are non-overlapping and continuous.

Fractionation of the habitat into microhabitats was observed but, as expected intuitively, it is not complete. On marine benches, the species present are non-randomly distributed across the bench platform from shore to seaward edge. The distributions of *Conus sponsalis* and *C. abbreviatus* are similar. Those of all other species differ significantly from each other at the 5% level of probability. The number and biomass of *Conus* decreases from a maximum near the landward edge to the seaward edge.

On reef platforms, the distribution of *Conus* is characterized by patchiness, or clumping, which is correlated with the nature of the substratum. Distribution is not related to distance from shore or breaker line, except that density is low at both extremes. Over most of the reef platform, the observed distribution reflects the uniformity of water movements across the reef.

Differential association with different kinds of substratum on marine benches serves to partially distinguish the microhabitats of *Conus sponsalis* and *C. abbreviatus,* the latter occurring more often in sandier regions. This is correlated with the observation that *C. abbreviatus* burrows in the substratum significantly more often than *C. sponsalis.* Although these two species commonly coexist, only *C. sponsalis* was found on benches or parts of benches exposed to dry air for long periods at low tide.

On Hawaiian coral reefs, most species of *Conus* are most often found associated with sand, the most prevalent type of substratum. Only one species, *C. pulicarius,* is probably entirely restricted to a sand substratum. The two most abundant species on the reefs, *C. flavidus* and *C. lividus,* occur characteristically on patches of sand among solid substratum. *C. abbreviatus,* which occupies sandier regions of marine benches, is also a sand dweller on the reefs. *C. pennaceus* characteristically occurs on or partly buried in sand under basalt or coral rocks during the day but crawls about on the surface of the sand at night. Other species are not commonly found under rocks.

Mechanical analyses of sand suggested no niche diversification with respect to particle size distribution of this moiety of the substratum.

Alimentary tracts of 1,930 specimens of 24 species of *Conus* collected in Hawaii were examined. From these, 1,073 prey organisms were identified, 879 of them to species. Three groups of species within the genus *Conus* may be distinguished on the basis of the nature of the food: most species feed exclusively on worms, mainly polychaetes. A second group feeds exclusively on other gastropods, and the third group feeds only on fishes. Eleven species, *C. sponsalis, C. abbreviatus, C. ebraeus, C. chaldaeus, C. rattus, C. distans, C. miles, C. imperialis, C. vexillum, C. vitulinus,* and *C. pertusus,* feed exclusively on polychaetes. Samples of the last three named species were rather small, however. Three species, *C. lividus, C. flavidus* and *C. quercinus,* feed both on polychaetes and on the enteropneust, *Ptychodera. C. leopardus* probably feeds on *Ptychodera. C. pulicarius* eats polychaetes and the echiuroid, *Thalassema. C. pennaceus, C. marmoreus,* and *C. textile* feed exclusively on other gastropods. *C. striatus, C. catus* and *C. obscurus* feed only on fishes.

On marine benches, the vermivorous species of *Conus* prey almost exclusively on members of the polychaete family Nereidae and superfamily Eunicea. At one station, the species *Perinereis helleri* was found to be the primary food of the three most abundant species of *Conus.* This polychaete was so abundant, however, that food cannot be said to be in short supply. Interspecific competition for food is therefore not indicated. Species of *Conus* common to both habitats feed proportionately more often on eunicids on reef platforms than they do on marine benches, for nereids are uncommon in the former habitat.

The typically reef-dwelling vermivorous species feed chiefly on polychaetes and enteropneusts associated with the sand moiety of the substratum. Of the two dominant species, *Conus flavidus* eats mainly Terebellidae, and *C. lividus* eats mainly *Ptychodera.*

On marine benches, the dominant species of *Conus* eat polychaetes which feed on algae, forming a three-step food chain. Calculation of a biomass pyramid for this food chain gave an herbivore/producer ratio of 4% and a carnivore/herbivore ratio of 6%.

Predation on *Conus* was difficult to measure. Certain fishes, other gastropods, octopi, crabs and starfishes are possible predators on adults. Predation on free-swimming larvae and newly-settled young is presumably of major ecological significance, but it was not amenable to investigation. Other factors which may also govern population density of the various species are also considered.

Ecological niches, ecological isolation, and interspecific competition in natural populations are discussed briefly. The comparative ecology of the more abundant species of *Conus* is summarized in the discussion, with emphasis on the extent of ecological isolation. Statistical analyses of certain dimensions of the niches of the ecologically most similar species are presented. Limitations of simple tests of significance of differences are noted. Degree of overlap is extremely important in niche diversification.

The adult ecological niche of each species of *Conus* studied differs significantly with respect to at least two of the following characteristics: nature of the food, nature of and relation to the substratum, and zonation or distribution pattern. The last is of particular importance only on marine benches. These differences are concluded to be the primary factors by which the ecological niches of species of *Conus* are differentiated. This is the mechanism which enables the maintenance of populations of large numbers of closely related, sympatric species of *Conus* in tropical regions.

LITERATURE CITED

Abbott, D. P. 1946. Some polychaetous annelids from a Hawaiian fish pond. Univ. Hawaii Res. Publ. No. **23**: 5-24.

Alpers, F. 1931. Zur kenntnis der anatomie von *Conus lividus* Brug., besonders des darmkanals. Jena. Zeitschr. Naturwiss. **65**: 587-658.

———. 1932a. Zur biologie des *Conus mediterraneus* Brug. Jena. Zeitschr. Naturwiss. **67**: 346-363.

———. 1932b. Ueber die nahrungsaufnahme von *Conus mediterraneus* Brug. eines toxoglossen prosobranchier. Pubbl. Staz. Zool. Napoli **11**: 426-445.

Andrewartha, H. G. & L. C. Birch. 1954. The Distribution and Abundance of Animals. Chicago: University of Chicago. 782 pp.

Banner, A. H. 1953. The Crangonidae, or snapping shrimp, of Hawaii. Pac. Sci. **7**: 3-144.

Beauchamp, R. S. A. & P. Ullyott. 1932. Competitive relationships between certain species of fresh-water triclads. Jour. Ecol. **20**: 200-208.

Bergh, R. 1896. Beiträge zur kenntnis der coniden. Nova Acta Ksl. Leop.-Carol. Akad. Naturf. **65**: 67-214.

Betts, M. M. 1955. The food of titmice in oak woodland. Jour. Anim. Ecol. **24**: 282-323.

Brown, F. A. Jr., M. Fingerman, M. I. Sandeen & H. M. Webb. 1953. Persistent diurnal and tidal rhythms of color change in the fiddler crab, *Uca pugnax*. Jour. Exp. Zool. **123**: 29-60.

Bryan, W. A. 1915. Natural History of Hawaii. Honolulu: Hawaiian Gazette Co. 596 pp.

Clements, F. E. & V. E. Shelford. 1939. Bio-Ecology. New York: John Wiley & Sons. 425 pp.

Clench, W. J. & Y. Kondo. 1943. The poison cone shell. Amer. Jour. Trop. Med. & Hyg. **23**: 105-121.

Cooper, D. M. & T. Dobzhansky. 1956. Studies on the ecology of *Drosophila* in the Yosemite region of California. I. The occurrence of species of *Drosophila* in different life zones and at different seasons. Ecology **37**: 526-533.

Crombie, A. C. 1947. Interspecific competition. Jour. Anim. Ecol. **16**: 44-73.

Da Cunha, A. B., T. Dobzhansky & A. Sokoloff. 1951. On food preferences of sympatric species of *Drosophila*. Evolution **5**: 97-101.:

Da Cunha, A. B., A. M. El-Tabey Shehata & W. de Oliveira. 1957. A study of the diet and nutritional preferences of tropical species of *Drosophila*. Ecology **38**: 98-106.

Daiber, F. C. 1956. A comparative analysis of the winter feeding habits of two benthic stream fishes. Copeia **1956**: 141-151.

D'Ancona, U. 1954. The struggle for existence. Leiden: W. J. Brill. 274 pp.

Darwin, C. 1859. On the origin of species by means of natural selection. London: J. Murray. 502 pp.

Dawes, B. 1930. Growth and maintenance in the plaice (*Pleuronectes platessa* L.). Part I. Jour. Mar. Biol. Assn. U. K. **17**: 103-147.

———. 1931. Growth and maintenance in the plaice (*P. platessa* L.) Part II. Jour. Mar. Biol. Assn. U. K. **17**: 877-947.

Dodge, H. 1953. A historical review of the mollusks of Linnaeus. Part 2. The class Cephalopoda and the genera *Conus* and *Cypraea* of the class Gastropoda. Bull. Amer. Mus. Nat. Hist. **103**: 1-134.

Dumas, P. C. 1956. The ecological relations of sympatry in *Plethodon dunni* and *Plethodon vehiculum*. Ecology **37**: 484-495.

Edmondson, C. H. 1928. The ecology of an Hawaiian coral reef. B. P. Bishop Mus. Bull. **45**. 64 pp.

———. 1946. Reef and shore fauna of Hawaii. B. P. Bishop Mus. Spec. Pub. **22**, 381 pp.

Eslick, A. 1940. An ecological study of *Patella* at Port St. Mary, Isle of Man. Proc. Linn. Soc. Lond. **152**: 45-59.

Fauvel, P. 1927. Polychètes Sédentaires. Faune de France, **16**. 494 pp.

Fischer-Piette, E. 1935. Les patelles d'Europe et d'Afrique du nord. Jour. Conchyl. **79**: 5-66.

Gause, G. F. 1934. The Struggle for Existence. Baltimore: Williams and Wilkins. 163 pp.

Gilbert, O., J. B. Reynoldson & J. Hobart. 1952. Gause's hypothesis: an examination. Jour. Anim. Ecol. **21**: 310-312.

Halstead, B. W. 1956. Animal phyla known to contain poisonous marine animals. Venoms (Amer. Assn. Adv. Sci.): 9-27.

Hartman, O. 1940. Polychaetous annelids. II. Chrysopetalidae to Goniadidae. Allan Hancock Pacific Expeditions **7**: 173-287.

———. 1944. Polychaetous annelids. V. Eunicea. Allan Hancock Pacific Expeditions **10**: 1-236.

———. 1948. The marine annelids erected by Kinberg, with notes on some other types in the Swedish State Museum. Ark. Zool. **42A**: 1-137.

———. 1954. Marine annelids from the northern Marshall Islands. Geol. Surv. Prof. Pap. **260-Q**: 619-644.

Helfrich, P. & A. J. Kohn. 1955. A survey to estimate the major biological effects of a dredging operation by the Lihue Plantation Co., Ltd. on North Kapaa Reef, Kapaa, Kauai. Preliminary Report. 31 pp. (Mimeographed. Available from the authors).

Hinegardner, R. T. 1957. The anatomy and histology of the venom apparatus in several gastropods of the genus *Conus*. M.S. Thesis, University of Southern California.

———. 1958. The venom apparatus of the cone shell. Hawaii Med. Jour. **17**: 533-536.

Holly, M. 1935. Polychaeta from Hawaii. B. P. Bishop Mus. Bull. **129**: 33 pp.

Holme, N. A. 1953. The biomass of the bottom fauna in the English Channel off Plymouth. Jour. Mar. Biol. Assn. U.K. **32**: 1-49.

———. 1954. The ecology of British species of *Ensis*. Jour. Mar. Biol. Assn. U.K. **33**: **145-172**.

Hutchinson, G. E. 1951. Copepodology for the ornithologist. Ecology **32**: 571-577.

———. 1953. The concept of pattern in ecology. Proc. Acad. Nat. Sci. Phil. **105**: 1-12.

Johnson, D. E. 1943. Systematic review of the chipmunks (genus *Eutamias*) of California. Univ. Calif. Pub. Zool. **48**: 63-148.

Koch, L. F. 1957. Index of biotal dispersity. Ecology **38**: 145-148.

Kohn, A. J. 1955. Studies on food and feeding of the cone shells, genus *Conus*. Ann. Rept. Amer. Malacol. Union, Bull. **22**: 31.

———. 1956a. The ecology collecting sack modified for marine organisms. Turtox News **34**: 33.

———. 1956b. Piscivorous gastropods of the genus *Conus*. Proc. Nat. Acad. Sci. **42**: 168-171.

———. 1959. The Hawaiian species of *Conus*. Pac. Sci. (In Press).

Kohn, A. J. & P. Helfrich. 1957. Primary organic productivity of a Hawaiian coral reef. Limnol. & Oceanogr. **2**: 241-251.

Lack, D. 1945. The ecology of closely related species with special reference to the cormorant (*Phalacrocorax carbo*) and shag (*P. aristotelis*). Jour. Anim. Ecol. **14**: 12-16.

———. 1947. Darwin's Finches. Cambridge: University Press. 208 pp.

MacArthur, R. 1955. Fluctuations of animal populations, and a measure of community stability. Ecology **36**: 533-536.

———. 1957. On the relative abundance of bird species. Proc. Nat. Acad. Sci. **43**: 293-295.

Margalef, R. 1956. Información y diversidad específica en las communidades de organismos. Inv. Pesq. Barcelona **3**: 99-106.

Mackenzie, J. M. D. 1950. Competition for nest-sites among hole-breeding species. Brit. Birds **43**: 184-185.

Mayr, E. 1948. The bearing of the new systematics on general problems: The nature of species. Adv. Gen. **2**: 205-237.

McCabe, T. T. & B. D. Blanchard. 1951. Three species of *Peromyscus*. Santa Barbara: Rood Associates. 136 pp.

McCaughey, V. 1918. A survey of the Hawaiian coral reefs. Amer. Nat. **52**: 409-438.

Nicholson, A. J. 1955. An outline of the dynamics of animal populations. Austral. Jour. Zool. **2**: 9-65.

Odum, E. P. 1953. Fundamentals of Ecology. Philadelphia: W. B. Saunders Co. 384 pp.

Odum, H. T. & E. P. Odum. 1955. Trophic structure and productivity of a windward coral reef community on Eniwetok Atoll. Ecol. Monogr. **25**: 291-320.

Ohba, S. 1952. Analysis of activity rhythm in the marine gastropod, *Nassarius festivus*, inhabiting the tide pool. I. On the effect of tide and food in the daytime rhythm of activity. Annot. Zool. Japon. **25**: 289-297.

Okuda, S. 1937. Polychaetous annelids from the Palau Islands and adjacent waters, the South Sea Islands. Bull. Biogeogr. Soc: Japan **7**: 257-316.

Ostergaard, J. M. 1950. Spawning and development of some Hawaiian marine gastropods. Pac. Sci. **4**: 75-115.

———. 1955. Some opisthobranchiate Mollusca from Hawaii. Pac. Sci. **9**: 110-136.

Park, T. 1954. Experimental studies of interspecies competition. II. Temperature, humidity, and competition in two species of *Tribolium*. Physiol. Zool. **27**: 177-238.

Pease, W. H. 1860. Descriptions of new species of Mollusca from the Sandwich Islands. Proc. Zool. Soc. Lond. Pt. 27. **1860**: 18-36.

Peile, A. J. 1939. Radula Notes VIII. 34. *Conus*. Proc. Malacol. Soc. Lond. **23**: 348-355.

Pilsbry, H. A. 1917. Marine Mollusks of Hawaii, I-III. Proc. Acad. Nat. Sci. Phil. **69**: 207-230.

———. 1920. Marine mollusks of Hawaii, XIV-XV. Proc. Acad. Nat. Sci. Phil. **72**: 360-382.

Pitelka, F. A. 1951. Ecologic overlap and interspecific strife in breeding populations of Anna and Allen humming birds. Ecology **32**: 641-661.

Prosser, C. L. 1955. Physiological variation in animals. Biol. Rev. **30**: 229-262.

Sandeen, M. I., G. C. Stephens & F. A. Brown, Jr. 1954. Persistent daily and tidal rhythms of oxygen consumption in two species of marine snails. Physiol. Zool. **27**: 350-356.

Smith, F. E. 1950. The benthos of Block Island Sound. Ph.D. Thesis, Yale University.

Strasburg, D. W. 1953. Comparative ecology of two salariin blennies. Ph.D. Thesis, University of Hawaii.

Svärdson, G. 1949a. Competition and habitat selection in birds. Oikos **1**: 157-174.

———. 1949b. Competition between trout and char (*Salmo trutta* and *S. alpinus*). Report, Inst. Freshw. Res., Drottninghom **29**: 108-111.

Takahashi, K. 1939. Polychaeta on coral reefs in Palau. Kagaku Nanyo (Science of the South Sea) **2**: 18-29. (in Japanese)

Test, A. R. 1945. Ecology of California Acmaea. Ecology **26**: 395-405.

Thorson, G. 1946. Reproduction and larval development of Danish marine bottom invertebrates. Medd. Komm. Havundersøg., Kbh., Plankton **4**: 523 pp.

———. 1956. Marine level-bottom communities of recent seas, their temperature adaptation and their "balance" between predators and food animals. Trans. N.Y. Acad. Sci. **18**: 693-700.

———. 1958. Parallel level bottom communities, their temperature adaptation and their "balance" between predators and food animals. *In* Perspectives in Marine Biology. Berkeley: University of California Press. 67-86.

Treadwell, A. L. 1906. Polychaetous annelids of the Hawaiian Islands collected by the steamer "Albatross" in 1902. U. S. Fish. Comm. Bull. **1903**: 1145-1181.

———. 1922. Leodicidae from Fiji and Samoa. Carnegie Inst. Wash. Pub. No. **312**: 127-170.

Turner, H. J. 1951. Fourth report on investigations of the shellfisheries of Massachusetts. State of Massachusetts. 21 pp.

Van Dongen, A. 1956. The preference of *Littorina obtusata* for Fucaceae. Arch. Néerl. Zool. **11**: 373-386.

Volterra, V. 1926. Variazioni e fluttuazioni del numero d'individui in specie animali conviventi. Mem. Accad. Lincei. Ser. 6. **2**: 31-113.

Wentworth, C. K. 1938. Marine bench-forming processes: Water-level benching. Jour. Geomorphol. **1**: 6-32.

———. 1939. Marine bench-forming processes. II, solution benching. Jour. Geomorphol. **2**: 3-25.

Wilson, D. P. 1952. The influence of the nature of the substratum on the metamorphosis of the larvae of marine animals, especially the larvae of *Ophelia bicornis* Savigny. Ann. Inst. Oceanogr. Monaco **27**: 49-156.

———. 1955. The role of micro-organisms in the settlement of *Ophelia bicornis* Savigny. Jour. Mar. Biol. Assn. U.K. **34**: 531-544.

THE STRUCTURE AND METABOLISM OF A PUERTO RICAN RED MANGROVE FOREST IN MAY[1]

Frank Golley,[2] Howard T. Odum[3] and Ronald F. Wilson[3]

Introduction

To develop a comparative science of world ecosystems, measurements of holistic properties are needed for all the important types of communities. Diverse ecosystems on land and water may differ widely in floristic and faunistic composition and in environmental conditions, but the basic function of communities may be placed on a comparable basis with measurements of photosynthesis, respiration, efficiency, biomass, and assimilation number. One major community type little studied from the functional viewpoint is the tropical mangrove swamp. According to a recent atlas of shore systems (McGill 1958), mangroves dominate about 75% of the world's coastlines between 25°N and 25°S latitude. This study reports measurements of structure and metabolism for a representative red mangrove community of the terrestrial type on the southern shores of Puerto Rico.

In the American tropics the mangrove swamp forest consists of a series of zones each dominated by one species of tree (Holdridge 1940, Davis 1940, Dansereau 1947). From open water and extending through the area which is covered by maximum high tides the red mangrove, *Rhizophora mangle* Roxb., is dominant. In this community the trees are supported by high, arching prop-roots, which make travel by an observer exceedingly difficult. Except for prop roots and the red mangrove seedlings, the forest floor is devoid of higher plant life. The next zone toward land is typically dominated by the black mangrove, *Avicennia tomentosa* Jacg., which characteristically sends up myriads of breathing roots 10-15 cm above the mud surface. The innermost zone is usually dominated by the white mangrove, *Laguncularia racemosa* Gaertn. Few herbaceous plants or epiphytes are associated with the mangrove trees in the first and second zones, but ferns and grasses may grow under the white mangroves. Animal life in the above-water parts of the mangrove forest is not abundant, but where the roots are submerged, as in tidal channels, massive epifauna are numerous, and include

[1] These studies were aided by a grant from the Rockefeller Foundation to H. T. Odum for studies in ecology at the Institute of Marine Science, The University of Texas, Port Aransas. Preparation of the manuscript was partly supported by the U.S. Atomic Energy Commission, Contract At(07-2)-10. A contribution from the Institute of Marine Biology, University of Puerto Rico, Mayaguez.

[2] Institute of Radiation Ecology, and Department of Zoology, University of Georgia, Athens, Georgia.

[3] Institute of Marine Science, The University of Texas, Port Aransas, Texas.

oysters, tunicates, and sponges. This underwater fauna does not derive food directly from the mangrove trees.

In January, 1958, May, 1959, and May, 1960, intensive studies over one to 2 wk periods were made in a red mangrove forest on the southern shore of Puerto Rico (18°N Lat, 67° W Long). Brief additional studies were made by Golley in June, 1961, while a visiting professor at the University of Puerto Rico at Mayaguez. The study area was on a peninsula of Magueyes Island, the location of the Institute of Marine Biology of the University of Puerto Rico. The peripheral red mangove forest occupied the greatest area on the peninsula, while a combination of red, black, white, and button (*Conocarpus erectus* Linn.) mangroves occupied a higher, central area. The study forest had not been disturbed since 1954 but evidences of earlier cutting were discovered. The forest is typical in this respect and is representative of other red mangrove forests in Puerto Rico as described by Holdridge (1940).

Methods

Environmental properties, vegetational structure, animal densities, and metabolism of principal components were investigated and a metabolic budget for one average day in May was compiled. The study of these quantities was facilitated by a boardwalk constructed through the swamp by the Marine Institute, which provided easy access to all zones of the forest. The area of the forest was surveyed by running measured lines from the boardwalk to the open water.

Environmental measurements.—Peat depth, temperature, light, wind movement, and depth and rate of tide movement were measured. To determine the depth of the peat 6 cores (5 cm in diameter) were taken on a transect from the mainland to the edge of open water. Temperature was measured on the mud surface and in the air above the mud at the study plots. Light measurements were taken vertically at one meter intervals from the mud surface to a point above the crown of the trees with a General Electric light meter, calibrated in foot-candles. The depth and cycle of the tides in May were measured and related to annual records obtained from Coker and Gonzalez (1960). Wind speed was measured with a Taylor anemometer at various strata within the center of the forest. Five min readings were taken at each level.

The biomass structure of the vegetation.—This was investigated in 2 plots in the center of the red mangrove community away from edge conditions. In 1959, on a 25m² quadrat all trees were harvested and the area and biomass of leaves, and number of roots were determined for meter strata from the mud to the crown. Leaf area was measured by tracing leaves on graph paper and then counting the number of cm² within the tracing. In 1960, on a 100 m² quadrat adjacent to the 25 m² quadrat, another method of estimating biomass was used. The diameter at breast height (dbh) was measured for all trees in the plot. Ten trees representing the most abundant diameter classes were harvested and the biomass of roots, trunks, branches, and leaves was determined (Tables I and II). Samples of fresh

TABLE I. Biomass of harvested red mangrove (in grams)

DBH Class cm	Height meters	Roots		Branches		Leaves		Trunk Dry Wt	Total Tree Dry Wt
		No.	Dry Wt	No.	Dry Wt	No.	Dry Wt		
1.2	3.6	5	159	9	136	143	109	374	778
1.5	3.7	4	131	8	136	123	99	409	775
1.5	3.7	5	134	9	187	179	123	438	882
1.7	3.6	6	293	9	239	173	133	596	1261
2.0	4.1	9	439	10	392	212	153	851	1835
2.4	6.0	8	622	15	438	243*	184	1255	2499
2.7	5.2	7	1379	17	851	529*	399	1856	4485
2.8	5.1	10	988	23	727	473*	357	1856	3928
3.7	6.0	6	2001	23	1056	691*	522	3575	7154
5.3	8.6	16	3514	40	4358	2375*	1608	6810	16290

* Number estimated from samples.

TABLE II. Biomass of red mangrove based on curves of biomass versus dbh. Average biomass per dbh class and per square-meter considered separately

DBH Class cm	Number of Trees	Roots		Leaves		Branches		Trunk	
		Per DBH Class	Per m²	Per DBH Class	Per m²	Per DBH Class	Per m²	Per DBH Class	Per m²
		Biomass in grams dry weight							
1- 2	67	180	121	110	74	180	121	440	295
2- 3	35	680	238	250	88	500	175	1250	438
3- 4	16	1700	272	510	82	1100	176	3100	496
4- 5	7	2700	189	900	63	2500	175	5200	364
5- 6	4	3840	154	1340	54	3900	156	7400	296
6- 7	3	4850	146	1760	53	5000	150	9500	285
7- 8	2	5840	117	2200	44	5950	119	11400	228
11-12	2	10000	200	3900	78	10100	202	19700	394
Total			1437		536		1274		2796

leaves, wood, and roots were dried at 100°C for 24 hrs in an oven to estimate dry weight biomass. The curves of biomass per dbh class from the harvested trees (Fig. 1) were used to estimate the biomass of the trees of known dbh on the entire 100 m² plot. The average biomass for the midpoint of a dbh class (1.5, 2.5, etc.) was read from the curves and multiplied by the number of trees in the class. The values of biomass of the dbh classes were summed to obtain the biomass of all trees on the plot. The few trees harvested in the 5+ cm dbh classes were insufficient to determine the shape of the curve in the larger size classes. The curves were extended in a straight-line through the 11-12 cm class (not shown in Fig. 1) to estimate the biomass of the few large

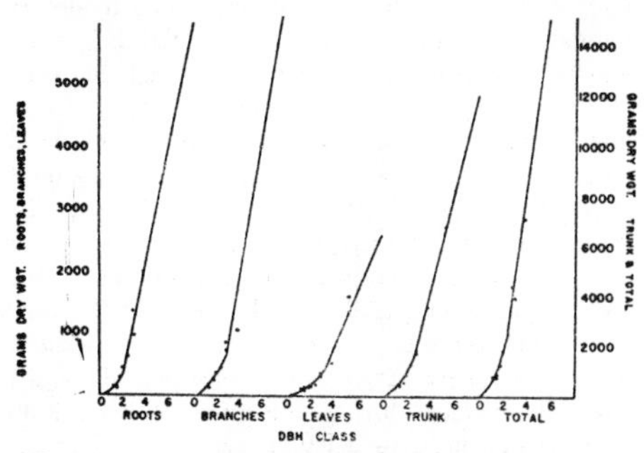

FIG. 1. Dry weight biomass in grams for roots, branches, leaves, trunks, and total trees from ten harvested red mangrove trees with dbh from 1 to 5 cm.

trees on the plot. Since Holdridge (1942) stated that red mangrove in Puerto Rico may reach 30 m in maximum height and 90 cm in maximum diameter, straight-line extension of the curves is probably reasonable.

Chlorophyll a was measured in sun and shade leaves from trees on the 25 m^2 plot in 1959. Values were multiplied by leaf area measurements to determine total chlorophyll. The method of chlorophyll extraction and analysis is described in Odum *et al.* (1958).

Growth of wood.—This was measured by determining the increase in the dbh of trees on the 100 m^2 study area from 1959 to 1960. From the 10 harvested trees a regression coefficient of wood biomass on dbh was calculated to be 3.39 x 10^3 g/cm. The increase in tree diameter was converted to grams of wood by multiplying the regression coefficient of wood biomass on dbh by the number of trees in each class and by the average dbh for each class. The wood deposition for each class was summed to obtain the total wood production during the year.

Photosynthesis and respiration.—These functions of the leaves were measured with a Liston-Becker infra-red CO_2 analyzer (Model 15A) operating in the field from a 500 watt Kohler gasoline generator in 1959 and from a 900 ft extension wire from the electrical supply of the Institute of Marine Biology in 1960. Initially, air was drawn over a bundle of leaves in a plastic bag at the rate of 2 l/min. The CO_2 readings were taken alternately from the bags and from free air outside the bags. After a series of measurements, the leaves were collected and their area and dry weight were determined.

In 1959, it was found that at high light intensities temperature increased in the bags during the mid-part of the day, affecting respiration and photosynthesis measurements. This was noticeable in the large discrepancy between respiration and gross photosynthesis of the forest, based on the 1959 balance sheet. Since respiration was about 6 times as great as photosynthesis, either photosynthesis was depressed or respiration was accelerated under the system of measurement.

In 1960, a plexiglass leaf chamber was constructed with a double wall. Ice water was circulated between the walls by an electric pump from a bucket containing fresh water and ice. Values were somewhat higher but still too small to match respiration data. During experiments with the cooled chamber CO_2 production was also influenced by the rate of air flow through the chamber. Measurements at 2 to 15 l/min flow rates showed CO_2 production increased to a flow rate between 10 and 15 l/min (Fig. 2). At these

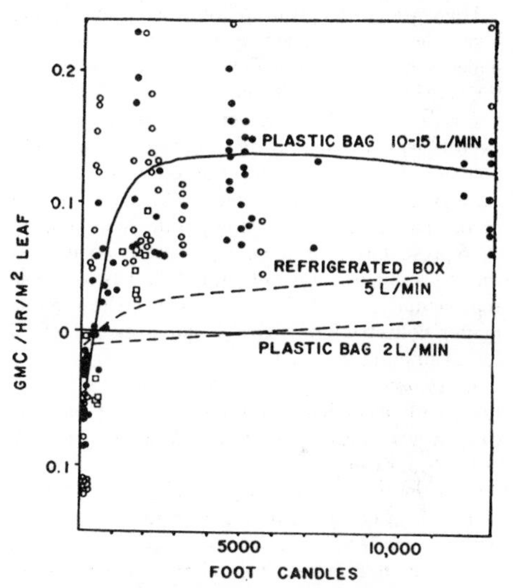

FIG. 2. Metabolism in gram carbon per hour per m^2 leaf surface of sun, shade, and seedling leaves of *Rhizophora mangle* measured at 10-15 l/m in one liter bags. Seedling data are indicated by dark circles, shade leaves by squares, sun leaves by open circles. Dashed lines summarize experiments at 2 and 5 l/m.

high rates there was no need for refrigeration of the plant chamber.

Earlier data were discarded and the data for calculation of metabolism were taken at flow rates between 10 to 15 l/min. In the field, the infra-red gas-analyzer was standardized by injection of CO_2 gas with a syringe into a a closed system of 20 l volume, previously cleared of CO_2 by absorption with NaOH.

Respiration of leaves was investigated at night by comparing CO_2 production of leaves in plastic bags with CO_2 content of free air. For budget computations we have assumed that night and day leaf respiration are equal. Respiration of prop roots was measured by enclosing the air exchange portion of the root (zone of pores) in plastic and drawing air over the root surface.

Respiration of the peat floor of the forest during low tide when the peat was exposed to air was measured by placing an aluminum sheet on the peat and drawing air under the sheet to an intake tube. The CO_2 production of the peat also varied with air flow, reading highest at the highest rates of flow available with field equipment (15 l/min). Thus soil respiration was a function of ground air flow. To calculate soil respiration it was necessary to know the rates of air movement over the forest floor.

Underwater respiration of the peat during high tide was measured by enclosing tidal water under a bell jar in the field and then measuring the O_2 change by the Winkler method.

Export of particulate organic matter.—Such export was estimated by pouring a measured volume of incoming and outgoing tidal water through a #10 plankton net. Water at the end of the boardwalk on incoming tide was sampled as representative of incoming water. Water on the study quadrat on outgoing tide was taken as representative of outflowing water. The organic matter collected was dried in an oven at 100°C for 24 hrs and weighed.

Biochemical oxygen demand.—Bottles were filled with incoming and outgoing water and the oxygen was measured by the Winkler method. After 6 days duplicate bottles were measured for oxygen content. The oxygen changes over 6 days were measured to estimate the magnitude of labile organic matter. Studies of decomposition of organic matter in sea water have shown that in 6 days between 60% and 80% of the organic matter is decomposed of that which will decompose over several months (See curves for sea waters in Sargent and Austin 1949, Fox, Oppenheimer, and Kittredge 1953, Rakestraw 1947). Thus the grams carbon in the labile organic matter can be estimated from 6 day oxygen decreases by dividing the grams O_2 consumed by .70.

Densities of animals.—Animal densities were determined by a variety of methods depending on the activity period and the habitat of the various taxa. There appeared to be 3 major habitats for animals in the forest: (1) areas containing mostly prop roots and seedlings, shaded by a few large trees; (2) dense thickets of small mangroves; and (3) areas of prop roots, seedlings, and a moderate number of medium sized trees (dbh 2-5 cm). Animal activity was partly influenced by time of day and partly by the tidal cycles. Square-meter quadrats were placed in each of the 3 types of habitats, twice during the day and once at night, to obtain estimates of animals living on the forest floor. In these surveys the observer sat motionless on the mangrove roots and observed the plot for one hour. All animals seen and the time each spent on the plot were recorded. Because the animals were active at different periods of the day, the census during the time of maximum activity was used to calculate densities. For instance, the snail, *Melampus coffeus,* was present throughout the day but ascended the roots and seedlings at high tide (occurring at night in May) (Golley 1960). Thus the night counts of these snails provided the most accurate estimates of density on the quadrats.

Many animals were censused on a 2 m wide transect beside the boardwalk (termed boardwalk census, Table III). There it was possible to move rapidly through the forest, with a minimum disturbance to the animal life. Traverses made by climbing over and under prop roots were mainly effective for these forms that could not move rapidly (such as the web spiders) or for going to a quadrat which would be under observation for a long time.

Flying insects were most difficult to census since it was not possible to use a sweep net or a similar device in the red mangrove forest. All insects flying across or along the boardwalk (in a cross-section area approximately 2 x 9 m) were recorded during day and night observation periods in 1959 and 1960. In 1960, volumes of air were examined visually at night by repeatedly flashing a spotlight into the forest for a few seconds, followed by a minute interval of darkness. Small flies (*Amphineuridae*) and gnats which remained on the mud or in the air above the m^2 quadrats for as long as 15 min were counted at 5 min intervals during the observation periods. The average count for the afternoon period (when these forms were most active) was used as an estimate of their density.

Leaf and branch dwellers were censused by carefully examining leaves, branches, and trunks of

TABLE III. Density and biomass of mangrove animals Abbreviations: weight, wt; individual, ind; quadrat, quad.

Taxa		Density per m²	Biomass per m²
SNAILS			
Melampus coffeus	6m² quad at night (1959: 23,35,41) (1960: 31,37,19) wt 0.0047/ind	31	0.15
Littorina anglepifera	4 in 100m² (0.04/m²) 4 per 20 tree stems (0.27/m²) wt 0.1025/ind	0.2	0.02
CRABS			
Aratus pisonii	8/20 trees, 7/21 trees, 7/9m² quad 1959, 5/9m² quad 1960. wt 0.3638/ind	0.6	0.218
Eurytium limosum	4 per 5m² quads; wt 0.948/ind	0.8	0.759
Uca mordax	4m² quad (6,9,3,1); wt 0.53/ind	5.0	2.65
Ucides cordatus and *Goniopsis eventata*	4 large crabs seen. Mangrove crabs within 8m of each other. Assume one per 16m² wt 24/ind.	0.06	1.44
INSECTS			
Aglaopteryx diaphana	6 night, m² quad (1959: 3,2,0) (1960: 1,1,0) ave 1.2/m² 15/21 trees, 9/20 trees, 1.36 trees/m²; wt 0.0025/ind	1.1	0.003
Gryllidae	18m² quad (1959: 0,2,3,0,0,4,0,1,4) (1960: 4,0,1,3,0,1,4,0,4) wt 0.003/ind	1.8	0.005
Amphinumidae	12m³ quad (1959: 2,1,1,1,1,0,0) (1960: 3,1,1,2,0,1) wt 0.001/ind	2.1	0.021
Sarcophagidae	12m² quad (1959: 2,1,1,1,1,0,0) (1960: 1,1,0,0,1,1) wt 0.003/ind	0.8	0.002
Unidentified forest floor insects	18m² quad (1959: 0,1,26,0,1,4,0,2,1) (1960: 0,1,0,4,1,2,0,0,1) est wt 0.002/ind	2.5	0.050
Rhagovelia plumbea	3m² night quad (6,6,0) est wt 0.0005/ind	4.0	0.002
Flying gnats	3m² quad (4,4,12) est wt 0.0001/ind	6.7	0.001
Flying insects	Boardwalk census, area 118 x 2m (0,9,14,16,13,8,11,2,11) est wt 0.003/ind	1.4	0.004
MISC. tree arthropods (incl crickets, ants, scorpions)	7/1031 leaves (1065 leaves/m²) 7.4/m² 3/20 trees, 3/21 trees 1.36 trees/m² est wt 0.002/ind	7.6	0.015
SPIDERS			
Olios antiguensis	6/100m²—0.06/m²		
Gasteracantha tetracantha	2/100m²—0.02/m²		
Tree spiders	2/20 trees, 3/21 trees —0.16/m²		
Spider webs	Boardwalk census, area 118 x 2m (1959: 71,58,48) (1960: 5,5)—.16/m² wt 0.003/ind	0.2	0.001
VERTEBRATES			
Fish	5m² night quad (1,1,2,0,0) est wt 1.3/ind	0.8	1.0
Anolis cristatellus	2/100m² area, 2 on 8 boardwalk censuses, est wt 0.60 gm	0.01	0.006
Birds (residents)			
Yellow warbler	1959: 4,3,0; 1960: 3,2,3.		
Grassquit	1959: 2,2,0; 1960: 3,2,3.		
Dove	1959: 1,2,2; 1960: 2,0,0.		
(Nonresidents)			
Rail	1959: 0,0,1; 1960: 1 1,2.		
Crow	1959: 0,0,0; 1960: 1,1,1.		
Hummingbird	1959: 1,0,0; 1960: 1,0,0.		
Vireo	1959: 0,0,0; 1960: 1,0,0.		
Green heron	1959: 0,0,0; 1960: 1,0,0.		
Unid	1959: 0,1,0; 1960: 2,2,1.		
	est 22 birds in forest or 0.003/m² est wt 10/ind	0.003	0.03
Total........		66.673	6.377

20-25 trees along the boardwalk and in the forest at night and during the day.

Birds were censused by making bird walks along the boardwalk at about 6:30AM 3 days in 1959 and 1960. Some of the dry land species on Magueyes Island utilized the edge of the mangrove forest for roosting and cover. Other birds, especially the larger water forms such as rails and green herons, mainly used the edge near open water and flew between several stands of mangroves. Actual time these transients spent in the forest was not estimated; actual counts of individuals were used in the density estimates. Resident birds were those seen or heard regularly and for which nests were located. These included the yellow warbler (*Dendroica petechia*), black-faced grass-quit (*Tiaris bicolor*) and the ground dove (*Columbigallina passerina*). Singing birds were counted as representing a pair of birds.

Specimens of the most abundant animals were collected, stored in formalin and later dried in an oven at 100°C for 24 hrs to determine dry weight biomass.

We have not attempted to obtain a complete species list of the fauna of the red mangrove forest. Our major objective was to outline the structure and function of the community, and this required identification of the dominant animals and determination of their density and biomass. Not all measurements are equally accurate; estimates of flying insects are probably least accurate, while estimates of mud dwellers, crabs and spiders are probably more accurate. Micro-organisms were not studied.

Animal respiration.—Respiration of a few medium-sized animals was measured in a simple variable respirometer (described in Teal 1959). Duplicate measurements were made at 26-27°C and the readings were corrected with a thermobarometer. Since the animals were quiet in the respirometer the metabolic rates should be considered minimal estimates of actual field respiration. Some approximations of oxygen consumption of forms too small or large to fit into the respirometer were taken from Spector (1957). Oxygen consumption of small arthropods was estimated as 1.0 ml O_2/g/hr and for vertebrates as 3.0 ml O_2/g/hr. The biomass estimates were multiplied by the oxygen consumption per gram dry weight to estimate the respiration of the animal population.

RESULTS

The Environment

The peninsula containing the mangrove study area was approximately 7400 m², with red mangrove occupying 4600 m², and a complex of red, black, white and button mangrove occupying 2800 m². The mangrove forest was protected from the open sea by a coral reef about 6 m from the forest, and separated from the forest by a narrow bed of *Thallasia testudinum* mixed with coral.

The swamp floor consisted of a layer of mangrove peat and roots about one meter thick in the center of the forest and graded to 1.2 m at the water edge and 0.8 m at the land edge. The peat rested on a coral platform. Samples of peat from the peat cores taken near the water edge were burned in a muffle furnace at 600°C for 6 hrs to determine the per cent ash. The first 3 sections (25 cm long) contained 54.7, 47.9 and 32.7% ash and the last section, from the coral base, contained 91% ash. These samples did not include sections of large roots. The limited data indicate that almost one-half of the mass of the peat is inorganic.

During May high tide occurred at night and reached a depth, on the study quadrats, of about 5-17 cm depending on wind conditions. On one night no tide reached the quadrat floor. According to the regime described by Coker and Gonzalez (1960), there is only one low and one high tide daily, each occurring at the same time of day over extended periods. High tide occurs at night in late spring and summer and at noon in winter. Changes in sea level between tides are small. The mean monthly annual changes in tide is 21.0 cm. Salinity of water flowing over the peat is 29.1 ‰.

Light intensity curves for different depths of the forest and times of day are plotted in Figure

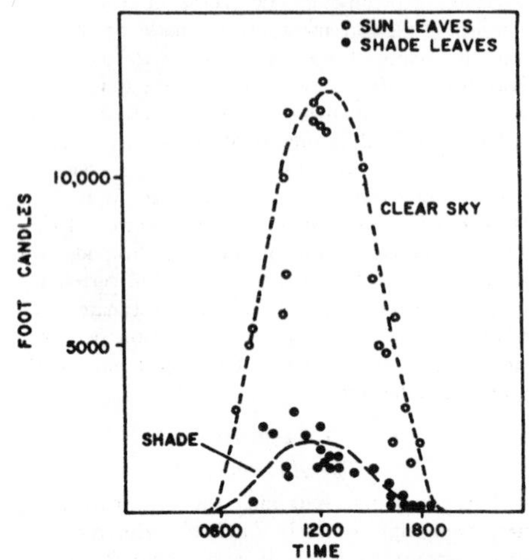

FIG. 3. The diurnal sequence of light intensity in the top of the forest (sun leaves) and under the forest canopy (shade leaves) in May, 1960.

3. On May 28, 1959, the temperatures on the mud surface in the shade, 2 cm in the mud, 60 cm in the air, and in the water during high tide were as follows:

	AM 8:50	9:30	10:40	12:25	PM 2:45	3:45	10:30
60 cm in air..	26	27	28	30	29	30	—
mud surface..	27	28	31	35	35	32	—
2 cm in mud..	27	28	30	31	35	32	—
water			no water				29

The temperature he mud surface, the habitat of many of the mangrove animals, reaches slightly higher levels than the other temperatures. On June 24, 1961, wind speed was measured on the study plots and at one meter above sea level in the ocean before the forest, at 9:00 and 11:00AM and 3:00PM. Average wind speeds in the direction of the prevailing wind in ft/min were: 1.5 at ground, 57.7 at 1m, 71.6 at 3.5m, 104 at 5.5m, and 736 at sea.

Structure of the Red Mangrove Forest

A vertical section through the forest shows distinct stratification (Fig. 4). In the lowest meter the arching roots (18 roots/m^2) form an intricate tangle. At the base of each root are conspicuous lenticels shown by Scholander to be used for oxygen respiration (Kramer and Kozlowski 1960). In dense profusion among the roots are seedlings (an average of 17/m^2 on 3 plots with 27, 4, and 19 seedlings) which sprout from their "viviparous embryos" dropped from the canopy into the soft mud. Above the lower meter of roots is the stratum of principal trunks with leaves less dense than above in the canopy or below in the seedling-root zone. Leaves in this stratum are of the shade type. Above the trunk zone at about 5 m is the area of maximum leaf biomass, with sun and shade leaves in profusion. All leaves in the upper 2 m exhibit the smaller size and thicker texture of sun leaves. Measurements of vertical structure are graphed in Figure 4.

The frequency distribution of dbh of red mangrove trees on the 100m^2 plot containing thickets, medium and large trees, is shown in Table II. Small trees from one to 4 cm dbh were most abundant. Dry weight biomass calculated from tree harvest (Table I) totaled 536 g in leaves, 1274 g in branches, 2796 g in trunks, and 1437 g above ground in prop roots per m^2 (Table II). Combining the leaf estimates made in 1959 and 1960 the average biomass of leaves is 778 g per m^2 for the forest. Underground portions of the vegetation, the peat and roots, totaled about 45,000 g dry weight/m^2. Small (0.5-1 cm diameter) and large (2+ cm diameter) roots were

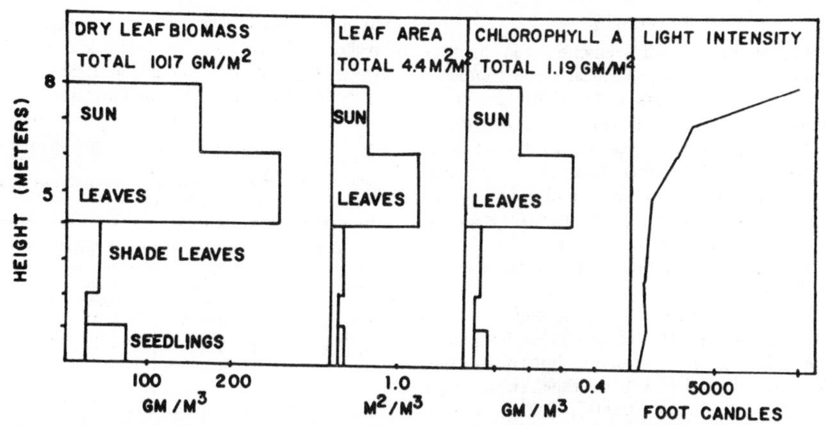

FIG. 4. Vertical distribution of leaf biomass, leaf area, chlorophyll *a*, and light intensity in the red mangrove forest, based on measurements and counts of all leaves on a 25m² quadrat in May, 1959.

taken from the core samples; large roots weighed 997 g and small roots 4000 g/m². A few large roots were also taken from the coral sections, indicating that mangrove roots extend into the coral substrate. These latter roots are not included in the estimates.

Chlorophyll *a*

Chlorophyll *a* was measured in sun and shade leaves of *Rhizophora mangle* and in the mangrove litter on the mud. In January, 1958, the following vertical sequence was measured in grams of Chlorophyll *a* per m² of leaf area: 6m, 0.25; 5m, 0.19; 4m, 0.32; 3m, 0.35; 2m, 0.23; 1m, 0.25; mud, 0.18 g/m² mud surface. In May, 1959, shade leaves contained 0.24 and 0.29 g/m² of leaf area; sun leaves contained 0.31 and 0.33 g/m² of leaf area in two different collections of leaves. Using mean values for shade and sun leaves and leaf area data in Figure 4, Chlorophyll *a* computed for each stratum is reported in Figure 4. Chlorophyll *a* for the whole quadrat was 1.19 g/m².

Animal Populations

Each stratum of the forest had its characteristic fauna. On the forest floor the dominant animals were the fiddler crab (*Uca mordax*) and the mud crab (*Eurytium limosum*), the snail (*Melampus coffeus*) and crickets (*Gryllidae*). On the trees the crab (*Aratus pisonii*), the snail (*Littorina angelifera*), roaches (*Aglaopteryx diaphana*), spiders (*Gasteracantha tetracantha*, and *Olios antiguensis*), lizards (*Anolis cristatellus*), and birds were encountered. Flying among the trees were various flies (Ephydridae and Sarcophagidae), gnats, butterflies, moths, and birds. When the forest was covered with water, fish, water striders, and large crabs emerged from burrows and entered from the sea edge.

The total fauna in May consisted of about 67 animals per m², which weighed a total of about 6.4 g dry weight (Table III). In terms of biomass the crabs were of greatest importance.

Photosynthesis of the Red Mangrove Forest

Rates of CO_2 uptake during the day determined for sun, shade, and seedling leaves of measured area at various light intensities are reported in Figure 2. A curve corresponding to the mean values of CO_2 output at the measured light intensities was fitted to the data by eye. Confidence limits of 95% were computed for sun, shade and seedling leaf data. At this level, the average confidence limits were 29% of the mean value for sun and seedling and 23% for shade leaves; this means that the confidence limits for the photosynthesis and respiration estimates based on the curve are within about 25% of the estimated value.

The number of hours at each 1000 ft-c light intensity during an average day was estimated from the light intensity curves for sun and shade leaves (Fig. 3). The CO_2 production per square meter of leaf for the daylight hours was calculated for seedlings, shade and sun leaves from the curve in Figure 2 and leaf area in Figure 4. Net daytime photosynthesis totaled 0.12 g C/m²/day for seedlings, 0.24 g C/m²/day for shade leaves, and 5.2 g C/m²/day for sun leaves.

Respiration

Based on night measurements, the average respiration per area of leaf surface was computed for sun leaves as 0.0449 g $C/m^2/hr$, for shade leaves as 0.0465 g $C/m^2/hr$, and for seedling leaves as 0.1470 g $C/m^2/hr$. Respiration rates for seedlings were slightly higher than those of other leaves. These estimates were multiplied by the area of leaves by strata (Fig. 4) to obtain the estimates of respiration by strata for 24 hrs (Fig. 5). Total leaf respiration of all strata was 5.4 g $C/m^2/day$.

The mean respiration of the lenticel zones of the prop-roots was 0.0046 g C/root/hr (based on the following measurements: horizontal zone, 0.0016; seedling, 0.0032; average sized roots, 0.0049, 0.0033, 0.0028, 0.0013, 0.0025, 0.0037, 0.0049, 0.0067, 0.0066; and large roots, 0.0098, 0.0085).

Measurement of the respiration of the forest floor was complicated since the mud was covered with water for about 10 hrs of each 24 hr period (at night). Based on the result of bell jar measurements (Table IV) made during high tide,

TABLE IV. Oxygen utilization of water covering forest floor by bell jar experiments

Unit	Initial Oxygen mg/l	Time Lapse hours	Final Oxygen mg/l	Mean Oxygen Change mg/l	Respiration g $O_2/m^2/hr$
	2.92		1.60		
A	2.75		1.40		
	2.82	3.3	1.50	−1.32	0.047
	5.43				
B	5.81		4.73		
	5.40		4.95		
	5.55	3.6	4.85	−0.70	0.039
	5.43				
C	5.81		5.00		
	5.40		5.15		
	5.55	3.7	5.07	−0.48	0.034
mean value					0.040

A Bell jar volume, 4.0 liters; area, 380cm².
B Bell jar volume, 3.54 liters; area, 177cm².
C Bell jar volume, 1.0 liters; area, 39cm².

respiration of the soil under water was about 0.04 g $O_2/m^2/hr$ or about 0.02 $C/m^2/hr$. The respiration of the soil during air exposure varied from 0.005 g $C/m^2/hr$ at air flow 0.1 cm/sec to 0.20 g $C/m^2/hr$ at air flow 3 cm/sec, as recorded in 41 measurements. Since the average wind velocity at ground level was 1.5 ft/min or .76 cm/sec, the soil respiration was about 0.012 g $C/m^2/hr$ or 0.168 g $C/m^2/14hr$. Whether peat was accumulating or decreasing during May is not known. The peat layer of a mangrove forest tends to develop a fairly constant inter-tidal equilibrium level (Chapman and Ronaldson 1958).

Respiration of the medium-sized animals (snails, crabs, and spiders) as measured in the field is presented in Table V, together with estimated values for insects and vertebrates. These values of oxygen consumption per gram body weight were multiplied by estimates of biomass (Table III) to obtain the total metabolism of the fauna—0.164 g $O_2/m^2/day$ or 0.082 gm $C/m^2/day$. These data indicate that the macrofauna account for only a small portion of the total consumption of the community.

TABLE V. Estimate of oxygen consumption of mangrove fauna

Taxa	Dry Wt/m²	ml O_2/g/hr	ml $O_2/m^2/hr$
Snails	0.170	1.49	0.253
Crabs	5.070	0.26	1.318
Insects	0.103	1.00	0.103
Spiders	0.001	7.67	0.008
Vertebrates	1.036	3.00	3.108
			4.790

Export

Every evening in May there was a gentle and gradual rise and fall of the tide. At night over a 10 hr period, 10 cm or 100 l/m² of water moved onto the quadrat and out again without perceptible turbulence or easily recognizable current. Data in Table VI indicate that the water moves out with 22.7 mg net dry particulate matter in each liter per day. Thus the particulate matter carried out in May is estimated at 2.27 g/m² or 1.14 g $C/m^2/day$.

TABLE VI. Estimate of export of particulate organic matter in tidal water[1]

Sample	Incoming Water mg/l	Outgoing Water mg/l	Change mg/l
1	1.36	38.20	36.4
2	1.87	10.60	8.7
average	1.62	24.40	22.7

[1] 100 liters of tidal water flow over one square-meter/day.

Data in Table VII indicate that the outgoing water carries more labile organic matter than the incoming water since there was 1.61 mgO_2/l water oxygen consumption in 6 days or 0.23 g $C/m^2/day$ labile organic matter exported, much less than the particulate value.

TABLE VII. Oxygen changes in biochemical oxygen demand bottles over a 6 day period in the dark at 28° C. Data as mg/1, oxygen

	Start	After 6 days	Change
Incoming Water	5.90 5.89	5.13 5.40	
	5.90	5.26	−0.64
Outgoing Water	4.88 4.53	2.40 2.50	
	4.70	2.45	−2.25
Difference	—	—	1.61 mg/1/6 days

Apparently the particulate matter includes a large proportion of non-labile organic matter. Bottles containing swamp water were still visibly full of particulate matter after one year. The only fraction of organic matter not included was the non-labile matter smaller than the net mesh.

Comparison of Gains and Losses

The various estimates of photosynthetic gain and loss due to respiratory consumption and export are included in one budget graph in Figure 5. The gains due to photosynthesis are plotted on the right. These include the observed daytime net photosynthesis plus an estimate of daytime photosynthesis that is consumed by concurrent daytime respiration. Thus the total length of bars to the right of the center line represents estimates of gross production. The bars to the left include day and night respiration and estimates of export as indicated. The graph in Figure 5 represents a synoptic view of the processes in the mangrove forest during an average day in May.

Apparently a large proportion of the gross photosynthesis is immediately used in plant respiration during the day and night. The estimates of gross photosynthesis and total attrition are sufficiently close to suggest that this forest is not in rapid succession. Without data from other

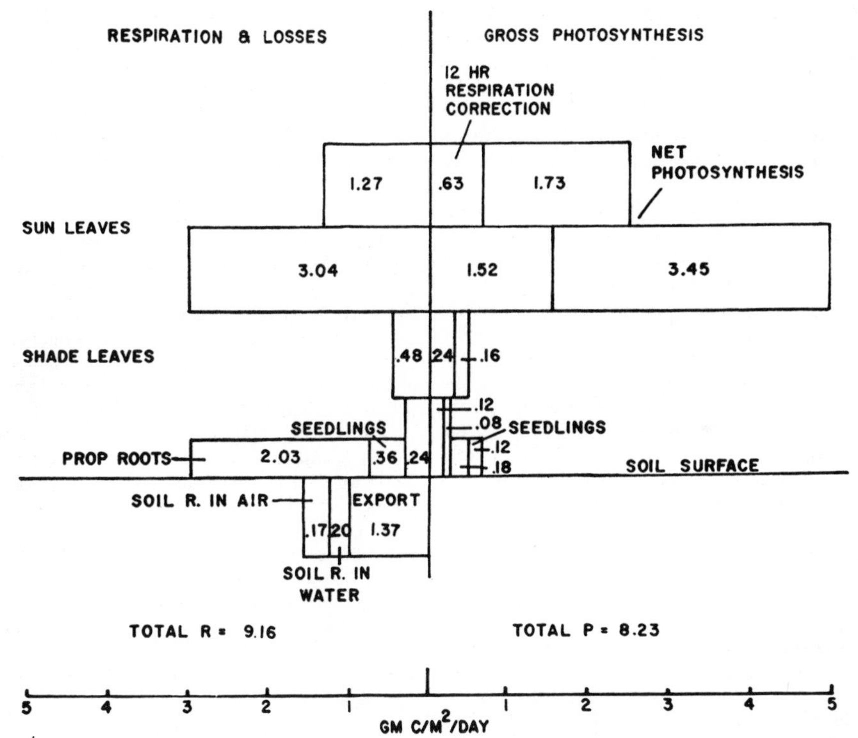

FIG. 5. Rates of photosynthesis, respiration, and export in the red mangrove forest components in gram carbon/m²/day in May.

seasons and other years, and further replications the results cannot be interpreted in too great detail.

Net estimates of annual growth of wood totaled 0.84 g/m²/day or about 0.42 g C/m²/day (Table VIII). Since there are some sites with an open

TABLE VIII. Net growth of wood of red mangrove on $100m^2$ plot

DBH Class cm	Number Trees 1959	DBH Increase cm/tree	Increase in Class* g/tree	Total Increase
0-1	4	.58	1965.0	7860
1-2	65	.08	271.0	17615
2-3	38	.00	0.0	
3-4	12	.00	0.0	
4-5	7	.19	643.7	4506
5+	10	.02	67.8	678
Totals	136			$30659/100m^2$

average growth of 306.6 $g/m^2/yr$ or .84 $g/m^2/day$

* Based on regression of 3388 g wood per cm dbh.

canopy in the forest and no trees which approach the maximum size for red mangrove (dbh, 90 cm), this forest may still be growing. Noakes (1955) in Malaya reported 130-140 cu ft/acre (14 g $C/m^2/day$) yield of timber based on mean annual increment over 25 yrs in growing red mangrove (*Rhizophora mucronata*) forest. Holdridge (1940) reported 2 in. diameter growth in 5 yrs on red mangrove plantations. These growth rates are much higher than those measured in our plot.

The organic matter supplying the soil falls as leaves from the euphotic zone above at a rate of about 1.3 $gm/m^2/day$ or 0.65 g $C/m^2/day$ (measured on a total of 24 m^2 in 4 replications as 1.4, 0.1, 1.9 and 2.8 g organic matter/m^2/day). Since export of particulate matter (1.1 g $C/m^2/day$) and soil respiration (0.37 g $C/m^2/day$) together are more than twice the estimates of the leaves, some other sources of organic matter to the mud may be present. Students of the field biology course at the University of Puerto Rico at Mayaguez set up a one hour plastic bell jar experiment to determine mangrove soil respiration in June, 1961. CO_2 was absorbed in KOH and each series contained a control, a clear plastic and a aluminum-foil covered box. Gross production of algae associated with the mud (calculated as the difference between covered and clear boxes) was 2.70 and 0.90 ml CO_2/hr or 1.134 and 0.378 g C/m^2/14 hours. Gains to the mud, the leaf fall plus the difference between gross production of the algae (using the highest value) and soil respiration in air, of 1.61 g $C/m^2/day$ were close to total losses to soil respiration and export of 1.74 g $C/m^2/day$. Although these are very preliminary data, they suggest that algae may be important producers in the mangrove community.

Comparison of Metabolism with Other Communities

With a total gross production and respiration exceeding 8 g $C/m^2/day$ or about 16 g organic matter/m^2/day, the red mangrove community is more fertile than most marine and terrestrial communities (Summarized by Odum and Odum 1959). The montane rain-forest, studied in the Luquillo Mountains, Puerto Rico, (Odum, Abbott, Selander, Golley, and Wilson, in manuscript; 17 g $C/m^2/day$) and the coral reefs, studied near Magueyes Island, Puerto Rico (Odum, Burkholder, and Rivero 1959); up to 22 g $C/m^2/day$) have greater gross production rates than the mangrove forest. The Puerto Rican red mangrove forest, although well adapted to survive on tropical shores, is not as efficient as the montane rainforest or the coral reef in the conversion of sunlight into organic matter in a similar light regime.

The ratio of gross photosynthesis per 12 hrs of day (0.68 g $C/m^2/hr$) to Chlorophyll a (1.19 g/m^2) is the assimilation ratio of the community (0.57 g C/g Chlorophyll a/hr or about 1.2 g O_2/g Chlorophyll a/hr). This ratio falls within the range (0.4 to 4.0 g O_2/g Chlorophyll a/hr) reported for other whole communities by Odum, McConnell and Abbott (1958).

SUMMARY

Measurements of structure and metabolism are reported from a stand of red mangrove (*Rhizophora mangle*) in southeastern Puerto Rico during May, 1959 and 1960 as follows: leaf biomass dry weight, 778 g/m^2; wood dry weight, 5507 g/m^2; peat and roots, 45,000 g/m^2; community chlorophyll *a*, 1.19 g/m^2; and animal biomass, 6.4 g/m^2. Total photosynthesis and leaf respiration were each estimated from measurements with a CO_2 analyzer and were about 8 g $C/m^2/day$. The forest respiration from air exchange holes in the prop root bases was 2.03 g $C/m^2/day$. Smaller magnitudes were found with estimates of leaf fall (1.3 g $C/m^2/day$), trunk growth (0.4 g $C/m^2/day$), tidal export of particulate matter (1.1 g $C/m^2/day$), underwater respiration of the soil (0.2 g $C/m^2/day$) and soil respiration in air (0.168 g $C/m^2/day$). For animals a small metabolism (0.082 g $C/m^2/day$) indicates a minor role in the ecosystem.

ACKNOWLEDGMENTS

We are grateful to Dr. Juan Rivero, Director, and members of the staff of the Institute of Marine Biology,

University of Puerto Rico, Mayaguez, for suggesting and sponsoring this study. Thanks are due to Dr. A. Smalley, Tulane University, for identification of the crabs, Dr. J. P. E. Morrison, U. S. National Museum for the snails, Dr. P. E. Hunter, University of Georgia, for the spiders, and Dr. J. Maldonado, University of Puerto Rico, Mayaguez, for the insects.

REFERENCES

Chapman, V. J. and J. W. Ronaldson. 1958. The mangrove and salt marsh flats of the Auckland isthmus. New Zealand Dept. of Scientific and Industrial Research Bull. **125:** 1-79

Coker, R. F. and J. G. Gonzalez. 1960. Limnetic copepod populations of Bahia Fosforescente and adjacent waters, Puerto Rico. J. Elisha Mitchell Sci. Soc. **76:** 8-28.

Dansereau, P. 1947. Zonation et succession sur la restinga de Rio de Janeiro. I. La halosere. Rev. Canad. Biol. **6:** 448-477.

Davis, J. H. 1940. The ecology and geologic role of mangroves in Florida. Pap. Tortugas Lab. **32:** 303-412.

Fox, D. L., C. H. Oppenheimer, and J. S. Kittredge. 1953. Microfiltration in oceanographic research. II. Retention of colloidal micelles by adsorptive filters and by filter-feeding invertebrates; proportion of dispersed organic to inorganic matter and to organic solutes. J. Marine Res., **12:** 233-243.

Golley, F. B. 1960. Ecologic notes on Puerto Rican mollusca. Nautilus **73:** 152-5.

Holdridge, L. R. 1940. Some notes on the mangrove swamps of Puerto Rico. Caribbean Forester **1:** 19-29.

———. 1942. Trees of Puerto Rico. USDA, Forest Service, Tropical Forest Exp. Stat. Occas. Papers **1.**

Kramer, P. J. and T. T. Kozlowski. 1960. Physiology of Trees. McGraw Hill, New York. 642 pp.

McGill, J. T. 1958. Coastal Landforms of the world. Map suppl. *in* Russell, R. J. 1959. Second coastal Geography Conf. Coastal Studies Institute, Louisiana State University. 472 pp.

Noakes, P. S. P. 1955. Methods of increasing growth and obtaining natural regeneration of the mangrove type in Malaya. Malayan Forester **18:** 23-30.

Odum, H. T. and E. P. Odum. 1959. Principles and concepts pertaining to energy in ecological systems. Chapter 3 *in* Odum, E. P., Fundamentals of Ecology. Saunders, Phila. 546 pp.

———**, W. McConnell, and W. Abbott.** 1958. Chlorophyll "A" of communities. Publ. Inst. Marine Sci. Texas **5:** 65-96.

———**, P. Burkholder, and J. Rivero.** 1959. Measurements of productivity of turtle grass flats, reefs, and the Bahia Fosforescente of southern Puerto Rico. Publ. Inst. Marine Sci. Texas **6:** 159-170.

———**, W. Abbott, R. K. Selander, F. B. Golley, and R F. Wilson.** 1960. Trophic structure and productivity of the lower montane rain-forest of Puerto Rico. In manuscript. 52 pp.

Rakestraw, N. 1947. Oxygen consumption in sea water over long periods. J. Marine Res. **6:** 259.

Sargent, M. C. and T. S. Austin. 1949. Organic productivity of an atoll. Trans. Amer. Geophys. Un. **30:** 245-249.

Spector, W. S. 1957. Handbook of Biological Data. Saunders. 584 pp.

Teal, J. M. 1959. Respiration of crabs in Georgia salt marshes and its relation to their ecology. Physiol. Zool. **32:** 1-14.

Concepts in Marine Ecology

THE AMERICAN NATURALIST

Vol. XCIV | March-April, 1960 | No. 875.

THE EVOLUTION OF STABILITY IN MARINE ENVIRONMENTS NATURAL SELECTION AT THE LEVEL OF THE ECOSYSTEM

M. J. DUNBAR

Department of Zoology, McGill University, Montreal, Canada

One of the most striking contrasts between the lower and the higher latitudes is manifested by the stability of the warm-adapted floras and faunas and the instability of the ecosystems of the cooler parts of the world. The argument developed here is that this contrast suggests, among other things, that selection may apply at the level of the ecosystem as well as at the levels of the individual and the specific population. Ecosystems can compete, and evolution of the stable ecosystem can be looked upon as a process of learning, analogous to the learning of regulated behavior in the nervous systems of animals.

OSCILLATIONS IN NUMBER

In this discussion, I am starting from the premise that oscillations are bad for any system and that violent oscillations are often lethal. The more violent the oscillation in specific numbers in any ecological situation, the greater the danger of extinction of species, at least of local extinction, causing serious disturbance of the community, possibly the extinction of the whole system, again locally.

Population oscillations of fairly wide amplitude are well known in terrestrial environments among rodents and their predators, game birds, rabbits, and insects, and much study has been devoted to them. Oscillations of this sort, in which a primary oscillator, a herbivore, causes sympathetic oscillations in carnivore populations which in turn take part (however small a part) in reducing the amplitude of the oscillations of the herbivore, appear only in relatively simple ecosystems, containing a limited, usually quite small, number of species. Such simplicity is found as a rule only in cool climates with marked seasonal variation, that is to say, in temperate and polar climates. (A special case might be made here for desert regions, in which ecosystems are also simple.) Oscillations of this sort are absent in tropical and subtropical environments, which foster much more complex ecosystems in which there is great multiplicity of energy paths along which overloadings can be released, with consequent great decrease or virtual elimination of the time-lag effect which is largely responsible for the oscillations in the simpler systems (Hutchinson, 1948; Riley, 1953; Odum, 1959; MacArthur, 1955).

The simplicity of the ecosystems in cool and cold climates is presumably partly or mainly a result of the novelty of the present polar conditions. Whether or not we accept the views of Zenkevitch (1949) on the ancient origin of the cold-adapted fauna or the more conservative estimates of the age of the present polar cooling of Barghoorn (1953) and Durham (1950), there can be little doubt that the conditions existing during the Pleistocene were special. The short million years since the end of the Pliocene, or the few million years of glacial climate possibly since some time in the Pliocene or Miocene, and the much shorter time available for colonization of areas actually glaciated, have not been long enough to allow the adaptation to the new conditions towards the poles of more than a relatively small number of species. Such adapted forms have by establishing themselves first, further lowered the chances of successful establishment by others, the more so because the number of possible niches and habitats in the colder regions is very much less than in warmer parts of the world.

SEASONAL OSCILLATIONS

What is true of fluctuations whose periods occupy a few years is also true of *seasonal* oscillations—they are most marked in temperate and polar regions, and absent or poorly developed in the tropics both on land and in the sea. In the plankton, for instance, upon whose numerical behavior everything else in the sea depends, production is continuous at a steady controlled level all year round in tropical waters, according to Steemann Nielsen (1957). Plant growth on land, in equatorial regions, has a similar year-round activity; individual species may develop periodicities of flowering and of new leaf growth, not necessarily annual periodicities, but the total mass and the total activity level remain fairly constant (Holttum, 1953). Steemann Nielsen describes extraordinary stability in the planktonic system in tropical seas. The phytoplankton does not at any time completely consume the available phosphate and nitrate supplies, which are continuously present at low levels; the zooplankton does not deplete the phytoplankton, and predation upon the zooplankton does not, apparently, cause oscillations. This means that the levels of production are delicately controlled in a hydrographic system which is itself highly stable. Except for certain areas of large-scale upwelling, which are among the most productive regions in the world, the warmer seas are very stable, and vertical movements of water are effectively and stubbornly resisted. The supply of nutrient salts is thus limited to (1) regeneration within the euphotic layer, and (2) such quantities as can enter any given area from outside by horizontal transport. These two sources have to maintain the equilibrium against constant loss from the system of salts contained in detritus which sinks out of the photosynthetic layer before it is mineralized. The quantities involved in this loss are not yet known.

The contrast between this tropical stability and the sharply oscillating annual cycles of standing crop in temperate and polar systems is obvious, and it is equally clear that the greatly differing climatic conditions are intimately associated with the contrast. The annual variation of light intensity and angle of incidence, the larger amplitude of annual temperature variation, and the unstable condition of the colder waters in winter, are all involved here. It may not be so obvious, however, that these annual oscillations in both marine and terrestrial ecosystems have something important in common with the "fur-bearer" type of oscillations already mentioned. Both types can be attributed to climate; the former as just described and the latter owing to the simplicity of the ecosystem, which is itself a result of climatic demands. Both groups of oscillations are to a high degree the result of the onset of the glacial climate. On the original premise, then, that oscillations are disadvantageous both to the individual species and to the ecosystem as a whole, I conclude that the steady systems of the tropics are the result of long evolution and that oscillations observed in the higher latitudes are systems of non-adaptation. Arctic and antarctic faunas are immature, still in the elementary stages of evolutionary "learning."

ADAPTATIONS TO COLD CLIMATES

The next thing to look for are signs of incipient adaptation, in this ecological sense, to life in cold climates. Adaptation in this case would be a matter of changes in the system which would tend to (1) increase the complexity of the ecosystem, that is, increase the number of species involved; (2) lower the rates of production from the maxima possible in both plants and animals; (3) spread the rate of grazing upon the plants as evenly as possible over the 12 months of the year, and (4) carry these processes up to higher trophic levels. This would involve control of the breeding rates of both plant and animal populations, and possibly also the regulation of the manner in which individuals in the herbivore and carnivore populations grow, and of their energy requirements. There is evidence that processes of this sort have in fact been evolved, or are in process of evolution.

1. The findings of Thorson (1950, 1952) can profitably be interpreted in this light. Thorson found that in marine shallow water benthos in temperate and arctic regions, the pelagic larva may or may not be retained. If it is retained, the spawning is restricted to a short period in the spring calculated to coincide with the abundance of the spring plankton; if it is not, the spawning period is much longer and could extend throughout the year. There is a direct relationship between the mean temperature of the environment and the proportion of species which retain the pelagic larva. The stabilizing effect of the loss of larva and production of larger eggs is thus best developed in the colder regions.

2. Steele (1959), working on the eastern Canadian arctic shallow water Amphipoda, concludes that whereas the littoral species have developed short seasonal breeding periods in the spring or summer, the benthonic forms as a rule breed all the year round. The immediate or proximate reason

for this may well be that the detrital or bacterial food of these species is present the year round anyway; nevertheless, the ecological effect is towards high biomass in winter.

3. Recent work on the marine fauna of the arctic regions has brought to light the fact that many species produce their young not simply at any time of year, but specifically in the dead of winter. In a situation in which the spring bloom of phytoplankton is sudden and violent, in which there may be no fall bloom at all and the winter stock of plant food tends to be quite small, so far as we know, one would have expected that all the animal species which depended on phytoplankton in the younger stages would have evolved a breeding cycle that would release the young at the most propitious moment, at some time between the beginning and the peak of the spring phytoplankton bloom. Many species of course do this, notably the copepods, but there are many that do not, and as the study of arctic breeding cycles advances, more such species appear. MacGinitie (1955) gives many examples of invertebrate species in the waters of the Point Barrow area, Alaska, which begin to develop their eggs in October and later, and many which produce ripe eggs at that time; some of the amphipods had hatched young in the marsupium. Dunbar (1957) records the dominant arctic pelagic amphipod *Themisto libellula* as maturing late in the autumn or in the early winter and carrying hatched young in December. *Gammarus setosus* breeds in the colder part of the year, and differs in this respect from its close relative *G. oceanicus* (MacIntyre, personal communication). And among the 114 species of amphipods taken in Ungava (Dunbar, 1954), making up a very large material collected over four seasons, only 55 species were found to include ovigerous or paedigerous females during the spring and summer months, from June to September. The material collected by the Russian drifting station expedition in the polar basin in 1950–1951 included several species recorded as spawning in winter (Brodskii and Nikitin, 1955). This winter breeding has the effect of increasing the number of species in the system. This is analogous to the adoption of nocturnal or diurnal habit, and it also spreads the rate of predation on the available food more evenly over the year.

4. There is something possibly significant in this regard in the large size and lowered growth rate of arctic poikilotherms. Harvey (1955), in discussing growth and metabolism in marine poikilotherms, writes: "In addition to this general inverse relation between age or size of animals and losses by respiration, there is an inverse relation between age and growth rate (the daily percentage increase in organic matter). It appears usual that growth rate decreases more rapidly with increasing size than respiration decreases.... In consequence, a greater proportion of food assimilated by young animals is built into new tissue than by old animals. Hence the same rate of plant production may permit a greater biomass of a stable community consisting mostly of aged, larger, slow-growing, slow-respiring animals than of one mainly composed of small quick-growing animals. The latter fauna, however, may synthesise more animal tissue yearly, the rate of turnover of living tissue in the animals being greater."

This, the development of large size and slow growth and metabolic rates, is precisely what happens in cold polar waters, which raises the question whether this situation may not be the result of adaptation to the total ecological condition, assuming that there is selection favoring stability.

5. The two-phase or alternating breeding cycle demonstrated in certain cold-water members of the zooplankton (for example, Dunbar, 1957), such as *Themisto libellula*, *Sagitta elegans*, *Thysanoessa* spp., may also be interpreted in this light, since it leads to stability of the species population and maintenance of large numbers of individuals under conditions of slow growth, slow maturation and probably limited food supply. In this type of cycle two broods coexist in the same body of water, the two broods being separated in time so that they may be partially or completely isolated from each other reproductively.

6. Finally, there is an example of stability which is as striking as that of the tropical plankton, and about which a little more is known. The families of oceanic birds are remarkable for the stability of their population numbers. The range of a species may increase or decrease over a period of years, as in the case of the Fulmar Petrel in the British Isles or of the Gannet in the North Atlantic, but oscillations of the sort discussed here do not appear to occur. They breed on small island groups or rock cliff shores, with a food supply that must be considered to be virtually without limit for practical purposes, and with great mobility. In spite of this plenty, the oceanic birds are most modest in their breeding rates. The numbers of eggs laid are small, often only one per pair, and breeding may not occur every year per individual, at least in some species. Why this frugality amid such plenty? I suggest that these are highly evolved stable populations which have in the past been subjected to the stress of oscillation in an oscillating system, and that they have responded to selection for this self-regulating character of a restricted breeding rate, tending toward stability. Baker (1938) quotes examples of many tropical (terrestrial) birds which lay only two eggs in a clutch, though they belong to genera in which five or six eggs are the rule among temperate representatives.

Some of the sea-bird populations mentioned above breed in polar or temperate regions. But if they do, they are migrants, and are present toward the poles only when the food supply is maximal for the year. As such they serve to cut down the animal marine populations when the latter are high, and depart for other parts when the food supply grows less; they do not prey upon animals during the low-level period in the oscillation. Moreover, birds, as homotherms, are less affected physiologically by the polar climates than are the poikilotherms—the problems of adaptation are much simpler. This is true also of the sea mammals, whose numbers also appear to be steady. Certain of the most numerous sea mammals that breed in the north are migratory; the phenomenon of migration itself may perhaps profitably be considered as the result of evolution at the ecological or ecosystem-level, as something which tends to stabilize both the systems of the breeding areas and those of the wintering areas.

The case made here, then, is that the stable non-oscillating system is the "ideal" and the product of a long process of evolution; and that the oscillating systems of regions affected by the present glacial climate of the world are in process of evolution towards better adjustment.

SELECTION OF ECOSYSTEMS

As to the mechanisms by which selection might take effect at this level, they are of the ordinary Darwinian sort except that the criterion for selection is survival of the system rather than of the individual or even the species. For instance, suppose an ecosystem, locally defined, begins to develop oscillations to a lethal degree, a degree such that one or more vital parts are not able to survive; the resulting empty environmental space, as in Cuvierian cataclysms, is available for occupation by communities from the adjacent regions; and these adjacent systems, as their survival suggests, are not of precisely the same constitution as the extinguished system. Perhaps a drift effect has occurred in certain species, or some other isolation effect. One or more of the specific elements will have growth rates, breeding potential and/or metabolic adjustment to temperature different from the former system; and if the difference is favorable to the continued survival of the system, its chances of survival are enhanced. In this way the system dominant in any geographic region changes, and changes (if the present assumptions are correct) in the direction of greater stability.

In this sort of selection, characters which may be of immediate advantage to the species are not necessarily selected into the stock in the long run, as may happen to the character which may be advantageous to the individual but ultimately lethal to the species. A breeding rate high enough to give one species an immediate advantage over another may ultimately prove to be lethal to the whole system. It thus becomes possible, in fact necessary, for selection to favor *slower* growth rate, *longer* time to reach maturity, *winter* spawning, and so on, under certain determining circumstances. It should be pointed out here that the model suggested in this paper has already been called for, as it were, by Hutchinson (1957) who writes (pp. 424-425): "So far little attention has been paid to the problem of changes in the properties of populations of the greatest demographic interest.... A more systematic study of evolutionary change in fecundity, mean life span, age and duration of reproductive activity and length of post-reproductive life is clearly needed. The most interesting models that might be devised would be those in which selection operated in favor of low fecundity, long pre-reproductive life and on any aspect of post-reproductive life." There is much more in the paper just quoted of direct relevance to the thesis put forward here. Slobodkin (1953), also, has pointed out that in certain circumstances a low rate of reproduction may have a greater selective advantage than a high rate. Utida (1957) comes to the conclusion from the mathematical approach that selection in favor of dampening of oscillations is to be expected, and points out that both Hutchinson (1954) and Slobodkin (1953) had come to similar conclusions from other directions. Chitty (1957) describes a noninfectious pathological condition which develops in *Microtus agrestis* under stress of crowding, and adds: "The main theoretical difficulty is to explain why such a deleterious condition, if controlled by an hereditary factor, has not been eliminated by natural selection." Finally, and in the contrary sense, Barnes (1958), describing the synchronization of spawning in *Balanus balanoides* with the spring phytoplankton outburst, regards such synchronization in arctic regions as essential; it is in fact essential only within the framework of classical selection theory, and it is clear that there are many forms in which no apparent effort at such synchronization is made.

There are probably strict limits to the degree to which an ecosystem can stabilize itself in an environment as variable (seasonally) as the climates of the higher latitudes. Perhaps stability comparable to the tropical stability was not achieved in higher latitudes even before the onset of the glacial climates. The thesis here is simply that there is a selection acting constantly in favor of greater stability, whether or not such selection meets a situation of equilibrium with an oscillating environment, beyond which greater stability of the system may become impossible.

SUMMARY

Starting from the premise that oscillations are dangerous for any system and that violent oscillations may be lethal, this paper contrasts the highly stable production systems of tropical waters with the seasonal and longer-term oscillations of temperate and polar waters. The differences are climatically determined, and since the present glacial type of climate is young in the climatic history of the earth, the ecological systems of the higher latitudes are considered as immature and at a low level of adaptation. That they may be in process of evolution toward greater stability is suggested by a number of phenomena, such as the development of large, slow-respiring, slow-growing individuals, and the production of the young in many arctic invertebrates in mid-winter or late fall. These and other observed peculiarities of high latitude fauna tend to make the most efficient use of the available plant food and to spread the cropping pressure over as much of the year as possible. Oceanic birds are cited as examples in which stable populations have been achieved by evolution of lower breeding rates, and the phosphate and nitrate cycles in the upper layers of tropical seas are discussed. It is emphasized that selection here is operating at the level of the ecosystem; competition is between systems rather than between individuals or specific populations.

LITERATURE CITED

Baker, J. R., 1938, The evolution of breeding seasons. *In* Evolution, ed. by G. R. de Beer. pp. 161-177. The Clarendon Press, Oxford.

Barghoorn, E. S., 1953, Evidence of climatic change in the geologic record of plant life. *In* Climatic change, ed. by H. Shapley. pp. 235-248. Harvard University Press, Cambridge.

Barnes, H., 1958, Processes of restoration and synchronization in marine ecology. The spring diatom increase and the "spawning" of the common barnacle, *Balanus balanoides* (L.). Coll. Intern. Biol. Mar. Stat. Biol. Roscoff (Ser. B, No. 24): 68-85.

Brodskii, K. A., and M. M. Nikitin, 1955, Hydrobiological work. The drift of the scientific station 1950-51, ed. by M. M. Somov. Sect. 4, Art. 7. (Air Force Cambridge Center Translation, 1956).

Chitty, D., 1959, Self-regulation of numbers through changes in viability. Cold Spring Harbor Symp. Quant. Biol. 22: 277-280.

Dunbar, M. J., 1954, The amphipod Crustacea of Ungava Bay, Canadian Eastern Arctic. J. Fish. Res. Bd. Canada 11 (6): 709-998.

1957, The determinants of production in northern seas: A study of the biology of *Themisto libellula* Mandt. Can. J. Zool. 35 (6): 797-819.

Durham, J. W., 1950, Cenozoic marine climates of the Pacific coast. Geol. Soc. Amer. Bull. 61: 1243-1264.

Harvey, H. W., 1955, The chemistry and fertility of sea waters. 224 pp. Cambridge University Press.

Holttum, R. E., 1953, Evolutionary trends in an equatorial climate. Symp. Soc. Exp. Biol. 7 (Evolution): 159-173. Cambridge University Press.

Hutchinson, G. E., 1948, Circular causal systems in ecology. Ann. N. Y. Acad. Sci. 50 (4): 221-246.

1954, Notes on oscillatory populations. J. Wildlife Management 18: 107-109.

1957, Concluding remarks. Cold Spring Harbor Symp. Quant. Biol. 22: 415-427.

MacArthur, R., 1955, Fluctuations of animal populations, and a measure of community stability. Ecology 36 (3): 533-536.

MacGinitie, G. E., 1955, Distribution and ecology of the marine invertebrates of Point Barrow, Alaska. Smithsonian Misc. Coll. 128 (9): 1-201.

Odum, E. P., 1959, Fundamentals of ecology. 546 pp. W. B. Saunders Co., Philadelphia.

Riley, G. A., 1953, Theory of growth and competition in natural populations. J. Fish. Res. Bd. Canada 10 (5): 211-223.

Scheffer, V. B., 1951, The rise and fall of a reindeer herd. Sci. Monthly 73 (6): 356-362.

Slobodkin, L. B., 1953, An algebra of population growth. Ecology 34: 513-517.

Steele, D. H., 1959, Marine Amphipoda of eastern and northern Canada. McGill University Thesis, M.S.

Steemann Nielsen, E., 1957, The balance between phytoplankton and zooplankton in the sea. J. du Conseil 23: 178-188.

Thorson, G., 1950, Reproductive and larval ecology of marine bottom invertebrates. Biol. Rev. 25: 1-45.

1952, Zur jetzigen Lage der marinen Bodentier-Ökologie. Verh. d. Deutschen Zool. Gesellsch. Wilhelmshaven 1951 (34): 276-327.

Utida, S., 1957, Population fluctuation, an experimental and theoretical approach. Cold Spring Harbor Symp. Quant. Biol. 22: 139-151.

Volterra, V., 1928, Variations and fluctuations of the number of individuals in animal species living together. J. du Conseil 3: 1-51.

Zenkevitch, I. A., 1949, Sur l'ancienneté de l'origine de la faune marine d'eau froide. XIII Congrès Internat. de Zool.: 550.

COMMUNICATION OF STRUCTURE IN PLANKTONIC POPULATIONS[1]

Ramon Margalef
Instituto Investigaciones Pesqueras, Barcelona, Spain

ABSTRACT

Two alternative approaches are usual in the study of ecosystems. One emphasizes biomass and production and deals with the system in terms of matter and energy; the other is more concerned with structure. Both approaches are complementary and for their synthesis we need quantitative expressions for structure and for structure transmission along time. Information theory can be helpful. Biotic diversity is an expression of the capacity for carrying information of the ecosystem (= "channel width") at the level of individuals. Diversity can be appreciated also at other levels, *e.g.*, at the biochemical level (pigment composition of phytoplankton). Many factors, in planktonic communities chiefly turbulence of water, introduce an amount of noise. Ecosystems can accumulate "history" or increase maturity—succession—only if "noise" remains relatively low. As changes in the energy flow across the ecosystem are related to diversity, this is an area where the study of relations between mass, energy and structure may be rewarding.

The scientific study of assemblages of organisms present in natural communities has been conducted in two different and separate ways. In the first, which is awakening active interest today, one considers assemblages in terms of matter (biomass) and energy flow (productivity) and is concerned with the forces that introduce changes in the system (true dynamics). In the second and more traditional approach, one analyzes the community as a complex of individuals belonging to different species with definite ecological requirements; changes are dealt with by so-called population dynamics, or, more correctly a population kinematics. Both approaches have developed with a certain independence and a large gap has been growing between them. To bridge this important gap seems to be a rewarding field of ecological research.

Lack of coordination between similar complementary aspects, one concerned with matter and energy and the other with structure, causes difficulties in many other areas of biology also. For instance in the speculations on the origin of life, the biochemical hypotheses are ingenious but are untenable until provision is made for a mechanism for isolating the products of reactions through some sort of spatial structure of the system. How did the primeval "soup" develop an organization?

In living systems spatial configurations may influence the future, but time is required to build an organization, accumulate biomass, and canalize energy. Just as there are rate limitations for construction of a living being through evolution, so the slower velocities at which organisms grow, travel, multiply, act, react, make tries, and select possibilities limit rates for evolution of ecological structure.

It is proposed that the speed at which organization can be built and transmitted involves specific problems of communication theory. The "information" in an ecological set may be measured (Margalef 1957). This note concerns the time change of informational content of ecological structure during the exercise of decisions or selection that influence future events, building up more organization in this or that way, and channelling energy along certain paths.

The static aspect of the structure: Biotic diversity or channel capacity

Ideas developed in the field of information theory prove helpful for developing quantitative expressions of structural properties of ecosystems. So far, the application of information theory to ecological prob-

[1] A paper presented to ASLO at Stillwater, Oklahoma, in August, 1960. The author wishes to express thanks to Dr. H. T. Odum, University of Texas, Institute of Marine Science, for both his stimulating discussions and his subsequent careful revision of the manuscript.

lems has been unimaginative and hardly qualifies as information theory or theory of communication. We have contented ourselves with expressing the degree of complexity of a biotic community as a function of the number of distinguishable states that can be assumed by a given set of individuals.

If $N_a, N_b \ldots N_s$ are the numbers of individuals of species $a, b, \ldots s$, and N is their sum, the expression for information

$$I = \frac{1}{N} \log_2 \frac{N!}{N_a!\, N_b! \ldots N_s!} \qquad (1)$$

represents an entropy and proves an advantageous substitute for the proposed indices of "biotic diversity." I is expressed in average number of bits per individual.

The actual structure of a population at a given time (t_1) represents a complex pattern. For the moment, consider that different individuals of the same species are indistinguishable and interchangeable. Such simplification discards a great mass of "information" of finer quality and amounts to the adoption of a *code*, for building a model that may be convenient in our preliminary approach. The biotic diversity (1) is a measure of the entropy of the system after coding and in an actual situation at time t_1 is the maximum amount of information.

The dynamic aspect of structure: Organization or integration

Communication theory concerns a certain amount of coded information put into a channel, the way it flows along the channel becoming more or less deteriorated through noise, and how any remaining information is recovered and decoded to be put into operation or use.

The initial structure of the community at time t_1 may also be considered as a message in the simple way described above. The comparison of species with the different letters or symbols of an alphabet is here pertinent. At a later time (t_2) the structure of the same community may be equal or different, but is interpretable in the terms of the same *code*. We can construct a model based on information theory to describe how populations change along time. However we have to remember that we are dealing only with a small fraction of the total amount of organization or "information": the limitation is fixed by our coding.

The diagram in Figure 1 indicates the coding and transmission of information in a mathematical or analogue electrical circuit parallel to the transmissions of information in the real ecological association.

Changes, if any, can effect: 1) the capacity of the channel, and 2) the degree of effective utilization of such capacity. Biotic diversity measures the maximum information that can be transmitted and, thus, the capacity of the channel. Biotic diversity rides on a "carrying wave" of births and deaths. The relative constancy in the proportions of the different species, that is a requisite for the conservation of channel capacity, has to be kept by negative or stabilizing feedback. In fact, the channel can widen or contract with time, owing to extinction of species, to introduction of new ones, or to differential multiplication of those already present.

It is customary to assign those species between which there is a negative feedback to separate ecological niches. Preservation of biotic diversity is, thus, related to the niche structure of the community. As the number of niches increases along succession up to a certain level, the carrying capacity of the channel widens. But the part of biotic diversity contributed by the possible existence of more than one species in a niche, drops as this event becomes less probable. Actually, with succession, biotic diversity can drop, as a result of the combination of both trends. Information carrying capacity of the channel, in terms of the selected code, can thus be reduced. But we have to remember that such changes are accompanied by the substitution of species by competition, and by a shift of "information" from the interindividual structure to the intraindividual level, where it is no longer apparent in our model.

Imagine that in the time elapsed from t_1 to t_2 no changes have occurred in the localization of species and that there has been an exact replacement of dead individ-

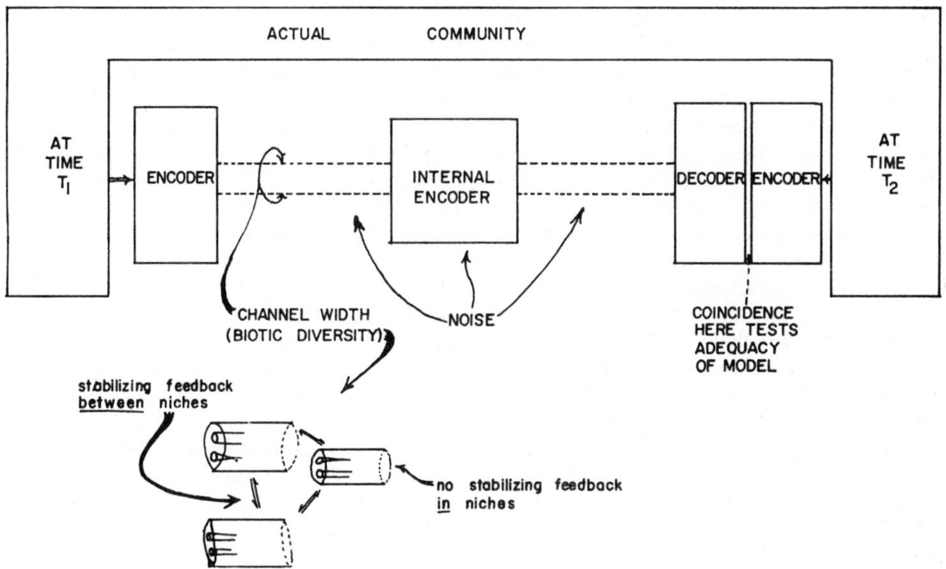

FIG. 1. Parallel between information transmission in the ecological community and in an analogue circuit.

uals. The pattern has been preserved and the information I_1 at time t_1 is completely recovered, then $I_2 = I_1$ at time t_2. Such may be actually the case in a community fixed on a solid substrate, but also in a community in which organisms have moved in the meantime in a regular and predictable way, so that their changes of place can be considered as a strictly determined encoding.

Ordinarily, a number of ecological factors (in planktonic communities mainly the turbulence of water) introduce randomness in the system or noise in terms of communication theory. Amount of noise, is highly relevant to the information to be transmitted. If the information contained in an initial pattern as gained through historical processes, is to influence future events, level of noise cannot be too high. The transmitted influence on future events may involve the preservation of the efficiency of the ecosystem. Here is a point where both complementary aspects of the community focus, *viz.*, the structural and the trophic-dynamic.

Only with a certain degree of stability (that is, with a low level of noise) can animals put into effective use the information stored by the species, such as migrating where food really is or mates await. In turbulent water, it is impossible to obtain profit from a great part of such information as, for example, indirect responses and internal rhythms. In a highly stimulating paper, Quastler (1959) sets forth the basis for a quantitative study of the relations between the degree of integration of an ecological frame (the reduction of noise) and the possibility of a reduction in the number of bearings that an organism has to take in such environment (feature sampling). It is obvious that in a 'primitive' environment that introduces a high level of noise, species tend to have direct responses to different factors and find no use for a capacity of storing 'information' effective in the sampling of a few features; these species can maintain themselves by a high rate of multiplication, with high power output, but with a low efficiency. Thus, dunes, creeks, the limnetic and pelagic spaces can be considered as environments with a level of noise higher than forests, the depths of the sea, or the coral reefs.

Succession is a result of storage and utilization of information and the effect of noise that reduces the amount of recoverable formation also defines the point where succession has to stop. Thus the maximum degree of maturity attainable by a community is controlled by environment.

It has been said previously that substitution of species, along succession, shifts information from the interindividual to the intraindividual level. The intraindividual channel involves the copy of genetic information and thermodynamically is cheaper or more efficient than the maintenance of structure by a feedback system such as is operating between species. A high content of information has to be coupled with a restricted feature sampling, only possible when the environment has a low noise level.

Migration and turbulence

Applications of this model are easy to grasp in connection with plankton communities, notwithstanding their general usefulness for all sort of communities. The transmission of information can also be used to analyze movements of planktonic organisms and also the part played by migration and turbulence in their movements.

Ashby (1957) offers an excellent starting point. With a large number of animals distributed in an environment, every individual in the place a at time t_1 has definite probabilities to be encountered in the places a,b,c . . . at time t_2. It is possible to write a matrix with the transition probabilities, and to compute the average state distribution and also the entropy or noise. The very fact that we have to substitute a deterministic model (such as including a statement that individual x will be met two hours later precisely at level b) by a system of probabilities, means that we have to contend with an amount of noise.

The information that fills the whole capacity of channel at the beginning (I_1) splits in two parts: information at the output (E) and noise (N). Perhaps the ratio E/I may be a good measure of what Bray calls "order or organization" and Quastler "integration." Quastler bases his measure on the ratio between the number of effective configurations and the number of possible ones.

If individuals remain fixed or change in a completely determinate way, the quotient $E/I = 1$. When movement is at random, the recovered information amounts only to the statement of number of species and number of individuals, and the quotient E/I is, of course, much lower. Such a quotient may be related to: 1) properties of the community, such as the swimming power of animals and the connections between organisms due to colony formation, epibiosis, etc., and 2) the turbulence of water. Now the problem is to put this into practical use.

Let us consider real populations subjected to births and deaths. It may be safely stated that when multiplication rate is low, average life long and fluctuations buffered, then the future situations are more predictable. Communities with such characteristics afford a channel with a lower noise level and may be considered as more organized.

With time, channel capacity may become larger, so that

$$I_1 < I_2 = E + N = I_1 + N'$$

and N' may be spoken of as a creative or "negative" noise.

Structural problems at the levels of specific and biochemical composition

We come again to our problem of bridging in some quantitative way the two aspects of a biotic community. Matter and energy flow can be expressed quantitatively in an obvious way. Information theory seems to offer a possibility to express quantitatively structural properties, at least such as are manifested at the level of assemblages of individuals. The translation of an energy budget into an information budget by simply converting calories into bits (Patten 1959) may give an idea of the amount of information to be expected in the whole organization of the community, but, for the moment, takes us no further.

When a mixed phytoplanktonic population is subjected to a sudden increase of the supporting capacity of the environment, the different species are led to take full advantage of their respectives rates of increase. As such rates differ widely, the result is a decrease of the diversity (I, in (1) above); and one or a few species manifest strong dominance. If ecological succession proceeds, diversity rises again to attain a stable value or drops slightly at the end.

We can consider the same population also in another way, either as a complex of photosynthetic mechanisms, or more simply, as an assemblage of different assimilatory pigments. An animal population could be simplified also to the form of a mixture of selected easily analyzable, chemical compounds. Now, the interesting fact, manifested in natural populations of marine plankton as well as in experimental populations artificially assembled in culture vessels, is that changes of the "diversity" in the specific level and in the biochemical level show a striking parallelism or analogy. In the case of a sudden increase of nutrients, chlorophyll *a* increases at a much higher rate than the other pigments present in the plankton. Thus if we have calculated a sort of diversity index based upon the kinds of molecules of different pigments (chlorophylls and carotenes), such diversity index drops, as does the diversity index based on distribution of individuals into species. More labile components and also those having a key position (as chlorophyll *a*) respond more quickly than stable components with a more complex biochemical dependence (carotenes). Other chlorophylls are in an intermediate position. The parallelism is maintained also along ecological succession up to a certain point. After the initial outburst with rapid unequal increase of the pigments, the differences of concentration between the different pigments diminishes with diversification of the photosynthetic mechanisms just as there is a completion and diversification of the niche structure of the community in succession.

These relations are also invaluable in the study of the spatial distribution of populations. For example, in the layers of contact and mixing between two water masses, two cases can be identified: 1) the two water masses supply complementary nutrients, then growth of the populations in the boundary receives a push, biotic diversity drops and pigment diversity also diminishes, 2) there is no complementary enrichment; populations simply mix, biotic diversity increases as a result of the mixture in the boundary, but pigment diversity does not change if both populations are in a similar physiological state.

Relations between structure and energy flow

We need to find a model to visualize the relations between the structural changes, the flow of energy and the biomass. Biomass in one aspect can be imagined as mass of different species, in others as mass of different biochemical subsystems cutting or overlapping across the limits of species.

The energy flow is distributed in any case through a number of subsystems (species, photosynthetic mechanisms). Subsystems, which convert energy more efficiently into information, accumulate biomass at a higher rate.

Noise, in the form of fluctuations with time, a sort of turbulence, may restrict possible assemblages of species to those that are able to follow the fluctuations and that work with a low efficiency.

I feel unable to draw a quantitative expression of the way variations in the possible energy flow are reflected in the relative biomass supported. No doubt someone will provide the ecologists with a coherent and workable model. With it, dynamics of planktonic populations including the related problems of migration, stratification, influence of water turbulence, and the practical problems of analysis of phytoplankton populations through cell counts or through pigment analysis will enter in a new phase of scientific study.

REFERENCES

ASHBY, W. R. 1957. An Introduction to Cybernetics. John Wiley & Sons, New York. 295 pp.

BRAY, J. R. 1958. Notes toward an ecologic theory. Ecology, **39**: 770–776.

MARGALEF, R. 1957. La teoria de la informacion en Ecologia. Mem. Real Acad. Ciencias Artes Barcelona, **32**: 373–449.

PATTEN, B. C. 1959. An introduction to the cybernetics of the ecosystem: the trophic-dynamic aspect. Ecology, **40**: 221–231.

QUASTLER, H. 1959. Information theory of biological integration. Amer. Natur., **93**: 245–254.

Vol. 100, No. 910 The American Naturalist January–February, 1966

FOOD WEB COMPLEXITY AND SPECIES DIVERSITY

ROBERT T. PAINE

Department of Zoology, University of Washington, Seattle, Washington

Though longitudinal or latitudinal gradients in species diversity tend to be well described in a zoogeographic sense, they also are poorly understood phenomena of major ecological interest. Their importance lies in the derived implication that biological processes may be fundamentally different in the tropics, typically the pinnacle of most gradients, than in temperate or arctic regions. The various hypotheses attempting to explain gradients have recently been reviewed by Fischer (1960), Simpson (1964), and Connell and Orias (1964), the latter authors additionally proposing a model which can account for the production and regulation of diversity in ecological systems. Understanding of the phenomenon suffers from both a specific lack of synecological data applied to particular, local situations and from the difficulty of inferring the underlying mechanism(s) solely from descriptions and comparisons of faunas on a zoogeographic scale. The positions taken in this paper are that an ultimate understanding of the underlying causal processes can only be arrived at by study of local situations, for instance the promising approach of MacArthur and MacArthur (1961), and that biological interactions such as those suggested by Hutchinson (1959) appear to constitute the most logical possibilities.

The hypothesis offered herein applies to local diversity patterns of rocky intertidal marine organisms, though it conceivably has wider applications. It may be stated as: **"Local species diversity is directly related to the efficiency with which predators prevent the monopolization of the major environmental requisites by one species."** The potential impact of this process is firmly based in ecological theory and practice. Gause (1934), Lack (1949), and Slobodkin (1961) among others have postulated that predation (or parasitism) is capable of preventing extinctions in competitive situations, and Slobodkin (1964) has demonstrated this experimentally. In the field, predation is known to ameliorate the intensity of competition for space by barnacles (Connell, 1961b), and, in the present study, predator removal has led to local extinctions of certain benthic invertebrates and algae. In addition, as a predictable extension of the hypothesis, the proportion of predatory species is known to be relatively greater in certain diverse situations. This is true for tropical vs. temperate fish faunas (Hiatt and Strasburg, 1960; Bakus, 1964), and is seen especially clearly in the comparison of shelf water zooplankton populations (81 species, 16% of which are carnivores) with those of the presumably less productive though more stable Sargasso Sea (268 species, 39% carnivores) (Grice and Hart, 1962).

In the discussion that follows no quantitative measures of local diversity are given, though they may be approximated by the number of species represented in Figs. 1 to 3. No distinctions have been drawn between species within certain food categories. Thus I have assumed that the probability of, say, a bivalve being eaten is proportional to its abundance, and that predators exercise no preference in their choice of any "bivalve" prey. This procedure simplifies the data presentation though it dodges the problem of taxonomic complexity. Wherever possible the data are presented as both number observed being eaten and their caloric equivalent. The latter is based on prey size recorded in the field and was converted by determining the caloric content of Mukkaw Bay material of the same or equivalent species. These caloric data will be given in greater detail elsewhere. The numbers in the food webs, unfortunately, cannot be related to rates of energy flow, although when viewed as calories they undoubtedly accurately suggest which pathways are emphasized.

Dr. Rudolf Stohler kindly identified the gastropod species. A. J. Kohn, J. H. Connell, C. E. King, and E. R. Pianka have provided invaluable criticism. The University of Washington, through the offices of the Organization for Tropical Studies, financed the trip to Costa Rica. The field work in Baja California, Mexico, and at Mukkaw Bay was supported by the National Science Foundation (GB-341).

THE STRUCTURE OF SELECTED FOOD WEBS

I have claimed that one of the more recognizable and workable units within the community nexus are subwebs, groups of organisms capped by a terminal carnivore and trophically interrelated in such a way that at higher levels there is little transfer of energy to co-occurring subwebs (Paine, 1963). In the marine rocky intertidal zone both the subwebs and their top carnivores appear to be particularly distinct, at least where macroscopic species are involved; and observations in the natural setting can be made on the quantity and composition of the component species' diets. Furthermore, the rocky intertidal zone is perhaps unique in that the major limiting factor of the majority of its primary consumers is living space, which can be directly observed, as the elegant studies on interspecific competition of Connell (1961a,b) have shown. The data given below were obtained by examining individual carnivores exposed by low tide, and recording prey, predator, their respective lengths, and any other relevant properties of the interaction.

A north temperate subweb

On rocky shores of the Pacific Coast of North America the community is dominated by a remarkably constant association of mussels, barnacles, and one starfish. Fig. 1 indicates the trophic relationships of this portion of the community as observed at Mukkaw Bay, near Neah Bay, Washington (ca. 49° N latitude). The data, presented as both numbers and total calories consumed by the two carnivorous species in the subweb, *Pisaster ochraceus*,

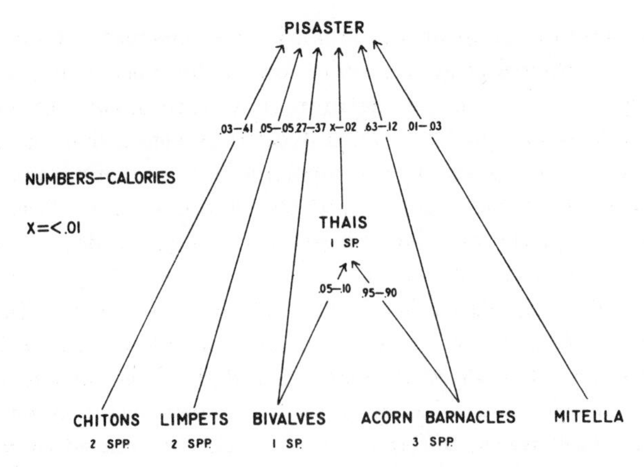

FIG. 1. The feeding relationships by numbers and calories of the *Pisaster* dominated subweb at Mukkaw Bay. *Pisaster*, N = 1049; *Thais*, N = 287. N is the number of food items observed eaten by the predators. The specific composition of each predator's diet is given as a pair of fractions; numbers on the left, calories on the right.

a starfish, and *Thais emarginata*, a small muricid gastropod, include the observational period November, 1963, to November, 1964. The composition of this subweb is limited to organisms which are normally intertidal in distribution and confined to a hard rock substrate. The diet of *Pisaster* is restricted in the sense that not all available local food types are eaten, although of six local starfish it is the most catholic in its tastes. Numerically its diet varies little from that reported by Feder (1959) for *Pisaster* observed along the central California coastline, especially since the gastropod *Tegula*, living on a softer bottom unsuitable to barnacles, has been omitted. *Thais* feeds primarily on the barnacle *Balanus glandula*, as also noted by Connell (1961b).

This food web (Fig. 1) appears to revolve on a barnacle economy with both major predators consuming them in quantity. However, note that on a nutritional (calorie) basis, barnacles are only about one-third as important to *Pisaster* as either *Mytilus californianus*, a bivalve, or the browsing chiton *Katherina tunicata*. Both these prey species dominate their respective food categories. The ratio of carnivore species to total species is 0.18. If *Tegula* and an additional bivalve are included on the basis that they are the most important sources of nourishment in adjacent areas, the ratio becomes 0.15. This number agrees closely with a ratio of 0.14 based on *Pisaster*, plus all prey species eaten more than once, in Feder's (1959) general compilation.

A subtropical subweb

In the Northern Gulf of California (ca. 31° N.) a subweb analogous to the one just described exists. Its top carnivore is a starfish (*Heliaster kubiniji*), the next two trophic levels are dominated by carnivorous gastropods, and the main prey are herbivorous gastropods, bivalves, and barnacles. I have collected there only in March or April of 1962–1964, but on both sides of the Gulf at San Felipe, Puertecitos, and Puerta Penasco. The resultant trophic arrangements (Fig. 2), though representative of springtime conditions and indicative of a much more stratified and complex community, are basically similar to those at Mukkaw Bay. Numerically the major food item in the diets of *Heliaster* and *Muricanthus nigritus* (a muricid gastropod), the two top-ranking carnivores, is barnacles; the major portion of these predators' nutrition is derived from other members of the community, primarily herbivorous mollusks. The increased trophic complexity presents certain graphical problems. If increased trophic height is indicated by a decreasing percentage of primary consumers in a species diet, *Acanthina tuberculata* is the highest carnivore due to its specialization on *A. angelica*, although it in turn is consumed by two other species. Because of this, and ignoring the percentages, both *Heliaster* and *Muricanthus* have been placed above *A. tuberculata*. Two species, *Hexaplex* and *Muricanthus* eventually become too large to be eaten by *Heliaster*, and thus through growth join it as top predators in the system. The taxonomically-difficult gastropod family Columbellidae, including both herbivorous and carnivorous species (Marcus and Marcus, 1962) have been placed in an intermediate position.

The Gulf of California situation is interesting on a number of counts. A new trophic level which has no counterpart at Mukkaw Bay is apparent, interposed between the top carnivore and the primary carnivore level. If higher level predation contributes materially to the maintenance of diversity, these species will have an effect on the community composition out of proportion to their abundance. In one of these species, *Muricanthus*,

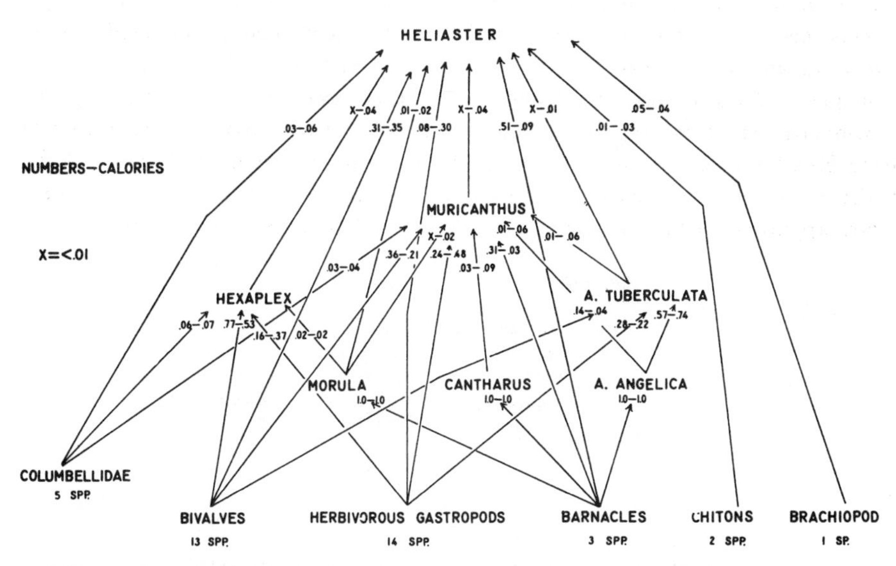

FIG. 2. The feeding relationships by numbers and calories of the *Heliaster* dominated subweb in the northern Gulf of California. *Heliaster*, N = 2245; *Muricanthus*, N = 113; *Hexaplex*, N = 62; *A. tuberculata*, N = 14; *A. angelica*, N = 432; *Morula*, N = 39; *Cantharus*, N = 8.

the larger members belong to a higher level than immature specimens (Paine, unpublished), a process tending to blur the food web but also potentially increasing diversity (Hutchinson, 1959). Finally, if predation operates to reduce competitive stresses, evidence for this reduction can be drawn by comparing the extent of niche diversification as a function of trophic level in a typical Eltonian pyramid. *Heliaster* consumes all other members of this subweb, and as such appears to have no major competitors of comparable status. The three large gastropods forming the subterminal level all may be distinguished by their major sources of nutrition: *Hexaplex*—bivalves (53%), *Muricanthus*—herbivorous gastropods (48%), and *A. tuberculata*—carnivorous gastropods (74%). No such obvious distinction characterizes the next level composed of three barnacle-feeding specialists which additionally share their resource with *Muricanthus* and *Heliaster*. Whether these species are more specialized (Klopfer and MacArthur, 1960) or whether they tolerate greater niche overlap (Klopfer and MacArthur, 1961) cannot be stated. The extent of niche diversification is subtle and trophic overlap is extensive.

The ratio of carnivore species to total species in Fig. 2 is 0.24 when the category Columbellidae is considered to be principally composed of one herbivorous (*Columbella*) and four carnivorous (*Pyrene*, *Anachis*, *Mitella*) species, based on the work of Marcus and Marcus (1962).

A tropical subweb

Results of five days of observation near Mate de Limon in the Golfo de Nocoya on the Pacific shore of Costa Rica (approx. 10° N.) are presented in Fig. 3. No secondary carnivore was present; rather the environmental resources were shared by two small muricid gastropods, *Acanthina brevidentata* and *Thais biserialis*. The fauna of this local area was relatively simple and completely dominated by a small mytilid and barnacles. The co-occupiers of the top level show relatively little trophic overlap despite the broad nutritional base of *Thais* which includes carrion and cannibalism. The relatively low number of feeding observations (187) precludes an accurate appraisal of the carnivore species to total web membership ratio.

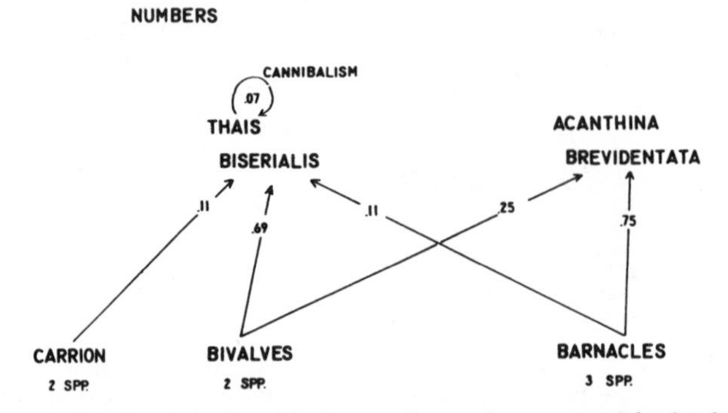

FIG. 3. The feeding relationship by numbers of a comparable food web in Costa Rica. *Thais*, N = 99; *Acanthina*, N = 80.

CHANGES RESULTING FROM THE REMOVAL OF THE TOP CARNIVORE

Since June, 1963, a "typical" piece of shoreline at Mukkaw Bay about eight meters long and two meters in vertical extent has been kept free of *Pisaster*. An adjacent control area has been allowed to pursue its natural course of events. Line transects across both areas have been taken irregularly and the number and density of resident macroinvertebrate and benthic algal species measured. The appearance of the control area has not altered. Adult *Mytilus californianus*, *Balanus cariosus*, and *Mitella polymerus* (a goose-necked barnacle) form a conspicuous band in the middle intertidal. The relatively stable position of the band is maintained by *Pisaster* predation (Paris, 1960; Paine, unpublished). At lower tidal levels the diversity increases abruptly and the macrofauna includes immature individuals of the above, *B. glandula* as scattered clumps, a few anemones of one species, two chiton species (browsers), two abundant limpets (browsers), four macroscopic benthic algae (*Porphyra*-an epiphyte, *Endocladia*, *Rhodomela*, and *Corallina*), and the sponge *Haliclona*, often browsed upon by *Anisodoris*, a nudibranch.

Following the removal of *Pisaster*, *B. glandula* set successfully throughout much of the area and by September had occupied from 60 to 80% of the available space. By the following June the *Balanus* themselves were being crowded out by small, rapidly growing *Mytilus* and *Mitella*. This process of successive replacement by more efficient occupiers of space is continuing, and eventually the experimental area will be dominated by *Mytilus*, its epifauna, and scattered clumps of adult *Mitella*. The benthic algae either have or are in the process of disappearing with the exception of the epiphyte, due to lack of appropriate space; the chitons and larger limpets have also emigrated, due to the absence of space and lack of appropriate food.

Despite the likelihood that many of these organisms are extremely long-lived and that these events have not reached an equilibrium, certain statements can be made. The removal of *Pisaster* has resulted in a pronounced *decrease* in diversity, as measured simply by counting species inhabiting this area, whether consumed by *Pisaster* or not, from a 15 to an eight-species system. The standing crop has been increased by this removal, and should continue to increase until the *Mytilus* achieve their maximum size. In general the area has become trophically simpler. With *Pisaster* artificially removed, the sponge-nudibranch food chain has been displaced, and the anemone population reduced in density. Neither of these carnivores nor the sponge is eaten by *Pisaster*, indicating that the number of food chains initiated on this limited space is strongly influenced by *Pisaster*, but by an indirect process. In contrast to Margalef's (1958) generalization about the tendency, with higher successional status towards "an ecosystem of more complex structure," these removal experiments demonstrate the opposite trend: in the absence of a complicating factor (predation), there is a "winner" in the competition for space, and the local system tends toward simplicity. Predation by this interpretation interrupts the successional process and, on a local basis, tends to increase local diversity.

No data are available on the microfaunal changes accompanying the gradual alteration of the substrate from a patchy algal mat to one comprised of the byssal threads of *Mytilus*.

INTERPRETATION

The differences in relative diversity of the subwebs diagrammed in Figs. 1-3 may be represented as Baja California (45 spp.) >> Mukkaw Bay (11 spp.) > Costa Rica (8 sp.), the number indicating the actual membership of the subwebs and not the number of local species. All three areas are characterized by systems in which one or two species are capable of monopolizing much of the space, a circumstance realized in nature only in Costa Rica. In the other two areas a top predator that derives its nourishment from other sources feeds in such a fashion that no space-consuming monopolies are formed. *Pisaster* and *Heliaster* eat masses of barnacles, and in so doing enhance the ability of other species to inhabit the area by keeping space open. When the top predator is artificially removed or naturally absent (i.e., predator removal area and Costa Rica, respectively), the systems converge toward simplicity. When space is available, other organisms settle or move in, and these, for instance chitons at Mukkaw Bay and herbivorous gastropods and pelecypods in Baja California, form the major portions of the predator's nutrition. Furthermore, *in situ* primary production is enhanced by the provision of space. This event makes the grazing moiety less dependent on the vagaries of phytoplankton production or distribution and lends stability to the association.

At the local level it appears that carnivorous gastropods which can penetrate only one barnacle at a time, although they might consume a few more per tidal interval, do not have the same effect as a starfish removing 20 to 60 barnacles simultaneously. Little compensation seems to be gained from snail density increases because snails do not clear large patches of space, and because the "husks" of barnacles remain after the animal portion has been consumed. In the predator removal area at Mukkaw Bay, the density of *Thais* increased 10- to 20-fold, with no apparent effect on diversity although the rate of *Mytilus* domination of the area was undoubtedly slowed. Clusters (density of 75–125/m^2) of *Thais* and *Acanthina* characterize certain rocks in Costa Rica, and diversity is still low. And, as a generality, wherever acorn barnacles or other space-utilizing forms potentially dominate the shore, diversity is reduced unless some predator can prevent the space monopoly. This occurs in Washington State where the shoreline, in the absence of *Pisaster*, is dominated by barnacles, a few mussels, and often two species of *Thais*. The same monopolistic tendencies characterize Connell's (1961a,b) study area in Scotland, the rocky intertidal of northern Japan (Hoshiai, 1960, 1961), and shell bags suitable for sponge settlement in North Carolina (Wells, Wells, and Gray, 1964).

Local diversity on intertidal rocky bottoms, then, appears directly related to predation intensity, though other potential factors are mentioned below. If one accepts the generalizations of Hedgpeth (1957) and Hall

(1964) that ambient temperature is the single most important factor influencing distribution or reproduction of marine invertebrates, then the potential role of climatic stability as measured by seasonal variations in water temperature can be examined. At Neah Bay the maximum range of annual values are 5.9 to 13.3 C (Rigg and Miller, 1949); in the northern Gulf of California, Roden and Groves (1959) recorded an annual range of 14.9 to 31.2 C; and in Costa Rica the maximum annual range is 26.1 to 31.7 C (Anon., 1952). Clearly the greatest benthic diversity, and one claimed by Parker (1963) on a regional basis to be among the most diverse known, is associated with the most variable (least stable) temperature regimen. Another influence on diversity could be exercised by environmental heterogeneity (Hutchinson, 1959). Subjectively, it appeared that both the Mukkaw Bay and Costa Rica stations were topographically more distorted than the northern Gulf localities. In any event, no topographic features were evident that could correlate with the pronounced differences in faunal diversity. Finally, Connell and Orias (1964) have developed a model for the organic enrichment of regions that depends to a great extent on the absolute amount of primary production and/or nutrient import, and hence energy flowing through the community web. Unfortunately, no productivity data are available for the two southern communities, and comparisons cannot yet be made.

PREDATION AND DIVERSITY GRADIENTS

To examine predation as a diversity-causing mechanism correlated with latitude, we must know why one environment contains higher order carnivores and why these are absent from others. These negative situations can be laid to three possibilities: (1) that through historical accident no higher carnivores have evolved in the region; (2) that the sample area cannot be occupied due to a particular combination of *local* hostile physiological effects; (3) that the system cannot support carnivores because the rate of energy transfer to a higher level is insufficient to sustain that higher level. The first possibility is unapproachable, the second will not apply on a geographic scale, and thus only the last would seem to have reality. Connell and Orias (1964) have based their hypothesis of the establishment and maintenance of diversity on varying rates of energy transfer, which are determined by various limiting factors and environmental stability. Without disagreeing with their model, two aspects of primary production deserve further consideration. The animal diversity of a given system will probably be higher if the production is apportioned more uniformly throughout the year rather than occurring as a single major bloom, because tendencies towards competitive displacement can be ameliorated by specialization on varying proportions of the resources (MacArthur and Levins, 1964). Both the predictability of production on a sustained annual basis and the causation of resource heterogeneity by predation will facilitate this mechanism. Thus, per production unit, greater stability of production should be correlated with greater diversity, other things being equal.

The realization of this potential, however, depends on more than simply the annual stability of carbon fixation. Rate of production and subsequent transfer to higher levels must also be important. Thus trophic structure of a community depends in part on the physical extent of the area (Darlington, 1957), or, in computer simulation models, on the amount of protoplasm in the system (Garfinkel and Sack, 1964). On the other hand, enriched aquatic environments often are characterized by decreased diversity. Williams (1964) has found that regions of high productivity are dominated by few diatom species. Less productive areas tended to have more species of equivalent rank, and hence a greater diversity. Obviously, the gross amount of energy fixed by itself is incapable of explaining diversity; and extrinsic factors probably are involved.

Given sufficient evolutionary time for increases in faunal complexity to occur, two independent mechanisms should work in a complementary fashion. When predation is capable of preventing resource monopolies, diversity should increase by positive feedback processes until some limit is reached. The argument of Fryer (1965) that predation facilitates speciation is germane here. The upper limit to local diversity, or, in the present context, the maximum number of species in a given subweb, is probably set by the combined stability and rate of primary production, which thus influences the number and variety of non-primary consumers in the subweb. Two aspects of predation must be evaluated before a generalized hypothesis based on predation effects can contribute to an understanding of differences in diversity between *any* comparable regions or faunistic groups. We must know if resource monopolies are actually less frequent in the diverse area than in comparable systems elsewhere, and, if so, why this is so. And we must learn something about the multiplicity of energy pathways in diverse systems, since predation-induced diversity could arise either from the presence of a variety of subwebs of equivalent rank, or from domination by one major one. The predation hypothesis readily predicts the apparent absence of monopolies in tropical (diverse) areas, a situation classically represented as "many species of reduced individual abundance." It also is in accord with the disproportionate increase in the number of carnivorous species that seems to accompany regional increases in animal diversity. In the present case in the two adequately sampled, structurally analagous, subwebs, general membership increases from 13 at Mukkaw Bay to 45 in the Gulf of California, a factor of 3.5, whereas the carnivore species increased from 2 to 11, a factor of 5.5.

SUMMARY

It is suggested that local animal species diversity is related to the number of predators in the system and their efficiency in preventing single species from monopolizing some important, limiting, requisite. In the marine rocky intertidal this requisite usually is space. Where predators capable of preventing monopolies are missing, or are experimentally removed, the systems become less diverse. On a local scale, no relationship between latitude (10° to 49° N.) and diversity was found. On a geographic scale, an increased stability of annual production may lead to an increased capacity for systems to support higher-level carnivores. Hence tropical, or other, ecosystems are more diverse, and are characterized by disproportionately more carnivores.

LITERATURE CITED

Anon. 1952. Surface water temperatures at tide stations. Pacific coast North and South America. Spec. Pub. No. 280: p. 1–59. U. S. Coast and Geodetic Survey.

Bakus, G. J. 1964. The effects of fish-grazing on invertebrate evolution in shallow tropical waters. Allan Hancock Found. Pub. 27: 1–29.

Connell, J. H. 1961a. Effect of competition, predation by *Thais lapillus*, and other factors on natural populations of the barnacle *Balanus balanoides*. Ecol. Monogr. 31: 61–104.

———. 1961b. The influence of interspecific competition and other factors on the distribution of the barnacle *Chthamalus stellatus*. Ecology 42: 710–723.

Connell, J. H., and E. Orias. 1964. The ecological regulation of species diversity. Amer. Natur. 98: 399–414.

Darlington, P. J. 1957. Zoogeography. Wiley, New York.

Feder, H. M. 1959. The food of the starfish, *Pisaster ochraceus*, along the California coast. Ecology 40: 721–724.

Fischer, A. G. 1960. Latitudinal variations in organic diversity. Evolution 14: 64–81.

Fryer, G. 1965. Predation and its effects on migration and speciation in African fishes: a comment. Proc. Zool. Soc. London 144: 301–310.

Garfinkel, D., and R. Sack. 1964. Digital computer simulation of an ecological system, based on a modified mass action law. Ecology 45: 502–507.

Gause, G. F. 1934. The struggle for existence. Williams and Wilkins Co., Baltimore.

Grice, G. D., and A. D. Hart. 1962. The abundance, seasonal occurrence, and distribution of the epizooplankton between New York and Bermuda. Ecol. Monogr. 32: 287–309.

Hall, C. A., Jr. 1964. Shallow-water marine climates and molluscan provinces. Ecology 45: 226–234.

Hedgpeth, J. W. 1957. Marine biogeography. Geol. Soc. Amer. Mem. 67, 1: 359–382.

Hiatt, R. W., and D. W. Strasburg. 1960. Ecological relationships of the fish fauna on coral reefs of the Marshall Islands. Ecol. Monogr. 30: 65–127.

Hoshiai, T. 1960. Synecological study on intertidal communities III. An analysis of interrelation among sedentary organisms on the artificially denuded rock surface. Bull. Marine Biol. Sta. Asamushi. 10: 49–56.

———. 1961. Synecological study on intertidal communities. IV. An ecological investigation on the zonation in Matsushima Bay concerning the so-called covering phenomenon. Bull. Marine Biol. Sta. Asamushi. 10: 203–211.

Hutchinson, G. E. 1959. Homage to Santa Rosalia or why are there so many kinds of animals? Amer. Natur. 93: 145–159.

Klopfer, P. H., and R. H. MacArthur. 1960. Niche size and faunal diversity. Amer. Natur. 94: 293–300.

______. 1961. On the causes of tropical species diversity: niche overlap. Amer. Natur. 95: 223–226.

Lack, D. 1949. The significance of ecological isolation, p. 299–308. *In* G. L. Jepsen, G. G. Simpson, and E. Mayr [eds.], Genetics, paleontology and evolution. Princeton Univ. Press, Princeton.

MacArthur, R., and R. Levins. 1964. Competition, habitat selection, and character displacement in a patchy environment. Proc. Nat. Acad. Sci. 51: 1207–1210.

MacArthur, R. H., and J. W. MacArthur. 1961. On bird species diversity. Ecology 42: 594–598.

Marcus, E., and E. Marcus. 1962. Studies on Columbellidae. Bol. Fac. Cienc. Letr. Univ. Sao Paulo 261: 335–402.

Margalef, R. 1958. Mode of evolution of species in relation to their place in ecological succession. XVth Int. Congr. Zool. Sect. 10, paper 17.

Paine, R. T. 1963. Trophic relationships of 8 sympatric predatory gastropods. Ecology 44: 63–73.

Paris, O. H. 1960. Some quantitative aspects of predation by muricid snails on mussels in Washington Sound. Veliger 2: 41–47.

Parker, R. H. 1963. Zoogeography and ecology of some macro-invertebrates, particularly mollusca in the Gulf of California and the continental slope off Mexico. Vidensk. Medd. Dansk. Natur. Foren., Copenh. 126: 1–178.

Rigg, G. B., and R. C. Miller. 1949. Intertidal plant and animal zonation in the vicinity of Neah Bay, Washington. Proc. Calif. Acad. Sci. 26: 323–351.

Roden, G. I., and G. W. Groves. 1959. Recent oceanographic investigations in the Gulf of California. J. Marine Res. 18: 10–35.

Simpson, G. G. 1964. Species density of North American recent mammals. Syst. Zool. 13: 57–73.

Slobodkin, L. B. 1961. Growth and regulation of Animal Populations. Holt, Rinehart, and Winston, New York.

______. 1964. Ecological populations of Hydrida. J. Anim. Ecol. 33 (Suppl.): 131–148.

Wells, H. W., M. J. Wells, and I. E. Gray. 1964. Ecology of sponges in Hatteras Harbor, North Carolina. Ecology 45: 752–767.

Williams, L. G. 1964. Possible relationships between plankton-diatom species numbers and water-quality estimates. Ecology 45: 809–823.

Reprinted for private circulation from THE AMERICAN NATURALIST
Vol. 102, No. 925, May-June 1968

MARINE BENTHIC DIVERSITY: A COMPARATIVE STUDY*

Howard L. Sanders

Woods Hole Oceanographic Institution, Woods Hole, Massachusetts 02543

INTRODUCTION

One of the major features of animal communities is their diversity, that is, the number of species present and their numerical composition. It has long been recognized that tropical regions, by and large, support a more diverse fauna than do regions of higher latitude. In the aquatic medium it is also evident that the marine habitats contain a greater wealth of species than do brackish regions. The reasons why certain environments harbor many kinds of organisms while others support a very limited number of species are still unclear. Various theories based on time (Fischer, 1960; Simpson, 1964), climatic stability (Klopfer, 1959; Fischer, 1960; Dunbar, 1960), spatial heterogeneity (Simpson, 1964), competition (Dobzhansky, 1950; Williams, 1964), predation (Paine, 1966), and productivity (Connell and Orias, 1964) have been proposed to explain these differences.

In the present paper, data collected from soft-bottom marine and estuarine environments of a number of differing regions will be used in a comparative study of within-habitat diversity. A new diversity measurement will be presented that is independent of sample size, and a hypothesis will be proposed to explain the observed patterns of diversity as well as to provide a framework for interpreting other diversity studies.

MATERIALS AND METHODS

The stations used in this study are as follows:

RH-14: Arabian Sea off Cochin, Kerala State, India (14 m).
RH-26: Bay of Bengal off Porto Novo, Madras State, India (20 m).
RH-28: Vellar River estuary, Porto Novo, Madras State, India (2 m).
RH-30: Bay of Bengal off Madras, India (15 m).
RH-33: Kakinada Bay, Andhra State, India (2 m).
RH-36: Bay of Bengal off Kakinada, Andhra State, India (37 m).
RH-41: Arabian Sea off Bombay, India (20 m).
RH-51: Indian Ocean off Hellville, Nossi Bé, Madagascar (18 m).
C#1: Outer continental shelf south of New England (40°27.2'N 70°47'W, 97 m).
S1.3: Upper continental slope south of New England (39°58.4'N 70°40.3'W, 300 m).
D#1: Upper continental slope south of New England (39°54.5'N 70°35'W, 487 m).
F#1: Lower continental slope south of New England (39°47'N 70°45'W, 1,500 m).

*Contribution No. 1959 from the Woods Hole Oceanographic Institution, Woods Hole, Massachusetts 02543.

G#1: Lower continental slope south of New England (39°42′N 70°39′W, 2,086 m).
GH#1: Abyssal rise south of New England (39°25.5′N 70°35′W, 2,500 m).
DR-12: Continental slope off northeast South America (07°09′S 34°25.5′W, 790 m).
DR-33: Continental slope off northeast South America (07°53.5′N 54°33.3′W, 535 m).
POC 1, 2, 3, 4: Pocasset River, Cape Cod, Massachusetts (0.5 m).
R Series: Buzzards Bay, Massachusetts (20 m).

All samples were collected with an Anchor dredge or a Higgins meiobenthic sled (the RH series of samples). The sediments were processed through a fine-meshed screen with 0.4 mm apertures, and the animals were carefully picked out and sorted in the laboratory. Sanders, Hessler, and Hampson (1965) gave details on the Anchor dredge and the methodology of processing.

THE RAREFACTION METHODOLOGY AND RESULTS

Since most diversity measurements are affected by sample size (see later discussion), it would be most useful to have a procedure which will allow one to compare directly samples of differing sizes. If this can be achieved, it may then be possible to perceive more clearly the factors influencing biological diversity. The rarefaction method, which permits each sample to generate a line, was developed to achieve this end. This methodology was applied to benthic marine samples collected from boreal estuary, boreal shallow marine, tropical estuary, tropical shallow marine, and deep-sea environments. The usual difficulty inherent in comparing samples of different sizes is that as sample size increases, individuals are added at a constant arithmetic rate but species accumulate at a decreasing logarithmic rate. The rarefaction method, instead, is dependent on the shape of the species abundance curve rather than the absolute number of specimens per sample. In all cases, the sediments were soft oozes and, therefore, comparable in regard to particle size.

The comparison was based on the polychaete-bivalve fraction of the samples rather than the entire fauna. Since these two groups comprise about 80% of the animals by number in most of the samples (Fig. 1), one can feel justified in generalizing from whatever results may be found. This study shows a systematic pattern of diversity that can be correlated with the variability of the physical environment.

With the method of diversity analysis developed for this study, samples with different numbers of specimens and from different regions of the world were compared directly. The procedure was to keep the percentage composition of the component species constant but reduce the sample size, that is, to artificially create the results that would have been obtained had smaller samples with the identical faunal composition been taken. Using this technique, the expected number of species present in populations of different sizes, that is, numbers of species per 10, 25, 50, 100, 200, . . ., 1,000, 2,000, etc., was determined.

In order to evaluate the validity of this method, one must understand how the values are obtained. The species are ranked by abundance, and

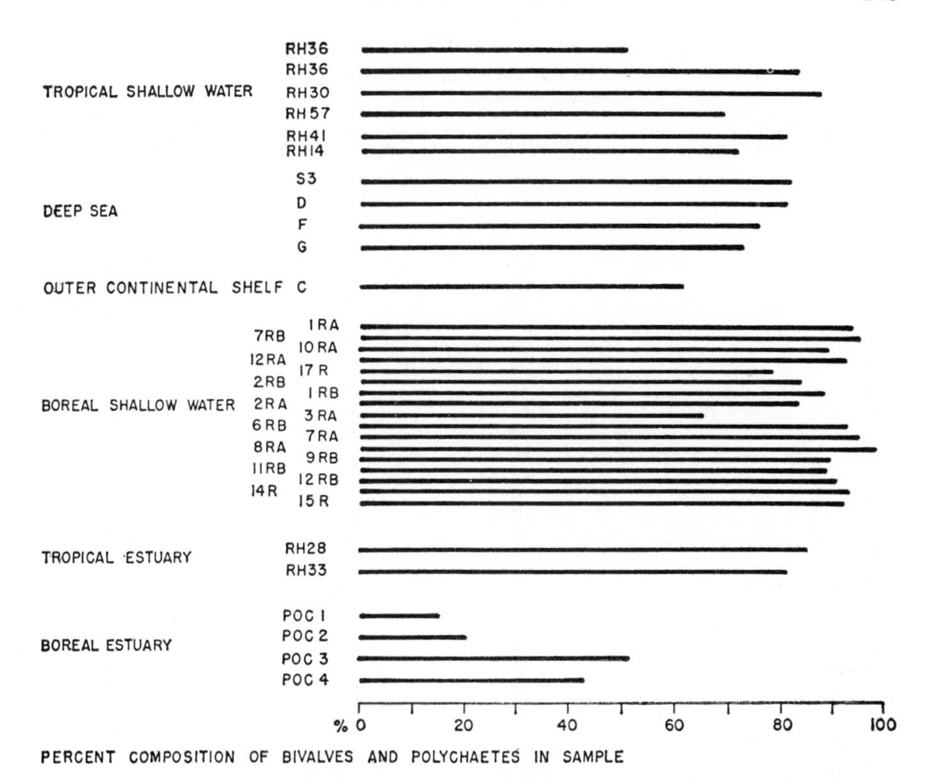

FIG. 1.—Percentage composition of the polychaete-bivalve fraction of the soft-bottom samples used in the analysis.

the percentage composition of each species and the cumulative percentage are plotted. In a hypothetical sample (Table 1) there are 1,000 individuals and 40 species. As an example, the number of species at the 25-individual level will be determined. The percentage composition is the same as in the original sample, but the number of individuals is reduced to 25. Since 25 specimens in this reduced sample represent 100% of the individuals present, then each individual specimen forms 4% of the sample. In the original sample, seven species each comprise 4% or more, and in total they compose 76% of the sample by number. Therefore, each of these seven species will be present in the reduced sample. This leaves a residue of 24% of the original sample comprising the remaining 33 species. Because none of these species forms more than 4% of the original sample, those species of this group that will appear in the reduced sample cannot be represented by more than one individual. Since one specimen comprises 4% of the reduced sample, therefore $24\%/4\% = 6$ species; $7 + 6 = 13$ species present per 25 individuals.

The determination of species per 100 individuals is as follows: (1) Since each individual represents 1% of the sample, then (2) 15 species in Table 1 each comprise $\geqq 1.0\%$ of the fauna and cumulatively $= 92.1\%$ of the

TABLE 1

HYPOTHETICAL SAMPLE WITH 1,000 INDIVIDUALS AND 40 SPECIES

Rank of Species by Abundance	Number of Individuals	% of Sample	Cumulative of Sample %
1	365	36.5	36.5
2	112	11.2	47.7
3	81	8.1	55.8
4	61	6.1	61.9
5	55	5.5	67.4
6	46	4.6	72.0
7	40	4.0	76.0
8	38	3.8	79.8
9	29	2.9	82.7
10	23	2.3	85.0
11	21	2.1	87.1
12	15	1.5	88.6
13	13	1.3	89.9
14	12	1.2	91.1
15	10	1.0	92.1
16	8	0.8	92.9
17	7	0.7	93.6
18	7	0.7	94.3
19	6	0.6	94.9
20	6	0.6	95.5
21	5	0.5	96.0
22	5	0.5	96.5
23	5	0.5	97.0
24	4	0.4	97.4
25	3	0.3	97.7
26	3	0.3	98.0
27	3	0.3	98.3
28–33	2 each	0.2 each	99.3
34–40	1 each	0.1 each	100.0
Total Number	1,000		

sample. (3) The residue = 7.9% of the sample; 7.9%/1.0% = 7.9 species. (4) 15 + 7.9 = 22.9 species per 100 individuals.

For species per 200 individuals: (1) Each individual forms 0.5% of the sample, and (2) 23 species each represent $\geqq$ 0.5% of the fauna and cumulatively = 97.0% of the sample. (3) The residue = 3.0%; 3.0%/0.5% = 6.0 species. (4) 23 + 6.0 = 29.0 species per 200 individuals.

Using this technique, we have made, in Figure 2, arithmetic plots of the number of species at different population levels up to the total number of individuals for samples from high latitude, low latitude, shallow water, deep sea, estuarine, and marine regions. The curvilinear nature of the lines is due to the fact that individuals are being added at a constant rate but the progressively rarer species are added at a continuously decreasing rate. The circles or the termination of the lines in Figure 2 give the actual number of individuals and species present in the samples. The curves themselves give the interpolated number of species at the different population levels.

What is significant is that each environment seems to have its own characteristic rate of species increment. Lowest diversity, that is, the fewest number of species per unit number of individuals, is found in the boreal

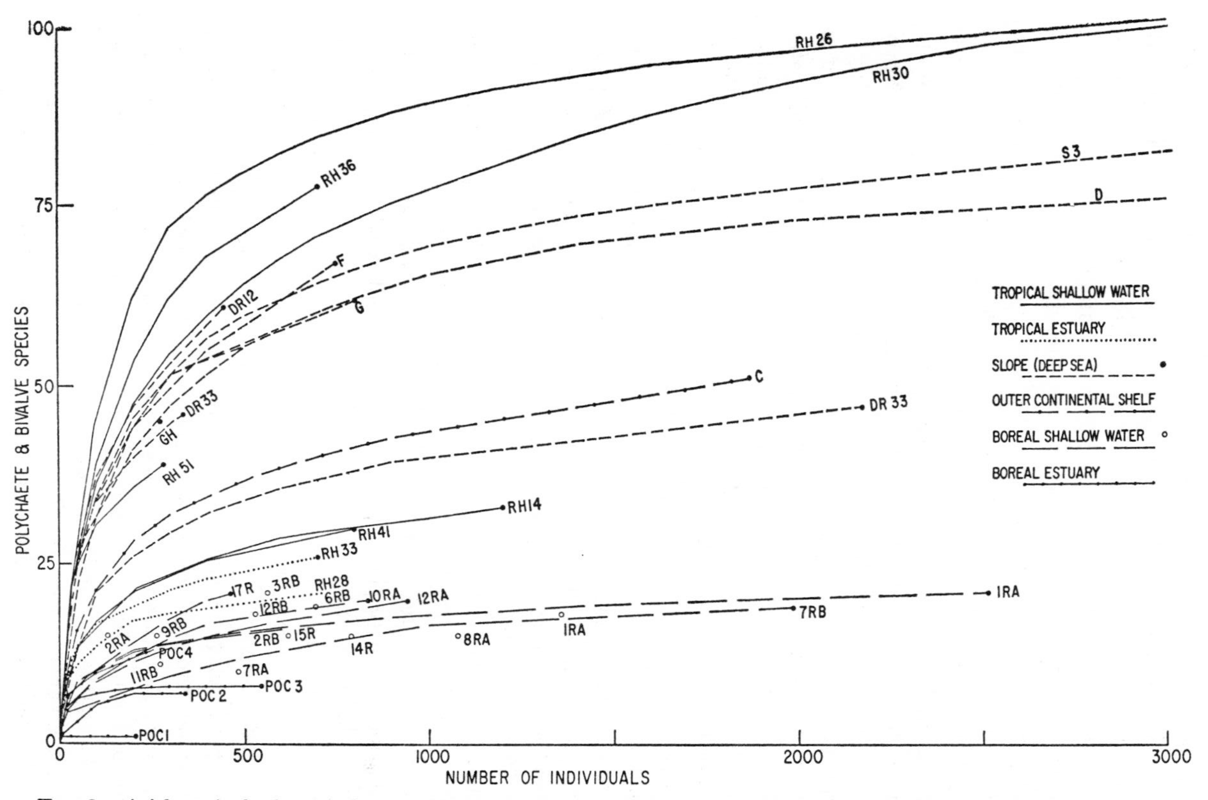

FIG. 2.—Arithmetical plot of the number of species at different population levels using the rarefaction methodology for stations from differing regions. The termination of a curve gives actual number of individuals and species found in the sample. The remainder of the curve is interpolated by the rarefaction methodology. The circles with station numbers are actual samples that have not been rarefied.

estuary as represented by the Pocasset River, Cape Cod, Massachusetts. The station highest up the estuary, POC 1 in Figure 2, with a mean sediment salinity of 7‰ and a range of 9.5‰ per tidal cycle, has the lowest diversity within the series. POC 2, next highest up the estuary, with a mean sediment salinity of 17‰ and with a 3‰ salinity range per tidal cycle, has the next lowest diversity. The diversity increases at POC 3, with a mean salinity of 20.7 and a tidal variation of 1.4‰. Still higher diversity is present at POC 4, where the mean salinity rises to 22.9‰ and the tidal variation is 1.8‰ (for details on faunal distribution in the Pocasset estuary and its relationship to salinity, see Sanders, Mangelsdorf, and Hampson, 1965). Besides the low and variable sediment salinities in the Pocasset River, there are, as well, large seasonal temperature changes.

Somewhat higher diversity values occur in tropical estuaries. These are represented by stations RH-28, the Vellar River estuary at Porto Novo, Madras State, India, and RH-33, at the mouth of the Godavari estuary at Kakinada, Andhra State, India. During the periods of heavy rainfall of October and November, the salinities in these shallow bodies of water are reduced to zero. Yet, in the dry season of May and June, the salinities are more than 34‰ (Jacob and Rangarajan, 1959). The probable reason for the greater diversity values in low- as compared with high-latitude estuaries is that it is easier to tolerate reduced salinities at high temperatures than at low temperatures (Panikkar, 1940). As a result, more marine forms are able to invade estuaries in the tropics than in higher latitudes.

Among the boreal shallow marine samples (the R-station series) diversity is modest. These samples were taken from Buzzards Bay, Massachusetts, in 20 m of water at all seasons of the year (Sanders, 1960). Here the annual temperature change is more than 23°C, with winter temperatures often less than −1.0°C and summer temperatures of more than 22°C. Such appreciable changes in annual temperature are as large as that found in any marine region of the world. This pronounced seasonal temperature variation probably accounts for the low diversity values. Note that within the R series of samples in Figure 2, the actual samples with lower densities, instead of being widely scattered throughout the graph, are clustered about the interpolated curves derived from the larger samples, thus verifying the validity of our methodology. This same clustering demonstrates that diversity values are sample-size independent when derived by the rarefaction technique. At the outer edge of the continental shelf, station C, the amplitude of temperature change has been reduced to 10°C and the faunal diversity has increased (Sanders, et al., 1965).

The most diverse values are found among the tropical shallow marine samples, although there is appreciable spread in the position of the curves. The highest values are from the three Bay of Bengal samples: RH-26, off Porto Novo, Madras State, India, in 20 m of water; RH-36, off Kakinada, Andhra State, India, in 37 m depth; and RH-30, off the city of Madras, in 15 m depth. All of these stations are too deep to be affected by the freshening of the surface water during the monsoons (LaFond, 1958;

Murty, 1958; Ramamurthy, 1953). Station RH-51, from a depth of 18 m off Nossi Bé, Madagascar, gives an intermediate value. Lowest values are from two stations in the Arabian Sea, RH-14 in 14 m off Cochin, India, and RH-41, in 20 m depth off Bombay, India. The probable cause for these modest diversity values is the low-oxygen minimum layer found throughout the northern Arabian Sea at the 100 to 200 meter depth. During the southwest monsoons, this low-oxygen water is pushed onto the continental shelf off India, creating a severe stress condition for the bottom fauna which is probably reflected in the reduced number of species present. Banse (1959) found, at almost precisely the site and depth of our Cochin station, oxygen values of only 5% saturation during the southwest monsoon, and Carruthers, Gogate, Naidu, and Laevastu (1959) obtained similar low-oxygen values at a location of equivalent depth near our Bombay station.

Our deepwater diversity curves, derived from stations Sl.3, D#1, F#1, and G#1 from the continental slope, station GH#1 on the abyssal rise (all south of New England), and stations DR-12 and DR-33 from the continental slope off northeastern South America, with but a single exception, are confined to a narrow sector of the graph. The physical factors in this environment are rigidly constant, with low temperatures, high salinity (see Sanders, et al., 1965), and high oxygen values. The single exception to our deepwater diversity pattern, DR-33, is due to the aggregation of a single polychaete species which forms more than 85% of the sample. If this species is arbitrarily excluded, the residual diversity, DR-33′, is similar to that found in other deep-sea samples. Thus the deep-sea benthic fauna appears to possess a relatively high diversity of the same general order as that present in tropical shallow seas.

It should be clearly pointed out here that this method of measuring diversity is valid when the fauna is randomly or evenly distributed but not aggregated. Even in cases of aggregation, it may be possible to uncover inherent diversity (DR-33′) by using this methodology. On the other hand, in samples with little aggregation, the diversity is only slightly increased by eliminating the most abundant species.

Applying confidence limits to the curves for certain of the environments is meaningless. In cases where the number of samples included is small, the confidence limits will be broad. The Bay of Bengal series, the pair of stations from the shallow depths of the Arabian Sea off India, and the two tropical estuarine stations suffer from this weakness.

The best that can be done is to represent the ranges in Figure 2 as bands of values (Fig. 3). The number of samples is given at each of the rarefied sample sizes from 100 individuals and larger. Environments and sample sizes with single samples, the Pocasset series, and aberrant station DR-33 are excluded. The clear separation of the environmental bands strongly indicate that these diversity differences are real.

Limitations of Methodology

The rarefaction method for measuring diversity must be used with dis-

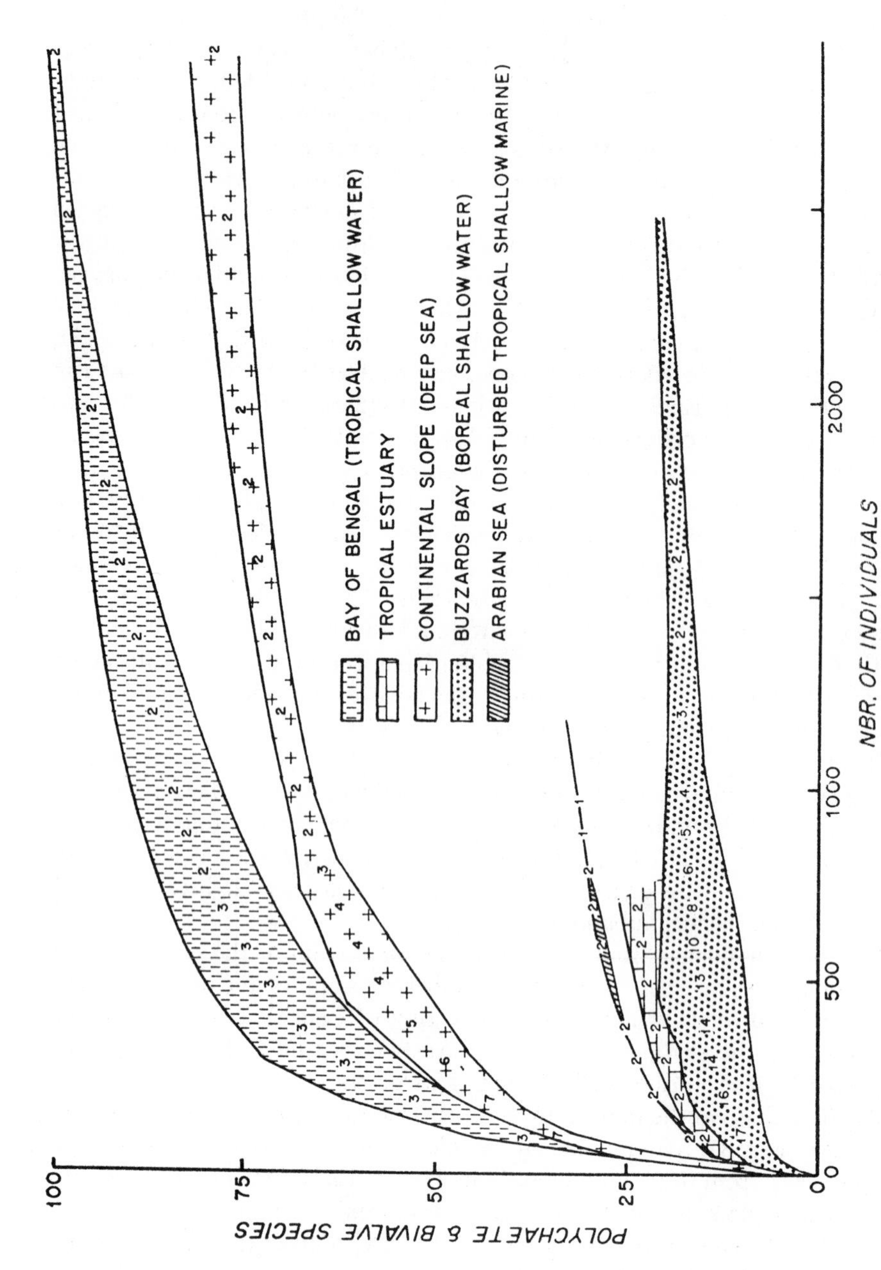

FIG. 3.—Range for diversity values found for a number of the regions included in this study.

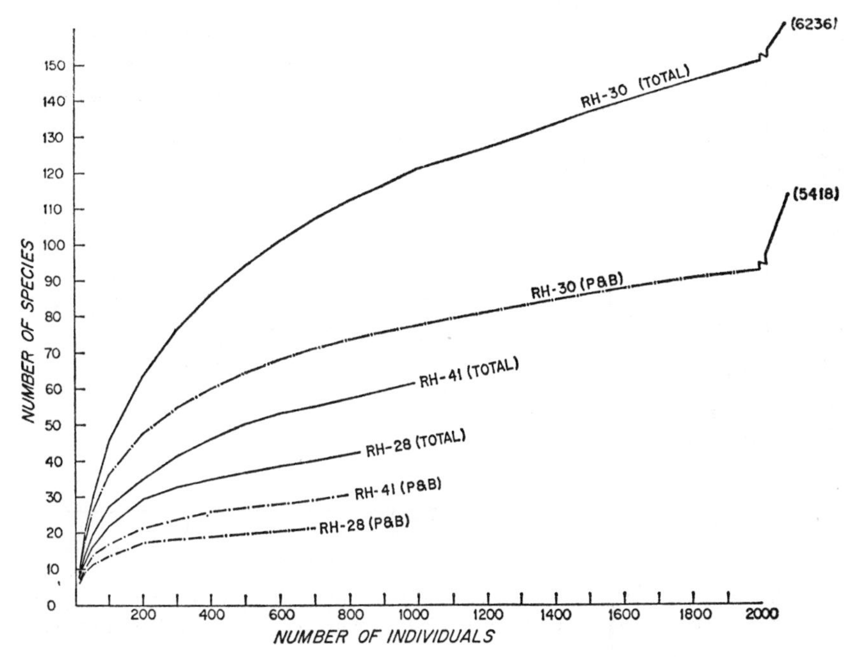

Fig. 4.—Diversity values for the total fauna compared with the polychaete-bivalve component of the same samples. Numbers in parentheses for station RH-30 are the actual numbers of specimens in the total sample and in the polychaete-bivalve fraction of that sample.

crimination to be meaningful. Such a technique is valid only when the same groups of organisms are compared and contrasted. With the inclusion of additional groups, the diversity values for a given faunal density increase. This phenomenon can be clearly observed with the few representative samples used in Figure 4. In each case, the diversity for the total fauna is decidedly higher than that of the polychaete-bivalve component of the sample.

Another requisite is that all the habitats sampled be similar; that is, the comparison must be made among a within-habitat (MacArthur, 1965) series of environments (in the present situation, the soft estuarine and marine oozes). Differing habitats from the same geographic region have differing diversity values (between-habitat comparison [MacArthur, 1965]). Thus the sand bottom fauna in Buzzards Bay is more diverse than the mud bottom fauna (Fig. 5). (Probably the fauna of stable sand bottoms will always be inherently more diverse because of the greater variety of microhabitats.)

In order to have the data comparable, it is necessary that the sampling procedures such as the type of gear used, the methodology utilized in processing the sample, and the screen size employed in washing the samples should be approximately similar.

Finally, this method does not specify which species taken from the residue will be present, and it can be used only to interpolate, not to extrapolate.

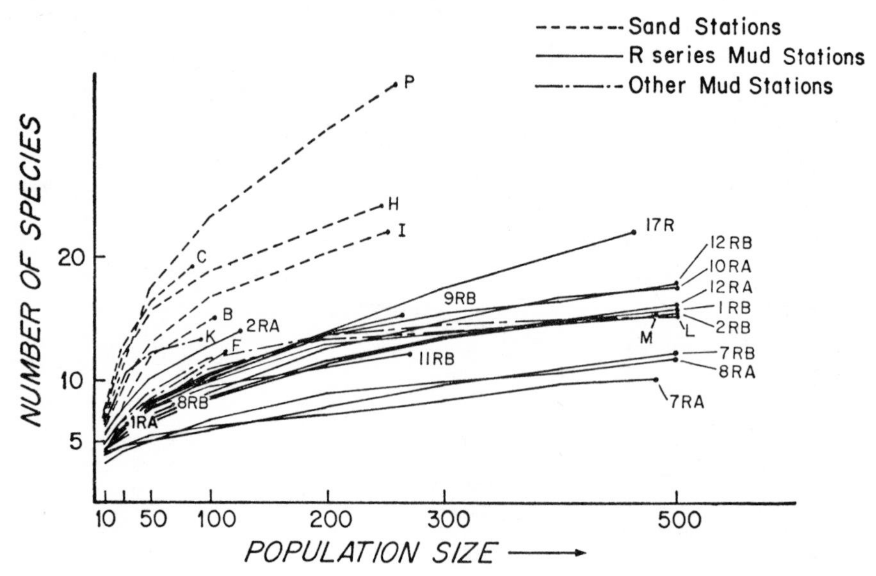

FIG. 5.—Comparison of diversity between mud- and sand-bottom samples from Buzzards Bay, Massachusetts, up to population sizes of 500 individuals.

THE PHYSICALLY CONTROLLED AND THE BIOLOGICALLY ACCOMMODATED COMMUNITIES

The interpretation of the curves in Figure 2 and the more general analyses of the total fauna might best be understood by describing two contrasting types of communities, both of which are abstractions. One can be called the "physically controlled community." In environments harboring this kind of community, the physical conditions fluctuate widely and the animals are exposed to severe physiological stress.

In the physically controlled community the adaptations are primarily to the physical environment. Examples of such communities are those found in hypersaline bays, high arctic terrestrial environments, and deserts. The physically controlled communities are always eurytopic and are characterized by a small number of species. A similar paucity of species occurs in environments of recent past history, such as most freshwater lakes.

The other extreme condition might be called the "biologically accommodated community." These communities are present where physical conditions are rather constant and uniform for long periods of time. Because of the historic constancy of the physical environment, physical conditions are not critical in controlling the success or failure of the species. With time, biological stress (intense competition, nonequilibrium conditions in prey-predator relationships, simple food web, etc.) is gradually mediated through biological interactions resulting in the evolution of biological accommodation. The resulting stable, complex, and buffered assemblages are always characterized by a large number of stenotopic species. The deep-sea regions, tropical shallow-water marine regions, and tropical rain forests best represent such conditions.

There is no such thing as a "pure" physically controlled or biologically accommodated community. All communities are the result of both their physical and biological components and are therefore somewhat intermediate between these extreme types. What determines the structure of any community is the relative proportions of these two parts.

In predominantly physically controlled communities there can be no close coupling of a species to its environment, as would be the case in the predominantly biologically accommodated communities. This is due to the variations in the amplitude of environmental factors; that is, there is no precise reproducibility from year to year. For example, one year the temperature may be slightly higher, so that one species is favored regarding breeding, which results in the "year-class" phenomenon, that is, a tremendous increase in the number of the new year class. (The year-class phenomenon probably is a characteristic feature of the predominantly physically controlled communities.) The next year, the temperature may be slightly lower, so that the same species is adversely affected, resulting in an unsuccessful breeding. At the same time, another species is favored in its breeding by the reduced temperature. The same would be true for the effects of other environmental variables, such as salinity and oxygen, on growth, breeding, metabolism, etc. Therefore, animal species of the physically controlled communities must adapt to a broad spectrum of physical fluctuations which does not allow the biological interrelationships to develop very far. (In some intertidal environments, the prey may be biologically controlled while its predator is physically controlled [Connell, 1961]).

From the concepts summarized above, it is possible to present the stability-time hypothesis in Figure 6. Where physiological stresses have been historically low, biologically accommodated communities have evolved. As the gradient of physiological stress increases, resulting from increasing physical fluctuations or by increasingly unfavorable physical conditions regardless of fluctuations, the nature of the community gradually changes

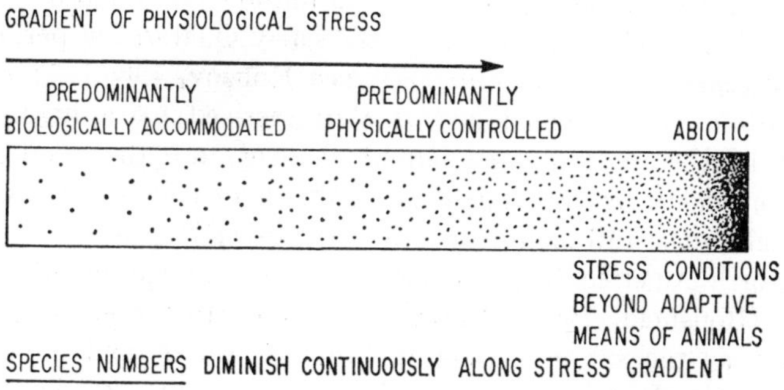

GRADIENT OF PHYSIOLOGICAL STRESS

PREDOMINANTLY BIOLOGICALLY ACCOMMODATED

PREDOMINANTLY PHYSICALLY CONTROLLED

ABIOTIC

STRESS CONDITIONS BEYOND ADAPTIVE MEANS OF ANIMALS

SPECIES NUMBERS DIMINISH CONTINUOUSLY ALONG STRESS GRADIENT

Fig. 6.—Bar graph representation of the stability-time hypothesis.

from a predominantly biologically accommodated to a predominantly physically controlled community. Finally, when the stress conditions become greater than the adaptive abilities of the organisms, an abiotic condition is reached. The number of species present diminishes continuously along the stress gradient.

When the stability-time hypothesis is applied to Figure 2, the closer the curves approach the ordinate, as shown with the shallow tropical marine and deep-sea samples, the nearer they approximate the biologically accommodated community. They describe assemblages in which there are large numbers of species per unit number of individuals (high diversity). In these environments, physical conditions are constant and have remained constant for a long period of time. The closer the curves approach the abscissa, as in the cases of the boreal estuary and boreal shallow marine samples, the greater are the physiological stress conditions imposed by the physical environment. In these assemblages there is a small number of species per unit number of individuals (low diversity). Here the physical conditions are highly variable and approach the idealized physically controlled community.

Thus each environment in Figure 2 appears to have its own unique family of curves. Such lack of randomness implies biological organization, with the nature of the organization or structure differing in different environments. Such organization is determined by the degree of stability of the physical environment and the past history of the physical environment, that is, to what degree an animal association is physically controlled and to what degree it is nonphysically regulated or biologically accommodated.

TIME

It requires appreciable time to evolve a highly diverse fauna, and the time component of our stability-time hypothesis is perhaps best illustrated with lakes. Most lakes are of a relatively transitory nature, or of recent geologic origin. It has been 10,000 years or less since the last glaciation, and the aquatic fauna from such recently glaciated regions shows limited diversification. However, there are a few ancient lakes—for example, the rift-valley lakes of Africa and Lake Baikal in Russian Siberia.

Lake Baikal was formed either about 30 million years ago in the middle Tertiary or at the end of the Tertiary and early Quaternary periods about one million years ago. (For references, see Kohzov, 1963.) This lake, in common with other ancient lakes, is characterized by a highly diverse fauna. One of the most diverse faunal elements are the gammarid amphipods, represented by 239 endemic and one nonendemic species.

To appreciate the full significance of such diversity, in all of what was glaciated North America there are no more than 28 species of gammarid amphipods (Bousfield, 1958). Certain of these crustaceans are confined to streams near the sea, can tolerate brackish water, and have recently evolved from closely related marine forms. Others are restricted to cave streams and springs. A few are limited to ponds. Still others occur in sloughs and temporary bodies of water. Only seven gammarid species are confined to lakes and rivers, and it is these few amphipods, distributed throughout vast areas of North America, which are the ecologic equivalents of the 240 species present in ancient Lake Baikal. (It should be mentioned that the diversity effects of time on a physically fluctuating environment of constant magnitude and periodicity remain unanswered.)

Lake Baikal has further implication to the concepts proposed in this paper. Two distinct and essentially separate faunas exist there. One element, broadly distributed through much of Siberia, is confined to the shal-

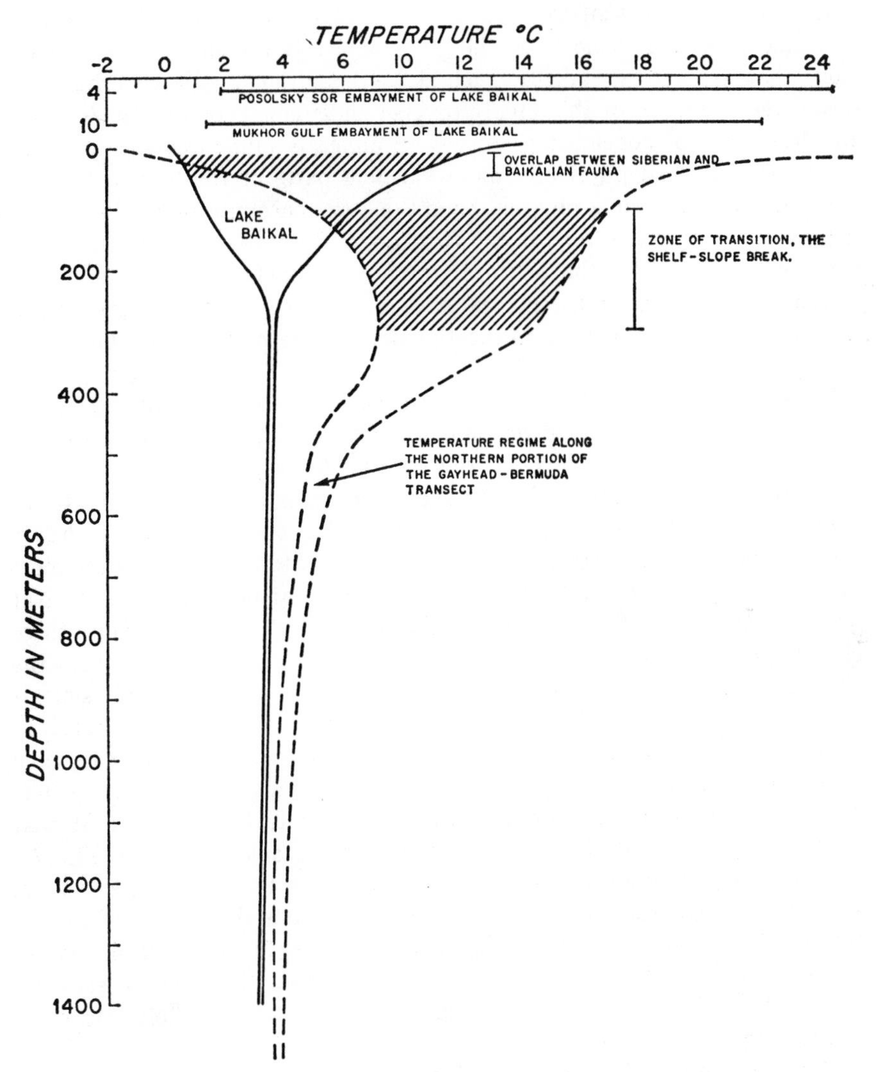

FIG. 7.—Temperature regime with depth in Lake Baikal and the northern part of the Gayhead-Bermuda transect.

low gulfs or bays and is not found deeper than 20 m in the lake. This group is represented by 137 species of nonendemic, free-living, benthic or pelagobenthic macroinvertebrates, 54% of which are insects. The other, an entirely endemic Baikalian fauna of 580 species of which are only 4% are insects, avoids the shallowed depths and embayments, and representative species are found down to 1,620 m, the maximum depth. (Lake Baikal is the deepest freshwater lake in the world.) The ecotone or region of overlap between these two faunas is very narrow (15 to 50 m depth), and but 22 species occur there (Fig. 7).

The Siberian eurytopic component exists in an environment of highly variable seasonal temperatures (see Fig. 7, Posolosky Sor and Mukhor Gulf) and low, fluctuating oxygen content (Kohzov, 1963). The deeper-dwelling Baikalian element, on the other hand, lives under physical conditions that are hardly varying. Such stable conditions, with time, have allowed the evolution of this highly diverse endemic stenotopic fauna, while the less diverse, nonendemic, eurytopic Siberian fauna remains confined to the shallow and physically more variable parts of the lake.

Similar rigidly stable physical conditions are encountered in the greater depths of Great Slave Lake in the Northwest Territories of Canada (Rawson, 1953). Yet, within the depth range from 200 to 600 m, only four species of macrofaunal benthic invertabrates were collected from this large "post-glacial" lake, again demonstrating the significance of the time component of diversity.

These findings from Lake Baikal are entirely analogous to the conditions occurring in the boreal region of the Gayhead-Bermuda transect. The continental shelf, particularly in the shoaler depths, harbors an impoverished fauna. The continental slope, however, supports a benthic fauna of high diversity. The region of very rapid and pronounced faunal change occurs somewhere between the depth range of 100 to 300 m and is the most marked zoogeographical boundary encountered along the transect. Not only are there specific and generic differences, but in some groups these changes are of familial and even ordinal significance. Conceivably, the changeover from the diverse stenotopic deep-sea fauna, which is physiologically adapted to constant temperature conditions (as well as other constant environmental factors), to that of the relatively depauperate eurytopic boreal littoral fauna, which exists under a varying temperature regime (other environmental factors tend also to fluctuate here), occurs at that depth where seasonal changes in temperature become large (see Fig. 7).

The same interpretation might be applied to the faunal changes in Lake Baikal, although the zone of rapid transition from the eurytopic assemblages of limited diversity to the highly diverse fauna of deeper water takes place at shallower depths (Fig. 7). Note that the range of seasonal temperature change in the transitional zone is approximately the same in both regions (Fig. 7).

Both limnetic Lake Baikal and the marine area of study south of New England are in boreal regions dominated by a continental climate.

In both situations, the pronounced seasonal changes in temperature are imposed on the shallow-water fauna while the benthos of greater depths are insulated from these changes (Fig. 7). With time, similar patterns of diverse stenotopic faunal assemblages have evolved in the physically stable deeper waters while the highly unstable shallow waters continue to support a rather impoverished fauna. Thus these two unrelated freshwater and marine faunas, molded by similar physical forces and time, have evolved and diversified in a parallel and analogous manner.

Such an interpretation is not at variance with the stability-time hypothesis as shown diagramatically in Figure 6. A continuous diminution in stress conditions certainly takes place from shallow to deep depths, but over a spatially restricted portion of this gradient the rate of change is very great. The region of abrupt change represents the transitional zone from the predominantly physically controlled to the predominantly biologically accommodated community.

Fischer (1960) concluded that the greater diversity in the tropics occurs not only because of the greater stability and longer history of that environment but also because the temperature is nearer the midpoint of the temperature range that protoplasm can endure. High temperature, per se, does not play a critical role in promoting diversity, for the highly diverse assemblages of the deep-sea and the endemic fauna of Lake Baikal evolved in relatively low-temperature environments. The critical factors appear to be time and environmental stability.

EXTRAPOLATIONS FROM THE STABILITY-TIME HYPOTHESIS

Our proposed hypothesis has another use. It allows us to predict. Hutchins (1947) pointed out that in the Northern Hemisphere a much greater seasonal change in water temperature takes place along the western edges of oceans at temperate latitudes than along the eastern edges. Such temperature conditions result from the prevailing west-to-east wind patterns in the middle latitudes. Thus the coastal boreal regions of eastern United States and parts of eastern Asia are dominated by a continental climate of high summer temperature and low winter temperature, while the outer European coasts and the western coast of North America are dominated by a maritime climate of appreciably less seasonal temperature change.

On this basis we can predict that two distinct types of boreal shallow-water communities exist. One can be termed the "continental climate boreal community" and would be exemplified by our Buzzards Bay series of samples. This community will be characterized by low faunal diversity. Furthermore, many, if not most, of the infaunal species will cease to grow and become inactive during the cold winter months (personal data and observations). The other boreal marine shallow-water community can be called the "maritime climate boreal community," characterized by greater faunal diversity and without the complete cessation of growth among the infauna during the winter months.

The model also suggests that the great upwelling regions of the oceans, such as the areas off southwest Africa and the Peruvian and Chilean coasts of South America, will typically show low benthic diversity. The abundant organic matter depletes the available oxygen as it sinks, so that the bottom water contains little or no oxygen. The stress condition resulting from the reduced available oxygen will be reflected in low diversities and, with further oxygen depletion, in low faunal densities or ultimately in abiotic conditions. Indeed, data by Gallardo (1963) from the upwelling areas off northern Chile provide impressive evidence for this interpretation. He found that the oxygen content of the bottom water was less than 5% saturated and that the sediments were reduced at water depths from 50 to 400 m. The benthic samples yielded few individuals, averaging only 6 to 7 per cubic meter.

SPATIAL VARIATION AND TEMPORAL VARIATION

Let us now consider a possible mechanism that would give the few though often numerically abundant species in predominantly physically unstable environments and the many species in the physically stable environments. We must consider two types of variation: *temporal variation,* which has already been discussed in some detail, and *spatial variation,* or habitat diversity.

When temporal variation is large (physiological stress conditions in physically unstable environments), it masks the effects of spatial variation (i.e., the wide range of habitats utilized by lemmings in the high latitudes of North America; see also MacArthur [1965] on within- and between-habitat avian diversities below). When temporal variation is small (minimal stress conditions in physically stable environments), the effects of spatial variations are realized, resulting in the progressive division of species with time. Thus with spatial variation occurring within the distributional range of a species, different selective forces will be acting on the species in different parts or habitats of its range. Initially, this process may result in the formation of separate subspecies and, if the gene flow is sufficiently attenuated, of separate species.

MacArthur (1965) pointed out that in a new environment (this may be comparable to environments of large temporal variation or high physiological stress), such as the initial stages in the colonization of an island by birds, few species are present but are distributed through a number of habitats. With time (this may be comparable to environments of decreasing temporal variation or reduced physiological stress) the number of species increases, but this enrichment is reflected in the *between-habitat* or β diversity of Whittaker (1965) (increase in the total number of species for all habitats) rather than the *within-habitat* or α diversity of Whittaker (1965) (number of species in a specific habitat remains constant).

A concept somewhat similar to MacArthur's constancy of within-habitat diversity is the earlier "parallel community hypothesis" postulated by

Thorson (1952) for the marine infauna. He contended that while there is a continuous gradient of species diversity from the arctic to the tropics for the epifauna, the number of infaunal species remains approximately the same. The findings in the present investigation, which is clearly a study of within-habitat diversity, give diametrically opposite results. Samples from historically stable environments of long duration and low physiological stress give high within-habitat diversity values, while samples from historically recent and/or variable environments yield low within-habitat diversity values.

Thorson, at the time he suggested his concept (1952), had only a very limited amount of data on the tropical marine infauna. At least one tropical locality upon which this interpretation is based, the Persian Gulf, with very high salinities and temperatures, represents a stress environment. Our own findings in the present study show that diversity can be quite variable in the tropics. Regions of low stress, such as the Bay of Bengal (RH-26, RH-30, and RH-36), support a very diverse infauna. Conversely, tropical areas of high stress, as exemplified by the shallow-water samples from the west coast of India (RH-14, RH-41), give much reduced values. Thorson (1966) recently found very high tropical infaunal diversities at shelf depths off the west coast of Thailand and he now feels that the applicability of the parallel community hypothesis to tropical environments should be carefully scrutinized.

There still appears to be an underlying difference between avian populations and benthic infaunal invertebrates. Among birds, with time and a physically stable environment, an increase in species occurs. This species enrichment takes place entirely as between-habitat diversity, while within-habitat diversity remains unchanged. With our infaunal benthic organisms, on the other hand, species enrichment is reflected both in the between- and the within-habitat diversities.

A lucid genetic interpretation for the relationship of environmental stability to diversity has been given by Grassle (1967). He pointed out that populations present in physically stressed and unpredictable environments show broad adaptations to these conditions by maintaining a high degree of genetic variability. Thus, even though the stress may be expressed in a variety of ways, a portion of the polymorphic population will probably survive. These genetically flexible species are opportunistic and cosmopolitan, and they have little tendency to speciate.

The price paid for this variability is "the genetic load or loss of fitness relative to the maximum in a more uniform environment." In stable environments "the expression of deleterious genes outweighs the advantages obtained from maintaining genetic flexibility." Therefore, in stable and predictable environments, such genetic variability will be selected against.

Diversity differences found in the present study between stable and unstable environments can be interpreted on the basis of genetic variability. The flexibility needed for survival in an unstable environment necessitates a larger utilization of the environment by each species. Thus diversity and genetic flexibility would be inversely related. (For a comprehensive discussion of the genetic basis of benthic diversity, see Grassle [1967].)

THE DEFINITIONS AND MEANINGS OF DIVERSITY

We might pause here and ask what we precisely mean by the word "diversity." It is apparent from looking through the literature that there are two definitions. This has resulted in some confusion.

One kind of diversity is the *numerical percentage composition* of the various species present in the sample. The more the constituent species are represented by equal numbers of individuals, the more diverse is the fauna. The less numerically equal the species are, the less diverse the sample is or, conversely, the greater is the dominance in the sample. This is a measure of how equally or unequally the species divide the sample, and the number of species involved is immaterial. Diversity measurements of this kind include the MacArthur "Broken Stick" model (1957), the Preston lognormal distribution (1948), and the Simpson index (1949). Such diversity might be designated after Whittaker (1965) as *dominance diversity.*

The other kind of diversity is determined by the *number of species.* The more species in a sample or the more species present in a species list for a given environment, the greater the diversity. Measurements of this sort are the α values of Fisher, Corbet, and Williams (1943), Margalef's d values (1957), the methodologies of Gleason (1922) and of Hessler and Sanders (1967), and the rarefaction technique used in the present paper. Such diversity can be designated after Whittaker (1965) as *species diversity.*

Since the number of species present and the relative dominance or lack of dominance in a sample are both measures of diversity, one might assume that they must be highly correlated with one another. Thus, a large number of species per unit number of individuals reflects low dominance; alternatively, a small number of species indicates high dominance. In Table 2 we will test this assumption.

Eighteen of the stations used in Figure 2 are included in the analysis. Each station is represented by a single sample, except station R, which is a composite of 15 samples taken from the same locality in Buzzards Bay. (The Pocasset series of samples are excluded because of the very few species present.)

The samples are ranked by dominance diversity from highest (lowest dominance) to lowest (highest dominance) diversity. These values are determined by plotting the percentage composition of the species along the ordinate and ranking the species by abundance along the abscissa (Fig. 8). The resultant cumulative frequency curve is used as a measure of dominance diversity.

Maximum diversification occurs when all the species in a sample are represented by exactly the same number of individuals, and the cumulative frequency curve, in this case, is a diagonal straight line that can be described by the formula $x = y$. Such a straight line forms the base line (Fig.

TABLE 2

MEASUREMENT OF THE CORRELATION BETWEEN SPECIES DIVERSITY AND DOMINANCE DIVERSITY AT DIFFERENT FAUNAL DIVERSITY LEVELS USING THE PEARSON PRODUCT MOMENT CORRELATION FOR 18 STATIONS INCLUDED IN THIS STUDY

Sample Size	*N*	*r* Value	Critical *r* Value at 5% Level	Critical *r* Value at 1% Level
Spp./10 ind.	18	.974	.456	.575
Spp./25 ind.	18	.948	.456	.575
Spp./50 ind.	18	.894	.456	.575
Spp./100 ind.	18	.766	.456	.575
Spp./200 ind.	18	.642	.456	.575
Spp./300 ind.	16	.636	.482	.606
Spp./400 ind.	15	.649	.479	.623
Spp./500 ind.	14	.622	.514	.641
Spp./600 ind.	14	.623	.514	.641
Spp./700 ind.	13	.748	.532	.661
Spp./800 ind.	9	.832	.632	.765
Spp./900 ind.	8	.769	.666	.798
Spp./1,000 ind.	8	.819	.666	.798
Spp./1,200 ind.	8	.817	.666	.798
Spp./1,400 ind.	7	.832	.707	.834
Spp./1,600 ind.	7	.833	.707	.834
Spp./1,800 ind.	7	.832	.707	.834
Spp./2,000 ind.	6	.801	.754	.874
Spp./2,500 ind.	4	.572	.878	.959
Spp./3,000 ind.	4	.519	.878	.959

8). What is measured is the deviations in percentage composition of a given sample from a hypothetical sample containing the same number of equally abundant species, that is, the degree of departure from the base line. The greater the departure, the greater the dominance and, conversely, the smaller the diversity. (See also Sanders, 1963, pp. 87 and 88.)

The stations are also ranked by species diversity from highest (most species per unit number of individuals) to lowest (least species per unit number of individuals) diversity at various population levels from 10 to 3,000 individuals. The dominance-diversity ranking is compared to the species-diversity ranking at each of the derived sample-size levels, using the rarefaction method.

This relationship is measured using the Pearson product moment correlation. The findings are given in Table 2. The square of the correlation coefficient, *r*, gives the approximate correlation, that is, the variance accounted for by the correlation.

This analysis reveals that species diversity and dominance diversity are indeed correlated at the 5% significance level, except with the largest samples. Using the 1% significance level, a consistent correlation exists only with the smaller-size samples (10 to 400 individuals). At intermediate sample sizes (500 to 1,800 individuals) the relationship is marginal, with the correlation values fluctuating around the critical correlation value. Among the largest sample sizes (2,000 to 3,000 individuals), species diversity and dominance diversity are not significantly correlated.

Thus a correlation, although not a particularly intimate one, exists between species diversity and dominance diversity. This conclusion is in agreement with the recent suggestion of Whittaker (1965) that the rela-

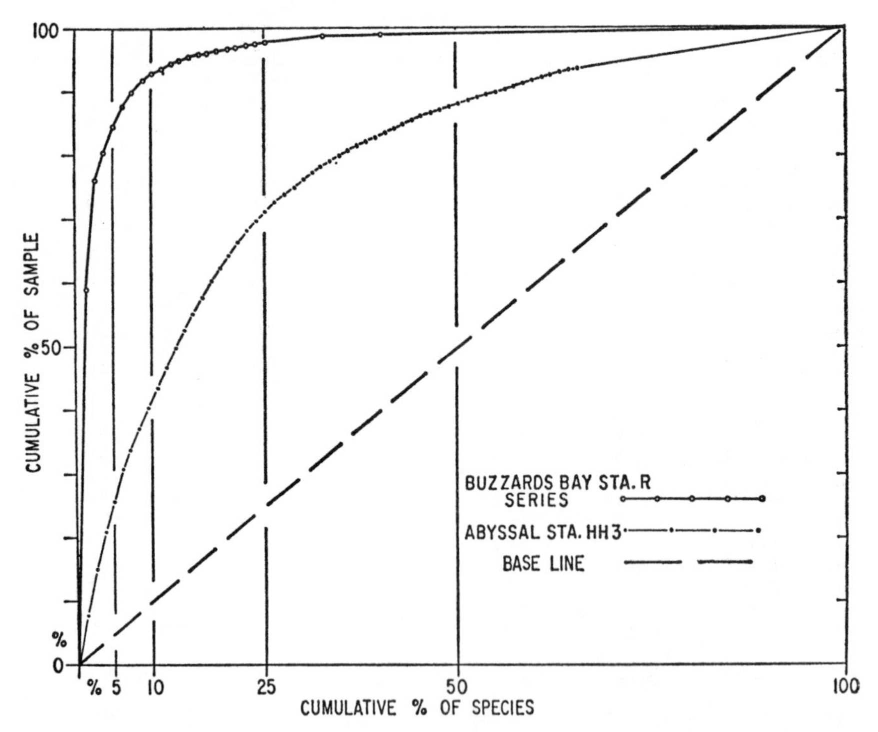

FIG. 8.—Degree of dominance of a sample related to numerical percentage composition of the included species plotted cumulatively. For explanation of figure, see text.

tionship between these two diversity measurements is weak. The degree of correlation appears to be sample-size dependent. A strong correlation is found with small sample sizes, a weaker correlation occurs among intermediate sample sizes, and either a weak or no significant relationship exists among the largest sample sizes.

Only at the smallest sample size does this correlation account for most of the variance (about 95% at the 10-individual level, about 90% per 25 individuals, and about 80% per 50 individuals). At such small population sizes only the most abundant species would normally be present, and one might expect a high level of correlation between dominance and species diversities. As sample size becomes larger, the less common species begin to appear and the percentage of variance accounted for by the correlation diminishes.

What, then, determines this relationship? As mentioned earlier, dominance diversity is independent of the number of species present. However, all species-diversity indexes are affected not only by the number of species but also by how a sample is divided among these species (percentage composition). There is always a dominance-diversity component in all species-diversity measurements because, while the height of the curve is determined by the number of species present, the shape of the curve is set

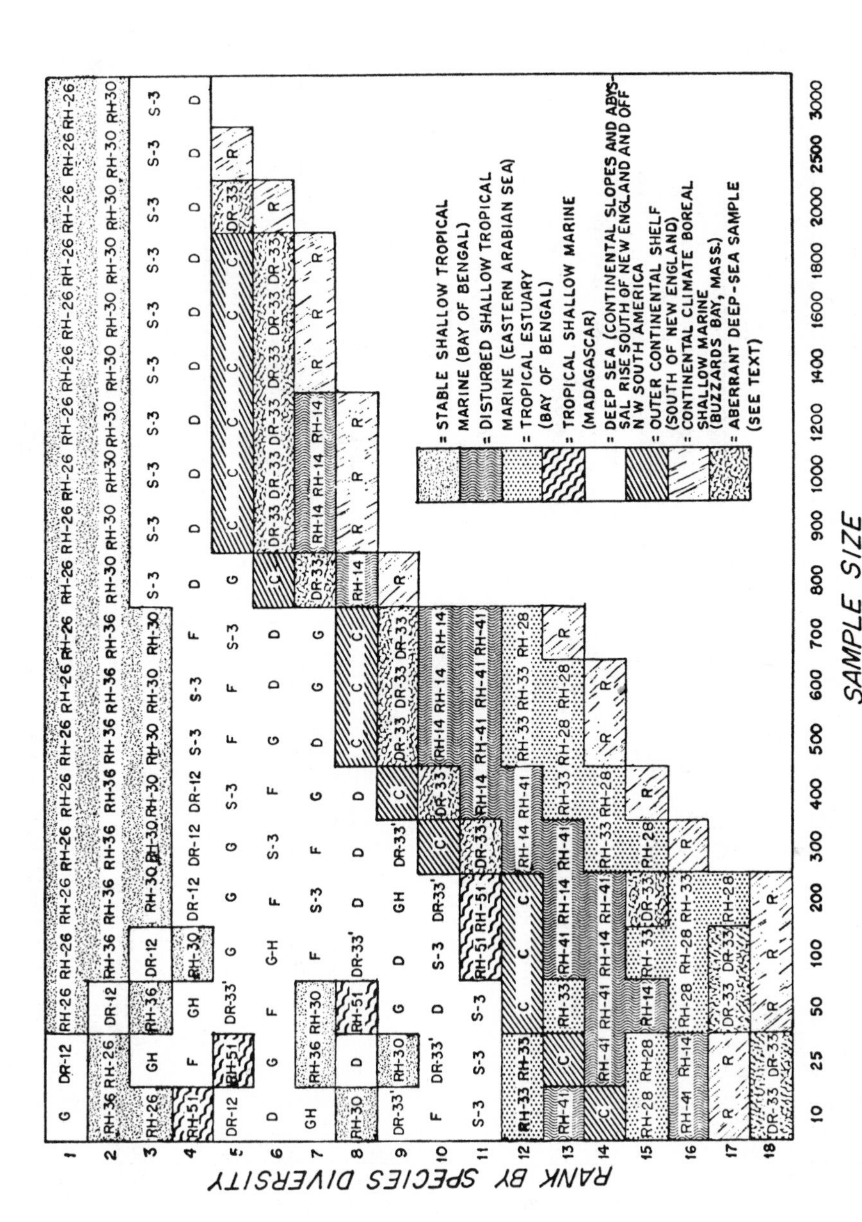

Fig. 9.—Ranking of stations by species diversity at different sample sizes.

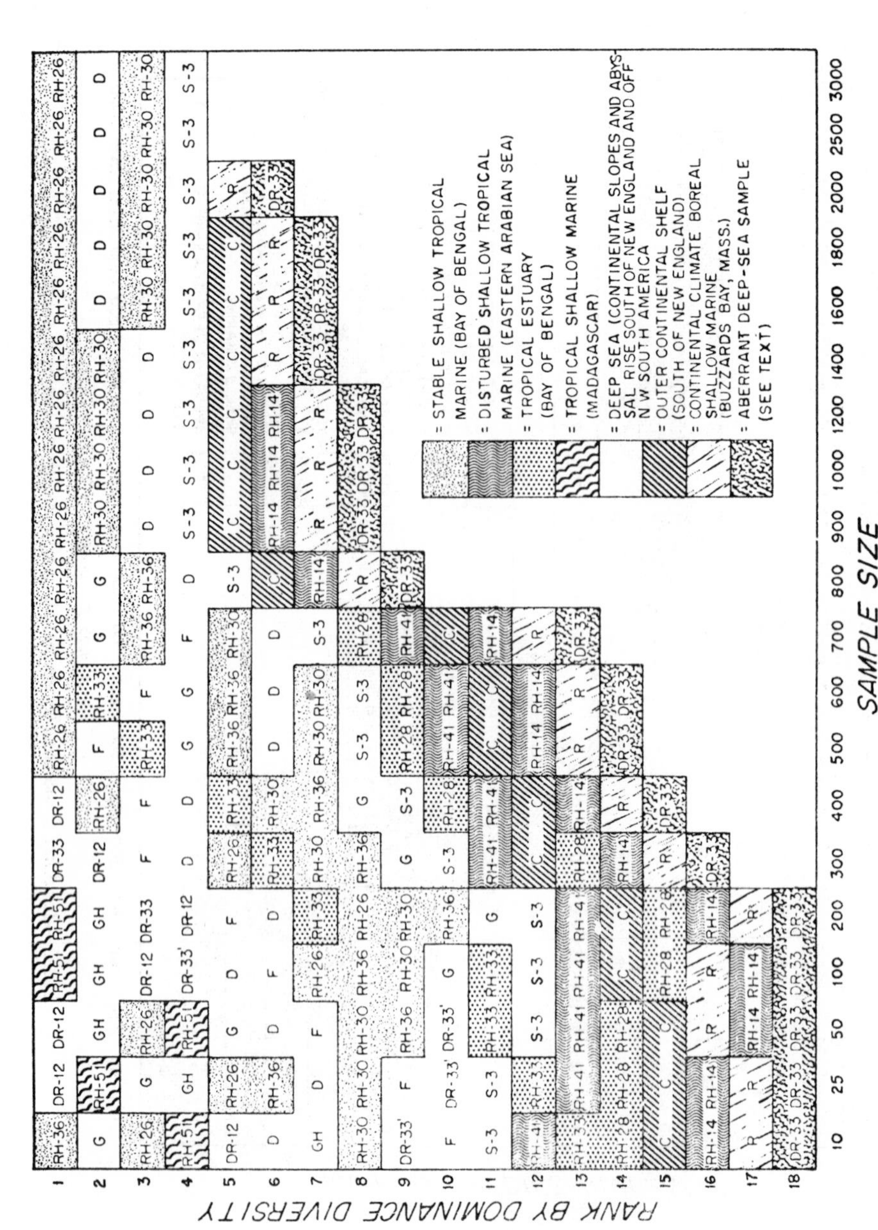

FIG. 10.—Ranking of stations by dominance diversity at different sample sizes.

by dominance. Probably it is the dominance dependency of species-diversity measurements that is primarily responsible for the percentage of variance accounted for by the correlation. Such an interpretation is consistent with the observation that the percentage of variance is very high with smallest sample sizes when only the numerically dominant species are present. Of critical importance, then, is the ecological significance of these two measures of diversity.

If these stations are ranked by both dominance diversity and species diversity, much better clustering of stations by environments is obtained within the species-diversity series (Figs. 9 and 10). Except at low densities, the species-diversity pattern remains stable by environment over the spectrum of population sizes. The stations with the highest diversity values are from the shallow marine depths of the Bay of Bengal (stations RH-26, RH-36, and RH-30). Then, except for aberrant station DR-33, there is a block of samples which includes all the deep-sea stations. They, in turn, are followed by the single shallow marine sample (RH-51) from Madagascar and the single station from the outer continental shelf (C). Next comes the previously mentioned atypical deep-sea station DR-33, the two stations from the stress shallow waters of the Arabian Sea off India (RH-14 and RH-41), the two tropical estuarine samples from India (RH-28 and RH-33), and, finally, the R series of samples from Buzzards Bay, Massachusetts.

Within the dominance-diversity series (Fig. 10), no such clear-cut groupings are present. Stations from a single environment often are widely separated. The Bay of Bengal stations (RH-26, RH-36, and RH-30) with the highest species-diversity values often show wide variability in ranking, both within a specific population size and among differing population sizes. The deep-sea stations do not form a solid block, but have stations from other environments interspersed among them. At those population sizes where the Madagascar station (RH-51) achieves first ranking in the dominance-diversity series, it does no better than eleventh rank using species-diversity criteria. The two tropical estuary stations, RH-33 and RH-28, have appreciably differing dominance-diversity values, yet their species-diversity values are almost identical. Station RH-33 gives intermediate to high dominance-diversity but low species-diversity values. The disturbed tropical stations, RH-14 and RH-41, show reasonable agreement in ranking when both diversity measurements are compared. The values are always low; yet the two stations are usually contiguous by species-diversity ranking and are always separated by dominance-diversity analyses. Outer continental shelf station C usually shows higher species- than dominance-diversity ranking. The R series of samples from Buzzards Bay give low values by both diversity methods. It occupies the lowest species-diversity rank over most of the sample-size spectrum and the next to lowest rank by dominance-diversity criteria. Aberrant deep-sea station DR-33 has last ranking in the dominance-diversity series, but its diversity values increase with sample size using species-diversity ranking. (The pronounced effect of the single numerically dominant species, forming 84.43% of this sample, is gradually mediated as the population size increases and the inherent species diversity begins to emerge. This effect is so overwhelming by dominance-diversity standards that station DR-33 is restricted to lowest ranking throughout the entire range of population sizes.) The only good agreement found between species- and dominance-diversity rankings occurs at the smallest population sizes, where only the most common species would be present.

In brief, then, the stations in the species-diversity series clearly and sharply sort themselves by environment and generally follow the gradient from the biologically accommodated to the physically controlled environmental situations. Within the dominance-diversity series, no such clear-cut groupings are present. Stations from a single environment often are widely separated. Further, there is often little agreement in the position of a station in one series as compared with the other.

These findings can only be interpreted to mean that the high level of agreement between environment and species diversity indicates that such a measure is a conservative and, therefore, ecologically powerful tool. On the other hand, the much poorer fit with dominance diversity suggests that this type of diversity is more variable in its relationship to the physical environment.

THE RELATIONSHIP OF OTHER DIVERSITY CONCEPTS TO THE STABILITY-TIME HYPOTHESIS

Pianka (1966) has summarized the various hypotheses advanced to explain the causes of latitudinal diversity gradients. He was able to separate them into six more or less distinct groupings, although most of the hypotheses contain components of more than one grouping. Presented in their most elemental form, they are:

a) *The time theory* (Simpson, 1964).—All communities tend to diversify with time. Older communities, therefore, are more diverse than younger communities.

b) *The theory of spatial heterogeneity* (Simpson, 1964).—The more heterogeneous and complex the physical (topographic) environment, the more complex and diverse its flora and fauna become.

c) *The competition theory* (Dobzhansky, 1950; Williams, 1964).—Natural selection in higher latitudes is controlled by the physical environment, while in low latitudes biological competition becomes paramount.

d) *The predation hyopthesis* (Paine, 1966).—There are more predators in the tropics who intensively crop the prey populations. As a result, competition among prey species is reduced, allowing more prey species to coexist.

e) *The theory of climatic stability* (Klopfer, 1959; Fischer, 1960; Dunbar, 1960; Connell and Orias, 1964).—Because of the greater constancy of resources, environments with stable climates have more species than environments of variable or erratic climates.

f) *The productivity theory* (Connell and Orias, 1964).—All other things being equal, the greater the productivity, the greater the diversity.

To fit better into the context of the present paper, two of these theories are rephrased as follows:

e') The *theory of climatic stability* is generalized to the *theory of environmental stability.* The moıe stable the environmental parameters—such as temperature, salinity, oxygen—the more species present.

c') The *competition theory* is altered to state that in environments of high physiological stress, selection is largely controlled by the physical variables, but in historically low stress environments, natural selection results in biologically accommodated communities derived from past biological interactions and competition.

How do the data from the marine benthic samples presented earlier in this paper and the derived *stability-time hypothesis* fit these theories? The *time theory* and the *theory of environmental stability* are most directly applicable to the *stability-time hypothesis.* They, in turn, determine the expression of the *competition theory* and the *theory of spatial heterogeneity.* Biologically accommodated communities resulting from past biological interactions (including competition) are realized in physically stable environments of long temporal continuity. Similiarly, the potentials of spatial heterogeneity can be achieved only under these same environmental conditions.

Neither the *predation theory* nor the *productivity theory* can readily be explained by the *stability-time hypothesis.* The *predation theory* was recently postulated by Paine (1966) for rocky intertidal marine organisms, although he feels it may have wider application. In the intertidal environment, the epibenthic animals experience alternating periods to exposure and immersion. Therefore, these organisms are subjected to desiccation, high salinity imposed by evaporation, exposure to freshwater rain, and air temperatures that are often significantly higher or lower than the seawater temperature. Thus, all rocky intertidal assemblages, independent of latitude, especially at the higher intertidal levels (see Connell, 1961), must be considered predominantly physically regulated communities, and the adaptations are primarily to the physical environment and the biological interactions are poorly developed.

Conceivably, the *productivity theory,* which says that the more food produced, the greater the diversity, may have some validity. Yet this effect is readily masked by numerous environmental variables (Hessler and Sanders, 1967). High productivity itself, from the sheer amount of organic matter produced, can create severe stress conditions and low diversity. For example, in some upwelling areas, high production is responsible for low oxygen content of the water on and just above the ocean floor. Similarly, the highly productive eutrophic lakes often have bottom water devoid of oxygen. In contrast, the high diversity values for the deep-sea benthos, shown by Hessler and Sanders (1967) and in the present paper, come from regions of low productivity.

COMPARISON USING CERTAIN OTHER FAUNAL INDEXES

Numerous indexes have been formulated to measure diversity. Odum, Cantlon, and Kornicker (1960) pointed out that in all types of presentation, logarithmic functions are involved. They recognize four categories. Three have pertinence to our paper:

1. *Cumulative species versus logarithm of abundance.*—This was exemplified by Gleason (1922), Fisher et al. (1943), and Margalef (1957). Such an index is obtained by determining the rate of species increase as additional samplings are made from the same population.

2. *Number of species of particular abundance versus logarithm of abundance.*—This method was formulated by Preston (1948) and is based on the premise that, if the presence of all species found in a given habitat can be revealed, the abundance distribution would follow a lognormal curve. Since such a complete revelation is usually impossible, the resulting curve is truncated at its rarer end. However, the shape of the curve allows one to approximate the total number of species in the habitat, including those as yet undiscovered.

3. *Abundance versus logarithm of rank.*—This type of index was proposed by MacArthur (1957). The observed abundances are compared with theoretical abundances derived from a model containing contiguous, non-overlapping niches.

One of the fundamental drawbacks of most diversity indices is that they are sample-size or density dependent. Hairston and Byers (1954), in an analysis of cumulative samples of soil arthropods from a singel habitat by both the logarithmic series of Fisher,. et al. (1943) and the lognormal distribution of Preston (1948), found that the results depended on the size of the total sample. Margalef (1957) pointed out that his diversity measurement, in common with other diversity indexes, increases with enlarged samples. A similar finding was reported by Williams (1964) for the Simpson diversity index (1949). Hairston, in a later paper (1959), demonstrated that the MacArthur model (1957) is also density dependent. From the analysis of his own data, he interpreted this phenomenon of increased diversity with increased sample size to mean that rare species are clumped. With repeated samples, there will be a greater likelihood of obtaining a new rare species than of obtaining a member of a rare species already collected. He concluded "that an increase in heterogeneity with an increase in sample size lies in the spatial distribution of the species concerned, and the inverse relationship between clumping and abundance."

In our study, single samples were collected from comparable sediment environments. Both small samples and large samples from the same environment (Fig. 2, boreal shallow marine and deep-sea) fall along the same diversity curve. If we apply the rarefaction method to a series of 15 samples of greatly differing sizes (35 to 2,514 individuals) which were carefully selected for sediment homogeneity and taken during the course of a 2-year period from a single locality in Buzzards Bay (Sanders, 1960), we find in the semilogarithmic plot in Fig. 11 no tendency for smaller samples to be

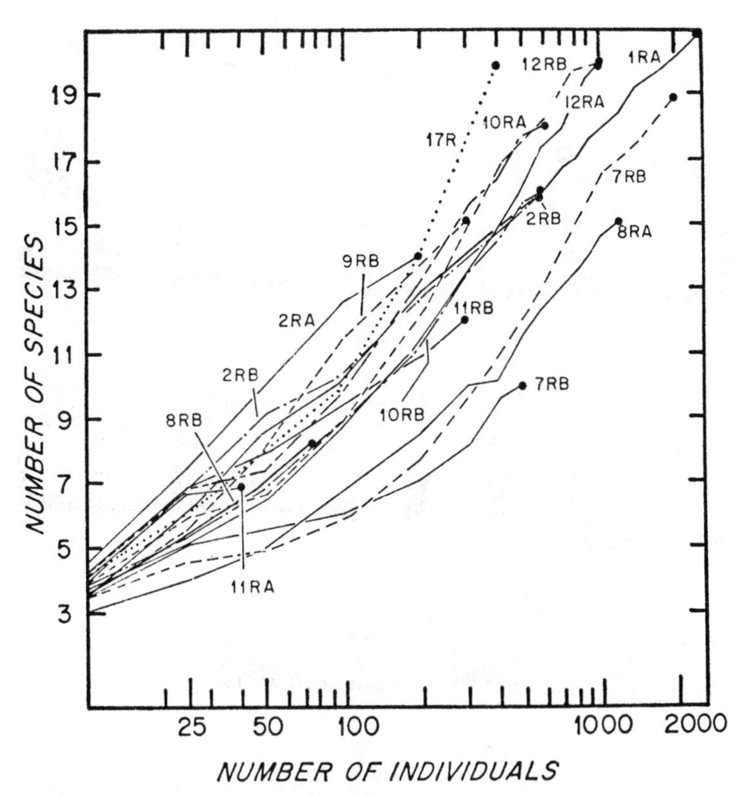

FIG. 11.—Rarefaction curves for the 15 station R samples from Buzzards Bay, Massachusetts.

less diverse than larger ones—that is, smaller-sized samples do not show a tendency to rise more steeply than larger-sized samples. Thus, both under the conditions of our sampling program and by using the rarefaction method of measuring diversity, no increase in heterogeneity takes place with increasing sample size. After all, what could be more homogeneous than a series of different-sized subsamples, each with the same percentage composition as the original samples? This, in essence, is what the rarefaction method does.

Now that we have demonstrated the effectiveness of the proposed rarefaction methodology in obtaining constancy of diversity at all population sizes in our samples, can similar stability be achieved when other diversity formulas are applied to the identical data whose internal homogeneity has been demonstrated over the entire range of sample size? We will attempt to answer this question by applying a number of diversity measurements to the data presented in Figure 2.

With the Preston truncated lognormal distribution analysis, the numbers of species with abundances of 1 to 2, 2 to 4, 4 to 8, etc., are plotted as points. The resulting curve is assumed to approximate a lognormal distribution. Such an estimate is essentially independent of sample size because a dou-

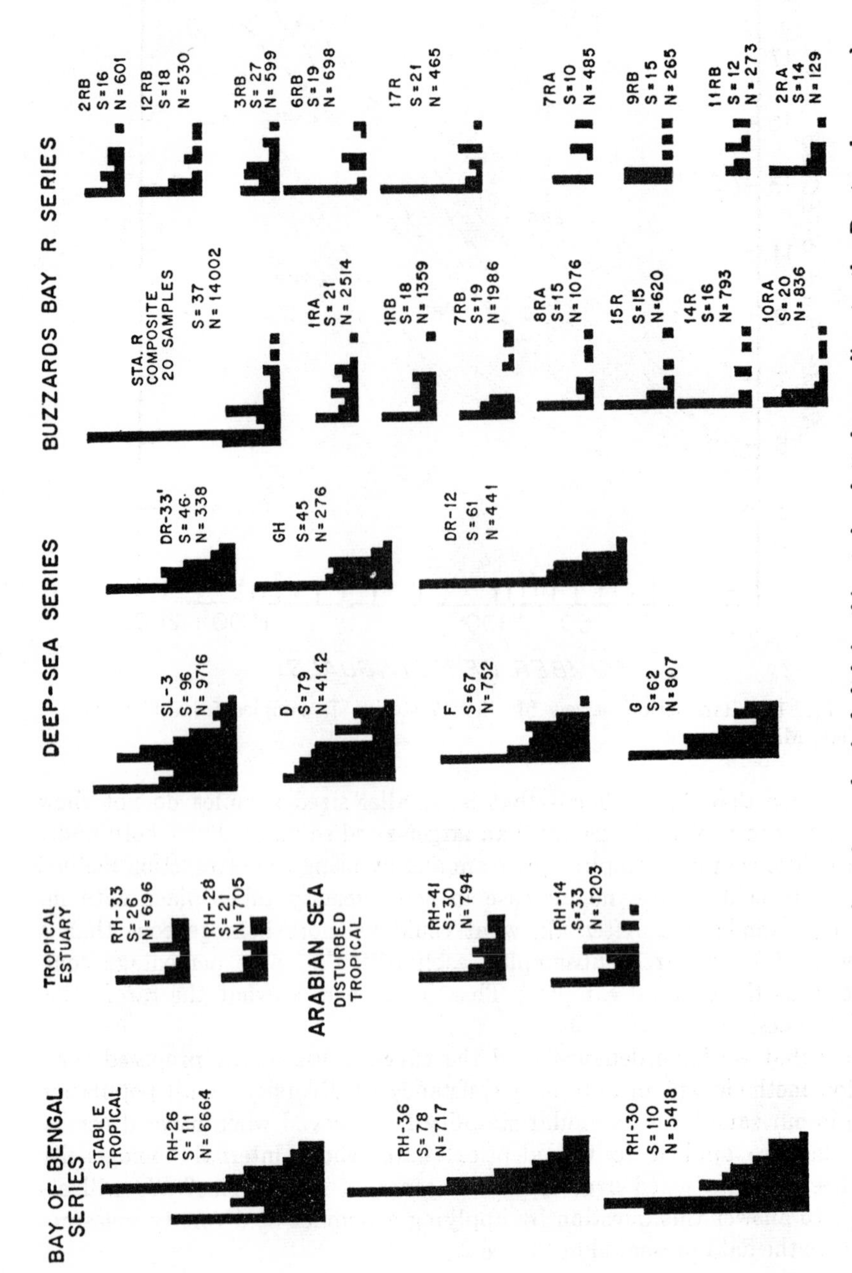

FIG. 12.—Histogram representation of samples included in this study plotted according to the Preston lognormal distribution.

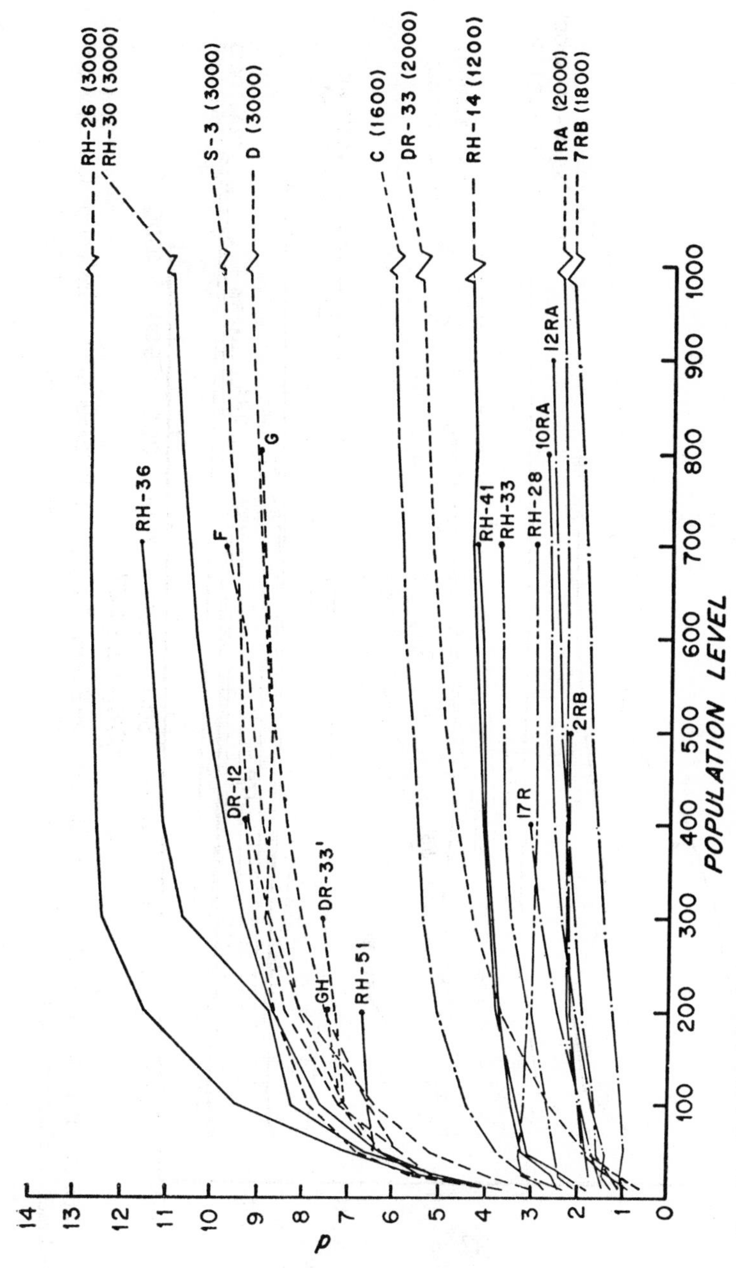

FIG. 13.—Relationship of the Margalef index to rarefaction data. Numbers in parentheses after the stations are the largest population levels to which the samples have been rarefied.

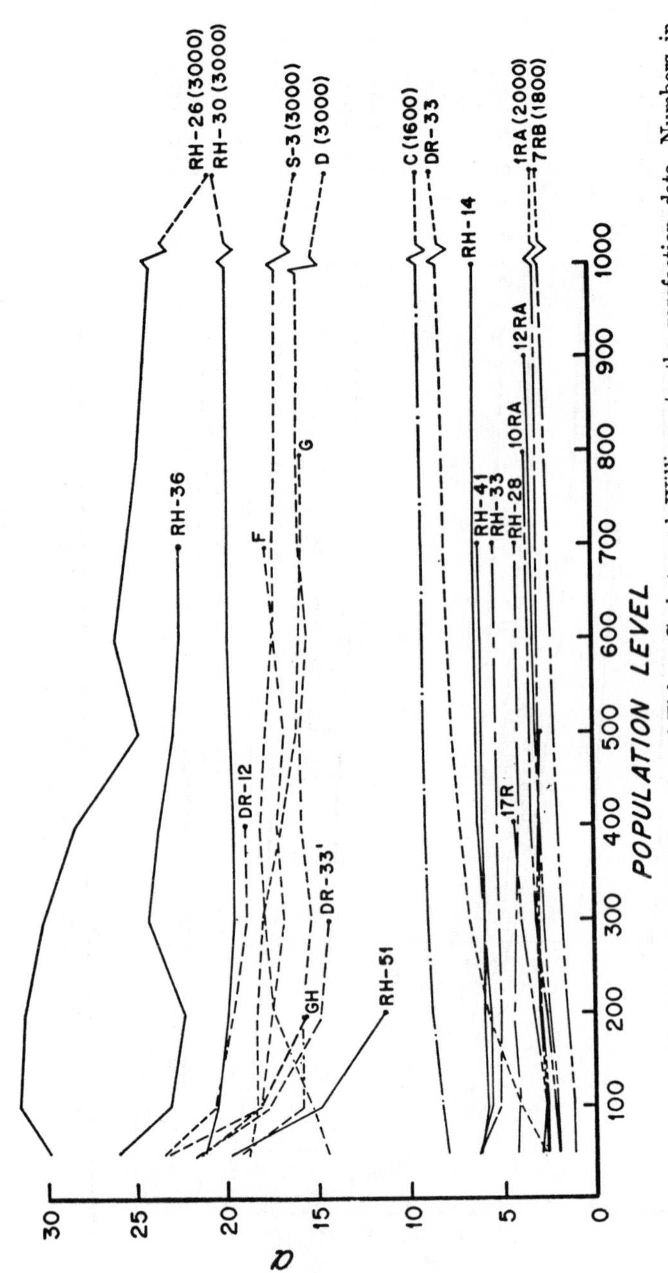

FIG. 14.—Relationship of the logarithmic series of Fisher, Corbet, and Williams to the rarefaction data. Numbers in brackets after the stations are the largest population levels to which the samples have been rarefied.

bling of individuals simply displaces the curve one unit to the right and adds a new unit to the left or rare end of the curve. When the entire suite of species is finally revealed, a lognormal rather than a truncated lognormal curve describes the situation.

Plotting our data by this method in Figure 12 gives histograms that are difficult to interpret. For example, the rarefaction curves for the two disturbed tropical shallow marine samples, RH-14 and RH-41, are essentially identical. The histograms derived for these same stations by using the Preston methodology are totally unlike (Fig. 11), and it would take a great amount of ingenuity to fit the histogram of station RH-14 to a truncated lognormal distribution. While the rarefaction curves for the two tropical shallow marine samples, RH-26 and RH-30, are somewhat alike, the Preston histograms are very different. The histogram for RH-26 also does not remotely fit a truncated lognormal pattern.

No attempt will be made to consider each of the histograms in Figure 12. It seems evident from the examples already chosen that the truncated lognormal distribution pattern cannot convincingly be made to fit these samples. Extrapolations of the total number of expected species made from such fitted truncated lognormal curves by adding to the left tail and converting them into normal distribution curves give results that are unrealistic. By such analyses, samples taken from the same environment and displaying similar rarefaction curves often have normal distribution curves containing appreciably differing numbers of species.

Using Margalef's (1957) index (which is essentially the same as Gleason's [1922] formulation), $d = (S - 1)/\ln N$, where S = number of species, N = number of individuals, and d = index of diversity, we find (Fig. 13) that d in all samples is initially low. As sample size becomes larger, the d value rapidly increases. At high densities (large sample sizes), the rate of increase gradually diminishes. This effect is most pronounced in samples with high diversities—the shallow waters of the Bay of Bengal and the deep sea. In high-stress environments—the Buzzards Bay series, tropical estuaries, and the disturbed tropical shallow marine—this effect is less pronounced. These data indicate that diversity indexes of the Margalef and Gleason types are influenced by sample size, even when such samples are internally homogeneous.

When the same samples are plotted using the α values of Fisher et al. (1943), $S = \alpha \ln (N/\alpha + 1)$, almost opposite results are obtained (Fig. 14). In the more diverse samples, the α values are highest at low densities, rapidly decrease as density increases, and then more slowly decrease until an approximate equilibrium is reached. With low-diversity samples, the tendency for higher α values at low density is either absent, poorly developed, or weakly opposed. When internally homogeneous samples are used, this diversity index also is not independent of sample size. In comparison with the d values, the α indexes tend to stabilize at lower faunal densities.

Application of the Simpson diversity index (1949), $C = (y/N)^2$, where

y = number of individuals in species N = total number of individuals, and C = measure of concentration of dominance, to these data does show an increase in diversity with sample size. The rate of change is decidedly less than when α or d indexes are used. This does not mean that the Simpson index is necessarily a more valid diversity measurement. The formula for calculating the diversity index, as shown by Williams (1964), greatly exaggerates the contributions of the few abundant species, while the influence of the many species with few individuals is insignificant. Thus there can be little increase in the diversity index as additional rare species are added. This is the explanation for the relatively low rate of diversity change with increasing sample size.

The degree of exaggeration by this methodology can be demonstrated by comparing the Simpson index per 100 individuals with the number of species for 100 individuals as determined by the rarefaction method for a number of the samples used in our study (Fig. 15). The maximum value by the rarefaction method is 4.93 times greater than the minimum value. Yet the maximum-to-minimum ratio, using the Simpson index, is 16.5:1.00.

We also compared the species numbers using the rarefaction method with the MacArthur model (1957):

$$\frac{N}{S} = \sum_{i=1}^{r} \frac{1}{(S - i + 1)},$$

where N = number of individuals, S = number of species, i = interval between successively ranked species and rarest, and r = rank in rareness. What we measured was the deviation in percentage composition of our actual and rarefied samples from the expected percentage composition derived from the MacArthur model for the same number of species. Like Hairston (1959) and others, we found the common species to be more common and the rare species to be rarer than expected from the model.

As shown in Table 3, there was a strong tendency within each environ-

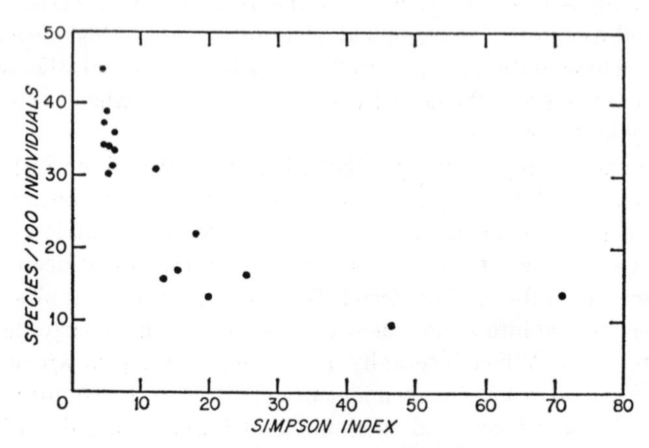

FIG. 15.—Measure of the exaggeration obtained by the Simpson index as compared with rarefaction methodology at the 100-individual level.

TABLE 3

PERCENTAGE DEVIATION OF THE RAREFACTION DATA FROM THE MACARTHUR MODEL FOR THE ENTIRE SAMPLE AND AT THE 100-INDIVIDUAL LEVEL FOR CERTAIN OF THE STATIONS

Station	Species Number	Number of Individuals	Deviation for Total Sample	Deviation/100 Individuals	Type of Environment
RH-26...	111	6,664	64.58	35.77	Shallow tropical marine
RH-30...	110	5,418	76.28	26.62	Shallow tropical marine
RH-36...	78	717	55.42	33.82	Shallow tropical marine
RH-51...	39	285	17.90	9.04	Shallow tropical marine
RH-14...	33	1,203	98.43	64.86	Stress shallow tropical marine
RH-41...	30	794	72.27	41.21	Stress shallow tropical marine
RH-28...	21	705	66.39	46.93	Tropical estuary
RH-33...	26	696	47.79	21.43	Tropical estuary
S1-3.....	96	9,716	89.32	49.23	Deep sea
D.......	79	4,142	63.12	18.51	Deep sea
DR-33*..	47	2,171	156.76	124.28	Deep sea
G.......	62	807	50.59	19.77	Deep sea
F.......	67	752	49.75	32.78	Deep sea
DR-12...	61	449	34.63	15.84	Deep sea
DR-33′...	46	338	36.87	29.07	Deep sea
GH......	45	276	21.67	17.14	Deep sea
C........	51	1,861	89.69	56.55	Outer continental shelf
1RA.....	21	2,514	119.61	97.27	Shallow boreal marine
7RB.....	19	1,976	109.05	44.41	Shallow boreal marine
12RA....	20	959	104.27	64.79	Shallow boreal marine
10RA....	20	836	103.35	63.70	Shallow boreal marine
2RB.....	16	465	98.17	83.20	Shallow boreal marine
17R.....	21	465	93.23	70.67	Shallow boreal marine

* The gross difference between this sample and the other deep-sea samples is caused by the pronounced aggregation of a single species (for further comments, see text).

ment for progressively smaller samples to show better agreement (less deviation) with the model. The rarefied samples containing 100 individuals always gave a better fit than the actual samples from which they were derived. Finally, when a large number of rarefied samples of different sample sizes from a common sample were compared (Fig. 16), the larger the sample size (except at very low numbers), the greater the departure from the model.

Thus even in homogeneous environments, the MacArthur model is markedly density dependent. For a given faunal density in Figure 16 and Table 3, samples from high-stress environments showed greater deviations from the model than samples from low-stress environments. Such effects are readily masked by sample size. In this regard, it is more than a coincidence that one of the few studies giving a good fit to the MacArthur model (Kohn, 1959) was based on small samples from a stable, low-stress environment. In all fairness to MacArthur, it should be pointed out that he has recently (1966) disavowed the validity of this index.

Lloyd and Ghelardi (1964) proposed the equitability concept as a measure of how a sample is apportioned among its constituent species. Because

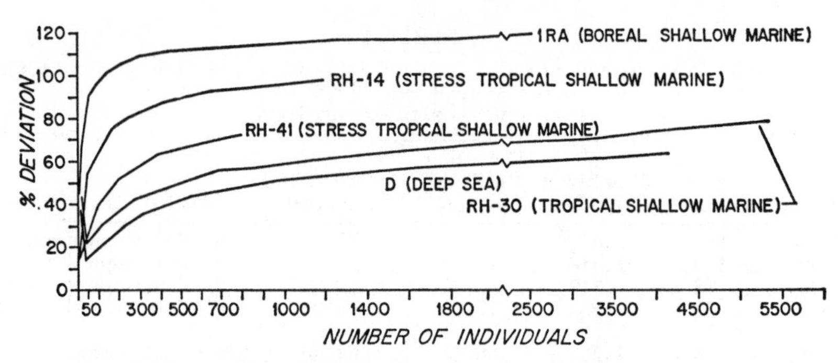

FIG. 16.—Deviation from the MacArthur model using the rarefaction method for certain stations of this study.

numerical equality is never achieved in practice, they suggested "equitability" rather than "evenness" as a more realistic standard. The Shannon-Wiener information function:

$$H(s) = -\sum_{r=1}^{s} p_r \log_2 p_r,$$

where s = total number of species and p_r = observed proportion of individuals that belong to the rth species ($r = 1, 2, \ldots, s$), provides the basis for this measure when combined with some theoretical distribution of abundances, in their case, the MacArthur model shown above.

From this model, given an s, a hypothetical diversity function can be calculated:

$$M(s) = -\sum_{r=1}^{s} \pi_r \log_2 \pi_r,$$

where π_r is the theoretical proportion of individuals in the rth species ranked in order of increasing abundance from $i = 1$ to s. By setting $M(s') = H(s)$, a calculation for the number of hypothetical "equitably distributed" species, s', is obtained. Equitability (ϵ) is the ratio of the "equitably distributed" species (s') to the actual number of observed species (s), $\epsilon = s'/s$. Higher values mean greater equitability; lower values, less equitable apportionment within the sample.

In Figure 17 we have plotted the equitability value (ϵ) for a range of population sizes for a number of our samples. Since in the rarefaction procedure the percentage composition of the original sample is unaltered, we should expect the equitability value to remain constant throughout the spectrum of population sizes. Figure 17 clearly shows that this is not the case. In every sample there is a continuous decrease in the equitability value with increasing sample size. This decrease is most pronounced at the smaller population sizes, particularly in samples from physically unstable environments. Thereafter, there is a more gradual reduction in the magnitude of decrease with increasing sample size. At larger population

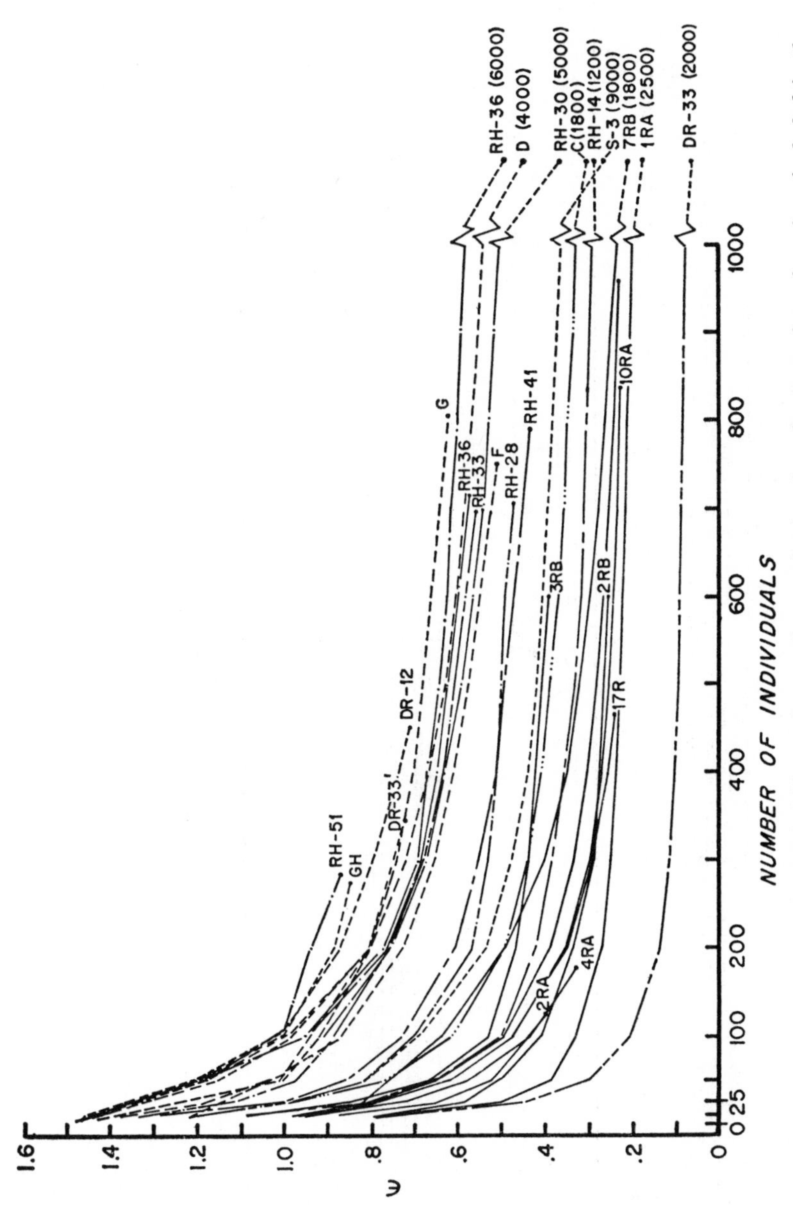

FIG. 17.—Relationship of the equitability value to the rarefaction data for a number of stations included in the study. Numbers in parentheses after the stations are the largest population levels to which the samples have been rarefied.

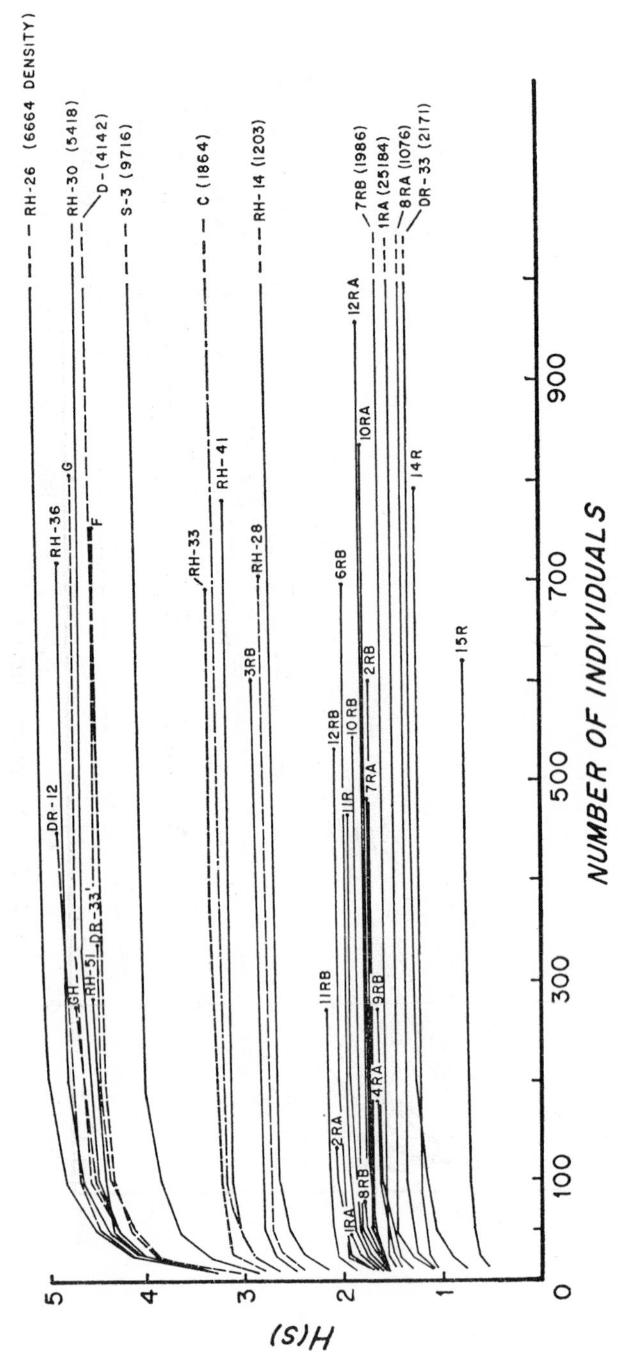

FIG. 18.—Relationship of the information function to the rarefaction data. Numbers in parentheses after the stations are the actual numbers of individuals obtained at those stations.

levels, the magnitude of decrease remains larger for samples from physically stable environments (shallow tropical marine and the deep sea).

Thus equitability is a measurement that is markedly sample-size dependent. This is not surprising when we remember that it is intimately related to the MacArthur model, which we have already shown to be highly sensitive to sample-size differences.

As a final diversity index, we will consider the Shannon-Wiener information function $H(s)$ itself. This function has the attribute of being influenced by both the number of species present and how evenly or unevenly the individuals are distributed among the constituent species. In other words, $H(s)$ is sensitive to both species and dominance diversities.

When the information function is plotted against the rarefaction data (Fig. 18), it very rapidly reaches a stable value and remains essentially constant over a broad spectrum of population sizes. Such stability is achieved at population sizes of about 200 individuals for the high-stress environments (boreal shallow marine, tropical estuary, and disturbed tropical shallow marine) and about 400 individuals for low-stress environments (tropical shallow marine and the deep sea).

Therefore, unlike the other diversity indexes tested, the information function is relatively sample-size independent, and samples of differing sizes, except at lowest faunal densities, can be directly compared.

In summary, when our series of samples were reduced to lower homogeneous population sizes by using the rarefaction method and then compared with various proposed diversity indexes, we found that most of these indexes were decidedly affected by sample size. On the other hand, the Shannon-Wiener information function, except when applied to small-sized samples, possesses the critical characteristic of a useful diversity index, that of being relatively sample-size independent.

SUMMARY

In this paper a methodology is presented for measuring diversity based on rarefaction of actual samples. By the use of this technique, a within-habitat analysis was made of the bivalve and polychaete components of soft-bottom marine faunas which differed in latitude, depth, temperature, and salinity. The resulting diversity values were highly correlated with the physical stability and past history of these environments. A stability-time hypothesis was invoked to fit these findings, and, with this hypothesis, predictions were made about the diversities present in certain other environments as yet unstudied. The two types of diversity, based on numerical percentage composition and on number of species, were compared and shown to be poorly correlated with each other. Our data indicated that species number is the more valid diversity measurement. The rarefaction methodology was compared with a number of diversity indexes using identical data. Many of these indexes were markedly influenced by sample size. Good agreement was found between the rarefaction methodology and the Shannon-Wiener information function.

ACKNOWLEDGMENTS

The ideas presented in this paper have been discussed with numerous individuals. I particularly would like to thank R. R. Hessler, J. H. Connell, and L. B. Slobodkin for their comments and criticisms. M. Rosenfeld and D. W. Spencer generously devoted many hours to both the statistical aspects of this paper and the application of appropriate computer programs. A General Electric 225 computer was used in the analyses.

Support for the acquisition of the various data used came from various sources. The tropical shallow water and estuarine samples from the Indian Ocean were collected as a result of support by the National Science Foundation as a part of the U.S. Program in Biology, International Indian Ocean Expedition. The boreal shallow-water samples from Buzzards Bay, Massachusetts, were collected during the period from 1956 to 1958 under grant NSF G-4812. Support for the collection of deep-sea samples was obtained under grants NSF G-15638, GB-3269, and GB-563. Grant NSF GB-563 also provided support for the collections of boreal estuarine samples from the Pocasset River, Massachusetts.

LITERATURE CITED

Banse, K. 1959. On upwelling and bottom-trawling off the southwest coast of India. J. Marine Biol. Ass. India 1:33–49.

Bousfield, E. L. 1958. Fresh-water amphipod crustaceans of glaciated North America. Can. Field-Natur. 72:55–113.

Carruthers, J. N., S. S. Gogate, J. R. Naidu, and T. Laevastu. 1959. Shoreward upslope of the layer of oxygen minimum off Bombay: Its influence on marine biology, especially fisheries. Nature 183:1084–1087.

Connell, J. H. 1961. Effects of competition, predation by *Thais lapillus*, and other factors on natural populations of the barnacle *Balanus balanoides*. Ecol. Monogr. 31:61–104.

Connell, J. H., and E. Orias. 1964. The ecological regulation of species diversity. Amer. Natur. 98:399–414.

Dobzhansky, T. 1950. Evolution in the tropics. Amer. Sci. 38:209–221.

Dunbar, M. J. 1960. The evolution of stability in marine environments. Natural selection at the level of the ecosystem. Amer. Natur. 94:129–136.

Fischer, A. G. 1960. Latitudinal variations in organic diversity. Evolution 14:64–81.

Fisher, R. A., A. S. Corbet, and C. B. Williams. 1943. The relation between the number of species and the number of individuals in a random sample of an animal population. J. Anim. Ecol. 12:42–58.

Gallardo, A. 1963. Notas sobre la densidad de la fauna bentonica en el sublitoral del norte de Chile. Gayana Zool. 10:3–15.

Gleason, H. A. 1922. On the relation between species and area. Ecology 3:158–162.

Grassle, J. F. 1967. Influence of environmental variation on species diversity in benthic communities on the continental shelf and slope. Unpublished Ph.D. dissertation. Duke Univ., Durham, N.C.

Hairston, N. G. 1959. Species abundance and community organization. Ecology 40:404–416.

Hairston, N. G., and G. W. Byers. 1954. The soil arthropods of a field in southern Michigan. A study in community ecology. Contrib. Lab. Vertebrate Biol. Univ. Michigan 64:1–37.

Hessler, R. R., and H. L. Sanders. 1967. Faunal diversity in the deep-sea. Deep-Sea Res. 14:65–78.

Hutchins, L. W. 1947. The basis for temperature zonation in geographical distribution. Ecol. Monogr. 17:325–335.

Jacob, J., and K. Rangarajan. 1959. Seasonal cycles of hydrological events in the Vellar estuary. First All-India Congr. Zool., Proc., Part 2, Scientific Papers, p. 329–350.

Klopfer, P. H. 1959. Environmental determinants of faunal diversity. Amer. Natur. 93: 337–342.

Kohn, A. J. 1959. The ecology of *Conus* in Hawaii. Ecol. Monogr. 29:47–90.

Kohzov, M. 1963. Lake Baikal and its life. Monogr. Biol. 11. 352 p.

LaFond, E. C. 1958. Seasonal cycle of the sea surface temperatures and salinities along the east coast of India. Andhra Univ. Mem. Oceanogr. 2: 12–21.

Lloyd, M., and R. J. Ghelardi. 1964. A table for calculating the "Equitability" component of species diversity. J. Anim. Ecol. 33:217–225.

MacArthur, R. H. 1957. On the relative abundance of bird species. Nat. Acad. Sci. Proc. 43:293–295.

———. 1965. Patterns of species diversity. Biol. Rev. 40:510–533.

———. 1966. Note on Mrs. Pielou's comments. Ecology 47:1074.

Margalef, R. 1957. La teoria de la informacion en ecologia. Memorias de la real academia de ciencias y artes (Barcelona) 33:373–449.

Murty, C. B. 1958. On the temperature and salinity structures of the Bay of Bengal. Current Sci. 27:249.

Odum, H. T., J. E. Cantlon, and L. S. Kornicker. 1960. An organizational hierarchy postulate for the interpretation of species individual distributions, species entropy, ecosystem evolution and the meaning of a species variety index. Ecology 41:395–399.

Paine, R. T. 1966. Food web complexity and species diversity. Amer. Natur. 100: 65–75.

Panikkar, N. K. 1940. Influence of temperature on osmotic behavior of some crustacea and its bearing on problems of animal distribution. Nature 146:366–367.

Pianka, E. R. 1966. Latitudinal gradients in species diversity: A review of concepts. Amer. Natur. 100:33–46.

Preston, F. W. 1948. The commonness and rarity of species. Ecology 29:254–283.

Ramamurthy, S. 1953. Hydrobiological studies in Madras coastal waters. J. Madras Univ., Series B. 23:148–163.

Rawson, D. S. 1953. The bottom fauna of Great Slave Lake. J. Fisheries Res. Board Can. 10:486–520.

Sanders, H. L. 1960. Benthic studies in Buzzards Bay. III. The structure of the soft-bottom community. Limnol. Oceanogr. 5:138–153.

———. 1963. Components of ecosystems, p. 86–91. *In* Gordon A. Riley [ed.] Marine Biology I. First Int. Interdisciplinary Conf., Proc. Port City Press, Baltimore.

Sanders, H. L., R. R. Hessler, and G. R. Hampson. 1965. An introduction to the study of the deep-sea benthic faunal assemblages along the Gay Head–Bermuda transect. Deep-Sea Res. 12:845–867.

Sanders, H. L., P. C. Mangelsdorf, Jr., and G. R. Hampson. 1965. Salinity and faunal distribution in the Pocasset River, Massachusetts. Limnol. Oceanogr. 10 (Suppl.):R216–R228.

Simpson, E. H. 1949. Measurement of diversity. Nature 163:688.

Simpson, G. G. 1964. Species density of North American recent mammals. Syst. Zool. 13:57–73.

Thorson, G. 1952. Zur jetzigen Lage der marinen Bodentier-Ökologie: Verhandlungen Deut. Zool. Ges. Wilhelmshaven 1952, p. 276–327.

———. 1966. Some factors influencing the recruitment and establishment of marine benthic communities. Neth. J. Sea Res. 3:267–293.

Whittaker, R. H. 1965. Dominance and diversity in land plant communities. Science 147:250–260.

Williams, C. B. 1964. Patterns in the balance of nature. Academic, New York. 324 p.

Potential Productivity of the Sea

Organic production by marine plankton algae is comparable to agricultural yields on land.

John H. Ryther

Under ideal conditions for photosynthesis and growth, what is the maximum potential rate of production of organic matter in the sea? Is this potential ever realized, or even approached? How does the sea compare with the land in this respect? These questions may be approached empirically with some measure of success but, aside from the time and effort required by this method, one can never be certain how close to the optimum a given environment may be and, hence, to what extent the biotic potential is realized.

However, we do know with some degree of certainty the maximum photosynthetic efficiency of plants under carefully controlled laboratory conditions; and there is a considerable literature concerning the effects of various environmental conditions on photosynthesis, respiration, and growth, particularly with respect to the unicellular algae. From such information it should be possible to estimate photosynthetic efficiencies and, for given amounts of solar radiation, organic production under natural conditions. This indirect and theoretical approach cannot be expected to provide exact values, but it does furnish a supplement to the empirically derived data which may help substantiate our concepts both of the environmental physiology of the plankton algae and the level of organic production in the sea.

An attempt has been made to use this joint approach for the marine environment in the following discussion. The only variable considered is light, and the assumption is made that virtually all of the light which enters the water (and remains) is absorbed by plants. Such situations are closely approximated in plankton blooms, dense stands of benthic algae, eelgrass, and other plants. For the rest, it is assumed that temperature, nutrients, and other factors are optimal, or at least as favorable as occur under ideal culture conditions. Given these conditions, I have attempted to calculate the organic yields which might be expected within the range of solar radiation incident to most of the earth. These data are then compared with maximal and mean observed values in the marine environment and elsewhere, and an attempt is made to explain discrepancies.

The calculations which appear below are based, for the most part, upon experimentally derived relationships between unicellular algae and the environment, and are therefore applicable only to this group. This must be kept in mind when, later in the discussion, comparisons are drawn between the theoretical yields and observed values of production by larger aquatic and terrestrial plants.

The values for the efficiency of photosynthesis under natural conditions are based on the utilization of the visible portion of the solar spectrum only (400 to 700 mμ), or roughly half of the total incident radiation. In converting these efficiencies to organic yields, it is assumed that the heat of combustion of the dry plant material is 5.5 kcal per gram, which closely approximates values for unicellular algae reported by Krogh and Berg (*1*), Ketchum and Redfield (*2*), Kok (*3*), Aach (*4*), Wassink *et al.* (*5*), and others.

The author is on the staff of the Woods Hole Oceanographic Institution, Woods Hole, Mass. This article is based on a paper presented by the author at the AAAS meeting in Washington, D.C., December 1958.

Reflection and Backscattering

Of the sunlight which strikes the surface of the ocean, a certain fraction is reflected from its surface and never enters the water. The remainder penetrates to depths which depend upon the concentration of absorbing and scattering particles or dissolved colored substances. While scattering may be as important as absorption in the vertical attenuation of the light, it makes little difference as far as the biological utilization of the radiation is concerned, since the scattered light is eventually absorbed, with the exception of a small fraction which is backscattered up out of the water. The combined reflected and backscattered light is lost to the aquatic system; the rest remains in the water, where, under the ideal conditions postulated, it is absorbed entirely by plants.

The fraction of the incident radiation which is reflected and backscattered has been studied by Powell and Clarke (*6*), Utterback and Jorgenson (*7*), and Hulburt (*8*). The two factors have been treated separately, but they may be considered together here. Their combined effect is rather small, ranging from about 3 to 6 percent, depending somewhat upon who made the measurements and the conditions under which the measurements were made. The highest values were observed when the sky was overcast. Sea states, ranging from flat calm to whitecap conditions, made surprisingly little difference. Reflection and backscattering were also found by Hulburt to be independent of the sun's angle, despite the fact that reflection increases greatly with the angle (from the zenith) of the incident light, particularly at angles above 60°. The explanation for this apparent contradiction lies in the fact that as the sun approaches the horizon, indirect sky light becomes increasingly important, and it eventually exceeds the intensities of the sun itself.

Hulburt's data also indicate that backscattering is not greatly influenced by the amount of particulate matter in the water, since his values in the clear Gulf Stream did not differ appreciably from those made in the turbid waters of Chesapeake Bay.

For the calculations which are made here, it is considered that an average of 5 percent of the incident radiation is lost through the combined effects of reflection and backscattering.

Photosynthesis and the Visible Spectrum

We first consider the efficiency of photosynthesis in sunlight at levels below the saturation intensity. Within this range, photosynthesis is directly proportional to the light intensity (or very nearly so), and the efficiency is therefore constant.

Despite the vast numbers of studies of quantum yield (that is, photosynthetic efficiency) in the literature, few data are available for the entire visible spectrum. Figure 1*A* shows two such series of measurements, one with the green alga *Chlorella* (Emerson and Lewis, *9*), the other with the diatom *Navicula minima* (Tanada, *10*). The ordinate is expressed as quantum requirement (the number of quanta required to reduce 1 mole of CO_2) rather than a reciprocal, quantum yield (moles of CO_2 reduced per quantum) as shown originally by the authors. Although the two organisms have strikingly different pigment complements, the curves are surprisingly similar, with minimal requirements in the red and yellow parts of the spectrum, maximal in the blue-green. *Navicula* appears to be somewhat more efficient than *Chlorella,* but the differences may not be significant.

Figure 1*B* illustrates the fact that the energy per quantum between 400 and 700 mμ decreases from a maximum of 71 kcal per mole quanta of blue light to 41 cal per mole quanta of red light. The heat of combustion of one reduced mole of CO_2 (reduced to CH_2O) is 112 kcal. A quantum requirement of 10 therefore represents an efficiency of $112/(41 \times 10)$, $= 27.3$ percent in red light and $112/(71 \times 10) = 15.7$ percent in blue light. Figure 1*C* shows the efficiencies of *Chlorella* and *Navicula* throughout the visible solar spectrum.

The spectral distribution of daylight varies with solar altitude and with the water vapor, carbon dioxide, and dust content of the atmosphere. Figure 2 shows the spectral distribution of daylight under average atmospheric conditions and with an air mass of 2 (solar angle = 30° from zenith) as given by Moon (*11*).

If the curves in Fig. 1*C* are averaged and the mean efficiency for the entire visible spectrum is calculated, weighing the mean for the average spectral distribution of sunlight as given in Fig. 2, this value turns out to be 18.4 percent. Taking into consideration a 5-percent reflection and backscattering loss, the efficiency of photosynthetic utilization of visible sunlight *below saturation intensity* incident to the water surface is 17.5 percent.

In extremely turbid waters and in those containing organic stains (the "yellow substance" described by Kalle, *12*), blue and green light may be selec-

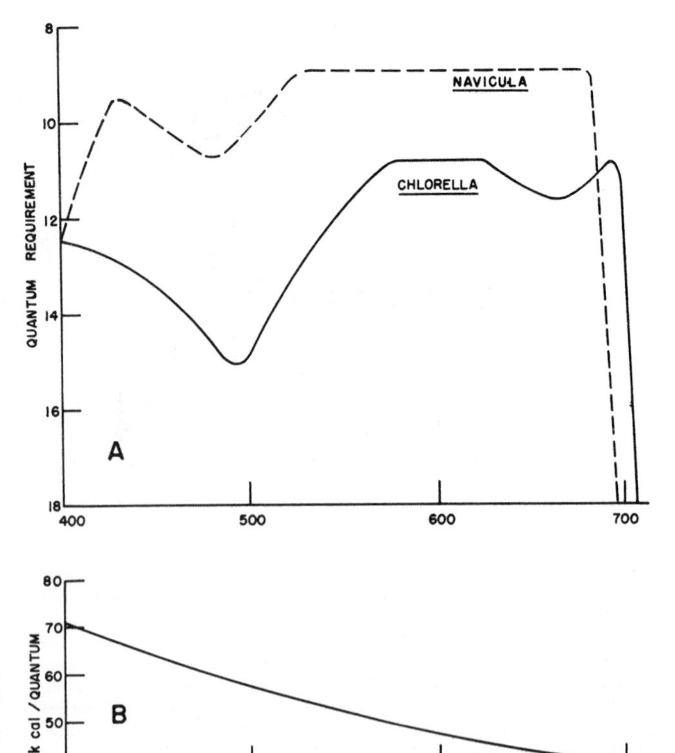

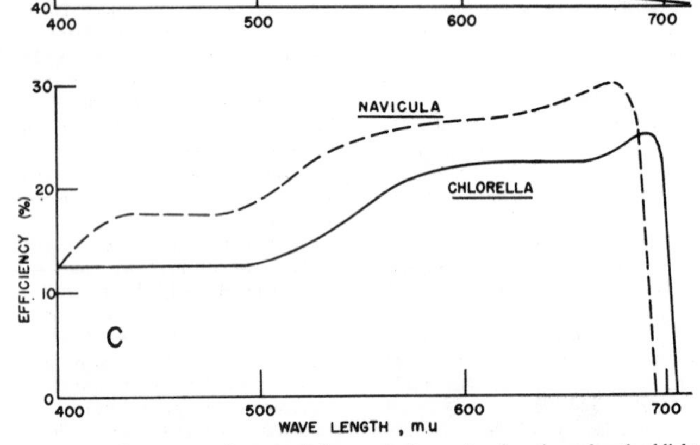

Fig. 1. (*A*) Quantum requirement of photosynthesis as a function of wavelength of light for *Chlorella* [after Emerson and Lewis, *9*] and for *Navicula* [after Tanada, *10*]. (*B*) Energy per mole quantum of light as a function of wavelength. (*C*) Efficiency of photosynthesis as a function of wavelength, calculated from (*A*) and (*B*).

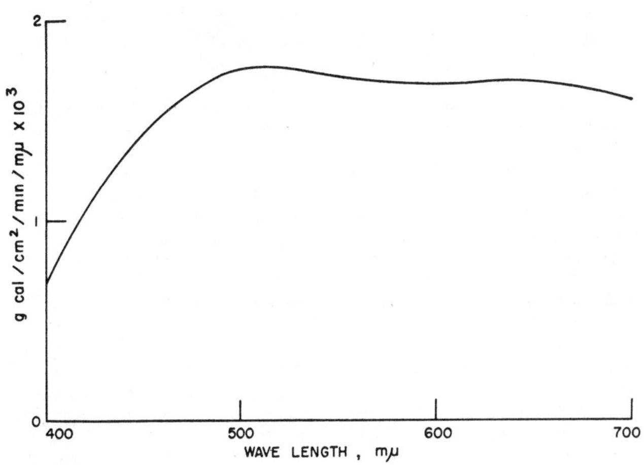

Fig. 2. The spectral distribution of daylight under average atmospheric conditions with air mass equal to 2. [After Moon, *11*]

tively absorbed, resulting in somewhat higher efficiencies in the utilization of the light penetrating to greater depths. On the other hand, in normal, clear oceanic water the red light is selectively absorbed by the water and blue-green light penetrates to the greatest depths, where it is used still less effectively than the average incident daylight considered above. These modifications are not considered in this article, since we are dealing with an idealized situation in which all of the light entering the water is absorbed by plants.

Intensity Effect

Above the saturation point, photosynthesis does not increase in proportion to light intensity, but remains constant or, at high intensities, is actually depressed, owing to photooxidation or other inhibitory processes.

Figure 3*A* shows a curve of photosynthesis by marine plankton algae as a function of light intensity, from Ryther (*13*). This is a mean curve of experiments with cultures of 14 species of organisms, preconditioned to a variety of different light regimes. Photosynthesis was measured by C^{14} uptake under solar radiation during the 4-hour period (10 A.M. to 2 P.M.) when the intensity is nearly constant and maximum. Graded intensities were obtained with neutral density filters. Almost identical curves were obtained by Steemann Nielsen and Jensen (*14*) for natural plankton populations.

Photosynthetic efficiencies remain constant, or nearly so, up to the saturation point, but then decline sharply at higher intensities. This decrease is illustrated by the difference between the actual photosynthesis curve in Fig. 3*A* and the dotted line, which is an extrapolation of the linear portion of the solid curve and represents photosynthesis if the efficiency remained constant. Figure 3*B* shows relative efficiencies as a function of light intensity, obtained from the ratio between the solid and dotted lines in Fig. 3*A*.

Using the data in Fig. 3*A*, Ryther (*13*) has calculated relative photosynthesis throughout the day and at various depths within the euphotic (illuminated) zone of the ocean for days with different values for total incident radiation. Several curves were thereby produced showing values for total daily photosynthesis at several depths within the euphotic zone relative to the hourly rate of photosynthesis at light saturation.

On extremely dull days, when the intensity never reaches the saturation region, photosynthesis is directly proportional to light intensity at all depths, and the curve of photosynthesis with depth shows an exponential decrease from the surface, as does that of light. On bright, sunny days, intensities at the surface exceed saturation and normally produce inhibition (which occurs at ⅓ or less the intensity of full sunlight). On such days, photosynthesis at the surface is less than that at intermediate depths. In all cases, photosynthesis at depths where the surface light is reduced to 10 percent or less is directly proportional to intensity, and in this region it decreases exponentially, following the light curve.

By extrapolating the lower, exponential portion of the photosynthesis curve to the surface, one may create a hypothetical curve of photosynthesis if the latter maintained the same efficiency at all depths. The ratio of the actual photosynthesis curve to this hypothetical exponential curve will then show the reduction in efficiency caused by light intensities above saturation in the upper waters. This has been done in Fig. 4 for a series of photosynthesis curves on days of varying incident radiation. Since photosynthesis at the various depths is a function of light intensity and not of depth per se, the units on the ordinate of Fig. 4 are natural logarithms of I_0/I and thus represent the depths to which given fractions of the incident radiation penetrate. The curve for the day with lowest radiation (20 gcal/cm² day) is exponential all the way to the surface, indicating that on such a day there is no reduction in photosynthetic efficiency from the effects of light intensity. On days of progressively higher light intensity, the photosynthesis curve departs more and more from the exponential curve illustrating the increasing reduction in efficiency.

If it is assumed that the maximum efficiency (with no intensity effect) is 17.5 percent, as calculated in the previous section, Fig. 5 shows the cumulative intensity effect with efficiencies plotted as a function of total daily incident radiation. The points were obtained from Fig. 4 from the ratio of the actual photosynthesis curves for each value of radiation to the exponential curve of maximum (17.5 percent efficiency. It may be seen that efficiencies decrease from 17.5 percent at low intensities to 6.5 percent on a day when 600 g cal/cm² reaches the earth's surface. It is noteworthy that the efficiency curve does not decrease in a regular way with increasing intensities, but that the rate of decrease becomes less at higher intensities. This is due to the fact that higher values of daily radiation are caused not only by higher intensities of sunlight but to an even greater extent by longer days including more hours of low intensity light.

We are now ready to calculate photosynthesis for different values of incident radiation from the efficiency curve shown in Fig. 5. This is done by multiplying the efficiency by one-half the appropriate values of radiation (that portion of the solar spectrum available for photosynthesis). This gives the amount of energy fixed in photosynthesis. Dividing this by 5.5 (the heat of combustion of a gram of average plant material, as discussed in the first section) we obtain a value which represents grams of organic matter produced per day beneath a square meter of water surface, provided that all the light entering this 1-meter-square column of water is effectively absorbed by plants. These values, shown as the upper broken line in Fig. 5, are equivalent to "real photosynthesis" or "gross production." They are hypothetical in the sense that they cannot be observed as a yield, since the plants must draw upon this organic matter to satisfy their own metabolic requirements. We must therefore subtract an amount of organic matter equivalent to the plants' respiration in order to calculate the amount of material available for harvest, the so-called "net production."

Respiratory Loss

Under conditions of active growth, photosynthesis at light saturation is some 10 to 20 times as great as dark respiration (see Ryther, *15*). Higher values have been reported, but it seems doubtful that they could represent steady-state conditions in natural populations. If we take a ratio of 15:1 as average for $P:R$ (photosynthesis: respiration) at optimal light, it is obvious that over a 24-hour period, half of which is dark, and within an entire plant community, of which many of the plants are in suboptimal light at all times, respiration must account for a much greater fraction of photosynthesis.

In calculating the ratio $P:R$ in natural communities, the oversimplified assumption will be made that respiration remains constant and independent of light and photosynthesis. While the literature pertaining to this subject is contradictory and in a state of great confusion (see, for example, Rabinowitch, *16*), there is mounting evidence that respiration and photosynthesis are not wholly independent processes. However, since there is no good quantitative formulation of a relationship between them which may be incorporated into our calculations, it must be neglected here.

As mentioned above, the data from Fig. 3*A* together with light intensity values for a group of days with varying total incident radiation have been used to calculate photosynthesis as a function of radiation. (See Ryther, *13*, for a full description of these calculations). The values given by this treatment represent photosynthesis per day beneath a square meter of surface relative to the value for photosynthesis per cubic meter per hour at light saturation. For example, a value of 100 would mean that daily photosynthesis beneath a 1-meter-square water column is 100 times as great as photosynthesis within a 1-cubic-meter aliquot of that water column for 1 hour at optimal light intensity (assuming that the plant population is evenly distributed within this water column).

Since respiration is 1/15 photosynthesis at light saturation and is also stipulated to be constant with respect to light, depth, and time of day, we may calculate total daily respiration in the same relative units as photosynthesis. The curves of photosynthesis and respiration as functions of radiation are shown in Fig. 6. They cross at 100 g cal/cm² × day, which may be considered the daily compensation level for an entire plant community. The value $(R/P) \times 100$ is the percentage of respiratory loss and is shown as the lower broken line in Fig. 6. It ranges from 100 percent at radiation values of 100 g cal/cm² day or less to 28 percent on extremely bright, long days.

Net Production

Returning to Fig. 5, gross production may be reduced by the respiratory loss (Fig. 6), giving the curve of net production, which begins at 100 g cal/cm² day and reaches a value of 25 g/m² day under radiation of 600 g cal/m² day (the lower broken line in Fig. 5).

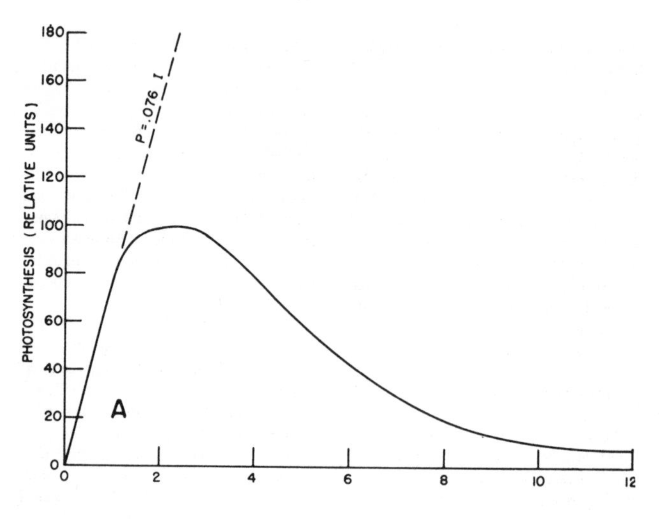

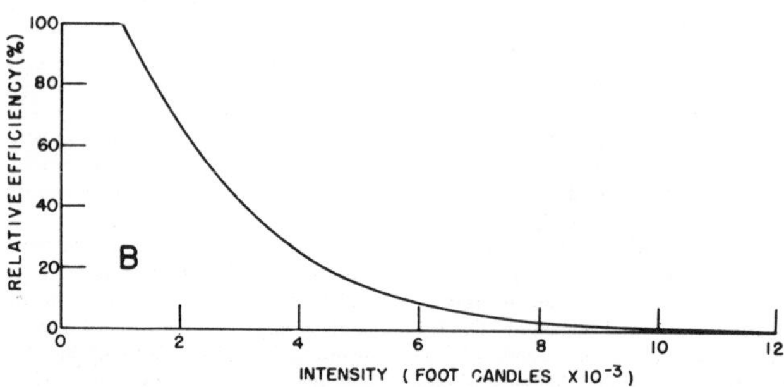

Fig. 3. (*A*) Photosynthesis of marine phytoplankton as a function of light intensity [after Ryther, *13*]. Broken line is the extrapolation of the linear portion of the solid line representing hypothetical sustained maximum photosynthetic efficiency. (*B*) Efficiency of photosynthesis as a function of light intensity, calculated from *A*.

Table 1. Gross and net organic production of various natural and cultivated systems in grams dry weight produced per square meter per day.

System	Gross	Net
A. Theoretical potential		
Average radiation (200 to 400 g cal/cm² day)	23–32	8–19
Maximum radiation (750 g cal/cm² day)	38	27
B. Mass outdoor Chlorella *culture (26)*		
Mean		12.4
Maximum		28.0
C. Land (maximum for entire growing seasons) (18)		
Sugar cane		18.4
Rice		9.1
Wheat		4.6
Spartina marsh		9.0
Pine forest (best growing years)		6.0
Tall prairie		3.0
Short prairie		0.5
Desert		0.2
D. Marine (maxima for single days)		
Coral reef (*27*)	24	(9.6)
Turtle grass flat (*28*)	20.5	(11.3)
Polluted estuary (*29*)	11.0	(8.0)
Grand Banks (Apr.) (*30*)	10.8	(6.5)
Walvis Bay (*23*)	7.6	
Continental Shelf (May) (*19*)	6.1	(3.7)
Sargasso Sea (Apr.) (*31*)	4.0	(2.8)
E. Marine (annual average)		
Long Island Sound (*32*)	2.1	0.9
Continental Shelf (*19*)	0.74	(0.40)
Sargasso Sea (*31*)	0.88	0.40

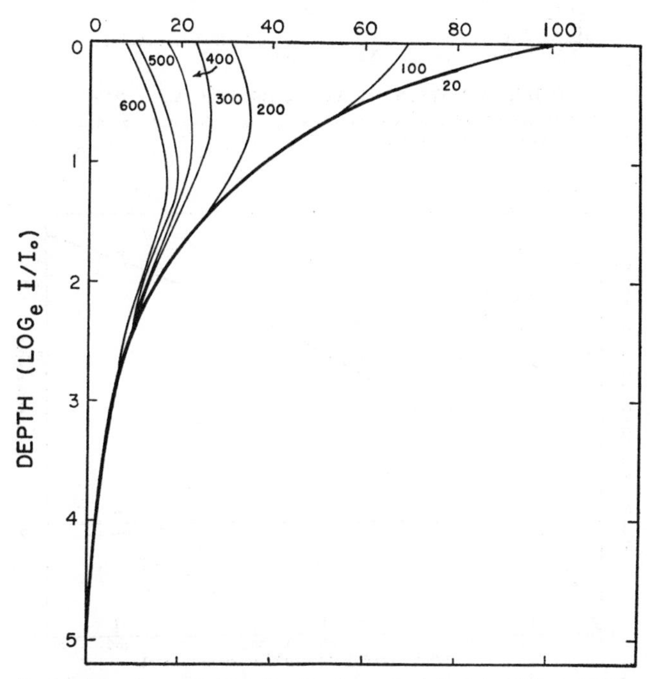

Fig. 4. Relative photosynthesis as a function of water depth for days of different incident radiation. Numbers beside curves show gram calories per square centimeter per day.

Although the annual range of daily incident radiation is extremely wide, even for a given latitude, this short-term variability is probably not very significant in affecting the general level of organic production of a given area. If one examines the tables compiled by Kimball (*17*) showing mean monthly radiation for different latitudes, it appears that over 80 percent of the data (including all latitudes and seasons) fall within a range of 200 to 400 g cal/cm² day. Thus, over most of the earth for most of the year a potential production of organic matter of some 10 to 20 g/m² day may be expected, while for shorter periods of fine summer weather, a net production of 25 g/m² day or slightly more may occur.

Comparison of Theoretical and Observed Production Rates

We may now compare the production rates which were calculated in the preceding sections with some values which have been observed empirically. Since the former are based on hypothetical situations in which all light entering the water is absorbed by plants, the observational data, to be comparable, must be restricted to natural environments in which these conditions are at least closely approximated (for example, in dense plankton blooms, thick stands of benthic algae and rooted plants). In addition to these maximal values, the theoretical potential may be contrasted with average oceanic productivity rates.

We may also extend this comparison to the terrestrial environment, including some of the better agricultural yields, bearing in mind, however, that the physiology and hence, perhaps, the biotic potential of land plants may differ significantly from those of algae.

Finally, we may include the yields of *Chlorella* grown in outdoor mass culture, drawing here upon the excellent, continuing studies of H. Tamiya and his collaborators. These are of particular interest, since the conditions of these experiments were as optimal as possible and since the physiology of *Chlorella* is identical or closely similar to that of the organisms upon which our calculations are based. Thus the *Chlorella* yields will serve as a check for the theoretical production rates.

It is important, in making these comparisons, to keep in mind the distinction between gross and net production as defined above. Some of the data refer to true photosynthesis measurements (gross production) while others, sucn as the *Chlorella* experiments and the agricultural yields, are based on the actual harvest of organic matter (net production). In those cases in which only gross production values are available and where radiation data are given, net production has been obtained from Figure 5 and is shown in parentheses.

The theoretical production potential for average and maximal radiation, and the observational data for both marine and terrestrial environments, are given in Table 1. In each case the original source is given, except for the land values, where reference is made to the recent compilation by Odum (*18*). The various methods by which the values were obtained will not be discussed here except in the case of the unpublished data, in which gross production was calculated from chlorophyll and light, according to the method of Ryther and Yentsch (*19*) and net production was measured by the C^{14} method, uncorrected for respiration as this method is interpreted by Ryther (*20*). Where gross production (photosynthesis) was originally reported as oxygen evolution, this has been converted to carbon assimilation, using an assimilatory quotient

$$\left(\Delta\frac{+O_2}{-CO_2}\right)$$

of 1.25 (see Ryther, *20*). Carbon uptake, in turn, has been converted to total organic production by assuming that the latter is 50 percent carbon by weight.

The maximal values for the marine environment represent the seven highest such values known to me. In addition to these, data are given for three regions (one inshore, one coastal, and one offshore) which have been studied over long enough periods of time to justify the calculation of annual means.

Discussion

The mean yield of *Chlorella* obtained by the Japanese workers is almost identical to the mean theoretical production for days of average radiation (12.4 versus 13.5 g/m² day). These yields of *Chlorella* were produced only during the warmer part of the year, presumably owing to the poor growth of *Chlorella* at low temperatures. The highest yields of *Chlorella* (up to 28 g/m² day) were, according to Tamiya, "obtained on fair days in the warmer months." This maximum is approximately the same as the theoretical net production for days of maximum radiation. Thus, the *Chlorella* yields agree very well with the theoretical productive potential of the sea.

The land values for net production quoted from Odum's tables range from 18.4 g/m² day for the highest yields of sugar cane to 0.2 g/m² day for deserts. The best agricultural yields are generally of the same order of magnitude as the theoretical net production of the sea, as are the values for the salt marsh and the pine forest (during its years of best growth). Uncultivated grasslands range from 3.0 for tall prairie to 0.2 for desert

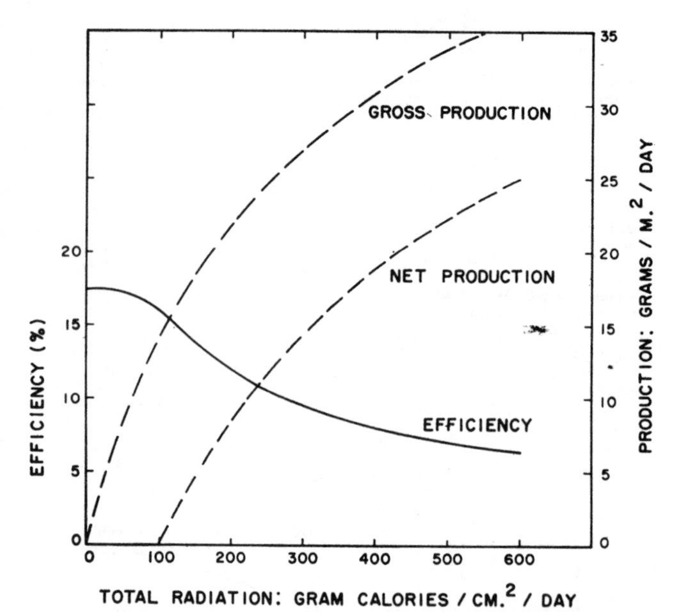

Fig. 5. Photosynthetic efficiency and theoretical maximum potential gross and net production as a function of incident radiation.

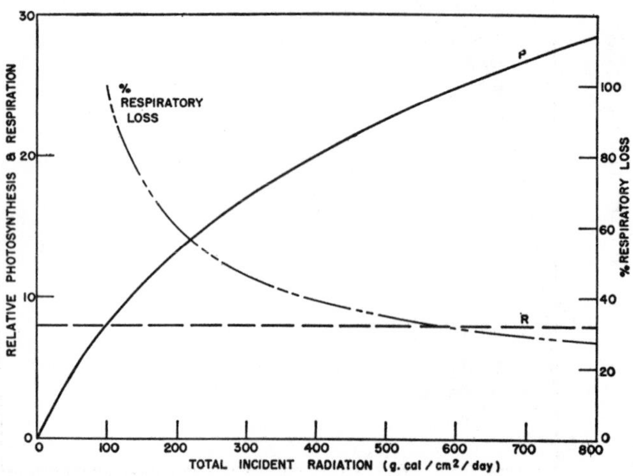

Fig. 6. Relative photosynthesis, respiration, and percentage of respiratory loss as a function of incident radiation.

conditions. Because of the extreme contrasts among terrestrial environments, mean values for the land as a whole are difficult to determine and would have little meaning. It is interesting, however, that Schroeder's estimate (*21*) of the annual production of all the land is equivalent to a mean daily production of 0.55 g/m^2, roughly the same as the value given in Table 1 for short prairie grass.

With regard to the marine data, it is perhaps surprising that net production rates differ by less than a factor of 2 in such diverse environments as a coral reef, a turtle grass flat, a polluted creek, and the Grand Banks. This alone would indicate that production in each case is limited by the same basic factor, the photosynthetic potential of the plants, and indeed these and the other high values in *D* in Table 1 all closely approach the theoretical potential.

Seasonal studies have been made of three marine areas, Long Island Sound, the continental shelf off New York, and the Sargasso Sea off Bermuda. In each case temporary rates of production were observed during the spring flowering which approached the theoretical maximum, but the annual means were more than an order of magnitude lower (*E* in Table 1). True, these regions do not, throughout the year, satisfy the postulated conditions necessary to obtain this maximum, namely, that all light entering the water be absorbed by plants. For example, in his Long Island Sound studies, Riley (*22*) found that no more than one-third of the incident radiation was utilized by plants, the remainder presumably being absorbed by nonliving particulate and dissolved materials. Using Riley's techniques, I estimated that only 25 to 40 percent of the light penetrating the continental shelf waters was absorbed by the phytoplankton. This alone, however, is insufficient to account for the discrepancy between observed and potential production rates. In the clear waters of the Sargasso Sea only 10 to 20 percent of the light is absorbed by the phytoplankton during most of the year. But there is little if any other particulate matter present; the remainder of the light is absorbed by the water itself. This is not a cause but an effect of low production. The underlying reason for low production rates here and in most parts of the ocean is the limitation of essential nutrients in the upper, euphotic layers and the inadequacy of vertical mixing processes in bringing deep, nutrient-rich water to the surface.

With the exception of the three planktonic communities which have been discussed, the seasonal cycles of marine production are largely unknown and can only be surmised. Probably high levels may be maintained throughout the year in benthic populations such as the coral reef, the turtle grass flats (see *D* in Table 1) and in thick beds of seaweeds, provided that seasonal temperature extremes do not impair growth. While the concentrations of nutrients in the surrounding waters may be very low, the fact that they are continually being replenished as the water moves over the plants probably prevents their ever being limiting. Plankton organisms, on the other hand, suspended as they are in their milieu, can probably never maintain high production rates in a given parcel of water, for their growth rapidly exhausts the nutrients from their surrounding environment and any mixing process which enriches the water must, at the same time, dilute the organisms. However, high plankton production may be sustained in a given geographic area (a polluted estuary, a region of permanent upwelling of deep water, and so forth), which is continually replenished with enriched water. In these situations, the productive capacity of the sea may be sustained for long periods, perhaps permanently.

For most of the ocean, as stated above, no such mechanism for nutrient replenishment is available. The combined meteorological and hydrographic conditions which produce the typical spring flowering of the phytoplankton over much of the oceans have been adequately described elsewhere and need not be discussed here. Suffice it to say that, in the oceans as a whole, as seasonal studies have demonstrated, high production approaching the theoretical maximum under optimal conditions is restricted to periods of a few days or, at most, weeks, per year.

Steemann Nielsen (*23*) has recently estimated the net production of the entire hydrosphere as 1.2 to 1.5×10^{10} tons of carbon per year, roughly one-tenth the earlier estimates made by Riley (*24*) and others, and about comparable to Schroeder's figure (*21*) for the land. Our production estimates are somewhat higher than those of Steemann Nielsen, the annual mean net production of organic matter for the Sargasso Sea (0.40 g/m^2 day) being about 6 times as great as his value for the same area, and twice his average for the oceans as a whole. This discrepancy appears to be largely due to the fact that Steemann Nielsen's values are based on single observations which probably seldom included seasonal maxima. His observations in the Sargasso Sea, for example, were made in June and did not differ greatly from our June values, which were the seasonal minima. If the Sargasso Sea is one of the less fertile parts of the ocean, as is generally believed, then our data would indicate that the seas are more than twice as productive as the land (*25*).

References and Notes

1. A. Krogh and K. Berg, *Intern. Rev. ges. Hydrobiol. Hydrog.* 25, 205 (1931).
2. B. H. Ketchum and A. C. Redfield, *J. Cellular Comp. Physiol.* 33, 281 (1949).
3. B. Kok, *Acta Botan. Neerl.* 1, 445 (1952).
4. H. G. Aach, *Arch. Mikrobiol.* 17, 213 (1952).
5. E. C. Wassink, B. Kok, J. L. P. van Oorschot, "The efficiency of light-energy conversion in *Chlorella* cultures as compared with higher plants," in "Algal Culture from Laboratory to Pilot Plant," *Carnegie Inst. Wash. Publ. No. 600* (1953), pp. 55–62.
6. W. M. Powell and G. L. Clarke, *J. Opt. Soc. Am.* 26, 111 (1936).
7. C. L. Utterback and W. Jorgensen, *ibid.* 26, 257 (1936).
8. E. O. Hulburt, *ibid.* 35, 698 (1945).
9. R. Emerson and C. M. Lewis, *Am. J. Botany* 30, 165 (1943).
10. T. Tanada, *ibid.* 39, 276 (1951).
11. P. Moon, *J. Franklin Inst.* 230, 583 (1940).
12. K. Kalle, *Ann. Hydrog. mar. Meteor.* 66, 1 (1938).
13. J. H. Ryther *et al.*, *Biol. Bull.* 115, 257 (1958).
14. E. Steemann Nielsen and E. A. Jensen, *Galathea Repts.* 1, 49 (1957).
15. J. H. Ryther, *Deep-Sea Research* 2, 134 (1954).
16. E. I. Rabinowitch, *Photosynthesis and Related Processes* (Interscience, New York, 1956), vol. 2, part 2, pp. 1925–1939.
17. H. H. Kimball, *Monthly Weather Rev.* 56, 393 (1928).
18. E. P. Odum, *Fundamentals of Ecology* (Saunders, Philadelphia, ed. 2, 1959).
19. J. H. Ryther and C. S. Yentsch, *Limnol. Oceanog.* 2, 281 (1957).
20. ———, *ibid.* 1, 72 (1956).
21. H. Schroeder, *Naturwissenschaften* 7, 8 (1919).
22. G. A. Riley, *Bull. Bingham Oceanog. Coll.* 15, 15 (1956).
23. E. Steemann Nielsen, *J. conseil, Conseil permanent intern. exploration mer* 19, 309 (1954).
24. G. A. Riley, *Bull. Bingham Oceanog. Coll.* 7, 1 (1941).

25. This paper is contribution No. 1016 of the Woods Hole Oceanographic Institution. The work was supported in part by research grant G-3234 from the National Science Foundation and under contract AT (30-1)-1918 with the Atomic Energy Commission.

26. H. Tamiya, *Ann. Rev. Plant Physiol.* 8, 309 (1957).
27. H. T. Odum and E. P. Odum, *Ecol. Monographs* 25, 291 (1955).
28. H. T. Odum, *Limnol. Oceanog.* 2, 85 (1957).
29. J. H. Ryther *et al.*, *Biol. Bull.* 115, 257 (1958).
30. J. H. Ryther and C. S. Yentsch, unpublished data.
31. J. H. Ryther and D. W. Menzel, unpublished data.
32. G. A. Riley, *Bull. Bingham Oceanog. Coll.* 15, 324 (1956).

Photosynthesis and Fish Production in the Sea

The production of organic matter and its conversion to higher forms of life vary throughout the world ocean.

John H. Ryther

Numerous attempts have been made to estimate the production in the sea of fish and other organisms of existing or potential food value to man (*1–4*). These exercises, for the most part, are based on estimates of primary (photosynthetic) organic production rates in the ocean (*5*) and various assumed trophic-dynamic relationships between the photosynthetic producers and the organisms of interest to man. Included in the latter are the number of steps or links in the food chains and the efficiency of conversion of organic matter from each trophic level or link in the food chain to the next. Different estimates result from different choices in the number of trophic levels and in the efficiencies, as illustrated in Table 1 (*2*).

Implicit in the above approach is the concept of the ocean as a single ecosystem in which the same food chains involving the same number of links and efficiencies apply throughout. However, the rate of primary production is known to be highly variable, differing by at least two full orders of magnitude from the richest to the most impoverished regions. This in itself would be expected to result in a highly irregular pattern of food production. In addition, the ecological conditions which determine the trophic dynamics of marine food chains also vary widely and in direct relationship to the absolute level of primary organic production. As is shown below, the two sets of variables—primary production and the associated food chain dynamics—may act additively to produce differences in fish production which are far more pronounced and dramatic than the observed variability of the individual causative factors.

The author is a member of the staff of the Woods Hole Oceanographic Institution, Woods Hole, Massachusetts.

Primary Productivity

Our knowledge of the primary organic productivity of the ocean began with the development of the C^{14}-tracer technique for *in situ* measurement of photosynthesis by marine plankton algae (*6*) and the application of the method on the 1950–52 *Galathea* expedition around the world (*5*). Despite obvious deficiencies in the coverage of the ocean by *Galathea* (the expedition made 194 observations, or an average of about one every 2 million square kilometers, most of which were made in the tropics or semitropics), our concept of the total productivity of the world ocean has changed little in the intervening years.

While there have been no more expeditions comparable to the *Galathea*, there have been numerous local or regional studies of productivity in many parts of the world. Most of these have been brought together by a group of Soviet scientists to provide up-to-date world coverage consisting of over 7000 productivity observations (*7*). The result has been modification of the estimate of primary production in the world ocean from 1.2 to 1.5×10^{10} tons of carbon fixed per year (*5*) to a new figure, 1.5 to 1.8×10^{10} tons.

Attempts have also been made by Steemann Nielsen and Jensen (*5*), Ryther (*8*), and Koblentz-Mishke *et al.* (*7*) to assign specific levels or ranges of productivity to different parts of the ocean. Although the approach was somewhat different in each case, in general the agreement between the three was good and, with appropriate condensation and combination, permit the following conclusions.

1) Annual primary production in the open sea varies, for the most part, between 25 and 75 grams of carbon fixed per square meter and averages about 50 grams of carbon per square meter per year. This is true for roughly 90 percent of the ocean, an area of 326×10^6 square kilometers.

2) Higher levels of primary production occur in shallow coastal waters, defined here as the area within the 100-fathom (180-meter) depth contour. The mean value for this region may be considered to be 100 grams of carbon fixed per square meter per year, and the area, according to Menard and Smith (*9*), is 7.5 percent of the total world ocean. In addition, certain offshore waters are influenced by divergences, fronts, and other hydrographic features which bring nutrient-rich subsurface water into the euphotic zone. The equatorial divergences are examples of such regions. The productivity of these offshore areas is comparable to that of the coastal zone. Their total area is difficult to assess, but is considered here to be 2.5 percent of the total ocean. Thus, the coastal zone and the offshore regions of comparably high productivity together represent 10 percent of the total area of the

oceans, or 36 × 10^6 square kilometers.

3) In a few restricted areas of the world, particularly along the west coasts of continents at subtropical latitudes where there are prevailing offshore winds and strong eastern boundary currents, surface waters are diverted offshore and are replaced by nutrient-rich deeper water. Such areas of coastal upwelling are biologically the richest parts of the ocean. They exist off Peru, California, northwest and southwest Africa, Somalia, and the Arabian coast, and in other more localized situations. Extensive coastal upwelling also is known to occur in various places around the continent of Antarctica, although its exact location and extent have not been well documented. During periods of active upwelling, primary production normally exceeds 1.0 and may exceed 10.0 grams of carbon per square meter per day. Some of the high values which have been reported from these locations are 3.9 grams for the southwest coast of Africa (*5*), 6.4 for the Arabian Sea (*10*), and 11.2 off Peru (*11*). However, the upwelling of subsurface water does not persist throughout the year in many of these places—for example, in the Arabian Sea, where the process is seasonal and related to the monsoon winds. In the Antarctic, high production is limited by solar radiation during half the year. For all these areas of coastal upwelling throughout the year, it is probably safe, if somewhat conservative, to assign an annual value of 300 grams of carbon per square meter. Their total area in the world is again difficult to assess. On the assumption that their total cumulative area is no greater than 10 times the well-documented upwelling area off Peru, this would amount to some 3.6 × 10^5 square kilometers, or 0.1 percent of the world ocean. These conclusions are summarized in Table 2.

Food Chains

Let us next examine the three provinces of the ocean which have been designated according to their differing levels of primary productivity from the standpoint of other possible major differences. These will include, in particular, differences which relate to the food chains and to trophic efficiencies involved in the transfer of organic matter from the photosynthetic organisms to fish and invertebrate species large and abundant enough to be of importance to man.

The first factor to be considered in this context is the size of the photosynthetic or producer organisms. It is generally agreed that, as one moves from coastal to offshore oceanic waters, the character of these organisms changes from large "microplankton" (100 microns or more in diameter) to the much smaller "nannoplankton" cells 5 to 25 microns in their largest dimensions (*12, 13*).

Since the size of an organism is an essential criterion of its potential usefulness to man, we have the following relationship: the larger the plant cells at the beginning of the food chain, the fewer the trophic levels that are required to convert the organic matter to a useful form. The oceanic nannoplankton cannot be effectively filtered from the water by most of the common zooplankton crustacea. For example, the euphausid *Euphausia pacifica*, which may function as a herbivore in the rich subarctic coastal waters of the Pacific, must turn to a carnivorous habit in the offshore waters where the phytoplankton become too small to be captured (*13*).

Intermediate between the nannoplankton and the carnivorous zooplankton are a group of herbivores, the microzooplankton, whose ecological significance is a subject of considerable current interest (*14, 15*). Representatives of this group include protozoans such as Radiolaria, Foraminifera, and Tintinnidae, and larval nuplii of microcrustaceans. These organisms, which may occur in concentrations of tens of thousands per cubic meter, are the primary herbivores of the open sea.

Feeding upon these tiny animals is a great host of carnivorous zooplankton, many of which have long been thought of as herbivores. Only by careful study of the mouthparts and feeding habits were Anraku and Omori (*16*) able to show that many common copepods are facultative if not obligate carnivores. Some of these predatory copepods may be no more than a millimeter or two in length.

Again, it is in the offshore environment that these small carnivorous zooplankton predominate. Grice and Hart (*17*) showed that the percentage of carnivorous species in the zooplankton increased from 16 to 39 percent in a transect from the coastal waters of the northeastern United States to the Sargasso Sea. Of very considerable importance in this group are the Chaetognatha. In terms of biomass, this group of animals, predominantly carnivorous, represents, on the average, 30 percent of the weight of copepods in the open sea (*17*). With such a distribution, it is clear that virtually all the copepods, many of which are themselves carnivores, must be preyed upon by chaetognaths.

Table 1. Estimates of potential yields (per year) at various trophic levels, in metric tons. [After Schaeffer (*2*)]

Trophic level	Ecological efficiency factor					
	10 percent		15 percent		20 percent	
	Carbon (tons)	Total weight (tons)	Carbon (tons)	Total weight (tons)	Carbon (tons)	Total weight (tons)
0. Phytoplankton (net particulate production)	1.9 × 10^{10}		1.9 × 10^{10}		1.9 × 10^{10}	
1. Herbivores	1.9 × 10^9	1.9 × 10^{10}	2.8 × 10^9	2.8 × 10^{10}	3.8 × 10^9	3.8 × 10^{10}
2. 1st stage carnivores	1.9 × 10^8	1.9 × 10^9	4.2 × 10^8	4.2 × 10^9	7.6 × 10^8	7.6 × 10^9
3. 2nd stage carnivores	1.9 × 10^7	1.9 × 10^8	6.4 × 10^7	6.4 × 10^8	15.2 × 10^7	15.2 × 10^8
4. 3rd stage carnivores	1.9 × 10^6	1.9 × 10^7	9.6 × 10^6	9.6 × 10^7	30.4 × 10^6	30.4 × 10^7

Table 2. Division of the ocean into provinces according to their level of primary organic production.

Province	Percentage of ocean	Area (km^2)	Mean productivity (grams of carbon/m^2/yr)	Total productivity (10^9 tons of carbon/yr)
Open ocean	90	326 × 10^6	50	16.3
Coastal zone*	9.9	36 × 10^6	100	3.6
Upwelling areas	0.1	3.6 × 10^5	300	0.1
Total				20.0

* Includes offshore areas of high productivity.

Table 3. Estimated fish production in the three ocean provinces defined in Table 2.

Province	Primary production [tons (organic carbon)]	Trophic levels	Efficiency (%)	Fish production [tons (fresh wt.)]
Oceanic	16.3×10^9	5	10	16×10^5
Coastal	3.6×10^9	3	15	12×10^7
Upwelling	0.1×10^9	1½	20	12×10^7
Total				24×10^7

The oceanic food chain thus far described involves three to four trophic levels from the photosynthetic nannoplankton to animals no more than 1 to 2 centimeters long. How many additional steps may be required to produce organisms of conceivable use to man is difficult to say, largely because there are so few known oceanic species large enough and (through schooling habits) abundant enough to fit this category. Familiar species such as the tunas, dolphins, and squid are all top carnivores which feed on fishes or invertebrates at least one, and probably two, trophic levels beyond such zooplankton as the chaetognaths. A food chain consisting of five trophic levels between photosynthetic organisms and man would therefore seem reasonable for the oceanic province.

As for the coastal zone, it has already been pointed out that the phytoplankton are quite commonly large enough to be filtered and consumed directly by the common crustacean zooplankton such as copepods and euphausids. However, the presence, in coastal waters, of protozoans and other microzooplankton in larger numbers and of greater biomass than those found in offshore waters (*15*) attests to the fact that much of the primary production here, too, passes through several steps of a microscopic food chain before reaching the macrozooplankton.

The larger animals of the coastal province (that is, those directly useful to man) are certainly the most diverse with respect to feeding type. Some (mollusks and some fishes) are herbivores. Many others, including most of the pelagic clupeoid fishes, feed on zooplankton. Another large group, the demersal fishes, feed on bottom fauna which may be anywhere from one to several steps removed from the phytoplankton.

If the herbivorous clupeoid fishes are excluded (since these occur predominantly in the upwelling provinces and are therefore considered separately), it is probably safe to assume that the average food organism from coastal waters represents the end of at least a three-step food chain between phytoplankton and man.

It is in the upwelling areas of the world that food chains are the shortest, or—to put it another way—that the organisms are large enough to be directly utilizable by man from trophic levels very near the primary producers. This, again, is due to the large size of the phytoplankton, but it is due also to the fact that many of these species are colonial in habit, forming large gelatinous masses or long filaments. The eight most abundant species of phytoplankton in the upwelling region off Peru, in the spring of 1966, were *Chaetoceros socialis*, *C. debilis*, *C. lorenzianus*, *Skeletonema costatum*, *Nitzschia seriata*, *N. delicatissima*, *Schroederella delicatula*, and *Asterionella japonica* (*11*, *18*). The first in this list, *C. socialis*, forms large gelatinous masses. The others all form long filamentous chains. *Thalossiosira subtilis*, another gelatinous colonial form like *Chaetoceros socialis*, occurs commonly off southwest Africa (*19*) and close to shore off the Azores (*20*). Hart (*21*) makes special mention of the colonial habit of all the most abundant species of phytoplankton in the Antarctic—*Fragiloriopsis antarctica*, *Encampia balaustrium*, *Rhizosalenia alata*, *R. antarctica*, *R. chunii*, *Thallosiothrix antarctica*, and *Phaeocystis brucei*.

Many of the above-mentioned species of phytoplankton form colonies several millimeters and, in some cases, several centimeters in diameter. Such aggregates of plant material can be readily eaten by large fishes without special feeding adaptation. In addition, however, many of the clupeoid fishes (sardines, anchovies, pilchards, menhaden, and so on) that are found most abundantly in upwelling areas and that make up the largest single component of the world's commercial fish landings, do have specially modified gill rakers for removing the larger species of phytoplankton from the water.

There seems little doubt that many of the fishes indigenous to upwelling regions are direct herbivores for at least most of their lives. There is some evidence that juveniles of the Peruvian anchovy (*Engraulis ringens*) may feed on zooplankton, but the adult is predominantly if not exclusively a herbivore (*22*). Small gobies (*Gobius bibarbatus*) found at mid-water in the coastal waters off southwest Africa had their stomachs filled with a large, chain-forming diatom of the genus *Fragilaria* (*23*). There is considerable interest at present in the possible commercial utilization of the large Antarctic krill, *Euphausia superba*, which feeds primarily on the colonial diatom *Fragilariopsis antarctica* (*24*).

In some of the upwelling regions of the world, such as the Arabian Sea, the species of fish are not well known, so it is not surprising that knowledge of their feeding habits and food chains is fragmentary. From what is known, however, the evidence would appear to be overwhelming that a one- or two-step food chain between phytoplankton and man is the rule. As a working compromise, let us assign the upwelling province a 1½-step food chain.

Efficiency

The growth (that is, the net organic production) of an organism is a function of the food assimilated less metabolic losses or respiration. This efficiency of growth or food utilization (the ratio of growth to assimilation) has been found, by a large number of investigators and with a great variety of organisms, to be about 30 percent in young, actively growing animals. The efficiency decreases as animals approach their full growth, and reaches zero in fully mature or senescent individuals (*25*). Thus a figure of 30 percent can be considered a biological potential which may be approached in nature, although the growth efficiency of a population of animals of mixed ages under steady-state conditions must be lower.

Since there must obviously be a "maintenance ration" which is just sufficient to accommodate an organism's basal metabolic requirement (*26*), it must also be true that growth efficiency is a function of the absolute rate of assimilation. The effects of this factor will be most pronounced at low feeding rates, near the "maintenance ration," and will tend to become negligible at high feeding rates. Food conversion (that is, growth efficiency) will therefore obviously be related to food availability, or to the concentration of prey organisms when the latter are sparsely distributed.

In addition, the more available the food and the greater the quantity consumed, the greater the amount of "internal work" the animal must perform to digest, assimilate, convert, and store the food. Conversely, the less available the food, the greater the amount of "external work" the animal must perform to hunt, locate, and capture its prey. These concepts are discussed in some detail by Ivlev (*27*) and reviewed by Ricker (*28*). The two metabolic costs thus work in opposite ways with respect to food availability, tending thereby toward a constant total effect. However, when food availability is low, the added costs of basal metabolism and external work relative to assimilation may have a pronounced effect on growth efficiency.

When one turns from consideration of the individual and its physiological growth efficiency to the "ecological efficiency" of food conversion from one trophic level to the next (*2*, *29*), there are additional losses to be taken into account. Any of the food consumed but not assimilated would be included here, though it is possible that undigested organic matter may be reassimilated by members of the same trophic level (*2*). Any other nonassimilatory losses, such as losses due to natural death, sedimentation, and emigration, will, if not otherwise accounted for, appear as a loss in trophic efficiency. In addition, when one considers a specific or selected part of a trophic level, such as a population of fish of use to man, the consumption of food by any other hidden member of the same trophic level will appear as a loss in efficiency. For example, the role of such animals as salps, medusae, and ctenophores in marine food chains is not well understood and is seldom even considered. Yet these animals may occur sporadically or periodically in swarms so dense that they dominate the plankton completely. Whether they represent a dead end or side branch in the normal food chain of the sea is not known, but their effect can hardly be negligible when they occur in abundance.

Finally, a further loss which may occur at any trophic level but is, again, of unknown or unpredictable magnitude is that of dissolved organic matter lost through excretion or other physiological processes by plants and animals. This has received particular attention at the level of primary production, some investigators concluding that 50 percent or more of the photoassimilated carbon may be released by phytoplankton into the water as dissolved compounds (*30*). There appears to be general agreement that the loss of dissolved organic matter is indirectly proportional to the absolute rate of organic production and is therefore most serious in the oligotrophic regions of the open sea (*11*, *31*).

All of the various factors discussed above will affect the efficiency or apparent efficiency of the transfer of organic matter between trophic levels. Since they cannot, in most cases, be quantitatively estimated individually, their total effect cannot be assessed. It is known only that the maximum potential growth efficiency is about 30 percent and that at least some of the factors which reduce this further are more pronounced in oligotrophic, low-productivity waters than in highly productive situations. Slobodkin (*29*) concludes that an ecological efficiency of about 10 percent is possible, and Schaeffer feels that the figure may be as high as 20 percent. Here, therefore, I assign efficiencies of 10, 15, and 20 percent, respectively, to the oceanic, the coastal, and the upwelling provinces, though it is quite possible that the actual values are considerably lower.

Conclusions and Discussion

With values assigned to the three marine provinces for primary productivity (Table 2), number of trophic levels, and efficiencies, it is now possible to calculate fish production in the three regions. The results are summarized in Table 3.

These calculations reveal several interesting features. The open sea—90 percent of the ocean and nearly three-fourths of the earth's surface—is essentially a biological desert. It produces a negligible fraction of the world's fish catch at present and has little or no potential for yielding more in the future.

Upwelling regions, totaling no more than about one-tenth of 1 percent of the ocean surface (an area roughly the size of California) produce about half the world's fish supply. The other half is produced in coastal waters and the few offshore regions of comparably high fertility.

One of the major uncertainties and possible sources of error in the calculation is the estimation of the areas of high, intermediate, and low productivity. This is particularly true of the upwelling area off the continent of Antarctica, an area which has never been well described or defined.

A figure of 360,000 square kilometers has been used for the total area of upwelling regions in the world (Table 2). If the upwelling regions off California, northwest and southwest Africa, and the Arabian Sea are of roughly the same area as that off the coast of Peru, these semitropical regions would total some 200,000 square kilometers. The remaining 160,000 square kilometers would represent about one-fourth the circumference of Antarctica seaward for a distance of 30 kilometers. This seems a not unreasonable inference. Certainly, the entire ocean south of the Antarctic Convergence is not highly productive, contrary to the estimates of El-Sayed (*32*). Extensive observations in this region by Saijo and Kawashima (*33*) yielded primary productivity values of 0.01 to 0.15 gram of carbon per square meter per day—a value no higher than the values used here for the open sea. Presumably, the discrepancy is the result of highly irregular, discontinuous, or "patchy" distribution of biological activity. In other words, the occurrence of extremely high productivity associated with upwelling conditions appears to be confined, in the Antarctic, as elsewhere, to restricted areas close to shore.

An area of 160,000 square kilometers of upwelling conditions with an annual productivity of 300 grams of carbon per square meter would result in the production of about 50×10^6 tons of "fish," if we follow the ground rules established above in making the estimate. Presumably these "fish" would consist for the most part of the Antarctic krill, which feeds directly upon phytoplankton, as noted above, and which is known to be extremely abundant in Antarctic waters. There have been numerous attempts to estimate the annual production of krill in the Antarctic, from the known number of whales at their peak of abundance and from various assumptions concerning their daily ration of krill. The evidence upon which such estimates are based is so tenuous that they are hardly worth discussing. It is interesting to note, however, that the more conservative of these estimates are rather close to figures derived independently by the method discussed here. For example, Moiseev (*34*) calculated krill production for 1967 to be 60.5×10^6 tons, while Kasahara (*3*) considered a range of 24 to 36×10^6 tons to be a minimal figure. I consider the figure 50×10^6 tons to be on the high side, as the estimated area of upwelling is probably generous, the average productivity value of 300 grams of carbon per square meter per year is high for a region where photosynthesis can occur during only half the year, and much of the primary production is probably diverted into smaller crustacean herbivores (*35*). Clearly, the Antarctic must receive much more intensive study before its productive capacity can be assessed with any accuracy.

In all, I estimate that some 240 million tons (fresh weight) of fish are produced annually in the sea. As this figure is rough and subject to numerous sources of error, it should not be considered significantly different from Schaeffer's (*2*) figure of 200 million tons.

Production, however, is not equivalent to potential harvest. In the first place, man must share the production with other top-level carnivores. It has been estimated, for example, that guano birds alone eat some 4 million tons of anchovies annually off the coast of Peru, while tunas, squid, sea lions, and other predators probably consume an equivalent amount (*22, 36*). This is nearly equal to the amount taken by man from this one highly productive fishery. In addition, man must take care to leave a large enough fraction of the annual production of fish to permit utilization of the resource at something close to its maximum sustainable yield, both to protect the fishery and to provide a sound economic basis for the industry.

When these various factors are taken into consideration, it seems unlikely that the potential sustained yield of fish to man is appreciably greater than 100 million tons. The total world fish landings for 1967 were just over 60 million tons (*37*), and this figure has been increasing at an average rate of about 8 percent per year for the past 25 years. It is clear that, while the yield can be still further increased, the resource is not vast. At the present rate, the industry can continue to expand for no more than a decade.

Most of the existing fisheries of the world are probably incapable of contributing significantly to this expansion. Many are already overexploited, and most of the rest are utilized at or near their maximum sustainable yield. Evidence of fishing pressure is usually determined directly from fishery statistics, but it is of some interest, in connection with the present discussion, to compare landings with fish production as estimated by the methods developed in this article. I will make this comparison for two quite dissimilar fisheries, that of the continental shelf of the northwest Atlantic and that of the Peruvian coastal region.

According to Edwards (*38*), the continental shelf between Hudson Canyon and the southern end of the Nova Scotian shelf includes an area of 110,000 square miles (2.9×10^{11} square meters). From the information in Tables 2 and 3, it may be calculated that approximately 1 million tons of fish are produced annually in this region. Commercial landings from the same area were slightly in excess of 1 million tons per year for the 3-year period 1963 to 1965 before going into a decline. The decline has become more serious each year, until it is now proposed to regulate the landings of at least the more valuable species such as cod and haddock, now clearly overexploited.

The coastal upwelling associated with the Peru Coastal Current gives rise to the world's most productive fishery, an annual harvest of some 10^7 metric tons of anchovies. The maximum sustainable yield is estimated at, or slightly below, this figure (*39*), and the fishery is carefully regulated. As mentioned above, mortality from other causes (such as predation from guano birds, bonito, squid, and so on) probably accounts for an additional 10^7 tons. This prodigious fishery is concentrated in an area no larger than about 800×30 miles (*36*), or 6×10^{10} square meters. By the methods developed in this article, it is estimated that such an upwelling area can be expected to produce 2×10^7 tons of fish, almost precisely the commercial yield as now regulated plus the amount attributed to natural mortality.

These are but two of the many recognized examples of well-developed commercial fisheries now being utilized at or above their levels of maximum sustainable yield. Any appreciable continued increase in the world's fish landings must clearly come from unexploited species and, for the most part, from undeveloped new fishing areas. Much of the potential expansion must consist of new products from remote regions, such as the Antarctic krill, for which no harvesting technology and no market yet exist.

References and Notes

1. H. W. Graham and R. L. Edwards, in *Fish and Nutrition* (Fishing News, London, 1962), pp. 3–8; W. K. Schmitt, *Ann. N.Y. Acad. Sci.* **118**, 645 (1965).
2. M. B. Schaeffer, *Trans. Amer. Fish. Soc.* **94**, 123 (1965).
3. H. Kasahara, in *Proceedings, 7th International Congress of Nutrition, Hamburg* (Pergamon, New York, 1966), vol. 4, p. 958.
4. W. M. Chapman, "Potential Resources of the Ocean" (Serial Publication 89–21, 89th Congress, first session, 1965) (Government Printing Office, Washington, D.C., 1965), pp. 132–156.
5. E. Steemann Nielsen and E. A. Jensen, *Galathea Report*, F. Bruun *et al.*, Eds. (Allen & Unwin, London, 1957), vol. 1, p. 49.
6. E. Steemann Nielsen, *J. Cons. Cons. Perma. Int. Explor. Mer* **18**, 117 (1952).
7. O. I. Koblentz-Mishke, V. V. Volkovinsky, J. G. Kobanova, in *Scientific Exploration of the South Pacific*, W. Wooster, Ed. (National Academy of Sciences, Washington, D.C., in press).
8. J. H. Ryther, in *The Sea*, M. N. Hill, Ed. (Interscience, London, 1963), pp. 347–380.
9. H. W. Menard and S. M. Smith, *J. Geophys. Res.* **71**, 4305 (1966).
10. J. H. Ryther and D. W. Menzel, *Deep-Sea Res.* **12**, 199 (1965).
11. ———, E. M. Hulburt, C. J. Lorenzen, N. Corwin, "The Production and Utilization of Organic Matter in the Peru Coastal Current" (Texas A & M Univ. Press, College Station, in press).
12. C. D. McAllister, T. R. Parsons, J. D. H. Strickland, *J. Cons. Cons. Perma. Int. Explor. Mer* **25**, 240 (1960); G. C. Anderson, *Limnol. Oceanogr.* **10**, 477 (1965).
13. T. R. Parsons and R. J. Le Brasseur, in "Symposium Marine Food Chains, Aarhus (1968)."
14. E. Steemann Nielsen, *J. Cons. Cons. Perma. Int. Explor. Mer* **23**, 178 (1958).
15. J. R. Beers and G. L. Stewart, *J. Fish. Res. Board Can.* **24**, 2053 (1967).
16. M. Anraku and M. Omori, *Limnol. Oceanogr.* **8**, 116 (1963).
17. G. D. Grice and H. D. Hart, *Ecol. Monogr.* **32**, 287 (1962).
18. M. R. Reeve, in "Symposium Marine Food Chains, Aarhus (1968)."
19. Personal observation; T. J. Hart and R. I. Currie, *Discovery Rep.* **31**, 123 (1960).
20. K. R. Gaarder, *Report on the Scientific Results of the "Michael Sars" North Atlantic Deep-Sea Expedition 1910* (Univ. of Bergen, Bergen, Norway).
21. T. J. Hart, *Discovery Rep.* **21**, 261 (1942).
22. R. J. E. Sanchez, in *Proceedings of the 18th Annual Session, Gulf and Caribbean Fisheries Institute, University of Miami Institute of Marine Science, 1966*, J. B. Higman, Ed. (Univ. of Miami Press, Coral Gables, Fla., 1966), pp. 84–93.
23. R. T. Barber and R. L. Haedrich, *Deep-Sea Res.* **16**, 415 (1952).
24. J. W. S. Marr, *Discovery Rep.* **32**, 34 (1962).
25. S. D. Gerking, *Physiol. Zool.* **25**, 358 (1952).
26. B. Dawes, *J. Mar. Biol. Ass. U.K.* **17**, 102 (1930–31); *ibid.*, p. 877.
27. V. S. Ivlev, *Zool. Zh.* **18**, 303 (1939).
28. W. E. Ricker, *Ecology* **16**, 373 (1946).
29. L. B. Slobodkin, *Growth and Regulation of Animal Populations* (Holt, Rinehart & Winston, New York, 1961), chap. 12.
30. G. E. Fogg, C. Nalewajko, W. D. Watt, *Proc. Roy. Soc. Ser B Biol. Sci.* **162**, 517 (1965).
31. G. E. Fogg and W. D. Watt, *Mem. Inst. Ital. Idrobiol. Dott. Marco de Marshi Pallanza Italy* **18**, suppl., 165 (1965).
32. S. Z. El-Sayed, in *Biology of the Antarctic Seas III*, G. Llano and W. Schmitt, Eds. (American Geophysical Union, Washington, D.C., 1968), pp. 15–47.
33. Y. Saijo and T. Kawashima, *J. Oceanogr. Soc. Japan* **19**, 190 (1964).
34. P. A. Moiseev, paper presented at the 2nd Symposium on Antarctic Ecology, Cambridge, England, 1968.
35. T. L. Hopkins, unpublished manuscript.
36. W. S. Wooster and J. L. Reid, Jr., in *The Sea*, M. N. Hill, Ed. (Interscience, London, 1963), vol. 2, p. 253.
37. *FAO Yearb. Fish. Statistics* **25** (1967).
38. R. L. Edwards, *Univ. Wash. Publ. Fish.* **4**, 52 (1968).
39. R. J. E. Sanchez, in *Proceedings, 18th Annual Session, Gulf and Caribbean Fisheries Institute, University of Miami Institute of Marine Science* (Univ. of Miami Press, Coral Gables, 1966), p. 84.

40. The work discussed here was supported by the Atomic Energy Commission, contract No. AT(30-1)-3862, Ref. No. NYO-3862-20. This article is contribution No. 2327 from the Woods Hole Oceanographic Institution.